FERROELECTRIC POLYMERS

PLASTICS ENGINEERING

Founding Editor

Donald E. Hudgin

Professor
Clemson University
Clemson, South Carolina

Additional Volumes in Preparation

FERROELECTRIC POLYMERS

CHEMISTRY, PHYSICS, AND APPLICATIONS

HARI SINGH NALWA

Hitachi Research Laboratory
Hitachi, Ltd.
Hitachi City, Ibaraki, Japan

CRC Press
Taylor & Francis Group
Boca Raton London New York

CRC Press is an imprint of the
Taylor & Francis Group, an **informa** business

CRC Press
Taylor & Francis Group
6000 Broken Sound Parkway NW, Suite 300
Boca Raton, FL 33487-2742

First issued in paperback 2019

© 1995 by Taylor & Francis Group, LLC
CRC Press is an imprint of Taylor & Francis Group, an Informa business

No claim to original U.S. Government works

ISBN-13: 978-0-8247-9468-2 (hbk)
ISBN-13: 978-0-367-40168-9 (pbk)

Visit the Taylor & Francis Web site at
http://www.taylorandfrancis.com

and the CRC Press Web site at
http://www.crcpress.com

to my father, Kadam Singh,

mother, Sukh Devi, and

daughters, Surya and Ravina

Preface

During the past decade, considerable progress has been made in the field of electronic polymers. Research activity on electroactive organic polymers has been focused mainly on three aspects: dielectric-ferroelectric, electrically conductive, and nonlinear optical properties. Pyroelectric, piezoelectric, and ferroelectric phenomena in inorganic and organic molecular materials have been studied for quite some time. In 1963, Kocharyan first reported on piezoelectricity for two organic polymers, poly(methyl methacrylate) and poly(vinyl chloride). It was not until 1969 that Kawai first demonstrated significant enhancement in the piezoelectric properties of poly(vinylidene fluoride) by electret formation. In 1971, the ferroelectric properties of poly(vinylidene fluoride) were also reported. These discoveries eventually led to the new field of ferroelectric polymer science, and, since then, poly(vinylidene flouride) PVDF has established itself as a "ferroelectric polymer" par excellence. Subsequently, there has been tremendous growth during the past decade in exploring the material science, physics, and technology of poly(vinylidene fluoride) and other fluorocarbon polymers. Concurrently, the search began for other classes of novel ferroelectric polymeric materials. From an electronic point of view, ferroelectric polymers are polar and possess a nonconjugated backbone; therefore, they are highly insulating materials. On the other hand, research activities on electrically conductive polymers began in earnest with the discovery of polyacetylene in 1977. Achieving conductivity as high as that of copper metal was a landmark discovery that added the term "conducting polymers" to the scientific literature and resulted in a new field of science. Currently, nonconjugated polar polymers (originally piezoelectrics) and π-electron-conjugated polymers are emerging as key materials in the study of nonlinear optical phenomena for future photonic technologies.

Poly(vinylidene fluoride) provides the best example of recent advances that have been made in the understanding and applications of a single organic ferroelectric polymer. The developments in poly(vinylidene fluoride), its copolymers, and blends as ferroelectric materials have been extraordinarily fast. The importance of this field can be seen by the enormous growth recently in research on novel ferroelectric polymers such as polyamides (odd nylons), cyanopolymers, copolymers of vinylidene fluoride and vinyl fluoride, polyureas, polythioureas, biopolymers such as polypeptides and cyanoethyl cellulose, ferroelectric liquid crystal polymers, blends, polymer-ceramic composites, and other polymeric materials. Because of this rapid expansion in such a short time, the large volume of experimental data on ferroelectric polymers remains scattered. It is timely, therefore, to consolidate the current knowledge on ferroelectric polymers into a single reference source. This book collates the chemistry, physics, and technology of the ferroelectric polymers by having experts in the area summarize and report on the extensive technical literature published over the last few decades—making it a valuable reference for those who are engaged in ferroelectric polymer research. I hope that all researchers interested in this area will find this book useful as a reference text for various aspects of ferroelectric polymers.

Each chapter in this book is essentially self-contained, but cross-references are provided to relate any given chemical structures to their properties. The chapters have been organized to create continuity and cohesiveness for this subject. My goal is to introduce readers to the remarkable changes that have occured in the material science of polar polymers due to advancements in both novel materials and industrial applications. The importance of the supermolecular structures in relation to the electrical properties is stressed since the industrial applications strongly depend on this relationship. The thermally stimulated depolarization (TSD) technique yields information on the charge storage and charge decay phenomena in electrets. This has proven to be one of the most important electro-analytical tools for characterizing polymers, copolymers, and composites. Chapter 1 gives details on TSD measurement techniques, theoretical approaches, and organic molecular and polymeric materials (including biopolymers) for which the electret technique has been used. The applied aspects of the electret technique in various fields of science are presented. Finally, it has been demonstrated that not only can thermoelectret formation significantly increase the magnitude of the pyroelectric, piezoelectric, and ferroelectric properties of polar polymers, but it can also be used to introduce such activities into specialty polar materials. Thermoelectret formation has been frequently employed and is the most useful technique in generating second-order nonlinear optical properties such as second-harmonic generation and electro-optic effects in organic polar polymers.

The chapters are divided according to the type of polar polymeric materials discussed, with each chapter covering chemical aspects, spectroscopic studies, and electrical properties of individual ferroelectric polymers. The book has been divided into two parts: Part I covers the ferroelectric polymers, while Part II discusses their applications in industries. The preparative methods, processing, spectroscopic characterization, chemical, physical, mechanical, dielectric, pyroelectric, piezoelectric, and ferroelectric properties of poly(vinylidene fluoride) PVDF, its copolymers, and blends are presented in Chapters 2 to 5. In a similar way, Chapters 6 to 11 highlight detailed, topical reviews of current developments on synthesis, processing, spectroscopy, and electrical properties of odd-numbered nylons, alicyclic and aromatic nylons, fluorinated nylons, polyurethanes, cyanopolymers (including polyacrylonitriles), vinylidene cyanide polymers and copolymers, polyureas, polythioureas, melamine–thiourea polymers, biopolymers (including polypeptides), cellulose derivatives, chitin, amylose, bone, collagen, keratin, proteins and deoxyribonucleic acid (DNA), ferroelectric liquid crystalline (FLC) polymers and copolymers, and polymer–ferroelectric ceramic composites.

Chapter 12 covers the broad area of various nonlinear optical (NLO) interactions in ferroelectric polymers. Because ferroelectric polymers can be used as nonlinear optical materials, experimental results are presented on second-order nonlinear optical susceptibility of PVDF, its copolymers and blends, polyureas and vinylidene cyanide copolymers with and without NLO chromophore side chains, and ferroelectric liquid crystalline polymers. Chapter 13 describes the dielectric properties of a wide variety of organic polymers including ferroelectric polymers. Ferroelectric polymers have already found widespread applications ranging from solid-state technology to biomedical engineering. Part II concludes the book with examples of applications of ferroelectric polymers in the field of electronics, photonics, and biomedical technology. Details of a wide variety of devices such as actuators, biomedical transducers, converters, hydrophones and arrays, detectors, diagnostic imaging, electro-optics, noninvasive cardiopulmonary sensors, optical data storage, photopyroelectric detection, vidicon tubes, robotics, sensors, transmitters, ultrasonic resonators and transducers, and underwater acoustic hydrophones are presented in Chapters 14 to 19. Piezoelectric polymers have also been used in the field of nonlinear optics for future optical communication and signal-processing technologies. The comprehensive, detailed description of ferroelectric polymers and their based technologies presented here benefit from the knowledge of tailor-made organic materials. This book brings together for the first time detailed coverage on the field of ferroelectric polymers in an organized sequence of chapters. I hope that this comprehensive text will be a focal point that proves stimulating for researchers working in chemistry, polymer science, physics, material science, electronics, optics, and biomedical engineering.

I am deeply indebted to Professor Padma Vasudevan of Indian Institute of Technology, New Delhi, for introducing ferroelectric polymers to me during my doctoral studies and for her excellent teaching and guidance. I would like to express my gratitude to Professor Seizo Miyata and Dr. Toshiyuki Watanabe of Tokyo University of Agriculture and Technology, Tokyo; and Drs. Akio Mukoh, Atsushi Kakuta, Akio Takahashi, Yasuo Imanishi, and other colleagues of Hitachi Research Laboratory, who have supported my efforts in completing this book. Special thanks are due to Professor Satya Vir Arya, my former teacher; and to Dr. Robert Lewis, Krishi Pal Raghuvanshi, Rakesh Misra, Yogesh, Dilbagh Tahlan, Dr. V. B. Reddy, Jagmer Singh, and the late Jaivir Singh, Ranvir Singh Chaudhary, Arvind Kumar, and Bhanwer Singh Chaudhary for their continuous encouragements. My wife, Dr. Beena Singh Nalwa, deserves the deepest appreciation for her consistent support, patience, and assistance that provided me strength throughout this work. I have the greatest appreciation for the authors of the chapters, who laid the true foundation of this book, for their timely efforts and for providing valuable contributions on this subject. Finally, I am very thankful to Russell Dekker and Rod Learmonth at Marcel Dekker, Inc., who promoted this unique project and provided guidance to bring this volume to a fruitful completion.

Hari Singh Nalwa

Contents

Part II: APPLICATIONS

Contributors

Munehiro Date   Biopolymer Physics Laboratory, The Institute of Physical and Chemical Research, Saitama, Japan

Danilo De Rossi   Centro "E. Piaggio," University of Pisa, Pisa, Italy

Claudio Domenici   C.N.R. Institute of Clinical Physiology, Pisa, Italy

Eiichi Fukada   Kobayashi Institute of Physical Research, Tokyo, Japan, and Institute for Super Materials, ULVAC Japan, Ltd., Tsukuba, Japan

Péter Hedvig   Plastics Research and Development, Ltd., Budapest, Hungary

Thomas R. Howarth   Naval Research Laboratory, Orlando, Florida

B.-J. Jungnickel   Physics Department, German Plastics Institute, Darmstadt, Germany

R. Glen Kepler   Sandia National Laboratories, Albuquerque, New Mexico

Rudolf Kiefer   Fraunhofer-Institut für Angewandte Festkörperphysik, Freiburg, Germany

Naokazu Koizumi   Institute for Chemical Research, Kyoto University, Kyoto, Japan

Karol Mazur   Department of Physics, Technical University, Zielona Góra, Poland

Seizo Miyata   Tokyo University of Agriculture and Technology, Tokyo, Japan

Hari Singh Nalwa   Hitachi Research Laboratory, Hitachi, Ltd., Hitachi City, Ibaraki, Japan

P. K. C. Pillai   Department of Physics, Indian Institute of Technology, New Delhi, India

Kurt M. Rittenmyer   Naval Research Laboratory, Orlando, Florida

G. Scherowsky   Institut für Organische Chemie, Technical University of Berlin, Berlin, Germany

Iwao Seo   Advanced Materials Laboratory, Mitsubishi Chemical Corporation, Ibaraki, Japan

Elisa Stussi   Centro "E. Piaggio," University of Pisa, Pisa, Italy

Shigeru Tasaka   Department of Materials Science, Shizuoka University, Shizuoka, Japan

Kohji Tashiro   Department of Macromolecular Science, Osaka University, Toyonaka, Osaka, Japan

Toshiyuki Watanabe   Department of Materials Systems Engineering, Tokyo University of Agriculture and Technology, Tokyo, Japan

Eiso Yamaka   Department of Computer Science, Tsukuba College of Technology, Tsukuba, Japan

Dechun Zou*   Advanced Materials Laboratory, Mitsubishi Chemical Corporation, Ibaraki, Japan

---

*Current affiliation: Tokyo University of Agriculture and Technology, Tokyo, Japan

1
Polymeric Electrets

P. K. C. Pillal
Indian Institute of Technology, New Delhi, India

I. INTRODUCTION

The electret effect in solid-state physics has attracted a great deal of attention on account of its numerous practical applications. The electret is popular for device applications because of its small size and weight, and also due to its very long lifetime of surface charge. An *electret* is a solid dielectric piece polarized by the simultaneous application of an electric field and heat. Electrets are usually prepared by cooling a heated dielectric in a strong electric field, following the technique of discoverer of the electret, Mototaro Eguchi [1–4]. The electrified sample exhibits electrical charges of opposite signs on its two sides. Electrets are considered as counterparts of magnets. They are metastable, and their polarization decays slowly with time after the termination of the polarization treatment. The persistance and the magnitude of polarization, however, depend on a number of factors, such as the material used for the preparation, the magnitude of the polarizing field, the polarizing temperature, polarization time, thickness of the sample, electrode material, etc. Michael Faraday was the first to outline the basic principles of the electret when he published *Experimental Researches in Electricity.* Oliver Heviside coined the word "electret" to describe dielectric bodies which retain their electric moments even after the externally applied electric field has been reduced to zero [3].

Eguchi [1] prepared the first electrets and studied their various characteristics. He melted equal parts of carnauba wax and rosin with the addition of some beeswax, and permitted the mixture to solidify in a strong DC field between two parallel electrodes. He explained his success in the following way:

> When an electric field is applied to a melted substance, the molecules or clusters of molecules (which supposedly contain electric doublets) orient themselves with their axes in the direction of the electric field—so that when the melted substance so-

lidifies, the molecules will retain their orientation in the immobile state, causing the substance to retain a permanent electric polarization [1].

Investigations of the electret state opened up a new technique for studying the mechanism of polarization and absorption of charges in amorphous and crystalline dielectrics. The earlier literature on the subject has been reviewed by Gutmann, Pillai et al., Sessler, and several other workers in the field [3,6–13,101,476,477].

II. TYPES OF ELECTRETS

Electrets exhibit a quasi-permanent electrical charge after the polarization treatment is terminated. As they are a source of electric field, they are particularly useful in instruments and devices where one requires a stable electrostatic field. Depending on the method of inducing polarization and the materials used for the preparation, electrets, in general, can be classified as: (a) thermoelectrets, (b) ceramic electrets, (c) photoelectrets, (d) thermophotoelectrets, (e) magnetoelectrets, (f) radioelectrets, etc. A brief review of various types of electrets is given below.

A. Thermoelectrets [1–19]

Thermoelectrets are usually prepared by the method adopted by the pioneer worker Eguchi [1]. Several other workers in the field have prepared electrets with modifications to improve their characteristics [17–25]. In general, the material from which an electret is to be made is melted or softened between two parallel high-voltage electrode assemblies, as shown in Figure 1. As the external field acts on this molten substance for a considerable time, the necessary charge redistribution is attained in it. It is then allowed to cool and slowly solidify under the same electric field. The charge distribution in the material is thus "frozen in" under these conditions. After solidification, the electric field is switched off. Upon removal of the electrodes, the solidified dielectric shows an electric field on either surface, usually positive on one side and negative on the other side. To retain the charge, the electret must be short-circuited and kept under controlled conditions of humidity and temperature. Electrets can also be prepared at low temperatures. However, the magnitude of the resultant charge and the permanency of the electret will be comparatively poorer than for electrets prepared with material in the molten state [20,21]. An electret prepared in this way, by the simultaneous application of temperature and electric field, is known as a *thermoelectret.*

An electret forming and measuring unit is shown in Figure 1 [26]. This instrument is used for the preparation and characterization of thermoelectrets of various types of dielectric materials. It consists of a cylindrical furnace with kanthal wire as heating element and a detachable lid. A jumbo contact thermometer and an electric relay are used to maintain the temperature of the cell as required within an accuracy of $\pm 0.5°C$. The electrical connections of the sample for high-voltage supply are taken through two UHF Teflon sockets fixed on the aluminum lid which is placed on the top of the apparatus. The sample is sandwiched between two polished aluminum electrodes and held tightly in the sample holder as shown in the figure. The sample is then connected to the high-voltage DC supply and heated to the required polarizing temperature under electric stress for a predetermined time. It is later allowed to cool down to room temperature

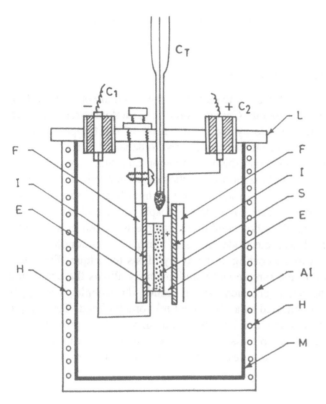

Figure 1   Electret forming and measuring unit [26]: I = Teflon insulation; E = Al electrode; H = heating coil; AI = asbestos cloth insulation; M = metal box; S = sample; C_T = contact thermometer; C_1, C_2 = Teflon-insulated connectors; F = brass frame of sample holder.

slowly under the same electric field. After this polarization treatment, the sample is taken out and short-circuited with an aluminum foil for further investigations.

B.   Fabrication of Ceramic Electrets [2,7,28,65–67]

The physical properties of low-melting-point organic materials are inadequate for certain types of applications. Hans Jaffe [27,28] showed that ceramic materials can be poled in the desired manner after they are fired and cooled if they are properly compounded and processed. This important work opened up a new field for electret research and applications.

Materials for electrified ceramics may be fabricated in a conventional way, that is, by pressing or extruding and slip-casting. Thereafter they usually are fired at high temperatures, in the range of 1200–1800°C. After the firing process, the fabricated ceramics are polarized with the application of high voltage in the usual way. Ceramic electrets can be prepared in any size and shape according to the desired application.

Gubkin and Skanavi [65] prepared stable ceramic electrets from various titanate materials, such as strontium-bismuth titanate, barium titanate, strontium titanate, calcium titanate, etc. They showed that ceramic electrets retain the majority of their charge without the need of shorting them with metal foils. Thomas A. Dickinson [66,67] showed that various kinds of ceramic and glass could be used for the preparation of stable electrets.

C. Preparation of Photoelectrets [29–50]

The photoelectret state in a photoconductor is achieved by the simultaneous application of an electric field and suitable illumination. Charge stored in a photoelectret is studied by depolarizing the photopolarized sample by radiation release of the trapped carriers. After the discovery of the photoelectret effect in sulfur by Nadzhakov [31,32], several other workers have made attempts to explain this phenomenon quantitatively [29–50].

To study the polarization and depolarization properties of photoconductors such as CdS, ZnS(Cu,Cl), ZnCdS, β-carotene, etc., an apparatus in which both preparation and characterization studies can be made under controlled atmospheric conditions has been used as shown in Figure 2 [30,50]. The photoconducting sample is placed between two electrodes A and B, where the lower electrode, B, is a transparent conducting glass electrode prepared by tin-indium oxide coating by spray pyrolysis technique. This electrode gives electrical connection to the sample and simultaneously allows the irradiation to pass through to generate the photogenerated carriers inside the sample. The upper electrode, A, is movable and is connected to an insulated micrometer head. The sample is surrounded by an electrostatically shielded heating arrangement, which provides uniform desired temperature inside the sample assembly. The entire system is mounted on a vacuum unit to provide vacuum or any desired gas, depending on the atmosphere required for the studies.

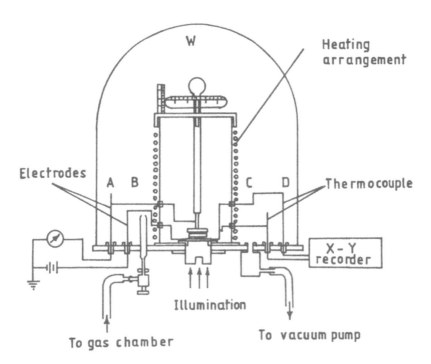

Figure 2 Apparatus for photopolarization and depolarization studies [30,50].

The photopolarization process in the sample has been studied by applying an electric field across the sample and simultaneously illuminating the sample in a perpendicular direction with the desired radiations. The phenomena of dark and photodepolarization can also be studied by using the same setup after switching off the field and illumination by observing the decay current from the sample—photoelectret—either by keeping it in dark or by reilluminating it.

D. Thermophotoelectret [30,50]

For the study of thermophotoelectret effect, the photoconducting sample is heated to a uniform higher temperature and then an electric field and appropriate illumination are applied for a known time using the apparatus in Figure 2. The polarization and depolarization currents are measured by means of an electrometer. The heated sample is cooled down to room temperature in the presence of the external field. The dark decay and photodecay characteristics are carried out on the thermopoled samples. The thermal depolarization currents of the electret, so formed, can also be observed upon reheating the sample at a uniform heating rate, in the absence of the field. The most important advantage of this apparatus is that the study of photopolarization and depolarization, as well as thermal polarization and depolarization currents, can be carried out without disturbing the sample. Some typical curves are shown in Figures 3, 4, and 5.

E. Magnetoelectrets [51–57]

Bhatnagar [51,55] obtained persistent polarization in certain types of organic materials by the simultaneous application of temperature and magnetic field and named this type

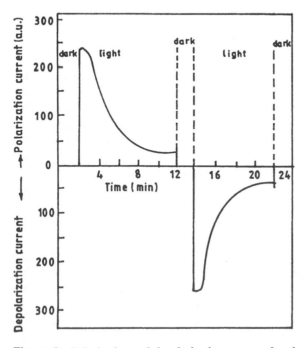

Figure 3   Polarization and depolarization curves of a photoelectret [50].

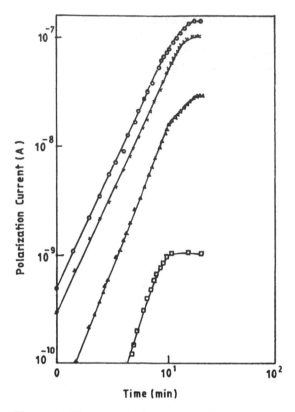

Figure 4 Photopolarization current in β-carotene at various forming field strengths [30,50]. □=1; △=1.8; ×=2.5; and ○=3.3 kV/cm.

of poled samples *magnetoelectrets*. This class of electrets retains the surface charge for a fairly long time, like thermoelectrets. Pillai and co-workers [52,56,57] also reported magnetoelectret effect in shellac wax. To confirm the influence of magnetic field, they prepared samples without the application of magnetic field under identical conditions. Although such samples showed some surface charge of low value due to the stripping and molding charge formation, it decayed to zero within a few days. However, samples prepared under thermal and magnetic field treatment showed much higher surface charge and also considerable persistence, similar to that of thermoelectrets. They also studied the effect of the poling magnetic field, polarizing time, and polarizing temperature on the formation and characteristics of the magnetoelectret. McMahan [53] showed that when certain organic compounds are solidified under the influence of a magnetic field, an anisotropic change of the dielectric constant takes place, causing it to be greater than normal in the direction of the applied magnetic field and lower at right angles to this direction. He postulated this mechanism of orientation of the molecules through diamagnetic coupling with the field. Diamagnetic anisotropy is observed in many aromatic ring compounds and, according to Selwood [54], the principal susceptibility is generally perpendicular to the plane of the ring. Hence, in a magnetic field such molecules should tend to orient themselves so that the plane of the rings is parallel to the field; the mol-

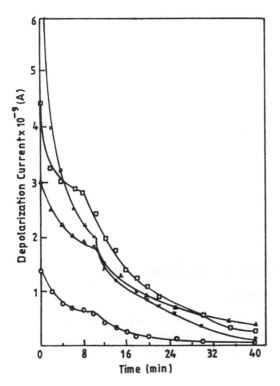

Figure 5 Photodepolarization decay curves of β-carotene photoelectrets [30,50], fabricated as in Figure 4.

ecules tend to orient with an axis of highest polarization in the direction of the magnetic field (Figs. 6a and 6b).

F. Electret Formation by Corona Charging [58–61]

A relatively new technique for the preparation of large-area electrets has been achieved by Pillai and co-workers. They developed a scorotron charging unit, shown in Figure 7, for corona charging polymer films and also for electrosensitizing photoconducting surfaces for electrophotographic applications. In this type of charging unit, three corona steel wires of 140 μm diameter were stretched across a perspex framework and the spacing between them was kept at 1 cm. A 15-wire grid in a curved configuration is used to provide uniform charging of the sample. The earthed metallic plate above the corona wires acts as a screen. The corona begins at about 8 kV, and the corona voltage can be varied up to 13 kV depending on the applications. A DC voltage is used for corona charging so that the same unit can be used for either the negative or positive mode of charging by applying positive or negative voltage, respectively, to the grid. Highly corona-charged layers can be obtained by varying the grid potential from 1 kV to 3 kV and optimizing the time of charging. The grid voltage and charging time were optimized to 1.5 kV and 60 s, respectively, keeping the corona voltage constant at 13 kV. A typical surface potential decay characteristics with time of a positively corona-charged Michler's ketone layer for both dark decay and light decay are shown in Figure 8.

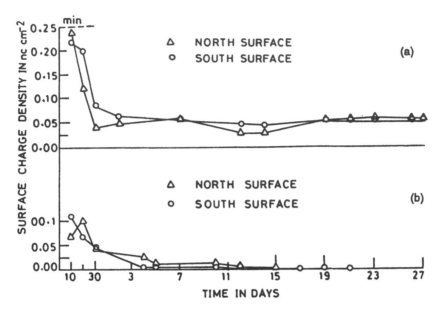

Figure 6 (a) Variation of surface charge density versus time for shellac wax magnetoelectret formed with 4.75 kG/mm magnetic field strength in 75-min polarization time [52]. (b) Variation of surface charge density versus time for shellac wax electret formed without any magnetic field in 75 min [52].

G. Thermoelectret Preparation by the Charge Injection Method

The conventional method of making thermoelectrets is usually very slow. It is also difficult to produce electrets of large area and complexity by this method. A novel method has been developed by Pillai and Shriver [62] for the rapid production of electrets of large area and any suitable size and shape. In this method, a dielectric sheet is continuously passed through an oven maintained at or near the molten state of the dielectric sheet or foil to be poled. When the heated sheet comes out of the oven, charging based on brush corona discharge technique is produced by applying a high-voltage DC potential to a metal wire brush or a series of thin wire electrodes kept at a short distance from the dielectric surface. This method permits charge injection at an elevated temperature. The moving dielectric with induced charges now passes through a grounded air or water-cooled metallic plate. The injected and the oriented charges in the dielectric are thus

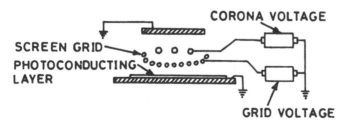

Figure 7   Corona charging unit [58,59].

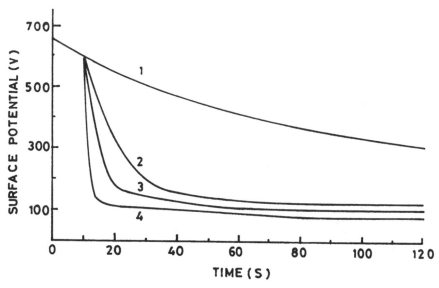

Figure 8 Discharge characteristics of positively charged Michler's Ketone layer: (1) dark decay; (2) decay under illumination of 0.245 mW/cm²; (3) 4.9 mW/cm²; (4) 85.75 mW/cm² [59].

"frozen in," and the resultant electret gives superior characteristics and stability of surface charge. Usually, electrets of opposing polarities may be produced on opposite sides of the dielectric sheet. Pillai and Shriver have prepared stable electrets of Teflon, polyethylene, polypropylene, Plexiglas, metal-impregnated kapton, mica, fused quartz, Mylar, quartz crystal, polycarbonate, Teflon-coated fiberglass, gadolinium molybdate single crystal, etc., by this technique. Typical examples of charge decay characteristics of brush corona-charged electrets of polyethylene and Teflon are shown in Figures 9 and 10. They have also obtained enhanced surface charge and better stability of charge on these samples by repeated brush corona charging of the same sample before the sample has been cooled. These electrets have been used in a number of applications, particularly in pollution studies [63]. They also produced polymeric electrets with excellent surface charge characteristics by means of Tesla-coil charging technique instead of brush corona charging [64]. The charge decay characteristics of Tesla-coil charged Teflon electrets are shown in Figure 11.

H.   Radioelectrets [69–75]

When certain types of solids are exposed to penetrating, high-energy radiations such as γ rays, β rays, X rays, etc., to a dose of about 10^6 rads, in the presence of an external polarizing field, they acquire charge storage during and after the termination of the field and irradiation. In many cases, these radiation-induced polarizations persist for a long time, and such polarized samples are popularly known as *radioelectrets*. Radioelectrets have been produced in a number of materials, such as carnauba wax, Teflon films, and ionic solids such as lithium fluoride and calcium fluoride. The charging is produced in the dielectric due to the electron–hole pairs created by the ionizing radiations, which drift in the applied electric field toward the electrodes and are trapped inside the material.

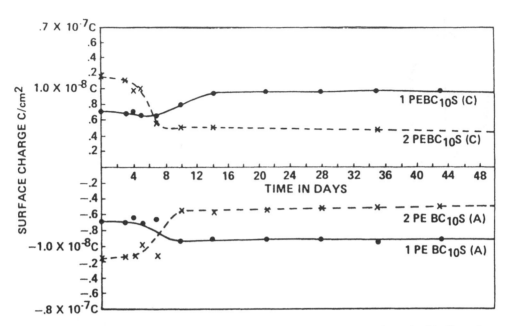

Figure 9 Charge decay characteristics of polyethylene thermoelectrets charged with 10 strokes of brush corona charge [62].

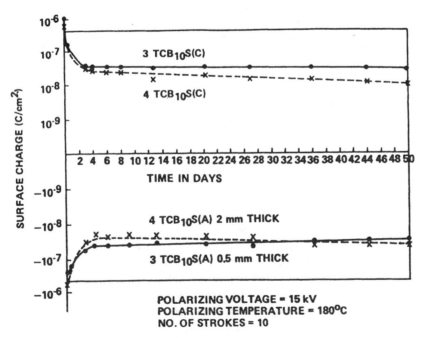

Figure 10 Charge decay characteristics of brush corona-charged Teflon electrets with 10 strokes [62].

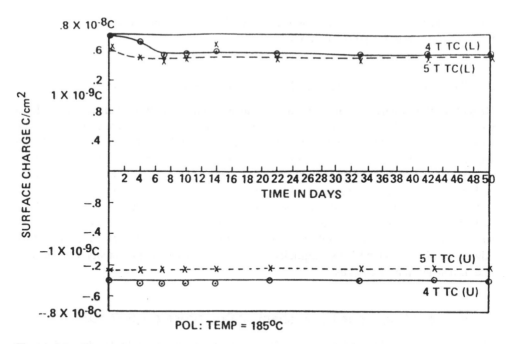

Figure 11 Charge decay characteristics of Tesla coil-charged Teflon electrets [62].

The radiation-stimulated polarization effects were studied by TSC thermograms and other methods usually followed for the study of thermoelectrets. The persistence of polarization in radioelectrets was found to depend on the magnitude of poling field, poling time, and irradiation dose, as well as on the type of materials used for the fabrication. These types of electrets are generally used for radiation dosimetry.

I. Electret Formation by Electron Beam Charging [80–86]

Polarization can be induced in dielectrics, particularly in polymer foils, by electron beam charging technique. Negative space charge layers are formed by injection of electrons on dielectrics. The electron beam charging is generally accomplished with nonpenetrating or partially penetrating electrons on nonmetalised polymer foils or on foils metalized on one surface. Polymer foils of polyethylene terephthalate (PET), polyfluoroethylene propylene (PTEP), polytetrafluoroethylene (TFE), and several other polymers have been poled by this technique. If the dielectric is irradiated with a low-energy beam, positive charging of the surface of incidence is also possible. Electron beams of the order of 0.5–1.0 MeV can be applied on dielectrics with thickness greater than 1.0 mm under atmospheric conditions. Low-energy beams are generally useful for charging thinner dielectric foils under vacuum condition. The electron beams are generated by means of an electron gun, glow cathodes, or radiofrequency discharges. Electrostatic beam focusing and scanning of the beam on the foil surfaces are employed to obtain uniformity of charge. Surface charge of about 10^{-7} C/cm^2 has been obtained by this method. The foil electret obtained in this manner has very good uniformity and stability of charges for several years at room temperature. The major advantages of electron beam charging are that it gives complete control over charge depth, lateral charge distribution, and charge density. Therefore, this technique is used extensively these days for the production of large-area electrets for practical applications.

J.   Electret Formation by the Liquid Contact Method [503–506]

Dielectric films metallized on one surface and in contact by its metallized side with a wet electrode such that a thin liquid layer resides between that contact and the liquid such as water or ethyl alcohol have been used for charging the films. A potential is applied between the electrode and the rear metallization of the film. Under these conditions, charge double layers will be formed at both solid–liquid interfaces. The interaction of electrostatic and molecular forces causes a charge transfer to the polymer film. A compensation charge of equal and opposite sign flows into the metallized rear surface. A large area of the film can be charged by gradually moving the electrode over the surface of the film. This method allows one to produce monocharge electrets. The main advantage of this method is its simplicity, control over initial charge density, and uniform lateral distribution of charges.

K.   Electret Formation by Mechanical Compression [507,508]

Stable polarization state in a dielectric can also be achieved by compression deformation. Electrets of polymethylmethacrylate by this technique were first reported by Novkov and Polovikov [507]. Miller and Murray also obtained persistent polarization in polystyrene and polymethylmethacrylate by cooling with flow under pressure [508].

III.   PRESERVATION OF ELECTRETS [87–92]

To obtain best results and permanency, the electret must be kept with its charged forces short-circuited, exactly as a magnet must be kept with a soft iron keeper. If it is kept open-circuited, the electret shows a decay of its charge, but it regains its full charge after having been short-circuited for some time. Electrets are exceedingly sensitive to humidity. It has been reported by Pillai and others [89] that the surface charge decreases considerably with time of exposure to humid atmosphere. They have observed a shift in the peak position of the thermally stimulated discharge current (TSDC) thermogram of polymer electrets from 350 K for dry samples to 334 K for wet samples. Therefore, instruments and devices using electret elements should be kept under controlled conditions of humidity, pressure, and temperature for better stability and performance, as these parameters have definite bearing on the permanency and decay of electrets. Wieder and Kaufman [91] reported that when electrets are immersed in water or exposed to humid air or highly ionized air, their charges are irreversibly neutralized. The effect of different types of vapors, both polar and nonpolar, on the behavior of electrets has been systematically studied by Beeler et al. [92]. They observed a decrease in the relaxation time and an increase in the magnitude of the volume polarization following an infusion of vapor into the dielectric. They also observed an increase in the transfer of charge by means of interfacial discharge at the dielectric–electrode gap. They have noted that infusion of a polar vapor into a nonpolar solid dielectric gives anomalously large values of polarization.

IV. EFFECT OF PRESSURE ON THE CHARACTERISTICS OF ELECTRETS [93,94]

A number of workers in the field have showed that the limiting surface charge density finally appearing on an electret depends on the air pressure surrounding the electret. The charge is found to be less if the electret kept under reduced pressure and is greater under pressures higher than atmospheric pressure. The charge density is shown to be proportional to pressure at low values of pressures; and a saturation effect exists at higher atmospheric pressures, as the polarizing field was insufficient to produce an electrification capable of yielding a surface potential gradient equal to the breakdown strength of air at that pressures.

V. ANALOGY BETWEEN ELECTRETS AND MAGNETS [1,2,90]

Eguchi [1,2] called the electret the electrical counterpart of a magnet because of its similarities to a permanent magnet. Just as the word "magnet" implies a material that exhibits a magnetic field at its two poles, an electret shows an electric field at its two opposing surfaces. If an electret is cut into two pieces, it gives two complete electrets. If the surface layers are removed one after the other, the remaining body remains an electret. However, reheating destroys the electret. John A. Eldridge [90] showed that the mechanical interaction between magnets is exactly the same as that between electrets. According to him, the two classes of polarized bodies may be called "polaret." In spite of these similarities, there are several striking differences between electrets and magnets. Magnets appear in nature, whereas electrets are always artificial. Magnets, if used properly, can be permanent, but the lifetime of electrets is limited due to slow decay of the induced polarization in the dielectric. Magnets can be used in air, but electrets, if exposed to humid air, soon lose their free charge. Moreover, in some cases, electrets show a charge reversal phenomenon, while magnets do not. Growth, decay, and reversal of charge are peculiar characteristics of electrets, whereas magnets do not show such behavior.

VI. CHARACTERISTICS OF ELECTRETS

A. Hetero- and Homocharge Formation in Electrets

Eguchi [1] found that freshly prepared electrets from wax mixtures, just after removal from the high-voltage electrode assembly, exhibited charges which had sign opposite to that of forming electrodes. That is, the surface which had been in contact with the anode showed a strong negative charge, and the cathode showed positive charge. These initial charges, positive toward the cathode and negative toward the anode, were found to decay within a few hours or days after the preparation of the electrets. Subsequently they passed through zero and assumed a sign equal to that of the corresponding forming electrode. This type of charge would reach a maximum and then gradually decay with time. This charge was found to be somewhat permanent for all practical purposes, lasting for several years when used for practical applications. Andrew Gemant [4] named the former type of charge "heterocharge" and the latter type of charge "homocharge." Later, Gross [475] gave the well-known "two-charge theory" to explain the behavior of an electret. Typical charge decay characteristics of a thermoelectret are shown in Figure 12.

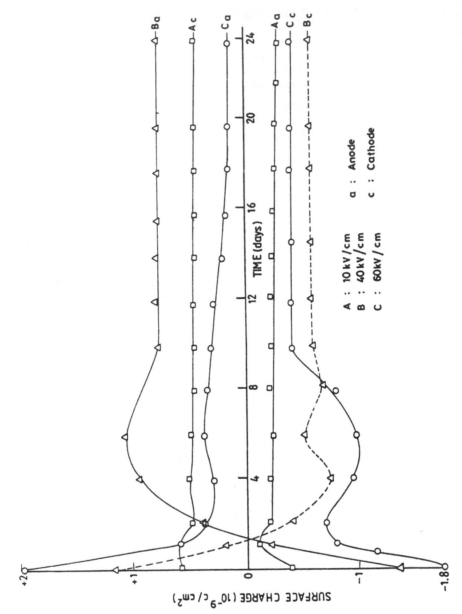

Figure 12 Decay of surface charge density with time for *Boswellia glabra* (plant resin) samples polarized at different polarizing fields [19].

B. Theories of Electrets

1. Gross's Two-Charge Theory [11,475,476,478]

Gross [11,475] treats the electret from the point of view of an absorptive dielectric. He describes the electrification to the enormous increase in the charging and discharging rates of the dielectric with rise of temperature. Further, he assumes a decay of electrification caused partly by external and partly by internal conduction within the dielectric itself. The dielectric absorption related to the movement of ions or the orientation of dipoles in the interior of the dielectric as shown in Figure 13a gives rise to heterocharge, while the conduction in the dielectric–electrode interface which produces the homocharge as illustrated in Figure 13c.

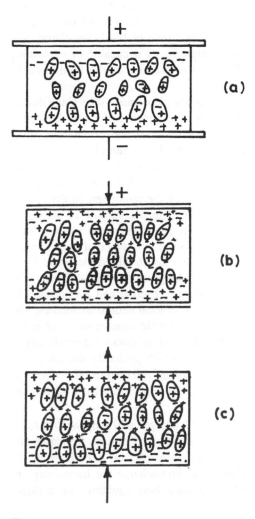

Figure 13 (a) Showing the displacement of ions and orientation of dipoles in the liquid state of the dielectric during the formation of electret. (b) Showing ions of the same sign of the high-voltage electrodes which are sprayed on the dielectric surface at the time of solidification. (c) Showing the possible nature of the solidified electret when it gives homocharge after the decay of initial heterocharge.

In polar materials, the formation of heterocharge is due mainly to the orientation of the dipoles. With sufficiently high field strengths, conduction currents surge into the interface and consequently electrons are fed into the dielectric or extend from its surface and are transferred to the electrodes. This facilitates the formation of homocharges, which appear in the form of surface charges, later spraying over a certain depth within the dielectric. Because of this process, a weakening of the field occurs which reduces the conduction currents. With short-circuit, a part of the dielectric polarization disappears immediately, and the rest follows more or less slowly. Due to this decay of the hetero-charge, the homocharge component begins to prevail in the resulting field. Again, the conduction currents play a role and cause the decay of homocharge, but there is no recombination of these two types of charges. Thus Gross concludes that there exist two types of charges of opposite signs in a short-circuited dielectric. The heterocharge is associated with dielectric absorption which is due to a homogeneous volume polarization of the dielectric. Homocharge is due mainly to the surface breakdown, which is purely an ionic surface charge. He has demonstrated the coexistence of the heterocharge and homocharge in a polarized sample. Thus the two-charge theory is capable of explaining most of the experimentally observed results. Many other workers in the field strongly supported Gross's theory.

According to Andrew Gemant [87], three independent phenomena take place during the formation of electret, and the observed effect is a superposition of these three phe-nomena. During the polarization process, a displacement of ions and orientation of di-poles take place in the liquid state as shown in Figure 13a. The accumulation of cations near the cathode and anions near the anode remains frozen-in while the liquid is being solidified with the external field on. Heterocharge is produced by the ionic displacement within the dielectric. While the dielectric solidifies, ions of the same sign as that of the high-voltage electrodes are sprayed over the respective dielectric surface as shown in Figure 13b, particularly during the time of solidification. Since the spray charges have the same sign as that of the forming electrodes, they explain the existence of homocharge. Simultaneously, the oriented dipoles form domains. When the solidified dielectric is re-moved from the electrode assembly, it shows a heterocharge as shown in Figure 13b. Then the electric forces originating from the dielectric are repelled and a migration of the charge carriers is started, until they finally recombine to form neutral molecules. The ions sprayed over the molten surface are in turn attracted by the dipoles oriented in the bulk of the sample owing to their sign and stay in a position as shown schematically in Figure 13c. At this state the electret has a homocharge which gradually decays as the dipolar orientation slowly disintegrates by taking up random orientation. However, the dipolar orientation in the electret material accounts for the remarkably long lifetime of the homocharge.

The relative contribution of these different types of charges will vary according to the chemical nature of the dielectric and also on the polarizing conditions. Wiseman and Feaster [96] extended the two-charge theory to explain some of their experimental ob-servations. They have shown that the heterocharge and homocharge are interacting en-tities, their interactions being given by the following static field equations for a three-layer capacitor:

$$E_1 = \frac{V - (d_2/K_2)(\sigma_\gamma - P_s)/\epsilon_0}{d_1 + (K_1 d_2/K_2) + (K_1 d_3/K_3)} \tag{1}$$

$$E_2 = \frac{V + (d_1/K_1) + (d_3/K_3)(\sigma_\gamma - P_s)/\epsilon_0}{(K_2 d_1/K_1) + d_2 + (K_2 d_3/K_3)} \tag{2}$$

and

$$E_3 = \frac{K_1 E_1}{K_3} \tag{3}$$

The electret's net charge density is given as

$$(\sigma_\gamma - P_s) = \epsilon_0 V\left(\frac{K_2}{d_2}\right) - \frac{Q}{A}\left[\left(\frac{K_2 d_3}{K_3 d_2}\right) + \left(\frac{K_2 d_1}{K_1 d_2}\right) + 1\right] \tag{4}$$

For a short-circuited system this equation reduces to

$$(\sigma_\gamma - P_s) = -\frac{Q}{A} \tag{5}$$

If V is not zero, the measured value $-Q/A$ contains an additional term $\epsilon_0 V(K_2/d_2)$ arising from nonpersistent polarization of the dielectric. The fields E_1 and E_2, which are responsible for the formation of homocharge and heterocharge, also depend on the gaps d_1 and d_3.

Perlman [10,479], Baumann and Wiseman [97], Wieder and Kaufman [91], Wild and Stranathan [480], Palm [481], and many other workers strongly support the modified two-charge theory.

Gubikin [482] gives a phenomenological theory of electret. Adams [88] and Swann [483] have developed theories for short-circuited electrets. Swann has given a mathematical theory based on (a) a distribution of surface and volume charges in electrets, which disappear according to ohmic conductivity having no relation to the decay of polarization, and (b) a distribution of polarization which decays with time. The theories of Swann and Gubikin agree only qualitatively with experiments.

Perlman [484] extended the theory by including the shielding effect with a uniform polarization. It has been found experimentally that at least two time constants are required to describe the polarization response. The superposition principle, the static field equations, and a relationship between the measured net surface charge and the charges on the dielectric are introduced and combined to yield an integral equation for internal field E:

$$E = \frac{4\pi \, \sigma \sum_{j=1}^{n+1} d_{2j-1}}{\left[K \sum_{j=1}^{n+1} d_{2j-1} + \sum_{j=1}^{n} d_{2j}\right]} \tag{6}$$

where

d's with odd subscripts represent thickness of air gaps
d's with even subscripts represent thickness of electret
K is the dielectric constant of electret
σ is the externally measured charge density

Piech and Handerek [485] express the effective charge of an electret in the form

$$\sigma_{\text{eff}} = \sigma_f + \sigma_\gamma^{\text{het}} - \sigma_\gamma^{\text{homo}} \tag{7}$$

where

σ_f is the density of bound charges
σ^{het} is the density of ionic component of real heterocharge
σ^{hom} is the density of real homocharge

Gross and Gultinger [486] considered an electret with its electrodes as a three-layer capacitor, as assumed by Wiseman and Feaster [96], and developed an expression for the surface charge in terms of conduction currents as

$$\sigma_{12} = G_1 E_1 - G_2 E_2 = - \sigma_{21} \tag{8}$$

where

G is the electrical conductance
E is the electric field
$G_1 E_1$ is the current through the surface layer
$G_2 E_2$ is the current through the main body of the sample
σ_{12} is the charge accumulated at the interface between the surface layer and the main body

Gross and Gultinger [486] applied the method of ionic thermal currents to a polarized electret which is discharged by the thermally stimulated discharge current technique and showed a method of calculation of dipole relaxation times and their activation energies. Perlman and Meunier [487], van Turnhout [488], and several others [489,490] have used it to analyze the electret effect quantitatively.

To explain the characteristics of plastic electrets, Pillai et al. [491] have suggested that the observed surface charge is due to the dielectric polarization as well as due to the various mechanisms responsible for charge transfer across the electret–electrode gap including that due to the stripping and molding charges. The effective surface charge may be written as

$$\sigma_{eff} = \sigma_D + \sigma_i + \sigma_{hom} + \sigma_M \tag{9}$$

where

σ_D is the charge due to dipolar orientation
σ_i is the charge contribution from ionic polarization produced by the displacement of positive and negative ions in the material
σ_{hom} is the charge contribution from external polarization such as the spray charge from the high-voltage electrode to the dielectric surface and also the charge due to the breakdown of the interspace between electrode and electret
σ_M is the charge contribution due to molding and stripping

The magnitude and nature of the effective charge depends on the relative contributions of each of these quantities and varies from material to material and also on the polarization parameters.

Fabel and Henisch [492] studied the characteristics of the electret by measuring the surface potential. The measured surface voltage is given by:

$$V = \frac{V_0 d_2}{d_1 + d_2} \tag{10}$$

where

V_0 is the forming voltage
d_1 and d_3 are the distances of charge layers from the positive and negative electrodes
d_2 is the thickness of the dielectric between the two positively and negatively charged
    layers

The polarization characteristics of the electrets have been studied extensively by the thermally stimulated discharge current (TSD) method; the details of some of these studies are given separately. Recently, a number of other methods utilizing heat pulses [492–494], pressure pulses [495,496], and electron beams [497–499] have been reported to investigate the charge distribution and polarizations in the thickness direction of polymer electrets. The results of these studies have extended the understanding of electret formation in dielectrics [501]. Recently, the laser intensity modulation method (LIMM) and the pressure propagation method (PWP) of measuring the polarization distributions in piezoelectric materials have been compared and critically analyzed by Alquie et al. [502], who concluded that these methods are complementary in many respects and the use of both methods on the same samples can yield additional information.

VII. EXPERIMENTAL METHODS TO INVESTIGATE POLARIZATION IN ELECTRETS

Of the various types of electrets, thermoelectrets are most important from the scientific and application points of view. Therefore, most studies have been carried out on thermoelectrets prepared from polymeric materials.

The following three different techniques are generally employed to investigate the characteristics of electrets.

A. Nonisothermal relaxation studies:
    (1) Thermally stimulated discharge (TSD) studies
    (2) Thermally stimulated polarization (TSP) studies
B. Isothermal methods:
    (1) DC step response method
    (2) Charge measurement
C. AC method: In this case dielectric parameters are measured over a range of frequencies

A. Nonisothermal Relaxation Studies

1. Thermally Stimulated Discharge (TSD) Studies [167–233]

In recent years, the thermally stimulated discharge current technique has become a powerful tool in the investigation of dielectric relaxations in polymers. The relaxation time, activation energy and distribution function of dipolar relaxation, attempt-to-escape frequency, and capture cross section have been successfully studied for many polymers.

In this method, the dielectric is initially polarized to form an electret. The electret is then heated at a constant rate and the depolarization current is recorded as a function of temperature.

The charge of an electret may be generated by various mechanisms, such as orientation of permanent dipoles, trapping of charges by structural defects and impurity centers, and building up of charges near heterogeneities such as the amorphous–crystalline interfaces in semicrystalline polymers and grain boundaries in polycrystalline materials. At room temperature, the charge decay measurements are rather time-consuming, because at such low temperatures, the dipoles and charges remain virtually immobile. In recent years, the thermally stimulated discharge current (TSD) technique has been used successfully to study the various mechanisms involved in electret formation and also the charge storage phenomena in them. In fact, TSD shortens the measurement time considerably. However, during TSD, the polarization of an electret is destroyed. This method is used extensively for the analysis of electrets, which in effect helps in the development of electrets having stable characteristics and long lifetimes. This is mainly because the TSD technique is characterized by a very low equivalent frequency (10^{-2}–10^{-4} Hz) as compared to the AC dielectric measurement (1–10^{+12} Hz) and consequently leads to a better resolution of the different relaxation processes [193].

The TSD method has its analogs in various branches of solid-state physics, as in thermally stimulated conductivity and thermoluminescence [194–198]. It was applied in a systematic way by Bucci and Fieschi to determine the dielectric relaxation in impurity-vacancy dipole complexes of ionic crystals. Recent studies of dielectric relaxations of various polymers by TSD [199–234] have shown that the total charge stored in a polymer electret is very sensitive to the structure of the forming material, and thus TSD study of polymers, polymer blends, and composites is expected to give valuable information about the molecular interactions and the extent of mixing between different components.

When an electret is reheated at a linear rate, a discharge current is produced in the external circuit. This thermally stimulated discharge current is recorded as a function of temperature. The resulting TSD spectrum shows maxima which corresponds to various decay processes. In polymer electrets, these decay modes correspond to various polymeric relaxations. Therefore, analysis of TSD spectra of polymer electrets provides very useful information on various aspects of charge storage mechanisms and relaxation processes.

Charge decay of the electret takes a long time at room temperature. Frei and Groetzinger [199] proposed the idea of thermal stimulation of decay. Von Altheim [200] also visualized that the charge released is controlled by temperature-dependent molecular motion. The technique based on thermal stimulation was later used by Murphy and Riberio [201], Gross [202,203], and Gubkin and Matsonashirli [204]. Bucci et al. [198,205] studied the relaxation of impurity dipoles in ionic crystals. They also developed a theory for a Debye type of relaxation. However, most polymeric materials show deviations form Debye relaxations. Gross [206] and van Turnhout [193] explained this deviation on the basis of distribution in relaxation times. Jonscher [207–210] also proposed a model to explain the non-Debye behavior. According to this model, flatness or broadening of the dielectric loss peak or TSD peaks can be interpreted on the basis of manybody interactions between dipoles or hopping of charge carriers, electronic as well as ionic, leading to mutual screening. The theoretical development of space charge or trapped charges was given by Perlman [211] and van Turnhout [214]. During the past few years, this technique has been used extensively to understand the electret formation mechanism and relaxation phenomena in polymers [212–234].

The TSD spectra of polymer electrets are, in general, very complex, consisting of several relaxation peaks. Special techniques such as peak cleaning and partial heating [235] have been developed to analyze the complex TSD spectra of polymers.

Sessler and West [236] studied the properties of electrets formed by electron beam irradiation using the TSD technique. Maeda et al. [237] applied this technique to study the mechanisms involved in electrets prepared by the application of AC fields. Pillai and co-workers [238,239] studied the properties of electrets prepared from polymer blends and showed that the observed polarization is due to induced dipole formation from deep trapping of charge carriers originating in the bulk during polarization. Ronarch et al. [240] employed this technique to study the phase separation in polypropylene block copolymers. Srivastava et al. [241] used it to investigate magnetoelectrets prepared from polymers.

TSD has also been carried out to study the effect of cross linking [228], stereo- and structural isomerism [242], doping [218], and co-polymerization [219] on the relaxation behavior of polymeric systems. TSD has also been carried out on several other organic and inorganic materials [231–233,235,243,259,260].

a. Factors Affecting TSD Spectra

Polarizing Field (E_p): In the formation of an electret, the dielectric is subjected to an electric field (E_p), known as the polarizing field. By varying the poling field, one can distinguish between the volume polarization and the space charge polarization. In TSD spectra, the peak due to dipolar or volume polarization increases linearly with E_p, whereas the space charge polarization varies nonlinearly with the polarizing field [190,193] (Fig. 14).

Polarization Temperature (T_p) [193]: The polarizing temperature, T_p, has a significant influence on the position and magnitude of the peak current due to various processes in the TSD spectra. For a process with a single relaxation time, the peak in the TSD spectrum remains unaltered, while its magnitude increases with increasing T_p. In a dis-

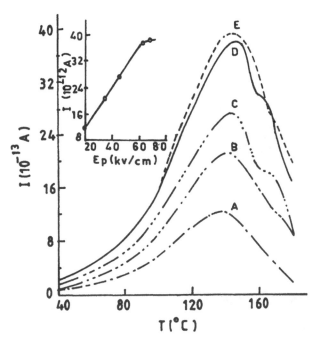

Figure 14   Effect of poling field E_p on TSD thermograms of polypropylene polycarbonate blends. T_p = 100°C, A = 20 kV/cm, B = 35.5 kV/cm, C = 47 kV/cm, D = 63.5 kV/cm, E = 70.6 kV/cm, t_p = 1 h. Inset shows variation of I_{max} with E_p [113,118].

tributed relaxation process, the position and magnitude of peak current depends on T_p, unless T_p is chosen so high so that the dipoles and/or charge carriers acquire sufficient mobility to achieve nonequilibrium excited states. Also, in a distributed process at low T_p, not all the slow dipoles are activated; only fast dipoles contribute to the TSD. As T_p is increased, more and more slow dipoles are activated and the magnitude of the peak current increases while its position shifts toward the higher-temperature side (Fig. 15).

Polarization Time (t_p) [193]:   The polarization time has significant effect on the position and magnitude of the peak current in TSD spectra. According to van Turnhout [193], t_p should change logarithmically to obtain charges of the same magnitude as with the charge of T_p.

Electrode Material [237–240]:   The effect of electrode material on the magnitude of charges acquired by the electret has been studied by a number of workers, and it has been shown that good correlation exists between the charge and the work function of the metal electrode used for the application of the poling field.

Electret Thickness [246]:   The charge of an electret has been found to depend on its thickness. The heterocharge released in a TSD has a linear relationship with the thickness of the specimen, leading to the existence of volume polarization as proposed by Gross.

Heating Rate: The TSD peaks shift to higher temperature with increasing heating rate. Similarly, the magnitudes of the peaks are also found to be increased. By varying the heating rate, several workers calculated the activation energy of the dielectric relation process.

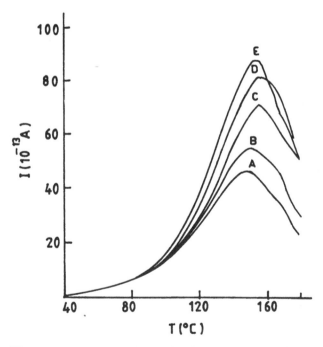

Figure 15   Effect of polarization temperature T_p on TSD thermogram of polypropylene polycarbonate blends. E_p = 47 kV/cm, A = 100°C, B = 120°C, C = 140°C, D = 150°C, E = 160°C, t_p = 1 h [113,118].

Humidity: If the electrets are exposed to water vapor or are immersed in water, the polarization is found to decrease; the effect is more pronounced for nonshorted electrets.

Other Factors: In addition to the above factors, the effect of pressure [252], gaseous vapors [253], and high-energy radiations [254] on the charge storage characteristics have been studied.

b. Mechanisms Responsible for TSD [193,255–265].   All the decay processes are thermally activated, and their relaxation times depend exponentially on temperature. Therefore, these processes are accelerated during TSD and will generate an appreciable discharge current. TSD peaks occur because of the release of the charge at a faster rate over a certain temperature range.

Several processes contribute to the discharge of electrets. In electrets made from polar materials, the disorientation of dipoles plays a major role. This tends to destroy the persistent dipolar polarization by redistributing all dipoles at random.

The activation energy is not same for all the dipoles. If the differences of various activation energies is not large, all the individual peaks overlap and merge into a broad peak, which is due to disorientation of polar side groups in polymers at low temperatures (e.g., the β peak). Another cause is due to the difference in rotational mass of the dipoles. The differences occur where dipoles are disoriented by the motion of main-chain segments. This is responsible for the appearance of α peaks which are located at T_g.

In addition to dipoles, the electret usually contains immobilized space charges, which are nonuniformly distributed, mostly near the electrodes. During heating, they are mobilized and neutralized either at the electrode or in the sample by recombination with charges of opposite sign due to the diffusion under the influence of the local fields. This space charge peak, identified as the ρ peak, appears at higher temperatures, as the neutralization of space charge requires them to move over many atomic distances.

At higher temperatures, neutralization of space charge also occurs due to the recombination with thermally generated carriers. These are uniformly distributed in the entire specimen and are responsible for the conductivity of the material. The loss of space charge owing to recombination with thermal carriers is not always noticeable.

Both dipole and space charge relaxations may occur in polar materials, only space charge can be active but in nonpolar materials. The dipole and space charge peak may be distinguished by observing their dependence on the poling field, which for the space charge peak is expected to be nonlinear. Several other differences have also been reported in the literature. If there is a distribution of relaxations, the shape and position of the current peaks depend on the forming and storage conditions of an electret.

c. Theoretical Background of TSD [266–281,288].   Bucci et al. [281] developed a theory for analyzing the ionic thermocurrents for dipoles with a single relaxation time. This theory has been generalized independently by Gross [282,283], van Turnhout [283–288], and Perlman [286,287,401] to study the polarization mechanisms of thermo-electrets having distributed relaxation times.

The general form of the TSD current is given as [283,288]

$$I(T) = A \exp\left[-\frac{E}{KT} - B \int_{T_0}^{T} \exp\left(-\frac{E}{KT} \right) dT \right] \tag{11}$$

where

$I(T)$ = externally measured discharge current
T = absolute temperature

T_0 = starting temperature of TSD
E = activation energy of dipoles/trap depth of charge carriers
K = Boltzmann constant

A and B are constants, which depend on the type of polarization and the heating rate. The values of A and B for trapping near the surfaces are

$$A = \frac{N_0 e \delta^2}{2 \epsilon \tau_0 d} \mu\tau$$

$$B = \frac{2}{\beta \tau_0} \tag{12}$$

and for dipolar decay these constants are

$$A = \frac{N p^2 E_f}{k \tau_0 T_f}$$

$$B = \frac{1}{\beta \tau_0} \tag{13}$$

where

N_0 = initial charge density in traps
$\mu\tau$ = charge mobility-free lifetime product
δ = penetration depth of charge
ϵ = dielectric permittivity
d = sample thickness
p = dipole moment
E_f = electret-forming field
T_f = electret-forming temperature
e = electronic charge
N = dipole concentration

The initial depolarization current, I_d, just above the initial temperature T_0 of Eq. (11), follows a relation of the type

$$\ln I_d(T) = \ln A - \frac{E}{kT} \tag{14}$$

The activation energy (E) responsible for the process can be calculated from the slope of the straight line obtained from $I_n(T)$ versus $1/T$ plotted as suggested by the initial rise method of Garlick and Gibson [289]. A low value of E is obtained when there is a distribution in relaxation. It requires separation of the individual peaks when different processes overlap and give a single TSD peak.

A more sophisticated method based on all the data has been advocated by Bucci et al. [281] by writing Eq. (11) in the following form:

$$\ln\left[\int_{t(T)}^{\infty} \frac{I_d(t)\, dt}{I(T)} = \ln \tau = \ln \tau_0 + \frac{E}{kT}\right] \tag{15}$$

where

$$\tau = \tau_0 \, \exp\!\left(-\frac{E}{kT}\right) \tag{16}$$

and

$$\ln \tau_0 + \frac{E}{kT} = \ln\left[\frac{Q_S(T)}{I(T)}\right] \tag{17}$$

where Q_S is the remaining charge at temperature T, $I(T)$ is the current at the same temperature T, E is the activation energy responsible for the release of charge carrier, and τ_0 is the relaxation constant. The parameter E can be calculated from the semilog plot of Q_S/I versus $1/T$. This method is known as Bucci's modified full-curve method. This method is applicable to the TSD of a uniform polarization.

From Eq. (11), the expression at which maximum TSD occurs is obtained by applying the condition

$$\frac{dI(T)}{dT} = 0$$

which yields

$$\tau_0 = \frac{kT_m^2}{\beta E \, \exp(E/kT)}$$

where T_m is the peak temperature, β is the heating rate, and the corresponding relaxation time at the peak temperature (T_m) can be calculated as

$$\tau_{T_m} = \frac{kT_m}{\beta E} \tag{18}$$

where E is the activation energy.

The activation energy can also be calculated by the charge ratio method suggested by van Turnhout [283]. It applies to the space charge-limited drift of the charge carriers:

$$\frac{Q_R(T)T_m^2}{Q_S(T)T^2} = \exp\!\left(-\frac{E}{KT} + \frac{E}{KT_m}\right) \tag{19}$$

where

Q_R = charge released up to temperature T
Q_S = remaining charge at T
T_m = peak temperature

E is calculated from a straight-line semilog plot of

$$\frac{Q_R(T)T_m^2}{Q_S(T)T^2} \quad \text{versus} \quad \frac{1}{T}$$

The linearity of the ratio of the released charge to the stored charge is typical of SCL drift. It occurs as it is the stored charge that determines the decay rate of retained charge. This is not true for other decay processes such as the dipolar reorientation.

Knowing E, the other parameters such as $\tau(T)$ and τ_0 can be calculated as follows. From the condition for maximum of Eq. (11) we have

$$\tau_0 = \frac{kT_m^2}{\beta E \, \exp(E/kT_m)} \tag{20}$$

where β is the heating rate in the TSD experiment.

Since $\tau(T) = T_0 \, \exp(E/KT)$, knowing τ_0, τ can be calculated at any temperature.

Several other methods have been reported in the literature for the evaluation of TSD data [261,262,277,278]. Hino et al. [290] utilized the shift of T_m with β for the calculation of β. Although this method gives an average value of E, it is not suitable for polymers, because large variation in β is not advisable, as both low and high values of β introduce errors. Nicolas and Woods [291] have summarized a number of methods which can be used to distinguish and evaluate different kinetics, but these methods make use of only one or two points on the curves, leading to rather large uncertainties. An iterative method has been suggested by Coswell and Woods [292], and a least-square fit by van Turnhout [283].

The initial charge stored in the electret due to homocharge can be calculated to a certain depth as follows [170]:

$$P_r = Q\left(1 - \frac{r}{2d}\right)^{-1} \tag{21}$$

where

$$r = 2d\left(\frac{Q}{Q + Q_0}\right) \tag{22}$$

P = polarization/unit volume
r = charge penetration depth
d = thickness of the sample
Q = charge released/unit area by TSD
Q_0 = surface charge density prior to the TSD experiment

An analytical method for calculating activation energy and frequency factor in case of complex TSD curves is based on the similarity between thermoluminescence and TSD. The complex glow peak is decomposed into a set of separate glow peaks analytically, and then trapping parameters are evaluated [293,294].

d. Determination of Capture Cross Section and Attempt-to-Escape Frequency. According to Frohlich's bistable method [295], the orientation of dipoles during the polarization and their randomization during TSD could be considered in terms of the transition of charge carriers across a potential hill (E_0). This theory holds good for the experimental TSD curves of dipolar materials, so a capture cross section (σ_D) and an attempt-to-escape frequency (ν_D) can be assigned to the induced dipoles and can be calculated using the Grossweiner's relation [295,296]:

$$\nu_D = \left\{\frac{3T'\beta}{2(T_m - T')T_m} \, \exp\frac{E}{kT_m}\right\} \, s^{-1} \tag{23}$$

and

$$\sigma_D = \frac{v_D}{2 \cdot 1 \times 10^{24} \times T_m^2} \text{ m}^2 \tag{24}$$

where T' is the half-width temperature on the lower-temperature side of the TSD curve and β is the heating rate.

2. Thermally Stimulated Polarization (TSP) Studies [297–313]

When a dielectric is charged while being heated linearly, polarization can be induced in it. Such thermally stimulated polarization (TSP) has many advantages, such as: (a) the resulting TSP current reveals how the orientation of the dipoles is proceeding in the material; (b) the search for the optimum poling temperature is eliminated; (c) unnecessary overheating of the sample is avoided; and (d) it also reveals the temperature at which the ohmic conduction becomes significant [297–313]. The charging current can be written as

$$i(t) = \alpha(T)P_e(T) \exp\left[-\int_0^t \alpha(T) \, dt \right] + g(T)E \tag{25}$$

where $P_e(T)$ is the equilibrium value of polarization P, which is given by

$$P_e(T) = \epsilon_0(\epsilon_S - \epsilon_\alpha)_T E \tag{26}$$

and

$$\epsilon_S - \epsilon_\alpha = \frac{Np^2}{3kT} \tag{27}$$

where

K is Boltzmann's constant
$\alpha(T)$ is the relaxation frequency
N is the number of dipoles/m^3
p is dipole moment
$g(T)$ is ohmic conductivity $= g_0 \exp(u/kT)$

The two contributions to the charging current behave differently as functions of temperature. The orientation of the dipoles is a transient process giving rise to a peak, whereas the conduction current, which derives from the motion of equilibrium carriers, increases steadily with temperature. The conduction current appears at a higher temperature than the dipole orientation current. In the temperature range of low conduction, the TSP and TSD currents are equal, but when the temperature eventually becomes high, the conduction current dominates in TSP.

To separate out the contributions of the dipole orientation and the conduction current, the sample is reheated for a second time in the same field. As no more dipoles will orient during a second heating, only the conduction current will be observed. Another way to separate the two contributions is to change the heating rate. This will alter the shape and position of the dipole peak, but the conduction current will remain the same.

B. Isothermal Methods

1.   DC Step Response Method

When a step-function voltage is applied to a dielectric, it results in a flow of current which initially decays with time before attaining a steady state. Conduction current gives information on various mechanisms involved in charge carrier generations, trapping, transport, and polarization in dielectrics [302,303].

Response of polymeric materials to step-function voltage has been investigated by a number of researchers [304–316]. Wintle [307] has given the theoretical treatment of the problem. A step function represents a wide Fourier spectrum, containing components of very low frequency. Therefore, all the information required to construct dielectric spectra in a wide range $(10^{-4}-10^{10}$ Hz) is contained in the recorded current. Fourier transformation technique was developed by Hamon, Williams, and Baird [304–306]. Absorption currents in several polymers find their origin from dipolar polarization processes at temperatures below T_g.

At temperatures above T_g, the absorption currents have been related to charge carrier accumulation near the electrodes—MWS polarization, generation of carriers in the bulk, and subsequent trapping and hopping of charge carriers between localized states.

Steady-state conduction currents have also been analyzed in various polymers. In most cases, the absorption current and steady-state conduction current in a polymer are controlled by different mechanisms. Schottky effect has been considered for steady-state conduction in poly(ethyline terphthalate), poly(tetrafluoroethylene) [317,318], and polycarbonate [319]. The effect of space charge on the Schottky barrier and hence on steady-state currents in polystyrene has been reported by Kamisako et al. [320].

Diffusion-limited Richardson-Schottky (R-S) field-assisted thermionic emission of holes dominates dark conductivity in poly(vinylcarbazole) when the electrode is of high-work-function material [321]. Poole-Frenkel (P-F) effect has been reported to be the dominant mechanism of conduction in polyvinylacetate films [131].

Charge transport in thin polypropylene films has been observed to be due to tunneling mechanism at low temperatures and Schottky emission at high temperatures. Space charge-limited currents have been observed to flow in polystyrene in the steady state [322].

2.   Charge Measurement

The properties of electrets have been studied mainly by the measurement of the surface charge. The charge measurement technique, therefore, has special importance for the study of electret effect. A number of methods, including the induction method, the dissectible capacitor method, the generating voltmeter method, the radiation discharge method, the Liechtenberg figure technique, the capacitive probe method, the dynamic capacitor technique, and the thermal pulse method, have been used for charge measurement by a number of workers in the field. Some of the important methods have been reviewed by Pillai et al. [328] and Sessler [329].

a. Induction Method.   The induction method was used by Eguchi [330], Gemant [331], and many other workers. An improved method based on the induction technique, as given by Pillai et al. [328], is described below. A special cylindrical measuring device was developed to enable the transfer of charge from the surface of the electret to a movable electrode and then to an electrometer. Figure 16 shows a schematic cross-sectional view of the apparatus. It consists of a fixed lower electrode on which the electret is placed to measure the surface charge and another, movable electrode, which is made to rest very close to the surface of the electret for a short time by a spring mechanism,

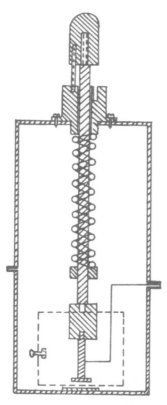

Figure 16   Charge measuring instrument [328].

as shown in the figure. When the movable electrode is released by lifting the upper electrode, a charge, equal in magnitude, but opposite in sign, is transferred immediately to an electrometer connected to this movable upper electrode. The typical surface charge versus time curve obtained by this method is shown in Figure 12.

b.   Vibrating Probe Method.   To measure surface charge and voltage, a cylindrical, double-walled vibrating probe (Fig. 17) was developed by Narula [332,333].

This probe consists of two coaxial aluminum cylinders, separated by Teflon insulation between them. The probe is made to vibrate at a predetermined frequency just above the electret surface. The voltage induced on the inner cylinder is monitored with the help of an AC microvoltmeter. The outer cylinder is grounded. The measured AC voltage is given by

$$V_{AC} = \frac{Q}{C_0} \, a \, \sin \, wt \tag{28}$$

where

$w = 2\pi f$
Q = charge on the surface
C_0 = average capacitance
f = driving frequency
a = vibrating constant

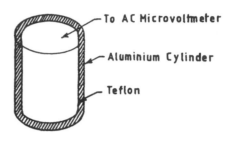

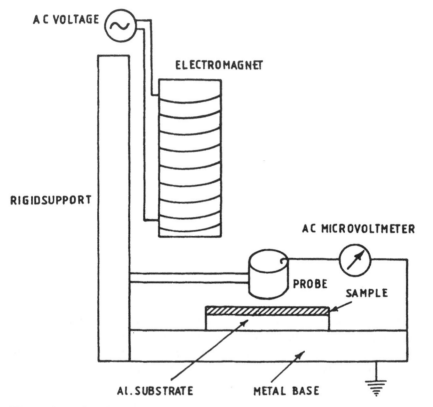

Figure 17   Vibrating-probe charge measuring assembly [333].

Keeping a, C_0, and f constant,

$$V_{AC} \propto Q$$

Reedyk and Pearlman [334] used another vibrating electrode method for the measurement of surface charge (Fig. 18). They calculated the electric field E_2 in the air gap between the electrode and the surface of the sample as

$$E_2 = \frac{kV - (\sigma_\gamma - \rho_s)/\epsilon_0}{d_1 + Kd_2} \tag{29}$$

where $(\sigma_\gamma - \rho_s)$ is the net surface charge density.

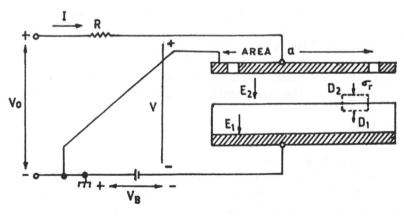

Figure 18 Vibrating-electrode method for surface charge measurement [334].

Improvements in the Induction Method: In this method an electrode is vibrated in the field of the electret to produce a signal that is displayed on an oscilloscope.

Kakiuchi [343] improved the induction method by avoiding the frictional charges due to surface contact. He made an apparatus with the upper electrode fixed and the lower electrode, with the sample, movable.

Wieder and Kaufman [344] measured the surface charge of plastic electrets by the induction method. Baumann and Wiseman [345] carried out both noise and charge measurements on the same electret. Kallmann and Rosenberg [346] used the radiation discharge method and the lifted-electrode method to measure the polarization in the electret. Wild and Stranathan [347] used equipment for measuring equivalent surface charge by using automatically operated measuring stations, a condenser, a DC amplifier with electrometer tube input, and a recorder. Each measuring station contained a dissectible condenser, the top plate of which served as a measuring electrode.

c. **Liechtenberg Figures Method.**   Devins and Reynolds [335] discussed two methods for studying the surface charge distribution of electrets. The first involves dusting the surface with dielectric powders which stick to the charged areas, forming the well-known Liechtenberg figures. While this method gives a detailed picture of the position of charges on the surface, it gives little information as to the magnitude of the charges. In their second method, a metallic probe in the form of a flat plate is placed close to the charged surface, and the charge on it is measured by means of an electrometer. This is quite similar to the lifted-electrode method, but the area of the electrode is small and it is allowed to move over the surface so as to scan continuously the charge distribution on the electret surface. Charge movement can be readily observed by this technique as a change in the oscilloscope pattern with time.

d. **Dissectible Capacitor Method.**   Gross [336,337] designed and developed the dissectible capacitor method for determining the effective surface charge density of an electret. This method has been widely used [338–340]. Gross used a lifted electrode, which was suspended on an iron piston and moved in the field of a coil. A special shielding device was used to shield the electret surface every time after lifting the electrode to avoid the influence of the polarized dielectric on the measuring electrode during the act of measuring. A fiber string electrometer was used as a measuring instrument. Heating arrangement was also used for measurements at elevated temperatures.

Gross [337] and Gemant [338] subsequently pointed out that there will be some error in this method due to the interference between the electret surface and the movable electrode. When the electrode is lifted, the interference will be increased. Treating the system as a two-layer capacitor, Gross calculated the electric field (E) in the air gap as

$$E = \frac{\epsilon V - K\sigma}{1 + \epsilon_{xd}/D} \text{ V/cm} \tag{30}$$

where σ is the real surface charge density, ϵ is the dielectric constant, D is the sample thickness, V is the applied voltage, and d is the thickness of the air gap. From this equation one can see that the electric field in the air gap of thickness d varies during movement of the electrode. It also produces some charge due to contact electrification between the metal electrode and the dielectric surface. However, this method shows some improvement over the ordinary induction method. Most of the drawbacks have been largely eliminated through modern electret research. Wiseman and Feaster [339] considered the electret–electrode combination as a three-layer capacitor as shown in Figure 19. The gaps d_1 and d_3 can usually be made negligibly small by molding the dielectric to the lower electrode or carefully machining the surfaces.

e. Continuous Measurement of Surface Charge.   Freedman and Rosenthal [341] developed a method that is capable of continuously measuring and recording electret strength. The apparatus consists of a mechanically driven capacitor in which the electret is mounted. The voltage output of the capacitor is passed on to the associated equipment to record the variation of surface charge continuously with time.

f. Generating Voltmeter Method.   Kojima and Kato [342] measured the surface charge without touching the surface of the electret, by a generating voltmeter. In this method, a grounded vane V roates with a constant speed between the surfaces of the electret E and the electrode A, from which an induced current is drawn out and measured.

g. Capacitive Probe Technique.   The capacitive probe method has been used by a number of workers [348–350] to measure surface charge and also as a "local charge meter" to map charge distributions on an electret surface. This method gives great ac-

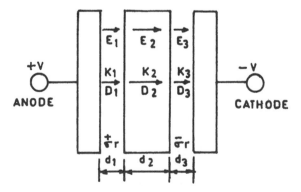

Figure 19   A three-layer capacitor. E_1, E_2, and E_3 are electric fields; K_1, K_2, and K_3 are dielectric constants; D_1, D_2, and D_3 are electric displacements $\pm\sigma_r$ charge densities and d_1, d_2, and d_3 are air gap distances.

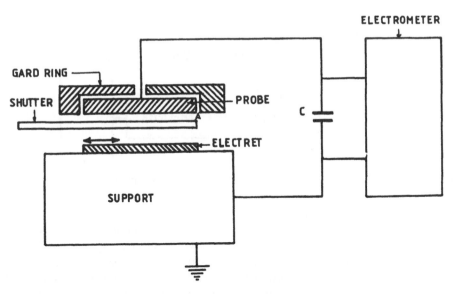

Figure 20 Schematic diagram of a capacitive probe setup for measuring charge densities [350].

curacy. It can be used to determine average charge density as well as charge distribution on the surface of an electret ranging from 5×10^{-11} to 10^{-6} C cm^{-2} (Fig. 20).

h. **Dynamic Capacitor Method.** In the dynamic capacitor method [351], the AC voltage induced on an electrode near an electret is measured. The AC voltage is generated as a result of variations of the electrostatic field of the electret, which are produced by vibrating an electrode in the vicinity of the electret. AC signals can also be produced by periodic shielding of the electrode/electret by means of a rotating shutter. By applying modern signal processing techniques, one can obtain fairly good results using this method.

In addition to the above methods, other techniques based on the charge compensation method [352,353], the thermal pulse method [354,355], polarization and depolarization current methods, etc. [256,257], have also been used by some workers in the field to study the characteristics of electrets.

i. **Thermal Pulse Method.** The thermal pulse method has been successfully used to find the charge centroids in electrets [358–360]. When the sample is irradiated with appropriate light pulses, absorption of the light takes place in the dielectric. As the heat diffuses in the sample it causes a corresponding change in the dielectric properties and consequently the potential on the surface of the electret. By measuring the voltage transient, one can find the charge centroid. Modification of this method by way of periodic heating of the sample electrode with a laser beam has been used by Lang and Das Gupta [361,362]. The nonuniform heating of the polarized sample generates pyroelectric currents which are found to be a function of the charge distribution in the electret.

j. **Pressure Pulse Method.** Another important method developed by Sessler, Secker, Laurenceau, Podgorsak, and others [363–368] to determine the charge distribution in an electret is based on the application of a step-function compressional wave through the poled sample. The charge distribution is obtained from the time derivative of the voltage

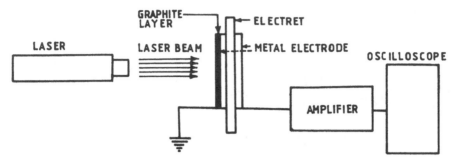

Figure 21 Setup for laser-induced pressure pulse method for charge measurement [367].

in the external circuit during propagation of the pressure step through the electret. This is a nondestructive method. Recently, Sessler and co-workers [367,368] have used laser-induced pressure pulses to explore the charge distribution in poled samples. An experimental setup for a laser-induced pressure pulse method is shown in Figure 21. Improved performance of this technique using shorter laser pulses and better coupling of the light with the electret surface and very high resolutions have been reported.

Pressure steps of the order of 1-ns rise time were obtained by using piezoelectric quartz [369,370] and electrically exciting it. It is possible to display the charge or field distribution directly, an advantage over other methods.

C. AC Method [310,314–316,323]

Study of dielectric parameters by the AC method is one of the earliest and commonly used techniques. The measurement of dielectric loss as a function of temperature at fixed frequencies or vice versa shows maxima at a characteristic relaxation temperature/ frequency of the polymer. The dielectric constant shows dispersion regions corresponding to various types of polarization processes. Glass transition phenomena in several polymers have been investigated by AC dielectric studies [326,327].

AC dielectric results are very often compared with the results of other techniques, such as mechanical measurements. They are also used to study the dielectric dispersion in single crystals of inorganic and organic materials [324,325].

VIII. STRUCTURE–PROPERTY CORRELATIONS

Since electret formation is a volume effect, the molecules/dipoles of the molten polar substances will realign in a regular manner while cooling under the influence of a high electric field. Gemant [371] stated that the normally disordered crystallites attain an orientation in the electric field. Microphotographs by Chandy [372], Nakata [373], and Gutman [374] taken parallel and perpendicular to the polarizing field in the electret, clearly showed the orientation effect. Characteristic spherulites were seen in the microphotographs of sections taken perpendicular to the polarizing field. Molecular orientation has also been studied using X-ray diffraction methods by Ewing [375], Kakiuchi [376], Good and Stranathan [377], Chandy and Bhawalkar [378,379], and Pillai [380]. X-ray diffraction techniques have also been used by many other workers to study the structural properties of waxes and other long-chain compounds [381–384,387].

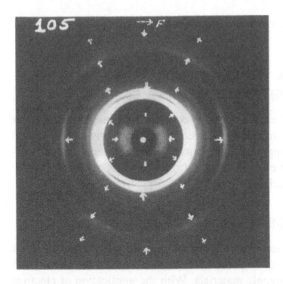

Figure 22 X-ray diffraction patterns of carnaubna wax electrets: X rays perpendicular to the field direction, E = 14 kV/cm, T = 85°C [380].

Kakiuchi [376] studied the X-ray diffraction patterns of a number of long-chain compounds and showed that the microcrystals of these compounds orient themselves in the direction of the electric field. Bennet [381] showed from his experimental investigations that the molecular group state of the liquid accentuates as the liquid solidifies, leading to a definite crystalline structure, which can be strongly oriented in the electric field. Such oriented crystalline structures exhibit different dielectric properties in different directions. Keller [386] has shown, by X-ray diffraction and by polarizing microscopic studies, that crystalline spherulites form and grow in an amorphous mass, according to the temperature of formation.

In polar dielectrics, the dipoles are found to be oriented in the direction of the field when it is kept under an electric stress. This preferred orientation effect has been extensively studied by Pillai [380] by X-ray diffraction techniques in a number of electret-forming materials. A definite structural difference between the poled and unpoled samples has been reported. Some typical X-ray diffraction photographs of carnauba wax thermoelectrets are shown in Figure 22. A very strong diffraction pattern with sufficient intensity for the first five side-spacing reflections could be obtained within 1 h of X-ray exposure. A considerable variation of intensity around the Debye rings has been obtained in photographs taken with the X-ray beam perpendicular to the forming field. On the other hand, variation of intensity is rather negligible in photographs taken with the X-ray beam parallel to the direction of the field. The magnitude of orientation of the crystallites is found to be greater for longer duration of polarization. Similarly, the variation of intensity and consequently the orientation of the crystallites is maximum with respect to the diffraction photographs of electrets prepared at higher temperatures, near the softening temperature of the material. Samples prepared at high temperatures without the application of the field do not show any variation of intensity around the Debye rings.

This establishes the influence of electric field and temperature on the formation of electret.

IX. SUBSTANCES THAT SHOW THE ELECTRET EFFECT
[50,95–98,101–102,110,389–393,517–530]

In the beginning, most investigators of electrets used waxes and wax mixtures and other organic compounds for the preparation and study of electrets. Later, Gemant, Wieder and Kaufman, and others emphasized the importance of preparing electrets using plastic substances. The stability and magnitude of charge obtained from this class of materials have been far superior to other organic materials.

X. POLYMERIC ELECTRETS [98–124]

In recent years, electrets prepared from various types of polymers have gained great scientific, technological, and commercial importance over the electrets prepared from waxes, resins, and other low-melting organic materials. With the application of electrets prepared from Mylar and Teflon in microphones, the study of the polarization properties of polymers has gained added significance. Polymers are preferred for electret preparation due to the facts that:

1. They exhibit very good charge storage and retention capacity, as they have very high concentrations of deep traps, besides being very good insulating materials.
2. They possess superior thermomechanical properties.
3. They can also be processed into thick and thin films or any other desired size and shape.

A. Polarization in Polymers

Polymers are of two types: (a) polar polymers such as poly(vinylidene fluoride) (PVDF) and (b) nonpolar polymers such as poly(ethylene) (PE) and poly(tetrafluoroethylene) (Teflon). When a polymeric material is subjected to an electric field, especially at elevated temperatures, the dipoles will be reoriented. In nonpolar polymers, however, there will be no dipolar alignment. Polymers are very good insulating as well as charge storage materials. When solid polymeric materials are subjected to an electric field, a net dipole moment over the bulk results. In nonpolar polymers, the dipole moment is due to the induced dipoles; whereas in a polar material, it is due to the orientation of the dipoles in the field direction. The total polarization, however, is due to the contributions from atomic, dipolar, and space charge polarizations, as shown in Figure 13. The electronic polarization is due to the displacement of electrons relative to the atomic nuclei. The displacement of electrons relative to the atomic nuclei in a molecule when its constituents are of different types gives rise to atomic polarization. Unequal sharing of electrons between atoms of a molecule may give rise to permanent dipoles. However, a molecule has a net dipole moment only if the polar groups or polar bonds are asymmetric. Orientational polarization results from the alignment of these dipoles in the direction of the field. These three types of polarizations are due to the charge carriers which are locally bound in atoms or molecules or structures of the dielectric. In general, free charge carriers also exist in dielectrics, which are mobile and can migrate at least some distance through the dielectric. In fact, these charge carriers give rise to space charge polarization or interfacial polarization when trapped in the material or at the interface or accumulated near the electrodes. The polarization of nonpolar materials is due mainly to the distortion of the electronic and atomic structures. Dipolar orientation in a nonpolar polymer may arise from the presence of impurities and also from the oxidation at polymer surface. Polar polymers exhibit dipolar orientation. The orientational polarization of these polymers depends very much on their chemical and physical nature.

Since most polymers contain both amorphous and crystalline regions, whose conductivities are different, Maxwell-Wagner-Sillars polarization—interfacial polarization— is also expected to contribute to the polarization of all polymers. In addition, space charge polarization, which is due to the accumulation of charge carriers near the electrodes and trapping of charge carriers at various localized states such as chain foldings, chain ends, etc., can also contribute to the total polarization.

The charge generation and retention properties of polymers also depend very much on their chemical structure as well as their physical nature, which limit their accessibility beyond a certain range. It is possible to modify the properties of polymers by doping, co-polymerization, substitution, or by blending two or more polymers together. These properties, in turn, affect the dielectric and charge formation characteristics which are determined by intra- and intermolecular interactions.

B. Conduction Mechanisms in Polymers [103,125,146–155]

Most polymers are insulators because of their low conductivity and low dielectric loss and they can, therefore, be charged electrically. The storage of charges in polymer electrets depends on a number of material parameters such as crystallinity, cross linking, additivities, irradiation history, stereoregularity, and water absorption [153].

The activation energy of polymers is temperature dependent, and a transition to an increased energy is observed at the T_g of polymers. A low temperature region of low activation energy is attributed to impurity conduction, and a high-temperature region with high activation energy arises due to a higher dissociation energy to form carriers for intrinsic conduction.

Another polymeric transition in conductivity corresponds to the temperature region $T_g < T < T_m$. This is due to decreased lifetimes of the molecular confirmations, favorable to the hopping of carriers and thermal disintegration of local ordered structure in the amorphous content. Since space charge polarization involves migration of charge carriers, it is significant only above T_g. Orientation of polarization can be observed below T_g also, if secondary relaxations involve motion of the polar groups. That is, polymers with flexible polar side chains contribute to dipolar polarization below T_g as well. In polymers in which polar side group is rigidly attached to the main chain itself, dipolar orientation cannot occur without cooperative motion of the main-chain segments. In general, total polarization of a polymer shows a marked increase at T_g, as temperature is varied from below T_g to higher temperatures.

Conduction in polymers is dependent on the charge carrier mobility, carrier concentration, and injunction of charges at the electrodes. In substances capable of electret formation, trapping sites are always present, so mobilities are trap-modulated and in turn affect the conduction processes [146]. The transport of charge within the macromolecule itself and the transport from molecule to molecule are the two main factors which greatly affect conduction in polymers. The intermolecular transport of charge is generally poor, as the interaction between polymer chains is weak. This can be improved with increase

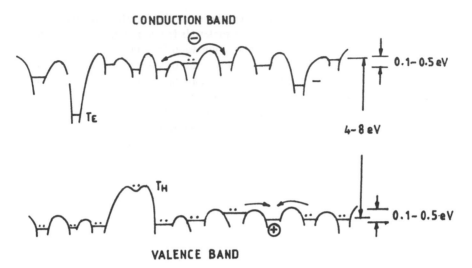

Figure 23   Energy-level diagram for polymers showing carrier transport by hopping, electron (T_E) and hole (T_H) traps [149].

in the degree of crystallinity. Polymeric materials are usually partially crystalline and, therefore, possess only short-range order. This gives rise to narrow conduction bands and low electron mobility [148]. Bauser [149] has given an energy band model (Fig. 23). The valence band and the conduction bands are narrow and may even split into individual energy levels spaced closely together. The energy differences are due to differences in the local environment of each molecule. Under such conditions, conduction may take place by hopping of electrons from one level to another. In addition, electrons can also make intermolecular jumps to molecules on neighboring chains. The spread in the conduction and valence levels is small compared to the width of the forbidden gap. Thus, a band scheme may be considered in polymers with actual transport of charges due to hopping.

Since the majority of polymers are semicrystalline, their electrical conduction mechanism is more complex than in inorganic ordered materials. Conduction in polymers may also depend on sample preparation and history. As a consequence, different conduction mechanisms have been reported for the same material [103,125,146,147]. The conductivity of polymers varies exponentially with temperatures and can be expressed as

$$\sigma = \sigma_0 \exp\left(-\frac{E}{kT}\right) \tag{31}$$

where

  E = activation energy
  k = Boltzmann's constant
  T = temperature

The conduction in polymers is normally explained by (a) intrinsic charge carrier generation and (b) charge carrier injunction from contacts at high fields.

Carrier mobility obeys the relation [129–132]

$$\mu = \mu_0 \exp\left(-\frac{E_\mu}{kT}\right) \tag{32}$$

where E_μ is the activation energy for charge carrier mobility. The carrier mobilities in polymers, both for electron and hole, are experimentally observed to be temperature activated, establishing the validity of the hopping model [129–132]. The magnitude of the carrier mobility generally lies in the range 10^{-8}–10^{-4} $m^2V^{-1}S^{-1}$. Low carrier mobility usually results from the absence of any long-range order in polymers. The carriers, therefore, cannot move over long distances without being captured or scattered. Steady-state conduction phenomena have been studied extensively in polymers by a number of workers.

The density of intrinsic carriers in polymers is very low at room temperature. The conduction mechanism in polymers cannot be explained adequately on the basis of band theory, as polymers are weakly bonded materials and the overlap of molecular orbitals is very small, and also due to the fact that they are heterogeneous in nature and their structure is nonperiodic. Therefore, their energy bands are very narrow and there is disorder in the energy levels, which produces localized states for the trapping of charge carriers. In addition, the localized states can also originate from chain foldings, chain ends, bulk and surface dipole states, crystalline–amorphous interfaces, and impurities present in the material [103].

The detrapping of charge crariers takes place from the traps depending on the trap depth and on the energy available for detrapping. It may be purely thermal and depends on the molecular motion and molecular environment. The local field may assist in the detrapping process. Because of the detrapping and subsequent trapping of charge carriers, the time required for charge transport is very high. The number of charge carriers that contribute to the conduction depends on the number of trapping sites between the conduction and valence bands and also on the trap levels.

C. Charge Carrier Injunction

The charge injunction or emission from the forming electrodes to the dielectric was established several years ago. Many investigators have invoked different types of mechanisms, such as (a) Schottky-Richardson emission and Poole-Frenkel effect, (b) space charge-limited current, (c) tunneling current, (d) ionic conduction, and (e) impurity conduction to explain the conduction mechanisms in polymers. Since these mechanisms are important in understanding the charge generation, trapping and transport of carriers in polymer electrets, brief explanation of these are given in the following sections.

D. Schottky-Richardson Emission and Poole-Frenkel Effect

In the presence of an applied electric field, the barrier profiles are altered. If the applied field is high enough and the electrode makes ohmic contact with the dielectric, then the charge carriers are injected into the insulator by lowering the barrier of metal–insulator interface. This effect is known as Schottky-Richardson (S-R) emission. The expression for the current due to S-R emission is given by

$$I = AST^2 \exp\left(-\frac{\phi_S}{kT} + \beta_{SR}V^{1/2}\right) \tag{33}$$

where β_{SR} is the Schottky field lowering constant

$$\beta_{SR} = \frac{1}{kT} \left(\frac{e^3}{4\pi\epsilon\epsilon_0 d} \right)^{1/2} \tag{34}$$

where e is electronic charge, ϵ_0 is the permittivity of free space, ϵ' is the high-frequency dielectric constant of the material, ϕ_S is the Schottky potential barrier, V is the applied field and S is the surface area of the sample.

The first term of Eq. (33) represents the Richardson equation,

$$I = AST^2 \exp\left(-\frac{\phi_S}{kT} \right) \tag{35}$$

It predicts a linear relationship between $\ln(I/T^2)$ and I/T whose slope yields the Schottky potential barrier ϕ_S. Equation (34), at a given temperature, predicts a linear relationship between $\ln I$ and $V^{1/2}$ with a slope β_{SR} which should vary inversely with temperature. If the charge carrier is trapped in a Coulombic potential well, then the charge carrier can be detrapped by lowering the trap depth upon the application of an electric field.

The current resulting from the detrapping of the carriers from these traps also has the same linear dependence on the applied voltage as the Schottky-Richardson emission. This effect is known as the Poole-Frenkel effect. The expression for this effect is given by

$$I = AST^2 \exp\left(-\frac{\phi_{PF}}{kT} + \beta_{PF}V^{1/2} \right) \tag{36}$$

where β_{PF} is field lowering constant and ϕ_{PF} is the trap depth. For a simple process,

$$\beta_{PF} = 2\beta_{SR} \tag{37}$$

According to some workers [142–145], β_{PF} becomes comparable to β_{SR}. The distinction between the two processes may be made by using different electrodes because S-R emission is electrode dependent whereas the P-F effect is not [155]. These two effects are important for understanding the electret effect in pure and doped polymeric systems [136–142].

E.   Space Charge-Limited Current (SCLC) [156–158]

If the electrode–insulator (polymer) contact is ohmic and the insulator is trap free, the accumulation of carriers near the electrode results in a space charge buildup. Mutual repulsion between individual charges limits the total charge injected into the sample, and the resulting current is said to be space charge-limited current (SCLC).

A complete mathematical analysis of the time-dependent space charge current is so complex that satisfactory explanation has not yet been obtained. Mott and Gurney were the first to emphasize the importance of an injecting contact between the metal and the insulator [156]. They provided an expression relating the current density J and applied voltage V, for a trap-free insulator, as follows:

$$J = \frac{9}{8} \epsilon_0 \epsilon \mu \cdot \frac{V^2}{d^3} \tag{38}$$

where ϵ_0 is the permittivity of free space, ϵ is the dielectric constant of the sample, μ is the dipole moment, d is the thickness of the sample, and V is the applied voltage.

If there are traps in the insulator, the space charge-limited current may be decreased by several orders. Rose [157] and Lamberts modified the theory and Rose has given the following equation for the current density:

$$J = \frac{9}{8} \, \epsilon_0 \epsilon \mu \theta \, \frac{V^2}{d^3} \tag{39}$$

where θ is the trap limiting factor, which is the ratio of the trapped charge to free charge.

Assuming that the carriers are trapped at shallow traps of average depth E_T, then θ is given by

$$\theta = \frac{n_{\text{eff}} \, \exp(-E_T/KT)}{N + n_{\text{eff}} \, \exp(-E_T/KT)} \tag{40}$$

where n_{eff} is the density of states in the conduction band and N is the density of traps.

Assuming that $N \gg n_{\text{eff}} \, \exp(-E_T/KT)$, then θ becomes

$$\theta = \frac{n_{\text{eff}} \, \exp(-E_T/KT)}{N} \tag{41}$$

Substituting for θ, the equation for J becomes

$$J = \frac{9}{8} \, n_{\text{eff}} \, \exp\left(- \frac{E_T}{KT}\right) \frac{\epsilon_0 \epsilon \mu}{N} \cdot \frac{V^2}{d^3} \tag{42}$$

The above equation suggests that SCLC in solids depends on the carrier transport and trapping, and is independent of carrier generation. Since only very few organic solids show ohmic conduction, the conduction mechanism in organic solids is mainly extrinsic in nature. The influence of electrode material and temperature has been the subject of many studies [159–161].

F. Tunneling Current

If the electrode contact is ohmic, the applied field is very high, and the thickness of the dielectric film is less than 100 Å, then the width of the potential barrier decreases to such an extent that the probability of finding an electron on the other side of the potential barrier by quantum mechanical tunneling increases as given by the Fowler-Nordheim equation [162]:

$$I = ASV^2 \, \exp\left(- \frac{\phi}{V}\right) \tag{43}$$

which describes the tunneling current for a sample case and predicts a linear relationship between $\ln(I/V^2)$ and $(1/V)$. The tunneling current is temperature independent.

G. Ionic Conduction

Ionic conduction occurs in polymers which contain ionic groups or to which ionic materials have been added. In these materials, absorption of water plays a dominant role because water may act as a source of ions, as a dielectric impurity, or as a local structure modifier. In amorphous polymers, it can also occur due to the drift of defects on application of an electric field. In various polymers, particularly polymers with halogens in their molecular structure, as in poly(vinylchloride), poly(vinyl fluoride), and poly(trifluoroethylene), electrical conduction has been found to be ionic [163,164]. Ionic conduction is characterized by its high activation energy in comparison to electronic activation energy and noticeable polarization effect under the action of a DC field. In addition, large transit time for ions is also observed. Activation energies are also found to be larger at higher temperatures than at lower temperatures [165].

H. Impurity Conduction [166]

The creation of electrons or holes in polymers can be achieved by introducing electron-acceptor or electron-donor materials into the polymer matrix to form charge-transfer complexes. The electrical properties can be controlled in a way similar to that for doped semiconductors with n- or p-type impurities.

Table 1 shows some of the important polymeric materials that have been used to study the electret effect and their applications in devices [110].

XI. NONLINEAR POLYMERS BY ELECTRET FORMATION

Polymers are the subject of extensive research and development for nonlinear optical applications because of their ability to tailor molecular structures which have inherently fast response times and large second- and third-order molecular susceptibilities. Polymers have excellent mechanical properties, environmental resistance, high laser damage thresholds, and excellent processing options. The nonlinearity in polymers and other organic materials can be considerably enhanced by electrically poling them. Such poled organic polymers—electrets—play a major role in optical signal processing, particularly as spacial light modulators, and have also found application in neural networks. Those polymers which show third-order nonlinear optical susceptibility ($\chi^{(3)}$) are found to be excellent candidates in optical signal processing systems as well as in tunable filters, degenerate four-wave mixing, phase conjugation, and sensors [445,446].

The conjugated organic polymers show largest $\chi^{(3)}$ values with response time in femtoseconds. The nonlinear properties of organic polymers are investigated extensively because of their subpicosecond response times, large nonresonant nonlinearities, low DC dielectric constants, low absorption, low switching energy, ease of processing and synthesis, room-temperature operation, good stability, and mechanical and structural integrity.

In polymers, control of orientation and symmetry is obtained by poling the polymer. The poled polymers show second-order nonlinear optical susceptibilities and electrooptical coefficients larger than well-known inorganic crystals such as lithium niobate. Spin-coating of poly(p-hydroxy styrene) with NLO chromophores on ITO-coated conducting glass substrate, followed by drying and poling above T_g, yields very good transparent films with high second-harmonic generation efficiencies [445].

Recently it has been found that nonlinear optical susceptibility can be considerably enhanced by chemical or electrochemical doping. In certain cases, doping leads to improved transparency in the visible region of the electromagnetic spectrum [448]. Griffin and Prasad have fabricated electrically poled Langmuir-Blodgett film from a co-polymer of $(NBSBV)_n$-$(C^x SBV)_m$ and studied its nonlinear optical properties. Polymer thin-film devices are found to be useful in integrated optics. Lytel [449,450] demonstrated the use of selective electrical poling of thin polymer films, in which active, poled channel wave

Table 1 Some Important Polymeric Materials Used for Electret Studies

Name of polymer	Abbreviation	References
Polymethyl methacrylate	PMMA	117,401,425,430,435,438
Polyvinyl acetate	PVAc	117,397
Polycarbonate	PC	113,118,420,433
Cellophane		274
Poly(vinylchloride)	PVC	271,396,426,429
Cellulose acetate		173
Poly(N-vinyl carbazole)	PVK	399,423
Polyethylene	PE	395,412,413,416,420,428
Poly(vinyl alcohol)	PVA	181
Polyacrylonitrile		172,399
Polystyrene	PS	119,400,407,411,420,421,425,438,443
Polyvinyl fluoride	PVF	402
Polyvinylidene fluoride	PVDF	403,409,410,413,417,431,439
Polypropylene	PP	118,404,419,420
Polytetrafluoroethylene (Teflon)	PTFE	405,418
Polyethylene terepthalate (Mylar)	PET	406,408,420,424
Polytrifluoroethylene		414
Nylon-6		415
Kapton		420,422
Plexiglass		420
Fibreglass		420
Poly(vinyl cinnamate)	PVCn	423
Polyimide		434
Polyacrylonitrile butadiene—styrene	ABS	438
Ethylcellulose	EC	438,440

guides are fabricated which are reported to have achieved 1.3-GHz modulation. The second-order nonlinear optical polymers are found to be useful in implementing neural networks for optical computing [451]. Optical implementation is particularly attractive, as optical devices are inherently parallel. Both second- and third-order nonlinear optical polymeric systems are prospective candidates for optical implementation of neural net architectures [452].

For optical processing based on the principle of light control by light, third-order nonlinear optical processes provide the main operations of optical logic, optical switching, optical memory storage, optical bistability, phase conjunction, photorefractivity, and third-harmonic generation. Molecular and polymeric materials have become an important class of material for nonlinear optical applications because of their excellent properties. Polymeric structures have the additional advantage of incorporating sturctural modifications not only in the main chain but also in the side chain. In addition, these materials offer flexibility of fabrication in various forms. A number of polymeric materials, such as poly-*p*-phenylene vinylene (PPV), polydiacetylene, polythiophene, polyacetylene, polysilane, and polygermane have been investigated for their nonlinear optical properties. Interesting and important results have been reported on oriented polymers. Oriented polydiacetylene gave the largest value of $\chi^{(3)}$ along the polymer chain. Uniaxially and biaxially oriented films of PPV showed the largest nonlinearity [453–457].

In poled polymers, the noncentrosymmetric alignment of NLO-active chromophores is achieved by electrically poling the material at the glass–rubber transition temperature. The alignment under these conditions is locked in by cooling the sample with the field still applied. The linear electrooptical coefficient for a poled polymer is given by [458,459]

$$\gamma \propto (1\mu \cdot \underline{\beta}/E_p)/5kT \tag{44}$$

where μ is the dipole moment, β is the first hyperpolarizability, E_p is the electric field used for poling, k is Boltzmann's constant, and T is the temperature during poling. $\mu \cdot \underline{\beta}$ is the scalar product of the vector μ and the tensor $\overline{\beta}$. From the measurement of LEO coefficient and μ, the vector part of β can be determined for molecules in a poled polymer sample.

Le Grange and co-workers have given a general discussion of the relation of three-dimensional order and the $\chi^{(2)}$ phenomenon. These concepts are applied to various crystals, liquid-crystal side-chain polymers, and Langmuir-Blodgett films [460–463]. The nonlinear properties of poly(vinylidene fluoride) and its copolymers with poly(trifluoroethylene) have been enhanced considerably by poling them [464]. Similar results have also been reported in alternating co-polymers of vinyl acetate and vinylidene cyanide [465]. Eich and co-workers at IBM have demonstrated second-harmonic generation (SHG) in corona-poled amorphous glasses containing p-nitroaniline side groups on a poly(ethylene) backbone [466]. Poled pure organic glass-forming materials also exhibit second-harmonic generation [467]. However, it has been reported that when the poling field is removed, the relaxation and mobility processes of the polymer release the dopants from its field-induced orientation, which in turn reduces the SHG intensity. Therefore, polymer mobility and the free-volume behavior and the effect of charge decay with time must be considered simultaneously when examining the temporal stability of poled polymer films. Doped PS and PMMA films poled at high temperature have been used for NLO studies by a number of workers [468–472]. The formation of single-mode passive wave guide structures have been demonstrated in poled PMMA films containing photochemically active nitrones [474]. Noncentrosymmetry required for second-order nonlinear optical effects has been achieved by photopolymerizing an acrylic matrix while simulataneously orienting NLO-active chromophores with a DC field. The intensity of all the poled and cross-linked films was unchanged for several days at room temperature, but gradually decayed to zero over many months due to the low T_g of the materials [474]. Therefore, it would be necessary to use permanently poled polymers for NLO applications.

XII. BIOPOLYMERS AND BIOELECTRETS [509–530]

Some biological materials and biopolymers are found to exhibit the polar uniaxial orientation of molecular dipoles in their structure and can be considered as bioelectret. Such materials show pyroelectricity and piezoelectricity. Mascarenhas [509] investigated TSDC properties after poling of a number of biological substances including collagen, gelatin, polysaccharides, polynucleotides (DNA, RNA), enzymes, and bound water in biopolymers. The electret state has been considered in various biophysical models as a basis for the understanding of membranes, neural signals, biological memory regeneration, electrical mediation in tissue growth, image formation in the retina, and other phenomena. Biocompatible polymeric materials such as Teflon, Dacron, etc., are now used extensively after proper polarization treatment for biomedical applications such as antithrombogenic surfaces and artificial membranes.

Fukada and co-workers [510,517–519,522] observed pyroelectricity and piezoelectricity in various kinds of uniaxially oriented biopolymers. Lang [521] also reported pyroelectricity of bone and tendon. The electrical stimulation of repair and growth of bone has attracted a great deal of interest in the field of orthopedics. Pyro- and piezoelectric studies in various types of biological systems showed the presence of natural polarity in the structure of various parts of animals and plants [520]. Bennis et al. [511] have shown that TSC technique is particularly adaptable for studying the characteristics of bioelectrets of apatites and similar biological materials.

Mascarenhas [509,513], after extensive investigations of the electret properties of various biological materials, indicated the possibilities for charge storage in bioelectrets and TSD of bioelectrets for the diagnosis of diseases. Reichle et al. [525], Pillai and Goel [523], Ledwith et al. [526], and Chatain et al. [524] have carried out TSD studies on biological materials including hemoglobin, β-carotene, dibenz(b,f)azepine derivatives, and poly-amino acid derivatives. After studying the TSDC spectra of gelatin, collagen, bound water, alpha-chitin, etc., Mascarenhas showed that TSDC spectra have proved to be a very sensitive tool for the characterization of bioelectret materials. Forhlich [527] proposed the use of quantum mechanics to study enzyme action. It was demonstrated by Ficat et al. [528,529] that biocompatible piezoelectric PVDF film electret implants have been shown to be effective with respect to stimulation of bone growth. Use of electret bandages has found to accelerate the wound-healing process, and charged Teflon has been found to be useful as a core-filling material in endodontics [530]. These studies reported by various workers thus show the growing importance of the use of bioelectrets in the biomedical field. However, further extensive research investigations are necessary for its full exploitation.

XIII. ELECTRETS OF POLYMER BLENDS
[113,117,118,436,437,522,531–542]

Polymer blends have attracted a great deal of attention in recent years. Most of the studies reported on these class of materials are based on their mechanical properties. Although persistent polarization and relaxation properties of various types of individual polymers have been studied by several researchers, not much work has been reported on the electrical and electret-forming properties of polymer blends.

Polymer blends are simple physical mixtures of the constituent polymers with no covalent bonds occurring between them. These blends may be homogeneous (single-phase) solid solutions or heterogeneous (multiphase) mixtures. In the last few years, this field has also attracted the attention of materials researchers, because of the possibility of modifying the electrical properties to a specific requirement by blending two or more polymers. In addition, blending involves only physical mixing of polymers, compared to the laborious chemical process involved in the formation of new long-chain molecules. Properties of polyblends vary considerably depending on the nature and miscibility of their constituents.

Recently, a few workers have reported the electret-forming properties of polymer blends (Table 2). Blends of poly(vinylidene fluoride) and poly(methylmethacrylate) offer a combination of toughness, self-extinguishing structural behavior, and resistance to chemicals along with very useful pyro-, piezo-, and dielectric properties [536–539]. Pillai et al. [113,539] studied the properties of electrets prepared from blends of poly(propylene) and poly(carbonate) and observed synergism in the total polarization. They have conducted thermally stimulated discharge current studies on compression-molded thin films of polypropylene-polycarbonate blends (of various weight ratios) and reported that polycarbonate dipoles are so constrained in the blends that they do not contribute much to the total polarization. They have also studied the effects of variation of field and temperature on polarization in different blend ratios. TSDC studies on the blends of poly(vinyl carbazole) and poly(carbonate) have shown that these blends can be used successfully for electrophotographic applications. Sekar et al. [117,531,534] have shown that polyblends of PMMA and PVAc gives a Schottky-Richardson type of conduction mechanism. Blending of these polymers has resulted in an increase in the electrical conductivity. The activation energy is initially decreased with decreasing PMMA content in the polyblends. They observed from the TSDC studies that the polarization is due to induced dipolar orientation resulting from the deep trapping of charge carriers during polarization. They also reported that the dipolar polarization is relatively small toward the total polarization. The peak temperature T_m of the TSDC spectra of the polyblends shifts toward the lower-temperature side with decreasing PMMA content. This result is in agreement with the results of electrical conductivity studies of the polyblends which have shown an increase in electrical conductivity and a decrease in activation energy with decreasing PMMA content in polyblends.

Figure 24 shows a typical comparison plot of polarization for given polarizing conditions for various weight ratios of PMMA:PVAc polyblends. It shows that polyblend electrets prepared from a 50:50 weight proportion of PMMA:PVAc yields maximum polarization. This may be due to the increased trapping of charges due to the increase in the trapping sites created by the polymer blend phase boundaries because of maximum heterogeneity in this mixed system. The activation energy of this composition is also found to be high when compared to other compositions. The peak temperature T_m is found to shift toward the lower-temperature side with decreasing PMMA content in the blends, due to the plasticization effect. The influence of poling temperature and poling field on the TSDC spectra of polyblends suggests that the observed polarization is due to the induced dipolar orientation resulting from the deep trapping of charge carriers [117]. The electrical conduction in this polyblend has also been studied as a function of voltage and temperature with three different electrode materials, and it has been found that the Schottky-Richardson emission type of conduction mechanism is found to be dominant in this polyblend. However, some of the injected charge carriers are trapped in the shallow trapping sites present in the polyblend, and their release is controlled by the Poole-Frenkel conduction mechanism, especially above 70°C [537].

In Figure 25, the effect of poling field E_p on the TSDC spectra of PMMA:PVAc polyblend prepared with 10:90 weight ratio is given. The variation of the α peak is proportional to E_p. However, the slope changes for the ρ peak at about 100 KV/cm field strength (Figure 24, inset).

Thermoelectret state in polyvinylchloride:polyethelene (50:50) polyblend has been investigated by Tripathi et al. [533], and the observed results are discussed in terms of

Table 2 Important Copolymers and Polymeric Blends Used for Electret Studies

Name of polymer	References
Celluloseacetate-polyvinyl acetate	436,437
Polycarbonate-polypropylene	238,239,273,312,333,542
Poly(vinyl carbazole):poly(carbonate)	113,539
Polymethyl methacrylate:polyvinyl acetate	117,531,534
Polyvinylchloride:polyethelene	533
Polyvinyliden fluoride:polymethylmethacrylate	532

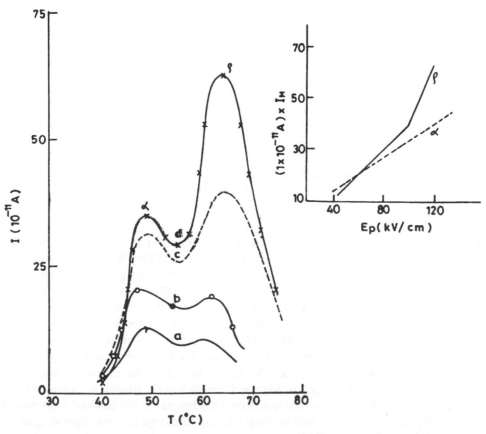

Figure 24 Comparison of TSDC spectra in different polyblends of PMMA:PVAc electrets, T_p = 80°C, E_p = 100 kV/cm: (a) PMMA:PVAc in the ratio 10:90 by weight; (b) PMMA:PVAc in the ratio 80:20 by weight; (c) PMMA:PVAc in the ratio 20:80 by weight; (d) PMMA:PVAc in the ratio 50:50 by weight [117,534]. *Inset*: Variation of α and ρ peaks with E_p [117,534].

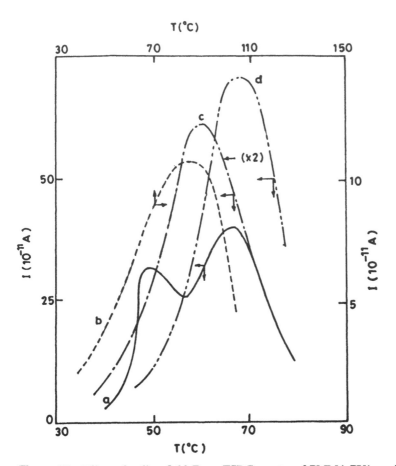

Figure 25 Effect of poling field E_p on TSDC spectra of PMMA:PVAc polyblend electrets (10: 90 weight ratio): (a) 40 kV/cm; (b) 60 kV/cm; (c) 100 kV/cm; (d) 120 kV/cm [117,534].

space charge that leads to induced dipole formation. Narula et al. [532] studied the effect of substrate on the heat of fusion and melting behavior of PVDF:PMMA polyblend. They observed that the substrate on which the polyblend samples are grown affects the morphology of PVDF in the polyblend and also has marked influence on the melting behavior and heat of fusion values. Pillai et al. have investigated the absorption currents in the polyblend of polycarbonate (PC) and polypropylene (PP) mixed in the weight ratio of 50:50, in the temperature range 50–170°C. They observed a relaxation peak in both the isochronal current and dielectric loss graphs. They also evaluated the activation energy of the relaxation process from the Arrhenius shift of the relaxation peak [542]. They have also carried out polarization and depolarization studies in polypropylene: polycarbonate polyblends. They explained the polarization on the basis of formation of induced dipoles which are formed by deep trapping of charge carriers originating in the polyblend during formation [192].

The above results on polyblend electrets suggest that by blending polymer components appropriately, one can tune electrical properties such as conductivity, dielectric parameters, total polarization obtained by electret formation, etc., for a desired application. However, further extensive research work is needed to explore the possibility of using these types of electrets for practical applications.

REFERENCES

1. M. Eguchi, *Phil. Mag. 49*:178 (1925).
2. M. Eguchi, *Jpn. J. Phys. 1*:10 (1922).
3. F. Gutmann, *Rev. Mod. Phys. 20*:457 (1948).
4. A. Gemant, *Phil. Mag. 20*:929 (1935).
5. P. K. C. Pillai, K. Jain, and V. K. Jain, *Phys. Sat. Solid. (a) 3*:341 (1972).
6. *JMS Rev. Macromol. Chem. Phys. C 31*(4):353 (1991).
7. F. Gutmann, *Rev. Mod. Phys. 20*:457 (1948).
8. G. M. Sessler, *Topics in Applied Physics—Electrets*, Springer-Verlag, Berlin, 1980, chaps. 1 and 2.
9. A. Gemant, *Phil. Mag. 20*(136):929 (1935).
10. M. M. Perlman, *J. ElectroChem. Soc. 6*(3–4):95 (1986).
11. B. Gross, *Charge Storage in Dielectrics*, Elsevier, Amsterdam, 1964.
12. V. A. Johnsons, Electrets, Pt. 1, A State of the Art Survey, Pt. II, A Bibliography, U.S. Government Res. Rep. TR-1045 and TR-1174, Harry Diamond Laboratories, Washington, DC, 1962.
13. G. M. Sessler, *Electrical Properties of Polymers*, Academic Press, New York, 1982.
14. M. Eguchi, *Proc. Phys. Math. Soc. Jpn. 1*:326 (1919).
15. J. Valasek, *Phys. Rev. 15*:537 (1920).
16. E. P. Adams, *J. Franklin Inst. 204*:469 (1927).
17. G. M. Sessler and J. E. West, *Appl. Phys. Lett. 17*:507 (1970).
18. P. K. C. Pillai and V. K. Jain, *Phys. Stat. Solid. 28*:415 (1968); *J. Appl. Phys. 40*:3487 (1969); *J. Phys. D. (Appl. Phys.) 3*:1 (1970).
19. P. K. C. Pillai, K. G. Balakrishnan, and V. K. Jain, *J. Appl. Phys. 42*:525 (1971).
20. P. K. C. Pillai, K. Jian, and V. K. Jain, *J. Electrochem. Soc. 118*:1675 (1971).
21. G. Groetzinger and H. Kretsch, *Z. Phys. 103*:337 (1936).
22. W. M. Good and J. D. Stranathan, *Phys. Rev. 56*:810 (1939).
23. N. P. Baumann and G. G. Wiseman, *J. Appl. Phys. 25*:1391 (1954).
24. H. Gelin and R. Stubbs, *J. Chem. Phys. 42*:967 (1965).
25. S. Mascarenhas and C. Arguello, *J. Electrochem. Soc. 115*:386 (1968).
26. G. K. Narula, Ph.D. thesis, Indian Institute of Technology, New Delhi, 1984.
27. H. Jaffe, *Ind. Eng. Chem. 2*:264 (1950).
28. H. Jaffe, *Phys. Rev. 73*:1261 (1948).
29. P. K. C. Pillai and M. Goel, *Phys. Stat. Solid. (a) 6*:9 (1971).
30. P. K. C. Pillai and M. Goel, *Polymer 16*:5 (1975).
31. G. Nadzhakov, *Phys. Z. 39*:226 (1938).
32. G. Nadzhakov, *Chem. Rev. 204*:1865 (1937).
33. P. K. C. Pillai, K. G. Balakrishnan, and V. K. Jain, *J. Appl. Phys. 42*:525 (1971).
34. P. K. C. Pillai and K. G. Balakrishnan, *Il Nuovo Cimento* 15B:2, 11 Giugno (1973).
35. H. Kallmann and B. Rosenberg, *Phys. Rev. 97*:1596 (1955).
36. H. Kallmann, J. R. Freeman, and M. Silver, *Rev. Mod. Phys. 33*:553 (1961).
37. V. M. Fridkin and I. S. Zheludev, *Photoelectrets and Electrophotographic Process*, Consultant Bureau Enterprises, Inc., New York, 1961, p. 140.
38. M. L. Chetkarov, *Sov. Phys. Solid State 3*:1594 (1962).
39. E. I. Adirovich, *Sov. Phys. Dokl. 6*:335 (1961).
40. N. A. Mankov, *Zurn. Eksp. Teor. Fiz. 6*:1031 (1936).
41. P. S. Tartakoviaskii, *Zurn. Eksp. Teor. Fiz. 10*:1025 (1940).
42. H. P. Kallmann and J. Rennert, *Electronics 32*:39 (1959).

43. H. Kallmann and J. R. Freeman, *Phys. Rev. 109*:1506 (1958).
44. V. M. Fridkin, *Sov. Phys. Dokl. 2*:552 (1957).
45. V. M. Fridkin, *Sov. Phys. Dokl. 4*:985 (1960).
46. B. Rosenberg, *J. Opt. Soc. Am. 48*:581 (1958).
47. B. Rosenberg, *J. Chem. Phys. 37*:61 (1961).
48. D. Chapman, R. J. Cherry, and A. Morrison, *Proc. R. Soc. (A)301*:173 (1967).
49. P. K. C. Pillai, K. G. Balakrishnan, and B. D. Sharma, *Indian J. Pure Appl. Phys. 9*:803 (1971).
50. M. Goel, Ph.D. thesis, Indian Institute of Technology, New Delhi, 1972.
51. C. S. Bhatnager, *Indian J. Pure Appl. Phys. 2*:331 (1964).
52. P. K. C. Pillai, K. Jian, and V. K. Jain, *J. Electrochem. Soc. Solid State Sci. 118*:10, 1675 (1971).
53. W. McMahon, *J. Am. Chem. Soc. 78*:3290 (1956); *J. Am. Chem. Soc. 113*:10, 1675 (1971).
54. P. W. Selwood, *Magnetochemistry*, Interscience, New York, 1956, chap. VII.
55. M. L. Khare and C. S. Bhatnager, *Indian J. Pure Appl. Phys. 7*:160 (1969).
56. P. K. C. Pillai and V. K. Jain, *J. Appl. Phys. 40*:3487 (1969).
57. P. K. C. Pillai and V. K. Jain, *J. Phys. D. 3*:829 (1970); *J. Phys. D. 116*:836 (1969).
58. S. Chandrasekhar, Ph.D. thesis, Indian Institute of Technology, New Delhi, 1979.
59. P. K. C. Pillai, N. Shroff, and A. K. Tripathi, *J. Electrostat. 17*:269 (1985).
60. S. K. Agarwal, Ph.D. thesis, Indian Institute of Technology, New Delhi, 1977.
61. P. K. C. Pillai and Navneet, *J. Electrostat. 15*:31 (1984).
62. P. K. C. Pillai and E. L. Shriver, NASA Tech. Rep. TR-R-457, 1975.
63. P. K. C. Pillai, C. S. Su, and E. L. Shriver, *Environmental Lett. 7*(3):261 (1974).
64. P. K. C. Pillai and E. L. Shriver, Paper No. 56, *Proc. 147th Meeting of the Electro-Chemical Society*, Toronto, May 11–16, 1975.
65. A. N. Gubkin and G. I. Skanavi, *Zheksperteor. Fiz. 32*(1):140 (1957).
66. T. A. Dickinson, *Ceramic Ind.*, p. 62 (1949).
67. T. A. Dickinson, *Ceramic Age 57*(3):17 (1951).
68. E. Schleicher, *Exp. Tech. Phys. 7*(4):168 (1959).
69. P. V. Murphy, *J. Chem. Phys. 38*:2400 (1963); *J. Chem. Solids 24*:329 (1963).
70. B. Gross, *Phys. Rev. 107*:368 (1957).
71. B. Gross, G. M. Sessler, and J. E. West, *J. Appl. Phys. 47*:968 (1976).
72. R. Andrecichin and B. T. Kolomiets, *Sov. Phys. Solid State 4*:598 (1962).
73. P. V. Murphy, S. Costa Ribeira, F. Milanez, and R. J. de Moraes, *J. Chem. Phys. 38*:2400 (1963).
74. R. C. Alger, Radiation effects in polymers, *Physics and Chemistry of the Organic Solid State, Vol. 2* (D. For, M. M. Labers, and A. Weissberger, eds.), Interscience, New York, 1965, chap. 9.
75. A. Charlly, Radiation effects in polymers, *Polymer Science, Vol. 2* (A. D. Jenkins, ed.), North-Holland, Amsterdam, 1972, chap. 23.
76. E. B. Podgorsak and P. R. Morem, *Phys. Rev. B8*:3405 (1973).
77. E. B. Podgorsak and P. R. Morem, *Appl. Phys. Lett. 24*:580 (1974).
78. B. Gross and R. J. de Moraes, *Phys. Rev. 126*:930 (1962).
79. G. M. Sessler and J. E. West, *Appl. Phys. Lett. 17*:509 (1970).
80. B. Gross, G. M. Sessler, and J. E. West, *J. Appl. Phys. 45*:2841 (1974).
81. B. Gross, *Phys. Rev. 107*:368 (1957).
82. G. M. Sessler and J. E. West, *J. Polymer Sci. Polymer Lett. Ed. 7*:367 (1969).
83. B. Gross and K. A. Wright, *Phys. Rev. 114*:725 (1959).
84. J. G. Trump, K. A. Wright, and A. M. Clarke, *J. Appl. Phys. 21*:345 (1950).
85. S. Matsuoka, H. Sunaga, R. Tanaka, H. Hagiwara, and K. Asaki, *IEEE Trans. NS-23*:1447 (1976).
86. D. A. Berkley, *J. Appl. Phys. 50*:3447 (1979).

87. A. Gemant, *Direct Current*, p. 145 (1953).
88. E. P. Adams, *J. Franklin Inst. 204*:469 (1927).
89. P. K. C. Pillai, K. Jain, and V. K. Jain, *Il Nuovo Cimento 11B*(2) (1972).
90. J. A. Eldridge, *Am. J. Phys. 16*(6):327 (1948).
91. H. H. Wieder and S. Kaufman, *J. Appl. Phys. 24*(2):156 (1953).
92. J. R. Beeler, Jr., J. D. Stranathan, and G. G. Wiseman, *J. Chem. Phys. 32*(2):442 (1960).
93. G. E. Sheppard and J. D. Stranathan, *Phys. Rev. 60*:360 (1941).
94. M. L. Miller and J. R. Murray, *J. Polymer Sci.*, Pt. 2-A, 4, 703 (1966).
95. A. Gemant, *Direct Current*, 145–147 (1953).
96. G. G. Wiseman and G. R. Feaster, *J. Chem. Phys. 26*(3):521 (1957).
97. N. P. Baumann and G. G. Wiseman, *J. Appl. Phys. 25*:1391 (1954).
98. H. H. Wieder and S. Kaufman, *J. Appl. Phys. 24*(2):156 (1953).
99. F. Binder, *Z. Naturforsch. 6-a*:714 (1951).
100. G. M. Sessler and J. E. West, *J. Accoust. Soc. Am. 40*:1433 (1966).
101. G. M. Sessler (ed.), *Electrets—Topics in Applied Phys.*, Vol. 3, Springer-Verlag, Berlin, (1980).
102. G. M. Sessler, Polymeric electrets, *Electrical Properties of Polymers* (D. A. Seanor, ed.), Academic Press, New York, 1982, chap. 6.
103. D. A. Seanor, Electrical conduction in polymers, *Electrical Properties of Polymers* (D. A. Seanor, ed.), Academic Press, New York, 1982, chap. 1.
104. N. F. Mott, *Phil. Mag. 19*:835 (1969).
105. K. W. Boer, *J. Non. Cryst. Sol. 4*:583 (1970).
106. J. J. O'Dwyer, *The Theory of Electrical Conduction and Breakdown in Solid Dielectrics*, Clarendon Press, Oxford, 1973.
107. K. C. Kao and W. Hwang, *Electrical Transport in Solids*, Pergamon Press, Oxford, 1981.
108. J. Mort and G. Pfister, *Electronic Properties of Polymers*, John Wiley, New York, 1982.
109. M. Ieda, *IEEE Trans. Electr. Insul. E1-19*:162 (1984).
110. Hari Singh Nalwa, *JMS Rev. Macromol. Chem. Phys. C31*(4):341 (1991).
111. Y. Wada, *IEEE Trans. Electr. Insul. E1-22*:255 (1987).
112. G. M. Sessler, Polymeric Electrets, *Electronic Properties of Polymers* (J. Mort and G. Pfister, eds.), John Wiley, New York, 1982, p. 59.
113. P. K. C. Pillai, G. K. Narula, A. K. Tripathi, and R. G. Mendiratta, *Phys. Rev. B-27*:2508 (1983).
114. H. von Seggern, *IEEE Trans. Elect. Insul. E1-21*(3):281 (1986).
115. S. S. Bamji, K. J. Kao, and M. M. Perlman, *J. ECC 6*:373 (1979).
116. S. B. Lang and D. K. Das Gupta, *Ferroelectrics 55*:151 (1984).
117. R. Sekar, Ph.D. thesis, Indian Institute of Technology, New Delhi, 1987.
118. G. K. Narula, Ph.D. thesis, Indian Institute of Technology, New Delhi, 1984.
119. Rashmi, Ph.D. thesis, Indian Institute of Technology, New Delhi, 1980.
120. C. D. Han (ed.), *Polymer Blends and Composites in Multiphase Systems*, Advances in Chemistry Series 206, American Chemical Society, Washington, DC, 1984.
121. C. D. Hang and H. K. Chuang, *J. Appl. Polymer Sci. 30*:4331 (1985).
122. P. K. C. Pillai, G. K. Narula, A. K. Tripathi, and R. G. Mendiratta, *J. Appl. Phys. 53*:6953 (1982).
123. W. Wollman and H. U. Poll, *Thin Solid Films 26*:201 (1975).
124. A. C. Lilly, Jr., and J. R. McDowell, *J. Appl. Phys. 39*:141 (1986).
125. R. Hogarth and T. Iqbal, *Phys. Stat. Solid (A) 65*:11 (1981).
126. N. Mott and R. W. Gurney, *Electronic Processes in Ionic Crystals*, Oxford University Press, London, 1957.
127. K. W. Boer, *J. Non.-Cryst. Solids 4*:583 (1970).
128. H. Bauser, *Kunstoffe 62*:192 (1972).
129. D. K. Davis, *J. Phys. D. Appl. Phys. 5*:162 (1972).

130. K. Masuda and M. Silver (eds.), *Energy and Charge Transfer in Organic Semiconductors*, Plenum Press, New York, 1974.
131. P. K. Nair, Ph.D. thesis, Indian Institute of Technology, New Delhi, 1976.
132. K. W. Boer, *J. Non.-Cryst. Solids 4*:583 (1970).
133. A. K. Jonscher, *Thin Solid Films 1*:213 (1961).
134. V. Adamec, *J. Phys. D. (G.B.) 14*(8):1487 (1981).
135. H. Sodolski, *Phys. Stat. Solid (a) 77*:749 (1983).
136. M. Ieda, G. Sawa, and S. Kato, *J. Appl. Phys. 42*:3737 (1971).
137. P. C. Arnett and N. Klein, *J. Appl. Phys. 46*:1399 (1975).
138. M. Chybieki, *Phys. Stat. Solid. (a) 39*:271 (1977).
139. D. M. Taylor and T. J. Lewis, *Phys. D. Appl. Phys. 4*:1346 (1971).
140. A. C. Lilly, Jr., and J. R. McDowell, *J. Appl. Phys. 39*:141 (1986).
141. W. Wollman and H. U. Poll, *Thin Solid Films 26*:201 (1975).
142. R. B. Hall, *Thin Solid Films 8*:263 (1971).
143. R. J. G. Simmons, *Phys. Rev. 155*:657 (1971).
144. J. R. Yeargan and H. L. Taylor, *J. Appl. Phys. 39*:5600 (1971).
145. A. K. Jonscher and A. Ansari, *Phil. Mag. 23*:205 (1971).
146. K. C. Kao and W. Hwong, *Electrical Transport in Solids*, Pergamon Press, New York, 1981.
147. H. G. Elias, *Macromolecules–I, Structure and Properties*, Plenum Press, New York, 1977, chap. 4.
148. J. Mort and A. I. Lakatos, *J. Non.-Cryst. Solids 4*:117 (1970).
149. H. Bauser, *Kunstoffe 62*:192 (1972).
150. F. H. Martin and J. Hirsch, *J. Appl. Phys. 43*(3):1001 (1972).
151. P. Khurana, Ph.D. thesis, Indian Institute of Technology, New Delhi, 1987.
152. M. M. Perlman and R. A. Creswell, *J. Appl. Phys. 42*:531 (1971).
153. A. Seanor Donald, *Electrical Properties of Polymers*, Academic Press, New York, 1982.
154. A. Lapu, M. Guirgea, T. Bal Tug, and P. Giluck, *J. Polymer Sci. Polymer Phys. Ed. 12*: 2399 (1974).
155. P. V. Wright, *J. Polymer Sci. Polymer Phys. Ed. 14*(5):955 (1976).
156. N. Mott and R. W. Gurney, *Electronic Processes in Ionic Crystals*, Oxford University Press, London, 1957.
157. M. Lampert, *Rep. Prog. Phys. 27*:329 (1964).
158. M. Lampert and P. Mark, *Current Injection in Solids*, Academic Press, New York, 1970.
159. M. Kryzewski and A. Zymonski, *J. Polymer Sci. 14*:245 (1970).
160. S. Sworakowski, *Acta Phys. Polontica 32*:1027 (1967).
161. J. S. Chutia and K. Bania, *Thin Solid Films 55*:387 (1978).
162. H. H. Fowler and L. Nordheim, *Proc. R. Soc.* (London) *A 119*:173 (1928).
163. A. Oster, *Z. Angew. Phys. 23*:120 (1970).
164. M. Kosaki, H. Oshima, and M. Ieda, *J. Phys. Soc. Jpn. 20*:1012 (1970).
165. S. Saito, H. Sasabe, T. Nakajima, and K. Yada, *J. Polymer Sci., A-26*:1297 (1968).
166. H. Block, Nature and application of electrical phenomena in polymers, *Electrical Phenomena in Polymer Sciences*, Advances in Polymer Sciences, Vol. 33 (H. J. Centow et al., eds.), Springer-Verlag, Berlin, 1979, pp. 93–167.
167. P. K. C. Pillai, K. Jain, and V. K. Jain, *Phys. Stat. Solid. (A) 13*:341 (1972).
168. T. Hinto, *J. Appl. Phys. 46*:1956 (1975).
169. P. K. C. Pillai, P. K. Nair, and R. Nath, *Polymer 17*:921 (1976).
170. J. van Turnhout, *Thermally Stimulated Discharge of Polymer Electrets*, Elsevier, New York, 1975.
171. M. L. Miller, *J. Polymer Sci. A-2*(4):685 (1966).
172. P. K. C. Pillai, T. C. Goel, and S. F. Xavier, *Eur. Polymer J. 15*:1149 (1979).
173. P. K. C. Pillai, B. K. Gupta, and S. Chandrasekhar, *J. Electrostat. 9*:315 (1981).
174. M. M. Perlman and R. A. Creswell, *J. Appl. Phys. 42*:531 (1971).

175. J. van Turnhout, *Polymer J.* 2:173 (1971).
176. P. K. C. Pillai, K. Jain, and V. K. Jain, *Phys. Stat. Solid. (a) 17*:221 (1973).
177. J. van Turnhout, *Top. Appl. Phys. 33*:81 (1980).
178. D. Muller, *Phys. Stat. Solid. A 23*:165 (1974).
179. P. C. Mehendru, K. Jain, V. K. Chopra, and P. Mehendru, *J. Phys. D 8*:305 (1975).
180. R. Sharma, L. V. Sud, and P. K. C. Pillai, *Polymer 21*:925 (1980).
181. P. C. Mehendru and K. Jain, *Phys. Lett. A 51*:287 (1975).
182. P. K. C. Pillai and R. C. Ahuja, *J. Polymer Sci. Polymer Phys. Ed. 12*:2465 (1974).
183. B. Gross, G. M. Sessler, and J. E. West, *J. Appl. Phys. 47*:468 (1976).
184. Y. Oka and N. Koizumi, *Jpn. J. Appl. Phys. 22*:L 281 (1983).
185. R. A. Creswell and M. M. Perlman, *J. Appl. Phys. 41*:2635 (1970).
186. C. Lacabanne, D. Chatain, J. Guillet, G. Setytve, and J. F. May, *J. Polymer Sci., Polymer Phys. Ed. 13*:445 (1975).
187. M. Goel and P. K. C. Pillai, *J. Macromol. Sci.—Phys. B16*:397 (1979).
188. J. van Turnhout, *Adv. Static Electr. 1*:56 (1971).
189. S. Ikeda and K. Matsuda, *Jpn. J. Appl. Phys. 20*:2319 (1981).
190. P. K. C. Pillai and M. Goel, *J. Electrochem. Soc. 120*:395 (1973).
191. H. S. Nalwa, L. R. Dalton, and P. Vasudevan, *Eur. Polymer J. 21*:943 (1985).
192. P. K. C. Pillai, G. K. Narula, A. K. Tripathi, and R. G. Mendiratta, *Phys. Rev. B 27*:2508 (1983).
193. J. van Turnhout, *Thermally Stimulated Discharge of Polymer Electrets*, Elsevier, Amsterdam, 1975, chap. 35.
194. P. Braunlinch, Thermoluminescence and thermally stimulated conductivity, *Thermoluminescence of Geological Materials* (D. J. McDougall, ed.), Academic Press, New York, 1968, p. 61.
195. R. H. Bube, *Photoconductivity of Solids*, John Wiley, New York, 1960, chap. 9.
196. D. Curie, *Luminescence in Crystals*, John Wiley, New York, 1963, chap. 6.
197. A. G. Milness, *Deep Impurities in Semiconductors*, Wiley-Interscience, New York, 1973, chap. 9.
198. C. Bucci, R. Fieschi, and G. Guidi, *Phys. Rev. 148*:816 (1966).
199. H. Frei and G. Groetzinger, *Phys. Z. 37*:720 (1936).
200. O. G. Von Altheim, *Ann. Phys. 35*:417 (1939).
201. P. V. Murphy and S. C. Riberio, *J. Appl. Phys. 34*:2061 (1963).
202. B. Gross, *Phys. Rev. 66*:26 (1944).
203. B. Gross, *J. Electrochem. Soc. 115*:376 (1968).
204. A. N. Gubkin and B. N. Matsonashirli, *Sov. Phys. Solid State 4*:878 (1962).
205. C. Bucci and R. Fieschi, *Phys. Rev. Lett. 12*:16 (1964).
206. B. Gross, *Appl. Opt. Suppl. 3*:176 (1969).
207. A. K. Jonscher, *J. Non.-Cryst. Solids 8–10*:293 (1972).
208. A. K. Jonscher, *J. Phys. C6*:1235 (1973).
209. A. K. Jonscher, *Nature* (London) *253*:717 (1973).
210. A. K. Jonscher, *J. Electrostat. 3*:53 (1977).
211. M. M. Perlman, *J. Electro.-Chem. Soc. 119*:892 (1972).
212. R. A. Creswell and M. M. Perlman, *J. Appl. Phys. 41*:2365 (1970).
213. M. M. Perlman, *J. Appl. Phys. 42*:2645 (1975).
214. J. van Turnhout, *Polymer J. 2*:183 (1971).
215. J. Vander Schurren and A. Linkens, *J. Appl. Phys. 51*:4697 (1980).
216. D. Ronarch, *J. Appl. Phys. Lett. 37*:707 (1980).
217. M. Kryzewski, M. Zielinski, and S. Sapieha, *Polymer 17*:212 (1976).
218. T. Mizutani, Y. Suzoki, and M. Hanai, *Jpn. J. Appl. Phys. Pt. I 21*:1639 (1982).
219. Y. Suzoki, G. Cai, T. Mizatani, and M. Ieda, *Jpn. J. Appl. Phys. Pt. I 21*:1759 (1982).
220. H. Sodoloski, *Phys. Stat. Solid. (A) 76*:303 (1983).

221. A. Mishra, *J. Appl. Polymer Sci.* *27*:1967 (1982).
222. J. Vanderschurren, M. Landang, and J. Nierette, *IEEE Trans. Elect. Insul.* *E1-17*:189 (1982).
223. J. P. Reboul and A. Tourelle, *J. Polymer Sci. Polymer Phys. Ed.* *22*:21 (1984).
224. M. Topic and Z. Katovic, *Polymer* *26*:1141 (1985).
225. M. Onoda, H. Nayakama, and K. Amakawa, *Jpn. J. Appl. Phys.* *24*:1375 (1985).
226. H. S. Nalwa and P. Vasudevan, *Makromol. Chem. Phys.* *186*:1255 (1985).
227. H. Von Seggern, *IEEE Trans. Elect. Insul.* *E1-21*:281 (1986).
228. P. K. C. Pillai, P. K. Nair, and R. Nath, *Polymer* *17*:925 (1976).
229. S. K. Sharma, A. K. Gupta, and P. K. C. Pillai, *J. Mat. Sci. Lett.* *5*:224 (1986).
230. P. K. C. Pillai, T. C. Goel, and S. F. Xavier, *Eur. Polymer J.* *15*:1152 (1979).
231. P. K. C. Pillai and M. Goel, *Polymer* *16*:5 (1975).
232. P. K. C. Pillai, A. K. Gupta, and M. Goel, *Macromol. Chem.* *181*:951 (1980).
233. P. K. C. Pillai, T. C. Goel, A. K. Tripathi, and R. Sekar, *J. Mat. Sci. Lett.* *4*:1131 (1985).
234. P. K. C. Pillai, G. K. Narula, and A. K. Tripathi, *J. Mat. Sci.* *17*:3017 (1982).
235. R. A. Cresswell and M. M. Perlman, *J. Appl. Phys.* *41*:2369 (1970).
236. G. M. Sessler and J. E. West, *J. Appl. Phys.* *47*:3480 (1976).
237. A. Maeda, K. Kojima, Y. Takai, and M. Ieda, *Jpn. J. Appl. Phys.* *23*:1260 (1980).
238. P. K. C. Pillai, G. K. Narula, A. K. Tripathi, and R. G. Mendiratta, *J. Appl. Phys.* *53*:6953 (1982).
239. P. K. C. Pillai, G. K. Narula, A. K. Tripathi, and R. G. Mendiratta, *Phys. Rev.* *B27*:2510 (1983).
240. D. Ronarch, P. Audren, and J. L. Moura, *J. Appl. Phys.* *58*:466 (1985) and *58*:474 (1985).
241. R. K. Srivastava, M. Quereshi, and C. S. Bhatnagar, *Jpn. J. Appl. Phys.* *17*:1537 (1978).
242. E. Marschal and H. B. Benoit, *J. Polymer Sci. Polymer Phys. Ed.* *16*:949 (1979).
243. T. Hashimoto, M. Shiraki, and T. Sakai, *J. Polymer Sci. Polymer Phys. Ed.* *13*:2401 (1975).
244. A. C. Lilly, R. H. Hendersson, P. C. Sharp, and L. L. Stewart, *J. Appl. Phys.* *41*:2002 (1970).
245. G. M. Sessler and J. E. West, *J. Appl. Phys.* *47*:3482 (1976).
246. J. Vanderschurren and A. Linkens, *J. Appl. Phys.* *51*:4697 (1980).
247. B. Gross, *Charge Storage in Dielectrics*, Elsevier, Amsterdam, 1964.
248. T. Hino, *Jpn. J. Appl. Phys.* *12*:611 (1973).
249. T. Hino, *J. Appl. Phys.* *46*:1956 (1975).
250. A. Gemant, *Phil. Mag.* *20*:929C (1935).
251. M. M. Perlman and R. A. Cresswell, *J. Appl. Phys.* *42*:2365 (1970).
252. B. Ai, C. P. Stoka, H. T. Giam, and P. Destoul, *J. Appl. Phys. Lett.* *34*:821 (1979).
253. B. Contaloube, C. Dreyfus, and J. Leunier, *J. Polymer Sci. Polymer Phys. Ed.* *17*:95 (1979).
254. C. Bowlt, *Contemp. Phys.* *17*:461 (1976).
255. T. Miyamoto and K. Shibahama, *J. Appl. Phys.* *44*:5372 (1973).
256. F. Gutmann and L. E. Lyons, *Organic Semiconductors*, John Wiley, New York, 1967.
257. G. M. Sessler and J. E. West, *Phys. Rev. B.* *10*:4488 (1974).
258. S. Radhakrishnan and S. Haridoss, *Phys. Stat. Solid. A.* *41*:649 (1977).
259. G. M. Sessler, J. E. West, D. A. Berkley, and G. Morgenstern, *Phys. Rev. Lett.* *38*:368 (1977).
260. P. Huo, P. Cebe, *J. Polymer Sci.* *30*:239 (1992).
261. M. Mourgues Martin, A. Bernes, C. Lucabanne, O. Nouvel, and G. Seutre. *IEEE Trans. Elect. Insul.* *27*(4):795 (1992).
262. H. Nunes da Cunha and R. A. Moreno, *IEEE Trans. Elect. Insul.* *27*(4):708 (1992).
263. E. Marchal, *IEEE Trans. Elect. Insul.* *E-1*(3):323 (1986).
264. A. Toureille and J. P. Febout, *IEEE Trans. Elect. Insul.* *E-1*(3):343 (1986).
265. J. A. Giacomitti and J. A. Malmonge, *IEEE Trans. Elect. Insul.* *E-1*(3):383 (1986).
266. M. Topic and Z. Katovic, *Polymer* *26*:1141 (1985).
267. D. Ronarch, P. Audren, and J. L. Moura, *J. Appl. Phys.* *58*(1):474 (1985).

268. J. Sworakowski, M. T. Figuelredo, G. F. Leal Ferreira, and M. Campos, *J. Appl. Phys.* 56(4): 1149 (1984).
269. J. P. Reboul and A. Toureille, *J. Polymer Sci. Polymer Phys. Ed.* 22:21 (1984).
270. M. Zielinski and M. Kryszewski, *Phys. Stat. Solid.* (a) 42:305 (1977).
271. P. K. C. Pillai, K. Jain, and V. K. Jain, *Indian J. Pure Appl. Phys.* 11:597 (1973).
272. P. K. C. Pillai and Rashmi, *Polymer* 20:1245 (1979).
273. P. K. C. Pillai, G. K. Narula, A. K. Tripathi, and R. G. Mendiratta, *Phys. Stat. Solid.* (a) 67:649 (1981).
274. P. K. C. Pillai and M. Mollah, *J. Macromol. Sci.—Phys.* B17(1):69 (1980).
275. P. K. C. Pillai, B. K. Gupta, and M. Goel, *J. Polymer Sci. Polymer Phys. Ed.* 19:1461 (1981).
276. A. Maeda, K. Kojima, Y. Takai, and M. Ieda, *Jpn. J. Appl. Phys.* 23(5):260 (1984).
277. V. V. Popik, V. N. Zhikharev, I. D. Seikorskii, *Sov. Phys. J.* 33(3):275 (1990).
278. W. Kohler, D. R. Robells, P. T. Dao, C. S. Willard, and O. S. Williams, *J. Chem. Phys.* 93(12):9157 (1990).
279. P. G. Karmazova, *Bulg. J. Phys.* 17(5):454 (1991).
280. V. Halperu, *J. Phys. D. Appl. Phys.* 26(2):307 (1993).
281. C. Bucci, R. Fieschi, and G. Guidi, *Phys. Rev.* 148:816 (1966).
282. B. Gross, *Appl. Opt. Suppl.* 3:176 (1969); *J. Electrochem. Soc.* 115:376 (1968).
283. J. van Turnhout, *Thermally Stimulated Discharge Currents of Electrets*, Elsevier, Amsterdam, 1975, chap. 2.
284. P. H. Ong and J. van Turnhout, *TSD of Polymer Electrets Having a Distributed Polarization*, Central Laboratorium, TNO (Delff) Publication, 1972, p. 72164.
285. R. A. Cresswell and M. M. Perlman, *J. Appl. Phys.* 41:2366 (1970).
286. M. M. Perlman and R. A. Cresswell, *J. Appl. Phys.* 42:537 (1972).
287. M. M. Perlman, *J. Appl. Phys.* 42:2645 (1975).
288. J. van Turnhout, in *Topics in Applied Physics—Electrets* (G. M. Sessler, ed.), Springer-Verlag, Berlin, New York, 1980.
289. G. F. J. Garlick and A. F. Gibson, *Proc. Phys. Soc. Lond.* 60:574 (1948).
290. T. Hino, K. Suzuki, and K. Yamashita, *Jpn. J. Appl. Phys.* 12:651 (1973).
291. K. H. Nicholas and J. Woods, *J. Appl. Phys.* 15:783 (1964).
292. T. A. T. Coswell and J. Woods, *Br. J. Appl. Phys.* 18:1045 (1967).
293. J. C. Renicar and R. J. Fleming, *J. Polymer Sci. Polymer Phys. Ed.* 10:1979 (1979).
294. R. P. Khare and J. D. Ranade, *Phys. Stat. Solid.* (a) 32:221 (1975).
295. L. I. Grossweiner, *J. Appl. Phys.* 24:1306 (1953).
296. V. K. Jain, C. L. Gupta, S. K. Agarwal, and R. C. Tyagi, *Thin Solid Films*, 30:245 (1975).
297. V. Adamec and E. Mateova, *Polymer* 16:166 (1975).
298. S. W. S. McKeever and D. M. Huges, *J. Phys. D* 8:1520 (1975).
299. T. W. Hiekmott, *J. Appl. Phys.* 46:2583 (1975).
300. P. Muller, in *Amorphous Semiconductors '76* (I. Kosa Smogyi, ed.), Akad. Kiado, Budapest, 1977.
301. J. van Turnhout, A. H. Van Rheenen, in *Proc. 4th Int. Conf. on Physics of Non-Crystalline Solids* (G. H. Frischat, ed.), Trans. Tech. Publ. Aedor., Mannsdorf, 1977.
302. D. W. McCall and E. W. Anderson, *J. Chem. Phys.* 32:237 (1960).
303. A. C. Lilly and J. R. McDowell, *J. Appl. Phys.* 39:141 (1968).
304. B. V. Hamon, *Proc. IEE 99*, Pt. IV:151 (1952).
305. G. Williams, *Polymer* 4:27 (1963).
306. M. E. Baird, *Rev. Mod. Phys.* 40:219 (1968).
307. H. J. Wintle, *J. Non-Cryst. Solids* 15:471 (1974).
308. H. J. Wintle, *Solid State Electron.* 18:1039 (1975).
309. V. Adamic, *Polymer* 237:219 (1970).
310. D. K. Das-Gupta and K. Joyner, *J. Phys. D. Appl. Phys.* 9:829 (1976).

311. P. K. C. Pillai and Rashmi, *Eur. Polymer J. 17*:611 (1980).
312. P. K. C. Pillai, G. K. Narula, A. K. Tripathi, and R. G. Mendiratta, *Phys. Stat. Solid. (a) 77*:693 (1983).
313. P. J. Atkins and R. J. Fleming, *J. Phys. D. 13*:625 (1980).
314. H. J. Wintle and T. C. Chapman, *J. Appl. Phys. 51*:3435 (1980).
315. T. C. Chapman and H. J. Wintle, *J. Appl. Phys. 53*:7425 (1982).
316. J. Lowell, *J. Phys. D. Appl. Phys. 16*:2223 (1983).
317. G. Lengyel, *J. Appl. Phys. 37*:807 (1966).
318. G. Sawa, M. Yada, and M. Ieda, *Jpn. J. Appl. Phys. 12*:475 (1975).
319. S. Chand, J. P. Aggarwal, and P. C. Mehendru, *Thin Solid Films 99*:351 (1983).
320. S. Kamisako, S. Akiyama, and S. Shinohara, *Jpn. J. Appl. Phys. 13*:1780 (1974).
321. P. J. Reucroft and S. K. Ghosh, *J. Non-Cryst. Solids 15*:399 (1974).
322. A. F. Burnester and V. J. Caldecourt, *J. Polymer Sci. Polymer Phys. Ed. 6*:1639 (1968).
323. T. W. Hickmott, *J. Appl. Phys. 46*:2583 (1975).
324. A. K. Batra, S. C. Mathur, and A. Man Singh, *Phys. Stat. Solid. (a) 77*:399 (1983).
325. J. S. Dhull and D. R. Sharma, *J. Phys. D. Appl. Phys. 15*:2307 (1982).
326. S. Saito, and T. Nakajinia, *J. Appl. Polymer Sci. 2*:93 (1959).
327. W. Reddish, *J. Polymer Sci. C 14*:123 (1966).
328. P. K. C. Pillai and V. K. Jain, *J. Sci. Ind. Res. 29*(6):270 (1970); P. K. C. Pillai and V. K. Jain, *J. Appl. Phys. 40*:3487 (1969).
329. G. M. Sessler, Physical principles of electrets, *Topics in Applied Physics Electrets*, Vol. 33, Springer-Verlag, Berlin, Heidelberg, New York, (1980).
330. M. Eguchi, *Phil. Mag. 49*:178 (1925).
331. A. Gemant, *Phil. Mag. 57* (No. 136 Suppl.):20 (1935).
332. P. Khurana, Ph.D. thesis, Indian Institute of Technology, New Delhi, 1987.
333. G. K. Narula, Ph.D. thesis, Indian Institute of Technology, New Delhi, 1984.
334. C. W. Reedyk and M. M. Pearlman, *J. Electrochem. Soc. 115*:49 (1968).
335. J. C. Devins and S. L. Reynolds, *Rev. Sci. Instrum. 26*:11 (1975).
336. B. Gross, *J. Chem. Phys. 17*(10):866 (1949).
337. B. Gross, *Br. J. Appl. Phys. 1*:254 (1950).
338. A. Gement, *Rev. Sci. Instrum. 11*:65 (1940).
339. G. G. Wiseman and G. R. Feaster, *J. Chem. Phys. 26*(3):521 (1957).
340. A. Chowdry and C. R. Westgate, *J. Phys. D. 7*:L149 (1974).
341. A. L. Freedman and L. Rosenthal, *Rev. Sci. Instrum. 21*:896 (1950).
342. S. Kojima and K. Kato, *J. Phys. Soc. Jpn. 6*:207 (1951).
343. Y. Kakiuchi, *J. Phys. Soc. Jpn. 6*:278 (1951).
344. H. H. Wieder and S. Kaufman, *J. Appl. Phys. 24*:156 (1953).
345. N. P. Baumann and G. G. Wiseman, *J. Appl. Phys. 25*:1391 (1954).
346. H. Kallmann and B. Rosenberg, *Phys. Rev. 6*:97, 1956 (1955).
347. J. W. Wild and J. D. Stranathan, *J. Chem. Phys. 27*:1055 (1957).
348. D. K. Davies, *J. Sci. Instrum. 44*:521 (1967).
349. T. R. Foord, *J. Phys. E. 3*:334 (1970).
350. G. M. Sessler and J. E. West, *Rev. Sci. Instrum. 42*:15 (1971).
351. L. A. Freedman and L. A. Rosenthal, *Rev. Sci. Instrum. 21*:896 (1950).
352. G. M. Sessler and J. E. West, *J. Electrochem. Soc. 115*:836 (1968).
353. D. Dreyfus and J. Lewiner, *J. Appl. Phys. 45*:721 (1974).
354. G. Morgenstern, *Appl. Phys. 46*:4357 (1975).
355. G. Dreyfus and J. Lewiner, *Phys. Rev. 14*:5451 (1976).
356. P. D. Southgate, *Appl. Phys. Lett. 28*:250 (1976).
357. D. K. Das Gupta and K. Doughty, *J. Phys. D 11*:2415 (1978).
358. R. E. Collins, *Appl. Phys. Lett. 26*:675 (1975); *J. Appl. Phys. 47*:4804 (1976); *Rev. Sci. Instrum. 48*:83 (1977).

359. H. von Seggern, *Appl. Phys. Lett. 33*:134 (1978).
360. R. E. Collins, *Ferroelectrics 33*:65 (1981).
361. S. B. Lang and D. K. Das Gupta, *Ferroelectrics 39*:1249 (1981).
362. S. B. Lang and D. K. Das Gupta, *J. Appl. Phys. 59*:2151 (1986).
363. G. M. Sessler and R. Gerhard-Multhaupt, *Rad. Phys. Chem. 23*:363 (1984).
364. P. E. Secker and J. N. Chubb, *J. Electrostat. 16*:1 (1984).
365. P. Laurenceau, G. Dreyfus, and J. Lewiner, *Phys. Rev. Lett. 38*:46 (1977).
366. E. B. Podgorsak and P. R. Moran, *Appl. Phys. Lett. 24*:580 (1974).
367. G. M. Sessler, J. E. West, and R. Gerhard, *Polymer Bull. 6*:109 (1981).
368. G. M. Sessler, J. E. West, and R. Gerhard, *Phys. Rev. Lett. 48*:563 (1982).
369. R. Gerhard-Multhaupt, *Phys. Rev. B 27*:2494 (1983).
370. W. E. Semmenger and M. Haardt, *Solid State Commun. 41*:917 (1982).
371. A. Gemant, *Direct Current*, p. 148 (1953).
372. K. C. Chandy, Ph.D. Thesis, University of Saugar, 1956.
373. K. Nakata, *Proc. Phys. Math. Soc. Jpn. 9*:179 (1927).
374. F. Gutman, *Rev. Mod. Phys. 20*:457 (1948).
375. M. Ewing, *Phys. Rev. 36*:378 (1930).
376. Y. Kakiuchi, *Sci. Pap. I.P.C.R. 1119-1120*:40 (1943).
377. W. M. Good and J. D. Stranathan, *Phys. Rev. 56*:810 (1939).
378. K. C. Chandy and D. R. Bhawalkar, *Proc. Natl. Acad. Sci.* (India), Allahabad, XXV, Section A, Part I (1956).
379. A. C. Mathew, Ph.D. thesis, University of Saugar, 1960.
380. P. K. C. Pillai, Ph.D. thesis, University of Saugar, 1962.
381. R. D. Bennet, *Phys. Rev. 36*:65 (1930).
382. C. S. Smith and R. L. Barrett, *J. Appl. Phys. 18*:177 (1947).
383. A. Muller, *Proc. R. Soc.* (Lond.) *127 A*:417 (1930).
384. A. van Hook and L. Silver, *J. Chem. Phys. 10*:686 (1942).
385. M. Ida, *J. Phys. Soc. Jpn. 10*(4):318 (1955).
386. A. Keller, *Nature* (Lond.) 169:913 (1952).
387. S. L. Khanna, *Indian J. Pure Appl. Phys. 6*:98 (1968).
388. S. Mikola, *Z. Phys. 32*:476 (1925).
389. P. K. C. Pillai and V. K. Jain, *J. Appl. Phys. 40*:3487 (1969).
390. K. G. Balakrishnan, Ph.D. thesis, Indian Institute of Technology, New Delhi, 1973.
391. Rashmi, Ph.D. thesis, Indian Institute of Technology, New Delhi, 1980.
392. G. M. Sessler and J. E. West, *J. Acoust. Soc. Am. 40*:1433 (1966).
393. P. K. C. Pillai, K. G. Balakrishnan, and T. N. R. Kurup, *J. Phys. D. Appl. Phys. 5*:1027 (1972).
394. F. Micheron, *Makromol. Chem. Macromol. Symp. 1*:173 (1986).
395. T. Hoshimoto, T. Sakai, and S. Miyata, *J. Polymer Sci. Polymer Phys. Ed. 16*:1965 (1978).
396. I. M. Talwar and D. L. Sharma, *J. Electrochem. Soc. 125*:434 (1978).
397. R. Sharma, L. V. Sud, and P. K. C. Pillai, *Polymer 21*:925 (1980).
398. R. J. Comstock, S. I. Stupp, and S. H. Carr, *J. Macromol. Sci. Phys. 13*:101 (1977).
399. P. K. C. Pillai and R. C. Ahuja, *J. Polymer Sci. Polymer Phys. Ed. 12*:2468 (1974).
400. A. Bui, H. Carchano, J. Guastavino, D. Chatain, P. Gauter, and C. Lacabanne, *Thin Solid Films 21*:313 (1974).
401. R. A. Creswell and M. M. Perlman, *J. Appl. Phys. 41*:2375 (1970).
402. J. P. Rearden and P. F. Waters, *Proc. Symp. Thermally Photo-Stimul. Currents*; Insul., Electrochem. Society, Princeton, NJ, 1976, p. 185.
403. N. Murayama and H. Hashizumo, *J. Polymer Sci. Polymer Phys. Ed. 14*:989 (1976).
404. X. Takamatsu and E. Fukada, *Kobimshi Kagatsu 29*:505 (1972).
405. B. Gross, G. M. Sessler, and J. E. West, *J. Appl. Phys. 22*:L281 (1983).
406. E. Marchal, H. Benoit, and O. Vogl, *J. Polymer Sci. Polymer Phys. Ed. 16*:949 (1978).

407. P. K. Watson, F. W. Schmidlin, and R. V. La Donna, *IEEE Trans. Elect. Insul.* 27(4):680 (1992).
408. Zhong-fu Xia, Guo-mao Yang, and Xi-min Sun, *IEEE Trans. Elect. Insul.* 23(4):702 (1992).
409. P. A. Ribeiro, J. A. Giacometti, M. Raposo, and J. N. Marat Mendes, *IEEE Trans. Elect. Insul.* 27(4):744 (1992).
410. G. Eberle and W. Eisenmenger, *IEEE Trans. Elect. Insul.* 27(4):768 (1993).
411. E. Marchal, *IEEE Trans. Elect. Insul. E* 1-21(3):323 (1986).
412. A. Toureille and J. P. Reboul, *IEEE Trans. Elect. Insul.* 1-21(3):343 (1986).
413. S. B. Lang and D. K. Das-Gupta, *IEEE Trans. Electr. Insul. E* 1-21(3):399 (1986).
414. Y. Oku and N. Koizumi, *Jpn. J. Appl. Phys.* 22:L281 (1983).
415. S. Ikeda and K. Matsuda, *Jpn. J. Appl. Phys.* 15:963 (1976).
416. D. K. Das-Gupta, A. Svatik, A. T. Bulinski, R. J. Denrby, S. Bamji, and D. J. Carlsson, *J. Phys. D. Appl. Phys. (U.K.)*, 23:12, 1599 (1990).
417. W. Stark, F. Harnisch, and W. Manthey, *J. Electrostat.* 25(3):277 (1990).
418. P. G. Kurmazova, G. A. Mekishev, *Bulg. J. Phys.* 17(5):454 (1991).
419. Lu Tingji and G. M. Sessler, *IEEE Trans. Elect. Insul.* 26(2):228 (1991).
420. P. K. C. Pillai and E. L. Shriver, NASA Technical Report, NASA TRR-457, 1973.
421. P. K. C. Pillai and Rashmi, *Eur. Polymer J.* 17:611 (1981).
422. B. L. Sharma and P. K. C. Pillai, *Phys. Stat. Solid. (a)* 71:583 (1982).
423. P. K. C. Pillai, P. K. Nair, and R. Nath, *Polymer* 17:924 (1976).
424. M. Goel, Ph.D. thesis, Indian Institute of Technology, New Delhi, 1972.
425. P. Khurana, Ph.D. thesis, Indian Institute of Technology, New Delhi, 1987.
426. I. M. Taylor and D. R. Bhawalkar, *Indian J. Pure Appl. Phys.* 7:685 (1969).
427. P. K. C. Pillai and V. K. Jain, *J. Phys. D. Appl. Phys.* 3:829 (1970).
428. J.-P. Reboul and A. Toureille, *J. Polymer Sci. Polymer Phys. Ed.* 22:21 (1984).
429. T. S. Gancheva and P. D. Dinev, *Eur. Polymer J.* 19:5 (1983).
430. J. Vanderschueren, *J. Polymer Sci. Polymer Phys. Ed.* 12:991 (1974).
431. D. K. Das-Gupta and K. Doughty, *Ferroelectrics* 60:51 (1981).
432. S. Eliasson, *J. Phys. D. Appl. Phys.* 19:1965 (1986).
433. Y. Aoki and J. O. Brittain, *J. Appl. Polymer Sci.* 20:2879 (1976).
434. J. K. Quamara, P. K. C. Pillai, and B. L. Sharma, *Acta Polymerica* 33:205 (1982).
435. H. Solunov and T. Vassilev, *J. Polymer Sci. Polymer Phys. Ed.* 12:1273 (1974).
436. P. K. C. Pillai, B. K. Gupta, and M. Goel, *J. Polymer Sci. Polymer Phys. Ed.* 19:1461 (1981).
437. P. Alexandrovich, F. E. Karasz, and U. J. Macknight, *J. Appl. Phys.* 47(10):4251 (1976).
438. A. K. Gupta, Ph.D. thesis, Indian Institute of Technololgy, New Delhi, 1985.
439. T. Kaura, Ph.D. thesis, Indian Institute of Technology, New Delhi, 1983.
440. P. K. C. Pillai and A. K. Gupta, *Makromol. Chem.* 181:951 (1980).
441. P. K. C. Pillai and A. K. Gupta, *J. Electrostatics* 16:79 (1984).
442. P. K. C. Pillai, A. K. Gupta, and S. K. Sharma, *Angew. Makromol. Chem.* 130:91 (1985).
443. P. K. C. Pillai and A. K. Gupta, *J. Mat. Sci. Lett.* 2:397 (1983).
444. P. K. C. Pillai, A. K. Gupta, and S. K. Sharma, *Acta Polymerica* 36(7):393 (1985).
445. D. R. Ulrich, in *Organic Materials for Non-Linear Optics*, Publication No. 69 (R. A. Hann and D. Bloor, eds.), Royal Society of Chemistry, London, 1989, p. 69.
446. S. R. Marder, J. E. Sohn, and G. D. Stucky (eds.), Materials for Nonlinear Optics: Chemical Perspectives, ACS Symposium Series 455, American Chemical Society, Washington, D.C., (1991).
447. A. J. Heeger, J. Orenstein, and D. R. Ulrich (eds.), *Non-Linear Optical Properties of Polymers*, Materials Research Society Symp. Proc., Pittsburgh, Vol. 109, 1987.
448. P. N. Prasad and D. R. Ulrich (eds.), *Non-Linear Optical and Electroactive Polymers*, Plenum Press, New York, 1988, p. 243.
449. R. Lytel, *Non-Linear Optics in Polymers*, NATO Workshop, NICE-Sophia Antipolis, 1988.

450. R. Lytel, Advances in organic integrated optic devices, Abstract 3.5, Int. Conf. on Organic Materials for Nonlinear Optics, June 29–30, 1988.

451. K. J. Malloy and C. L. Giles, *Non-Linear Optical Properties of Polymers*, Materials Research Society Proc., Vol. 109 (A. J. Heeger, J. Orenstein, and D. R. Ulrich, eds.), Material Research Society, Pittsburgh, PA, 1988, p. 177.

452. D. S. Chemla and J. Zyss (eds.), Nonlinear Optical Properties of Organic Molecules and Crystals, Academic Press, New York, 1987.

452a. J. Zyss, Molecular Nonlinear Optics, Academic Press, New York, 1994.

453. H. S. Nalwa, Organic materials for third-order nonlinear optics, *Adv. Mater.* 5:341 (1993).

454. C. Sauterret, J. P. Hermann, R. Frey, F. Pralene, J. Duciuing, R. H. Banghman, and R. R. Chance, *Phys. Rev. Lett. 36*:956 (1976).

455. D. N. Rao, J. Swiatkiewiez, P. Chopra, S. K. Ghosal, and P. N. Prasad, *Appl. Phys. Lett. 48*:1187 (1986).

456. H. S. Nalwa, Organometalic materials for nonlinear optics, *Appl. Organometal. Chem. 5*: 349 (1993).

457. F. Kajzar, J. Messier, and C. Rosilio, *J. Appl. Phys. 60*:3040 (1986).

458. J. W. Perry, Nonlinear optical properties of molecules and materials, in *Materials for Nonlinear Optics*, (S. R. Marder, J. E. Sohn, and G. D. Stucky, eds.), ACS Symp. Ser. 455, Washington, DC, 1991, Chap. 4, p. 67.

459. K. D. Singer, M. G. Kuzyk, and J. E. Sohn, *J. Opt. Soc. Am. B 4*:968 (1987).

460. H. S. Nalwa, T. Watanabe, and S. Miyata, Optical second harmonic generation in organic molecular and polymeric materials: measurements, techniques, and materials. In: Progress in Photochemistry and Photophysics, J. F. Rabek (eds.), CRC Press, Boca Raton, FL, 1992, chap. 4, pp. 103–185.

461. G. R. Meredith, J. G. Van DuSen, and D. Williams, *J. Macromol. 15*:1385 (1982).

462. K. D. Singer, S. E. Sohn, and S. Lalama, *J. Appl. Phys. Lett. 49*:248 (1986).

463. J. D. Le Grange, M. G. Kuzyk, and K. D. Singer, *Mol. Cryst. Liq. Cryst. 250 b*:567 (1987).

464. B. Berge, A. Wicker, J. Lajenwiez, and J. F. Legrand, *Europhys. Lett. 9*:657 (1989).

465. M. Kishinoto, M. Sato, and H. Gano, *Springer Proc. Phys. 36*:196 (1988).

466. M. Eich, A. Sen, H. Looser, G. C. Bjorklund, J. D. Swalen, R. Jweing, and D. Y. Yoon, *J. Appl. Phys. 66*:2559 (1989).

467. M. Eich, H. Looser, D. Y. Yoon, R. J. Tweing, G. C. Bjorklund, and J. C. Baumert, *J. Opt. Soc. Am. B 6*:1590 (1989).

468. H. L. Hampsch, G. K. Wong, J. M. Torkelson, S. J. Bethke, and S. G. Grubb, *Proc. SPIE 1104*:267 (1989).

469. H. L. Hampsch, J. M. Torkelson, S. J. Bethke, and S. G. Grubb, *J. Appl. Phys. 67*:1037 (1990).

470. C. Ye, T. J. Marks, J. Yang, and G. K. Wong, *Macromolecules 20*:2322 (1987).

471. E. M. Williams, *The Physics and Technology of Xerographic Processes*, Wiley-Interscience, New York, 1984.

472. D. A. Seanor, in *Electrical Properties of Polymers* (D. A. Seanor, ed.), Academic Press, New York, 1982.

473. K. W. Beesen, K. A. Horn, M. McFarland, A. Nahata, C. Wu, and J. T. Yardley, in *Materials for Nonlinear Optics* (S. R. Marder, J. E. Sohn, and G. D. Stucky, eds.), ACS Symp. Ser. 455, Washington, DC, 1991, chap. 20.

474. D. R. Robello, C. S. Willand, M. Seozzafava, A. Ulman, and D. J. Williams, in *Materials for Nonlinear Optics* (S. R. Marder, J. E. Sohn, and G. D. Stucky, eds.), ACS Symp. Ser. 455, Washington, DC, 1991, chap. 18.

475. B. Gross, *Phys. Rev. 66*(1 & 2):26 (1944).

476. B. Gross, *Proc. 5th Int. Symp.—Electrets*, Heidelberg, 1985, p. 9; *IEEE Trans. Elect. Insul. E 1-21*:249 (1986).

477. R. Gerhard-Mullhaupt, *IEEE Trans. Elect. Insul. E 1-22*(5):531 (1987).

478. B. Gross, *Endeavour 30*:115 (1971).

479. M. M. Perlman, *J. Appl. Phys. 31*(2):356 (1960).

480. J. W. Wild and J. D. Stranathan, *J. Chem. Phys. 27*(5):1055 (1957).

481. K. Palm, *Exp. Tech. der Phys. 4*(6):253 (1956).

482. A. N. Gubikin, *Zh. Tekh. Fiz. 27*(9):1954 (1957).

483. W. F. G. Swann, *J. Franklin Inst. 250*:219 (1950).

484. M. M. Perlman, *J. Electochem. Tech. 6*:95 (1968).

485. T. Piech and J. Handerek, *Phys. Stat. Solid. 9*:361 (1965).

486. B. Gross and W. Gultinger, *J. Appl. Sci. Res B 16*:189 (1956).

487. M. M. Perlman and J. L. Meunier, *J. Appl. Phys. 36*(2):420 (1965); *31*:356 (1960).

488. J. van Turnhout, *Polymer J. 2*:173 (1971).

489. C. W. Reedyk and M. M. Perlman, *J. Electrochem. Soc. 115*:49 (1968).

490. V. F. Zolotarov, D. G. Semak, and D. V. Chepur, *Phys. Stat. Solid. 21*:437 (1967).

491. P. K. C. Pillai, K. Jain, and V. K. Jain, *Phys. Stat. Solid. (a) 14*:K 29 (1971); *Nuovo Cimento 3 B*:225 (1971).

492. G. W. Fabel and H. K. Henisch, *Phys. Stat. Sol(a)*, 535 (1971).

493. P. I. Mopsik and A. S. De Reggi, *J. Appl. Phys. 53*:4333 (1982).

494. S. B. Lang and D. K. Das-Gupta, *Ferroelectrics 60*:23 (1984).

495. G. M. Sessler, J. E. West, and R. Gerhard, *Polymer Bull. 6*:109 (1981).

496. G. M. Sessler, J. E. West, and R. Gerhard, *Phys. Rev. Lett. 48*:563 (1982).

497. W. Eisenmenger and M. Haardt, *Solid State Commun. 41*:917 (1982).

498. G. M. Sessler, J. E. West, D. A. Berkley, and G. Morgenstern, *Phys. Rev. Lett. 38*:368 (1977).

499. D. W. Tong, *IEEE Trans. Elect. Insul. EI-17*:377 (1982).

500. G. M. Sessler, J. E. West, and H. Von Seggern, *J. Appl. Phys. 53*:4320 (1982).

501. G. M. Sessler and R. Gerhard-Multhaupt, *IEEE Trans. Elect. Insul. EI-21*(3):411 (1986).

502. C. Alquie, C. Laburthe Tolra, J. Lewiner, and S. B. Lang, *IEEE Trans. Elect. Insul. 27*(4):751 (1992).

503. R. E. Collins, *AWA Tech. Rev. 15*:53 (1973).

504. P. W. Chudleigh, *Appl. Phys. Lett. 21*:547 (1972); *J. Appl. Phys. 47*:4475 (1976).

505. P. V. Murphy, *J. Phys. Chem. Solids 24*:329 (1963).

506. B. Gross, *Z. Phys. 155*:479 (1959); *IEEE Trans. NS-25*:1048 (1978).

507. Yu. N. Novkov, F. I. Polovikov, *Sov. Phys. Solid State 8*:1240 (1966).

508. M. L. Miller and J. R. Murray, *J. Polymer Sci. Pt. 2 A*(4):697 (1966).

509. S. Mascarenhas, Bioelectrets: Electrets in biomaterials and biopolymers, in *Topics in Applied Physics, Vol. 23, Electrets* (G. M. Sessler, ed.), Springer-Verlag, Berlin, 1987, chap. 6.

510. E. Fukada, *IEEE Trans. Elect. Insul. 27*(4):813 (1992).

511. A. Bennis, F. Miskane, N. Hitmi, M. Vignoles, M. Heughebaert, A. Lamure, and C. Lacabanne, *IEEE Trans. Elect. Insul. 27*(4):826 (1992).

512. E. Menefee, Thermocurrent from alpha-helix disordering in Keratin, in *Electrets* (M. M. Perlman, ed.), Electrochemical Society, Princeton, NJ, 1973, p. 661.

513. S. Mascarenhas, *J. Electrostat. 1*:141 (1975).

514. E. Murphy and S. Merchant, in *Electrets* (M. M. Perlman, ed.), Electrochemical Society, New York, 1973, p. 627.

515. C. Linder and I. Miller, *J. Phys. Chem. 76*:3434 (1972).

516. M. Goel, S. Meera, and P. Pillai, In Abstr., Int. Workshop on Electrical Charges in Dielectrics, Kyoto, Japan (F. Kudada, ed.), 1978, p. 64.

517. E. Fukada and I. Yasuda, *Jpn. J. Appl. Phys. 3*:117 (1964).

518. E. Fukada, *Ferroelectrics 60*:285 (1984).

519. E. Fukada, *Rep. Prog. Polymer Phys. Jpn. 33*:269 (1991).

520. H. Athenstaedt, *Z. Anat. Entwickl. Gesch. 136*:249 (1972).

521. S. B. Lang, *Nature 212*:704 (1966).
522. E. Fukada, Piezoelectric Properties of Biological Polymers, *Quat. Rev. Biophys. 11*:59 (1983).
523. P. K. C. Pillai and M. Goel, *Polymer 16*:5 (1975).
524. D. Chatain, C. LaCabanne, M. Maitrot, G. Seytre, and J. F. May, *Phys. Stat. Solid. A 16*: 225 (1973).
525. M. Reichle, T. Nedetzka, A. Mayer, and H. Vogel, *J. Phys. Chem. 74*:2659 (1970).
526. A. Ledwith, K. C. Smith, and S. M. Walker, *Polymer 19*:51 (1978).
527. H. Frohlich, *Int. J. Quantum Chem.* 2:641 (1968); *Phys. Lett. 44 A*:385 (1973); *Proc. Natl. Acad. Sci. USA* 72:4211 (1975).
528. J. J. Ficat, G. Escourrou, M. J. Fauran, R. Durraux, P. Ficat, C. Lacabanne, and F. Micheron, *Ferroelectrics 51*:121 (1983).
529. J. J. Ficat, T. Thiechart, P. Ficat, C. Laeabanne, F. Micheron, and I. Bab, *Ferroelectrics 60*: 313 (1984).
530. N. M. West, J. E. West, J. H. Revere, and M. C. England, *J. Endodont.* 5:208 (1979).
531. R. Sekar, A. K. Tripathi, T. C. Goel, and P. K. C. Pillai, *J. Appl. Phys. 62*(10):4, 196 (1987).
532. G. K. Narula, P. K. C. Pillai, *J. Mat. Sci. Lett.* 8:627 (1989).
533. A. Tripathi, A. K. Tripathi, and P. K. C. Pillai, *J. Appl. Phys. 64*(4):2301 (1988).
534. R. Sekar, A. Tripathi, T. C. Goel, and P. K. C. Pillai, *J. Mat. Sci.* 22:3353 (1987).
535. N. K. Kalfoglou, *J. Appl. Polymer Sci. 32*:5247 (1986).
536. B. Hahn and V. J. Breety, *Macromolecules 18*:718 (1985).
537. C. Domenici, D. De Rossi, R. Nannini, and R. Verui, *Ferroelectrics (G.B.) 60*:61 (1984).
538. W. Medycki, B. Hilezer, J. K. Kruger, and A. Marx, *Polymer Bull. 11*:429 (1984).
539. P. K. C. Pillai, G. K. Narula, A. K. Tripathi, and R. G. Mendiratta, *J. Appl. Phys. 53*:6455 (1982).
540. J. Wanski and M. Kryzewski, The phase structure of poly(N-vinyl carbazole) and poly(carbonate) as studied by thermally stimulated current and thermo-optic analysis, in *Polymer Blends, Processing, Morphology and Properties* (M. Martuscelli, R. Palumbo, and M. Kryzewski, eds.), Plenum Press, New York, 1980.
541. H. S. Nalwa and P. Vasudevan, *Macromol. Chem. Makromol. Chem. Phys. 186*:1255 (1985).
542. P. K. C. Pillai, G. K. Narula, A. K. Tripathi, and R. G. Mendiratta, *Phys. Stat. Solid. (a)*: 77:695 (1983).

2
Crystal Structure and Phase Transition of PVDF and Related Copolymers

Kohji Tashiro
Osaka University, Toyonaka, Osaka, Japan

I. INTRODUCTION

Poly(vinylidene fluoride) (PVDF) exhibits a variety of characteristic mechanical and electric properties, such as piezoelectricity (the largest among the synthetic polymers), pyroelectricity, nonlinear optical property, etc. In a relatively early stage of the history of this polymer, a possibility of using it as a ferroelectric material was pointed out on the basis of crystal structure analysis by X-ray diffraction technique. If this were so, PVDF would be the first ferroelectric polymer, making it unique among many inorganic and organic ferroelectric substances. But it took a long time—about 10 years—until the ferroelectricity of PVDF was proved experimentally.

PVDF has a very simple chemical formula, $-CH_2-CF_2-$, intermediate between polyethylene (PE) $-CH_2-CH_2-$ and polytetrafluoroethylene (PTFE) $-CF_2-CF_2-$. This simplicity of chemical structure gives both high flexibility (as much as PE) and some sterochemical constraint (as seen in PTFE) to the main-chain structure of PVDF. Because of these structural characteristics, therefore, PVDF takes many types of molecular and crystal structures, which change depending on the preparation conditions of the samples. The crystal form which exhibits ferroelectricity, form I or β, is only of these crystalline modifications. Depending on the molecular conformation and the chain packing in the unit cell, the mechanical and electric properties are affected sensitively. These crystalline modifications can be transformed to each other reversibly or irreversibly under suitable external conditions. The transition mechanism is complicated but can be understood qualitatively on the basis of trans–gauche conformational exchange.

Although the chemical structure of PVDF is relatively simple, the monomeric unit has a directionality of CH_2(head)–CF_2(tail). Thus a structural irregularity of head-to-head and tail-to-tail linkage of monomers is introduced more or less into the skeletal chain. This type of irregularity affects the structural stability of the crystal modifications significantly. In other words, introduction of some structural irregularity or other kind of unit as a comonomer may give us a good (or bad ?) chance to modify the sterochemical structure of PVDF and to improve the physical properties. One of the typical examples with good success is a synthesis of the copolymer between VDF and trifluoroethylene (TrFE). That is, as explained in a later section, the measurement of the temperature dependence of dielectric constant, etc., revealed the existence of a ferroelectric phase transition between the ferro-/and paraelectric crystalline phases. This discovery of ferroelectric phase transition was the first example in synthetic polymers. By investigating the characteristic properties of the copolymers, the essence of the ferroelectricity of PVDF can be understood thoroughly. The structure and transition behavior of these copolymers are affected also by the sample preparation conditions as well as by the VDF/TrFE content. The existence of TrFE monomer plays an important role in a clear observation of ferroelectric phase transition. A change of TrFE comonomer into another type of monomer—tetrafluoroethylene (TFE), vinyl fluoride (VF), etc.—modifies the phase-transitional behavior remarkably.

Application of the concepts obtained in the study of PVDF and its copolymers has been made for other types of novel polymers with similar polar chemical structure. For example, poly(vinylidene cyanide) (PVDCN) and its copolymer with vinyl n-fatty acid ester have chemical structures close to that of PVDF; F atoms are displaced by CN groups. These polymers are also attracting attention because of their characteristic electric properties. Odd-numbered members of nylon (nylon 7, nylon 11, etc.) and ferroelectric liquid-crystalline polymers (FLCP) are also the candidates for ferroelectric polymers. These substances are still new, and their structural study has not yet been developed extensively compared with the study of PVDF and its copolymers.

In this chapter a detailed description of the structure and phase-transition behavior of PVDF and its copolymers with TrFE, TFE, etc., will be given on the basis of experimental results of X-ray diffraction, infrared and Raman spectroscopy, neutron scattering, NMR, thermal analysis, electric measurements, etc. The phase-transition mechanism will be discussed experimentally and theoretically from various points of view. The ferroelectric behavior of other types of polymers, i.e., odd nylons, VDCN copolymers, FLCP, etc. will also be discussed based on experimental data of X-ray diffraction and vibrational spectroscopy.

II. MOLECULAR AND CRYSTAL STRUCTURE OF PVDF

A. Outline

As mentioned in the introductory section, the uniqueness of PVDF may originate from its chemical structure just intermediate between PE and PTFE. Because of stereochemical repulsion between neighboring CF_2 and CF_2 groups, PTFE takes a helical conformation [1]. Although the torsional angle around the C—C bond (ca. 169°) deviates only slightly from the trans-rotational state (180°) of the planar-zigzag PE chain, the resultant skeletal chain of PTFE is appreciably rigid compared with that of PE, in which the trans and gauche bonds are easily exchangeable in the melt state. In fact, the critical ratio of PTFE random coil (= $\langle r^2 \rangle / n l^2$: $\langle r^2 \rangle$ is a mean-squared end-to-end distance of a finite chain with C—C bond length l and degree of polymerization n) was calculated to be ca. 30, appreciably higher than the value 7 of PE coil [2]. In the case of PE, on the other hand, the all-trans planar-zigzag conformation is the most stable, as observed predominantly in the crystalline state. The gauche bonds are also energetically stable but can appear only partly so at a temperature close to the melting state [3]. In PVDF, the CH_2 group plays a role of diluting the direct CF_2-CF_2 interactions by locating at an intermediate position between the adjacent CF_2 groups. By taking a suitable conformation, therefore, both the trans and gauche isomers are stably existent with a statistical probability similar to each other. This may be one of the main reasons why PVDF (and its copolymers) can take a variety of molecular conformation as well as crystal structures, even in the crystalline state.

As is already known, PVDF can take at least three types of molecular conformations, TGT$\bar{\text{G}}$, TTTT, and TTTGTTT$\bar{\text{G}}$, as explained below. Concerning the packing modes of these molecular chains into the unit cell, four types of crystalline modifications have been clarified to exist (another two types of crystalline modifications have been additionally proposed). These four crystalline modifications are named forms α, β, γ, and δ (or, correspondingly, forms II, I, III, and IV (or polar form II; II_p) in order of discovery in the history of PVDF. In Figures 1 and 2 are shown the molecular and crystal structures of these four modifications. In Table 1 are listed the unit cell parameters of these forms. In Figure 3 is summarized the interrelation between these modifications which can be transferred relatively easily but in a complicated manner by changing an external condition.

B. Detailed Description of Each Crystalline Modification

1. Form I

a. Preparation. Crystalline form I is prepared most typically by stretching at room temperature the unoriented form II which is obtained by cooling the melt at a normal

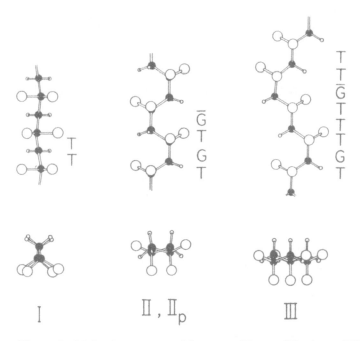

Figure 1 Molecular structure of four crystalline modifications of PVDF [252].

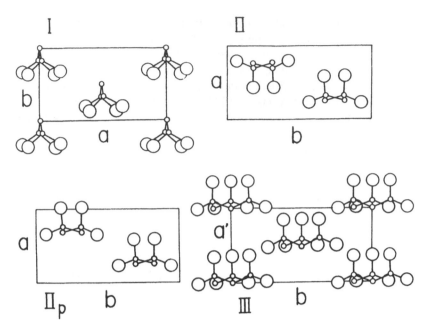

Figure 2 Crystal structure of four crystalline modifications of PVDF [252].

rate (e.g., 10–20°C/min or higher). Slow cooling of the melt does not give form II but mostly gives the form III spherulite; some form I spherulites are also obtained although the form I spherulites grow four to five times more slowly than the form III [4–6]. The stretching of form III also gives form I. But these mechanical processes give only (uniaxially) oriented sample. Rolling the form II film at room temperature results in the same situation, although some degree of preferential orientation may occur (usually the *b* axis orients preferentially in parallel with the film surface [7]). (If you try to get a doubly oriented form I sample, you may roll the uniaxially oriented rod at room temperature along the original draw direction.) Unoriented form I film can be obtained by

Table 1 Crystallographic Data of Crystalline Modifications of PVDF

	Unit cell parameters	Space group	Molecular chain	N^a
Form I	$a = 8.58$ Å, $b = 4.91$ Å, c (f.a) $= 2.56$ Å	$Cm2m$-C_{2v}^{14}	Slightly twisted Planar-zigzag	2
Form II	$a = 4.96$ Å, $b = 9.64$ Å, c (f.a) $= 4.96$ Å, $\beta = 90°$	$P2_1/c$-C_{2h}^{5}	TGT$\overline{\text{G}}$	2
Form II$_p$	Essentially the same as form II	$P2_1cn$ (?)	TGT$\overline{\text{G}}$	2
Form III	$a = 4.96$ Å, $b = 9.58$ Å c (f.a) $= 9.23$ Å, $\beta = 92.9°$	Cc-C_{s}^{4}	TTTGTTT$\overline{\text{G}}$	2

$^a N$ = number of chains in a unit cell; f.a. = fiber axis.

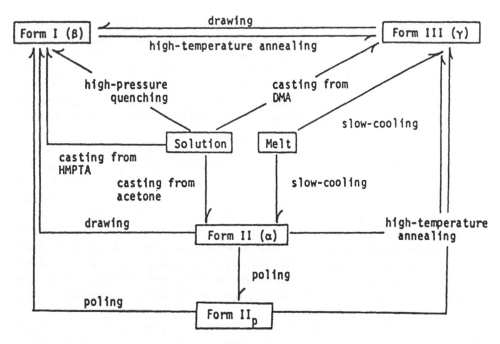

Figure 3   Interconversion between crystal modifications [252].

casting from a solution of PVDF dissolved into strongly polar hexamethylphosphoric-triamide (HMPTA) [8] or by precipitating a dimethylacetamide (DMA) solution into acetic acid aqueous solution [4,5]. Crystallization from a 0.02% boiling cyclohexanone solution at 80°C is reported to give form I single crystal [9]. Unoriented form I sample can also be obtained by extremely rapid quenching of the molten sample into a liquid nitrogen reservoir; the sample thickness should be as thin as possible so as to increase the thermal conductivity [10,11]. High-pressure and high-temperature annealing of the sample (ca. 400 MPa, 570 K) may also give the form I crystal, in which the molecular chains crystallize into extended-chain crystal (ECC) [12]. High-pressure quenching of the melt gives form I crystal [13]. Application of high voltage to the thin film results in a phase transition of form II to polar form II and finally to form I with the polar *b* axis oriented preferably along the normal to the film surface [14–16].

b.   X-Ray Structure Analysis.   The crystal structure of PVDF form I was proposed by several authors. Lando et al. reported a crystal structure essentially equivalent to that shown in Fig. 1 [17]. The molecular conformation is all-trans planar zigzag. The unit cell is of orthorhombic type with the space group $Cm2m-C_{2v}{}^{14}$, in which the two chains are packed with the CF_2 groups parallel to the *b* axis. Judging from this parallel packing of the polar CF_2 groups, Lando et al. predicted even at this early stage the ferroelectricity of this crystalline form. After that, the structure was refined by Gal'perin et al. [18] and by Hasegawa et al. [19]. Their point of refinement was that, as seen in PTFE, the CF_2-CF_2 steric hindrance should make the perfectly planar zigzag structure unstable and the CF_2 groups might be deflected right and left (statistically or regularly) from the zigzag plane. As one possible model, Hasegawa et al. employed the chain conformation model with regularly alternate deflections through the internal rotations around the C—C bonds:

the deflection angle σ is defined as the deviation angle of two neighboring C—C bonds projected along the chain axis (refer to Fig. 1). In order to satisfy the observed fiber period of 2.56 Å, the statistical packing of a deflected chain (repeating period is 5.12 Å) and its mirror image was introduced on the basis of the required space group symmetry; on average, the repeating period becomes 2.56 Å. The deflection angle σ was obtained so that the observed X-ray reflection intensities were reproduced as well as possible; the final σ obtained is 7°, as shown in Figure 1. This deflection angle corresponds to the internal rotational angle of the skeletal CC bond, 171.6°.

The deflected chain structure is supported by the energy calculation. Sometimes the energy calculations are performed with an assumption of perfectly planar zig-zag chain conformation [20]. Even recently, Karasawa and Goddard carried out a packing energy minimization of form I crystal and still obtained the perfectly planar-zigzag chain conformation as the lowest-energy structure [21]. By modifying the F $\cdots$ F nonbonded interatomic interaction potential energy function suitably so as to satisfy the vibrational spectroscopic data (discussed in a later section), the deflected chain conformation is found to be the lowest stable energy structure. Calculation by Hasegawa et al. gave the deflection angle 7° as the lowest energy state for the altnerately deflected chain model, coincident with the X-ray-analyzed structure [22]. Recently Tashiro et al. supported this result by energy minimization [23].

As has been pointed out, the PVDF chain is highly flexible. Therefore the zigzag chain may include various types of conformational defects, as proposed for the planar-zigzag chain of n-paraffin and PE [24]. Takahashi et al. found that the X-ray fiber diagram of form I contains some diffuse streaks [25,26]. The first detected streak is along the line connecting the 001 and the 110, 200 Bragg reflections [25], although the amorphous halo overlaps partly and makes it difficult to see the whole streak. Takahashi et al. assigned this streak to the TGTG̅-type kink structure contained in the all-trans chain; ... TTTTTTTTTTGTG̅TGTG̅TTTTTTTT.... The TGTG̅ form is the conformation taken by crystalline form II. This type of kink might be generated by shear stress applied to the sample. Another type of streak was also found [26], which appears between the 110–110 and 020–220 Bragg reflections and between the 110–200 and 111–201 reflections, suggesting disorder due to slippage of the structural units along the b axis and along the 110 plane, respectively. Although quantitative interpretation is not yet complete, these structural disorders might correlate with the above-mentioned intramolecular kink defect.

c. Vibrational Spectra. In order to clarify the details of the molecular and crystal structures of form I, the infrared and Raman vibrational spectroscopic measurements are also useful. In Figures 4 and 5 are reproduced the infrared and Raman spectra of form I [7,8,27,28]. Essentially, the spectra can be interpreted reasonably on the basis of the parallel packing of the simple planar-zigzag chains. The vibrational analysis for an isolated single chain is approximately enough for the interpretation of the spectra in the fingerprint region of 400–4000 cm^{-1} [7,8,29–31]. The factor group of the planar-zigzag form I chain is isomorphous to the point group C_{2v}. The vibrational modes can be classified into four symmetry species as follows.

$$\Gamma_m = 5A_1(\mu_b', \alpha_{bb}', \alpha_{cc}') + 2A_2(\alpha_{ac}') + 3B_1(\mu_c', \alpha_{bc}') + 4B_2(\mu_a', \alpha_{ab}')$$

where μ_i' and α_{ij}' are the transition dipole moment and the transition polarizability component, respectively. For example, 12 internal vibrational bands are expected to ap-

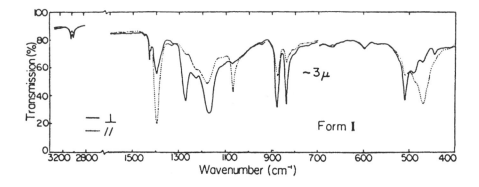

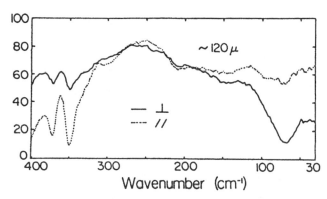

Figure 4 Polarized infrared spectra of PVDF form I [8].

pear in the infrared spectra: five A_1 bands with perpendicular polarization ($\mu'//b$), three B_1 bands with parallel polarization ($\mu'//c$), and four B_2 bands with perpendicular polarization ($\mu'//a$), where "perpendicular" and "parallel" mean the transition dipole points into the direction perpendicular and parallel to the chain axis (c), respectively. On the other hand, if we want to interpret the spectra in the whole frequency region, factor group analysis of the cyrstal structure has to be carried out [8]. The unit cell of form I contains two chains as shown in Figure 2, but an optically active (i.e., spectroscopically observable) lattice vibrational mode is predicted to be only one, a librational mode around the c axis as illustrated in Figure 6. The chains located at corner and center positions of the cell are crystallographically identical to each other because the Bravais lattice is of the C-centered type (the space group Cm2m). Therefore the true primitive unit cell is that constructed by connecting the corner and center chains and so contains only one chain. In general, the spectroscopically active mode is the mode in which the identical molecules in the neighboring unit cells vibrate in phase. In the case of form I crystal, this corresponds to the mode with the two chains at corner and center lattice sites librating in phase as shown in Figure 6a. This librational mode belongs to the B_2 symmetry species and can be observed as a perpendicular band in the far-infrared spectrum. The band at ca. 70

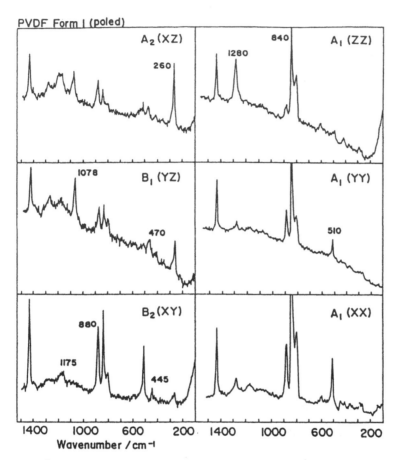

PVDF Form I (poled)

Figure 5 Polarized Raman spectra of the highly poled PVDF form I [7]. The coordinate system is as follows: $Z//$ draw axis and $Y//$ normal to the film surface.

cm^{-1} corresponds to it (Fig. 7a). This band is very broad, suggesting a disorder in the chain packing in the lattice.

The bands observed in Figures 4 and 5 can be assigned more definitely by carrying out normal mode calculation for the single chain and the crystal. The details may be found in Refs. 7, 8, 30, and 31. The band assignments thus obtained are in Table 2. Through a detailed search of the spectra, we may notice that the number of the observed bands is much more than that predicted from the factor group analysis of the crystal structure with simple planar-zigzag chains. As stated already, the X-ray analysis and energy calculation require the alternately deflected chain conformation. Therefore the monomeric units contained in the fiber repeating period are two, not one, indicating that not only in-phase but also out-of-phase modes should be vibrationally active. A deflection of the chain was tentatively assumed to be regularly alternate, but the statistical deflection might also be possible in some cases, because the X-ray analysis cannot say which model is better, regular or random. The energy difference between the regularly deflected chain

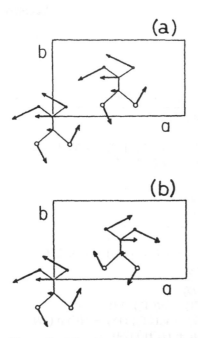

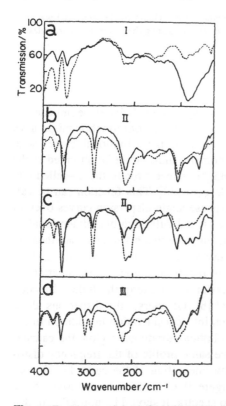

Figure 6 Librational lattice mode of PVDF form I crystal: (a) in phase; (b) out of phase [8]. the spectroscopically active mode is (a).

Figure 7 Polarized far-infrared spectra of PVDF cystalline forms I, II, II$_p$, and III measured at liquid nitrogen temperature [76]. The electric vector of the incident infrared beam is perpendicular (————) or parallel (————) to the draw direction of the sample.

Table 2   Vibrational Modes of PVDF From I Crystal [7]

| Species | Wavenumber (cm^{-1}) | | Potential energy distribution (%)^b |
	Obs.	Calc.^{a}	
A$_1$	2980	2975	v_s(CH$_2$) (99)
	1428	1434	δ(CH$_2$) (81)
	1273	1283	v_s(CH$_2$) (40) $-$ v_s(CC) (22) + δ(CCC) (26)
	840	844	v_s(CF$_2$) (59) + v_s(CC) (17)
	508	513	δ(CF$_2$) (98)
A$_2$	980	980	t(CH$_2$) (100)
	260	265	t(CF$_2$) (100)
B$_1$	1398	1408	w(CH$_2$) (58) $-$ v_s(CC) (34)
	1071	1074	v_a(CC) (53) + w(CH$_2$) (25) $-$ w(CF$_2$) (22)
	468	471	w(CF$_2$) (90)
B$_2$	3022	3024	v_a(CH$_2$) (99)
	1177	1177	v_a(CF$_2$) (71) $-$ r(CF$_2$) (18)
	880	883	r(CH$_2$) (62) $-$ v_a(CF$_2$) (18) $-$ r(CF$_2$) (19)
	442	444	r(CF$_2$) (70) + r(CH$_2$) (24)
	70	72	Librational lattice mode

^{a}Calculation for the planar-zigzag chain structure.
^{b}Symmetry coordinates: (v_a) antisymmetric stretching; (v_s) symmetric stretching; (δ) bending; (w) wagging; (t) twisting; (r) rocking. The sign (+) or ($-$) denotes the phase relation among the symmetry coordinates.

and the statistically deflected chain may not be so large as to definitely eliminate the possibility of statistical deflection. In an extreme case, the chain conformation might have no translational symmetry along the chain axis. Then all the vibrational modes with various phase angles between the neighboring monomeric units could be observed in the spectra. This means that the vibrational spectra might reflect directly the so-called frequency distribution function or the density of state $g(v)$ [32]. This function can be calculated based on the three-dimensional frequency-phase angle dispersion curves over all the directions of the crystal. In the case of polymer crystal with high one-dimensionality, the intramolecular interactions along the chain direction are much stronger than those in the lateral directions. Therefore the dispersion curves calculated for an isolated chain may represent the density of state in the fingerprint region to a good approximation. In Figure 8 is shown the calculated frequency dispersion curve along the (00δ_c) direction in the first Brillouin zone of PVDF form I crystal [8]. Of course, the dispersons along the other directions (δ_a, δ_b, δ_c) should also be taken into consideration in addition to the (00δ_c) direction, but except for the low-frequency lattice vibrational region, the calculation along the (00δ_c) direction gives essentially the same profile of the frequency distribution function $g(v)$. The $g(v)$ can be obtained by integration of the curve length contained in a constant span of a frequency. In Figure 9 is made a comparison of the calculated $g(v)$ with the actually observed infrared spectra. It should be noticed here that direct comparison of the spectra for both the intensity and band position cannot be made because the spectra are, in general, expressed as a product of $g(v)$ and the squared

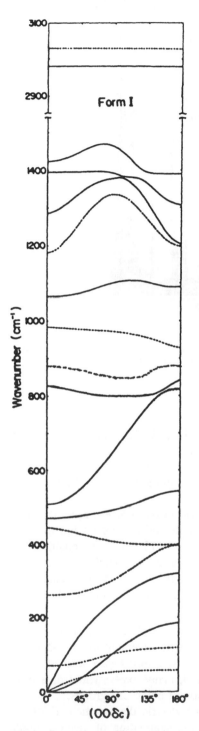

Figure 8 Vibrational frequency–phase angle dispersion curve calculated for PVDF form I crystal [8]. The solid and broken lines denote the in-plane and out-of-plane modes, respectively.

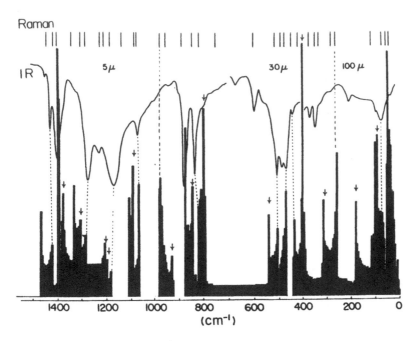

Figure 9 Frequency distribution function calculated for PVDF form I [8]. The solid curves represent the observed infrared spectra and the vertical lines at the top denote the positions of the Raman lines.

transition dipole moment μ': $I(\nu) \approx g(\nu) * (\mu')^2$. In Figure 9, the bands connected by broken lines to the peaks of the $g(\nu)$ originate from the optically active in-phase modes of the planar zigzag chain structure. Other peaks also correspond to the observed additional bands: for example, the shoulders of the bands at 890 and 840 cm^{-1} correspond to the double peaks in this frequency region.

Some bands still cannot be explained by these $g(\nu)$ peaks. For example, the 600- and 1230-cm^{-1} bands are observed for the molten sample and are considered to come from the amorphous phase. The bands at 1450, 1330, and 678 cm^{-1} are considered to be due to the head-to-head and tail-to-tail irregular linkages contained in the skeletal chain. This assignment was confirmed by comparing the spectra with those measured for alternating copolymer of CH_2CH_2 and CF_2CF_2, where the corresponding bands can be observed intensely [8,33,34].

d. Ferroelectricity of Crystal Form I. A ferroelectric is defined as a substance in which the electric dipoles can reorient reversibly following inversion of an externally applied electric field vector *without any change in the structure before and after the reorientation* [35]. In the case of PVDF form I crystal, many experimental data have been reported to show the electrically induced inversion of polar CF_2 groups.

D-E Hysteresis Loop and Inversion Current: the electric displacement D or the polarization P of the crystal is determined by the total summation of electric dipole moments in a unit volume of the lattice. Therefore the change in dipole orientation should reflect on the D (and P) explicitly. The time dependence of electric dipole reorientation is equivalent to the time evolution of charges, i.e., the current i. A clear D-E hysteresis loop and the corresponding inversion current were measured for PVDF form I [36,37]. It is important to notice that the *D-E* hysteresis loop can be obtained even at low temperature—below the glass transition point, ca. $-40°C$—suggesting that the dipole inversion may occur in the crystalline region.

Infrared and Raman Spectra: The infrared absorption intensity I is proportional to the square of the inner product between the transition dipole μ' and the electric field vector E;

$$I \propto (\mu'E)^2 = (\mu')^2 E^2 \cos^2 \phi \tag{1}$$

where ϕ is an angle between the μ' and E. As predicted from the factor group analysis of form I crystal, the transition dipole of the A_1 vibrational modes is parallel to the polar b axis and that of the B_2 modes is along the a axis [7,8]. Therefore, if the CF_2 dipoles change their orientation around the chain axis under the application of an external electric field, the A_1 and B_2 bands should change their relative intensities (see Fig. 10). The poling induces preferential orientation of the b axis along the film normal. Therefore, when the infrared beam is incident along the normal to the film surface, the transition dipoles μ_b' and μ_a' become, respectively, perpendicular and parallel to the incident electric vector. That is, the A_1 band is decreased and the B_2 band is increased after poling. In fact, as seen in Figure 10, the A_1 and B_2 bands behave in such a manner as predicted above, while the B_1 bands with the transition dipole parallel to the chain axis almost do not change their intensity, even after poling [7].

In-situ observation of dipole inversion was made by infrared measurement under an electric field [38,39]. Corresponding to the change in the *D-E* hysteresis loop, the infrared band intensity (for example, of the A_1 band) exhibits a clear hysteresis loop.

Poling-induced reorientation of dipoles can also be detected through polarized Raman spectral measurement. For example, Figure 5 shows polarized Raman spectra measured before and after poling treatment of a rolled form I sample [7]. In this figure the notation (XZ) or (YZ) indicates the Raman scattering geometry: the incident laser beam has an electric vector along the X or Y axis, respectively, and the scattered signal has an electric vector along the Z axis, where the Z axis is parallel to the draw axis of the chain and the Y axis is normal to the film surface. Before poling, the A_2 band at 260 cm^{-1} (with Raman polarizability tensor component α_{ac}') is observed relatively intensely for the (YZ) spectrum than for the (XZ) spectrum, while the B_1 band at 1078 cm^{-1} (α_{bc}') has an inverse tendency, indicating that the a axis orients preferably along the Y axis or the normal to the film surface. After poling, the relative intensity changes. The B_1 band is observed more intensely in the (YZ) spectrum and the A_2 band is in the (XZ) spectrum, indicating that the b axis is induced to orient into the Y direction from the X direction, although the degree of reorientation is not very high.

X-Ray Diffraction: The change in orientation of the unit cell axes should also be clearly detected in X-ray diffraction measurements. Kepler and Anderson [40] and Odajima and co-workers [41–45] evaluated the degree of b-axis orientation by measuring the change in the X-ray reflectional profile. The rolled sample was poled under high electric field and the $(311 + 021)$ reflectional intensity was measured as a function of azimuthal angle φ, which was defined as an angle between the polar b axis and the poling direction. The biaxial orientation distribution function $q(\varphi)$ is given as

$$q(\varphi) = \frac{I(\varphi)}{\int I(\varphi) \, d\varphi} \tag{2}$$

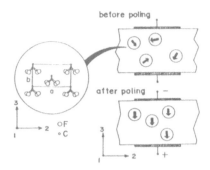

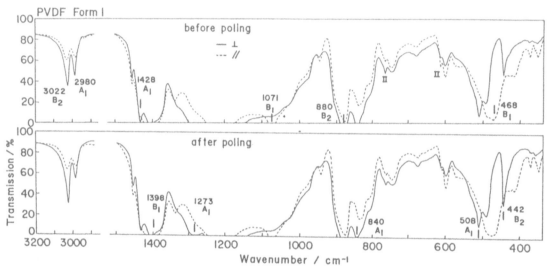

Figure 10 Structural change caused by poling of PVDF form I crystal and the associated infrared spectral change [7].

where $I(\varphi)$ is an intensity totally integrated over the reflectional width concerning the chain axial orientation. $q(\varphi)$ is approximately expanded in the following form:

$$q(\varphi) = q_0 + \langle\cos 2\varphi\rangle * \cos 2\varphi + \langle\cos 4\varphi\rangle * \cos 4\varphi + \langle\cos 6\varphi\rangle * \cos 6\varphi \qquad (3)$$

where $\langle\cos 2\varphi\rangle$ is a measure of the b axial orientation. This coefficient changes from -0.10 to 0.10 after poling, indicating that an a-axial orientation in the unpoled sample changes into the b-axial orientation after poling, although the degree of reorientation is not very high. Other coefficients change as follows: $\langle\cos 4\varphi\rangle = 0.04 \rightarrow 0.00$ and $\langle\cos 6\varphi\rangle = -0.05 \rightarrow -0.04$. That is, only the coefficient $\langle\cos 6\varphi\rangle$ does not change, even after poling. The coefficient $\langle\cos(m\varphi)\rangle$ is an average among the various orientations of the b axis induced during the poling process:

$$\langle\cos(m\varphi)\rangle = \int \cos(m\varphi) * g(\varphi)\, d\varphi$$

where $g(\varphi)$ is a distribution function of the b axis. An observed invariance of $\langle\cos 6\varphi\rangle$ can be interpreted reasonably by assuming that the $g(\varphi)$ might be nonzero only at the angle $\varphi = \varphi_0 + (\pi/3)n$, where φ_0 is an angle in the initial b-axis direction before poling and $n = 1,2, \ldots$. In other words, the b axis may be assumed to reorient in a 60°-step rotation. In this case, cos $6\varphi(m = 6$ in the above integration) becomes cos $6(\varphi_0 + \pi/3)$ and cos $6(\varphi_0 + 2\pi/3)$, which are both equal to cos $6\varphi_0$, resulting in an invariance of the coefficient $\langle \cos 6\varphi \rangle$.

Another application of X-ray diffraction technique to prove the reorientation behavior of the polar b axis of form I crystal is use of the Bijvoet pair. Let us assume that the b axis orients as shown in Figure 11a after enough poling treatment. By changing the sign of the applied electric field, the b axis will reorient as shown in Figure 11b. Thus the film surface is parallel to the (110) plane in case a and to the ($\overline{1}\overline{1}0$) plane in case b. For polar crystal, in general, the X-ray reflection intensity of the $hk0$ plane is not equal to that of the ($\overline{h}\overline{k}0$) plane, due to the effect of anomalous dispersion term included in an atomic scattering factor. Takahashi et al. measured the X-ray intensity from the {110} planes by alternately changing the sign of the electric field applied to the form I film

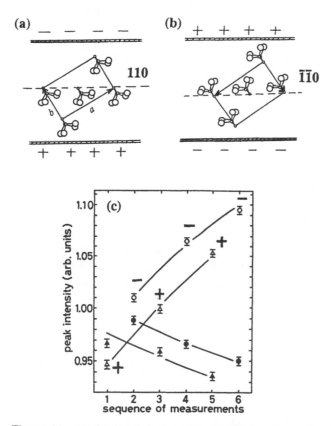

Figure 11 (a), (b) Reorientation of the PVDF form I crystal under the applied electric field, and (c) the X-ray intensity change induced reversibly with a change in the sign of the applied electric field of 280 MV/m [46]. Measurement was made by changing the sign of the electric field alternately. The initial film includes both the form I and form II crystals. The gradual intensity increment is due to the poling-induced transformation of form II to form II_p and to form I. The open symbols are for crystal form I, and solid symbols are for crystal form II_p.

[46]. As shown in Figure 11c, the intensity changes reversibly between I^+ and I^-, corresponding to the change in the sign of the field, indicating a reorientation behavior of the b axis the manner illustrated in Figures 11a and 11b.

NMR: From the half-width of proton free-induction decay function (FID) during the relaxation of magnetic spin orientation after an impulse by a strong magnetic field H_0, the spin-spin relaxation time T_2 was evaluated for various angles γ between the H_0 and the poling direction of PVDF form I film. The anisotropy in the T_2 thus obtained was analyzed as a function of the angle γ, from which the orientation function of the polar b axis was evaluated [47–49]. The γ dependence of T_2 cannot be explained by using a model of 180° inversion of the b axis, but it can on the basis of the 60° rotation model which is essentially equivalent to that described in the X-ray diffraction experiment.

The thus clarified reorientation behavior of the CF_2 dipoles under an electric field was interpreted theoretically on the basis of the idea of propagation of soliton (kink) along the chain axis by Hopfinger and co-workers [50,51]. Broadhurst and Davis [52] derived a thermodynamical equation to reproduce the *D-E* hysteresis loop of PVDF form I for the model that the rigid-rod segments can rotate about their chain axis between the six sites of different orientation directions.

e. Long-Range Interactions in PVDF Form I Crystal.   In ferroelectric substances, in general, long-range dipole–dipole interactions play an important role in their physical properties and ferroelectric phase-transition behavior. It is difficult, however, to estimate this dipole–dipole interaction quantitatively. One useful method is the measurement of the LO–TO splitting of polar phonons in the Raman spectra, where LO and TO denote the longitudinal and transverse optical modes, respectively [53]. In the LO mode, the direction of the transition dipole **M** generated by the vibrations of the charged atoms is parallel to the propagation direction of phonon **q**, while the moment **M** of the TO mode is perpendicular to **q**. The phonon accompanying such a dipole change is called a polar phonon. The transition dipoles generate a macroscopic electric field that couples in feedback fashion with the LO modes of the polar phonon. Due to the effect of these self-generated electrostatic forces, the vibrational frequency of the LO mode is shifted toward higher frequencies than that of the TO mode, the frequency of which is determined by the short-range interaction in the crystal lattice. This phonomenon is called "LO–TO splitting." Since the vibrational mode accompanying the change of dipole moment **M** is infrared active, the LO–TO splitting is expected to be observed in the Raman spectra of the "infrared-active" polar phonons.

In order to provide a more concrete image, some equations are developed here [54]. The three-dimensional displacement vector of an atom is denoted by $\mathbf{u}_k(\mathbf{r})$. The macroscopic polarization and electric field are denoted by $\mathbf{P}(\mathbf{r})$ and $\mathbf{E}(\mathbf{r})$, respectively. Under the harmonic, adiabatic, and electrostatic approximations, these quantities are coupled together in the following manner:

$$\left(\frac{m_k d^2\mathbf{u}_k(\mathbf{r})}{dt^2}\right) = \Sigma_{k'}\mathbf{R}_{k,k'}\mathbf{u}_{k'}(\mathbf{r}) - \mathbf{Q}\mathbf{E}(\mathbf{r}) \tag{4}$$

$$\mathbf{P}(\mathbf{r}) = \Sigma_k \mathbf{Q}_k\mathbf{u}_k(\mathbf{r}) + \chi\mathbf{E}(\mathbf{r}) \tag{5}$$

$$\mathbf{E}(\mathbf{r}) = -4\pi\mathbf{q}^0[\mathbf{q}^0\mathbf{P}(\mathbf{r})] \tag{6}$$

where m_k is the mass of the kth atom. $\mathbf{R}_{k,k'}$ is a force constant connecting the k and k' atoms, and **Q** is a matrix representing the apparent charges Q_k of the atoms. χ is the high-frequency electronic susceptibility. $\mathbf{q}^0 = \mathbf{q}/|\mathbf{q}|$ is a unit vector in the direction of $\mathbf{q}$. Equations (5) and (6) show, respectively, the generation of polarization and electrostatic field induced by atomic vibrations $\mathbf{u}$. This electrostatic field affects the vibrational mode $\mathbf{u}$ in a form $\mathbf{QE}$ in Eq. (4). In these equations the length of the phonon vector q is assumed as

$$\frac{\omega}{c} \ll q \ll \frac{1}{a}$$

where ω is the vibrational frequency of the atoms and a is the order of the unit cell. The $\mathbf{u}$, $\mathbf{P}$ and $\mathbf{E}$ change in an exponential form as

$$\mathbf{u}_k(\mathbf{r}) = \mathbf{u}_k^0 \exp[i(\mathbf{qr} - \omega t)]$$
$$\mathbf{P}(\mathbf{r}) = \mathbf{P} \exp[i(\mathbf{qr} - \omega t)] \tag{7}$$
$$\mathbf{E}(\mathbf{r}) = \mathbf{E} \exp[i(\mathbf{qr} - \omega t)]$$

By substituting Eq. (7) into Eqs. (4)–(6), we get finally the following dynamical equation,

$$|(\Omega_j^2 - \omega^2)\delta_{jj'} + (4\pi/\epsilon^\infty(\mathbf{q}^0))(\mathbf{q}^0 \cdot \mathbf{M}_j)(\mathbf{q}^0 \cdot \mathbf{M}_{j'})| = 0 \tag{8}$$

Ω_j is an eigenvalue of the jth mode and is determined by a short-range interaction $\mathbf{R}$ in the following equation:

$$|\mathbf{m}^{-1/2}\mathbf{R}\mathbf{m}^{-1/2} - \Omega^2| = 0 \tag{9}$$

where $\mathbf{m}$ is a matrix consisting of an atomic mass m_k and $\mathbf{R}$ is a matrix including $R_{k,k'}$. $\epsilon^\infty(\mathbf{q}^0)$ is a dielectric constant at infinitely high frequency and is given by

$$\epsilon^\infty(\mathbf{q}^0) = \Sigma \, \epsilon_i^\infty \times (q_i^0)^2 \qquad (i = 1, 2, 3) \tag{10}$$

q_i^0 is a component of $\mathbf{q}^0$ and ϵ_i^∞ is given as follows using a refractive index n_i:

$$\epsilon_i^\infty = n_i^2 \tag{11}$$

In Eq. (8), the transition dipole $\mathbf{M}_j$ of the jth mode is given by

$$\mathbf{M}_j = \frac{\Sigma_k Q_k e_j}{\sqrt{m_k}}$$

where e_j is an eigenvector of the jth mode determined by Eq. (9). The term $(\mathbf{q}^0 \cdot \mathbf{M}_j)$ in Eq. (8) is an inner product between the vectors $\mathbf{q}^0$ and $\mathbf{M}_j$. Therefore, when the phonon q propagates in the direction of angles θ and θ' from the transition moments $\mathbf{M}_j$ and $\mathbf{M}_{j'}$, respectively, the vibrational frequency ω is dependent on the product $\cos \theta * \cos \theta'$.

As an example, we will consider a case of triglycine sulfate (TGS) single crystal [55], which is one of the few organic ferroelectric substances and is often used as an infrared beam detector. This crystal belongs to the space group $P2_1$–C_2^2 and the normal vibrational modes are classified into two symmetry species, A and B. The A modes have transition moments M along the b axis and the B modes have moment M within the $a'c$ plane. The crystal is set on an optical holder as illustrated in Figure 12. The laser beam is irradiated in the direction $\mathbf{k}_i$, and the 90° scattered beam $(\mathbf{k}_s)$ is collected into the spectrometer. The crystal axis b or c is set at an angle θ from the phonon vector $\mathbf{q}$, which

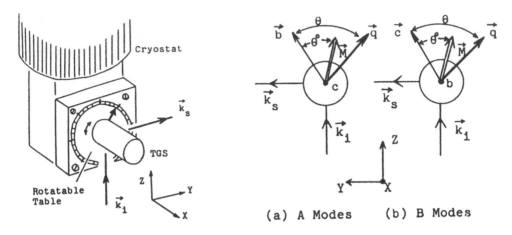

Figure 12 (a) Sample mounting system with a rotatable holder within the cryostat for Raman spectral measurement and (b) scattering geometries for the LO–TO measurements: (a) A modes and (b) B modes [55].

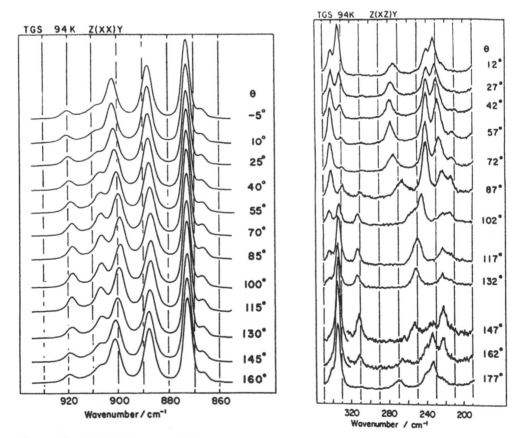

Figure 13 LO–TO frequency shifts measured for the A modes of TGS crystal at liquid nitrogen temperature [55]. The angle θ is between the *b* axis and the phonon vector (refer to Fig. 12).

is determined as $\mathbf{q} = \mathbf{k}_i - \mathbf{k}_s$ (see Figs. 12a and 12b). The angle θ is changed and the Raman spectra are measured. As shown in Figure 13, the band position shifts in a complicated manner with a change of θ. In Figure 14 is plotted the θ dependencies of the Raman band frequencies thus observed. Referring to Figures 12a and 12b, $\mathbf{q}^0$ is expressed as

$$
\begin{aligned}
&\text{A modes:} && \mathbf{q}^0 = (\sin\theta, \cos\theta, 0) && \theta = b \wedge \mathbf{q} && (12)\\
&\text{B modes:} && \mathbf{q}^0 = (\sin\theta, 0, \cos\theta) && \theta = c \wedge \mathbf{q}
\end{aligned}
$$

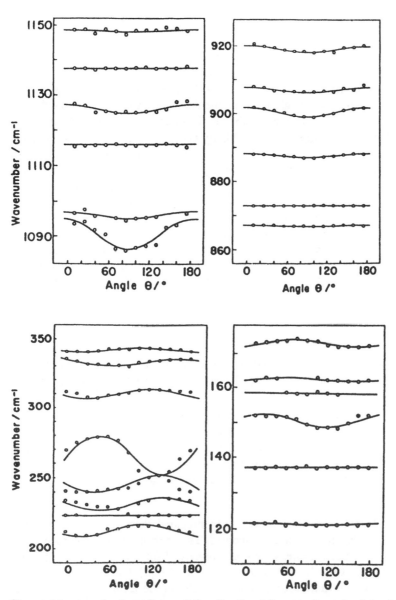

Figure 14 Angular dependence of the vibrational frequency measured for the A (upper) and B modes of TGS single crystal [55]. The angle θ is referred to in Figure 12.

From Eqs. (8), (10), (11), and (12), we get the following equations:

A modes: $$\omega_{LO}^2 = \frac{\Omega_{TO}^2 + 4\pi M^2 \cos^2(\theta - \theta_0)}{n_{a'}^2 \sin^2 \theta + n_b^2 \cos^2 \theta} \tag{13}$$

B modes: $$\omega_{LO}^2 = \frac{\Omega_{TO}^2 + 4\pi M^2 \cos^2(\theta - \theta_0)}{n_c^2} \tag{14}$$

where $n_{a'}$, n_b, and n_c are refractive indices in the a', b, and c directions, respectively. θ_0 is an angle between the M and the b or c axis: $\theta_0 = 0$ for the A modes; θ_0 of the B species is changeable in the $a'c$ plane depending on the type of the vibrational mode. The curves shown in Figure 14 can be reproduced quite well on the basis of Eqs. (13) and (14) by adjusting the parameters Ω_{TO} and M, where the Ω_{TO} corresponds to the frequency at $\theta = 90°$ in Eq. (13) for the A modes and can be checked by the infrared spectra. The solid curves in Figure 14 are the results obtained.

This method was applied to PVDF form I [7]. Different from the case of above-mentioned low-molecular-weight single crystals, the detection of the LO-TO band splitting is difficult because of the broad bandwidth and disturbance by the amorphous bands. The sample used for this experiment was the highly poled form I film. Let us consider the bands of the totally symmetric A_1 modes, which have transition dipoles parallel to the polar b axis. In Figure 15 is shown the setting geometry of the sample. In the case of A_1 species, the TO bands can be measured by setting the b axis (parallel to the normal to the film plane) with an angle $90°$ from the $\mathbf{q}$ vector $(= \mathbf{k}_i - \mathbf{k}_s)$. The LO band should be measured by rotating the film around the c axis so that the b axis is parallel to the $\mathbf{q}$. Figures 16a and 16b show the measured Raman spectra as a function of the rotation angle of the sample (the angle is between the $\mathbf{k}_i$ and the normal to the film.). As predicted from Eq. (8), the LO–TO band splittings $\Delta\omega$ are detected, which are in the range of 0–6 cm^{-1} depending on the vibrational mode. Starting from Eq. (8), the splitting width $\Delta\omega$ is approximately expressed by

$$\Delta\omega = \omega_{LO} - \omega_{TO} \approx \frac{2\pi M^2}{\epsilon^\infty \omega_{TO}} \tag{15}$$

Since the M^2 is proportional to the infrared absorbance A, the LO–TO splitting width is proportional to the oscillator strength A/ω_{TO}. In fact, the observed bands were found to satisfy this relation reasonably.

f. Domain Structure. The ferroelectrics possess a domain structure with the dipoles oriented in various directions. Domains are bounded by the domain walls. High-voltage poling induces a movement and fusion of domain walls, resulting finally in a single-domain structure. A possibility of domain structure in PVDF form I was reported by several authors [56], but detailed discussion will be referred to a later section.

2. Forms II and II$_p$

a. Preparation. Crystal form II is more easily prepared than the other crystalline forms as far as the unoriented sample is concerned. The molten sample is cooled down to room temperature at a normal rate, giving an unoriented form II. An extremely slow cooling of the melt gives form III and also form I in part [4,5]. On the other hand, an extremely high cooling rate (ultrafast quenching) results in the formation of form I, as stated in the previous section. Casting from acetone at room temperature gives the unoriented form II (strictly speaking, some preferential orientation occurs in the film plane).

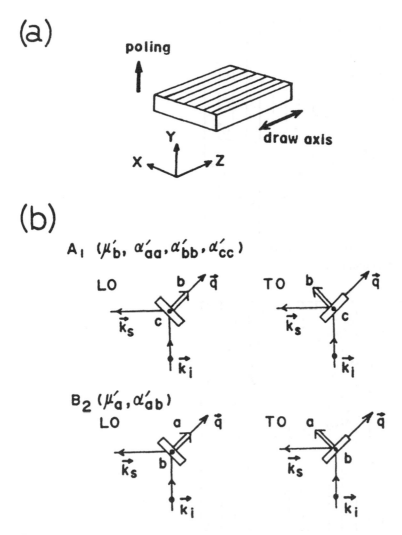

Figure 15 (a) Definition of the coordinate system fixed on the sample film. (b) Setting geometries for the measurement of the LO and TO bands [7]. Illustrations are for the A_1 and B_2 modes of PVDF form I crystal.

In order to get the oriented form II sample, the drawing of the unoriented sample must be made at comparatively high temperature, above 130–160°C, so as to avoid the transformation into form I; a highly oriented and pure form II sample is difficult to obtain. A direct stretching of the melt at room temperature during crystallization also gives uniaxially oriented form II sample. But this sample, after being annealed at 150°C for 24 h with the ends clamped by a metal holder, exhibits streak reflections in the X-ray fiber diagram, as discussed later [57]. Longer annealing of this sample causes gradual transformation from form II to form III [58,59]. Gohil and Petermann obtained this type of oriented form II sample [60], which shows streak electron diffraction, by stretching the ultrathin film cast from dimethyl formamide (DMF) solution at 135–160°C.

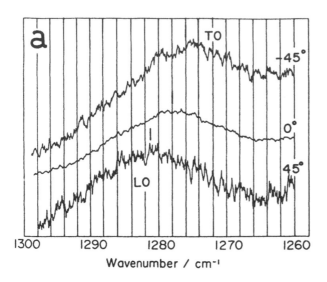

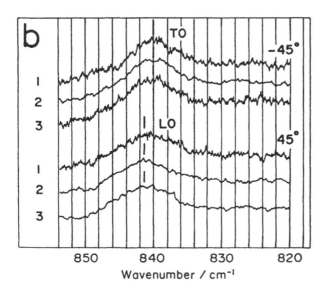

Figure 16 Raman spectra measured for PVDF form I crystal in the frequency regions of 1280 cm^{-1} [7]. The angle is between the Y axis of the sample and the incident photon vector k_i. Spectra 1, 2, and 3 are for the checking of the reproducibility of the band position.

Crystal form II_p, a polar modification of form II, can be obtained by applying high voltage to form II sample at room temperature [14–16]. Transformation behavior is dependent on the applied voltage: 1 MV/cm gives polar form II, but 5 MV/cm causes further transition from II_p to I [14,61,62]. The form II_p film is not unoriented, but the polar a axis orients preferentially in the direction of the normal to the film plane.

b. X-Ray Structural Analysis

Form II: The molecular conformation of form II was proposed on the basis of X-ray diffraction and infrared spectral data. Two models were presented, a glide-type

(TGT$\overline{\text{G}}$) and a 2_1 helical type. The infrared and Raman data supported the TGT$\overline{\text{G}}$ conformation [29,30,262]. After some trials, Doll and Lando proposed two types of crystal structural models: two chains with conformation TGT$\overline{\text{G}}$ packed in a monoclinic unit cell (space group P2$_1$–C$_2^2$), or a triclinic unit cell (space group P1-C$_1^1$) with the angle γ almost equal to 90° [63]. The glide plane of the chain is not parallel to the a axis but makes an angle of ca. 7°. Concerning the direction of the chains along the c axis, the monoclinic cell contains two antiparallel chains (because the b axis is taken as a unique 2_1 axis), while the triclinic cell contains parallel chains. Hasegawa et al. proposed another type of crystal packing structure [19]. The two TGT$\overline{\text{G}}$ chains are packed in a pseudo-orthorhombic (monoclinic) unit cell of γ = 90° with a space group P2$_1$/c-C$_{2h}^5$. The TGT$\overline{\text{G}}$ chains are packed with the glide plane coincident with that of the cell. The adjacent chains are connected by a point of symmetry and thus in an antiparallel fashion along the chain direction. Since both the structural models proposed by Doll and Lando and by Hasegawa et al. gave reliability factors of almost the same order, further investigations were required. For example, Farmer et al. carried out the packing energy calculation and stated that a contamination of head-to-head and tail-to-tail irregular linkage might cause the different chain packings [20]. In the course of structural investigation, Lando and co-workers analyzed the crystal structures of form II$_p$ and III [64,65], as will be mentioned below, in both of which the chains were proposed to be packed in statistical up and down modes. Forms II$_p$ and III samples are obtained by the treatment of form II sample. Therefore form II should also have such a statistical type of chain packing along the c-axis direction. By taking into consideration this structural connection, Bachmann and Lando [66] reanalyzed the X-ray diffraction data of form II and proposed a new packing structural model: The TGT$\overline{\text{G}}$ chains are packed in an orthorhombic unit cell (P2cm-C$_{2v}^4$) with statistically equal probability of up and down directions at the same site. Takahashi et al. prepared a form II sample which gives a "tilting" X-ray diffraction pattern; that is, the chain axis tilts from the draw direction by ca. 20° along the b axis and therefore some reflections are not arrayed in a straight line but displace up and down from the horizontal line [67]. This sample was useful to distinguish the apparently equivalent pairs of reflections such as (111 and $\overline{1}$11), (121 and $\overline{1}$21), etc. These pairs of reflections were found to give different X-ray intensities, indicating that the monoclinic unit cell should be employed because these reflections are equivalent for an orthorhombic type of unit cell. They reanalyzed the X-ray data again [68], based on the monoclinic unit cell proposed by Hasegawa et al. and concluded that (a) the crystal structure of P2$_1$/c-C$_{2h}^5$ gives the best fit between the observed and calculated X-ray reflection intensities, but (b) more sophisticated interpretation of the data can be made on the basis of the statistical packing of chains with four different orientations; the chain is packed upward (C) or downward ($\overline{\text{C}}$) along the c axis and also in the positive (A) or negative ($\overline{\text{A}}$) direction of the a axis [For example, the structure shown in Figure 2 corresponds to the packing model of the two chains with AC (left) and $\overline{\text{A}}\overline{\text{C}}$ (right) orientation. The form II$_p$ is assigned to (AC + A$\overline{\text{C}}$).] The existence probabilities determined at one site are 56% for AC, 27% for A$\overline{\text{C}}$, 10% for $\overline{\text{A}}$C, and 7% for $\overline{\text{A}}\overline{\text{C}}$ models. Since the chains need to be packed in a line with the same orientation along the a axis because of the energetical advantage, the evaluated probabilities of different molecular orientation allow us to speculate a coexistence of domains. For example, the layers consisting of the same sense (A or $\overline{\text{A}}$) may be arrayed along the b axis as illustrated in Figure 17 and the domains of opposite a-axial orientation are aggregated side by side (antidomain structural model) [69]. Detailed analysis of the X-ray reflectional width supported this idea of domains

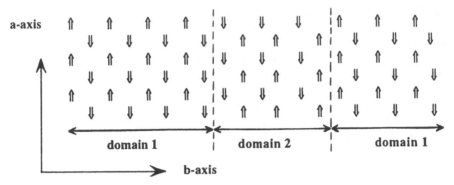

Figure 17 An illustration of antiphase domain structure of PVDF form II crystal [69]

[70,71]. Annealing induces the ordering of these domains into a single-domain-like structure [70].

Another type of disorder is also contained in the crystal structure of form II. Takahashi et al. [57], Weinhold et al. [72], Lovinger [73], and Gohil and Petermann [60] found that form II sample prepared under a special condition (refer to the previous section) shows streaks in the X-ray and electron diffraction patterns. The streak from 110 reflection, e.g., runs along the c^* axis and that of the 002 reflection runs along the $hk0$ direction. Gohil and Petermann [60] pointed out that these streaks can be observed only when forms I and II coexist in the sample. Takahashi et al. analyzed the streak pattern on the basis of the layer stacking disorder theory and proposed a kink model, in which some long trans sequence is included in a regular ... TGT$\overline{\text{G}}$ ... chain structure (... TGT$\overline{\text{G}}$TGT$\overline{\text{G}}$TTTGT$\overline{\text{G}}$TGT$\overline{\text{G}}$TG ...) [57]. As explained in a later section, this kink structure accelerates the conformational transformation from form II (TGT$\overline{\text{G}}$) to form III (TTTGTTT$\overline{\text{G}}$); in other words, the streak form II may be assumed as a transitory state and *nodus operandi* of the conformational change [73].

Form II$_p$:   Crystal structure of polar form II (II$_p$) was first proposed by Davis et al. based on the combination of X-ray diffraction and infrared spectral data [14]. The infrared spectra are not changed, except for the relative intensity change, even after transformation from form II to II$_p$, indicating that the molecular conformation is essentially the same as that of form II. The cell constants evaluated from X-ray diffraction are also the same, but the relative intensities are different from those of form II: The remarkable intensity reduction is observed for the reflections of 100, 120, and 210. Based on these intensity changes, a new chain-packing structure was proposed, in which the TGT$\overline{\text{G}}$ chains are packed together so that the chain dipoles become parallel to each other along the "polar" a axis by rotating around the chain axis. Concerning the direction of chains, several propositions were made. The two chains in the cell are packed in an antiparallel fashion along the c axis (up and down) [15,16], in a parallel fashion (up and up) [14], or in a statistical up- and down-mode packing [64]. Bachmann et al. analyzed the X-ray diffraction data of form II$_p$ and proposed a statistical packing of the up and down chains at the same site in the orthorhombic cell with the space group P2$_1$cn [64]. Comparing the accuracy of the structural analysis made for the other crystalline modifications, however, further refinement may be required for the structure of II$_p$ because the sample used in their analysis contains nonnegligible contamination of reflections from the antipolar form II and form I.

c. *Vibrational Spectra*

Form II: Measurement of infrared and Raman spectra of form II was made by several authors before the proposition of X-ray analyzed crystal structure. Typical polarized infrared spectra of form II are shown in Figure 18 in comparison with the Raman spectra. The infrared spectra were found to be very similar in pattern with those of poly(vinylidene chloride) [74] and could be reasonably interpreted by using a molecular chain conformation of $T\overline{G}TG$ [29–31,262]. The normal mode calculations made for the $T\overline{G}TG$ chain model supported this consideration [8,30,31]. As stated in the previous section, several models had been given for the chain packing in the form II unit cell. The factor group analysis made for the crystal structure may give us some useful information concerning the chain packing [8]. In the case of the structure $P2_1/c\text{-}C_{2h}^5$, the glide symmetry of an isolated $T\overline{G}TG$ molecular chain is still preserved in the unit cell. The correlation in the vibrational modes between the single chain and the unit cell may be described as follows:

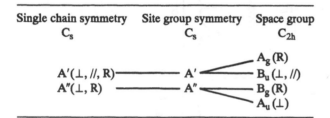

This correlation table predicts that the molecular vibrations of $T\overline{G}TG$ chain will split into two due to the intermolecular interactions; one is infrared active and the other is Raman active. Therefore we may not expect any band splitting in the infrared spectra as well as in the Raman spectra, but mutual exclusion should be observed as a slight band gap between the frequencies of the infrared and Raman bands. On the other hand, if the space group is $P2_1\text{-}C_2^2$ or $P1\text{-}C_1^1$, then all the intramolecular vibrations should be split into two bands due to the coupling effect of a symmetry reduction in the molecular chain at a site (site symmetry C1) and intermolecular correlation of the adjacent two chains. The correlation table should become as follows.

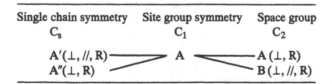

A close view of the infrared and Raman spectra measured at room temperature (Figure 18) and at liquid nitrogen temperature does not give any detectable amount of band splitting in each spectrum, inconsistent with the above-mentioned prediction for space group $P2_1$ or P1. On the other hand, according to the lattice vibrational calculation made for the structural model of $P2_1/c$, the frequency gap between the infrared and Raman bands of the corresponding vibrational modes is too small, 2–3 cm^{-1} at most, to detect in the observed spectral data beyond the experimental error. Therefore it may be difficult

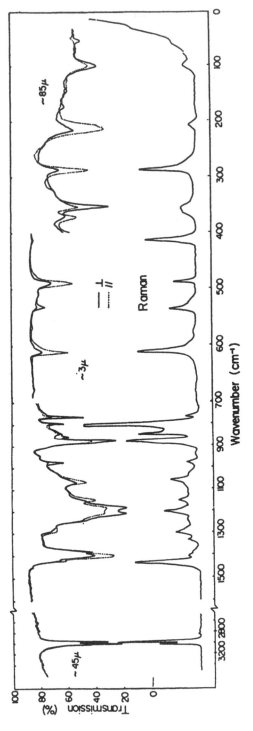

Figure 18 Infrared and Raman spectra of PVDF form II [8].

to judge which structural model is the best, but the observed spectra seem more consistent with the prediction made for the structural model of $P2_1/c$.

For the $P2_1/c$ model, five lattice vibrational modes are predicted by the normal mode calculation, which should appear at 84 cm^{-1} (translational lattice mode along the b axis, $L(T_b)$, B_g), 62 cm^{-1} (in-phase librational mode around the c axis, $L(R_c^0)$, B_g), 59 cm^{-1} (translational lattice mode along the a axis, $L(T_a)$, A_g), 51 cm^{-1} (out-of-phase librational lattice mode around the chain axis, $L(R_c'')$, A_u), and 11 cm^{-1} (translational lattice mode along the c axis, $L(T_c)$, A_g). As seen in Figure 7b, the far-infrared band was observed at 53 cm^{-1} with perpendicular polarization (the electric vector of the incident infrared beam $\perp$ the drawn direction of the film), which shifts to 60 cm^{-1} at liquid nitrogen temperature [8,75]. The band could be assigned to $L(R_c'')$ of the A_u symmetry species. The Raman bands observed at 99, 52, and 29 cm^{-1} were assigned to $L(T_b)$, $L(T_a)$, and $L(T_c)$, respectively.

Form II_p: The infrared spectra of form II_p are essentially the same with those of form II, as reported by Davis et al. [14]. Group theoretical consideration helps us to interpret the change in relative intensity of the TGTḠ bands caused by the preferential orientation of the polar a axis in the direction of the normal to the poled film surface. In the transmission spectra, the electric vector of the incident infrared beam is parallel to the film surface and therefore the vibrational modes with the transition dipoles parallel to the a axis (B_u species) are expected to decrease in intensity and those with dipoles parallel to the b axis (A_u species) should increase the intensity instead. In Figure 19 is shown a series of infrared spectra measured before and after poling of form II film [14,76–78]. The bands at 795, 612, and 489 cm^{-1} decrease the relative intensity remarkably, while the band at 766 cm^{-1} rather increases the intensity, although the whole of the bands intrinsic of form II decrease in absolute intensity because of partial transformation to crystalline form I. The tendency of these intensity changes can be confirmed more clearly in Figure 19c; the preferential orientation of the a axis was enhanced by annealing at 170°C. The bands at 612 and 489 cm^{-1} were assigned to B_u species and

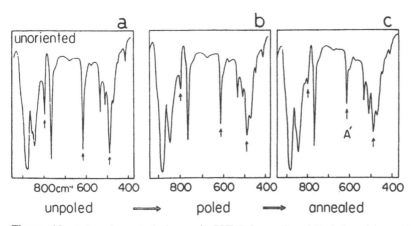

Figure 19 Infrared spectral change in PVDF forms II and II_p induced by poling and anneling (170°C) [112]. The arrows show the bands of the A′ symmetric species with the transition dipole moment along the polar a axis. As the degree of orientation of the a axis is increased, the A′ bands decrease in intensity.

that at 766 cm^{-1} to A$_u$. The poling-induced behavior of these bands is consistent with the prediction. The band at 795 cm^{-1} was assigned to A$_u$, but the intensity decreases after poling (and annealing) and should be reassigned to the B$_u$ mode.

Although the infrared spectra of form II$_p$ in the middle frequency region are essentially the same with those of II except for the remarkable change in relative intensity, the bands in the lower frequency region are different for II and II$_p$. For example Figure 20 shows the far-infrared spectral change measured at liquid nitrogen temperature in the poling process of form II → polar form II$_p$ → form I [76]. As the poling proceeds, the bands intrinsic to form II (e.g., 60 cm^{-1}) begin to decrease in intensity and the new band appears at 73 cm^{-1} (at liquid nitrogen temperature), which may be assigned to the lattice vibrational band of form II$_p$. This band increases in intensity gradually as the poling proceeds. At the same time, the band of form I at ca. 80 cm^{-1} also begins to increase in intensity. In the spectra of form II$_p$, splitting is observed for the band at ca. 220 cm^{-1}: The band splits into the two components with parallel and perpendicular polarization characters (with respect to the chain direction). If the structure proposed by Davies and Singh is adopted [15], the two glide-type chains are included in the unit cell but no

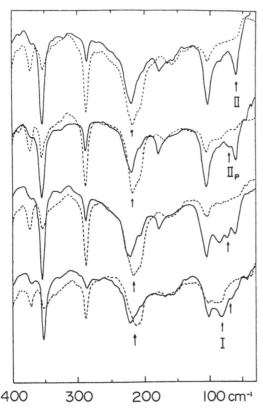

Figure 20 Polarized far-infrared spectral changes induced by poling treatment of the uniaxially oriented PVDF form II film measured at liquid nitrogen temperature [76]: ———, electric vector of incident infrared beam perpendicular to the draw direction; −−−−, electric vector of incident infrared beam parallel to the draw direction.

symmetrical relation exists between these two chains. If the structure of Bachmann et al. is employed [64], then the two chains are related by *n*-glide symmetry. In the former case, the bands may be split into two. In the latter case, too, the bands should split into two, but the following correlation exists:

Molecular chain	Space group
A′($\perp$, //) ————————	A′($\perp$, //)
A″ ($\perp$) ————————	A″ ($\perp$)

Therefore the original A′ bands of an isolated chain may split into a pair of bands with $\perp$ and // polarization characters [($\perp$ and $\perp$) or (// and //) are also possible]. The band near 210 cm^{-1} is originally a parallel band in the crystal form II. This band splits into one parallel and one perpendicular band in the form II$_p$, possibly because of such a correlation effect of the two chains.

Structural Defects and Vibrational Spectra: The infrared spectra of forms II and II$_p$ contain various additional bands which cannot be assigned on the basis of the molecular and crystal structure mentioned in the previous section. In the vibrational spectra of form I, the bands characteristic of the head-to-head and tail-to-tail linkages were assigned to 1450, 1330, and 678 cm^{-1}. These bands can also be commonly detected in the infrared spectra of forms II, II$_p$ (and III) [8,76]. In the far-infrared region, as seen in Figure 7, form I shows several parallel bands at 350 and 372 cm^{-1} in addition to the broad lattice vibrational band at 70 cm^{-1}. These bands may be assigned to the modes originating from the breakdown of symmetry of the zigzag chain, as discussed in the previous section. These bands can also be detected in the far-infrared spectra of forms II and II$_p$, suggesting the presence of some trans sequences within the gauche-type chain as a kind of structural defect (e.g., a kink structure) [76].

3. Form III

a. Preparation. Identification of crystal form III had been confused for a long time. One of the most typical methods of preparing form III is to cast from dimethylacetamide (DMA) or dimethylforamide (DMF) solution at ca. 60°C (slight warming of the solution is helpful to get a relatively transparent film) [8]. Isothermal crystallization of the molten sample at a temperature just below the melting point gives form III [4,5]. High-pressure and high-temperature crystallization gives also form III [79]. High-temperature annealing of the crystalline form II gives form III, but in this case, some conditions are needed to cause the solid-state transition from II to III, as mentioned in detail by Lovinger [80]. At first a mixture of large spherulite of form II and small-size spherulite of form III is prepared by slowly cooling the melt. This sample is annealed below 152°C, giving no transformation to form III. Annealing at 155–162°C induces the generation of form III spherulite at the boundary between the II and III spherulites. That is, the form III can be obtained through the solid-state transformation from form II for the first time when the form II sample contains some structural defect (such as a kink), heterogeneity, or form III lamellae at the first stage. This can also be said with regard to the preparation of the oriented form III sample. The oriented form III sample is difficult to prepare by a simple stretching of the solution-cast form III film, because it changes easily to form I by drawing. Stretching at high temperature gives an oriented sample of form III, but the degree of orientation is not very high [8,81]. An effective method is a use of oriented form II sample. The uniaxially oriented form II sample prepared by stretching the melt-quenched sample under a usual condition (140–150°C) is not transferred into form III, even annealing at high temperature (175°C) for 25 days [73]. The form II sample showing X-ray streaks in the fiber diagram or containing some kink defect, which is prepared by stretching the melt at room temperature during crystallization, transforms to form III slowly by annealing at 150°C for 24 h with the ends fixed [57]. Another way is first to prepare oriented form II_p by charging the oriented form II sample and then heat this II_p sample at a temperature close to the melting point, giving a well-oriented form III [82–84]. Heating of oriented form I sample at high temperature immediately below the melting point gives also oriented form III sample; this transition may be a ferroelectric-to-paraelectric phase transition, as discussed later, but it requires careful control of temperature [85,86]. Some investigators assign this transition rather to the phenomenon of melt and recrystallization [87].

b. X-Ray Structure Analysis. Cortili and Zerbi measured the infrared spectra of form III film cast from DMA solution (unannealed) and pointed out a similarity of the spectra between forms I and III [29,88]. They speculated that form III might be a crystalline modification with the molecular chain conformation essentially the same as that of form I (trans-zigzag) but with some degree of conformational disordering, although they cast some doubt on the existence of the form III itself. A similarity of the infrared spectra was also pointed out by Prest and Luca [4,5]. Hasegawa et al. measured the X-ray diffraction pattern for solution-cast and unoriented form III sample and indexed the observed reflections on the basis of a unit cell slightly modified from that of form I: The two zigzag chains are packed in the monoclinic unit cell with $\beta = 97°$ and the two chains are different in relative height along the chain axis [19]. Kobayashi et al. first interpreted the infrared and Raman spectra of as-cast form III film by this chain packing [8], but even at that time there remained some unassignable bands in the spectra of form III. After that, many investigators [89] pointed out that the X-ray diffraction data measured for well-annealed form III sample could not necessarily be interpreted by using the cell model proposed by Hasegawa et al. Weinhold et al. obtained relatively well-oriented form III sample and measured the X-ray fiber diagram [81]. The repeating period along the chain axis was 9.18 Å, which cannot be explained by the planar-zigzag conformation of form I. They proposed several conformational models, including $TTTGTTT\overline{G}$, $TGTGT\overline{G}T\overline{G}$, etc. Their proposition was supported by energy calculation [90] and normal mode calculation [91,92], although at that time a definite solution of molecular and crystal structure had not yet been obtained. Afterwards, in their second paper [65], Weinhold et al. analyzed the X-ray data in detail and proposed an orthorhombic unit cell of $a = 4.97$ Å, $b = 9.66$ Å, and c (fiber axis) $= 9.18$ Å, in which two $TTTGTTT\overline{G}$ chains are packed with their CF_2 dipoles parallel along the a axis. At almost the same time, Takahashi and Tadokoro [58] obtained a highly oriented form III sample from form II with streak X-ray reflections as stated above, and reported a monoclinic unit cell, not an orthorhombic cell, of $a = 4.96$ Å, $b = 9.58$ Å, c (f.a) $= 9.23$ Å, and $\beta = 92.9°$ in which two $TTTGTTT\overline{G}$ chains are packed. The sample which shows the tilting X-ray diagram could support this monoclinic cell on the basis of unequivalence of the reflections hkl and $\overline{h}kl$ [67]. The monoclinic cell was also supported by the analysis of electron diffraction [89]. The chains are packed in the cell in a statistical up and down mode (Lando et al.) or in a parallel mode (Takahashi and Tadokoro) along the chain direction.

As seen in Figure 2, the $TTTGTTT\overline{G}$ chains are packed in the unit cell with their dipoles parallel to the a axis and therefore the form III is polar crystal. Lovinger annealed polar form II_p at high temperature and found that the 100 and 120 reflections reappear in addition to the main reflections characteristic of the polar form III [93]. An observation of 121 reflection, which cannot be indexed using the form III structure, lead him to propose another crystal modification of form III in which the TTTGTTTḠ chains are packed in an antiparallel fashion, but this antipolar form III (form V) was included only in a minority population in the annealed form II_p sample. Karasawa and Goddard carried out the lattice energy minimization and supported a possibility of this antipolar form III [21].

c. Vibrational Spectra. As stated above, the infrared spectra of the as-cast form III are apparently similar to those of form I. This is checked in Figure 21, where the infrared spectra of forms I, II, as-cast III, and annealed III are compared with each other [92]. Cortili and Zerbi compared the spectra of form III sample obtained by resolidification of the melt with those of form I and speculated that form III should consist of a slightly distorted zigzag chain, although they did not give a clear answer for the existence of form III itself [29,88]. Kobayashi et al. [8] measured the infrared and Raman spectra of as-cast form III and interpreted them by adopting the crystal structure proposed by Has-

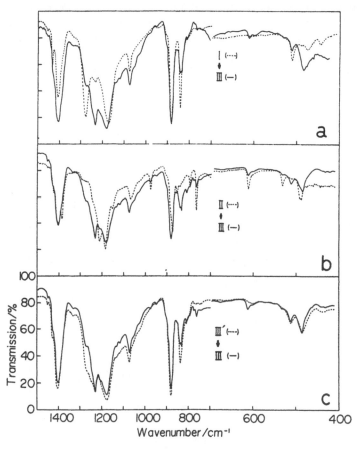

Figure 21   Infrared spectra (broken lines) of PVDF form I, form II, and as-cast form III in comparison with those of form III (solid lines), which is obtained by annealing forms I, II, and as-cast III at high temperature [92].

egawa et al. [19]. Although the main of the spectra could be explained well, many additional bands were still in question. After that, based on the molecular conformational model TTTGTTTḠ, Bachmann et al. [91] and Tashiro et al. [92] could assign the bands reasonably through the normal coordinates treatment. The bands can be classified roughly into three groups characteristic of T_m ($m \geq 3$), TTTG, and TG sequences. For example, the 840-cm^{-1} band, which is commonly observed for forms I and III, is characteristic of T_m, the 811- and 766-cm^{-1} bands are intrinsic of form III and assigned to the TTTG sequence, and the 614-cm^{-1} band, common to forms II and III, is assigned to the TG sequence.

At this stage we may naturally question why the vibrational spectra of the as-cast and unannealed form III films are similar to those of form I and why the spectra of annealed form III samples are so different from those of as-cast form III. In the infrared spectra of as-cast form III, the bands characteristic of long trans sequences are observed with appreciably high intensities: for example, the 1275-cm^{-1} band is characteristic of a longer trans sequence T_m ($m \geq 4$). These trans bands were found to decrease in intensity as the heat treatment proceeded at higher temperature. In Figure 22 are shown a series of infrared spectra of form III film annealed at various temperatures [92]. As the annealing temperature is increased, the trans bands (1275, 840 cm^{-1}, etc.) decrease in intensity and the gauche bands (811, 614 cm^{-1} etc.) increase instead. As a result, the whole spectral pattern changes gradually from the form I-like pattern to the pattern intrinsic of form III. In addition to this annealing-induced spectral change of as-cast form III, the DSC curve and far-infrared low-frequency bands, and the X-ray diffraction curve, change greatly. For example, the melting point of form III crystal shifts to higher tem-

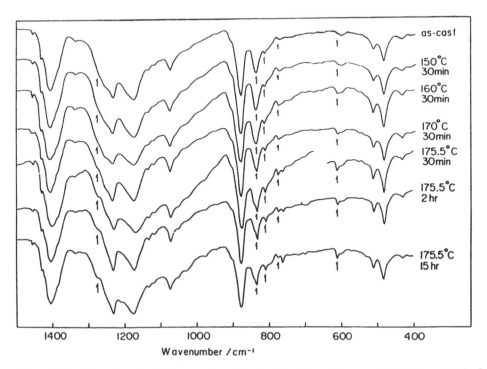

Figure 22 Infrared spectral change induced by heat treatment of as-cast PVDF form III [92].

perature by ca. 10°C, from 174°C to 186°C, when the as-cast film is annealed. At the same time the endothermic peak becomes sharper. In the far-infrared spectra the bands characteristic of form III that are well recognized at 299 and 90 cm^{-1} increase in intensity as the heat treatment proceeds. On the basis of all these experimental results, the difference in structural feature between as-cast form III and crystalline form III may be specified clearly. As understood from the intense 1275-cm^{-1} band, as-cast form III takes a more or less disordered molecular conformation containing a fairly long trans unit in the basic TTTGTTT$\overline{\text{G}}$ sequence. The low degree of the order in chain packing is reflected in the broad melting behavior at a comparatively low temperature, in the broadness of far-infrared spectra and X-ray reflections. Heat treatment of as-cast form III results in interchange between the trans and gauche isomers, as revealed by the infrared and Raman spectral measurements, and the molecular chain becomes more regular, with a TTTGTTT$\overline{\text{G}}$ repeat unit:

$$\ldots\text{TTTGTTT}\overline{\text{G}}\text{TTTTTTTGTTT}\overline{\text{G}}\ldots$$

$$\downarrow$$

$$\ldots\text{TTTGTTT}\overline{\text{G}}\text{TTTGTTT}\overline{\text{G}}\text{TTTG}\ldots$$

As a result of such a disorder-to-order transition in the moelcular conformation, the regularity of the lattice increases and reflects on the high-temperature shift and sharpening of the DSC melting and the sharpening of the X-ray reflections.

d. Conformational Change and Phase Transition Among Forms II, II$_p$, and III. As mentioned above, forms II and II$_p$ are transformed to form III by annealing at high temperature for a long time. The molecular conformation of forms II (II$_p$) and III are relatively close to each other:

Forms II and II$_p$ TGT$\overline{\text{G}}$TGT$\overline{\text{G}}$
Form III   TTTGTTT$\overline{\text{G}}$

As seen here, the transformation from II to III should be accompanied by a conformation change btween TG and TT. How can this conformational change occur? For the transformation II$_p$ → III, Servet et al. proposed an occurrence of flipflop motion of TGT$\overline{\text{G}}$ chain between TGT and T$\overline{\text{G}}$T [82–84]. For II → II$_p$, Lovinger [94] proposed a mechanism of TGT$\overline{\text{G}}$ → TCTC → T$\overline{\text{G}}$TG based on the calculated energy barrier between them and the possible existence of intermediate conformation of TCTC. Drey-Aharon et al. proposed an inversion mechanism through the propagation of kink along the skeletal chain [95]. In order to answer the question more exactly, however, we also have to take into consideration another factor: the orientation of the chain with respect to up and down along the chain direction and the parallel and antiparallel packing of the CF$_2$ groups along the a axis. According to the crystal structural models of Bachmann and Lando [66], in all the crystal forms of II, II$_p$, and III the chains are packed in a statistically disordered fashion of up and down along the chain axis. On the other hand, according to the models of Takahashi and co-workers [68], form II contains essentially alternately up and down chains along the c axis and the CF$_2$ groups are oriented in an antipolar mode along the a axis. Such regular packing is limited within a domain. These domains are assumed to gather together to form a crystallite at a weighted probability. The chains in form III are packed in polar (along the a axis) and parallel (along the c axis) fashion. Their proposed transition mechanism is as follows [59,71,96]. As pointed above already, the transformation from II to III is easily induced when the form II contains some structural defect such as a kink (TT sequence) along the chain. Such a defect is considered to generate as a joint between $T G T \overline{G}$ and $T \overline{G} T G$ sequences through the segmental flipflop motion:

$$T G T \overline{G} T G T \overline{G} T G T \overline{G} T G \rightarrow T G T \overline{G} T G T \underline{T T} \overline{G} T G T \overline{G}$$

This motion may act as a trigger for a generation of the TTTG sequence of form III. Another possible motion is an inversion motion which causes the simultaneous reverse of molecular orientation both along the c and a axes:

$$T G T \overline{G} T G T \overline{G} T G T \overline{G} T G \rightarrow T \overline{G} T G T \overline{G} T G T \overline{G} T G T \overline{G}$$

By repeating the cooperative flipflop motion coupled with an inversion motion, the antiparallel and antipolar form II can be changed into the parallel and polar form III. The transformation from II to II_p may be caused only through an inversion motion. If this assumption is correct, then the relatively easy transformation from II_p to III is induced through the flipflop motion. A similar mechanism was also proposed by Livinger [73]. This type of motion is related to the dielectric dispersion of PVDF form II [97,98]:

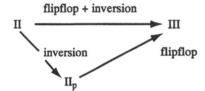

4. Factors Affecting the Crystalline Forms

a. Kinetic and Thermodynamic Factors.   As stated in the preparation sections, the preparation condition of the crystalline form is intrinsic to each form, but a slight change in the crystallization condition affects the resultant crystal modifications. Detailed investigation describes the kinetic and thermodynamic superiority between these crystalline forms as follows [99].

Solution Cast: Figure 21a shows the infrared spectra measured for PVDF form I film cast from a very polar HMPTA solution. Casting at room temperature for several months gives infrared spectra of highly crystalline and pure form I. When the evaporation rate of the solvent is increased by using a vacuum pump, crystal form III is obtained at room temperature. Casting on a hot stage increases the evaporation rate further, resulting in the infrared pattern of crystal form II. A similar situation can also be seen for casting from DMA solution: a slow evaporation for form III and a fast evaporation for form II. By extrapolating this tendency, we might obtain form I crystal by evaporating the DMA much more slowly. Casting from acetone solution at room temperature gives form II, but slow casting on a glass slide cooled by Dry Ice gives form III. Thus we cannot say simply that the crystal form obtainable is determined only by the polarity of the solvent used [100]. The result indicates the importance of controlling the evaporation rate of the solvent: As the evaporation rate is lowered, the crystalline form with higher trans fraction appears more predominantly, regardless of the solvent.

Crystallization from the Melt: A rapid cooling of the melt down to room temperature results in the crystallization of form II, while a slow cooling gives the highly crystalline form III. Much slower cooling at high temperature immediately below the melting point is reported to give spherulites of form I which crystallize four to five times more slowly than form III spherulites. Ultraquenching of the thin film at liquid nitrogen temperature followed by heat treatment at $-30°C$ gives form I crystal, although the reason has not yet been clarified.

Phenomenological factors affecting crystal modification or polymorphism may be classified mainly into two types: thermodynamic and kinetic effects. The crystalline form actually obtained is determined by a delicate balance between these two effects. Therefore, even the thermodynamically unstable crystalline phase might grow dominantly because of the kinetic effect. The experimental results stated above reveal that the crystalline form with the molecular conformation of higher trans content (I > III > II) is obtained more predominantly as the crystallization rate is lowered in both solution-casting and melt-cooling. It may be said that the crystalline form I is thermodynamically the most stable at room temperature and the form II is the most advantageous kinetically among all the crystalline forms. The crystallization behavior is sensitive to changes in chemical structure, for example, an introduction of the head-to-head and tail-to-tail linkage, as will be mentioned below.

b. Effect of Head-to-Head and Tail-to-Tail Linkages. The melting point of PVDF changes to some extent depending on the source supplied to the measurement, even when the same and pure crystalline form is prepared. For example, form II of KF (Kureha Chemicals, Japan) melts at 178°C, higher than Kynar (Pennwalt Chem., USA) at 161°C. High-resolution NMR revealed the existence of head-to-head and tail-to-tail (HHTT) linkage in PVDF chain: 5–6% in a Kynar sample and 3–4% in a KF sample [101–105]. It is possible that this difference in HHTT content gives the difference in the melting point. (Chain branchings may also affect the physical properties of this polymer, but no detailed study has been reported.) Lando and Doll investigated the effect of HHTT structure on crystal modification by synthesizing a series of copolymers consisting of VDF and a small content of TFE or TrFE (0–20%) [106]. Cooling from the melt at a normal rate leads to the crystallization of form II in the case of PVDF homopolymer and the copolymer containing 9mol% TrFE, but copolymer samples with 7mol% TFE or 17mol% TrFE comonomer crystallize into unoriented form I. These results are supported by a study of VDF-TrFE and VDF-TFE copolymer samples with a wider range of monomer ratio, as discussed in detail in a later section. Lovinger et al. [107] investigated the structure and phase-transition behavior of PVDF samples containing various degrees of HHTT sequences [105]. When the samples were crystallized from the melt, the form II was obtained at room temperature for samples containing HHTT content less than 12%, while form I crystal appeared for samples of higher HHTT. These observations are consistent with the energetical consideration that the stereochemical repulsion between nonbonded fluorine atoms in the HHTT part stabilizes the trans-zigzag structure rather than the gauche-type conformation. More quantitatively, Farmer et al. carried out the energy calculation in order to investigate the effect of HHTT on the relative stability of forms I and II [20]. They calculated the potential energy of 20-mers and 60-mers of PVDF chains with various contents of HHTT sequences. For a chain without HHTT linkage, $TG T\overline{G}$ conformation is more stable than all-trans form I chain, while the relation is reversed when the HHTT content exceeds ca. 11%. This value of HHTT is quite well coincident with the above-mentioned experimental value reported by Lovinger et al. [107].

Farmer et al. calculated the crystal lattice energy [20]. In the case of form I crystal, introduction of HHTT linkage increases the size of the unit cell, but the relative position and orientation of the zigzag chains are not changed from the original structure without HHTT defect. In the case of form II, HHTT defect modifies the packing mode of $TG T\overline{G}$ chains to an appreciable extent. That is, introduction of HHTT defect changes the relative height of one chain along the c axis by ca. 1 Å and rotates the direction of the glide plane of the chain from the a axis by several degrees. Farmer et al. pointed out a possibility of having two types of "realistic" crystal structure for form II: If X-ray structure analysis is done for samples with different HHTT content, they may give a crystal structure that is slightly but significantly different in some details of conformation and packing of the chains, as actually seen in the difference of the crystal structural models proposed by Doll and Lando [63] and Hasegawa et al. [19]. Both models, however, were corrected more or less in papers published later [66,68].

As stated above, commercially available samples are, strictly speaking, not an ideal (or isogeric) PVDF but contain HHTT linkages to some extent. How does the crystallization behavior change if this ideal PVDF specimen is obtained? Cais and Kometani succeeded in synthesizing a PVDF sample with HHTT content of only 0.2% by reductive dechlorination of the precursor polymer, poly(1,1-dichloro-2,2-difluoroethylene) [105]. As mentioned above, Lovinger et al. measured the X-ray diffraction of unoriented sample and confirmed the crystallization of form II attended with minor form I [107]. Unfortunately, as Cais and Kometani stated, this material could not be obtained with a molecular weight of sufficient magnitude to impart typical polymeric properties, thus obviating a meaningful comparison with commercial PVDF samples [105]. In fact, as stated later, the PVDF oligomer behaves in a different manner from that of PVDF. After that, they synthesized a sample with high enough molecular weight with the lowest HHTT content, ca. 2.8%. (Exactly, this is not the normal PVDF but the deuterated PVDF-d_2) [108].) This HHTT value is the lowest compared with those of the commercially available samples; for example, Kureha 1000 contains ca. 3.8% HHTT, Solef 1008 contains ca. 4.1%, and Kynar 900 contains ca. 5.2%. It is somewhat difficult to compare the properties of these low-HHTT samples directly with those of other samples because the former are deuterated species and so the unnegligible isotopic effect has to be taken into consideration. But the melting point, ca. 188°C, and density, 1.86 g/cm^3 (1.80 g/cm^3 corrected for the mass of deuterium) of this sample are definitely higher than the other data (e.g., the sample with HHTT 3.5% shows melting point 180°C and density 1.78 g/cm^3). If the difference comes from the difference in regularity of chemical structure, only a small reduction in HHTT content is considered to have remarkably large effect on the structure and physical properties of this polymer in the high-VDF-content region.

c. Effects of External Field. As already mentioned, an external field such as high pressure, high DC voltage, etc., induces the phase transition between the crystalline forms of PVDF. What happens in the crystal structure when the sample is subjected to an external field? It is not easy to investigate the structure in situ, but some reports have appeared.

High Pressure: Direct observation of phase transition from form II to form I during heating under constant pressure was made by Matsushige et al. [12]. A combined measurement of differential thermal analysis (DTA) and X-ray diffraction under 400 MPa pressure revealed that the initial form II crystal melts at ca. 280°C and the folded-chain crystal (FCC) and extended-chain crystal (ECC) of form I crystallize around 297°C. The FCC of form I melts at ca. 305°C and the ECC melts around 330°C. The ECC is a banded structure of 1500–2000 Å width. In this experiment the temperature was scanned at a heating rate of 5°C/min. If the temperature was kept constant at 278°C under the same pressure, form II sample gradually transferred to form I, even below the melting point of form II.

High Electric Field: Measurements of infrared and Raman spectra of PVDF under high DC voltage were reported by several authors, as mentioned in the section on form I. For example, a hysteresis loop of the infrared intensity versus electric field was observed, showing an inversion of CF_2 dipoles under electric field [38,39]. Reorientation behavior of the crystal lattice under high electric field was also measured by the X-ray diffraction method. Takahashi et al. showed an occurrence of reorientation of polar form I crystal by measuring a slight but definite difference in X-ray intensity between the 110 and $\overline{1}10$ reflections, which was ascribed to the effect of an anomalous dispersion term of the atomic scattering factor [46]. As already mentioned, Kepler and Anderson [40] and Takahashi et al. [41–45] estimated the distribution function of the polar b axis around the draw axis for the poled form I sample, but the distribution function evaluated by using the {111} intensity is largely different from that by {311 + 021} reflection. This difference was ascribed by the latter authors to a difference in the relative intensity of (111) and ($\overline{1}11$) reflections [109]. That is, for the othorhombic-type packing, the relative intensity of (111) and ($\overline{1}11$) reflections is equal to each other, whereas the intensity may be changed if the crystal structure of form I changes into a monoclinic system because of the effect of high electric field. Their model is that the zigzag chains rotate around the chain direction by ca. $7°$ from the b axis in the same direction due to the torque induced by the electric field.

III. FERROELECTRIC PHASE TRANSITION IN VDF-TrFE COPOLYMERS [110–113]

A. Discovery of Phase Transition

As stated in the introductory section [34], the ferroelectric phase-transition phenomenon of VDF-TrFE copolymers was discovered for the first time in a synthetic polymer in 1980 by Japanese research groups [114–116]. At first a curious thermal behavior in the dielectric constant was detected for a copolymer of VDF 55mol% sample. As shown in Figure 23 [115,117], in the vicinity of ca. 65°C, the dielectric constant showed a maximum and was found to satisfy the so-called Curie-Weiss law [$\epsilon = \epsilon_0 + C/(T - T_0)$; $C = 3600$ K for $T > T_0$ and $C = 500$ K for $T < T_0$, where $T_0 = 343$ K]. At the same time, the electric polarization of the poled film decreased down to zero in this temperature region. The piezoelectric constant also disappeared at this temperature. Anomalous heat capacity was detected [118]. The D-E hysteresis loop was clearly observed even at $-100°C$, indicating the ferroelectricity of this copolymer. The loop was not detected above the transition point. These phenomena observed for the bulk sample were interpreted as results of a ferroelectric phase transition occurring within a crystalline region. Tajitsu et al. measured the temperature dependence of the crystal lattice spacing for the X-ray reflection of 2θ ca. $19°$ [119]. The lattice spacing changed apparently continuously but remarkably in the vicinity of the transition point. Evaluating the lattice spacing change Δd between the ferro- and paraelectric phases, they found a relationship of

$$\Delta d \propto P_r^2$$

where P_r is the remnant polarization measured at each temperature. A phenomenological theory of second-order transition can reproduce this relation as an electrostriction effect in the ferroelectric crystal. Because of the long-range electrostatic interaction caused by spontaneous polarization P, the crystal lattice is deformed additionally by the quantity Δd from the normal lattice size determined by short-range interactions alone. Tajitsu et al. assumed that the molecular chains of planar-zigzag conformation rotate around the

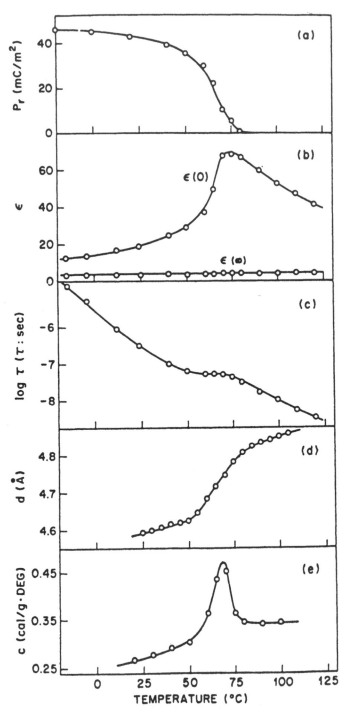

Figure 23 Temperature dependence of remnant polarization P_r, dielectric constants $\epsilon(0)$ and $\epsilon(\infty)$, logarithm of relaxation time τ, lattice spacing d, and specific heat c measured for VDF 55%-TrFE copolymer [263].

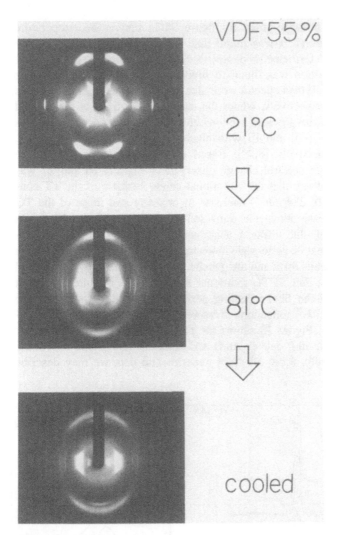

Figure 24 Temperatue dependence of X-ray fiber diagram taken for uniaxially oriented VDF 55%-TrFE copolymer [126].

chain axis *rigidly* above the phase-transition temperature. Such a rigid rotation of polar zigzag chains was considered to result in the disappearance of the polarity of the whole crystal. But Yamada et al. found the change of infrared spectra in the transition, suggesting that the transition may occur between the crystal forms corresponding to forms I and III of PVDF homopolymer [115]. At that time, unfortunately, the crystal structure of form III was misunderstood to be essentially equivalent to that of form I.

Tashiro et al. [120] and Lovinger et al. [121,122] measured the temperature dependence of X-ray fiber pattern as a function of temperature. In Figure 24 are shown the fiber diagrams of VDF 55% copolymer taken at 21, 81, and again at 21°C. The diagram at 21°C is similar to that of PVDF form I, while that taken at 81°C is rather close to the pattern of PVDF forms II and III, although the reflections are more diffuse than those of forms II and III of PVDF homopolymer. In Figure 25 is shown the temperature dependence of the (001) reflections. It was found that the repeating period along the chain axis changes from 2.55 Å (ferroelectric phase) to 4.60 Å (paraelectric phase). That is, this ferroelectric phase transition was found to involve a change in the molecular conformation. The infrared and Raman spectra were also found to change drastically, as shown in Figures 26 and 27, respectively, where the case of VDF 55% is reproduced [120,123–125]. As explained in the previous sections, the detailed band assignment was already made for PVDF forms I, II, and III with characteristic conformations of TT, TGTḠ, and TTTGTTTḠ, respectively [8,92]. Based on this information about the conformation-sensitive bands, the spectral change observed in Figures 26 and 27 was interpreted reasonably in such a way that the vibrational bands intrinsic of the TT conformational sequence (1288, 850, 884 cm^{-1}) decrease in intensity and those of the TG form (614 and 870 cm^{-1}) increase above the transition temperature. Figure 28 shows the temperature dependence of the infrared absorbance evaluated for the various conformation-sensitive bands. That is, large-scale molecular conformational changes were found to occur between the all-trans form and the gauche form constructed by a statistical combination of TG, TḠ, TTTG, and TTTḠ rotational isomers. This is quite consistent with the temperature change in the fiber-repeating period. In Figure 29 is shown the temperature dependence of the (hk0) reflections measured for unoriented VDF 55% copolymer sample [121–123,126]. Figure 30 shows the plot of the thus estimated lattice spacing against temperature. The unit cell expands and contracts above and below the phase-transition point, respectively. From all these experimental data we may describe

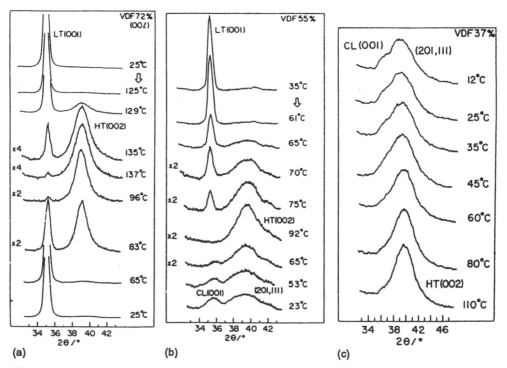

Figure 25 Temperature dependence of the meridional 001 reflection measured for VDF-TrFE copolymers with VDF content of (a) 72%, (b) 55%, and (c) 37% [124].

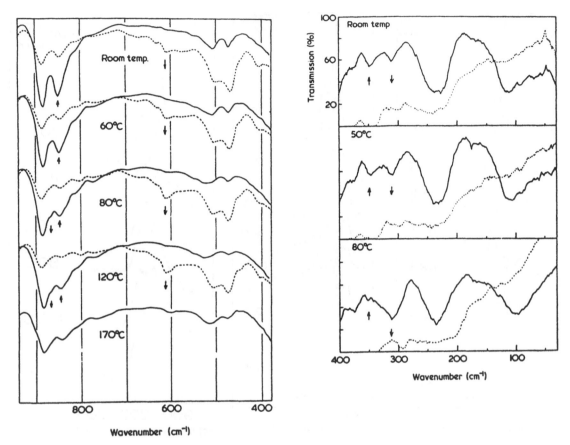

Figure 26 Temperature dependence of polarized infrared spectra measured for VDF 55%-TrFE copolymer [123].

the structural change in this ferroelectric phase transition of the copolymer as follows [123]. The low-temperature ferroelectric phase consists of the polar packing of zigzag chains, similar to the structure of PVDF form I, and the high-temperature paraelectric phase is constructed by an arrangement of gauche-type chains with a sequence of TG, T$\overline{G}$, TTTG, and TTT$\overline{G}$ isomers. Because of this conformational change, the unit cell size in the lateral direction also expands largely in the high-temperature phase. This type of structural change in the ferroelectric phase transition is quite unusual and different from that in low-molecular-weight ionic compounds, in which small ionic groups change their dipole orientation by rotation and/or displacement [35]. In the polymer system the CF_2 dipoles are combined together by strong covalent bonds and so the orientational change in the dipole moments require cooperative motion of the neighboring CF_2 groups through the large-scale trans-gauche conformational change.

B. Structural Analysis of Crystalline Phases

In Figure 24, we may notice that the fiber diagrams measured at room temperature before and after heating above the high-temperature phase are different from each other

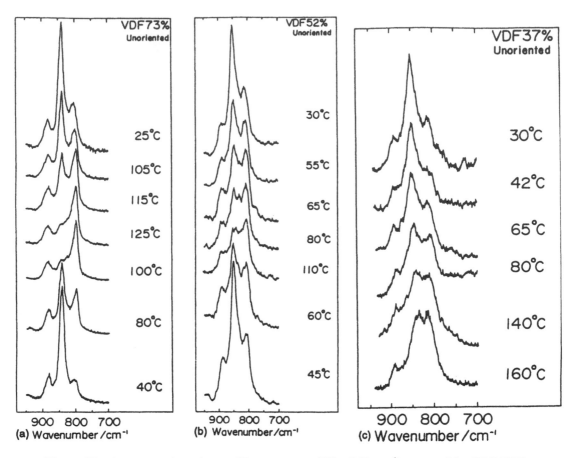

Figure 27 Temperatue dependence of Raman specra (700–1000 cm^{-1}) measured for VDF-TrFE copolymers with VDF content of (a) 73%, (b) 52%, and (c) 37% [124].

[123,126]. That is, the sample before heating exhibits a diagram very similar to that of PVDF form I, although the details are different. The sample obtained after cooling down to room temperature shows an appreciably different pattern, in which the equatorial reflections are not in a horizontal equatorial line but shift up or down from the line and the first layer reflections overlap with each other and form a very long and apparently diffuse arc. This pattern can be interpreted in terms of a tilting phenomenon of the crystallites; the chain axis is tilted from the draw axis by a constant angle [32]. The crystalline form giving this characteristic X-ray pattern is named the "cooled" (CL) phase. The details of the structure are mentioned in a later section. The sample showing the X-ray pattern of Figure 24a was prepared by stretching the CL sample at room temperature by several percentages of the original length. The crystalline phase thus prepared, with a very sharp and clear X-ray pattern, is called the "low-temperature" (LT) phase. When the LT phase is heated up, it changes into the CL phase on the way to the phase transition to the high-temperature (HT) phase. The lattice spacing, for example, changes discontinuously between the LT and CL phases, as shown in Figures 29 and 30. When it is cooled from the HT phase, on the other hand, the sample transfers

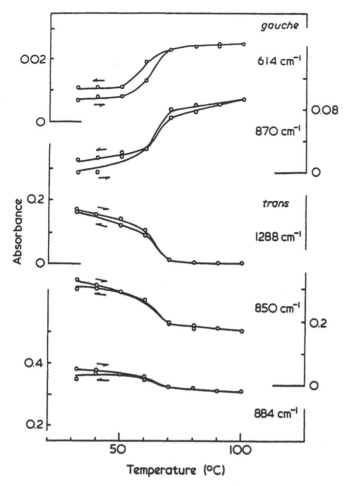

Figure 28 Temperature dependence of infrared relative absorbance of unoriented VDF 55%-TrFE copolymer [123].

apparently continuously to the CL phase. The transition scheme may be illustrated in the following way.:

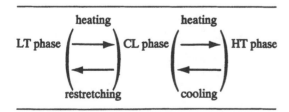

1. Crystal Structure of the Low-Temperature Phase

As pointed out above, the X-ray fiber diagram of the LT phase is essentially the same with that of PVDF form I. The innermost reflection (201) + (111) of the first layer line,

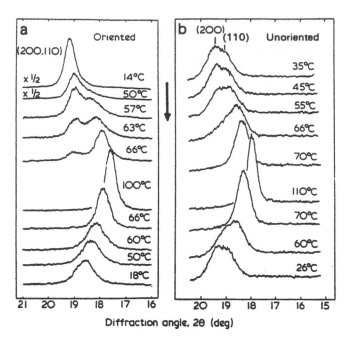

Figure 29 Temperature dependence of the X-ray (200,110) reflection measured for oriented and unoriented VDF 55%-TrFE copolymer samples [123].

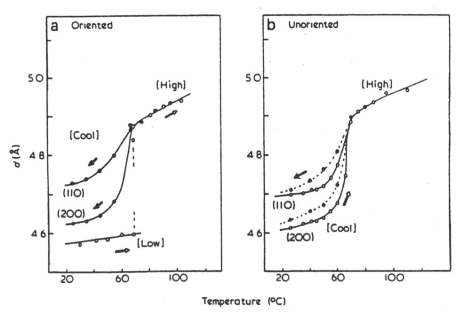

Figure 30 Temperature dependence of the lattice spacings $d(200, 110)$ measured for oriented and unoriented VDF 55%-TrFE copolymer samples [123].

however, is found to be composed of several reflections having slightly different lattice spacings [123]. In addition, these reflections do not lie on a straight layer line but shift up and down from the line. A similar situation is seen also for the equatorial reflections. That is, a slight degree of tilting phenomenon is implied from these data. As stated above, a more noticeable tilting is observed for the CL phase, which can be analyzed comparatively easily. The LT phase is obtained by stretching the CL phase by ca. 5% of strain at room temperature. Therefore, if the tilt line is assumed to be common to these two phases and only the tilting angle is assumed to be different, the analyzed results of the tilting phenomenon for the CL phase will supply useful information for indexing the reflections of the LT phase. The unit cell parameters and the tilting information of the LT phase thus obtained are as follows.

$$a = 9.12 \text{ Å} \qquad b = 5.25 \text{ Å} \qquad c \text{ (f.a.)} = 2.55 \text{ Å} \qquad \beta = 93°$$
$$\text{tilt angle } \phi = 3° \text{ in the tilt plane [130]}$$

The systematic absences of $h + k = 2n$ for hkl reflections give a plausible space group $C121\text{-}C_2^3$; the two-dimensional space group of the ab plane is equal to that of the space group $Cm2m\text{-}C_{2v}^{14}$ of PVDF form I. The X-ray intensities are calculated and compared with the observed data on the basis of the crystal structure of PVDF form I as a starting model. In the case of copolymers, however, the cocrystallization of VDF and TrFE monomers should be taken into consideration. That is, the weight content of VDF/TrFE monomers, the head-to-head and tail-to-tail linkage in VDF and TrFE sequence, tacticity in TrFE chain segments, etc., must be included into the calculation of intensities [127]. The alternative deflection of CF_2 groups along the chain axis, as likely as in PVDF form I, is also taken into consideration. The minimal R factor was obtained for the deflection angle $\sigma = 7°$, the value essentially the same with that in PVDF form I. In Figure 31 is shown the crystal structure of the LT phase of VDF 55% sample thus determined. In order to satisfy the relative intensities of the observed reflections more quantitatively, a coexistence of 60°-step domains, in which the b axis is distributed into $\pm$ 60° and $\pm$ 120° directions from the original b axis, was also introduced. The minimal R factor, 14%, was obtained assuming the statistical weight of the 60° domains by ca. 15%. In this way, the structural situation of the VDF 55% LT phase is very similar to that of PVDF form I.

2. The High-Temperature Phase

In the HT phase, the X-ray equatorial reflections are very sharp but the layer reflections are diffuse, making it difficult to carry out detailed structural analysis. An organic combination of the X-ray data with vibrational spectral data made it easier. The unit cell parameters are determined as follows (at 80°C):

$$a = 9.75 \text{ Å} \qquad b = 5.63 \text{ Å} \qquad c \text{ (f. a.)} = 4.60 \text{ Å} \qquad \frac{a}{b} = \sqrt{3}$$

As stated above, the infrared and Raman data indicate a molecular conformation consisting of random sequences of TG, T$\overline{\text{G}}$, TTTG, and TTT$\overline{\text{G}}$ rotational isomers, the averaged fiber period of which is calculated to be 4.62 Å on the basis of the chain structural models of PVDF forms II and III. It should be noticed that the model has a glide-type symmetry [123], not a helical one [121,122], as likely as the molecular symmetry of PVDF forms II and III. As suggested from the dielectric relaxation measurement [37,128,129] and the NMR measurement [130–134], the thermal rotational motion of

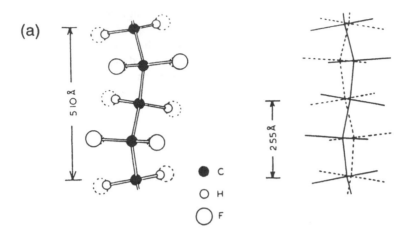

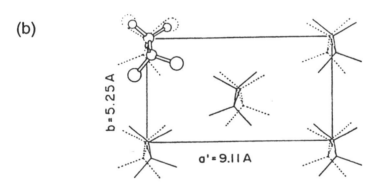

Figure 31 (a) Molecular and (b) crystal structure of the LT phase of VDF 55% copolymer [123]. In (a), an alternately deflected zigzag chain model with repeating period of 5.10 Å (left) and the statistical positioning of a deflected chain (solid line) and its mirror image (broken line) are illustrated. In the latter case the averaged repeating period is 2.55 Å.

chains around the c axis occurs violently at high temperature. Therefore a time-averaged molecular structure may be approximated by a model constructed by a collection of rings, which are the loci generated by rotating all the atoms about the chain axis, as illustrated in Figure 32. Fourier transformation of this model (molecular transform) gives the X-ray intensity distribution, which is in comparatively good agreement with the observed intensity for equatorial reflections as well as for layer reflections. Therefore we may assume that the HT phase is a kind of rotator phase in which the gauche-type chains rotate around the chain axis and therefore the total dipole moment becomes zero or a nonpolar crystal. As for the crystal structure of the HT phase, there remains an unresolved problem concerning the up and down directions of the chains along the c axis.

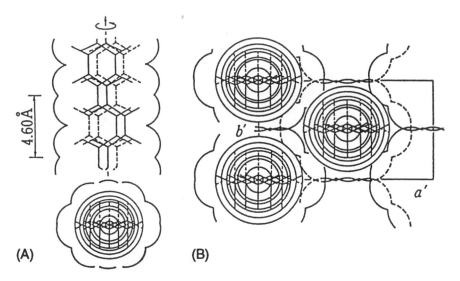

Figure 32 (a) Molecular and (b) crystal structures of the HT phase of VDF 55%-TrFE copolymer [123,126].

3. The Cooled Phase

The CL phase is obtained by cooling the HT phase down to room temperature. The positional shift of the equatorial and layer reflections from the horizontal lines can be interpreted by the tilting phenomenon. Based on the method proposed by Bunn et al., the tilt diagram can be reproduced well by assuming the following parameters:

$a = 9.16$ Å $b = 5.43$ Å c (f.a.) $= 2.528$ Å $\beta = 93°$ (assumed)

tilt angle $\phi = 18°$ in the tilt plane [130]

In Figure 33 is shown a comparison between the observed and calculated tilting X-ray diagrams taken by a cylindrical camera and a Weissenberg camera [for (001) reflections]. The ratio of the a and b axes deviates a little from the exact $\sqrt{3}$ of hexagonal unit cell, resulting in a slight difference in the lattice spacings of the (200) and (110) reflections, as actually seen in the X-ray profile of the equatorial line (Figure 29), where these reflections are observed apparently to split into at least two. Some authors assigned the CL phase to the mixture of the LT phase and the HT phase with the disordered chain conformation which is frozen during cooling from the high temperature to the room temperature [121,122,135]. We should not focus our attention on only one reflection (200 + 110) but rather interpret the X-ray pattern as a whole by using as many reflections of higher angles as possible, as shown in Figure 33. The X-ray pattern can be interpreted reasonably to originate from one crystalline phase [136], but the orderliness of the CL phase is not so high as the LT phase, as represented by the broadness of the reflections. In Figure 30 the lattice spacing of the HT phase changes apparently continuously into that of the CL phase, but the change occurs steeply, in a relatively narrow temperature range. The transition temperature from the CL to the HT phase is lower than that from the LT to HT phase. The LT phase is obtained from the CL phase by stretching at room temperature or by poling under high DC electric field, indicating that the stretching and

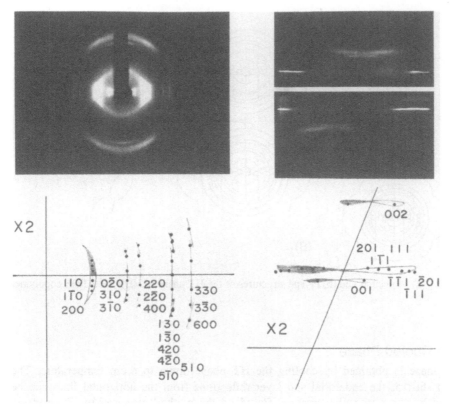

Figure 33 X-ray fiber diagram and Weissenberg photograph of the CL phase of VDF 55%-TrFE copolymer taken at room temperature, in comparison with the calculated reflection positions [123,126].

poling induce a structural regularization, possibly through the transformation of some disordered part into the regular trans form. From these experimental data, therefore, we may reasonably assume that the CL phase may contain some contribution of the gauche-type chain segments in the trans-zigzag crystalline region. Inclusion of some gauche segments may result in a shorter fiber-repeating period (2.528 Å compared to 2.55 Å of the LT phase) and a larger unit cell size. The characteristic tilting phenomenon and the existence of 60° domains must be also taken into acocunt in this discussion. Calculation of the X-ray intensities with the existence of the 60° domain taken into consideration gives the lowest R factor for the structure with the 60° domains of statistical weight of ca. 30%.

All these experimental facts could be interpreted reasonably by adopting a model for the CL phase shown in Figure 34: The long trans-zigzag segments are connected to each other with skew bond. The parallel array of these chains results necessarily in the tilting of trans segments in the (130) plane by ca. 18° and the CF_2 dipoles of the neighboring trans segments make an angle 60° to each other, leading to a natural formation of 60° domain structure in a statistical weight of about $\frac{1}{3}$. Of course we do not need to adhere to the skew model, but a combination of trans and gauche isomers may be more reasonable, as a defect equivalent to the skew bond (A possibility of the skew bond may not be ignored energeticlly, as pointed out by Farmer et al ([20]).

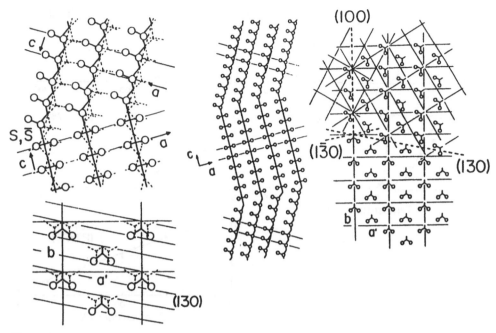

Figure 34 Crystal structural model of the CL phase [112,123,126].

In this way the CL phase may be assumed as a kind of superlattice consisting of the domains of long trans chains linked together with the boundaries of disordered trans-gauche bonds. Therefore the volume of the crystal, for example, cannot be calculated simply based on the unit cell parameters. The actually observed temperature dependence of the specific volume may be calculated by taking into consideration both the contributions of the trans segments and the gauche irregular linkages [137]. By assuming that the trans segments have repeating period of 2.53 Å along the chain axis and the irregular gauche linkage has an effective period of 2.30 Å along this direction, the averaged c axis length along the draw direction may be evaluated as follows: $\langle c \rangle = 2.53X + 2.30(1 - X)$, where X is the content of the trans segments and is evaluated from the intensity change of the X-ray (001) reflection as shown in Figure 35. Then the volume V is calculated by using the equation $V = ab\langle c \rangle \sin \beta$, where the a- and b-axis lengths are calculated from the temperature dependence of the observed lattice spacings d_{200} and d_{110}. The specific volume of the VDF 55% copolymer thus calculated is plotted against temperature in comparison with the observed temperature dependence of the macroscopic specific volume of the bulk sample [138], as shown in Figure 36 [137]. This figure indicates that the crystalline phase transition reflects explicitly on the thermal expansion curve of the bulk copolymer sample.

C. VDF Content Dependence of the Transition Behavior [126,139,140]

1. VDF 55% Copolymer

As discussed above, the VDF 55% copolymer exhibits crystalline phase transitions among the three phases: the LT, HT, and CL phases. The transition process has been summarized

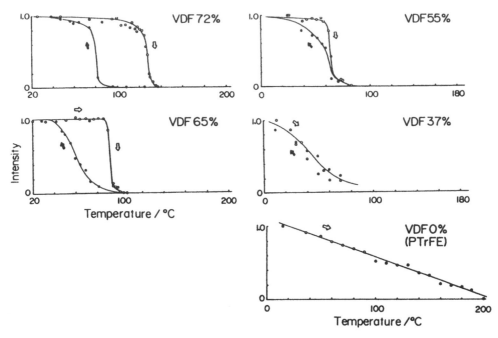

Figure 35 Temperature dependence of the integrated intensity of the (001) reflection evaluated for the various VDF-TrFE copolymers [126].

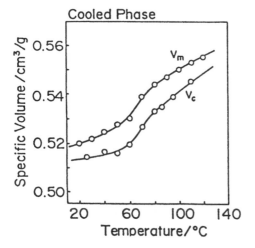

Figure 36 Temperature dependence of the specific volume calculated for VDF 55%-TrFE CL phase [137] in comparison with the observed specific volume of the bulk unoriented sample [138].

in the previous section. In the case of unoriented sample, the transition occurs reversibly between the CL and HT phases.

2.   VDF 70–80% Copolymers

In contrast to VDF 55% sample, the copolymer with VDF content 70–80% experiences the transition between the LT and HT phases predominantly even for unoriented samples. As shown in Figures 37 and 38, the X-ray reflections of the two phases coexist on the way of the transition, indicating that the transition is a typical first-order type. The thermal hysteresis between the heating and cooling processes is very large, supporting this transition type. Strictly speaking, the lattice spacings, for example, change differently between highly oriented samples and unoriented samples. As shown in Figure 38, for the oriented sample the (200, 110) interplanar spacing changes discontinuously between the LT and HT phases, with almost linear thermal expansion in each crystalline phase. On the other hand, in the case of unoriented sample, the curve of the HT phase changes in a slightly different manner from the former case: The spacing does not change in a linear manner, but apparently exhibits an S-shaped curve, suggesting a partial transition between the CL and HT phases as seen in the VDF 55% sample.

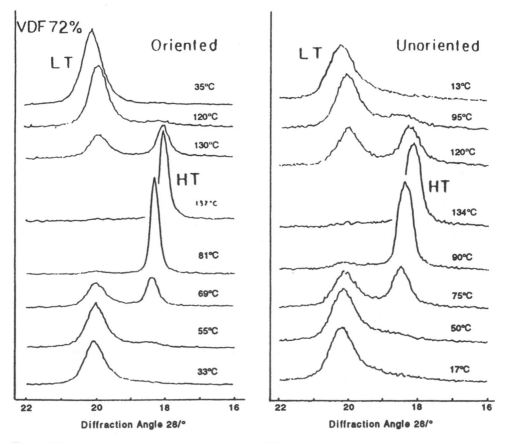

Figure 37   Temperature dependence of the X-ray (200, 110) reflection measured for oriented and unoriented VDF 72%-TrFE copolymer samples [126].

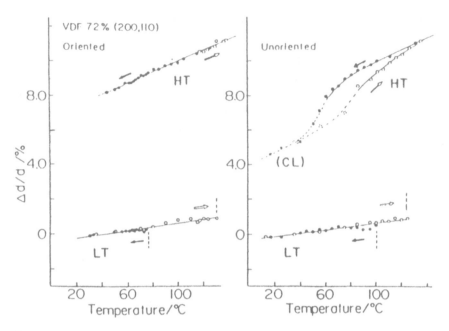

Figure 38 Temperature dependence of the lattice spacings $d(200, 110)$ measured for oriented and unoriented VDF 72%-TrFE copolymer samples [126].

3. VDF 65% Copolymer

This tendency can be detected more clearly for a VDF 65% sample [126]. As seen in Figures 39 and 40, the starting sample is the LT phase obtained by restretching the sample after heat treatment above the transition temperature. This phase transfers clearly and discontinuously to the HT phase. In the cooling process the HT phase transfers partly to the LT phase in a first-order transition manner but partly to the CL phase in a continuous way. Therefore, at room temperature, after cooling from the HT phase, the sample shows an overlap of the X-ray fiber diagram of the LT and CL phases with tilting phenomenon. The temperature dependencies of the X-ray pattern and the lattice spacing indicate this tendency clearly. This behavior is observed for both oriented and unoriented samples, but the tendency of the transition between the HT and CL phases is clearer for the unoriented sample. This characteristic behavior can also be detected through the measurement of the (001) reflection as a function of temperature. As shown in Figure 25, the 001 reflection changes sharply in the case of oriented VDF 72% sample. In the case of VDF 65% sample, the heating gives a sharp transition between the LT and HT phases, while the cooling causes rather broad change from the HT to both the LT and CL phases (Figure 35).

In this way the contribution of the crystalline phases in the transition changes systematically for copolymer samples of VDF 55, 65, and 72%. In the highly regularized sample of the LT phase, the transition occurs discontinuously between the LT and HT phases, but in the cooling process the contribution from the CL phase can be detected more clearly for the copolymer with lower VDF content. The unoriented sample exhibits this tendency more clearly:

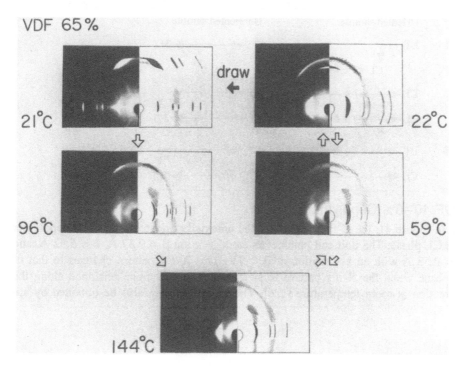

Figure 39 Temperature dependence of the X-ray fiber diagram taken for uniaxially oriented VDF 65%-TrFE copolymer [126].

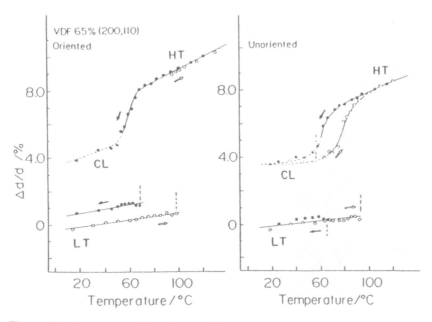

Figure 40 Temperature dependence of the lattice spacings d(200, 110) measured for oriented and unoriented VDF 65%-TrFE copolymer samples [126].

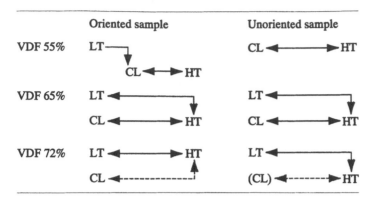

4. VDF 40–0% Copolymers

As seen in Figure 41, the X-ray diagram of the uniaxially oriented VDF 37% copolymer is of the CL phase. The unit cell parameters are $a' = a \sin \beta = 9.37$ Å, $b = 5.52$ Å, and c (f.a) = 2.53 Å with an assumption of $\beta = 93°$. This X-ray pattern changes to that of the LT phase when the X-ray pattern is measured for the sample tensioned along the draw direction at room temperature [126]. The LT pattern can also be obtained by ap-

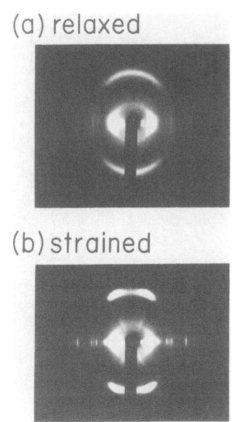

Figure 41   X-ray fiber diagrams of uniaxially oriented VDF37%-TrFE copolymer measured at room temperature under relaxed and strained conditions [126].

plying a high DC voltage [141]. In contrast to the case of a VDF 55% sample, the LT phase transforms back to the original CL pattern when the tension is removed. The (200) and (110) reflections observed separately at room temperature change apparently continuously and gradually into one as the temperature is increased from room temperature to 60°C, above which the reflection becomes sharp, as typically observed for the HT phase (see Figure 42). During this transition the trans–gauche conformational change can be detected, but the hysteresis between the heating and cooling processes is almost not seen, as shown in Figure 43. In Figure 44 is shown the temperature dependence of the Raman band intensity evaluated from Figure 27 for the VDF 73, 52, and 37% samples [124]. The 840-cm^{-1} trans band decreases in intensity and the 800-cm^{-1} gauche band increases in intensity as the temperature approaches the transition point, but the trans band does not disappear perfectly even above the transition point, indicating that the trans–gauche conformational change is not perfect—some trans portion remains even above the transition point. This tendency becomes clearer as the VDF content decreases to 13% and finally to 0% or PTrFE.

In the case of PTrFE, the X-ray pattern at room temperature is that of the CL phase [126], in which the equatorial reflections split up and down because of the tilting phe-

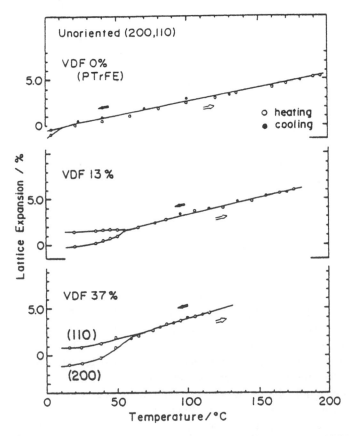

Figure 42 Temperature dependence of the lattice spacings $d(200, 110)$ measured for oriented VDF-TrFE copolymer samples with the VDF content of 37, 13, and 0% [126].

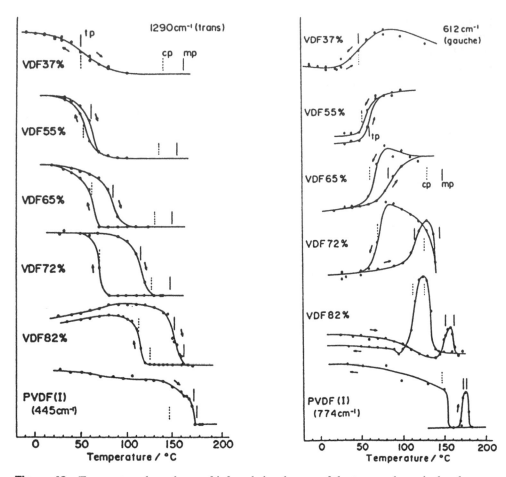

Figure 43 Temperature dependence of infrared absorbances of the trans and gauche bands measured for a series of VDF-TrFE copolymers [112]. mp, melting point; tp, transition point; cp, crystallization point.

nomenon and the first layer line is very diffuse. The (001) reflection cannot be detected, but the very broad meridional reflection with 2.30 Å spacing is observed instead. Lovinger and Cais assigned this reflection to the 002 reflection of the HT phase [142], but we should interpret the whole X-ray pattern, i.e., both the equatorial and layer reflections, consistently. The behavior of the (001) and (002) reflectional intensity as a function of VDF content should also be taken into account, as shown in Figure 45 [126]. The (001) intensity decreases down to zero as the VDF content decreases from 55% to 0%, just when the (002) reflection can still be observed with spacing $d = 1.26$ Å for all samples of VDF 55, 37, 13, and 0%. The (001) reflection should never be confused with the layer reflections of {201, 111}, although the former is observed only as a shoulder of the latter reflections and both the reflections appear on the meridional line because the chain axis of the sample is tilted from the draw direction in the CL phase. An explanation of these behaviors using the repeating period 2.30 Å or 4.60 Å is difficult to make. A simple equation can be developed to express the structure factors of the (001) and (002) reflec-

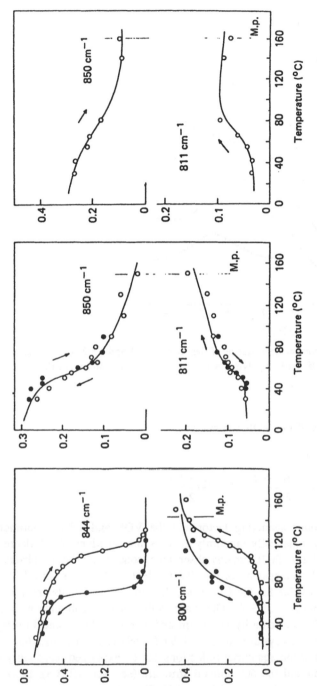

Figure 44 Temperature dependence of Raman scattering intensity of the trans (850 cm^{-1}) and gauche (800 and 811 cm^{-1}) bands for VDF-TrFE copolymers with VDF 73, 52, and 37% content [124].

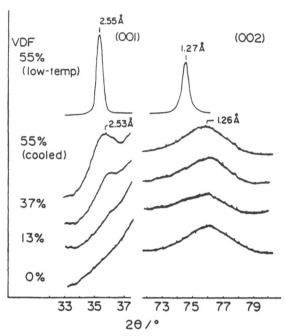

Figure 45 VDF content dependence of the X-ray (001) and (002) reflections measured for uniaxially oriented VDF-TrFE copolymer samples at room temperature [126].

tions as a function of the VDF content X:

$$F(001) \propto (f_F + f_H)X \tag{16}$$

$$F(002) \propto \frac{(4f_C + 3f_F + f_H)}{2} + \frac{(f_H - f_F)X}{2}$$

where f_i is an atomic scattering factor and the VDF and TrFE monomeric units are assumed to locate statistically randomly at the same positions. As X decreases to zero, $F(001)$ decreases but $F(002)$ remains observable, consistent with the observed tendency. Judging from the lattice spacing of the (002) reflection, the fiber period of PTrFE is ca. 2.53 Å, a common value for the CL phase; the unit cell constants are $a * \sin \beta = 9.68$ Å, $b = 5.62$ Å, and $c = 2.53$ Å. The c-axis value is in accordance with that reported from electron diffraction data by Gal'perin and Strogalin [143] and Lovinger and Cais [142]. Similarly to the crystal structure of VDF 55% CL phase, the displacements and relative intensities of the equatorial reflections can be explained reasonably by using the tilting phenomenon and the 60° domain structure (see Sec. III.B.3). Below room temperature the equatorial reflections (200) and (110) split apparently into two and collapse into one as the temperature is increased (see Figure 42).

In this way, the VDF-TrFE copolymers with VDF 0–40% exhibit the X-ray pattern of the CL phase and the transition between the CL and HT phases occurs in a wide temperature region with almost no detectable hysteresis. By measuring the inversion current, Oka and Koizumi suggested that PTrFE behaves like a ferroelectric substance

[141]; it is reasonably understood on the basis of the polar domain structure in the crystallites of the CL phase.

5. VDF 80–90% Copolymers

As stated above, copolymers with VDF content higher than 70% take the regular all-trans chain conformation of the LT phase, and the transformation between the LT and HT phases can occur in first-order fashion. When the VDF content is beyond 80%, the crystallization and transitional behavior change remarkably depending on the sample preparation conditions. For example, for copolymer samples of VDF 82 and 95%, melting and quenching into ice water give a crystalline form almost identical to PVDF form II [99]. However, when the sample is cooled slowly from the melt, VDF 82% copolymer crystallizes into the trans-zigzag form I type [144]. Such a situation can also be seen for the case of solvent casting. For example, VDF 95% copolymer crystallizes into form II when cast rapidly from methyl ethyl ketone solution, but into form III when the evaporation rate of the solvent is lowered. For VDF 82% samples, slow casting from acetone solution gives form I. Rapid casting gives form II, which transfers into form I when annealed at high temperature. That is, form II of VDF 82% copolymer is not as stable as form I. The crystallization behavior may be summarized in the following way:

	TGTG⁻ (form II)	TTTGTTTG⁻ (form III)	TT (form I)
Melt cooling			
VDF 100%	--→		slow cooling
95%	--------------------------------------→		
82%	---→		
72%		------------------→	
65%		------------------→	
Acetone cast			
VDF 100%	------------------------------→	slow cooling	
95%	-------------------------------→		
82%	--→		
72%		--------------------→	
65%		--------------------→	

That is, copolymer samples with VDF content higher than 80% are predominantly crystallized into form III (and form I) when cast slowly. Upon rapid casting, form II is mostly obtained. Samples with lower VDF content (<80%) crystallize in form I for any preparation condition. The situation is the same for the slow cooling of the melt. In this way copolymers of VDF content of ca. 80% may be a boundary at which the crystallization behavior changes sensitively depending on a slight change of the condition. From these experimental results we may deduce the relative stability of the crystalline forms of TT, TTTGTTTG̅, and TGTG̅, as discussed in the section on PVDF homopolymer.

In such a sense, therefore, a detailed description of the structure and transition behavior of the VDF 82% sample may be useful for understanding the essential features of the copolymers [144]. For example, Figure 46 shows the infrared absorbance of the trans (1276-cm^{-1}) and gauche (766-cm^{-1}) bands and those of the (200, 110) X-ray intensities of the LT and HT phases measured as a function of temperature for VDF 82% copolymer prepared by slow cooling from the melt. As stated above, this sample contains dominantly the all-trans LT phase together with a small content of unstable form II

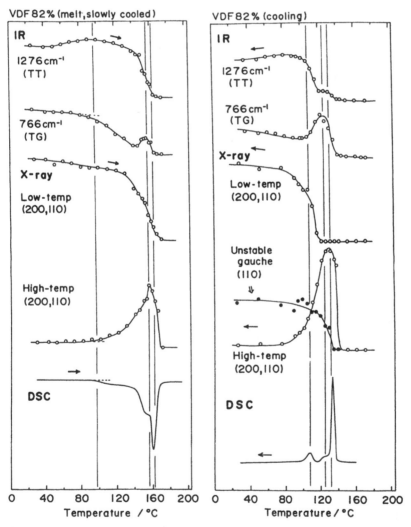

Figure 46 Temperature dependencies of the infrard absorbances of the trans and gauche bands, the X-ray (200, 110) intensities of the LT and HT phases in comparison with the DSC thermogram measured for the melt and slowly cooled VDF 82%-TrFE copolymer [144]. Left, heating process; right, cooling process.

crystal. As the temperature is increased, the gauche band characteristic of form II crystal begins to decrease in intensity around 100°C, where a small endothermic change is observed as a shoulder of the DSC curve. At the same time, the X-ray intensity of the HT phase begins to increase, during which the LT phase does not show as large a change as the former. At about 150°C the LT phase begins to transfer into the HT phase. The HT phase disappears at the melting point of ca. 162°C. In the cooling process the HT phase begins to appear around 135°C and then the unstable form II crystal appears around 125°C, though the relative amount is small. At 110°C the HT phase almost disappears and the LT phase increases in population, reflecting on the exothermic peak in the DSC thermogram. The unstable gauche form is saturated near 110°C and down to room temperature. On the other hand, in the case of melt-quenched VDF 82% sample, heating starts from the unstable form II. As the temperature increases, the form II decreases rapidly and transforms to the HT phase around 125°C (more exactly, in the region of 100–140°C), which is consistent with the transition observed for the unstable gauche form contained in the slowly cooled sample, as stated above. The HT phase thus generated melts around 145°C, which is rather lower than the melting point (162°C) of the HT phase observed for the slowly cooled sample. The unit cell of form II is larger by ca. 5.5% than that of PVDF form II, indicating a looser packing of the TGT$\overline{\text{G}}$ chains in the cell. At the same time, the crystallite size of this form II is considered to be rather small compared with that of the LT phase obtained by slow cooling from the melt. The transition behavior of VDF 82% copolymer may be summarized as follows:

LT phase (trans) ⇌ (150°C / 110°C) HT phase (gauche) ⇌ (162°C / 135°C) melt

Unstable gauche phase ⇌ (110–140°C / 125°C)

Unstable gauche phase → (125°C) HT phase → (145°C) melt

6. PVDF

In Figure 43 are plotted the infrared absorbances against temperature measured for a series of VDF-TrFE copolymers in both the heating and cooling processes [113]. In Figure 47 are shown the diagrams of transition and melting (or crystallization) temperatures plotted against VDF content for the case of unoriented samples [113]. The transition point increases gradually as the VDF content increases and at the same time the hysteresis between the heating and cooling processes increases. We may have a question: Where is the Curie transition temperature of PVDF form I? As predicted from the polar crystal structure, PVDF form I may experience a ferroelectric-to-paraelectric phase transition. Nakamura and Wada [145], Glass et al. [146], and Micheron [147] estimated the T_c of PVDF form I crystal as being located closely below the melting point on the basis of electric property measurements. Herchenröder et al. [148] assigned ca. 140°C to T_c, but this was doubted because piezoelectric activity can be observed still above this temperature. Lovinger et al. predicted that the T_c of PVDF form I crystal should be around 205°C by carrying out a straight-line extrapolation of the plot of T_c versus VDF content of VDF-TFE copolymers to the point of VDF 100% [149]. Ohigashi et al. also extra-

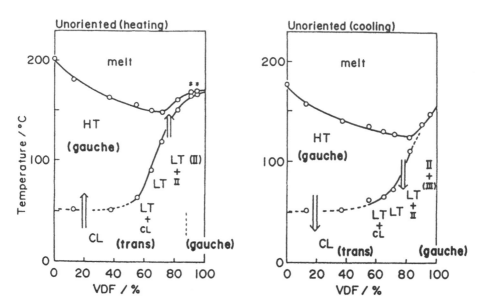

Figure 47 Transition behavior of unoriented VDF-TrFE copolymer samples with various VDF contents [112].

polated linearly a similar plot to the point of VDF 100% and stated that the T_c of PVDF should be above the melting point [150,151]. But why could they do a straight-line extrapolation? Tashiro et al. measured the temperature dependencies of infrared absorbance and X-ray diffraction intensity during heating of PVDF form I and found an occurrence of trans-to-gauche conformational exchange immediately below the melting point (ca. 172°C), as seen in Figure 43, where the gauche band increases and then decreases because of melting of the crystallites as observed in VDF 82% and 72% copolymers [85,86]. The infrared spectra of the generated HT phase are highly polarized, and the X-ray reflections also show a clear orientation, suggesting an oriented crystal-to-crystal transition. The infrared spectra of the HT phase is very similar to those of $T\overline{T}TGTTT\overline{G}$ conformation (form III), but the X-ray data showed the presence of (100), (120) reflections originally forbidden for form III, suggesting that the HT phase might be antipolar or nonpolar form III crystal; the structure of the latter is essentially the same as that proposed by Lovinger as form V [93]. As has been pointed out in several papers, the data presented by Tashiro et al. might be interpreted in another way: that the "center" of T_c is located above the melting temperature and only the beginning of the transition can be detected because a broad transition might overlap even partly with a broad melting [149]; or the melt of form I crystal and recrystallization to form III crystal might occur in the temperature region close to the melting point [87]. In any case, however, a possibility of ferroelectric phase transition of PVDF form I was presented, and T_c was proposed to be in the temperature region covering the melting phenomenon.

D. Molecular Motion

As revealed by X-ray diffraction and vibrational spectroscopic measurements, in the ferroelectric phase transition of VDF-TrFE copolymers a large conformational change occurs between trans and gauche forms attendant with a large rotational motion in the HT phase. Experiments with NMR [130–134,152,153] and quasi-elastic neutron scattering [154–156] clarified the molecular motion occurring in this transition.

By analyzing the bandwidth and the nuclear spin-lattice relaxation time T_1 of VDF 52–72% copolymers as a function of temperature, Ishii et al. [130,133,134] found that (a) in the low-temperature region of 30–80°C, the fraction of the mobile part in the sample (mobile fraction) is ca. 20% at most, originating from the flipflop motion of TrFE groups between two sites 180° apart; (b) between 80 and 114°C below the T_c, the mobile fraction increases because of an oscillation motion of VDF groups of trans-zigzag chains with an amplitude of 10°; (c) as the temperature approaches the transition point, the mobile fraction increases up to 100%, indicating an active chain motion in the crystalline region through the trans–gauche conformational exchange. T_1 becomes minimal at the transitional point, being interpreted in terms of one-dimensional diffusion motion of the conformational defects in the paraelectric phase. The activation energy of this diffusion motion was evaluated to be ca. 8.5 kcal/mol. The activation energy does not depend on the VDF content, but T_1 and the correlation time of the spin pair modulated by this diffusive motion of the defect are largely different between the copolymers. Hirschinger et al. [152,153] measured the free-induction decay signal $M(t)$ in addition to the T_1 of proton and fluorine nuclei and pointed out that the data of the HT phase could be interpreted better by using, not a flipflop kink mode of 0, 60, 120, or 180° reorientational motion, but a three-body crankshaft motion such as TGT$\overline{\text{G}}$T $\longleftrightarrow$ T$\overline{\text{G}}$TGT.

The trans–gauche conformational change should be correlated well with the repeating period along the chain axis. The c axial length is shortened drastically when the crystal transforms from the trans to gauche phases. It should be noticed, however, that the low-temperature trans chain itself contracts to some extent during the phase transition (by ca. 0.5%) [124]. This contraction is higher than the amount predicted for a simple negative thermal expansion of the chain at high temperature. As pointed out above, the gauche defect is considered to move along the chain axis through trans–gauche conformational change. This defect motion should affect the trans chain conformation more or less by intruding into the LT crystalline region, resulting in a contraction of the trans chain.

Legrand et al. analyzed experimental data of elastic and quasi-elastic incoherent neutron scatterings for VDF 70% copolymer [154–156]. The elastic scattering data were interpreted by assuming the model that the hydrogen atoms experience a diffusion motion within a cylinder of finite size. In the ferroelectric phase, the radius R and height H of the cylinder are 1.5 and 2.5 Å, respectively, and ca. 60% of the protons are spatially frozen in the sample. In the paraelectric phase, R and H increase to 2.0 and 3.3 Å, respectively, and the mobile protons are 87%. The residence time of a given conformation is 0.2 ns in the paraelectric phase. Quasi-elastic scattering data taken for the doubly oriented sample revealed that the molecular chains rotate around the chain axis combined with a translational motion along the chain axis.

E. Transition Mechanism and Phonon–Phonon Coupling

In general, the driving force of the phase transition is a soft mode [157]. A vibrational lattice mode with a given phase difference between the unit cells increases its amplitude, and the vibrational frequency shifts toward zero as the temperature approaches the transition point. This is considered to be caused by the softening of intermolecular interactions as a result of the anharmonic balance between the short-range force and the long-range electrostatic force. The intermolecular force becomes weaker and finally zero, just when the restoring force is zero and the atoms are frozen at the positions displaced from the original equilibrium positions; that is, a new structure is generated. After the phase transition is completed, the lattice mode of a new phase becomes harder as the temperature increases further. Such a soft mode is frequently detected in the second-order phase transition but is not limited to it. In the case of the first-order type also, the phonon is softened in the transition region. However, before the frequency decreases completely down to zero, the other type of phonon begins to appear at the transition temperature. The free energies of the two phases are equalized, and the structure changes suddenly and discontinuously from one form to the other. The mode which can work as a soft mode is dependent on the type of the crystal. In the case of VDF-TrFE copolymers, too, it is important to clarify what type of phonon plays a significent role in the ferroelectric phase transition and what kinds of interaction exist between the phonons.

For example, Figure 48 shows the temperature dependence of the far-infrared band frequency of the skeletal torsional mode of VDF 52% copolymer [76]. The 110-cm^{-1} band characteristic of the trans sequence appears to shift continuously to the lower-frequency side and transfer gradually to the gauche band position at 108 cm^{-1}. One interpretation of this spectral change may be as follows. This 110-cm^{-1} trans band is originally infrared inactive but could be observed because of the symmetry breakdown due to the loss of translational symmetry along the chain axis of the copolymer. According to the normal mode analysis, this band is considered to be a torsional mode with phase angle π between the neighboring monomeric units along the chain axis [7,8]. As the temperature approaches the transition point, anharmonicity of the torsional mode becomes larger, the vibrational amplitude is increased, and the frequency is softened. The amplitude increases further and finally reaches the level of the gauche position of the CF$_2$ groups, just when the molecular chain is trapped by a potential well of the gauche form and stabilized as a thermodynamically stable gauche crystalline phase. Figure 49 illustrates this situation schematically.

Figure 50 shows the ultrasonic velocity measured for unoriented copolymers with various VDF contents [76]. In the vicinity of the transition point, the anomalous minimum

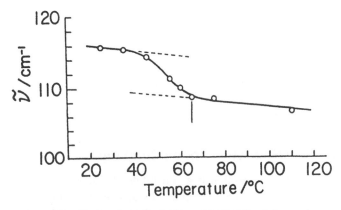

Figure 48 Temperature dependence of vibrational frequency measured for the skeletal torsional mode of VDF 52%-TrFE copolymer sample [76].

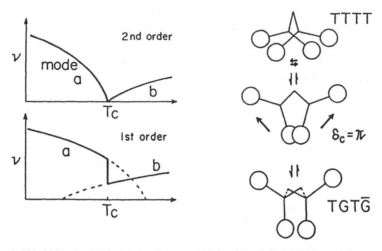

Figure 49 Schematic illustration of the structural change between the trans and gauche conformations (right) and the temperature dependence of the soft mode [112].

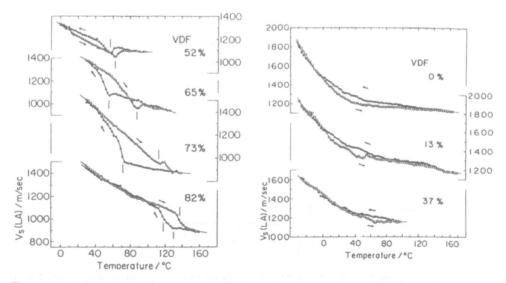

Figure 50 Temperature dependence of the ultrasonic velocity measured for unoriented VDF-TrFE copolymers [76].

where $a_2 = \alpha(T - T_c)$, c^P is the elastic modulus under a constant polarization, and β, γ, and δ are the coupling constants between P and ϵ. The linear coupling $\beta P \epsilon$ represents the piezoelectric effect on the elastic constant (β corresponds to the piezoelectric constant); the coupling $\gamma P^2 \epsilon$ is due to the electrostrictive effect. The elastic constant c^E under a constant electric field is given by

$$c^E = c^P - \left(\frac{\partial^2 A}{\partial \epsilon \partial P}\right)^2 \chi^\epsilon \tag{18}$$

Here χ^ϵ is an electric susceptibility [= $(\partial E/\partial P)^{-1}$]. The c^E changes with temperature in a fashion depending on the type of interaction between P and ϵ: for example, for a linear coupling $\beta P \epsilon$, c^E is given as

$$c^E = c^P - \beta^2 \chi^\epsilon = \begin{cases} c^P - \beta^2/[\alpha(T - T_c)] & T > T_c \\ c^P - \beta^2/[2\alpha(T - T_c)] & T < T_c \end{cases} \tag{19}$$

In Figure 51 is illustrated the calculated result for the different couplings between the P and ϵ for the second ($a_4 > 0$, $a_6 = 0$) and first ($a_4 < 0$, $a_6 = 0$) types of transition. The is observed for the velocity. Such an observation was made by several authors [158–160]. The Young's modulus of the sample was also found to show a minimum [161–164]. Such an anomaly in the mechanical property can be interpreted phenomenologically in terms of coupling between the electric polarization P and mechanical strain ϵ [76]. A Helmholtz free energy A of the polar crystal is expressed as [165]

$$A = A_0 + \tfrac{1}{2}c^P \epsilon^2 + \tfrac{1}{2}a_2 P^2 + \cdots + \beta P \epsilon + \gamma P^2 \epsilon + \delta P \epsilon^2 \tag{17}$$

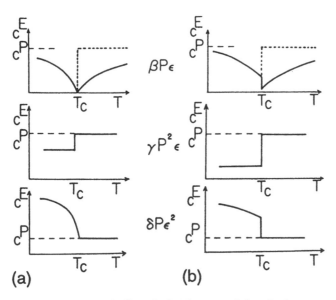

Figure 51 Theoretically calculated curve of the elastic constant c^E versus temperature for (a) second and (b) first-order transitions [76]. The solid and broken lines in the top figures are for $\beta \neq 0$ and $\beta = 0$ above the transition point, respectively.

anomalous change in the elastic modulus is reproduced qualitatively well by these calculations.

The far-infrared spectra suggested that the internal rotational mode with increasing amplitude has a possibility of soft mode in the transition, which may cause a large change in the CF_2 dipole or the spontaneous polarization P of the crystal. In other words, the temperature change in the polarization P may originate from the thermal behavior of the optical phonon. The strain ϵ is related with the acoustic phonon. Therefore, the anomalous thermal behavior of the ultrasonic velocity may be interpreted in such a way that the optical phonon is coupled with the acoustic phonon through the piezoelectric interaction ($\beta P\epsilon$), electrostrictive interaction ($\gamma P^2\epsilon$), and so on.

F. Factors Affecting the Transition Behavior

Ferroelectric phase transition of VDF-TrFE copolymers is affected by many factors.

1. Crystallite Size or Domain Size

As in general synthetic polymers, the melting point and the phase-transition point are affected largely by the crystallite size of the copolymer [166–169]. For example, the DSC curves were measured for various samples of VDF 73% copolymer which were melt-quenched in liquid nitrogen and then annealed isothermally at different temperatures (T_a) for 2 h [169]. The T_c and melting point change largely depending on the T_a: T_c increases at first with increasing T_a. When the sample is annealed above the transition temperature, the height of the endothermic peak decreases gradually and a new endothermic peak begins to appear at low-temperature side and shifts toward the lower temperature with further increase of T_a. The melting point increases during this process of annealing. Odajima et al. derived an equation describing an effect of the crystallite boundary on the phase-transition temperature in a one-dimensional ferroelectric system [170]. According to their result, the T_c increases for larger crystallite size: This corresponds to the behavior observed for samples annealed below the transition temperature. A lowering of transition temperature as observed for samples annealed above T_c cannot be explained. We must also consider other type of boundary effect. Ohigashi et al. interpreted the change of T_c and melting point of annealed and/or poled VDF 75% samples on the basis of idea of domains in the crystallite [171]. Including an effect of domain wall energy, T_c is approximately expressed by $T_c \approx T_0 - 2\epsilon^*d/t + \ldots$, where d and t are the width and thickness of a domain and ϵ^* is a constant that depends on the dielectric constant. Annealing and/or poling may change d and t, resulting in a decrease (or increase) of T_c. In their interpretation, however, no consideration was made for the chain conformation. The transition is accompanied by a trans–gauche conformational change. In the previous section it was pointed out that the CL phase including a gauche linkage as a boundary of domains transforms to the HT phase at lower temperature than the LT–HT phase transition; this is because of the existence of gauche irregular structure, which may act as a kind of nucleus for the trans-to-gauche conformational change. Therefore, if the annealing above T_c increases the number of such gauche boundaries, then the transition point might be shifted to the lower-temperature side.

Existence of domains has not yet been directly proved experimentally. The temperature dependence of the integrated half-width of the X-ray reflection suggests its existence [169]. Figure 52 shows the integrated half-width of the X-ray (200, 110) reflections measured for the LT and HT phases of unoriented VDF 73% copolymer samples as a function of temperature (heating process). The half-width of the LT phase is almost

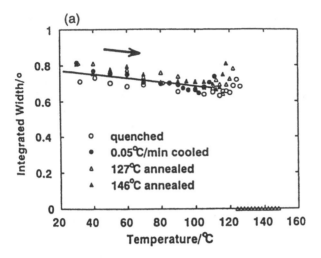

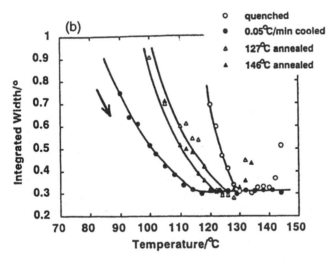

Figure 52 Temperature dependence of the integrated half-width of X-ray (200, 110) reflections evaluated for unoriented VDF 73%-TrFE copolymer samples [169]. The samples were quenched from the melt to liquid nitrogen temperature and then annealed at the various temperatures or cooled slowly from the melt to room temperature at a constant rate of 0.05°C min. (a) LT phase; (b) HT phase.

constant over a wide temperature region and is not much different between the samples treated under the various conditions, but the half-width of the HT phase changes drastically during the phase transition: At the initial stage the half-width is very large and decreases steeply as the transition proceeds and finally becomes very sharp, which is common to all the samples. Since the half-width may correspond to the size of the X-ray coherent crystalline region in an inversed relation, this experimental result can be interpreted in the following way (refer to Fig. 53). The LT phase of the unoriented VDF 73% sample consists of a collection of domains, the size of which reflect on the half-

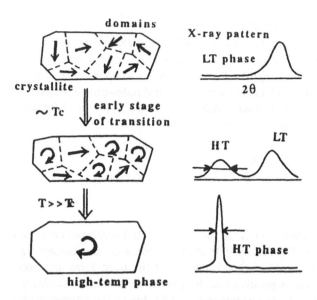

Figure 53 Illustration of structural change and corresponding X-ray profiles speculated for ferroelectric phase transition of VDF 73%-TrFE copolymer [169].

width of the X-ray reflection. As the temperature increases, the HT phase begins to appear at each domain and coexists with the LT phase domains. The domains of the HT phase are fused into one larger domain, because the large-amplitude rotation of the chains erases the difference between the domains. In this way the domain wall is gradually taken away as the transition proceeds further, and finally the whole crystallite is occupied by a large single domain of the HT phase in which the chains are packed in a hexagonal mode and rotate violently.

2. VDF Regular Sequence

An analysis of ^{19}F NMR spectra and infrared absorption spectra revealed a distribution of VDF monomeric units in a copolymer chain [127]. The VDF regular sequence is increased with an increase of VDF content following Markov chain statistics. This regularity is related to the cooperativity in the phase transition [126]. For the chain with VDF content, a probability of side-by-side arrangement of regular VDF segments is high. On the other hand, if the VDF content is low, this probability is low and the short VDF segments are isolated from each other. Therefore, even when some parts of the VDF segments experience a trans-to-gauche conformational change, this does not work as a trigger for the cooperative phase transition but the structural change occurs point-to-point gradually in a broad temperature region. For the sample with longer VDF sequences, the structural change occurring in a segment is propagated cooperatively to the neighboring VDF segments and a large-scale phase transition can be induced in a narrow temperature region. The regularity in the VDF segments along the chain axis may reflect the sharpness of the X-ray (001) reflection. In fact, the (001) reflection width is sharper for the sample with higher VDF content [124,126]. Besides, it is sharper for the LT phase than for the CL phase of the same VDF content.

3. Trans–Gauche Energy Difference

In Figure 54 is plotted the value of ln (D_T/D_G) against $1/T$, where D_T and D_G denote the relative infrared absorbances of trans (840-cm^{-1}) and gauche (612-cm^{-1}) bands, respectively, and T is absolute temperature [126]. The line slope in the temperature region of the paraelectric HT phase, where the long-range intermolecular dipole–dipole interaction may be diminished, gives a trans–gauche energy difference Δu (= $u_{TT} - u_{TG}$) intrinsic of a polymer chain. The energy difference thus estimated is

VDF	37%	55%	65%	72%	
Δu	−0.56	−1.52	±0.0	+0.69	kcal/mol monomeric unit
	(−0.55)	(−1.5)	(0)	(1.1)	

The figures in parentheses are the values corrected for the sequential constraint condition of the T and G rotational isomers [172]. As the VDF content decreases, Δu changes sign from positive to negative; i.e., the conformational stability of the trans form increases. For VDF 72% copolymer, where Δu is positive and the gauche form is more stable, the appearance of the LT trans phase at room temperature may be due to the intermolecular dipole–dipole interaction as well as the nonbonded van der Waals interaction. When such a trans-stable LT phase is heated and once a trans bond changes to a gauche bond, the transformation to the latently stable gauche chain conformation will occur drastically and cooperatively, resulting in a sharp first-order transition. For a trans-stable copolymer of VDF 37–55%, the conformational transition may be difficult. In fact, the Raman meas-

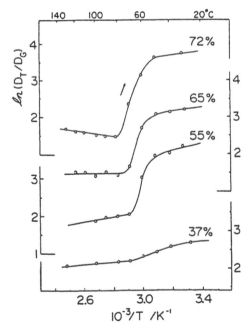

Figure 54 Plot of ln($D_t D_g$) versus $1/T$, where D_t and D_g denote the relative infrared absorbances for the trans (840 cm^{-1}) and gauche (612 cm^{-1}) bands, respectively [126].

urement indicated an incomplete conformational change even above the transition point for these copolymers with low VDF content [124].

4. Role of TrFE Unit

Similarly to PTFE, TrFE is stabilized into the trans form when the chain configuration is atactic and contains many head-to-head linkages [173]. An introduction of bulk TrFE monomer into the VDF chain increases the unit cell size of the ferroelectric phase, as is known from X-ray diffraction measurements [126,139,140]. This volume increment of the cell results in weaker intermolecular interaction between the VDF units because of an increase of the interchain distance. The dipole moment of VDF monomer unit is canceled in part by the introduction of TrFE monomer. Thus, not only the van der Waals interaction but also the dipole−dipole long-range interaction will change depending on the VDF content. Correspondingly, the stability of the trans crystal form, the transition temperature, etc., will be affected significantly. As described above, the TrFE units may experience some molecular motion even below the transition point [130,133,134,174]. This mobility of the TrFE unit might also play an important role in initiating the conformational change in the VDF segments.

All these factors described here must be coupled together in an organized fashion when any theoretical treatment is undertaken for the ferroelectric phase transition of VDF-TrFE copolymers. The details will be mentioned in a later section.

G. Structural Change Induced by External Field

In addition to the factors discussed above, an external field such as a high electric field, high pressure, and so on influences the transition behavior of the copolymers very much.

1. Poling Treatment

As stated above, a characteristic feature of ferroelectric material is an inversion of the dipole under application of an electric field. In the case of VDF-TrFE copolymers, too, an inversion of CF_2 dipoles has been observed with many measurements, including X-ray diffraction [46,175,176], infrared spectra [177,178], and NMR spectra [179].

An organized combination of infrared transmission and reflection spectral data gives us useful information concerning the reorientation behavior of CF_2 dipoles induced by poling treatment [178]. When the infrared beam is incident along the normal to the sample film and the transmission absorption spectra (TRS) are measured, the electric field E_0 of the infrared beam interacts with the transition dipoles μ' included within the film plane [refer to Eq. (1)]. When the infrared beam is incident on the metal surface with high incident angle (ca. 80° from the normal to the surface), the electric field pointing in the direction of the normal is enhanced remarkably, and thus only the vibrational modes with the transition dipoles normal to the film surface can be observed selectively (see Fig. 55). The spectra thus measured are called reflection-absorption spectra (RAS). By combining these two methods of TRS and RAS, we can deduce a concrete image of dipole orientation in the film. For example, Figure 56 shows the TRS and RAS of VDF 75% copolymer sample (LT phase) taken before and after poling treatment. In these figures the bands denoted by symmetry species A_1, B_1, etc., have essentially the same polarization characters as PVDF form I crystal discussed in the previous section: The A_1, B_1, and B_2 species have transition dipoles parallel to the b, c, and a axes, respectively. In the TRS, the bands of the B_2 species (// a) increase in intensity after poling of the film, and those of the A_1 species (// b) decrease instead. The B_1 mode at ca. 1400 cm^{-1}

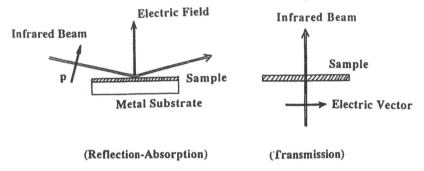

Figure 55 Illustration of the infrared spectral measurement in a transmission (TRS) and reflection-absorption (RAS) modes.

(// *c*) keeps the intensity unchanged. This B_1 band is intense in the TRS but weak in the RAS, indicating that the chain axis is included within the film plane. The A_1 and B_2 bands in the RAS change their intensities in an opposite manner to that of the TRS: The intensity of the B_2 bands decreases and that of the A_1 bands increases in the RAS after poling treatment. Therefore we may speculate that the applied electric field induces re-

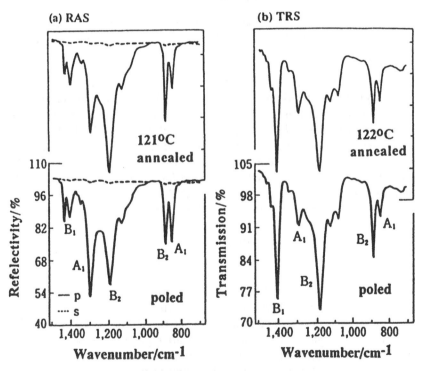

Figure 56 Transmission (TRS) and reflection-absorption (RAS) spectral changes taken before and after poling of VDF 73%-TrFE copolymer film [178]. The solid and broken curves correspond to measurement with the electric vector of the incident infrared beam parallel (*p* wave) and perpendicular (*s* wave) to the reflection plane (refer to Fig. 55).

orientation of the CF$_2$ dipoles (or the polar b axis) into the direction of the electric field while keeping the chain axis unchanged.

The poling not only induces a reorientation of dipoles but also works to increase the regularity of the crystalline state [140,171,175]. Figure 57 shows X-ray diagrams of VDF 52% copolymer taken for the LT phase (a), the CL phase (b), and the sample obtained by poling the CL phase (c and d) [175]. In Figure 57c the incident X-ray beams are normal to the film surface, i.e., parallel to the electric field E, and in Figure 57d they

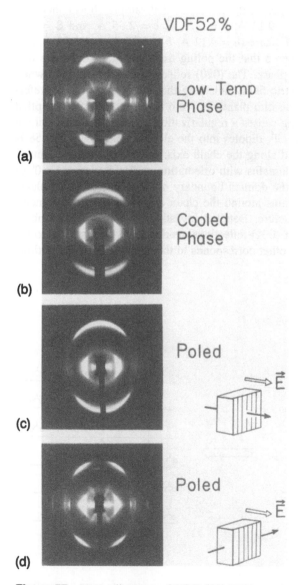

Figure 57 X-ray diagrams of VDF 52%-TrFE copolymer samples of various crystalline states: (a) low-temperature phase; (b) cooled phase; (c, d) poled sample of CL phase [175]. The draw axis is vertical.

are perpendicular to E. The characteristic features of the X-ray diffraction pattern of the CL phase are the tilting phenomenon and the broadness of reflections. By stretching the CL phase at room temperature along the draw axis, an LT phase is obtained in which the tilting phenomenon disappears almost completely and the layer-line reflections are remarkably sharp. In these uniaxially oriented samples of the LT and CL phases, the X-ray pattern is almost isotropic around the draw axis. On the other hand, the X-ray patterns are very different between the two scattering geometries for the poling-treated sample, as seen in Figures 57c and 57d. For example, the equatorial reflections are different in relative intensities between these two cases. The first-layer reflections of (201, 111) are also different. Compared with the CL phase, the poling-treated sample shows sharp reflections. The unit cell parameters (a = 9.12 Å, b = 5.22 Å, c = 2.55 Å, and β = 93°) are almost the same as those of the LT phase (a = 9.12 Å, b = 5.25 Å, c = 2.55 Å, and β = 93°). Therefore it may be considered that the poling treatment transforms the disordered CL phase into the regular LT phase. The (020) reflection has a maximum when the X-ray beam is parallel to the electric field, indicating that the b axis orients preferentially into the direction normal to the film plane, i.e., into the direction of the applied electric field. That is, to say, the poling causes a regularization of the crystalline state at the same time as the inversion of the CF_2 dipoles into the electric field. In Figure 58 is illustrated the structural change viewed along the chain axis: the CL phase has a multidomain structure consisting of small domains with orientation angles deviating by 0° or $\pm$ 60° from the standard b axis with the domain boundary of the (130) or ($1\bar{3}0$) plane. The poling induces rotation of the chains around the chain axis, and the multidomains disappear into the single domain structure. Detailed investigation of the Weissenberg photographs shows that the two sets of (001) reflections overlap: One corresponds to the crystallites of the tilted chains and the other corresponds to the crystallites of the untilted

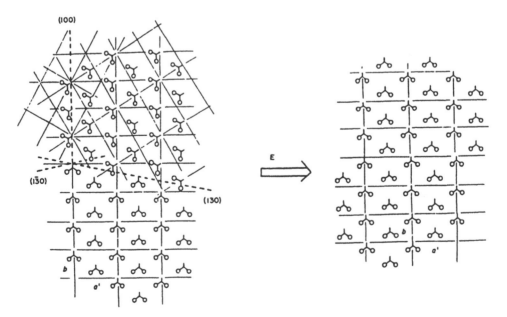

Figure 58 Dipole reorientation induced by poling of the cooled phase. The (100), (130), and ($1\bar{3}0$) planes are domain walls between the different dipole orientations [175].

chains along the draw axis. Before poling the sample contains only the tilted crystallites. Therefore we may consider that parts of the tilted crystallites change into crystallites of untilted chains. For example, as illustrated in Figure 59, in a crystallite with dipoles closely parallel to the applied electric field, the dipole reorientation may occur relatively easily while the tilting character remains unchanged. On the other hand, in a crystallite where the CF_2 dipoles are arrayed almost perpendicular to the electric field, the tilted segments may rise up vertically as the dipole reorientation is induced by the electric field. As discussed in the section on PVDF, it may be difficult to rotate the whole chain segment rigidly when the dipoles are reoriented toward the electric field. It is plausible to assume a mechanism in which a solitonlike motion of defects (such as gauche linkage) induce a simultaneous occurrence of CF_2 dipole rotation [50,51] and the standing up of tilted segments into the vertical direction.

Poled sample transforms into the HT phase at higher temperature than the untreated CL phase does. This is because the poled sample is close to the LT phase in structural regularity, although the degree of regularity depends on the effectiveness of the poling process.

An effect of electric field on the phase transition was investigated by measuring the infrared spectra in situ under high DC voltage. The transition point was found to shift to the higher-temperature side with increasing applied electric field. The rate of shift

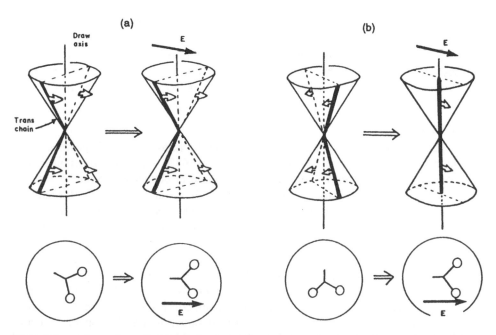

Figure 59 A model for the structural change in the cooled phase induced by an external electric field [175]. The bold lines represent the trans-zigzag segments, with the dipoles denoted by arrows. The circular cones represent the distribution of the tilted trans chains around the draw axis. In (a), where the electric field vector is parallel to the dipoles, no change occurs for the tilting angle. In (b), with the electric field vector perpendicular to the dipoles, the tilted chain segments stand up into the vertical direction.

dT_c/dE is estimated as 0.36°Cm/MV (VDF 74% [180]), 0.03°Cm/MV (heating) and 0.69°Cm/MV (cooling) (VDF 75%, [177]), and 0.01°Cm/MV (heating)(VDF 52% [181]).

The measurement of inversion current as a function of temperature is another method for investigating the effect of electric field on the transition. The inversion current was measured by applying an AC voltage of triangular shape to a VDF 75% sample in the course of heating and cooling processes [182]. The coercive electric field E_c evaluated from the voltage giving the maximal current was plotted against temperature as shown in Figure 60. E_c decreases drastically in the vicinity of the transition. After entering the HT paraelectric phase, E_c increases again to some extent. This change is considered to correspond to the transition from the paraelectric to ferroelectric phase induced under high electric field. As the temperature increases further, the curve of electric current versus electric field is that characteristic of the paraelectric phase and the E_c is not known in this region. The free energy G is developed as a function of P as

$$G = G_0 + \tfrac{1}{2}xP^2 + \tfrac{1}{4}yP^4 + \tfrac{1}{6}zP^6 \tag{20}$$

The electric field E is given as

$$E = \left(\frac{\partial G}{\partial P}\right)_T = xP + yP^3 + zP^5 \tag{21}$$

where $x = (T - T_0)/C$ is a linear electric susceptibility, and T_0 and C are constants. y and z are nonlinear electric susceptibilities, independent on temperature, and $y < 0$ and $z > 0$ for the first-order transition. From the relation between P and E, we may estimate the coercive electric field E_c as a function of temperature. The solid curve in Figure 60 shows the curve thus calculated, which is in good agreement with the observed one. The P versus E curve shows double hysteresis when the temperature is just above the transition point T_c. The transient increase of E_c in Figure 60 corresponds to this double hysteresis loop.

2. Tensile Force

The LT phase is obtained through the stress-induced transition of the CL phase at room temperature. In the case of copolymers with VDF 50–65%, this transformation is irreversible. For the VDF 0–40% samples, the LT phase is obtained only under tension. Such a disorder–order transformation occurs through a trans–gauche conformational change, as already discussed in previous sections. In this sense, therefore, we may speculate that the transition behavior should be affected by application of external tensile force. For example, Figure 61 shows the thermal contraction and expansion curves of a uniaxially oriented VDF65% sample measured in the heating and cooling processes, respectively, under the application of a constant tensile stress of 0.0–1.1 MPa [183]. If the transition temperature is defined as the point with the steepest slope, then the transition temperature is found to shift to the lower-temperature side as the tensile stress is increased. When the stress exceeds a certain critical value, the transition point increases to the higher-temperature side. This critical stress is considered to correspond to the point where the CL phase is forced to transfer to the LT phase. The increase of the transition temperature of the LT–HT phase is easily interpreted because the trans-to-gauche conformational change is suppressed by the application of tensile stress along the chain axis. A lower shift of the transition point in the low-stress region comes from a more complicated situation of mechanical distribution in the polymer material; the balance of

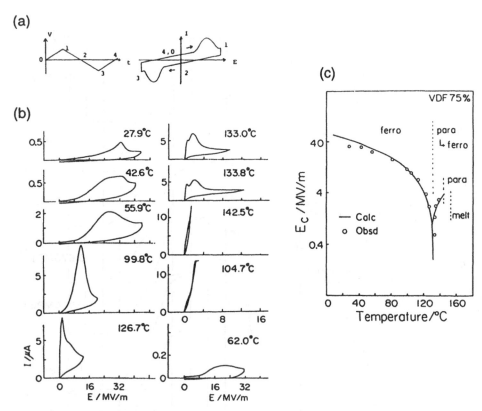

Figure 60 (a) A triangular-type high-voltage V applied to the film (left) and the corresponding polarization inversion current I (right) as a function of time. (b) Temperature dependence of I observed for VDF 75%-TrFE copolymer. (c) Comparison of the calculated coercive electric field E_c with the observed value as a function of temperature (heating process) [182].

the strain between the crystalline and amorphous regions must be taken into account in the analysis of the data. From the thermodynamic equilibrium between the LT (or CL) and HT phases, a Clausius-Clapeyron equation is derived with the three-dimensional compliance components taken into account; the temperature shift is expressed in an approximation by

$$\frac{dT_c}{d\sigma_3} = \frac{-k(\Delta l/l_0)}{\Delta S} \tag{22}$$

where σ_3 is a stress along the draw axis, ΔS is an entropy change during the transition ($= -12.4 \times 10^4$ J/Km^3), and $\Delta l/l_0$ is a change in the fiber period in the transition ($= 9.8\%$). The parameter k is determined by a ratio of Young's modulus of the crystalline and amorphous regions and the Poisson ratio of the crystalline and amorphous phases. The calculated value of $dT_c/d\sigma_3$ is $-5.7°$C/MPa, in good agreement with the actually observed value of $-5°$C/MPa. Such an interpretation can be made also for the case of tension along the lateral direction [184]. In this case the sample expands in both the heating and cooling processes and the application of tensile stress induces a shift of the

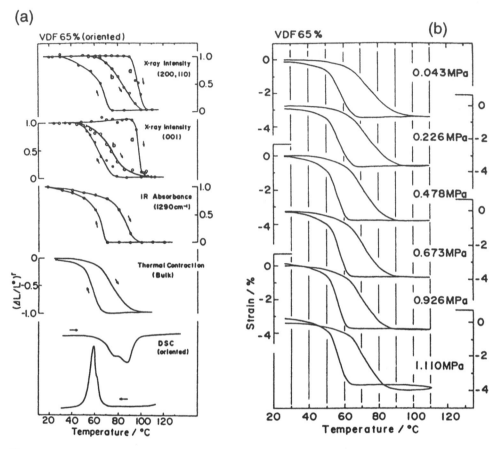

Figure 61 (a) Comparison of the temperature dependence of X-ray diffraction intensity and infrared absorbance of the LT phase with that of thermal contraction of the bulk sample and the DSC thermogram measured for VDF 65%-TrFE copolymer [183]. (b) Stress dependence of the thermal contraction curve of VDF 65% copolymer.

transition temperature as follows: the observed $dT_c/d\sigma_1 = -1°C/MPa$ and the calculated value is $-0.5°C/MPa$.

Sekimura et al. [185] measured the dielectric constant of a cast and annealed VDF52% copolymer film with evaporated gold electrodes on the surfaces, under tensile stress [172]. The lower shift of T_c for the tensile stress in the lateral direction is consistent with the above-mentioned result of the thermal expansion experiment, whereas that along the drawn axis is opposite to the above result. Many factors, such as the VDF content, the degree of orientation, the relative content of the LT and CL phases, the stress distribution associated with the aspect ratio of the used samples, the film–metal contact, etc., might affect the result, although the true reason for the discrepancy has not yet been clarified.

3. High Pressure

An application of hydrostatic pressure to VDF-TrFE copolymers causes regularization of the crystal structure as in the cases of tensile stress and electric field [12,163,164]. In the region of relatively low pressure p, the transition temperature increases at a rate

dT_c/dp = 0.314°C/MPa for VDF 53%, dT_c/dp = 0.383 (heating) and 0.401 (cooling) for VDF 65%, and dT_c/dp = 0.33 for VDF 73% polymer. For the VDF 65% case, an application of hydrostatic pressure higher than 150 MPa induces the transformation of crystalline state similar to that of the VDF 73% sample with typical first-order phase transition. For a VDF 54% unoriented sample, Matsushige et al. observed an enlargement of transition hysteresis above 200 MPa, ascribed to a transformation from an unstable crystal structure including gauche defects to a stable crystal (this transition is considered to correspond to that from the CL to LT phases, judging from the behavior) [12].

4. Electron Beam

During the observation of a thin copolymer film by electron microscopy, Lovinger noticed that the electron diffraction pattern changes even at room temperature, depending on the electron dose [186]. For example, a single crystal of VDF 73% copolymer shows an electron diffraction pattern typical of the LT phase at room temperate, but it changes gradually though discontinuously to the pattern of the HT phase as the electron irradiation increases. Odajima et al. also observed such an electron beam-induced phase transition by using γ rays as a source [187]. Recently, Macchi et al. also investigated this effect [188,189]. When the γ-ray dose is lower than 300 Mrad, the VDF 73% remains a crystalline material, but it becomes amorphous beyond 500 Mrad. This amorphous sample shows an apparently ferroelectric transition behavior as seen in the existence of maximal dielectric constant around 80°C, suggesting that this transition occurs between paraelectric phase and ferroelectric spin glass phase. In the case of VDF 65% sample, the γ-ray irradiation causes a transition to the HT phase at room temperature. The LT phase of a VDF 52% sample changes into the CL phase when the irradiation is relatively low (ca. 40 Mrad).

IV. FERROELECTRIC PHASE TRANSITIONS IN VDF-TFE COPOLYMERS

A. VDF Content Dependence

Vinylidene fluoride-tetrafluoroethylene (VDF-TFE) copolymers exhibit similar ferroelectric phase transitions as do VDF-TrFE copolymers, but the behavior is rather different depending on the VDF content [190–195].

1. Uniaxially oriented VDF 81%-TFE shows a transition between LT and HT phases, but the transition temperature is ca. 120°C (heating) and 115°C (cooling), which is appreciably lower than that of the corresponding VDF 82%-TrFE copolymer. The hysteresis between the heating and cooling processes is also small.

2. In the case of uniaxially oriented VDF 75%-TFE, the starting LT phase changes into the CL phase discontinuously and then transfers to the HT phase apparently continuously. Cooling back to the room temperature gives a mixture of CL and LT phases. These behaviors are very similar to those of the VDF 65%-TrFE copolymer.

3. Uniaxially oriented VDF 65%-TFE shows the tilted X-ray fiber pattern typical of the CL phase. This CL phase is transferred reversibly to the LT phase only under tension along the draw axis. The conformational change between the trans and gauche forms is limited, and the long trans sequences spread over the VDF and TFE segments, as seen in the observation of infrared and Raman spectra as a function of temperature. The X-

ray pattern remains that of the CL phase even at high temperature, although the layer lines become more diffuse because of a conformational disordering along the chain axis.

4. In the case of uniaxially oriented VDF 41%-TFE, the LT phase is obtained at room temperature even under relaxed condition, but this phase is easily transferred to the more stable CL phase by annealing above 110°C. In this temperature region, the initial LT phase transfers to the CL phase and then goes to the HT phase at higher temperature. This behavior is very similar to that of the VDF 55%-TrFE copolymer, as stated in the previous section. Around 110°C the infrared gauche band begins to increase its intensity, and the trans bands show some deflection in their temperature dependence, just when the LT-to-CL phase transition is observed. This is an important experimental correspondence between the X-ray and infrared data, because, as discussed in the section on VDF-TrFE copolymers, the CL phase has been considered to generate via the trans–gauche conformational change as a defect between the long trans segments. It is also important that the tilting phenomenon characteristic of the CL phase may be considered to relate to the generation of gauche bonds within the trans-zigzag skeletal chains. The TFE segmental parts remain in a trans form even at high temperature below the melting point.

5. VDF 23%-TFE shows the X-ray pattern of the LT phase for a uniaxially oriented sample at room temperature, but the pattern includes diffuse reflections at positions corresponding to the second, seventh, and eighth layer lines of polytetrafluoroethylene (PTFE) [196]. In the case of PTFE homopolymer, these X-ray reflections change to spots upon cooling the sample to low temperature via the several stages of the phase transition. PTFE experiences a large molecular motion attendant with large conformational change between the right- and left-handed helices via the trans-zigzag conformation because of the low energy barrier between the right- and left-handed helices. The trans-zigzag part behaves as a boundary between the right- and left-handed helical conformations, and the relative amount of this boundary region decreases with a drop in temperature. Such a structural change in the TFE segments is also observed in VDF 23%-TFE: the infrared band at 642 cm^{-1} characteristic of the helical form increases gradually over a temperature region of 0 to −100°C, and the 625-cm^{-1} trans band decreases simultaneously. Corresponding to this conformational change, the X-ray diffraction pattern changes greatly when the sample is cooled down to −70°C, for example (see Fig. 62). That is, the above-mentioned diffuse reflections change into spots. The innermost intense reflection of the first layer line changes to a group of three sharp reflections, which correspond to the reflections of the first layer line of the planar-zigzag chain and the seventh and eighth layer lines of the uniform PTFE helical chain. Therefore, it may be said that the conformational change occurs in the TFE segments from trans to helix, keeping the trans form of the VDF sequence. Such structural transformation in the TFE sequence detected for this copolymer is not found for the VDF-TFE copolymers with VDF contents of > 41%. The short TFE segments are considered to be stabilized in the trans form by a restraining effect of neighboring VDF trans sequences.

B. Comparison Between VDF-TrFE and VDF-TFE Copolymers

We may compare the above-mentioned transition behavior of VDF-TFE copolymers with that of VDF-TrFE copolymers, as illustrated in Figure 63 [194].

1. Crystalline Form at Room Temperature

The LT phase is detected for VDF > 70% in VDF-TrFE samples and VDF > 75% in VDF-TFE samples. The CL phase is for VDF 0–50% in VDF-TrFE samples and for

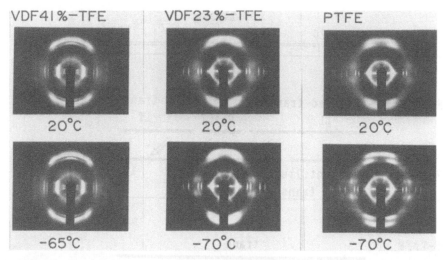

Figure 62 X-ray fiber diagrams of VDF-TFE copolymers with VDF 41, 23, and 0% measured at 20 and −70°C [194]. VDF 41%-TFE shows the pattern of the CL phase. VDF 23%-TFE shows the pattern of the LT phase with the component of the PTFE diagram overlapped.

VDF 40–65% in VDF-TFE samples. The VDF 65% sample may be typical in showing the large difference between these two types of copolymers. Uniaxially oriented VDF 65%-TrFE exists as a mixture of the LT and CL phases, while the corresponding VDF-TFE copolymer exists as the CL phase. The LT phase is present only under tension, the behavior of which is very close to that of VDF 37%-TrFE copolymer. In contrast to the case of VDF-TrFE copolymers, as the TFE content increases further, the CL phase is difficult to obtain, as is typically observed for samples of VDF 20–40% content. They exist as the untilted trans structure at room temperature, but they change easily to the CL phase irreversibly when the sample is heated above the transition point. In the case of VDF 23%-TFE, the trans structure is stable even above 200°C.

2. Local Structure

In the low range of TrFE or TFE content, these monomeric units take essentially the trans conformation. But as the TFE sequential length becomes longer, the characteristic behavior begins to appear. Just as in PTFE, the long TFE segments experience thermal motion around the chain axis through internal rotation via the trans form. Such thermal motion is "frozen in" at low temperature and then the more stable helical form is attained in the long TFE sequences, as observed for the VDF 23%-TFE sample and PTFE. In the case of PTrFE, rotational motion of the crystalline chain is considered to occur near 0°C, but the drastic conformational change is not detected as clearly as for PTFE.

3. Phase-Transition Behavior

The discontinuous first-order phase transition is observed for VDF 70–100% of VDF-TrFE copolymers and VDF 75–100% of VDF-TFE copolymers. In the latter case, however, the transition point is close to the melting point and the thermal hysteresis between the heating and cooling processes is narrower. For VDF-TrFE copolymers with VDF content < 50%, a broad and apparently continuous phase transition of the second-order type occurs with almost negligibly small thermal hysteresis. Such a transition is observed

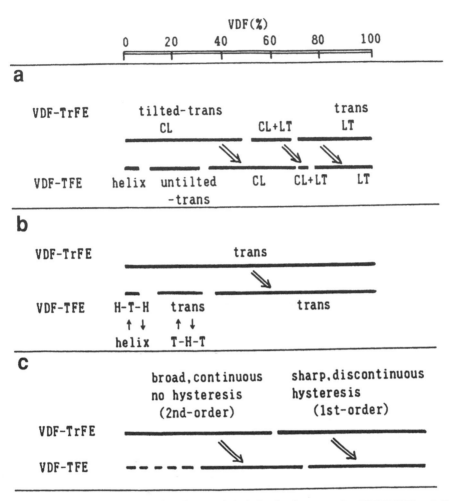

Figure 63 Comparison of the characteristic behavior between the VDF-TrFE and VDF-TFE copolymers: (a) crystal form at room temperature; (b) structure of commonomer parts; (c) phase transition [194].

in VDF-TFE samples with VDF contents of 40–65%. The conformational change in the VDF segments becomes imperfect for VDF-TrFE copolymers with lower VDF content. Such a phenomenon can also be seen in the case of the TFE copolymer system, with a much stronger influence. In fact, the VDF and TFE segments take the trans form preferentially in the case of VDF 64%, and the generation of gauche linkage is small in the population even at high temperature.

In this way the cooperativity and the completeness of the conformational transition are reduced by cutting off the continuous VDF arrays by the introduction of comonomer units into the chain, just when the degree of influence is different between TrFE and TFE components [190–194]. As a whole, the TFE unit may stabilize the trans conformation of the copolymer chain to a higher extent than the TrFE unit and make it more difficult for the chain to experience the trans-to-gauche conformational change at high temperature. In fact, the transition temperature of the trans-to-gauche change is very close to the melting point in the case of VDF-TFE copolymers compared with VDF-TrFE copolymers of similar VDF content. Effects of comonomer on chain conformational stability has been investigated by Farmer et al. [20] for PVDF homopolymer chains containing a given amount of head-to-head and tail-to-tail linkage consisting of TrFE or TFE units: The effect of the TFE unit to stabilize the trans conformation of VDF chain is larger by ca. twice that of the TrFE unit.

In addition to such an energetical consideration, we also need to take into account the thermal mobility of the monomeric units. The TFE unit is considered to be thermally more mobile than TrFE. This may be one of the reasons why VDF 64%-TFE copolymer takes the CL phase and VDF 65%-TrFE copolymer takes the LT phase under free tension. A smaller thermal hysteresis in the VDF-TFE copolymers is also considered to support this idea: The thermal motion in the comonomer units adjacent to the VDF segment makes the trans–gauche conformational exchange easier to occur. The lower transition point of VDF-TFE copolymers may also be an illustration of this factor. The smaller dipole moment in the VDF-TFE chain also plays as an important factor in reducing the transition temperature. In order to clarify the difference in role of comonomeric units in the phase transition, it may be necessary to carry out a computer simulation of the phase transition at a molecular level, for example.

C. Vibrational Spectra of Copolymers

As discussed in the previous sections, vibrational spectra have been playing an important role in interpretation of the essential features of the ferroelectric phase transition of VDF-TrFE and VDF-TFE copolymers. In particular, an observation of gauche bands and an exchange of trans and gauche band intensities at high temperature were some of the keys in clarifying the structural change in the transition. The infrared and Raman spectra of the copolymers are, however, not so simple as expected from the chemical structure, but require us to accumulate various types of information necessary for the band assignment. For example, a comparison of the spectra between the various crystalline modifications of PVDF homopolymer and the band assignment based on the normal modes calculation lead to a reasonable assignment of the bands intrinsic of the trans and gauche conformers (Sec. II.B).

More detailed investigation requires us to analyze the spectra through more sophisticated methods. Some methods have been reported to analyze the complicated infrared spectra of the copolymers, for example, normal mode calculation for model chains with particular monomeric sequences [8,195], a negative eigenvalue theorem for disordered structure [197], a utilization of Green function [198], and so on. Some of them will be mentioned here.

1. Band Intensity and VDF Sequence

Figure 64 shows infrared spectra taken for a series of unoriented VDF-TrFE copolymers cast from polar HMPTA solution at room temperature, which are composed almost purely of all-trans crystalline form [126]. In the section on PVDF, the sensitivity of the band intensity to the VDF sequential length was pointed out. For example, the band at ca. 1280 cm^{-1} needs the all-trans VDF sequential length to be longer than 4 in order to show a detectable intensity in the infrared spectra, and the 840-cm^{-1} band for the VDF trans sequence needs to be longer than 3. This relationship between band intensity and VDF sequential length can be checked quantitatively by analyzing the spectra of the copolymers shown in Figure 64. A polymer chain is assumed to be synthesized by

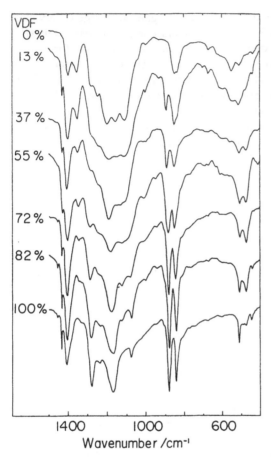

Figure 64 Infrared spectra of a series of VDF-TrFE copolymers cast from HMPTA (room temperature) [126].

random copolymerization of monomers A and B. The probability of the sequence ... BA_pB ... is given by

$$f(p) = (1 - X)^2 X^{p-1}$$

where X is the molar content of monomer A and p is an integer [199]. The total weight fraction of the regular sequential length longer than A_m is given by

$$F(m) = \frac{\displaystyle\sum_{p=m}^{\infty} pf(p)}{\displaystyle\sum_{p=1}^{\infty} pf(p)} = X^{m-1}[m + (1 - m)X] \tag{23}$$

If we assume that a certain mode of vibration can give its infrared-active intensity only for the regular monomer sequence longer than A_m, then the infrared absorbance $R(X)$, reduced by the sample thickness, is expressed as follows by utilizing the Lambert-Beer law (absorbance = molar extinction coefficient · concentration · thickness).

$$\frac{R(X)}{R(1)} = XF(m) = X^m[m + (1 - m)X] \tag{24}$$

In Figure 65 is shown $R(X)$ as a function of VDF content X for the various infrared bands. Solid curves represent Eq. (24) for the various m values. For example, the band intensity at 880 cm^{-1} is linearly proportional to X; that is, $m = 1$. This band can appear in the infrared spectra only if one isolated monomer A (= VDF) is present within the copolymer chain. The trans band at 840 cm^{-1} exhibits its infrared intensity for $m = 2$–3; it can be infrared active if the A monomers are arranged in a regular sequence longer than AA or AAA—in other words, TT or TTT. The band at 1285 cm^{-1} begins to appear for $m = 3$–4 or TTT-TTTT. The 1285-cm^{-1} band disappears above the ferroelectric phase transition, while the bands at 840 and 880 cm^{-1} remain observed. The result shows that the chain conformation changes from the long all-trans form into the gauche-type form consisting of TG (T$\overline{\text{G}}$) and TTTG (TTT$\overline{\text{G}}$) sequences, as already explained in the previous section.

2. Characteristic Bands of VDF and TFE Segments

When we search the polarized infrared spectra taken at room temperature for a series of uniaxially oriented VDF-TFE copolymers [195,200], the spectra are found to be reproducible in most parts by simple addition of two parent spectra of PVDF and PTFE with relative weight taken into account. That is, the bands intrinsic to VDF and TFE monomeric units are relatively easily identified: For example, the bands at 1400–1450 cm^{-1} are from VDF sequences, and the bands at 600–650 cm^{-1} are from TFE segments. Some of the bands, however, cannot be interpreted in such a simple manner [195]. For example, the bands in the 800- to 900-cm^{-1} region behave in a somewhat peculiar fashion, different from the bands in the other frequency regions. In order to interpret them, the normal mode calculation was made for the following copolymer models:

$$-[-(CH_2CF_2)_m-(CF_2CF_2)_n-]-$$

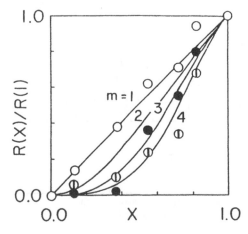

Figure 65 Dependence of reduced infrared absorbance $R(X)/(R(1)$ on VDF molar content X [126]. The curve with a fixed value of m includes a contribution from all the VDF regular sequences longer than (VDF)$_m$. ○ 880 cm^{-1}, ● 840 cm^{-1}, and ◐ 1285 cm^{-1}.

where $m:n$ is $\infty:0$ (PVDF), 8:2 (VDF 80%), 5:5 (VDF 50%), 2:8 (VDF 20%), and 0:∞ (PTFE). As discussed above, the TFE segments may be assumed to take the trans conformation in the copolymer chain as long as the TFE content is less than ca. 20%. Therefore, in this normal mode calculation the whole chains are assumed to be in the all-trans conformation, even for PTFE homopolymer (This assignment is considered not so unreasonable because the trans conformation of PTFE is reported to show very similar vibrational frequencies to those of (13/6) or (15/7) helical form, roughly speaking) [201]. In the frequency region 800–900 cm^{-1} the vibrational modes characteristic of CF$_2$ groups of VDF unit can appear. Based on the calculated vibrational frequencies and modes, the bands are classified into two types in this region: one is the mode with the phonon wave extending over the VDF segment only, and the other is the mode in which the phonon is located at the boundary between the VDF and TFE segments. From these calculations a state of density is constructed in this frequency region, as shown in Figure 66b. The vibrational frequencies of symmetric CF$_2$ stretching modes are not so dispersed but are concentrated in a comparatively narrow frequency region regardless of the relative con-

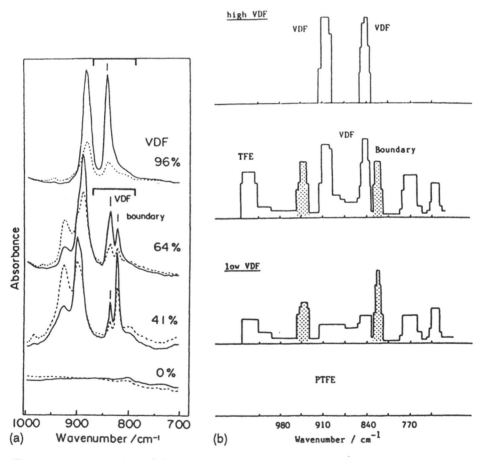

Figure 66 A comparison of observed polarized infrared spectra of VDF-TFE copolymers with the frequency distribution function calculated for the model structure —[—(CH$_2$CF$_2$)$_m$—(CF$_2$CF$_2$)$_m$—]— [195].

tent of VDF and TFE units. However, the number of the CF_2 modes or the vibrational density gradually decreases with increasing TFE monomers. On the other hand, the mode localized in the VDF-TFE boundary is located at almost constant frequency (ca. 830 cm^{-1}). Therefore we can expect an interchange of relative intensity between these two modes depending on the TFE content. Of course, the absorption intensity should be expressed as a product of the frequency distribution function and the transition dipole moments, as already explained, but the vibrational mode density corresponds qualitatively to the actually observed spectral change shown in Figure 66a.

V. PHASE TRANSITION IN THIN FILMS

Thin films are important for various reasons. (a) Effects of ferroelectric domain size and interface on the chain aggregation structure may be clarified through comparison with the behavior of three-dimensional bulk samples. (b) Poling treatment or dipole reorientation under external electric field can be made effectively and therefore detailed investigation is possible for the structurally homogeneous system. (c) By decreasing the film thickness down to the molecular dimension, the two-dimensional ferroelectric system may be realized as a limiting morphology and the three-dimensional ferroelectric system can be understood more deeply through clarification of the behavior of the two-dimensional system, which should be more easily treatable theoretically.

Methods to investigate the crystal structure of thin films have been developed. Some of the typical methods are X-ray diffraction measurement for Langmuir-Blodget films [202], infrared and Raman spectral measurement by utilizing reflection on the surface, and so on. As discussed in Section (III.G), a combination of two techniques, TRS and RAS, in infrared measurement is a useful technique for the purpose of obtaining information concerning the orientation and conformation of the chains in thin films, as explained below [178,203–206].

A. Morphology of Vacuum-Evaporated Thin Film

Vacuum evaporation is an interesting method of preparing thin films. The sample is put in a molybdenum boat, which is mildly heated under vacuum. The molecular chains evaporate and deposit on the surface of the substrate, which is set a certain distance above the heater. The substrates used may be alkali halide (KBr, NaCl, etc.), glass, quartz, silicon, germanium, and so on, depending on the purpose of the measurement. Alkali halide is especially useful for the measurement of transmission spectra. The RAS spectral measurement is made by using an aluminum-coated glass plate as a substrate. Epitaxial crystallization of chains on alkali halide was reviewed by Mauritz et al. for polyethylene, nylon 6, polyoxymethylene, polytetrafluoroethylene, etc. [207].

Regarding PVDF, a detailed study was reported by Takeno et al. investigating the orientation and crystalline form of the chains under varying conditions such as substrate temperature and evaporation rate [205,206]. For example, under the condition of substrate temperature −100°C and evaporation rate 1–3 Å/s, form I crystal is deposited with the chain axis included in a horizontal plane of the substrate. As the substrate temperature increases, form II chain is deposited. The chain orientation changes from horizontal to vertical as the evaporation rate is decreased. More mild condition (125°C and slow evaporation) leads to the formation of form III with vertical orientation. These behaviors can be easily understood on the basis of thermodynamic and kinetic advantage between forms I, II, and III (Sec. II.B.4).

VDF 73%-TrFE copolymer was evaporated onto KBr and aluminum-coated glass, for which the TRS and RAS were measured, respectively [178,203,204]. By adjusting the evaporation condition, the morphology formed on different substrates of KBr and aluminum-coated glass could be controlled to become identical to each other. For example, Figure 67 compares the TRS and RAS measured for vacuum-evaporated VDF 73%-TrFE thin film for a short evaporation time (film A). In Figure 68 are shown the spectra for a film evaporated for a longer time (film B), for which the spectra were measured before and after annealing at high temperature. As has already been mentioned, the infrared spectra of VDF 73%-TrFE copolymer may be interpreted to a good approximation on the basis of structure and symmetry of all-trans zigzag chain of PVDF form I. The A_1 mode possesses the transition dipole along the b axis, the B_1 mode along the c axis, the B_2 mode along the a axis, and the A_2 mode is infrared inactive. In the spectra of film A, the B_1 bands (// c) observed in the RAS are absent in the TRS, while the A_1 (// b) and B_2 (// a) bands are detected relatively intensely in the TRS. In film B, on the other hand, the TRS taken before annealing exhibits all the bands of the A_1, B_1, and B_2 species with relatively strong intensity as likely as in the RAS. After annealing film B, the B_1 transmission band intensity is reduced remarkably. This is consistent with a large increase in the corresponding RAS band intensity. In parallel, the intensities of the A_1 and B_2 bands increase in the TRS and decrease in the RAS. From these spectral data combined with the symmetry consideration, the structural changes illustrated in Figure 69 may be reasonably deduced. In an early stage of evaporation, i.e., in film A, the molecular chains crystallize with the c axis almost normal to the film surface. The a and b axes are considered to be isotropically oriented around the normal axis. The thickness of film is estimated to be about 1000 Å, from the infrared intensity. The average molec-

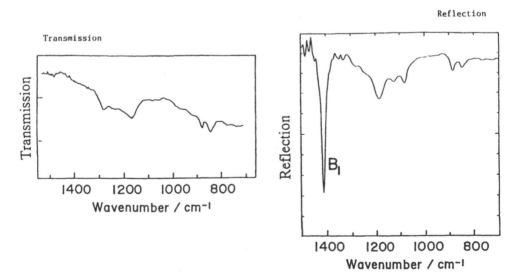

Figure 67 FTIR spectra of VDF 73%-TrFE copolymer thin film prepared by vacuum evaporation for a short time [178]. Left, transmission spectrum, right, reflection-absorption spectrum.

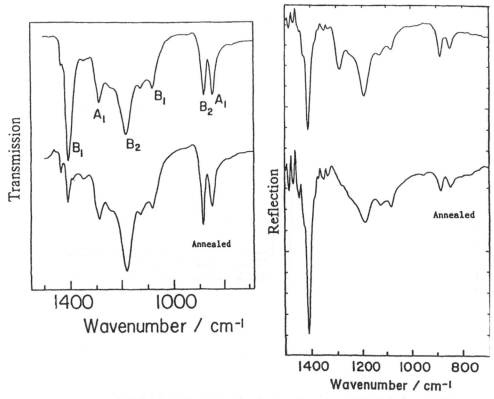

Figure 68 FTIR spectral change before and after heat teatment of VDF 73%-TrFE copolymer thin film prepared by vacuum evaporation for a long time [178]. Left, transmission spectrum; right, reflection-absorption spectrum.

ular weight of the vacuum-evaporated sample is about 4000, as measured by GPC, corresponding to a chain length of about 160 Å if the fully extended trans conformation is assumed. Therefore it may be considered that film A consists of about six layers of all-trans chains. The chain orientation remains unchanged after heat treatment, possibly because of a strong interaction between the chains and the substrate surface. In the relatively thick film B, or film prepared via a longer evaporation period, the additionally evaporated chain molecules are deposited over this thin layer of film A; the chain orientation is gradually disordered and at last the chain axis becomes almost parallel to the film surface. Heat treatment in the high-temperature region above the melting point induces rearrangement of the newly overlaid chains into a structure identical to that of the initial stage of evaporation. The infrared intensity increment suggests additional stacking of ca. 600 Å or four trans chain layers on film A. It may be emphasized here that the newly deposited molecular chains change the orientation by about 90° after heat treatment. As one plausible mechanism, the initially adsorbed molecular layer might serve as a nucleus for this large structural change.

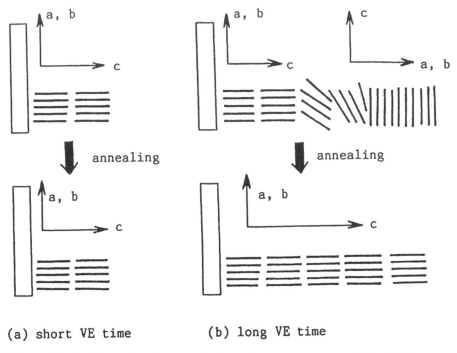

(a) short VE time   (b) long VE time

Figure 69   An illustration of structural change in vacuum-evaporated films [178].

B.   Phase Transition in Vacuum-Evaporated Thin Film

In Figure 70 are shown the temperature dependencies of the relative intensities of the trans (1280-cm^{-1}) and gauche (610-cm^{-1}) bands measured for vacuum-evaporated and annealed VDF 73% copolymer film in comparison with those measured for bulk VDF 73% sample [178,203,204]. In the bulk film the trans–gauche conformational change occurs around 120°C in the heating process and around 70°C in the cooling process. The trans–gauche conformational change is also observed for the vacuum-evaporated thin film, but the temperature region is much lower than the original one—about 70°C in the heating process and 50°C in the cooling process. The transition region is very broad compared with the bulk one. The low transition temperature is considered to come from the low molecular weight of the evaporated sample: The bulk sample has an average molecular weight of 10^5–10^6, which is reduced to the order of 10^3 in the evaporated sample. In the previous section the large reorientational motion of the chain axis from horizontal to vertical was pointed out as occurring by annealing at high temperature. This large structural transformation does not occur in the trans–gauche conformational transition region but occurs for the first time once the sample is heated above the melting temperature and cooled down again.

VI.   PHASE-TRANSITION THEORY

In order to interpret the phase-transition behavior of fluorine polymers including PVDF, VDF-TrFE copolymers, and VDF-TFE copolymers, several theories have been reported. Some of them will be reviewed here.

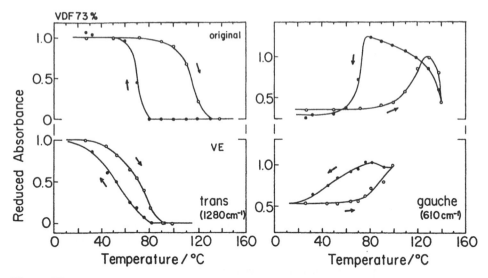

Figure 70 Temperature dependence of infrared absorbances of trans (left) and gauche (right) bands measured for vacuum-evaporated VDF 73%-TrFE copolymer in comparison with that of bulk sample [178].

A. Phenomenological Theory (Thermodynamics)

Ferroelectric phase transition has been interpreted phenomenologically by developing the free energy as a function of temperature T and polarization P [35]. Some applications have been indicated in previous several sections. Let us review this theory here again briefly. Gibb's free energy G is expressed as

$$G = G_0 + \tfrac{1}{2}\chi P^2 + \tfrac{1}{4}\xi P^4 + \tfrac{1}{6}\zeta P^6 \tag{25}$$

where $\chi = (T - T_0)/C$ is a linear electric susceptibility, and T_0 and C are constants. By setting the electric field as

$$E = \left(\frac{\partial G}{\partial P}\right)_T = \chi P + \xi P^3 + \zeta P^5 = 0 \tag{26}$$

the spontaneous polarization P_s is solved as

$$P_s^2 = \frac{-\xi \pm \sqrt{\xi^2 - 4\chi\zeta}}{2\zeta} \tag{27}$$

By substituting $\chi = (T - T_0)/C$, P_s is expressed as a function of temperature T. Depending on the sign of ξ, the curve of G versus P_s changes as shown in Figure 71.

In the case $\xi > 0$, the free energy G changes continuously from a curve with a single minimum at $P_s = 0$ to a curve with a double minimum at $P_s \neq 0$ as the temperature is decreased. That is, a second-order phase transition occurs from the paraelectric to the ferroelectric phase. On the other hand, for $\xi < 0$, starting from the single-minimum curve, there appears a state with two (three) minima at $P_s = 0$ and $\neq 0$ points at temperature T_c, and then the curve becomes one with double minima at $P_s \neq 0$. This corresponds to

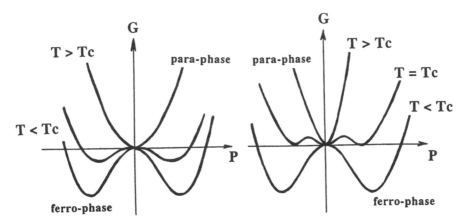

Figure 71 Free energy G and spontaneous polarization P_s.

the transition from the paraelectric to the ferroelectric phase with an intermediate state of coexistence of the para- and ferroelectric phases at T_c, that is, first-order transition. The phase-transition temperature T_c is expressed as

$$T_c = T_0 \qquad \text{for the second-order transition}$$

$$T_c = T_0 + \frac{3C\xi^2}{16\zeta} \qquad \text{for the first-order transition}$$

The hysteresis loop between P_s and E, the temperature dependence of the dielectric constant $\epsilon = (\partial P/\partial E)_{E=0}$, and so on, can be expressed in terms of the parameters ξ, ζ, etc. By fitting these theoretical curves with the actually observed data, the parameters can be determined (see Sec. III.G.1) [37,182].

If both mechanical work and electrical work are applied to the system, then the Helmholtz free energy A should be expressed in a more complicated fashion:

$$A = A_0 + \tfrac{1}{2}c_{ij}{}^P x_i x_j + \tfrac{1}{2}x_{ij}{}^x P_i P_j + \tfrac{1}{2}q_{ijk}x_i P_j P_k + \tfrac{1}{4}\xi_{ijkl}{}^x P_i P_j P_k P_l + \dots \qquad (28)$$

where x_i is the strain component and P_i is a component of polarization. If no external stress is applied, the stress X_i is given as

$$X_i = \left(\frac{\partial A}{\partial x_i}\right)_{P=0} \qquad (29)$$

From Eq. (28), we have

$$c_{ij}{}^P x_j + \tfrac{1}{2}q_{ijk}P_j P_k = 0$$

In a simpler expression, the strain x is given bv

$$x = QP_s^2 \qquad (30)$$

That is, the spontaneous polarization induces the strain of the crystal lattice, or a spontaneous deformation. This relation was observed for VDF 55% copolymer [119], although it might be apparent, as mentioned in Section III.A.

B. Mean-Field Theory

Taylor and co-workers developed a mean-field theory and applied it to PVDF and VDF-TFE copolymers [208,209]. A chain is assumed to be a linear array of torsional angles, ...θ-trans-θ'-trans.... The distribution of θ is expressed as $n_A(\theta)$ and that of θ' as $n_B(\theta')$. The intrachain energy V_1 is given by

$$V_1 \propto \int U(\theta, \theta')[n_{AB}(\theta, \theta') + n_{BA}(\theta, \theta')] \, d\theta \, d\theta' \tag{31}$$

where $U(\theta, \theta')$ is the potential energy related to the torsional angles θ and θ'. $n_{AB}(\theta, \theta')$ is a probability that the adjacent torsional angles A and B are θ and θ', respectively. The interchain interaction energy V_2 is

$$V_2 \propto \int W(\theta_1, \theta_1', \theta_2, \theta_2')n_A(\theta_1)n_B(\theta_1')n_A(\theta_2)n_B(\theta_2') \, d\theta_1 \, d\theta_1' \, d\theta_2 \, d\theta_2' \tag{32}$$

where W is the potential energy of a monomeric unit in a chain segment 1 of torsional angles θ_1 and θ_1' in the potential of another chain segment 2 with angles θ_2 and θ_2'. The mean-field approximation simplifies the complicated interchain interaction V_2 only as a functional of n_A and n_B and neglects the correlation between the chains. That is, the effective intrachain Hamiltonian is given as

$$H_{\text{eff}} \propto V_1 + \tfrac{1}{2}\left[\int\int [h_A(\theta) + h_B(\theta')]n_{AB}(\theta, \theta') \, d\theta \, d\theta' \right.$$
$$\left. + \int [h_A(\theta') + h_B(\theta)]n_{BA}(\theta, \theta') \, d\theta \, d\theta' \right] \tag{33}$$

where $h_A(\theta)$ and $h_B(\theta')$ are mean fields representing the interchain interactions: $h(\theta) = -\delta V_2(\theta)/\delta n(\theta)$. A transfer integral equation based on this Hamiltonian is solved, from which the distributions $n_A(\theta)$ and $n_B(\theta')$ are obtained. Then the free energy is calculated as a function of $n_A(\theta)$ and $n_B(\theta')$, from which the structure of minimal free energy is sought. $U(\theta, \theta')$ and $W(\theta_1, \theta_1', \theta_2, \theta_2')$ are computed numerically for the concrete chain structure. For example, in the case of PVDF, four structures give the free-energy minima, corresponding to forms II, I, III, and another conformation (TGTG) in order of low energy. Taylor and co-workers applied this method to VDF-TFE copolymer systems with various VDF content, and obtained good correspondence between the calculated and observed transition behavior of a series of copolymers.

C. Weiss Theory [35]

Let us consider two states corresponding to two opposite dipole orientations (upward and downward). By applying external electric field E along the direction of the upward dipoles, the energy level of the upward dipole is lowered by $-\mu E$ and that of the downward dipole by $+\mu E$. The number of upward dipoles is N_+ and that of downward dipoles is

N_- (total number of dipoles $N = N_+ + N_-$). The arrangement of these dipoles in a space gives a spontaneous polarization $P = N_+\mu - N_-\mu$. This polarization generates a coercive electric field around the dipoles by βP, and therefore the effective electric field working to the dipoles is not E but $F = E + \beta P$. By considering the Boltzmann distribution of dipoles into the two states, we may have the following equations, where k is a Boltzmann constant and T is a temperature:

$$N_+ = N * \frac{\exp(-\mu F/kT)}{\exp(\mu F/kT) + \exp(-\mu F/kT)}$$

$$N_- = N * \frac{\exp(\mu F/kT)}{\exp(\mu F/kT) + \exp(-\mu F/kT)}$$

$$P = N_+\mu - N_-\mu = N\mu * \tanh\left[\frac{\mu(E + \beta P)}{kT}\right]$$

Under the condition of $E = 0$, we may solve the relation between spontaneous polarization P_s and T. From the above equation, E can be expressed as a series of P as in Eq. (26), and the parameter ξ is given as

$$\xi = \frac{kT}{3\mu^4 N^3} > 0$$

Therefore, in this treatment of the equation, the Weiss theory gives the second-order transition. The transition point T_c is given by

$$T_c = \frac{N\mu^2\beta}{k} \tag{34}$$

This theory was applied to the copolymer case with trans–gauche conformational change [210]. In the development of the equations, the trans–gauche energy difference has to be introduced, which is essential in a treatment of the copolymer system.

The trans monomer groups are assumed to take two states of upward and downward dipoles, and the dipoles of gauche groups are assumed to orient in the direction 60° deviating from the trans dipole; i.e., gauche groups are assumed to have $\mu/2$ dipoles along the trans direction (refer to Fig. 72). In a simpler manner, the gauche groups may be assumed to be nonpolar as understood from the structure of the HT phase. The number of each state is given as follows:

$$N_{t+} = N * \frac{\exp[-(u_t - \mu F)/kT]}{A} \tag{35}$$

$$N_{t-} = N * \frac{\exp[-(u_t + \mu F)/kT]}{A} \tag{36}$$

$$N_{g+} = f * N * \frac{\exp[-(u_g - \mu F/2)/kT]}{A} \rightarrow f * N \frac{\exp[-u_g/kT]}{A} \tag{37}$$

$$N_{g-} = f * N * \frac{\exp[-(u_g + \mu F/2)/kT]}{A} \rightarrow f * N * \frac{\exp[-u_g/kT]}{A} \tag{38}$$

$$A = \exp\left[\frac{-(u_t - \mu F)}{kT}\right] + \exp\left[\frac{-(u_t + \mu F)}{kT}\right] + 2 * f * \exp\left[\frac{-u_g}{kT}\right] \tag{39}$$

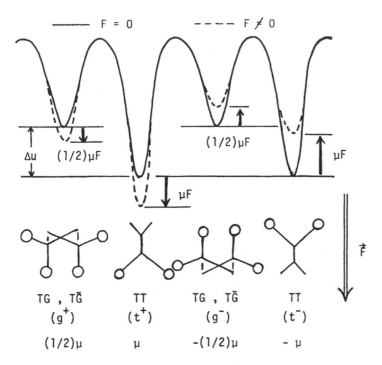

Figure 72 Potential-energy curve for trans and gauche isomers under the application of an external electric field [210].

where the coefficient f (= 2) in N_g means a statistical weight of the gauche state (TG and TḠ). For a volume V of a crystallite, the polarization P is expressed by

$$P = \frac{(N_{t+} - N_{t-})\mu}{V} \tag{40}$$

Substitution of Eqs. (35)–(38) into Eq. (40) gives P in the form which follows:

$$P = \frac{(\mu N/V)\{\exp[-(u_t - \mu F)/kT] - \exp[-(u_t + \mu F)/kT]\}}{A}$$

Putting $\exp(\mu F/kT) = x$, $\mu N/V = b$, $\Delta u = u_t - u_g$, and $4 * \exp(\Delta u/kT) = a$, we have

$$P = \frac{b[x - (1/x)]}{x + (1/x) + a} \tag{41}$$

Under the condition of $E = 0$, Eq. (41) gives the following relation concerning the transition point T_c:

$$T_c = \frac{(N\mu^2\beta/V_ck)}{[1 + 2 \exp(\Delta u/kT_c)]} \tag{42}$$

where V_c is the volume at T_c. If Δu is infinitely negative, i.e., the trans form is overwhelmingly stable, then T_c is equal to that given by the above-mentioned Weiss theory. Equation (42) predicts qualitatively that the transition temperature T_c will be higher for the polymer with more stable trans chain ($\Delta u < 0$), as long as the dipole moment μ, volume V_c, and internal effective field or β are constant (in the actual case, V_c, μ, and so on, are modified for copolymers with different VDF content).

By using the relation that the infrared absorbance (D_t and D_g of the trans and gauche conformations, respectively), is proportional to the number of each conformer, the following equation is written ($E = 0$):

$$D_t/D_g \propto \frac{N_{t+} + N_{t-}}{N_{g+} + N_{g-}} = \frac{\exp(-\Delta u/kT)[\exp(\mu\beta P/kT) + \exp(-\mu\beta P/kT)]}{4} \tag{43}$$

Therefore we have the approximate equation

$$\frac{\partial[\ln(D_t/D_g)]}{\partial(1/T)} = \frac{-\Delta u}{k} + \left(\frac{\mu\beta}{k}\right)^2 \left[P^2/T + P\left(\frac{\partial P}{\partial T}\right)\right] \tag{44}$$

Therefore, for the paraelectric state $P = 0$, the plot of $\ln(D_t/D_g)$ against $1/T$ gives the trans–gauche energy difference Δu, as was shown in the previous section.

D. Application of Ising Model (Bethe Approximation)

A verity of theories have been published for general ferromagnetic and ferroelectric phase transitions on the basis of the Ising model of a one-, two- or three-dimensional system. One of the most basic and essential theories which is considered to be useful for the polymer system may be the Bethe-Peierls approximation for a two-dimensional spin system [211,212]. It takes into consideration a two-dimensional arrangement of spins and an interaction between a spin (or dipole) and the neighboring spins (dipoles) of a certain coordination number. Takagi [213] and Fowler and Guggenheim [214] proposed approximation methods for calculating the number of possible spin arrangements, from which a partition function Z is derived comparatively easily with good approximation. Odajima applied this method to two-dimensional VDF-TrFE copolymer systems by introducing an energy difference between the trans and gauche conformers [215]. Ikeda and Sekimura derived another equation using essentially the same approach as Odajima [216].

At first a linear lattice of N points is considered. Each lattice point is denoted by $+$ or $-$, the number of which is N_+ and N_-, respectively ($N_+ + N_- = N$). A pair of $++$ and $--$ represents the trans (T) form with a dipole moment μ, while $+-$ and $-+$ represent the gauche form (G or $\overline{G}$) with zero dipole. Numbers of these pairs (Q_{++}, Q_{--}, Q_{+-}, and Q_{-+}) are calculated under periodic boundary condition and related with N_+ or N_- as follows:

$$N_+ = Q_{++} + \frac{Q_{+-} + Q_{-+}}{2} = Q_{++} + \frac{Q_0}{2}$$

$$N_- = Q_{--} + \frac{Q_{+-} + Q_{-+}}{2} = Q_{--} + \frac{Q_0}{2} \tag{45}$$

The total number of different configurations of the lattice is evaluated:

$$g(N_+, Q_0) = \frac{N_+! N_-!}{(Q_0/2)! 2 (N_+ - Q_0/2)! (N_- - Q_0/2)!} \tag{46}$$

A crystal is considered to be composed of ν molecular chains, each of which possesses N lattice points. The configurational energy $W(N_+, Q_0)$ with both the intra- and intermolecular interactions taken into account is expressed by

$$W(N_+, Q_0) = -\nu\left\{\frac{J(N - Q_0)}{2} - \frac{JQ_0}{2} + Lz\left[-\frac{N}{4} + N_+ \cdot \frac{(N - N_+)}{N}\right]\right\} - PE \qquad (47)$$

where P is the polarization of the crystal, given by

$$P = \mu\nu(Q_{++} - Q_{--}) = \mu\nu(N_+ - N_-)$$

and E is an electric field. J is an intrachain energy difference between the trans and gauche bondings and L is the interchain energy difference. The partition function $Z(N_+, Q_0)$ is written as

$$Z(N_+, Q_0) = g(N_+, Q_0)^\nu \exp\left[\frac{-W(N_+, Q_0)}{kT}\right] \qquad (48)$$

Free energy is obtained from $\log Z(N_+, Q_0)$ and the equilibrium values of N_+ and Q_0 are calculated by minimizing the free energy:

$$\frac{N_+(N_- - Q_0/2)}{N^-(N_+ - Q_0/2)} = \exp\left(\frac{-2F}{kT}\right) \qquad (49)$$

$$\frac{(2N_+ - Q_0)(2N_- - Q_0)}{Q_0^{\,2}} = \exp\left(\frac{2J}{kT}\right) \qquad (50)$$

where $F = zL(N_+ - N_-)/2N + \mu E$.

Substituting Eq. (49) into Eq. (50), we have the spontaneous polarization P as

$$P = \frac{N\nu\mu \, \exp(J/kT) \, \sinh(F/kT)}{\sqrt{1 + \exp(2J/kT) \, \sinh^2(F/kT)}} \qquad (51)$$

P is calculated against temperature T. Depending on the sign and magnitude of J or the trans–gauche energy difference, P changes with temperature in a second-order manner (for $J \geq 0$) or in a first-order manner (for $J < 0$). $J < 0$ means that the gauche form is more stable than the trans form for an isolated chain: This is the case of VDF 72%-TrFE copolymer with first-order phase transition. On the other hand, VDF 55%-TrFE apparently shows second-order transition, corresponding to the case of $J \geq 0$, because the trans form is more stable. In this way, Odajima's theory can reproduce the characteristic features of the phase transition in the copolymers qualitatively.

VII. OTHER RELATED POLYMERS

A. Ethylene-Tetrafluoroethylene Alternating Copolymer

An alternating copolymer of ethylene and tetrafluroethylene (ETFE) has a repeating structure of $-CH_2CH_2CF_2CF_2-$, assumed to be a model polymer consisting of only head-to-head and tail-to-tail linkages of VDF units [8,33]. At room temperature the trans conformation is overwhelmingly stable. The chains are packed in an orthorhombic unit cell with a packing mode similar to that of orthorhombic PE (herringbone type, see Fig. 73) [217–219]. Near 100°C, this copolymer shows a rotational order-disorder transition between orthorhombic and hexagonal crystal phases: the a and b axial lengths of the unit cell change greatly [220]. The zigzag chains are said to rotate rigidly around the chain axis. In this temperature region, however, the crytalline trans bands of infrared and Raman

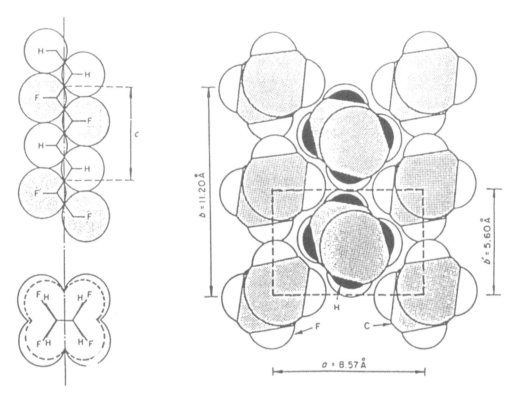

Figure 73   Crystal structure of ETFE copolymer [218].

spectra decrease in intensity with the deflection point of the intensity–temperature curve, and the gauche bands are detected though weak [34]. Correspondingly, the X-ray (004) reflection is found to decrease in intensity in this temperature region. That is, the molecular chain is considered to experience some conformational change between the trans and gauche bonds, although the degree of the change is not very high.

$$\ldots-CF_2-CF_2-CH_2-CH_2-CF_2-CF_2-CH_2-CH_2-CF_2-CF_2-CH_2-CH_2-CF_2-CF_2-CH_2-CH_2-$$

T   T   T   T   T   T   T   T   T   T   T   T   T   T   T   T

↓

T   T   T   T   T   T   G   T   T   T   T   T   T $\overline{G}$ T

In this way the ease of conformational change is dependent on the type of comonomer, and the trans stability increases in the order of VDF-TrFE < VDF-TFE < ETFE or in order of increasing amount of TFE structural units. That is, the intramolecular nonbonded interactomic interactions, especially a steric repulsion between nonbonded fluorine atomic pairs, is one of the most important factors governing the transition behavior of fluorine polymers.

B. VDF-VF and VF-TrFE Copolymers

Poly(vinyl fluoride) (PVF) has a polar crystal structure in which the all-trans chains are packed with their CF dipoles parallel along the b axis [221]. Since the side atom F is bonded in atactic configuration to the skeletal chain, the averaged molecular structure is quite similar to that of PVDF form I, when viewed along the chain axis. Therefore this polymer may also show ferroelectric behavior, although at the present stage no confirmation has not yet been made. Guerra et al. synthesized [222] a series of copolymers of vinylidene fluoride with vinyl fluoride (VDF-VF). For copolymers with VF content higher than 84%, the X-ray pattern of PVDF form I is obtained. No crystalline phase transition was detected up to the melting temperature. Maeda et al. synthesized a random copolymer between VF and TrFE monomers and found that the dielectric constant obeys the Curie-Weiss law around 143°C [223]. The transition is considered to be second-order type. The X-ray lattice spacing shows a discontinuity in the thermal expansivity in the transition region, supporting it.

C. Vinylidene Cyanide Copolymers

Cyanide group is highly polar. Poly(vinylidene cyanide) (PVDCN), which has a chemical structure of $-[-CH_2C(CN)_2-]-$, is another candidate to show the ferroelectric property. This homopolymer is not stable, and so the copolymers between VDCN and n-alkyl fatty acid ester have been synthesized [224,225].

$$-[-CH_2C(CN)_2-CH_2-CH-]-$$
$$|$$
$$OCO(CH_2)_{n-2}CH_3$$

These copolymers with different alkyl chain length n were reported to show characteristic electric behavior [226–228]. In particular, an alternating copolymer of $n = 2$ or VDCN-vinyl acetate (VDCN-VAc) copolymer [229–231] attracts attention because of its high piezoelectricity comparable to that of PVDF form I. Structural investigation was made for a series of the copolymers by using X-ray diffraction and infrared spectral measurements. The X-ray fiber pattern as a whole is very broad and appears at a glance, to be totally amorphous in the case of VDF-VAc copolymer [232]. But as the side-chain length or the number of n increases, the X-ray pattern shows characteristic crystalline reflections [224,225,233,234]. On the equatorial line, for example, a very intense and sharp reflection is observed and is assigned to the interchain spacing corresponding to the distance between two layers constructed by the parallel arrangement of chains with the side alkyl groups sticking out of the main chain. This spacing increases almost linearly with an increment of the side-chain length. In order to clarify the packing mode of the chains in the crystal (strictly speaking, it might be called paracrystalline state), packing-energy calculation was made for the various models; the two energetically possible structures are shown in Figure 74. The X-ray diffraction and infrared spectral data tell us that the main chain takes almost fully extended zigzag structure [fiber period is about 4.8 Å (or 9.6 Å) on average], and the alkyl side groups also take the all-trans conformation at low temperature. Because of the spatial repulsion between the adjacent CN groups and also between bulk ester groups, the main chain takes a slightly deflected conformation. In one of the two models, the chains with long arms are arranged side by side so that the CN groups and the alkyl groups are arrayed in parallel along the a axis. Another model is that these chains are packed in an antiparallel fashion with respect to the direction of CN and alkyl groups. Energetically, the antiparallel packing is overwhelmingly stable and can reproduce the experimental data better than another model can do. If the antiparallel packing is employed, this copolymer might be antiferroelectric.

st-P(VDCN-VOc)

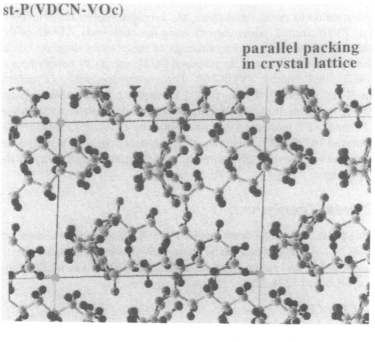

parallel packing
in crystal lattice

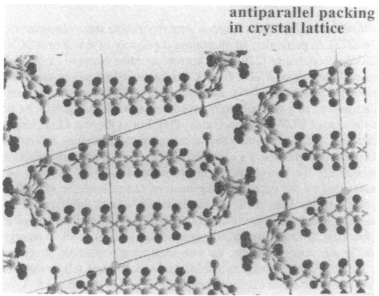

antiparallel packing
in crystal lattice

Figure 74   Crystal structural models of VDCN copolymers [234].

Atactic poly(acrylonitrile) [PAN, $-CH_2-CH(CN)-$] was proposed to have a polar crystal structure based on the energy calculation [235], from which a possibility of ferroelectricity might be deduced although no definite experimental data have been presented. The D-E hysteresis loop was measured for copolymers of PAN with allylcyanide [$-CH_2-CH(CH_2CN)-$] [236].

D. Odd Nylons

Since the crystal structure of odd nylon (e.g., nylon 11) was proposed by X-ray diffraction analysis [237], a possibility of ferroelectricity of this crystal has been speculated by many investigators. As shown in Figure 75, essentially all-trans zigzag chains are packed in a triclinic unit cell with the amide C=O groups parallel to each other. Piezoelectric constant, dielectric constant, and so on, have been measured for a series of odd nylons [238–240]. Until recently, however, no clear proof of ferroelectric behavior of odd nylons was presented. Scheinbeim and co-workers prepared films by melt-quenching and drawing at room temperature, which were found to show a clear D-E hysteresis loop [241–

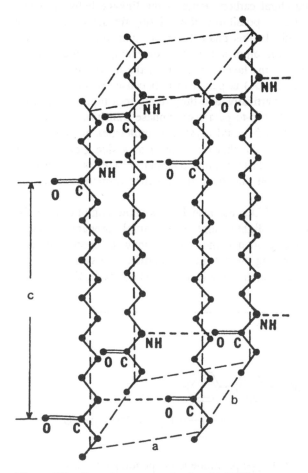

Figure 75 Crystal structure of nylon 11 [237].

243]. The spontaneous polarization was estimated to be ca. 48 mC/m^2 for nylon 11 and 86 mC/m^2 for nylon 7, which are larger than that of PVDF form I. Piezoelectric constant (d_{31} = 14 pC/N for nylon 11) and electric-mechanical coupling constant (k_{31} = 0.049 for nylon 11) were measured to remain active in the temperature region up to 200°C, which is also higher than PVDF form I, ca. 175°C. The D-E hysteresis loop indicates that odd nylons may be ferroelectric. Annealing of the films makes occurrence of dipole inversion difficult [242]. This was interpreted on the basis of shrinkage of the crystal lattice, particularly a shortening of intersheet distance: Hydrogen bonds in the sheets may play as a restraining factor for the inversion of amide groups into the electric field direction. From the experimental result that only the poorly crystalline sample shows a clear D-E hysteresis loop and a high piezoelectric effect, it may be suggested, as another possibility, that the chains in the strongly connected sheet structure cannot reorient, but the chains in the disordered crystallite with weakly hydrogen bonded sheets [242,244] or, in an extreme case, the chains in the noncrystalline phase, can rotate under the application of external electric field.

E. Ferroelectric Liquid-Crystalline Polymers

Liquid crystalline molecules having chiral carbon atoms as the linkage between a mesogenic group and end alkyl chains have a possibility of exhibiting ferroelectric behavior in the smectic C phase (Sm C*) [245]. In this phase the molecular axis tilts from the normal to the layer plane and the electric dipoles of C=O, for example, direct into the common direction to each other; that is, the layers are polar in the lateral direction. On the other hand, the smectic A phase (Sm A) is paraelectric because the molecules are packed in the layer with their axes set vertically, and they rotate in a liquidlike fashion around the molecular axis. The ferroelectric-to-paraelectric phase transition is observed between Sm C* and Sm A phases. If such liquid crystalline molecules are introduced into a polymer system as a component of the skeletal chain or as a side group sticking out of the main chain, a ferroelectric liquid crystalline polymer (FLCP) is obtained. In fact, many FLCPs have been synthesized and their ferroelectric properties measured [246].

Not only the low-molecular-weight FLCs but also FLCPs show complicated polymorphism and polytype, depending on a slight change of the sample preparation conditions [247]. Polymorphism is defined as a phenomenon that the inner-layer structure and/or layer-stacking mode are different among the samples. In particular, a variation in the stacking mode of layers with the same innerstructure is defined as polytype. For example, the FLCP with the chemical structure

was found for the first time to show a polytype phenomenon depending on a slight change in cooling rate from the isotropic state to room temperature [248,249]. In Figure 76 is

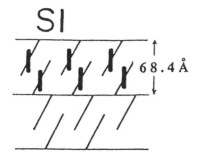

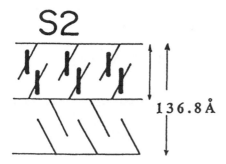

Figure 76 Polytype structure of ferroelectric liquid crystalline polymer [248]. A bold bar represents a mesogenic group and a thin line represents an alkyl side chain.

illustrated a possible stacking mode of layers in this FLCP. In model S1 the layers are stacked together with a period of one layer, while in model S2 the double layers are repeated with opposite-tilting directions of the side groups. These two structures are obtained by changing the cooling rate only slightly, from 0.2°C/min (for model S1) to 0.1°C/min (for model S2). If one layer is assumed to be polar, then S1 may be ferroelectric and S2 anti-ferroelectric as a possibility.

VIII. MOLECULAR THEORY OF PIEZOELECTRICITY AND PYROELECTRICITY IN PVDF

A. Piezoelectric Phenomena in Bulk Samples

The ferroelectric property of PVDF form I (and its copolymers) reflects directly on the highest piezo- and pyroelectric effects among the synthetic polymers. Piezoelectric strain or stress constant tensor, **d** or **e**, are defined, respectively, as follows for the polarization change ΔP caused by stress σ or strain ε under the condition of constant temperature and zero electric field:

$$
\begin{bmatrix} \Delta P_1 \\ \Delta P_2 \\ \Delta P_3 \end{bmatrix} = \begin{bmatrix} d_{11} & d_{12} & \cdots & d_{16} \\ d_{21} & d_{22} & \cdots & d_{26} \\ d_{31} & d_{32} & \cdots & d_{36} \end{bmatrix} \begin{bmatrix} \sigma_1 \\ \sigma_2 \\ \vdots \\ \sigma_6 \end{bmatrix} = \begin{bmatrix} e_{11} & e_{12} & \cdots & e_{16} \\ e_{21} & e_{22} & \cdots & e_{26} \\ e_{31} & e_{32} & \cdots & e_{36} \end{bmatrix} \begin{bmatrix} \epsilon_1 \\ \epsilon_2 \\ \vdots \\ \epsilon_6 \end{bmatrix}
$$

or

$$\Delta P = d\sigma = \epsilon\epsilon \tag{52}$$

Since the discovery of piezoelectricity in PVDF film, many trials have been made in order to clarify the origin of the effects [250,251]. In such a trial, we must take into account the following experimental facts concerning the piezoelectric effect in PVDF samples [252]. Piezoelectric constants are proportional to (a) the volume fraction of crystal form I, (b) the spontaneous polarization of the poled sample, (c) the electrostriction constant, (d) the degree of the b-axis orientation of crystallites, and (e) piezoelectric constant $d_{31}{}^M$ (M denotes the macroscopic quantity), ca. 10^{-11} C/N at room temperature, decreases largely by ca. one or two digits below the glass transition temperature ($-40°C$), while $d_{33}{}^M$ of ca. -10^{-11} C/N shows no appreciable change.

In order to reproduce these characteristic behaviors of PVDF form I bulk sample, it was necessary to evaluate to what extent the piezoelectric effect intrinsic of crystalline region may contribute to the bulk constant. Several theories have been proposed to evaluate the crystalline as well as the bulk piezoelectric constants [21,253–256].

1. Fluctuation of Dipoles

Broadhurst et al. assumed that the dipoles fluctuate within the crystalline lamella [253]. The total dipole moment $\mathbf{m}_c$, contributed from both the fluctuation of crystalline dipoles and the space charges trapped at the interfacial part between the crystallite and amorphous region, is expressed approximately by

$$\mathbf{m}_c = \left[\frac{n(\epsilon_c + 2)\mu_0\langle\cos\phi\rangle}{3} + ql_c \right]\mathbf{u} \tag{53}$$

where μ_0 is a dipole moment in a vacuum, n is the number of crystalline dipoles, $\langle\cos\phi\rangle$ is a term representing a fluctuation of dipole with an angle ϕ from the polar axis of crystallite, $\mathbf{u}$ is a unit vector along this polar axis, and ϵ_c is the dielectric constant of the crystallite. q is a charge on the interfacial surface, and l_c is the thickness of the crystallite. The direction of vector $\mathbf{u}$ of a crystallite deviates from the normal to the film surface by an averaged angle of θ_0 and fluctuates around the averaged angle ϕ_0. The total dipole moment M_s of the sample containing many crystalline lamellae is given as follows:

$$M_s = \left[\frac{N(\epsilon_c + 2)\mu_0 J_0(\phi)}{3} + (\Sigma q)l_c \right]\langle\cos\theta_0\rangle \tag{54}$$

where N is the total number of the dipoles in the sample, and J_0 is a zero-th Bessel function. Hydrostatic piezoelectric constant d_p is calculated as follows:

$$d_p = \left(\frac{1}{A}\right)\left[\frac{\partial(M_s/l_sA)}{\partial p}\right] = P_0\beta_c\left[\frac{(\epsilon_c - 1)}{3} + \frac{\phi_0^2\gamma}{2} + \frac{\partial(\ln l_c)}{\partial(\ln v_c)}\right] \tag{55}$$

where p is the hydrostatic pressure, A is the area of the sample surface, v_c is the crystal volume, β_c is the volume compressibility of the crystal [$\beta_c = -\partial \ln(v_c)/\partial p$], and γ is the Grüneisen coefficient for dipole librational frequency ω [$\gamma = -\partial(\ln \omega)/\partial(\ln v_c)$]. Suitable numerical parameters were substituted into this equation and the contribution of each term in the d_p was obtained as follows.

Electrostriction effect (the first term), 37%
Dipole fluctuation (the second term), 10%
Dimensional change (the third term), 28%

Dvey-Aharon et al. derived an equation for the calculation of $d_{31}{}^c$ of crystal form I on the basis of librational motion of dipoles in the mean field determined by the surrounding chains [254]. Thus calculated $d_{31}{}^c$ is ca. 0.02 pC/N, which is too small compared with the bulk value of ca. 20 pC/N. In order to reproduce this significant difference, they assumed that the crystalline region grows reversibly at the expense of amorphous material by the application of external electric field; the total dipole moment of the sample may become larger reversibly.

2. Lattice Dynamical Theory

In order to interpret the piezoelectricity from the molecular structural level, Tashiro et al. derived an equation to calculate the three-dimensional piezoelectric tensor components based on the lattice dynamic theory [255]. In a point-charge model, a polarization $\mathbf{P}_0$ of the crystal is given as follows:

$$\mathbf{P}_0 = \left(\frac{1}{Nv}\right) \sum_{kmi}\sum\sum Q_i \, \mathbf{X}_i \, (i, \, m, \, \mathbf{k}) \tag{56}$$

where $\mathbf{X}_i \, (i, \, m, \, \mathbf{k})$ is the position vector of the ith atom of the unit m in the kth unit cell, Q_i is the charge of the atom, v is the unit cell volume, and N is a number of unit cells in a crystal. The polarization change $\Delta\mathbf{P}$ induced by the unit cell deformation is written as

$$\Delta\mathbf{P} = \Delta\left(\frac{\sum\sum\sum Q_i \mathbf{X}_i}{Nv}\right) = \left(\frac{1}{Nv}\right) \sum\sum\sum Q_i \, \Delta\mathbf{X}_i - \mathbf{P}_0\left(\frac{\Delta v}{v}\right) = \Delta\mathbf{P}_1 + \Delta\mathbf{P}_2 \tag{57}$$

The first term $\Delta\mathbf{P}_1$ is a polarization change originating from the atomic displacement $\Delta\mathbf{X}_i$. $\Delta\mathbf{X}_i$, derived lattice dynamically under the condition of minimization of strain energy, is expressed as a combination of inner strain $\boldsymbol{\rho}_i$ and external strain $\mathbf{W}_i \, (i, \, m, \, \mathbf{k})\boldsymbol{\epsilon}$, where $\mathbf{W}_i$ is a matrix constructed by the atomic coordinates $\mathbf{X} \, (i, \, m, \, \mathbf{k})$ and $\boldsymbol{\epsilon}$ is a strain: $\Delta\mathbf{X}_i = \boldsymbol{\rho}_i + \mathbf{W} \, (i, \, m, \, \mathbf{k})\boldsymbol{\epsilon}$ [257]. $\boldsymbol{\rho}_i$ is given as follows by using the force constant matrix $\mathbf{F}_R$: $\boldsymbol{\rho} = -(\mathbf{F}_\rho)^{-1}\mathbf{F}_{\rho\epsilon}\boldsymbol{\epsilon}$, where $\mathbf{F}_{\rho\epsilon} = \hat{\mathbf{B}}\mathbf{F}_R\mathbf{B}\mathbf{W}$ and $\mathbf{F}_\rho = \hat{\mathbf{B}}\mathbf{F}_R\mathbf{B}$. $\mathbf{B}$ is the matrix connecting the internal displacement vector $\Delta\mathbf{R}$ (such as the bond length change, bond angle deformation, etc.) with $\Delta\mathbf{X}$ as $\Delta\mathbf{R} = \mathbf{B} \, \Delta\mathbf{X}$. Then $\Delta\mathbf{P}_1$ is expressed as

$$\Delta\mathbf{P}_1 = \left(\frac{1}{Nv}\right) \sum\sum\sum Q_i \, \Delta\mathbf{X}_i = \left(\frac{1}{Nv}\right) \sum\sum\sum Q_i[\boldsymbol{\rho}_i + \mathbf{W}(i, \, m, \, \mathbf{k})\boldsymbol{\epsilon}]$$

$$= \left(\frac{1}{v}\right)[\sum Q_i\boldsymbol{\rho}_i] = \left(\frac{1}{v}\right) \mathbf{Q}\boldsymbol{\rho} = -\left(\frac{1}{v}\right)\mathbf{Q}(\mathbf{F}_\rho)^{-1} \, \mathbf{F}_{\rho\epsilon}\boldsymbol{\epsilon} \tag{58}$$

$$= -\left(\frac{1}{v}\right) \mathbf{Q}(\mathbf{F}_\rho)^{-1} \, \mathbf{F}_{\rho\epsilon} \, \mathbf{s}\boldsymbol{\sigma}$$

Here $\mathbf{Q}$ is a matrix consisted of the charge Q_i and $\mathbf{s}$ is a compliance tensor ($\boldsymbol{\epsilon} = \mathbf{s}\boldsymbol{\sigma}$). The second term $\Delta\mathbf{P}_2$ indicates the polarization change due to the volume change and is expressed as

$$\Delta\mathbf{P}_2 = -\mathbf{P}_0 \frac{\Delta v}{v} = -\mathbf{P}_0\mathbf{s}'\boldsymbol{\sigma} \tag{59}$$

where

$$\mathbf{s}' = (s_{11} + s_{21} + s_{31}, s_{12} + s_{22} + s_{32}, \ldots, s_{16} + s_{26} + s_{36})$$

Therefore we have

$$\Delta \mathbf{P} = \Delta \mathbf{P}_1 + \Delta \mathbf{P}_2 = -\left[\left(\frac{1}{v}\right) \mathbf{Q}(\mathbf{F}_\rho)^{-1} \mathbf{F}_{\rho\epsilon} \, \mathbf{s} + \mathbf{P}_0 \mathbf{s}'\right]\sigma \tag{60}$$

By comparing this equation with the original definition [Eq. (56)], we have

$$\mathbf{d} = -\left[\left(\frac{1}{v}\right) \mathbf{Q}(\mathbf{F}_\rho)^{-1} \mathbf{F}_{\rho\epsilon} \, \mathbf{s} + \mathbf{P}_0 \mathbf{s}'\right] \tag{61}$$

Applying this equation to PVDF form I crystal, we obtain

$$\mathbf{d} = \begin{bmatrix} 0 & 0 & 0 & 0 & -54.82 & 0 \\ 0 & 0 & 0 & -7.64 & 0 & 0 \\ -0.45 & -7.23 & -44.98 & 0 & 0 & 0 \end{bmatrix} \text{pC/N}$$

where the Cartesian coordinates are defined as follows: 1 // c, 2 // a, and 3 // b (Fig. 15). The piezoelectric strain constant of the crystalline region was measured by the X-ray diffraction method under application of an electric field [258,259]; the crytsalline strain ϵ_3 along the b axis, for example, is induced by the electric field E_3 as $\epsilon_3 = d_{33}{}^c E_3$. The results are $d_{31}{}^c = \pm 0.1$ pC/N and $d_{33}{}^c = -20$pC/N. The observed values are scattered and do not have high accuracy, but the sign and order are comparable to the calculated values.

Using these calculated three-dimensional d values, we can calculate the macroscopic piezoelectric constants of the bulk PVDF film on the basis of a complex model consisting of crystalline and amorphous phases [256,260]. The macroscopic piezoelectric stress constant $e_{3j}{}^M$ is expressed as

$$e_{3j}{}^M = P_s^M[(\epsilon^c/2\epsilon^a + \epsilon^c)(\kappa_{3j}{}^a/\epsilon^a) + m_{3j}{}^M +$$
$$\Sigma\beta_i{}^c c_{ij}'^M] + (3\epsilon^a/2\epsilon^a + \epsilon^c)\phi\chi\Sigma d_{3i}{}^c c_{ij}'^M \tag{62}$$

where ϵ^a and ϵ^c are dielectric constants of the amorphous and crystalline regions, respectively, and P_s^M is a macroscopic polarization of the sample, which is expressed in terms of ϵ^a, ϵ^c, P_s^0, and so on. κ is an electrostriction constant (i.e., a change of dielectric constant by strain), $m_{3j}{}^M$ is a bulk Poisson's ratio, $\beta_i{}^c$ is a linear compressibility of the crystal, $c_{ij}'^M$ is a function of elastic constants, and ϕ is a volume fraction of the crystalline phase. χ is a measure of the degree of orientation of polarization vector of the crystal. Substituting a set of suitable numerical values into all these parameters, we can reproduce the actually observed macroscopic piezoelectric constants including their temperature dependencies above and below the glass transition temperature:

	Room temperature		Below T_g	
	Obs.	Calc.	Obs.	Calc.
$d_{31}{}^M$	20 pC/N	25.3	1	2.8
$d_{32}{}^M$	2	7.0	1	0.8
$d_{33}{}^M$	−30	−35.4	−8	−6.5
$e_{31}{}^M$	40 mC/m^2	28.7	10	11.0
$e_{32}{}^M$	3	7.2	6	3.6
$e_{33}{}^M$	−48	−40.7	−40	−41.6

Equation (62) consists of the four terms, the contribution percentages of which are evaluated as follows for $d_{31}{}^M$, for example.

	Room temperature	Below T_g
Electrostriction effect (κ)	5.4%	0.0%
Poisson's ratio (m)	87.9%	74.8%
Volume change ($\beta_c c'^M$)	−6.6%	−27.7%
Crystalline piezo effect (d_c, e_c)	13.4%	52.9%

Among the four terms Poisson's ratio contributes overwhelmingly to an appearance of macroscopic piezoelectric effects [261]. An intrinsic piezoelectric effect of the crystalline phase becomes important below the glass transition temperature. Equation (62) is found to satisfy all the requirements (a)–(e) as important factors of the piezoelectric effect in PVDF form I sample.

On the basis of these calculations, we can estimate the limiting values of the macroscopic piezoelectric effects of PVDF form I [256]. The results are

$$d_{31}{}^M = 144.9 \text{ pC/N} \qquad d_{32}{}^M = 16.7 \text{ pC/N} \qquad d_{33}{}^M = 186.3 \text{ pC/N}$$

$$e_{31}{}^M = 161.1 \text{ mC/m}^2 \qquad e_{32}{}^M = 17.8 \text{ mC/m}^2 \qquad e_{33}{}^M = -202.5 \text{ mC/m}^2$$

The limiting value $d_{31}{}^M = 145$ pC/N and the corresponding electromechanical coupling constant $k = 63\%$, a measure of transformation efficiency between the electrical and mechanical energies, are comparable to those of inorganic piezoelectric materials such as Rochell salt (275 pC/N and 66%) and PZT (110 pC/N and 31%). The actually reported maximal d constant of PVDF form I sample is ca. 40 pC/N.

B. Pyroelectricity in PVDF Sample

The pyroelectric effect of PVDF form I sample was also interpreted in a similar way with that discussed in the above section. For example, Broadhurst, et al. presented the equation of pyroelectric constant by exchanging the effect of pressure used in the evaluation of d_p with that of temperature T [253]:

$$p_y = \left(\frac{1}{A}\right)\left[\frac{\partial(M_s/l_s A)}{\partial T}\right] = P_0 \alpha_c \left[\frac{(\epsilon_c - 1)}{3} + \frac{\phi_0{}^2 \gamma}{2} + \frac{\phi_0{}^2}{T\alpha_c} + \frac{\partial(\ln l_c)}{\partial(\ln v_c)}\right] \qquad (63)$$

where α_c is the thermal expansion coefficient of the crytsal.

Wada and Hayakawa derived a phenomenological equation to represent the pyroelectric constant for the system consisting of spherical crystalline particles and an amorphous phase [260]:

$$p_y = P_s{}^M \left[\left(\frac{\epsilon^c}{2\epsilon^a + \epsilon^c}\right)\left(\frac{\lambda^a}{\epsilon^a} - \frac{\lambda^c}{\epsilon^c}\right) - \alpha\right] + \left(\frac{3\epsilon^a}{2\epsilon^a + \epsilon^c}\right)P_{sc}\left(\alpha_c + \frac{\partial \ln P_{sc}}{\partial T}\right) \qquad (64)$$

where λ is the temperature coefficient of dielectric constant ϵ, α is the linear thermal expansion coefficient of the film thickness, and P_{sc} is the spontaneous polarization of the spherical particle. The p_y was calculated as -2.74×10^{-5} C/m^2 K, which is in good agreement with the experiment value, -2.7×10^{-5} C/m^2 K.

IX. CONCLUDING REMARKS

In this chapter the molecular and crystal structures of PVDF and its related polymers have been reviewed. The chain conformation changes quite sensitively depending on a slight change in the environmental conditions or the sample preparation conditions. In particular, PVDF, VDF-TrFE, and VDF-TFE copolymers exhibit clear ferroelectric phase transitions in which the molecular conformation changes drastically between the trans and gauche structures, resulting in the characteristic behavior of the various physical properties including thermal expansion, temperature dependence of spontaneous polarization, and so on. In order to solve the various problems concerning the structural characteristics of PVDF, etc., many kinds of techniques have been developed, as explained in this chapter. The information obtained through these structural research is considered to be quite useful when we encounter various unsolved problems of other polymer materials concerning structure and relationships with physical properties. There are still many unsolved problems of PVDF crystal structure and phase transitional behavior. We are still at the threshold of research to obtain a final solution for these problems at the molecular structural level.

REFERENCES

1. P. De Santis, E. Gigkio, A. M. Liquori, and A. Ripamonti, Stability of helical conformations of simple linear polymers, *J. Polymer Sci. A1*:1383 (1963).
2. A. E. Tonelli, Conformational characteristics of poly(vinyl fluoride), poly(fluoromethylene), and poly(trifluoroethylene), *Macromolecules 13*:734 (1980).
3. Y. Kim. H. L. Strauss, and R. G. Snyder, Conformational disorder in crystalline *n*-alkanes prior to melting, *J. Phys. Chem. 93*:7520 (1989).
4. W. M. Prest, Jr., and D. J. Luca, The morphology and thermal response of high-temperature-crystallized poly(vinylidene fluoride), *J. Appl. Phys. 46*:4136 (1975).
5. W. M. Prest, Jr., and D. J. Luca, The formation of the γ phase from the α and β polymorphs of polyvinylidene fluoride, *J. Appl. Phys. 49*:5042 (1978).
6. S. Osaki and Y. Ishida, Effects of annealing and isothermal crystallization upon crystalline forms of poly(vinylidene fluoride), *J. Polymer Sci. Polymer Phys. Ed. 13*:1071 (1975).
7. K. Tashiro, Y. Itoh, M. Kobayashi, and H. Tadokoro, Polarized Raman spectra and LO-TO splitting of poly(vinylidene fluoride) crystal form I, *Macromolecules 18*:2600 (1985).
8. M. Kobayashi, K. Tashiro, and H. Tadokoro, Molecular vibrations of three crystal forms of poly(vindylidene fluoride), *Macromolecules 8*:158 (1975).
9. R. L. Miller and J. Raisoni, Single crystals of poly(vinylidene fluoride), *J. Polymer Sci., Polymer Phys. Ed 14*:2325 (1976).
10. C. C. Hsu and P. H. Geil, Morphology-structure-property relationships in ultraquenched poly(vinylidene fluoride), *J. Appl. Phys. 56*:2404 (1984).
11. Y. Oka and N. Koizumi, Formation of unoriented form I poly(vinylidene fluoride) by high-rate quenching and its electrical properties, *Bull. Int. Chem. Res. Kyoto Univ. 63*:192 (1985).
12. K. Matsushige, Pressure effect on phase transition in Ferroelectric polymers, *Phase Transitions 18*:247 (1989).
13. J. Scheinbeim, C. Nakafuku, B. A. Newman, and K. D. Pae, High-pressure crystallization of poly(vinylidene fluoride), *J. Appl. Phys 50*:4399 (1979).
14. G. T. Davis, J. E. McKinney, M. G. Broadhurst, and S. C. Roth, Electric-field-induced phase changes in poly(vinylidene fluoride), *J. Appl. Phys. 49*:4998 (1978).
15. G. R. Davies and H. Singh, Evidence for a new crystal phase in conventionally poled samples of poly(vinyldene fluoride) in crystal form II, *Polymer 20*:772 (1979).

16. D. Naegele, D. Y. Yoon, and M. G. Broadhurst, Formation of a new crystal form (α_p) of poly(vinylidene fluoride) under electric field, *Macromolecules 11*:1297 (1978).
17. J. B. Lando, H. G. Olf, and A. Peterlin, NMR and X-ray determination of the structure of poly(vinylidene fluroide), *J. Polymer Sci. Part A-1 4*:941 (1966).
18. Ye. L. Gal'perin and B. P. Kosmynin, The Crystalline structure of polyvinylidenefluoride, *Vysokomolekul. Soedin. 11*:1432 (1969).
19. R. Hasegawa, Y. Takahashi, Y. Chatani, and H. Tadokoro, Crystal structure of three crystalline forms of poly(vinylidene fluoride), *Polymer J. 3*:600 (1972).
20. B. L. Farmer, A. J. Hopfinger, and J. B. Lando, Polymorphism of poly(vinylidene fluoride): Potential energy calculation of the effects of head-to-head units on the chain conformation and packing of poly(vinylidene fluoride), *J. Appl. Phys. 43*:4293 (1972).
21. N. Karasawa and W. A. Goddard III, Force fields, structures, and properties of poly(vinylidene fluoride) crystals, *Macromolecules, 25*:7268 (1992).
22. R. Hasegawa, M. Kobayashi, and H. Tadokoro, Molecular conformation and packing of poly(vinylidene fluoride). Stability of three crystalline forms and the effect of high pressure, *Polymer J. 3*:591 (1972).
23. K. Tashiro, Y. Abe, and M. Kobayashi, Computer simultion of structure and ferroelectric phase transition of vinylidenefluoride copolymers (1) VDF content dependent of the crystal structure, *Ferroelectrics*, in press.
24. M. Maroncelli, H. L. Strauss, and R. G. Snyder, The distribution of conformational disorder in the high-temperature phases of the crystalline n-alkanes, *J. Chem. Phys. 82*:2811 (1985).
25. Y. Takahashi, H. Tadokoro, and A. Odajima, Kink bands in form I of poly(vinylidene fluoride), *Macromolecules 13*:1320 (1980).
26. Y. Takahashi, Structural disorder of poly(vinylidene fluoride) form I: Glides, *Macromolecules 26*:1471 (1993).
27. L. Lauchlan and J. F. Rabolt, Polarized Raman measurements of structural anisotropy in uniaxially oriented poly(vinylidene fluoride) (Form I), *Macromolecules 19*:1049 (1986).
28. G. L. Cessac and J. G. Curro, Raman scattering in uniaxially oriented samples of planar zigzag poly(vinylidene fluroide), *J. Polymer Sci., Polymer Phys. Ed. 12*:695 (1974).
29. G. Cortili and G. Zerbi, Chain conformations of poly(vinylidene fluoride) as derived from its vibrational spectrum, *Spectrochim. Acta 23A*:285 (1967).
30. S. Enomoto, Y. Kawai, and M. Sugita, Infrared spectrum of poly(vinylidene fluoride), *J. Polymer Sci. Part A-2 6*:861 (1968).
31. F. J. Boerio and J. L. Koenig, Vibrational analysis of poly(vinylidene fluoride), *J. Polymer Sci. Part A-2 9*:1517 (1971).
32. H. Tadokoro, *Structure of Crystalline Polymers*, Wiley Interscience, New York, 1979.
33. K. Zabel, N. E. Schlotter, and J. F. Rabolt, Structural characterization of an ethylene-tetrafluoroethylene alternating copolymer by polarized Raman scattering, *Macromolecules 16*:446 (1983).
34. K. Tashiro and M. Kobayashi, Thermally-induced structural phase transition of fluoride copolymers, *Polymner Prepr. Jpn. 36*:2339 (1987).
35. M. E. Lines and A. M. Glass, *Principles and Applications of Ferroelectrics and Related Materials*, Oxford Univ. Press, London, 1977.
36. T. Furukawa and G. E. Johnson, Measurement of ferroelectric switching characteristics in polyvinylidene fluoride, *Appl. Phys. Lett 38*:1027 (1981).
37. T. Furukawa, Ferroelectric properties of vinylidene fluoride copolymers, *Phase Transitions 18*:143 (1989).
38. D. Naegele and D. Y. Yoon, Orientation of crystalline dipoles in poly(vinylidene fluoride) films under electric field, *Appl. Phys. Lett. 33*:132 (1978).
39. T. Takahashi, M. Date, and E. Fukada, Dielectric hysteresis and rotation of dipoles in polyvinylidene fluoride, *Appl. Phys. Lett. 37*:791 (1980).

40. R. G. Kepler and R. A. Anderson, Ferroelectricity in polyvinylidene fluoride, *J. Appl. Phys.* 49:1232 (1978).

41. N. Takahashi and A. Odajima, Orientation behaviors in polyvinylidene fluoride film under electric field, *Jpn. J. Appl. Phys. 20*:L59 (1981).

42. N. Takahashi and A. Odajima, Fine structure of drawn polyvinylidene fluoride and its poling effects, *Charge Storage, Charge Transport and Electrostatics with Their Apolications* (Y. Wada, ed.), Elsevier, Amsterdam, 1979, p. 148.

43. N. Takahashi and A. Odajima, Ferroelectric reorientation of crystallites in polyvinylidene fluoride, *Ferroelectrics 32*:49 (1981).

44. N. Takahashi and A. Odajima, Fine structures of stretched poly(vinylidene fluoride) and their poling effects, *Rep. Progr. Polymer Phys. Jpn 21*:141 (1978).

45. A. Odajima, N. Takahashi, and H. Fujii, Biaxial orientation and piezoelectric constant of drawn poly(vinylidene fluoride), *Rep. Progr. Polymer Phys. Jpn. 23*:387 (1980).

46. Y. Takahashi, Y. Nakagawa, H. Miyaji, and K. Asai, Direct evidence for ferroelectric switching in poly(vinylidene fluoride) and poly(vinylidene fluoride-trifluoroethylene) crystals, *J. Polymer Sci. Part C: Polymer Lett. 25*:153 (1987).

47. D. C. Douglass, V. J. McBriety, and T. T. Wang, The use of NMR linewidths to study *b*-axis distributions in poled and unpoled PVDF, *J. Chem. Phys. 77*:5826 (1982).

48. V. J. McBrierty, D. C. Douglass, and T. T. Wang, Correlation between the piezoelectric constant d_{31} and *b*-axis distribution in polyvinylidene fluoride, *Appl. Phys. Lett. 41*:1051 (1982).

49. J. Clements, G. R. Davies, and I. M. Ward, A broad line NMR study of oriented poly(vinylidene fluoride), *Polymer 26*:208 (1985).

50. A. J. Hopfinger, A. J. Lewanski, T. J. Sluckin, and P. L. Taylor, Solitary wave propagation as a model for poling in PVF2, *Solitons and Condensed-Matter Physics* (A. R. Bishop and T. Schneider, eds.), Springer Verlag, New York, 1979, p. 330.

51. H. Dvey-Aharon, T. J. Sluckin, P. L. Taylor, and A. J. Hopfinger, Kink propagation as a model for poling in poly(vinylidene fluoride), *Phys. Rev. B 21*:3700 (1980).

52. M. G. Broadhurst and G. T. Davis, Ferroelectric polarization in polymers, *Ferroelectrics 32*:177 (1981).

53. W. Hayes and R. Loudon, *Scattering of Light by Crystals*, John Wiley, New York, 1978.

54. S. M. Shapiro and J. D. Axe, Raman scattering from polarphonons, *Phys. Rev. B6*:2420 (1972).

55. K. Tashiro, N. Yagi, M. Kobayashi, and T. Kawaguchi, Determination of the LO-TO splittings in the Raman spectra of ferroelectric triglycine sulfate (TGS), *Jpn. J. Appl. Phys. 26*: 699 (1987).

56. N. Takahashi and A. Odajima, A possibility of domain structure in crystallites of poly(vinylidene fluoride) form-I, *Rep. Progr. Polymer Phys. Jpn. 23*:161 (1980).

57. Y. Takahashi and H. Tadokoro, Formation mechanism of kink bands in modification II of poly(vinylidene fluoride). Evidence for flip-flop motion between TGTG and TḠTḠ conformations, *Macromolecules 13*:1316 (1980).

58. Y. Takahashi and H. Tadokoro, Crystal structure of form III of poly(vinylidene fluoride), *Macromolecules 13*:1317 (1980).

59. Y. Takahashi, Y. Matsubara, and H. Tadokoro, Mechanisms for crystal phase transformation by heat treatment and molecular motion in poly(vinylidene fluoride), *Macromolecules 15*: 334 (1982).

60. R. M. Gohil and J. Petermann, Chain conformational defects in polyvinylidene fluoride, *Polymer 22*:1612 (1981).

61. J. I. Scheinbeim, K. T. Chung, K. D. Pae, and B. A. Newman, The dependence of the piezoelectric response of poly(vinylidene fluoride) on phase-I volume fraction, *J. Appl. Phys. 50*:6101 (1979).

62. J. I. Scheinbeim, C. H. Yoon, K. D. Pae, and B. A. Newman, Ferroelectric hysteresis effects in poly(vinylidene fluoride) films, *J. Appl. Phys. 51*:5156 (1980).

63. W. W. Doll and J. B. Lando, Polymorphism of poly(vinylidene fluoride). III. The crystal structure of phase II, *J. Macromol. Sci.-Phys. B4*:309 (1970).

64. M. Bachmann, W. L. Gordon, S. Weinhold, and J. B. Lando, The crystal structure of phase IV of poly(vinylidene fluoride), *J. Appl. Phys. 51*:5095 (1980).

65. S. Weinhold, M. H. Litt, and J. B. Lando, The crystal structure of the γ phase of poly(vinylidene fluoride), *Macromolecules 13*:1178 (1980).

66. M. A. Bachmann and J. B. Lando, A Reexamination of the crystal structure of phase II of poly(vinylidene fluoride), *Macromolecules 14*:40 (1981).

67. Y. Takahashi, M. Kohyama, Y. Matsubara, H. Iwane, and H. Tadokoro, Tilting phenomena in forms II and III of poly(vinylidene fluoride): Evidence for monoclinic structures, *Macromolecules 14*:1841 (1981).

68. Y. Takahashi, Y. Matsubara, and H. Tadokoro, Crystal structure of form II of poly(vinylidene fluoride), *Macromolecules 16*:1588 (1983).

69. Y. Takahashi and H. Tadokoro, Structure and disorder of poly(vinylidene fluoride), *Ferroelectrics 57*:187 (1984).

70. Y. Takahashi and H. Tadokoro, Short-range order in form II of poly(vinylidene fluoride): Anti-phase domain structures, *Macromolecules 16*:1880 (1983).

71. Y. Takahashi, Packing disorder in form II of poly(vinylidene fluoride): Influence of elongation and annealing temperature, *Polymer J. 15*:733 (1983).

72. S. Weinhold, M. A. Bachmann, M. H. Litt, and J. B. Lando, Orthorhombic vs. Monodinic structures for the α and γ phases of poly(vinylidene fluoride): An analysis, *Macromolecules, 15*:1631 (1982).

73. A. J. Lovinger, Conformational defects and associated molecular motions in crystalline poly(vinylidene fluoride), *J. Appl. Phys. 52*:5934 (1981).

74. M. M. Coleman, M. S. Wu, I. R. Harrison, and P. C. Painter, Vibrational spectra and conformation of poly(vinylidene chloride), *J. Macromol. Sci.-Phys. B15*:463 (1978).

75. J. F. Rabolt and K. W. Johnson, Low frequency vibrations in polyvinylidene fluoride (form II), *J. Chem. Phys. 59*:3710 (1973).

76. K. Tashiro, Y. Itoh, S. Nishimura, and M. Kobayashi, Vibrational spectroscopic study on ferroelectric phase transition of vinylidene fluoride-trifluoroethylene copolymers: 2. Temperature dependences of the far-infrared absorption spectra and ultrasonic velocity, *Polymer 32*:1017 (1991).

77. P. D. Southgate, Room-temperature poling and morphology changes in pyroelectric polyvinylidene fluoride, *Appl. Phys. Lett. 28*:250 (1976).

78. M. Latour, Infrared analysis of poly(vinylidene fluoride) thermoelectrets, *Polymer 18*:278 (1977).

79. W. W. Doll and J. B. Lando, The polymorphism of poly(vinylidene fluoride) IV. The structure of high-pressure-crystallized poly(vinylidene fluoride), *J. Macromol. Sci.-Phys. B4*:889 (1970).

80. A. J. Lovinger, Crystalline transformations in spherulites of poly(vinylidene fluoride), *Polymer 21*:1317 (1980).

81. S. Weinhold, M. H. Litt, and J. B. Lando, Oriented phase III poly(vinylidene fluoride), *J. Polymer Sci., Polymer Lett. Ed. 17*:585 (1979).

82. B. Servet, D. Broussoux, F. Micheron, R. Bisaro, S. Ries, and P. Merenda, Oriented crystalline phases of PVDF and their piezoelectric properties, *Rev. Tech. Thomson-CSF 12*:761 (1980).

83. D. Broussoux, B. Servet, and F. Micheron, Morphology and structure of PVDF induced by mechanical and electrical orientation, *Rev. Tech. Thomson-CSF 12*:795 (1980).

84. B. Servet and J. Rault, Polymorphism of poly(vinylidene fluoride) induced by poling and annealing, *J. Phys. France 40*:1145 (1979).

85. K. Tashiro, K. Takano, M. Kobayashi, Y. Chatani, and H. Tadokoro, Phase transition at a temperature immediately below the melting point of poly(vinylidene fluoride) form I: A proposition for the ferroelectric Curie point, *Polymer 24*:199 (1983).

86. K. Tashiro, K. Takano, M. Kobayashi, Y. Chatani, and H. Tadokoro, A preliminary X-ray study on ferroelectric phase transition of poly(vinylidene fluoride) crystal form I, *Polymer Bull 10*:464 (1983).

87. Y. Takahashi and N. Miyamoto, Kink motion in poly(vinylidene fluoride) form I, *J. Polymer Sci., Polymer Phys. Ed. 23*:2505 (1985).

88. G. Cortili and G. Zerbi, Further infrared data on polyvinylidene fluoride, *Spectrochim. Acta 23A*:2216 (1967).

89. A. J. Lovinger and H. D. Keith, Electron diffraction investigation of a high-temperature form of poly(vinylidene fluoride), *Macromolecules 12*:919 (1979).

90. S. K. Tripathy, R. Potenzone, Jr., A. J. Hopfinger, N. C. Banik, and P. L. Taylor, Predicted chain conformation for a possible phase III form of poly(vinylidene fluoride), *Macromolecules 12*:656 (1979).

91. M. A. Bachmann, W. L. Gordon, J. L. Koenig, and J. B. Lando, An infrared study of phase-III poly(vinylidene fluoride), *J. Appl. Phys., 50*:6106 (1979).

92. K. Tashiro, M. Kobayashi, and H. Tadokoro, Vibrational spectra and disorder-order transition of poly(vinylidene fluoride) form III, *Macromolecules 14*:1757 (1981).

93. A. J. Lovinger, Annealing of poly(vinylidene Fluoride) and formation of a fifth phase, *Macromolecules 15*:40 (1982).

94. A. J. Lovinger, Molecular mechanism for $\alpha \rightarrow \delta$ transformation in electrically poled poly(vinylidene fluoride), *Macromolecules 14*:225 (1981).

95. H. Dvey-Aharon, P. L. Taylor, and A. J. Hopfinger, Dynamics of the field-induced transition to the polar *a* phase of poly(vinylidene fluoride), *J. Appl. Phys. 51*:5184 (1980).

96. Y. Takahashi and K. Miyaji, Long-range order parameters of form II of poly(vinylidene fluoride) and molecular motion in the α_c relaxation, *Macromolecules 16*:1789 (1983).

97. Y. Miyamoto, H. Miyaji, and K. Asai, Anisotropy of dielectric relaxation in crystal form II of poly(vinylidene fluoride), *J. Polymer Sci., Polymer Phys. Ed. 18*:597 (1980).

98. J. D. Clark and P. L. Taylor, Longitudinal polarization of *a*-poly(vinylidene fluoride) by kink propagation, *J. Appl. Phys. 52*:5903 (1981).

99. K. Tashiro and M. Kobayashi, Crystalline modifications of poly(vinylidene fluoride) and its derivatives exhibited at room temperature: The dependence on the crystallization conditions and VDF molar content of copolymers, *Rep. Progr. Polymer Phys. Jpn. 30*:119 (1987).

100. Y. Toida and R. Chujo, High-resolution NMR spectra and conformation of poly(vinylidene fluoride) and their relation with crystal modifications, *Polymer J. 6*:191 (1974).

101. C. S. Wilson III, NMR study of molecular chain structure of polyvinylidene fluoride, *J. Polymer Sci. Part A 1*:1305 (1963).

102. A. Tonelli, Conformational characteristics of poly(vinylidene fluoride), *Macromolecules 9*: 547 (1976).

103. A. E. Tonelli, F. C. Schilling, and R. E. Cais, 19F NMR chemical shifts and the microstructure of fluoro polymers, *Macromolecules 15*:849 (1982).

104. A. E. Tonelli, F. C. Schilling, and R. E. Cais, Carbon-13 nuclear magnetic resonance chemical shifts and the microstructures of the fluoro polymers poly(vinylidene fluoride), poly(fluoromethylene), poly(vinyl fluoride), and poly(trifluoroethylene), *Macromolecules 14*:560 (1981).

105. R. E. Cais and J. M. Kometani, Synthesis and two-dimensional 19F NMR of highly aregic poly(vinylidene fluoride), *Macromolecules 18*:1354 (1985).

106. J. B. Lando and W. W. Doll, The polymorphism of poly(vinylidene fluoride). I. The effect of head-to-head structure, *J. Macromol. Sci.-Phys. B2*:205 (1968).

107. A. J. Lovinger, D. D. Davis, R. E. Cais, and J. M. Kometani, The role of molecular defects on the structure and phase transitions of poly(vinylidene fluoride), *Polymer 28*:617 (1987).

108. R. E. Cais and J. M. Kometani, Polymerization of vinylidene-d_2 fluoride. Minimal regiosequence and branch defects and assignment of preferred chain-growth direction from the deuterium effect, *Macromolecules 17*:1887 (1984).

109. N. Takahashi and A. Odajima, On the structure of poly(vinylidene fluoride) under a high electric field, *Ferroelectrics 57*:221 (1984).

110. A. Odajima and K. Tashiro, Structures and phase transitions of ferroelectric fluoropolymers (in Japanese), *J. Cryst. Soc. Jpn. 26*:103 (1984).

111. A. J. Lovinger, Recent development in the structure, properties, and applications of ferroelectric polymers, *Jpn. J. Appl. Phys. Suppl. 24-2*:18 (1985).

112. K. Tashiro and M. Kobayashi, Structural phase transition in ferroelectric fluorine polymers: X-ray diffraction and infrared/Raman spectroscopic study, *Phase Transitions 18*:213 (1989).

113. K. Tashiro and M. Kobayashi, Structural phase transition in ferroelectric vinylidene fluoride-trifluoroethylene copolymers, *Jpn. J. Appl. Phys. Suppl. 24-2*:873 (1985).

114. T. Furukawa, M. Date, E. Fukada, Y. Tajitsu, and A. Chiba, Ferroelectric behavior in the copolymer of vinylidene fluoride and trifluoroethylene, *Jpn. J. Appl. Phys. 19*:L109 (1980).

115. T. Yamada, T. Ueda, and T. Kitayama, Ferroelectric-to-paraelectric phase transition of vinylidene fluoride-trifluoroethylene copolymer, *J. Appl. Phys. 52*:948 (1981).

116. Y. Higashihata, J. Sako, and T. Yagi, Piezoelectricity of vinylidene fluoride-trifluoroethylene copolymers, *Ferroelectrics 32*:85 (1981).

117. T. Furukawa, G. E. Johnson, and H. E. Bair, Ferroelectric phase transition in a copolymer of vinylidene fluoride and trifluoroethylene, *Ferroelectrics 32*:61 (1981).

118. H. Mizuno, Y. Nagano, K. Tashiro, and M. Kobayashi, A study of the ferroelectric phase transition of vinylidene fluoride-trifluoroethylene copolymers by AC calorimetry, *J. Chem. Phys. 96*:3234 (1992).

119. Y. Tajitsu, A. Chiba, T. Furukawa, M. Date, and E. Fukada, Crystalline phase transition in the copolymer of vinylidene fluoride and trifluoroethylene, *Appl. Phys. Lett. 36*:286 (1980).

120. K. Tashiro, K. Takano, M. Kobayashi, Y. Chatani, and H. Tadokoro, Structural study on ferroelectric phase transition of vinylidene fluoride-trifluoroethylene random copolymer, *Polymer 22*:1312 (1981).

121. A. J. Lovinger, G. T. Davis, T. Furukawa, and M. G. Broadhurst, Crystalline forms in a copolymer of vinylidene fluoride and trifluoroethylene (52/48 mol%), *Macromolecules 15*: 323 (1982).

122. G. T. Davis, T. Furukawa, A. J. Lovinger, and M. G. Broadhurst, Structural and dielectric investigation on the nature of the transition in a copolymer of vinylidene fluoride and trifluoroethylene (52/48 mol%), *Macromolecules 15*:329 (1982).

123. K. Tashiro, K. Takano, M. Kobayashi, Y. Chatani, and H. Tadokoro, Structure and ferroelectric phase transition of vinylidene fluoride-trifluoroethylene copolymers: 2. VDF 55% copolymer, *Polymer 25*:195 (1984).

124. K. Tashiro and M. Kobayashi, Vibrational spectroscopic study of the ferroelectric phase transition in vinylidene fluoride-trifluoroethylene copolymers: 1. Temperature dependence of the Raman spectra, *Polymer 29*:4429 (1988).

125. J. S. Green, J. P. Rabe, and J. F. Rabolt, Studies of chain conformation above the Curie point in a vinylidene fluoride/trifluoroethylene random copolymer, *Macromolecules 19*:1725 (1986).

126. K. Tashiro, K. Takano, M. Kobayashi, Y. Chatani, and H. Tadokoro, Structural study on ferroelectric phase transition of vinylidene fluoride-trifluoroethylene copolymers. (III) Dependence of transitional behavior on VDF molar content, *Ferroelectrics 57*:297 (1984).

127. T. Yagi and M. Tatemoto, A fluorine-19 NMR study of the microstructure of vinylidene fluoride-trifluoroethylene copolymers, *Polymer J. 11*:429 (1979).

128. T. Furukawa and G. E. Johnson, Dielectric relaxations in a copolymer of vinylidene fluoride and trifluoroethylene, *J. Appl. Phys. 52*:940 (1981).

129. N. Koizumi, N. Haikawa, and H. Habuka, Dielectric behavior and ferroelectric transition of copolymers of vinylidene fluoride and trifluoroethylene, *Ferroelectrics 57*:99 (1984).

130. F. Ishii, A. Odajima, and H. Ohigashi, Ferroelectric transition in vinylidene fluoride-trifluoroethylene copolymer studied by nuclear magnetic resonance method, *Polymer J. 15*: 875 (1983).

131. V. J. McBrierty, D. C. Douglass, and T. Furukawa, Magnetic resonance and relaxation in a vinylidene fluoride/trifluoroethylene copolymer, *Macromolecules 15*:1063 (1982).

132. V. J. McBrierty, D. C. Douglass, and T. Furukawa, A nuclear magnetic resonance study of poled vinylidene fluoride/trifluoroethylene copolymer, *Macromolecules 17*:1136 (1984).

133. F. Ishii and A. Odajima, Proton spin lattice relaxation in vinylidene fluoride/trifluoroethylene copolymer. I. Vinylidene fluoride 72 mol% copolymer, *Polymer J. 18*:539 (1986).

134. F. Ishii and A. Odajima, Proton spin lattice relaxation in vinylidene fluoride/trifluoroethylene copolymer. II. Effects of vinylidene fluoride content upon spin relaxation processes, *Polymer J. 18*:547 (1986).

135. T. Horiuchi, K. Matsushige, and T. Takemura, Intermediate structure at the ferro-to-paraelectric phase transition of VDF/TrFE copolymer (54/46 mol%), *Jpn. J. Appl. Phys. 25*:L465 (1986).

136. M. V. Fernandez, A. Suzuki, and A. Chiba, Study of annealing effects on the structure of vinylidene fluoride-trifluoroethylene copolymers using WAXS and SAXS, *Macromolecules 20*:1806 (1987).

137. K. Tashiro and M. Kobayashi, Ferroelectric phase transition and specific volume change in vinylidene fluoride-trifluoroethylene copolymers, *Rep. Progr. Polymer Phys. Jpn. 29*:169 (1986).

138. K. Matsushige, T. Horiuchi, S. Taki, T. Takemura, and K. Tagashira, Pressure effect on ferroelectric phase transition of vinylidene fluoride/trifluoroethylene copolymer, *Polymer Prepr. Jpn. 34*:3005 (1985).

139. A. J. Lovinger, T. Furukawa, G. T. Davis, and M. G. Broadhurst, Crystallographic changes characterizing the Curie transition in three ferroelectric copolymers of vinylidene fluoride and trifluoroethylene: 1. As-crystallized samples, *Polymer 24*:1225 (1983).

140. A. J. Lovinger, T. Furukawa, G. T. Davis, and M. G. Broadhurst, Crystallographic changes characterizing the Curie transition in three ferroelectric copolymers of vinylidene fluoride and trifluoroethylene: 2. Oriented or poled samples, *Polymer 24*:1233 (1983).

141. Y. Oka and N. Koizumi, Ferroelectric order and phase transition in polytrifluoroethylene, *J. Polymer Sci. Part B, Polymer Phys. 243*:2059 (1986).

142. A. J. Lovinger and R. E. Cais, Structure and morphology of poly(trifluoroethylene), *Macromolecules 17*:1939 (1984).

143. Y. L. Gal'perin and Y. V. Strogalin, The symmetry and dimension of the unit cell of polytrifluoroethylene, *Vysokomol. Soyed. 7*:16 (1965).

144. K. Tashiro and M. Kobayashi, Structure and ferroelectric phase transition of poly(vinylidene fluoride) form I and vinylidene fluoride-trifluoroethylene copolymers [IX] VDF 82% copolymer, *Polymer Prepr. Jpn. 34*:2241 (1985).

145. K. Nakamura and Y. Wada, Pizoelectricity, pyroelectricity, and the electrostriction constant of poly(vinylidene fluoride), *J. Polymer Sci. Part A-2 9*:161 (1971).

146. A. M. Glass, J. H. McFee, and J. G. Bergman, Jr., Pyroelectric properties of polyvinylidene fluoride and its use for infrared detection, *J. Appl. Phys. 42*:5219 (1971).

147. F. Micheron, Thermodynamic model of PVF2 and applications, *Ferroelectrics 28*:395 (1980).

148. P. Herchenröder, Y. Segui, D. Horne, and D. Y. Yoon, Ferroelectricity of poly(vinylidene fluoride): Transition temperature, *Phys. Rev. Lett. 45*:2135 (1980).

149. A. J. Lovinger, D. D. Davis, R. E. Cais, and J. M. Kometani, On the Curie temperature of poly(vinylidene fluoride), *Macromolecules 19*:1491 (1986).

150. K. Koga, N. Minato, and H. Ohigashi, Composition dependence of Piezoelectric and ferroelectric properties of vinylidene fluoride-trifluoroethylene copolymers, *Polymer Prepr. Jpn. 34*:953 (1985).

151. H. Ohigashi, Piezoelectric polymers—Materials and manufacture, *Jpn. J. Appl. Phys. Suppl. 24-2*:23 (1985).

152. J. Hirschinger, B. Meurer, and G. Weill, Nuclear spin relaxation in a vinylidene fluoride and trifluoroethylene copolymer 70/30 I. Dynamics and morphology in the ferroelectric phase, *J. Phys. France 50*:563 (1989).

153. J. Hirschinger, B. Meurer, and G. Weill, Nuclear spin relaxation in a vinylidene fluoride and trifluoroethylene copolymer 70/30 II. The one-dimensional fluctuations in the paraelectric phase, *J. Phys. France 50*:583 (1989).

154. J. F. Legrand, B. Frick, B. Meurer, V. H. Schmidt, M. Bee, and J. Lajzerowicz, Neutron scattering studies of the ferroelectric transition in P(VDF-TrFE) copolymers, *Ferroelectrics 109*:321 (1990).

155. J. F. Legrand, P. Delzenne, A. J. Dianoux, M. Bee, C. Poinsignon, D. Broussoux, and V. H. Schmidt, Conformationally disordered crystalline phase in VF_2-VF_3 copolymers, *Springer Proc. Phys. 29*:59 (1988).

156. J. F. Legrand, Structure and ferroelectric properties of P(VDF-TrFE) copolymers, *Ferroelectrics 91*:303 (1989).

157. J. F. Scott, Soft-mode spectroscopy: Experimental studies of structural phase transitions, *Rev. Mod. Phys. 46*:83 (1974).

158. K. Matsushige, K. Tagashira, T. Horiuchi, S. Taki, and T. Takemura, Ferroelectric phase transition of VDF/TrFE copolymer at high pressure, *Jpn. J. Appl. Phys. Suppl. 24-2*:868 (1985).

159. H. Tanaka, H. Yukawa, and T. Nishi, Dynamics of ferroelectric phase transition in vinylidene fluoride/trifluoroethylene (VF_2/F_3E) copolymers. I. acoustic study, *J. Chem. Phys. 90*:6730 (1989).

160. J. F. Krüger, J. Petzelt, and J. F. Legrand, Brillouin spectroscopic investigations of the ferroelectric phase transition in a polyvinylidenefluoride-trifluoroethylene copolymer, *Coll. Polymer Sci. 264*:791 (1986).

161. Y. Murata and N. Koizumi, Dynamic mechanical disperson in copolymers of vinylidene fluoride and trifluoroethylene, *Polymer J. 17*:385 (1985).

162. Y. Murata and N. Koizumi, Elastic modulus for copolymer of vinylidene fluoride and tetra-fluoroethylene, *Rep. Progr. Polymer Phys. Jpn. 30*:363 (1987).

163. N. Koizumi, Y. Murata, and N. Haikawa, Pressure dependence of the ferroelectric transition in copolymers of vinylidene fluoride and trifluoroethylene, *Jpn. J. Appl. Phys. Suppl. 24-2*: 862 (1985).

164. N. Koizumi, Y. Murata, and Y. Oka, Pressure dependence of ferroelectric transition and anomaly in bulk modulus in copolymer of vinylidene fluoride and trifluoroethylene, *Jpn. J. Appl. Phys. 23*:L324 (1984).

165. W. Rehwald, The study of structural phase transitions by means of ultrasonic experiments, *Adv. Phys. 22*:721 (1973).

166. J. E. Green, B. L. Farmer, and J. F. Rabolt, Effect of thermal and solution history on the Curie point of VF_2-TrFE random copolymers, *J. Appl. Phys. 60*:2690 (1986).

167. G. M. Stack and R. Y. Ting, Thermodynamic and morphological studies of the solid-state transition in copolymers of vinylidene fluoride and trifluoroethylene, *J. Polymer Sci. Part B, Polymer Phys. 26*:55 (1988).

168. H. Tanaka, H. Yukawa, and T. Nishi, Effect on crystallization condition on the ferroelectric phase transition in vinylidene fluoride trifluoroethylene (VF_2/F_3E) copolymers, *Macromolecules 21*:2469 (1988).

169. K. Tashiro, K. Ushitora, and M. Kobayashi, Annealing effects on ferroelectric phase transition of vinylidene fluoride-trifluoroethylene copolymers, *Polymer Prepr. Jpn. 42*:4498 (1993).

170. A. Odajima, Y. Takase, and K. Yuasa, Boudary effects on phase transitions in polymer ferroelectrics, *Polymer Prepr. Jpn. 34*:950 (1985).

171. G-R. Li, N. Kagami, and H. Ohigashi, The possibility of formation of large ferroelectric domains in a copolymer of vinylidene fluoride and trifluoroethylene, *J. Appl. Phys. 72*:1056 (1992).

172. A. Odajima and G. R. Mitchell, Trans-gauche energy difference in moelcular chains of vinylidene fluoride/trifluoroethylene copolymers in the paraelectric phase, *Polymer J. 16*: 587 (1984).

173. R. R. Kolda and J. B. Lando, The effect of hydrogen-fluorine defects on the conformational energy of polytrifluoroethylene chains, *J. Macromol. Sci.-Phys. B11*:21 (1975).

174. T. Yagi, Transition and relaxation in poly(trifluoroethylene), *Polymer J. 11*:711 (1979).

175. K. Tashiro and M. Kobayashi, Structural study of the ferroelectric phase transition of vinylidene fluoride-trifluoroethylene copolymers: 4. Poling effect on structure and phase transition, *Polymer 27*:667 (1986).

176. J. A. Day, E. L. V. Lewis, and G. R. Davis, X-ray structural study of oriented vinylidene fluoride/trifluoroethylene copolymers, *Polymer 33*:1571 (1992).

177. K. J. Kim, N. M. Reynolds, and S. L. Hsu, Spectroscopic studies on the effect of field strength upon the Curie transition of a VDF/TrFE copolymer, *J. Polymer Sci. Part B, Polymer Phys. 31*:1555 (1993).

178. K. Tashiro and M. Kobayashi, FTIR study on molecular orientation and ferroelectric phase transition in vacuum-evaporated and solution-cast films of vinylidene fluoride-trifluoroethylene copolymers: Effects of heat treatment and high-voltage poling, *Spectrochim. Acta., 50A*:1573 (1994).

179. J. Clements, G. R. Davies, and I. M. Ward, A broad-line nuclear magnetic resonance study of a vinylidene fluoride/trifluoroethylene copolymer, *Polymer 33*:1623 (1992).

180. K. Kimura and H. Ohigashi, Ferroelectric properties of thin films of vinylidene fluoride and trifluoroethylene copolymers (I), *Polymer Prepr. Jpn. 31*:717 (1982).

181. K. Tashiro, T. Ishioka, and M. Kobayashi, Electric field effect on ferroelectric phase transition of vinylidene fluoride-trifluoroethylene copolymers studied by infrared spectroscopy, unpublished data (1985).

182. K. Tashiro, M. Nakamura, M. Kobayashi, Y. Chatani, and H. Tadokoro, Polarization inversion current and ferroelectric phase transition of vinylidene fluoride-trifluoroethylene copolymer, *Macromolecules 17*:1452 (1984).

183. K. Tashiro, S. Nishimura, and M. Kobayashi, Thermal contraction and ferroelectric phase transition in vinylidene fluoride-trifluoroethylene copolymers. 1. An effect of tensile stress along the chain axis, *Macromolecules 21*:2463 (1988).

184. K. Tashiro, S. Nishimura, and M. Kobayashi, Thermal contraction and ferroelectric phase transition in vinylidene fluoride-trifluoroethylene copolymers. 2. An effect of tensile stress applied in the direction perpendicular to the chain axis, *Macromolecules 23*:2802 (1990).

185. T. Sekimura, A. Iwado, and S. Ikeda, Tensile stress dependence of dielectric constant and ferroelectric transition in vinylidene fluoride and trifluoroethylene copolymers, *Polymer Prepr. Jpn. 40*:4412 (1991).

186. A. J. Lovinger, Polymorphic transformations in ferroelectric copolymers of vinylidene fluoride induced by electron irradiation, *Macromolecules 18*:910 (1985).

187. Y. Takase, T. Ishibashi, and A. Odajima, Ferroelectric phase transition of vinylidene fluoride and trifluoroethylene copolymers and its high energy irradiation effects, *Polymer Prepr. Jpn. 33*:2567 (1984).

188. F. Macchi, B. Daudin, and J. F. Legrand, Electron irradiation induced structural modifications in ferroelectric P(VDF-TrFE) copolymers, *Nuclear Instr. Meth. Phys. Res. B46*:324 (1990).

189. F. Macchi, B. Daudin, and J. F. Legrand, Effect of electron irradiation on the ferroelectric transition of P(VDF-TrFE) copolymers, *Ferroelectrics 109*:303 (1990).

190. A. J. Lovinger, Ferroelectric transition in a copolymer of vinylidene fluoride and tetrafluoroethylene, *Macromolecules 16*:1529 (1983).

191. A. J. Lovinger, G. E. Johnson, H. E. Bair, and E. W. Anderson, Structural, dielectric and thermal investigation of the Curie transition in a tetrafluoroethylene copolymer of vinylidene fluoride, *J. Appl. Phys. 56*:2412 (1984).

192. Y. Murata and N. Koizumi, Curie transition in copolymers of vinylidene fluoride and tetrafluoroethylene, *Polymer J. 17*:1071 (1985).

193. A. J. Lovinger, D. D. Davis, R. E. Cais, and J. M. Kometani, Compositional variation of the structure and solid-state transformation of vinylidene fluoride/tetrafluoroethylene copolymers, *Macromolecules 21*:78 (1988).

194. K. Tashiro, H. Kaito, and M. Kobayashi, Structural changes in ferroelectric phase transitions of vinylidene fluoride-tetrafluoroethylene copolymers: 1. Vinylidene fluoride content dependence of the transition behaviour, *Polymer 33*:2915 (1992).

195. K. Tashiro, H. Kaito, and M. Kobayashi, Structural changes in ferroelectric phase transitions of vinylidene fluoride-tetrafluoroethylene copolymers: 2. normal-modes analysis of the infrared and Raman spectra at room temperature, *Polymer 33*:2929 (1992).

196. E. S. Clark and L. T. Muus, Partial disordering and crystal transitions in polytetrafluoroethylene, *Z. Kristallogr. 117*:119 (1962).

197. G. Zerbi, L. Piseri, and F. Labassi, Vibrational spectrum of chain molecules with conformational disorder. Polyethylene, *Mol. Phys. 22*:241 (1971).

198. V. N. Kozyrenko, I. V. Kumparenko, and I. D. Mikhailov, *J. Polymer Sci. 15*:1721 (1977).

199. M. Kobayashi, K. Akita, and H. Tadokoro, Infrared spectra and regular sequence length in isotactic polymer chains, *Makromol. Chem. 188*:324 (1968).

200. J. Green and J. F. Rabolt, Identification of a curie transition in vinylidene fluoride/tetrafluoroethylene random copolymers by spectroscopic methods, *Macromolecules 20*:456 (1987).

201. G. Zerbi and M. Sacchi, Dynamics of polymers as structurally disordered systems. Vibrational spectrum and stucture of poly(tetrafluoroethylene), *Macromolecules 6*:692, 700 (1973).

202. T. Horiuchi, K. Fakao and K. Matsushige, New evaluation method of evaporated organic thin films by energy dispersive x-ray diffractometer, *Jpn. J. Appl. Phys., 26*:L1839 (1987).

203. K. Tashiro, K. Yamamoto, M. Kobayashi, and O. Phaovibul, A preliminary study on the structure and ferroelectric phase transition of vacuum-evaporated thin films of vinylidene fluoride-trifluoroethylene copolymers, *Rep. Progr. Polymer Phys. Jpn. 30*:123 (1987).

204. K. Tashiro and M. Kobayashi, FTIR-study on molecular orientation in ultra-thin films of vinylidene fluoride-trifluoroethylene copolymers, *Rep. Progr. Polymer Phys. Jpn. 31*:183 (1988).

205. A. Takeno, N. Okui, T. Hiruma, T. Kitoh, M. Muraoka, S. Umemoto, and T. Sakai, Pyrolysis and deposition mechanism of poly(vinylidene fluoride) in vacuum deposition process, *Kobunshi Ronbunshu 48*:399 (1991).

206. A. Takeno, N. Okui, T. Hiruma, M. Muraoka, S. Umemoto, T. Sakai, Crystal form and molecular orientation of poly(vinylidene fluoride) thin films prepared by vapor deposition, *Koubunshi Ronbunshu 48*:405 (1991).

207. K. A. Mauritz, E. Baer, and A. J. Hopfinger, The epitaxial crystallization of macromolecules, *J. Polymer Sci. Macromol. Rev. 13*:1 (1978).

208. N. C. Banik, F. P. Boyle, T. J. Sluckin, P. L. Taylor, S. K. Tripathy, and A. J. Hopfinger, Theory of structural phase transitions in crystalline poly(vinylidene fluoride), *J. Chem. Phys. 72*:3191 (1980).

209. R. Zhang and P. L. Taylor, Theory of ferroelectric-paraelectric transitions in VF_2/F_4E random copolymers, *J. Appl. Phys. 73*:1395 (1993).

210. K. Tashiro, K. Takano, M. Kobayashi, Y. Chatani, and H. Tadokoro, Structure and ferroelectric phase transition of poly(vinylidene fluoride) form I and vinylidene fluoride-trifluoroethylene copolymers [II], *Polymer Prepr. Jpn. 30*:1882 (1981).

211. H. A. Bethe, Statistical theory of superlattice, *Proc. R. Soc. A150*:552 (1935).

212. R. Peierls, Statistical theory of superlattices with unequal concentration of the components *Proc. Camb. Phil. Soc. 32*:477 (1936).

213. Y. Takagi, *J. Phys. Soc. Jpn. 4*:99 (1949).

214. R. H. Fowler and E. A. Guggenheim, Statistical thermodynamics of superlattices, *Proc. R. Soc. A174*:189 (1940).

215. A. Odajima, A statistical theory of ferroelectric phase transition of vinylidene fluoride trifluoroethylene copolymers, *Ferroelectrics 57*:159 (1984).

216. S. Ikeda and T. Sekimura, Theory of phase transition of VDF-TrFE copolymers, *Polymer Prepr. Jpn. 42*:4501 (1993).

217. F. C. Wilson and H. W. Starkweather, Jr., Crystal structure of an alternating copolymer of ethylene and tetrafluoroethylene, *J. Polymer Sci. Polymer Phys. Ed. 11*:919 (1973).

218. T. Tanigami, K. Yamaura, S. Matsuzawa, M. Ishikawa, K. Mizoguchi, and K. Miyasaka, Structural studies on ethylene-tetrafluoroethylene copolymer 1. Crystal structure, *Polymer* 27:999 (1986).

219. B. L. Farmer and J. B. Lando, Conformational and packing analysis of the alternating copolymer of ethylene and tetrafluoroethylene, *J. Macromol. Sci.-Phys. B11*:89 (1975).

220. T. Tanigami, K. Yamaura, S. Matsuzawa, M. Ishikawa, K. Mizoguchi, and K. Miyasaka, Structural studies on ethylene-tetrafluoroethylene copolymer 2. Transition from crystal phase to mesophase, *Polymer* 27:1521 (1986).

221. G. Natta, I. W. Bassi, and G. Allegra, Crystal structure of atactic polyvinylfluoride, *Lincei-Rend. Sci. Fis. Mat. e Nat. 31*:350 (1961).

222. G. Guerra, G. D. Dino, R. Centore, V. Petraccone, J. Obrzut, F. E. Karasz, and J. MacKnight, Structural characterization of vinylidene fluoride/vinyl fluoride copolymers, *Makromol. Chem. 190*:2203 (1989).

223. K. Maeda, S. Tasaka, and N. Inagaki, Ferroelectric behavior of vinylfluoride-trifluoroethylene copolymer, *Jpn. J. Appl. Phys. 30*:L2107 (1991).

224. S. Tasaka, N. Inagaki, T. Okutani, and S. Miyata, Structure and properties of amorphous piezoelectric vinylidene cyanide copolymers, *Polymer 30*:1639 (1989).

225. M. Kishimoto, K. Nakajima, and I. Seo, Structure and physical properties of vinylidene cyanide-straight chain fatty acid vinyl ester copolymers, *Polymer Prepr. Jpn. 37*:2249 (1988).

226. D. Zou, S. Iwasaki, T. Tsutsui, S. Saito, M. Kishimoto, and I. Seo, Anomaly in dielectric relaxation in alternating copolymers of vinylidene cyanide and fatty acid ester, *Polymer 31*:1888 (1990).

227. T. T. Wang and Y. Takase, Ferroelectric dielectric behavior in the piezoelectric amorphous copolymer of vinylidenecyanide and vinyl acetate, *J. Appl. Phys. 62*:3466 (1987).

228. I. Seo, M. Kishimoto, H. Sato, and H. Gamo, Alternating copolymer of vinylidene cyanide and vinyl acetate: A new nonlinear optical material, *Nonlinear Optics of Organics and Semiconductors* (T. Kobayashi, ed.), Springer-Verlag, Berlin, 1989, p. 196.

229. Y. Inoue, K. Kawaguchi, Y. Maruyama, Y. S. Jo, and R. Chujo,[13] C n.m.r. studies on the microstructure of piezoelectric copolymers of vinylidene cyanide, *Polymer 30*:698 (1989).

230. P. A. Mirau and S. A. Heffner, Chain conformation in poly(vinylidene cyanide-vinyl acetate): Solid state and solution 2D and 3D n.m.r. studies, *Polymer 33*:1156 (1992).

231. S. Tasaka, K. Miyasato, M. Yoshikawa, S. Miyata, and M. Ko, *Ferroelectrics 57*:267 (1984).

232. S. Kurihara, Y. Takahashi, H. Miyaji, and I. Seo, Structural change on poling in a piezoelectric copolymer of vinylidene cyanide with vinyl acetate, *Jpn. J. Appl. Phys. 28*:L686 (1989).

233. K. Tashiro, M. Kobayashi, M. Kishimoto, and I. Seo, Structure of copolymers of vinylidene cyanide and *n*-fatty acid vinyl ester, *Polymer Prepr. Jpn. 41*:1273 (1992).

234. K. Tashiro, M. Kobayashi, M. Kishimoto, and I. Seo, Crystal structure of vinylidene cyanide copolymers: Energy minimization and computer simulation of X-ray diffraction pattern, *Polymer Prepr. Jpn. 42*:1442 (1993).

235. R. J. Hobson and A. H. Windle, Crystalline structure of atactic poly(acrylonitrile), *Macromolecules 26*:6903 (1993).

236. T. Nakamura, S. Tasaka, and N. Inagaki, Ferroelectricity in acrylonitrile copolymers, *Polymer Prepr. Jpn. 41*:4565 (1992).

237. R. Hasegawa, K. Kimoto, Y. Chatani, H. Tadokoro, and A. Sekiguchi, *Polymer Prepr. Jpn.*, p. 713 (1974).

238. G. Wu, O. Yano, and T. Soen, Dielectric and piezoelectric properties of nylon 9 and nylon 11, *Polymer J. 18*:51 (1986).

239. B. A. Newman, P. Chen, K. D. Pae, and J. I. Scheinbeim, Piezoelectricity in nylon 11, *J. Appl. Phys. 51*:5161 (1980).

240. J. I. Scheinbeim, Piezoelectricity in γ-form nylon 11, *J. Appl. Phys. 52*:5939 (1981).

241. J. W. Lee, Y. Takase, B. A. Newman, and J. I. Scheinbeim, Ferroelectric polarization switching in nylon-11, *J. Polymer Sci. Part B, Polymer Phys. 29*:273 (1991).

242. J. W. Lee, Y. Takase, B. A. Newman, and J. I. Scheinbeim, Effect of annealing on the ferroelectric behavior of nylon-11 and nylon-7, *J. Polymer Sci. Part B, Polymer Phys. 29*: 279 (1991).

243. Y. Takase, J. W. Lee, J. I. Scheinbeim, and B. A. Newman, High-temperature characterization of nylon-11 and nylon-7 piezoelectrics, *Macromolecules 24*:6644 (1991).

244. T. Furukawa, Y. Takahashi, D. Suzuki, and M. Kutani, Ferroelectric switching in odd-nylons, *Polymer Prepr. Jpn. 41*:4562 (1992).

245. R. B. Meyer, L. Liebert, L. Strzelecki, P. Keller, Ferroelectric liquid crystals, *J. Phys. (Paris), 36*:L69 (1975).

246. R. W. Cahn, E. A. Davis, I. M. Ward, A. M. Donald, A. H. Windle, Liquid crystalline polymers, Cambridge Univ. Press, London (1992).

247. J. Hou, K. Tashiro, and M. Kobayashi, Effect of chiral alkyl groups on the structural phase transition of ferroelectric liquid crystals investigated by differential scanning calorimetry, X-ray diffraction, and infrared/Raman spectroscopic methods, *J. Phys. Chem. 96*:2729 (1992).

248. K. Tashiro, J. Hou, and M. Kobayashi, First observation of the polytype phenomenon of layer stacking structure in a side-chain type ferroelectric liquid-crystalline polymer, *Macromolecules, 27*:3912 (1994).

249. K. Tashiro and M Kobayashi, Order-disorder transitions in crystalline polymers with characteristic mechanical and electric properties, *Ordering in Macromolecular Systems* (A. Teramoto, ed.), Springer-Verlag, Berlin, pp. 17–34, (1994).

250. R. Hayakawa and Y. Wada, Piezoelectricity and related properties of polymer films, *Progr. Polymer Sci. 11*:1 (1973).

251. Y. Wada and R. Hayakawa, Piezoelectricity and pyroelectricity of polymers, *Jpn. J. Appl. Phys. 15*:2041 (1976).

252. K. Tashiro, H. Tadokoro, and M. Kobayashi, Structure and piezoelectricity of poly(vinylidene fluoride), *Ferroelectrics 32*:167 (1981).

253. M. G. Broadhurst, G. T. Davis, J. E. McKinney, and R. E. Collins, Piezoelectricity and pyroelectricity in polyvinylidene fluoride—A model, *J. Appl. Phys. 49*:4992 (1978).

254. H. Dvey-Aharon, T. J. Sluckin, and P. L. Taylor, Theory of reversible crystallization and electric effects in PVF$_2$, *Ferroelectrics 32*:25 (1981).

255. K. Tashiro, M. Kobayashi, H. Tadokoro, and E. Fukada, Calculation of elastic and piezoelectric constants of polymer crystals by a point charge model: Application to poly(vinylidene fluoride) form I, *Macromolecules 13*:691 (1980).

256. K. Tashiro and H. Tadokoro, Estimating the limiting values of the macroscopic piezoelectric constants of poly(vinylidene fluoride) form I, *Macromolecules 16*:961 (1983).

257. K. Tashiro, Molecular theory of mechanical properties of crystalline polymers, *Prog. Polymer Sci. 18*:377 (1993).

258. N. Takahashi, A. Odajima, K. Nakamura, and N. Murayama, *Polymer Prepr. Jpn. 27*:364 (1978).

259. N. Takahashi, T. Ishibashi, and A. Odajima, Poling effect on fine structure of PVDF β crystal, *Polymer Prepr. Jpn. 27*:1788 (1978).

260. Y. Wada and R. Hayakawa, A model theory of piezo- and pyroelectricity of poly(vinylidene fluoride) electret, *Ferroelectrics 32*:115 (1981).

261. S. Tasaka and S. Miyata, Electrostriction and piezoelectric effects in poly(vinylidene fluoride), *Koubunshi Ronbunshu 36*:689 (1979).

262. F. J. Boerio and J. L. Koenig, Roman scattering in nonplanar poly(vinylidene fluoride), *J. Polymer Sci. Part A-2 7*:1489 (1971).

263. T. Furukawa, G. E. Johnson, H. E. Bair, Y. Tajitsu, A. Chiba, and E. Fukada, Ferroelectric phase transition in a copolymer of vinylidene fluoride and trifluoroethylene, *Ferroelectrics 32*:61 (1981).

3
Ferroelectric, Pyroelectric, and Piezoelectric Properties of Poly(vinylidene Fluoride)

R. Glen Kepler
Sandia National Laboratories, Albuquerque, New Mexico

I. INTRODUCTION

In the early 1970s it was discovered that poly(vinylidene fluoride) (PVDF) is ferroelectric. The structure of the repeating unit of PVDF is $-CH_2-CF_2-$. It is not a new material, having been first polymerized in the 1940s; and at the time of the discovery of its ferroelectric properties, it was produced commercially in very large quantities for a variety of applications. The major high-volume commercial applications include use as a base for durable long-life coatings for exterior finishes, as a chemically inert material for use in chemical processing equipment, and as an electrical insulator. PVDF is a crystalline polymer with a crystalline fraction of about 50% and high mechanical and impact strength, particularly in oriented fibers and films. It is resistant to creep under load and to fatigue after repeated flexure.

The discovery of the fact that PVDF is ferroelectric resulted from studies of polymer electrets, materials in which an effective macroscopic separation of positive and negative charges can be produced by the application of an electric field either by orienting permanent dipoles or creating a space charge by injecting free charges. In 1969, Kawai [1] found that PVDF exhibited an unusually large piezoelectric effect after poling, the process of inducing an effective macroscopic separation of charge by the application of an electric field. A typical poling process consists of heating a polymer film to some high temperature, applying an electric field, and then lowering the temperature to ambient temperature with the field applied. This observation led Bergmann and co-workers [2–4] to study the pyroelectric and nonlinear optical properties of PVDF films and to be the first to speculate about the possibility that PVDF is ferroelectric.

Since there are inconsistencies in the usage of the term "electret," particularly relative to polymer ferroelectrics, and many workers in the field have referred to PVDF as an electret, a brief discussion of the definitions used in this chapter is provided here.

An electret is a material into which a polarization can be induced by the application of a high electric field. Sessler [5] defines an electret as ''a piece of dielectric material exhibiting a quasi-permanent electrical charge. The term 'quasi-permanent' means that the time constants characteristic for the decay of the charge are much longer than the time periods over which studies are performed with the electret.'' Lines and Glass [6] state that ''Generally speaking, electret is used to describe any material in which a polarization persists after the application of an electric field.'' We believe that the more restrictive definition of Sessler is more appropriate, since it implies a material which is not in thermal equilibrium and thus excludes ferroelectrics. Heaviside [7] introduced the term ''electret'' in 1892, by analogy to ''magnet,'' when very little was known about the origin of polarization in solids; and studies of polarization in materials which were referred to as electrets have been done almost exclusively on materials in which the polarization is not in thermal equilibrium. Electrets are very important technological materials which play a major role in microphones. The lifetime of the polarization can be very long even though it is not in thermal equilibrium.

Ferroelectrics, on the other hand, are defined as crystalline materials in which the unit cell is polar and the direction of polarization can be changed by the application of an electric field [8]. The polarization is, therefore, in thermal equilibrium. The discovery of the very interesting properties of PVDF resulted from studies of polymer electrets, and even though it is now well established that it is ferroelectic, as we will show below, some workers continue to refer to it as an electret. Although some consider ferroelectrics to be a subset of electrets, by our definitions they are not.

The rest of this chapter is divided into six major sections: ''the Polymer,'' ''Ferroelectricity,'' ''Pyroelectricity,'' ''Piezoelectricity,'' ''Nonlinear Optical Properties,'' and ''Summary.'' The next section, titled ''The Polymer,'' consists of a brief discussion of the physical properties of PVDF and of some applications that are not related to its ferroelectric properties. It is divided into three subsections: one on the chemical structure, one on the solid-state structure, and on physical properties and applications. The following major section is titled ''Ferroelectricity.'' Since it is very unusual to find a ferroelectric polymer and some readers may be somewhat skeptical, the first subsection consists of a discussion of the experimental results which prove that PVDF is, indeed, ferroelectric. The following subsections are titled ''Theory of Chain Reorientation,'' ''Kinetics of Poling,'' ''Remanent Polarization—Experiments,'' ''Remanent Polarization—Theory,'' and ''Ferroelectric Phase Transition.'' The pyroelectricity exhibited by PVDF films is the next major subject. It is divided into four subsections: ''Definition and Measurements,'' ''Primary and Secondary Pyroelectricity,'' ''Reversible Changes in Crystallinity,'' and ''Theoretical Models.'' The title of the next major section is ''Piezoelectricity.'' It is divided into four subsections. ''Definitions,'' ''Measurement Techniques,'' ''Experimental Results,'' and ''Theoretical Models.'' The last two major sections are quite short and are not divided into subsections. The first of these is a discussion of some work on the nonlinear optical properties of PVDF, and the final major section is a brief summary of the most significant results presented in this chapter.

II. THE POLYMER

A. Chemical Structure

The chemical structure of PVDF is quite simple. The monomer unit consists of two atoms each of carbon hydrogen and fluorine, $-CH_2-CF_2-$. Polymerization was first disclosed in 1944 [9]. In this chapter we will not attempt to discuss the chemistry of PVDF and refer the reader to other chapters in this book and to a review by Lovinger [10] if interested. Typical molecular weights of commercial materials is on the order of 4×10^6 or 60,000 monomer units.

A common defect in the polymer chain is created during polymerization when, like carbons, e.g., the carbons bonded to two fluorine atoms, of the monomer bond to one another creating head-to-head (HH), ($-CF_2-CF_2-$), or tail-to-tail (TT), ($-CH_2-CH_2-$), defects. Such defects generally occur in pairs, a TT defect immediately following a HH defect, so the concentration of the two types are generally the same. The concentration of these defect pairs in commercial materials is typically 4–5%, and the defect concentration can be varied from about 3.5% to 6% under practical synthesis conditions. Polymers with defect concentrations ranging from 0.2% to 23.5% have been synthesized [11,12] and in the next section we will discuss the role of these defects in the physical properties of PVDF.

B. The Solid State

In the solid state, PVDF is crystalline with a crystallinity of approximately 50%. Four polymorphs are well established, referred to in this chapter as α, β, γ, and δ. There have been a large number of studies of the crystal structures of PVDF, and considerable nomenclature confusion has resulted. Many refer to the four polymorphs α, β, γ, and δ as phases II, I, III, and IV, respectively. We will follow the recommendations of Lovinger [10] and use the Greek alphabet symbols. When PVDF is cooled from the melt, the crystalline phase formed is the nonpolar α phase, the crystal structure of which is shown in Figure 1. The molecules are in a distorted trans-gauche-trans-gauche' (TGT$\overline{\text{G}}$) con-

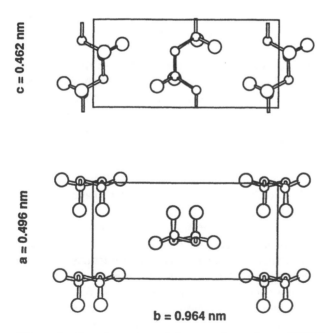

Figure 1 Crystal structure of the nonpolar α phase of PVDF. (Redrawn from Ref. 67.)

formation. The (TGTḠ) conformation is distorted because of steric hindrance between the fluorine atoms. A drawing of a space-filling model is shown in Figure 2. Since there is a large dipole moment associated with the carbon–fluorine bond, it is readily seen that in this conformation there is a net dipole moment perpendicular to the chain axis. In the crystal the unit cell is nonpolar, however, because there are two chains associated with each unit cell, and the dipole moments of the two chains are directed in opposite directions, as is readily seen in the bottom half of Figure 1.

The crystalline phase of most interest for ferroelectricity is the polar β phase shown in Figure 3. The polymer chains in this phase are in a distorted, planar zigzag, all-trans (TT) conformation. The planar zigzag conformation is distorted because the fluorine atoms are too large to allow a simple all-trans conformation, and it is believed that the atoms are statistically offset as indicated by the dotted lines in Figure 3. A drawing of a space-filling model of a segment of a polymer chain in this conformation is shown in Figure 4. As can be seen from Figures 3 and 4, this crystal structure results in a polar unit cell with a large dipole moment.

Since the crystalline phase formed when the polymer is cooled from the melt is the α phase, the lowest-energy molecular conformation is (TGTḠ). As mentioned in the previous section, the polymer, as usually synthesized, contains on the order of 5% HHTT defects, and the role of these defects in the physical properties of PVDF, particularly their effect on the energy of the different molecular conformations, has been of considerable interest. Farmer et al. [13] conducted a theoretical investigation of the effect of the density of these defects on the energy of the all-trans and the (TGTḠ) conformations. Both steric and electrostatic effects were included in the calculations. The steric potential energy was calculated numerically using the potential energy functions proposed by DeSantis et al. [14], and the electrostatic energy was calculated by assuming that each atom was charged and summing pairwise over the atoms. The calculated potential energies of the two chain conformations as a function of defect density are shown in Figure 5. At low defect concentrations this theory predicts that the (TGTḠ) is the most stable, as is observed experimentally, but that at high defect concentrations the all-trans conformation is most stable.

In 1966 Lando and Doll [15] speculated that the presence of these defects would have an effect on the polymorphism of PVDF and studied the crystal structures obtained in a variety of conditions with polymers containing varying amounts of either trifluoroethylene or tetrafluoroethylene units to simulate the effect of higher HHTT defect concentrations. They observed that the introduction of these units into the PVDF polymer chain increased the tendency of the polymer to crystallize in the β phase. The β phase was obtained by cooling from the melt when the concentration of tetrafluoroethylene units was above about 7 mol%. As discussed elsewhere in this book, these copolymers of vinylidene fluoride with either trifluoroethylene or tetrafluoroethylene have now been found to have very interesting and useful ferroelectric properties.

Cais and Komentani [11] have synthesized PVDF molecules with defect concentrations ranging from 0.2 to 23.5 mol%, and the effect of the defects on the crystal structures obtained has been studied [12]. It was found, in agreement with the predictions of Farmer et al. [13], that below a defect concentration of about 11 mol% the α phase was obtained by quenching from the melt, while above that concentration the β phase was obtained.

In addition to the α and β phases, two other polymorphs are well established, γ and δ. Their crystal structures are shown in Figure 6. Both of these structures have polar unit cells, but the dipole moments are smaller than that in the β phase. In the γ phase the

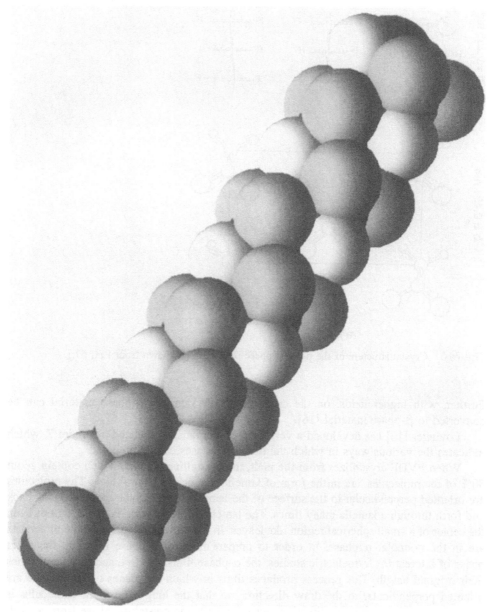

Figure 2   A space-filling model of a segment of a PVDF molecule in the α phase.

molecule is in a $(T_3GT_3\overline{G})$ conformation, and in the δ phase the molecular conformation is the same as that in the α phase. The major difference between the α and δ phases is the orientation of the dipole moments of the two molecules in the unit cell. In the δ phase they are parallel and in the α phase they are antiparallel. It is an interesting fact that, as we will discuss later in this chapter, α-phase material can be converted into δ-phase material simply by applying a high electric field [16], on the order of 100 MV/m.

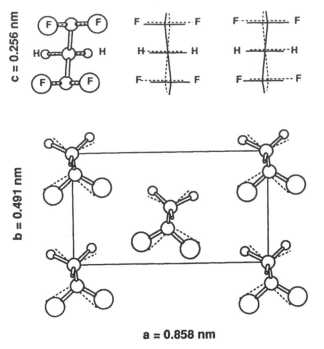

Figure 3   Crystal structure of the polar β phase of PVDF. (Redrawn from Ref. 67.)

Further, with higher fields, on the order of 500 MV/m, the δ-phase material can be converted to β-phase material [16].

Lovinger [10] has developed a very helpful diagram, reproduced in Figure 7, which indicates the various ways in which the different phases can be prepared.

When PVDF crystallizes from the melt, the crystalline regions, which contain about 50% of the molecules, are in the form of lamellae, typically 10 nm thick. The molecules are oriented perpendicular to the surface of the lamellae and a given molecule runs back and forth through a lamella many times. The lamella form spherulites, radiating out from the center of a small spherical region like leaves. In films prepared this way, the molecules are in the nonpolar α phase. In order to prepare films of β-phase material, the polar phase of interest for ferroelectric studies, the α-phase films are stretched to several times their original length. This process produces films in which the planes of the lamella are oriented perpendicular to the draw direction, so that the long axis of the molecules is oriented parallel to the draw direction and it converts the crystalline phase from α to β.

Films of even greater order can be prepared by rolling. X-ray pole figures show that this process can produce β-phase films in which the crystallites are oriented with the polymer-molecule axis, the *c* axis, parallel to the rolling direction. In addition, the dipole moments of the unit cells, which are parallel to the *b* axis, are in the plane of the film, and the *a* axis is perpendicular to the plane of the film [17].

C.   Physical Properties and Applications

As mentioned in the introduction, PVDF is used in many applications which are completely unrelated to its ferroelectric properties. It is very durable, with high mechanical

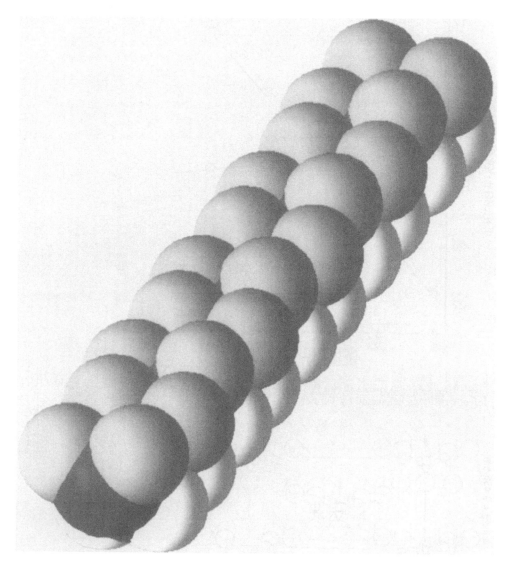

Figure 4   A space-filling model of a segment of a PVDF molecule in the β phase.

and high impact strength. In oriented films and fibers it exhibits exceptional mechanical strength. It exhibits excellent resistance to abrasion, superior resistance to deformation under load, and exceptional resistance to repeated flexure or fatigue. PVDF can be processed by essentially all the normal methods known for thermoplastics and is machinable; it is resistant to chemical attack and is essentially unaffected by sunlight. This combination of properties has led to a wide variety of applications. The fact that it is unaffected by sunlight and is tough has led to its use in durable, long-life finishes for exterior metal siding. Its resistance to creep and the fact that it has adequate electrical properties has led to many applications as an electrical insulator, particularly as a wire coating. Its resistance to chemical attack and processibility has led to many applications in chemical

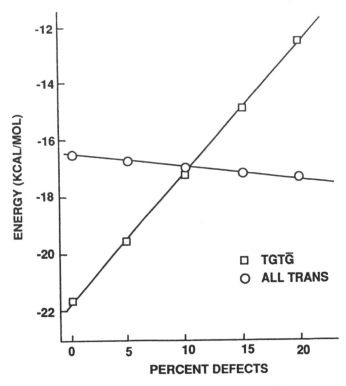

Figure 5 Potential energy of the all-trans and (TGTḠ) chain conformations plotted against HHTT defect density as calculated by Farmer et al. [13].

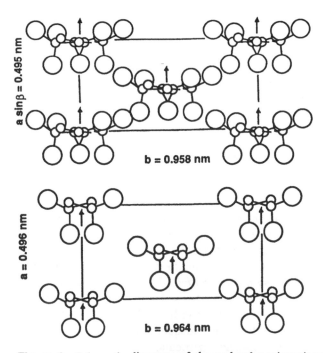

Figure 6 Schematic diagrams of the molecular orientations in the γ phase (upper) and the δ phase (lower). The arrows indicate the orientation of the dipole moments of the molecules.

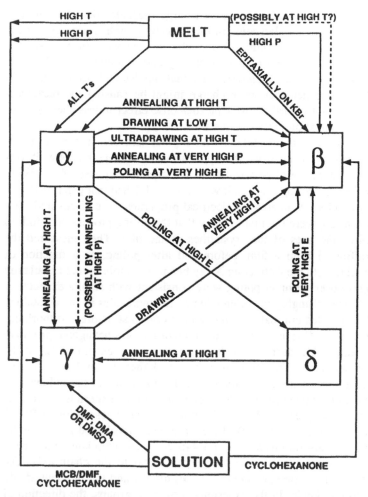

Figure 7 The interrelations among the four well-established phases of PVDF. DMF, DMA, DMSO, and MCB stand for dimethylformamide, dimethylacetamide, dimethyl sulfoxide, and monochlorobenzene, respectively, which along with cyclohexanone are solvents which are used for solvent casting. (From Ref. 10.)

processing equipment. Many of the physical properties, processing techniques, and applications of PVDF have been summarized by Dohany et al. [18]

III. FERROELECTRICITY

A. Proof of Ferroelectricity

We will begin our discussion of the ferroelectric properties of PVDF by first presenting the evidence that it is, in fact, ferroelectric. The early work by Kawai [1] involved poling thin films by applying a high voltage between electrodes on the two surfaces at an elevated temperature and then lowering the temperature to room temperature with the field applied. The fact that he was able to observe a large piezoelectric effect in such films showed that he had produced polarization in the films by orienting dipoles, but the experiment provided no evidence regarding whether the dipoles were in a crystalline phase or not. Similarly, the observation of pyroelectricity and nonlinear optical effects in similarly prepared PVDF films [2–4] showed only that dipoles had been oriented in the films. Some workers suggested that space charge might be causing the piezoelectricity, but space charge cannot contribute to a piezoelectric effect unless there are inhomogeneities in the mechanical properties of the film.

Since a ferroelectric material is crystalline, with a polar unit cell in which the direction of polarization can be changed by the application of an electric field, it was necessary to show that the direction of polarization of the crystalline phase of the PVDF films was being changed by the poling process. It was well established that β-phase PVDF has a polar unit cell long before its interesting electrical properties were discovered [19].

Kepler and Anderson [20] were the first to show that the poling process does indeed change the orientation of the axes of the crystallites in the film. They measured the intensity of X rays diffracted from a film, before and after poling, as a function of orientation of the film relative to the diffracting X-ray beam. A simple model calculation showed that the changes observed upon poling were consistent with results expected if the applied electric field was changing the orientation of the molecules in the crystalline region of the films. They pointed out that even though the crystal structure is orthorhombic, the lower-symmetry unit cell results from a small distortion of a hexagonal primitive lattice, the distortion being a 1% decrease in the separation between molecules in the direction of the dipole moments. Therefore, they assumed that the molecules in the crystallites could rotate in increments of 60° about their long axes under the influence of the applied electric field and obtained reasonable agreement with the experimental results.

Tamura et al. [21] and Naegele and Yoon [22] studied infrared absorption in PVDF films and concluded that the dipoles in the β phase were being reoriented reversibly by the poling process. Infrared absorptions at 512 and 446 cm^{-1} are associated with the β phase, and their transition moments are oriented perpendicular to the chain axis. The transition moment at 512 cm^{-1} is oriented along the CF_2 dipole, and the moment at 446 cm^{-1} is perpendicular to the dipole. In the experiments by both groups, the direction of travel of the incident light was parallel to the applied field, so that when dipoles are aligned in the direction of the field, the absorption at 512 cm^{-1} is smaller than the absorption when the dipoles are perpendicular to the applied field, while that at 446 cm^{-1} is larger. Tamura et al. compared unpoled, poled, and thermally depoled absorption spectra and correctly interpreted their results as showing that the direction of polarization of the β-phase crystals was being changed by the application of the electric field but also concluded that "the β crystals surrounded by the amorphous region exhibit ferroelectric-like behavior; the β crystal is not thought to be a ferroelectric crystal." They did not make it clear why they thought the β phase is not ferroelectric.

Naegele and Yoon [22] measured the intensity of absorption at the two wavelengths as they varied the applied field through more than a complete hysteresis loop. They also found that the electric field was reorienting the dipoles in the crystalline phase. In addition, the fact that they were able to observe changes in transmission as they reversed the direction of polarization showed that the polarization process involved intermediate steps as suggested by the 60° increment model of Kepler and Anderson [20]. Their results ruled out polarization reversal by rotation in a 180° step as hypothesized by some. Their results are shown in Figure 8.

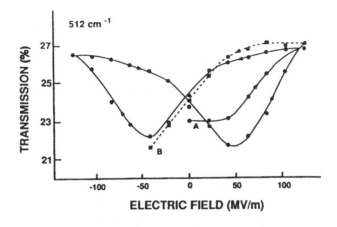

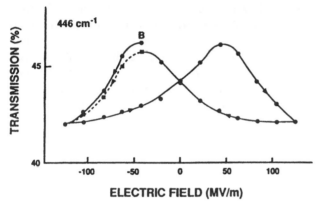

Figure 8 Dependence of infrared transmission at 512 cm⁻¹ (top) and 446 cm⁻¹ (bottom). The experiment started with an unpoled sample at point A in the top part of the figure. The applied electric field vas varied in increments of 21 MV/m in the direction indicated by the arrows. Point B in the top half of the figure is the same as point B in the bottom half and is the point in the experiment at which the wavelength was changed (From Ref. 22.)

Takahashi and Okajima have carried out a much more thorough investigation of the film orientation dependence of diffracted X-ray intensity on poled and unpoled samples than the investigation of Kepler and Anderson [20]. Among other things, they observed more diffraction peaks and corrected their data for absorption. They concluded that their data were consistent with the 60° increment model and also that the poling process led to a growth of the crystallites parallel to the film surface. Takahashi et al. [23] have also observed the few percent change in intensity of the diffracted X rays expected in a polar crystal when the direction of polarization is changed by 180°, providing direct evidence for polarization by an applied field.

The most convincing evidence for dipole reorientation and the 60° increment model was provided by Bur et al. [17]. They measured the intensity of X rays diffracted from a 1-mm-diameter cylinder machined from a PVDF 1-mm-thick film which had been stretched by rolling and poled at the same time. Films produced by rolling have a β-phase, single-crystal-like texture. The chain axis (c) is oriented in the rolling direction, as it is in stretched film, and the b axis (parallel to the dipoles) tends to be oriented in the plane of the film. Bur et al. [17] measured the intensity of X rays diffracted from the 1-mm-diameter cylinder at $2\theta = 20.7°$, the angle at which X rays are diffracted from the 110 and 200 planes in the β-phase PVDF crystals, as a function of the angle of rotation of the cylinder about its axis. The results expected for an unpoled sample and for unit cells rotated by 60° during poling, as predicted by the 60° increment poling model, are shown in the center of Figure 9. The direction corresponding to $\phi = 0$ in Figure 9 is parallel to the original plane of the film.

The experimental results obtained by Bur et al. [17] for a poled and an unpoled sample are shown in the bottom and top, respectively, of Figure 9. The unpoled sample was obtained by heating an initially poled sample to 152°C, holding it at that temperature for 10 min, then cooling to room temperature. It is easy to see that the data for the unpoled sample (top of Figure 9) correspond to the results expected for single crystal texture film depicted in the center of Figure 9 by the curve labeled (a). The orientation of the unit cells in an unpoled sample is shown schematically in the drawing to the right of curve (a). Apparently there are constraints of some sort that force the molecular chains to return to the orientation induced by the rolling process when the sample is thermally depoled. It is not obvious that would happen, but these results clearly show that it does.

The solid line through the data obtained on a poled sample at the bottom of Figure 9 is a least-squares fit of a mathematical model to the experimental data. In the model it was assumed that the intensity of the diffracted X rays varied with ϕ as

$$I(\phi) = CD(\rho) \sum A_i \exp\{\rho[\cos(\phi - \phi_i) - 1]\} \tag{1}$$

where

$$\phi_i = \phi_o + 60_i - 210$$

$$A_1 = A_3 = A_4 = A_6 = \frac{P}{4} + f\left(1 - \frac{P}{2}\right)$$

and

$$A_2 = A_5 = \frac{(1 - P)}{2} + fP$$

The quantity P represents the fraction of (200) plane normals that are aligned with the piezoelectrically active ($i = 1, 3, 4, 6$) directions in the specimen, and $f = [S_{110}/S_{200}]^2$, where S_{hkl} is the structure factor for the hkl reflection. $D(\rho)$ is a normalization constant chosen so that

$$1 = D(\rho) \int_{-\pi}^{\pi} \exp[\rho(\cos \phi - 1)] \, d\phi$$

and C is a scaling factor.

The fit to the data indicates that the full width at half-maximum of the distribution about each maximum is about 60°, reflecting the considerable overlap between the peaks observed in the data. This is not surprising, considering the way the original orientation of the crystallite axes was produced. In spite of the wide distribution, it is quite clear from these experimental data that the polymer chains reorient in an electric field by rotating in 60° increments about the chain axes in the crystallites and that PVDF is, therefore, ferroelectric.

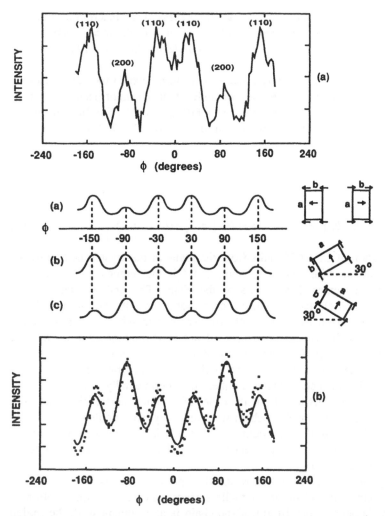

Figure 9 Intensity of X-rays diffracted from a cylindrical sample machined from a rolled film of PVDF 1 mm thick: (top) data obtained on a sample which had been initially poled and subsequently heated to 152°C for 10 min to unpole it; (bottom) data obtained on a poled sample. In the middle, between the two sets of experimental data, is a schematic diagram of the diffracted X-ray intensity expected versus angle of rotation for (a) an unpoled sample with a single-crystal texture, (b) a sample in which the unit-cell *b* axis has been rotated 60° clockwise and a sample in which the unit cell axis has been rotated 60° counterclockwise. A poled sample would consist of a sum of the (b) and (c) orientations in the 60° increment model. The unit cell orientations for each of the curves *a*, *b*, and *c* are depicted to their right. (Adapted from Ref. 17.)

B. Theory of Chain Reorientation

Aslaksen [24] was the first to develop a theoretical model of polymer chain reorientation in PVDF. He assumed that the temperature dependence of the polarization, the pyroelectric effect that had been recently reported, arises from the torsional oscillations of the macromolecules. Three torques were considered: a phenomenological one due to nearest-neighbor interaction, one due to the torsional stiffness of the macromolecule, and one due to the interaction of the permanent dipole moment with the local electric field. He concluded that molecular reorientation in an applied electric field could take place in 180° increments.

Dvey-Aharon et al. [25] subsequently undertook a more detailed theoretical examination of polarization reversal in PVDF and concluded that it cannot occur by a 180° rotation of the chain segments but that it can occur in 60° increments. For the 180° model they started with a phenomenological Hamiltonian of the form $H = T + U$, where

$$T = \frac{1}{2} I \sum \dot{\theta}_i^2 \tag{2}$$

and

$$U = \sum_i \left\{ A_1 [1 - \cos(\theta_i)] + A_2 [1 - \cos(2\theta_i)] + \frac{1}{2} k(\theta_i - \theta_{i+1})^2 \right\} \tag{3}$$

θ_i is the angle of rotation of CF_2–CH_2 units from the plane of the crystalline b axis, the dot signifies differentiation with respect to time, and I is the moment of inertia of a monomer unit about the center-of-mass axis of the chain. The first two terms in U represent the interchain and electric field forces on a monomer unit, and the third is the torsional rigidity of the chain. A_1 consists of two terms, one which depends on the orientation of neighboring chains Δ and one which depends on the product of the applied field E and the dipole moment p of the monomer unit. For it to be energetically favorable for a chain to rotate by 180° in this model, the applied field must be greater than Δ/p. Dvey-Ahron et al. [25] estimated, using a number of experimentally determined constants, that it would take an electric field of the order of 10 times greater than that used experimentally to meet this criterion for a chain surrounded by like-oriented chains. They suggested that it would be possible to reorient only those chains at a domain walls where half the nearest neighbors are oriented parallel to the applied field and half are antiparallel. In that environment, $\Delta = 0$.

The mechanism for chain reorientation which they proposed involves thermally activated formation of kinks in a chain at a lamella surface followed by rapid motion of the kink through the lamella, about 10 nm, if the chain is at a domain wall, the region in a crystallite where the chains on one side are oriented in one direction, and those on the other side are directed in the opposite direction. They showed that the motion of the kink, a soliton, through the lamella would be rapid compared with formation time of kinks. For the 180° model, they concluded that the energy required to form a 180° kink was so large that the waiting time for the formation of kinks was too long to make that model viable.

As an alternative, they considered the 60° model and modified their equations accordingly. In this case they found that there is an intermediate minimum in the potential at 30° which has a major impact on the ease with which chains can reorient. A domain boundary between two differently oriented regions is shown schematically in Figure 10, and the potential energy of the chain labeled 1 in the middle of the dashed hexagon in Figure 10 is plotted as a function of the chain rotation angle θ in Figure 11. The intermediate minimum in the potential in this model allows the formation of kinks and rotated segments in the middle of chains such as the one labeled 1 in Figure 10, and the kinks at each end of a rotated segment can rapidly propagate along the chain to produce a 60° chain rotation.

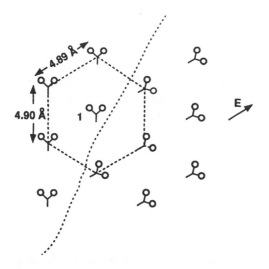

Figure 10 Domain boundary between regions of β-phase PVDF differing in the direction of polarization by 60°. (From Ref. 25.)

C. Kinetics of Poling

In early work on poling it was thought that the process was quite slow. It was typically achieved by heating samples to some high temperature, usually above 80°C, applying a high electric field, and then cooling to room temperature with the field applied. The

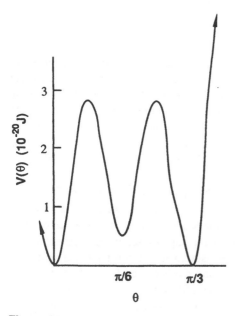

Figure 11 Potential energy versus angle of rotation θ for the PVDF molecule labeled 1 in the dashed hexagon in Figure 10. The zero for θ is the vertical direction. (From Ref. 25.)

effects of poling time, electric field, and poling temperature were investigated by a number of groups. [26–31]. It was found that the magnitude of the induced polarization and of the piezoelectric coefficient increases with increasing voltage and temperature, about as was expected within the electret hypothesis which tended to drive these studies.

Studies of the kinetics of poling, by measuring the time dependence of polarization reversal currents or of the pyroelectric coefficient after the application of a step voltage, have proved to be much more fruitful. Buchman [32] measured polarization-reversal switching currents at temperatures above 110°C and fields up to about 14 MV/m. The switching currents he observed initially increased for a time after a reversal of the applied voltage, went through a maximum, and then decreased to a low stable value. He found that he could fit the electric field E and temperature T dependence of the maximum switching current observed, J_{max}, with an empirical formula of the form

$$J_{max} = J_0 E^n \exp\left(-\frac{\alpha}{k_B T}\right) \tag{4}$$

with n varying from 0.7 to 1.6, $\alpha = 0.65$ eV, and k_B the Boltzmann constant. The reciprocal of the time at which the maximum current occurred τ_{max}^{-1} varied as

$$\tau_{max}^{-1} = \tau_0^{-1} E^{n'} \exp\left(\frac{b}{k_B T}\right) \tag{5}$$

where n' was on the order of or greater than n and β varied from 1.1 to 1.5 eV. At 140°C and an electric field of 14 MV/m, τ_{max} was about 30 s.

Blevin [31] studied the kinetics of the poling process by measuring the pyroelectric coefficient as a function of time during poling at a field of 100 MV/m. He reported poling times that ranged from 30 s for a poling temperature of 100°C to 5.5 days at a poling temperature of 20°C. Southgate [33] was the first to report that poling could take place at room temperature as well as at lower temperatures. He used a corona discharge to apply very high fields and reported that complete poling could be achieved in less than 1 s at room temperature. Hicks and Jones [34] also attempted to measure the poling time and reported room-temperature times of a few seconds.

In view of the very long poling times observed experimentally, Dvey-Aharon et al. [25] were somewhat dismayed when they estimated the poling time they would expect based on their 60° increment chain-rotation model and found that it predicted times on the order of 1 ms. Their estimate was based on the assumption that the approximately 10^7 chains in a crystallite would have to reorient roughly sequentially, because at any given time only a very few chains could be at a domain boundary. They rationalized their results by pointing out that their model was highly idealized and, thus, likely to provide only a lower bound to the actual poling time.

However, shortly after Dvey-Aharon et al.'s theoretical paper appeared, Furukawa and Johnson [35] published a very careful experimental study of switching time which shed new light on the polarization process. They measured the time dependence of electric displacement in a sample after the application of a step voltage using a circuit which allowed them to apply higher fields and make measurements at shorter times than previous studies. They used fields up to 200 MV/m and studied the effect of sample temperature from −100°C to 20°C. The circuit they used allowed them to measure displacement charge versus time following the application of a very-fast-rising step voltage down to times on the order of a microsecond. They observed a switching time of 4 μs at 20°C and 200 MV/m for films 7 μm thick.

These observations prompted Clark and Taylor [36] to reexamine the kinetic predictions of the 60° increment kink model of Dvey-Aharon et al. [25]. They assumed that the N polymer chain segments forming a crystallite reoriented sequentially in an electric field E and that the rate-limiting process was the formation of kinks. The energy barrier to kink formation was taken to be $U-\lambda E$, where U is the zero field barrier and λ is a constant of proportionality having the units of a dipole moment, and thus the rate of kink formation q was assumed to be

$$q = q_0 \exp\left[\frac{-(U - \lambda E)}{k_B T}\right] \tag{6}$$

The rate of change with time of the probability that n and only n chain segments have reoriented, $p_n(t)$, is equal to the rate at which regions with n chain segments reoriented are created minus the rate with which regions with n chain segments reoriented are lost:

$$\frac{\partial p_n}{\partial t} = q[p_{n-1}(t) - p_n(t)] \tag{7}$$

The solution to this equation is

$$p_n(t) = \frac{(qt)^n \exp(-qt)}{n!} \tag{8}$$

and if N chain segments with dipole moment μ must reorient to pole a certain crystallite, the dipole moment of that crystallite or crystalline region increases monotonically from 0 to $N\mu$. The average moment of a crystallite of N segments at time t, $P_N(t)$, is thus

$$P_N(t) = \sum_{n=0}^{N} n\mu p_n(t) + \sum_{n=N+1}^{\infty} N\mu p_n(t) \tag{9}$$

The total dipole moment of the sample $P(t)$ is then the sum of P_N over the distribution $Q(N)$ of N for all the crystallites:

$$P(t) = \Sigma_N Q(N) P_N(t) \tag{10}$$

Clark and Taylor [36] evaluated this expression for a variety of $Q(N)$ and obtained results which were all in qualitative agreement with the experimental results of Furukawa and Johnson [35] when the time dependence of the dielectric constant of PVF$_2$ is taken into account. The experimental results yielded 0.7 for the width at half-height for $\partial D(t)/\partial$ $(\log_{10} t)$, where $D(t)$ is the experimentally measured displacement,

$$D(t) = \kappa_0 \kappa(t^{-1}) E + P(t) \tag{11}$$

with κ_0 being the permittivity of free space and $\kappa(t^{-1})$ the real part of the frequency-dependent dielectric constant. The theoretical predictions range from 0.5 for very sharp distributions $Q(N)$ to 1.0 for very broad distributions. The agreement between the experimental results and the theoretical predictions for the intermediate width distribution

$$Q(N) \propto N \exp\left(\frac{-N}{N_0}\right) \tag{12}$$

are shown in Figure 12.

This theory predicts that the logarithm of the time t at which $\partial D(t)/\partial(\log_{10} t)$ is a maximum should vary linearly with the inverse of temperature. The appropriate plot of

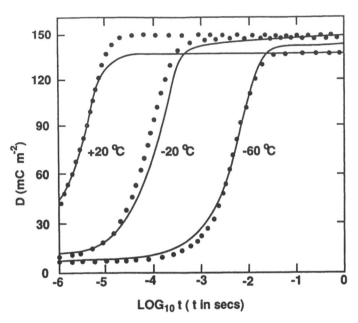

Figure 12 A comparison of the theory of Clark and Taylor (solid dots) with some of the experimental results of Furukawa and Johnson. (From Ref. 36.)

some of Furukawa and Johnson's data is shown in Figure 13, and it can be seen that this is approximately the case. The slopes of these lines give experimental values for U and λ of 18 kcal mol and $(230 \pm 50) \times 10^{-30}$ Cm, respectively. The intercept of the lines with the axis at infinite temperature gives N_0/q_0 as approximately 50 fs, and that value was used to calculate the theoretical points plotted in Figure 12.

Dvey-Aharon et al. [25] estimated that a kink in a chain is three or four monomer units long, and that for a 60° twist the interchain energy barrier is approximately 4 kcal/mol per monomer unit. The intrachain energy for a 60° twist was estimated to be about 1 kcal/mol, giving a total twist energy of about 13 kcal/mol. This value is sufficiently close to the experimental value of 18 kcal/mol to add significant support to the model.

The parameter λ has the units of a dipole moment and represents the reduction per unit electric field of the energy necessary to create a kink. The dipole moment of the monomer unit is about 30×10^{-30} Cm (this value will be discussed in more detail in the next section), considerably smaller than the experimentally determined value of $\lambda = 250 \times 10^{-30}$ Cm. Clark and Taylor [36] argue, however, that it is only one contribution to λ. Another is a reduction of steric hindrance to poling due to the slight twisting by the applied field of the unpoled nearest-neighbor molecules. Since steric hindrance is given by a Lennard-Jones potential, it is very sensitive to position. Clark and Taylor suggest that this effect will make a large contribution to λ.

Studies of many inorganic ferroelectrics have shown [6] that in many materials the switching time t_s follows either an exponential law,

$$t_s \sim \exp\left(\frac{\alpha}{E}\right) \tag{13}$$

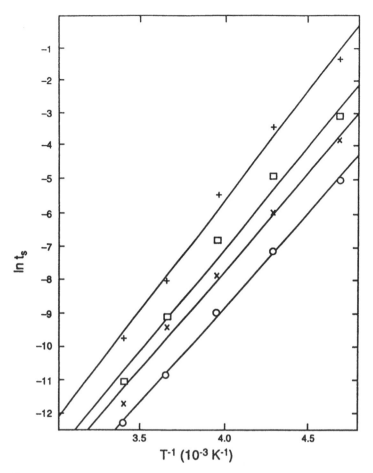

Figure 13 The logarithm of the switching time t_s, which is the time at which $\partial D(t)/\partial(\log_{10} t)$ is a maximum, versus the inverse of the temperature T. Circles are data taken with the electric field $E = 140$ MV/m; crosses, 160 MV/m; squares, $E = 180$ MV/m; plus signs, $E = 200$ MV/m. (From Ref. 36.)

or a power law,

$$t_s \sim E^{-n} \tag{14}$$

and phenomenological models have been developed, assuming nucleation-and-growth processes, which give these forms. Furukawa and Johnson [35,37] have compared their experimental data with these empirical relations, and the values of α and n were found to be larger than those for better-known ferroelectrics, $\alpha = (-2) \times 10^3$ MV/m and $n = 8-10$ for PVF$_2$ versus 1 MV/m and 1.5 for BaTiO$_3$ [6,38–40].

The kink propagation model implies that log t_s varies linearly with E rather than with $1/E$ as implied by Eq. (13). The logarithm of the switching time data of Furukawa and Johnson [35] are plotted versus $1/E$ in the top of Figure 14 and versus E in the bottom of Figure 14. Superficially the data agree equally well with either functional form. Clark and Taylor [36] have carried out a statistical analysis of the data and concluded

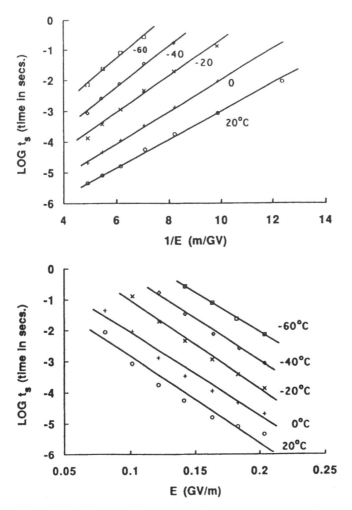

Figure 14 The logarithm of the switching time t_s at several temperatures versus (top) the reciprocal of the applied field E and (bottom) the applied field E. (Data from Ref. 35.)

that the formulas are equally good for fields from 120 to 200 MV/m; but for lower fields, down to 80 MV/m, both formulas break down, with the $1/E$ dependence breaking down more slowly.

Odajima et al. [41] have investigated the nucleation and growth hypothesis in more detail, attempting to combine it with aspects of the kink propagation model [25]. Al-Jishi and Taylor [42], however, have calculated the local electric field and concluded that it is sufficiently strong to pose difficulties for any interpretation of switching in terms of a macroscopic model of nucleation and growth. Takase and Odajima [43] and Takase et al. [44] have investigated the effect of large doses of radiation on the switching process and suggested that two nucleation processes are involved.

Polarization and polarization switching in ferroelectrics is a very complicated process and is not clearly understood in any material. Very large electric fields are involved. If two adjacent crystals of PVDF are stacked one on top of the other, with their polarizations parallel and vertical, and the polarization of one crystal is suddenly reversed, the change in the electric field at the interface of the two crystals would be on the order of 10^4 MV/m, clearly greater than the dielectric strength of PVDF. The effect of such fields have to be taken into account, but it is not clear how to do so. Many studies have been conducted on domains in more familiar ferroelectrics [6], but none has been reported for PVDF. Spiking can be observed in the poling current [45], which is reminiscent of Barkhausen pulses [6], but detailed investigations have not been undertaken.

In summary, the poling process in PVDF, is similar in many ways to that in more conventional ferroelectrics, but the kink propagation process envisioned by Dvey-Aharon et al. [25] is almost certainly involved. Many detailed aspects of the predictions of that model are in surprisingly good agreement with experimental results.

D. Remanent Polarization—Experiments

The magnitude of the saturation remanent polarization of PVDF, the polarization remaining after it has been poled and the applied field turned off, has been measured, primarily by measuring hysteresis loops, by many groups [46–50], only a few of which are referenced. At room temperature the conductivity of PVDF is quite low, and hysteresis loops can be obtained simply and at very low frequencies. Typical results are shown in Figure 15. The data shown were obtained on a poled sample by sweeping the electric field through one cycle in a symmetrical triangular wave pattern with a peak height of 280 MV/m and a period of 30 s. The data shown were obtained from one of a series of cycles. In this study, measurements were made on several samples of both uniaxially

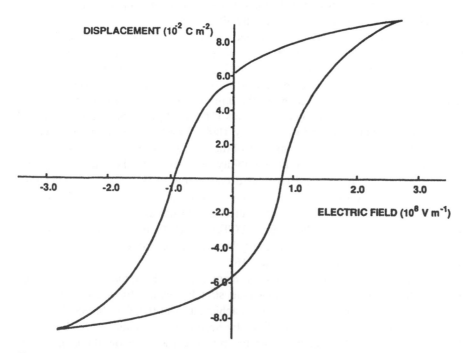

Figure 15 Hysteresis loop obtained on a 25-μm-thick film of PVDF. The data shown were obtained from one of a series of loops. (From Ref. 47.)

oriented and biaxially oriented films. The uniaxially oriented films are stretched in just one direction during manufacture, and the films produced consist of all β-phase material with the polymer molecules oriented predominantly in the stretch direction. The biaxially oriented films are stretched in two perpendicular directions, and the films produced consist of approximately half β-phase and half α-phase material with the long axis of the molecules aligned predominantly in the plane of the film. The remanent polariation, found from measurements of hysteresis loops on 10 uniaxially oriented samples [47], was 0.058 ± 0.003 C/m^2 and for 10 biaxially oriented samples, 0.063 ± 0.002 C/m^2. The coercive field, the electric field required to reduce the polarization to zero, was found to be about 90 MV/m for both types of sample.

The remanent polarization of the same 20 samples was also measured by measuring the amount of charge released by the sample when it was suddenly heated to above the melting temperature, about 175°C. The remanent polarization of the 10 uniaxially oriented samples obtained this way was 0.045 ± 0.003 C/m^2, and for the 10 biaxially oriented samples it was 0.057 ± 0.001 C/m^2.

The fact that the remanent polarization of the biaxially oriented samples was as large as or even larger than that of the uniaxially oriented samples was initially very surprising. The crystal structure of the α phase was known to be nonpolar, so it could not contribute to the polarization. There were other observations that something unusual was happening. Luongo [51] reported in 1972 that fields of 30 MV/m could convert α-phase material to β phase, and Southgate [33] and Latour [52] reported that, based on infrared spectra, something happened to α-phase material but what was happening was not clear. Das-Gupta and Doughty [53,54] reported changes in X-ray diffraction peaks which tended to indicate that the amount of α-phase material was reduced by the poling process, and Sussner et al. [55] showed that relatively high pyroelectric coefficients are exhibited by samples which, prior to poling, were essentially all α phase.

Davis et al. [16] clarified the situation when they showed, using X-ray diffraction and infrared absorption, that α-phase material was converted to δ-phase material by electric fields of the order of 120 MV/m, and that δ-phase material was converted further to β phase at higher fields. The conversion of α to δ phase involves somehow reversing the orientation of one of the two chains in the unit cell so that the dipole moments of the two chains are parallel. A number of workers [16,56–58] suggested that this could happen by simply flipping one of the chains by 180°. However, that seems unlikely in view of some of the results of the investigation of the poling mechanism discussed earlier. Dvey-Aharon et al. [59] investigated the possibility that a form of the kink propagation model might explain the results and concluded that it might. However, Lovinger [60] has suggested a mechanism which is simpler and requires no chain rotation. He pointed out that small intramolecular rotations about all G and $\overline{G}$ bonds of every second chain can cause the conformation to be altered from TGT$\overline{G}$ to T$\overline{G}$TG, thus reversing the dipole moment of every second chain but maintaining the overall chain direction. The proposed mechanism is shown schematically in Figure 16.

Annealing can have a large effect on the magnitude of the remanent polarization. Takase et al. [61] have reported that annealing stretch-oriented films close to 180°C increases the remanent polarization that can be achieved from 0.056 to 0.085 C/m^2; and Ohigashi and Hattori [62] have prepared films by heating them to above 280°C under a pressure of 2–5 kbar, with silicone oil as the pressure transmission fluid, and have achieved a remanent polarization of 0.1 C/m^2. Under these high-pressure and high-temperature conditions, lamellar crystals about 0.2 μm thick and 10 μm wide grew in

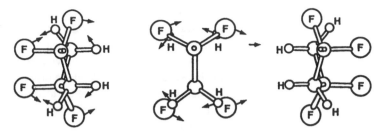

Figure 16 Schematic representation of the intramolecular rotations required in every second polymer chain to transform from the α to the δ phase. (From Ref. 60.)

the films. Ohigashi and Hattori speculate that thick β-phase crystals grow in the paraelectric phase, the phase just above the ferroelectric Curie temperature, which can be reached under high pressure.

Takase et al. [63] have found that remanent polarizations of 0.1 C/m^2 can be achieved in deuterated PVDF, and Bauer [64] reported the same for biaxially oriented films. Bauer and co-workers have developed a standard stress gauge using PVDF and in the process have demonstrated [65] that they can achieve a remanent polarization of 0.092 ± 0.002 C/m^2 using a patented poling process.

The remanent polarization is only weakly dependent on temperature from −100°C to 20°C, decreasing slightly with increasing temperature. The coercive field, on the other hand, is a strong function of temperature, decreasing from 180 MV/m at −100°C to 30 MV/m at +100°C [48].

It is very difficult to measure the remanent polarization much above room temperature using standard techniques, because samples become increasingly conductive. Bur et al. [17] worked around this problem by measuring the intensity of X rays diffracted from a cylinder of PVDF which had been machined from a 1-mm-thick poled film, as described earlier in the section titled "Proof of Ferroelectricity," after annealing at various elevated temperatures. At room temperature, they estimated that the remanent polarization of their sample was 0.043 C/m^2 by fitting their experimental data to the model described previously. As expected, this value is smaller than that obtained from samples that have not been rolled. Rolled samples start from a single-crystal-like texture in which the dipole moments are perpendicular to the poling electric field. In samples that have not been rolled but have been stretch oriented, the dipole moments are oriented randomly relative to the poling field. Using this technique, Bur et al. measured the remanent polarization as a function of temperatures above room temperature by measuring the intensity of X rays diffracted from the sample after heating it to an elevated temperature, holding it at that temperature for 10 min, and then lowering the temperature back to room temperature. Within the accuracy of their measurements, they found that the remanent polarization was stable up to 75°C and then decreased linearly with temperature up to 152°C, at which point about one-half the initial polarization was left.

Bur et al. [17] observed similar behaviors in the pyroelectric and hydrostatic piezoelectric coefficients in measurements made on the same material undergoing the same heat treatments. The decreases in the remanent polarization and in the pyroelectric and piezoelectric coefficients were attributed to a broad range of Curie temperatures resulting from a broad range of crystallite size. It was shown that the crystallinity at room temperature was not changing and other phases were not appearing. However, it appears that the depolarization might also, or perhaps even more logically, be explained by crystal melting. As Bur et al. [17] point out, thermodynamic transformations which occur over broad temperature ranges are common in polymers. If the phase transition involved were a Curie transition, some constraint on the polymer chains would have to orient the dipoles preferentially in the plane of the film as the sample temperature was lowered to room temperature. It appears to be at least as probable that, if crystallite melting is involved, the local stresses developing as the temperature is lowered would lead to the same crystal phase and orientation as was initially produced by the processing.

E. Remanent Polarization—Theory

A first-order estimate of what the magnitude of the remanent polarization might be can be obtained by assuming that local fields are not important and simply taking a sum of the monomer dipole moments. The dipole moment of a monomer unit in free space is 7.0×10^{-30} Cm [47]. This value is obtained by taking a vector sum of the bond dipole moments for the C—F and C—H bonds [66], 4.7×10^{-30} and 1.3×10^{-30} Cm, respectively, keeping in mind the fact that the carbon end is positive in the C—F bond and negative in the C—H bond. There are two monomer units per orthorhombic unit cell of dimensions $a = 8.58$ Å, $b = 4.91$ Å, and c (chain axis) $= 2.56$ Å, with their dipole moments parallel to the b axis [67], and the F—C—F and H—C—H bond angles are 108° and 112°, respectively [67]. Therefore, if it is assumed that the monomer units are rigid dipoles of dipole moment 7.0×10^{-30} Cm and that these dipoles are rigidly aligned in crystals, the polarization of a single β-phase crystal of PVF_2 would be 0.131 C/m². Since PVDF samples are typically about 50% crystalline, this result suggests that the remanent polarization might be 0.065 C/m², surprisingly close to the approximately 0.06 C/m² obtained experimentally. In a sample in which the c axes of the crystallites are aligned parallel to the draw direction and the b axes are randomly oriented in the plane perpendicular to the c axes, the 60° model [20] lowers this estimate to $3P_0/\pi = 0.062$ C/m². For the rolled samples used by Bur et al. [17], the estimate would be reduced to 0.056 C/m² if the initial orientation were perfect, because the dipoles would all be oriented 30° from parallel to the normal to the film surface in the poled condition.

These calculations neglect the effect of the local field on the polarizable molecules. Broadhurst et al. [68] were the first to recognize that this effect is important, and they used an Onsager cavity approach, which was originally developed for molecules in liquids, to calculate the remanent polarization. This approach is equivalent to assuming that the electric field of the dipoles inside a hypothetical cavity is negligible at the site of the molecule of interest, which is at the center of the cavity, and that the only contribution to the electric field from the oriented dipoles of the material in which the molecule is embedded is the Lorentz field. The Lorentz field is the electric field which results from the surface charge on the hypothetical spherical cavity. The magnitude of this field is $P/3\kappa_0$, where P is the polarization and κ_0 is the permittivity of free space.

To calculate the remanent polarization of PVDF, Broadhurst et al. assumed that the polarizability of the dipoles was given by the Clausius-Mossotti relation,

$$\alpha = \frac{3\kappa_0}{N} \frac{\kappa_\infty - 1}{\kappa_\infty + 2} \tag{15}$$

where N is the density of dipoles and κ_∞ is the optical frequency dielectric constant. By adding the permanent and polarizable components of the dipole moments, the total polarization is found to be

$$P = \frac{\kappa_\infty + 2}{3} P_0 \qquad (16)$$

with $P_0 = N\mu_0$ and $\mu_0 = 7 \times 10^{-30}$ Cm. Mopsik and Broadhurst [69] and Kakutani [70] used $\kappa_\infty = 3$, and this model then predicts [71] that the remanent polarization of PVDF is 0.22 C/m^2 rather than 0.13 C/m^2 as predicted by the rigid dipole model. For a 50% crystalline material, this model therefore predicts a remanent polarization of 0.11 C/m^2, significantly disagreeing with the experimental value of about 0.06 C/m^2.

The assumption of a negligible contribution to the local field by nearby dipoles, as was made in the above calculation, is good for a cubic lattice, but the crystal structure of PVDF is orthorhombic, not cubic. Purvis and Taylor [72,73] calculated the dipole field sums for the orthorhombic PVDF lattice, assuming point dipoles, and obtained results quite different from those obtained in the Lorentz field approximation. In fact, the point dipole approximation led to the conclusion that the local field decreased the dipole moment of the dipoles and that the polarization of a PVDF crystal would be 0.086 C/m^2, lower than the value suggested by the rigid dipole model. Subsequently, Al-Jishi and Taylor [74,75] recalculated the local field taking into account the fact that the dipoles are extended rather than points. For dipoles consisting of ± 0.43 electronic charges separated by 0.10 nm, chosen by considering the size and dipole moment of the CF$_2$ unit, they found that the crystal remanent polarization would be 0.127 C/m^2, very close to the value of 0.131 C/m^2 obtained by assuming rigid dipoles, and thus in good agreement with the experimental results.

Ogura and Chiba [76] have carried out calculations very similar to those of Taylor and co-workers and extended them to PVDF copolymers with trifluoroethylene. The results were essentially the same.

These calculations show that the polarization of the crystalline phase is very sensitive to the local field and to the structure of the lattice. It will therefore be very important to consider changes in the local field when calculating piezoelectric and pyroelectric coefficients, as we will discuss later.

F. Ferroelectric Phase Transition

No evidence is seen for a ferroelectric phase transition below the β-phase melting point, 175–180°C, in standard samples of PVDF. However, in PVDF which has been specially synthesized to contain 13.5–15.5 mol% HHTT defects [12], in vinylidene fluoride-tetrafluoroethylene copolymers with 72–82 mol% vinylidene fluoride [77], and in vinylidene fluoride-trifluoroethylene copolymers containing 54–90 mol% vinylidene fluoride [78], clear evidence for ferroelectric transitions is observed. Lovinger et al. [79] found that for vinylidene fluoride-tetrafluoroethylene copolymers containing 64–81 mol% vinylidene fluoride, the ferroelectric transition temperature varied linearly with vinylidene fluoride content. By extrapolating the data to 100% vinylidene fluoride, they estimated that the ferroelectric phase-transition temperature of PVDF is 195–197°C.

IV. PYROELECTRICITY

A. Definition and Measurements

Pyroelectricity is defined as the reversible change in polarization with temperature, and the pyroelectric coefficients p_i are defined by

$$\Delta P_i = p_i \, \Delta T \tag{17}$$

where ΔP_i is the reversible changes in polarization in the i direction and ΔT is the change in temperature. In thin-film PVDF samples the convention used to label sample direction is usually 1 for in the plane of the film and parallel to the draw direction and the chain axes, 2 for in the plane of the film and perpendicular to the draw direction, and 3 for perpendicular to the plane of the film. In PVDF p_3 is the only pyroelectric coefficient that has been measured, and it is usually measured by measuring the total charge displaced between the electrodes on opposite sides of a film when the temperature is changed a known amount. The surface charge on a poled dielectric sample is equal to the component of sample polarization perpendicular to the surface. Thus, the two electrodes on opposite sides of a poled sample become charged with equal amounts, assuming their areas are equal, of opposite sign charge when they are shorted together, and the amount of charge is equal to the sample polarization times the area of the electrodes. The pyroelectric coefficient can, therefore, be measured by simply measuring the amount of charge displaced between the two electrodes per unit area and per unit change in temperature. The electrodes used are typically very thin evaporated metal, which expands and contracts with the sample without constraining it when the temperature is changed. To obtain the true change in polarization the experimental data must be corrected for the fact that the electrode area has changed by subtracting the change in the charge on the electrodes due to this effect:

$$\frac{P}{A} \frac{\partial A}{\partial T} \tag{18}$$

If the electrodes are sufficiently stiff to impede the expansion and contraction of the sample, errors are introduced.

Bergman and co-workers [2–4] were the first to report measurements of the pyroelectric coefficient of PVDF. They reported values of -2.4×10^{-5} C/m^{2}K for biaxially oriented films (films stretched in two perpendicular directions in the plane of the film during manufacture) and about -0.5×10^{-5} C/m^{2}K for uniaxially oriented films.

Initially there were questions raised about how reversible the pyroelectric coefficient was in PVDF [26,80], because it was found that the magnitude of the coefficient depended on how high the sample had been heated after poling. The confusion was caused by the fact that in PVDF irreversible changes in polarization occur the first time a sample is heated to a specific temperature after poling, possibly by small crystallite melting with subsequent reorientation of the dipole moment upon recrystallization. Careful measurements of the pyroelectric and piezoelectric coefficients and of the remanent polarization by X-ray diffraction on samples processed in the same way show that the coefficients measured at room temperature are proportional to the remanent polarization [17].

The magnitudes of the pyroelectric coefficient at room temperature reported by various groups differ, presumably because of different sample preparation techniques, but they tend to be near the value reported by Bergman et al. [2].

On the other hand, there seems to be little agreement on the temperature dependence. Burkhard and Pfister [80] reported a monotonically increasing magnitude from about -1.5×10^{-6} C/m^{2}K at $-100°$C to -8×10^{-6} or -9×10^{-6} C/m^{2}K at 75°C for samples which had been poled at 112°C in an electric field of 30 MV/m and then annealed at 112°C. These measurements were made before much was known about the poling process. It is now recognized that poling at such a low field and subsequent annealing at 112°C would result in a quite low remanent polarization, as the reported magnitude of the pyroelectric coefficient indicates. Since the sample was annealed at 112°C, it is reasonable to assume that changes in the remanent polarization with temperature up to 75°C were reversible, and measurements made during both a heating and a cooling cycle gave the same results. The experimental technique used involved measuring the current flow between electrodes on opposite sides of the sample as a function of temperature while heating or cooling the sample at a constant rate.

Buchman [32] and Tamura et al. [21] used a pulse heating technique. Periodic heat pulses were applied to the samples by chopping a light beam which heated the sample as the temperature was slowly varied. Only the alternating current at the frequency of the heat pules was measured, so irreversible changes in the remanent polarization could not contribute to the pyroelectric signal.

Buchman [32] found that the pyroelectric coefficient at $-180°$C was about equal to that at 100°C and that it went through a minimum near the glass transition temperature at about $-50°$C. The minimum was about one-fifth of the maximum, and the magnitude was not determined. Tamura et al. [21] reported that the pyroelectric coefficient is approximately constant from $-80°$C to 50°C.

B. Primary and Secondary Pyroelectricity

It is frequently useful to separate pyroelectricity into two types, primary and secondary. Primary pyroelectricity is defined as the change in polarization with temperature observed when the sample dimensions are held constant. Secondary pyroelectricity is the additional contribution observed when the sample dimensions are allowed to relax to their equilibrium values. In both cases, primary and secondary pyroelectricity, the quantity of interest is the sample polarization. Secondary pyroelectricity should not be confused with the correction term given in Eq. (18), which must be applied to experimental data to obtain accurate measurements of the change in polarization with temperature.

Secondary pyroelectricity results from piezoelectricity, to be discussed in some detail later in this chapter, and can be determined from the piezoelectric coefficients. The equation is

$$p_i^s = d_{im} c_{mn} \alpha_n \tag{19}$$

where p_i^s is the i_{th} component of secondary pyroelectricity, d_{im} are components of the piezoelectric coefficient matrix, c_{mn} are components of the elastic stiffness matrix, α_n is the thermal expansion coefficient, and summation over repeated indices is indicated. This notation and the coefficients will be discussed in more detail in the section on piezoelectricity later in this chapter.

Kepler and Anderson [81,82] have conducted a series of experiments to determine the relative contribution of primary and secondary pyroelectricity in PVDF, and thus, to gain insight into the fundamental mechanisms involved. The piezoelectric, thermal expansion, and elastic stiffness coefficients of uniaxially and biaxially oriented PVDF films were measured and used to calculate the secondary pyroelectric coefficients from Eq. (19), and the results were compared with the experimentally measured pyroelectric coefficients [81]. The quantities determined in these experiments are shown in Table 1. Secondary pyroelectricity accounted for approximately half of the pyroelectricity observed.

Since these results suggested that half of the pyroelectricity was primary pyroelectricity, an experiment was undertaken to actually measure it [82]. The experimental concept was to change the temperature of a sample on a very short time scale, 200 ps, and to measure the change in polarization observed before the sample dimensions could change. Lightly dyed samples were heated suddenly with a 200-ps laser light pulse and the time dependence of the appearance of the pyroelectric charge measured [82]. Since inertial effects would clamp the sample dimensions, it was anticipated that the charge from primary pyroelectricity would appear as rapidly as the sample was heated, and that the charge from secondary pyroelectricity would appear as the sample dimensions relaxed to new equilibrium dimensions. No evidence for a significant primary contribution was observed, and subsequent experiments [83] led to the conclusion that the primary contribution to the pyroelectric coefficient can be no greater than 15%.

In these pulse heating experiments, the signal resulting from the sudden changes in temperature was observed for times up to 1 μs. The secondary pyroelectric signal resulting from oscillations in the sample dimensions could be readily observed. Just as with the calculation from the piezoelectric coefficients, the results could account for only about half of the previously measured coefficient [81]. In the earlier experiment the pyroelectric coefficient was determined by measuring the amount of charge displaced

Table 1 Piezoelectric, Pyroelectric, and Thermal Expansion Coefficients and Mechanical Properties of PVDF Thin Films

Material property	Coefficient	Biaxially oriented film	Uniaxially oriented film
Piezoelectric coefficients (pC/N)	d_{31}	4.34	21.4
	d_{32}	4.36	2.3
	d_{33}	-12.4	-31.5
	d_h	-4.8	-9.6
	$d_{33}{}^{\text{a}}$	-13.5	-33.3
Pyroelectric coefficients (10^{-5} C/m^{2}K)	p_3	-1.25	-2.74
	$p_3{}^{\text{a}}$ (calculated)	-0.44	-1.48
Thermal expansion coefficients (10^{-4} K^{-1})	α_1	1.24	0.13
	α_2	1.00	1.45
Mechanical properties	E (10^9 Pa)	2.5	2.5
	K (10^{-10} Pa^{-1})	2.6	2.6
	ν	0.392	0.392
	s_{11} (10^{-10} Pa^{-1})	4.0	4.0
	s_{12} (10^{-10} Pa^{-1})	-1.57	-1.57
	c_{11} (10^9 Pa)	5.04	5.04
	c_{12} (10^9 Pa)	3.25	3.25

[a]Calculated from the hydrostatic piezoelectric coefficients d_{31} and d_{32}.
Source: From Ref. 81.

when a sample was transferred from one oil bath to another with an accurately known temperature difference.

Nix et al. [84] have also conducted a series of experiments which show that not all the pyroelectric effect can be accounted for on the basis of secondary pyroelectricity. They measured the piezoelectric coefficients and calculated the secondary pyroelectric coefficient using Eq. (19). They found that secondary pyroelectricity could account for only 10% to 60% of the total pyroelectric effect. The variations in the experimental results were attributed to variations in sample processing, and they concluded that the primary contribution varied from 90% to 40% depending on the method of sample preparation.

In view of the pulse heating experimental results, Kepler and Anderson [82] suggested that a new effect, reversible changes in crystallinity with temperature, could account for most, if not all, of the discrepancy between the measurements of secondary pyroelectricity and the total pyroelectric coefficient.

C. Reversible Changes in Crystallinity

Since PVDF is a crystalline polymer with a crystallinity χ_c of about 0.5, any change in crystallinity would lead to a change in remanent polarization since the dipole moments of the polymer segments going into or out of the crystalline phase would either add to or subtract from the original polarization. This observation led Kepler and Anderson [82] to speculate that reversible changes in crystallinity might account for the missing contribution to the pyroelectric coefficient. Kavesh and Schultz [85], in a study of polyethylene, had concluded that its crystallinity decreases reversibly from 67% at 25°C to 57% at 110°C, and an effect of that magnitude in PVDF would by itself result in a pyroelectric coefficient of -5.9×10^{-5} C/m²K, more than twice the experimentally observed value.

A number of experiments were subsequently conducted which confirm that indeed reversible changes in crystallinity with temperature are contributing to the pyroelectric coefficient in PVDF. Schultz et al. [86] made temperature-dependent small-angle X-ray scattering measurements on PVDF, and the results were interpreted as evidence for reversible changes in crystallinity with temperature. Anderson et al. [87] measured the time dependence of the appearance of pyroelectric charge after a sudden change in temperature which resulted from absorption of a 200-ps laser light pulse. A gradual increase in charge, which occurs between about 1 ms and 1 s and accounts for 25% of the total pyroelectric charge, was attributed to reversible changes in crystallinity. In this experiment it was not possible to measure the time dependence of the appearance of pyroelectric charge for times beyond 1 s because of thermal equilibration effects.

Kepler and co-workers [83,88,89] pointed out that, if indeed the crystallinity changes reversibly with temperature, thermodynamic relations indicate that it should also change with an applied electric field. Since the Gibbs free energy G, given by

$$G = u + pv - sT - PE \tag{20}$$

does not change crossing a phase boundary, changes in G along the boundary are equal in the two phases. In this equation p is the pressure, T is the temperature, v and s are the molar volume and entropy, respectively, and P is the component of the polarization parallel to the electric field E. If we consider the amorphous-crystalline phase boundary, then $dG_a = dG_c$ and, since [90]

$$du = T\, ds - p\, dv + E\, dP \tag{21}$$

then

$$dG = v\,dp - s\,dT - P\,dE \tag{22}$$

and, since p is constant,

$$s_c\,dT + P_c\,dE = s_a\,dT + P_a\,dE \tag{23}$$

where the subscripts a and c denote the amorphous and crystalline phases, respectively. Therefore

$$(s_c - s_a)\,dT = (P_c - P_a)\,dE \tag{24}$$

and, since $(s_c - s_a)T = l$, where l is the molar latent heat of melting, the change dT_m in the melting temperature resulting from a change in the applied electric field is

$$dT_m = \frac{TP_c\,dE}{l} \tag{25}$$

under the simplifying assumption that $P_a = 0$.

If it is assumed that a fraction γ of the pyroelectric coefficient p results from reversible changes $d\chi_c/dT$ in crystallinity with temperature, then, with the approximation of perfect alignment of the crystallites,

$$\frac{d\chi_c}{dT} = \frac{\gamma p}{P_c} \tag{26}$$

The magnitude of the change in crystallinity produced by the application of an electric field becomes, by combining Eqs. (25) and (26),

$$\frac{d\chi_c}{dE} = -\frac{d\chi_c}{dT}\frac{dT_m}{dE} = -\frac{\gamma p}{l} \tag{27}$$

In deriving Eq. (27) it was assumed that a field-induced change in the melting temperature results in the same amount of melting or crystallization at constant temperature as an actual change in temperature of the same amount but of the opposite sign.

To test the reversible crystallinity hypothesis, Kepler and co-workers [83,88,89] measured the change in crystallinity induced by an electric field by measuring the change in intensity of X rays diffracted from the 200 and 110 planes in PVDF as a function of an applied electric field. Thin films were glued to an aluminum plate which served as one electrode, and a thin aluminum film evaporated on top of the film served as the second electrode. The incident and diffracted X rays passed through this electrode, and the intensity of X rays diffracted from a film was determined by integrating the total number of X-ray photons counted during an automatic scan of 2θ at $\frac{1}{2}°$ min^{-1} from $19.25°$ to $21.75°$. The experimental results are shown in Figure 17.

On the basis of measurements of the heat of melting, an estimated crystallinity of 0.5, and measurements of heat capacities of the amorphous and crystalline phases, the latent heat of melting of the crystalline phase at 298 K was estimated to be 55 J/g. Therefore, the molar latent heat l is 3.52×10^3 J/mol, using 64 g, the molecular weight of the monomer or repeat unit, as the molar unit. Using -2.7×10^{-5} C/m^{2}K for the pyroelectric coefficient [81] and an average density for PVDF [10] of 1.76 g/cm^3, the molar pyroelectric coefficient was determined to be -9.8×10^{-10}. With these values and the assumption that one-third of the pyroelectric coefficient results from reversible

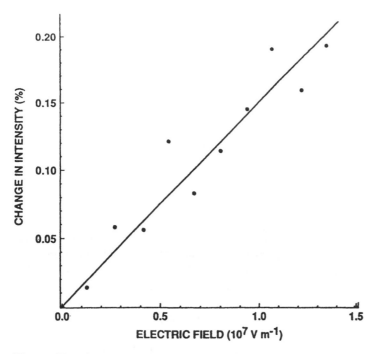

Figure 17 The percent change in intensity of the diffracted X-ray beam as a function of the applied electric field. The solid line is a least-squares fit to the data which was constrained to pass through zero. (From Ref. 88.)

changes in crystallinity, that is, $\gamma = 0.33$, it is predicted from Eq. (27) that the field dependence of crystallinity should be 2.76×10^{-11} m/V. The experimentally observed value, obtained from the data in Figure 17, is 1.8×10^{-10} m/V, more than six times larger. The experimental results, therefore, clearly support the hypothesis that reversible changes in crystallinity contribute significantly to the pyroelectric coefficient. Two possible reasons that these results predict that the contribution to the pyroelectric coefficient of reversible changes in crystallinity is much larger than observed experimentally have been suggested. One is that the estimated value of the latent heat of melting at 298 K is too large. The other is that the melting of crystallites may not result in complete disorientation of the dipole moments of the monomer units.

Nuclear magnetic resonance (NMR) has been used by Clements et al. [91] to measure the reversible changes in crystallinity between 20 and 60°C in PVDF. In broad-line NMR measurements the spectra show two component lines with distinctly different linewidths. The broad component was assigned to the crystalline regions of the polymer and the narrow component to an oriented noncrystalline phase. By measuring the intensity of the two lines as a function of temperature, they found that the crystalline fraction dropped from 0.503 to 0.427 over the temperature range 20–60°C for a uniaxially oriented sample. If the dipole moments of all the material which leaves the crystalline phase become randomly oriented, these results would predict a pyroelectric coefficient of -25×10^{-5} C/m²K assuming that the polarization of the crystalline phase is 0.013 C/m². Since the experimentally measured value of the pyroelectric coefficient is about 10 times smaller, reversible changes in crystallinity are again predicted to produce a much larger pyroelectric coefficient than is observed experimentally.

Dvey-Aharon et al. [92] have developed a theoretical model of reversible changes in crystallinity, and this model also predicts a contribution to the pyroelectric coefficient much larger than that observed.

The experimental and theoretical evidence firmly establish reversible changes in crystallinity as a major contributor to the pyroelectric coefficient of PVDF. The problem now is to explain why the contribution from reversible changes in crystallinity is so small. The most likely reason seems to be that dipoles leaving the crystalline phase remain oriented in their original direction to a large extent.

D. Theoretical Models

Aslaksen [24] was the first to attempt to develop a theoretical model for the pyroelectric effect in PVDF. His model was based on the fact that the average dipole moment depends on the amplitude of libration of the monomer dipoles and that the amplitude of libration increases with increasing temperature. Hayakawa and Wada [93,94] and Wada [95,96] proposed that trapped charge is an important contributor and have investigated that hypothesis in some detail. In general, trapped charge will contribute only if mechanical inhomogeneities are present, and it is now thought that any contribution from trapped charge is small.

Broadhurst et al. [97] pointed out that rigid dipoles would produce pyroelectricity simply because the number of dipoles per unit volume would change with temperature, and Mopsik and Broadhurst [69] improved on this model by including two additional effects. The dipoles were allowed to be polarizable so that the dipole moment depended on the local field, and the dipoles were assumed to librate as in Aslaksen's model [24]. This paper is the one referenced earlier in this chapter in the section on remanent polarization and is the one in which an Onsager cavity approach was used to calculate the local field. The Clausius-Mossotti relation, which is derived using the Lorentz field approximation, was used to determine the polarizability of the dipoles. The mean-squared amplitude of libration of the dipoles was assumed to be that for a classical harmonic oscillator,

$$\langle \phi_0{}^2 \rangle = \frac{2k_B T}{I\omega^2} \tag{28}$$

where I is the moment of inertia and ω is the frequency. The average contribution of the dipoles $\langle \cos \theta_0 \rangle$, with θ_0 being the angle between the film normal and the orientation of the dipole moment to the sample moment, was treated as an adjustable parameter.

Broadhurst et al.'s model was initially developed for homogeneous dielectrics and was subsequently extended to be applicable to crystalline polymers such as PVDF. The effects of compensating charge at the crystallite surfaces within the films, the lamellar morphology, and the presence of the amorphous phase were all included. Broadhurst et al. concluded that the agreement between their theory and experiment was good to within 20% and that the largest contribution to the pryoelectric coefficient, 50%, arises from the bulk dimensional changes, compared with 23% from changes in molecular dipole fluctuations, and 27% from changes in the local field. This model was based on the Lorentz local field approximation, as discussed in the remanent polarization section, but the very high remanent polarization for a single crystal, 0.022 C/m², which it predicts did not enter into the comparison of the predictions of this model with experiment. The experimentally determined remanent polarization was used and the difference ascribed to the distribution of the dipole moments of the crystalline regions, which was treated as an adjustable parameter.

The fact that this model predicts that 23% of the pyroelectric response results from dipole libration is notable in the light of the experimental results [83] which led to the conclusion that no evidence could be seen for a primary pyroelectric coefficient, and that if one did exist, it would have to be less than 15% of the total. In Broadhurst et al.'s theory there are two separable parts to the dipole libration contribution, one which results simply from the change in temperature without any other changes taking place and another which results from the changes in sample dimensions when the temperature is changed. The first contribution is primary pyroelectricity and is predicted to be 15% of the total. This prediction suggests that it may be worthwhile to try to do the pulse heating experiment with greater accuracy to test the prediction that dipole libration is contributing significantly to the pyroelectric coefficient.

Purvis and Taylor [73,98] and Al-Jishi and Taylor [75] have calculated the pyroelectric coefficient predicted by their models in which the local field was calculated taking the orthorhombic crystal structure into account using a point dipole and an extended dipole approximation, respectively. They assumed that the mechanical properties of stretch-oriented PVDF determined by Ohigashi [99] at temperatures well below the glass transition temperature were the properties of single crystals and that the only role of the amorphous phase at this temperature was to transfer stress and change dimensions with temperature. With these approximations they found the very significant result that the contribution to the pyroelectric coefficient resulting from changes in the monomer dipole moments, which in turn resulted from changes in the local field as the dimensions of the unit cell change with temperature, was larger than the contribution from the increase in the concentration of dipoles. Even though their model is quite simple, the predicted pyroelectric coefficient is in reasonable agreement with experiment and they did not include a contribution from dipole libration.

V. PIEZOELECTRICITY

A. Definitions

In some materials an applied force changes the polarization of the material, and this phenomenon is referred to as the direct piezoelectric effect. In order to discuss this phenomenon in PVDF, it will be useful to define a few terms first.

The force per unit area applied to a material is called the stress. For the discussion in this chapter it will be assumed that the state of stress is homogeneous throughout the material, there are no body forces or body torques, and all parts of the material are in static equilibrium. If we imagine a unit cube within the material with edges parallel to the x, y, and z axes, a force will be transmitted across each face and this force may be resolved into three components. If we consider the face of the cube which is perpendicular to the x axis and which intersects the axis on the positive side of the origin, the origin being considered to be at the center of the cube, the stress components are σ_{1j}, with 1 representing the x direction and j taking on the value 1, 2, or 3, corresponding to the x, y, and z directions. The stress perpendicular to this face of the cube in the positive x direction is referred to as σ_{11}. Stresses parallel to this face are referred to as σ_{12} and

σ_{13}, corresponding to forces parallel to the y and z axes, respectively. The assumption that the stress in the material is homogeneous requires that the stress normal to opposite faces of the cube be equal and in opposite directions, and the assumption of no body torques requires that $\sigma_{ij} = \sigma_{ji}$. It should be noted that the sign convention used here leads to tensile stresses for positive values of σ_{ii}.

Since the polarization in a material can have three mutually perpendicular components, it is represented by a vector with the three components P_i, where $i = 1, 2$, and 3. The change in the polarization induced by stress is then

$$\Delta P_i = d_{ijk}\sigma_{jk} \tag{29}$$

where repeated subscripts imply summation. The 27 d_{ijk} form a third-rank tensor and are called the piezoelectric strain coefficients.

A more compact notation is made possible by the symmetry in the stress tensor, $\sigma_{ij} = \sigma_{jk}$, required by the assumption of no body torques, an assumption clearly appropriate for the piezoelectric effect. When the terms of Eq. (29) are written out, terms arising from shear stress may be written in pairs such as

$$d_{312}\sigma_{12} + d_{321}\sigma_{21} = (d_{312} + d_{321})\sigma_{12} \tag{30}$$

Since d_{ijk} can now be set equal to d_{ikj}, it becomes possible to combine the jk subscripts into a single subscript m, allowing a simpler, matrix notation:

$$\Delta P_i = d_{im}\sigma_m \tag{31}$$

where $m = 1\text{--}6$. The components of d_{im} are equal in value to the corresponding d_{ijk} if $m = 1, 2$, or 3. Otherwise d_{im} is twice the corresponding tensor component.

The 18 coefficients d_{im} can usually be reduced by symmetry and in the case of PVDF, where the symmetry is $mm2$ and the three directions in the sample are defined as 1 being parallel to the draw direction, 2 perpendicular to the draw direction and in the plane of the film, and 3 perpendicular to the plane of the film, there are only five independent coefficients and those coefficients are d_{31}, d_{32}, d_{33}, d_{15}, and d_{24}.

The piezoelectric coefficients discussed so far are the direct piezoelectric coefficients. When an electric field is applied it can produce a strain, and the coefficients which describe the strain are referred to as converse piezoelectric coefficients.

The strain in a solid can be specified from the change in a vector x connecting two points within the solid when the solid is deformed. The tensor quantities δ_{ij} are defined as the dimensionless ratios of components of Δx (the change in the vector x) to the components of x:

$$\delta_{ij} = \lim_{x \to 0}\left(\frac{\Delta x_i}{x_j}\right) \tag{32}$$

When the subscripts are equal, the quantities δ_{ij} measure the increase in length per unit length parallel to the i axis. Otherwise, δ_{ij} is equal, for strains small compared with unity, to the angular rotation toward the i axis of a line parallel to the j axis before the deformation.

With these definitions it is clear that, if $\delta_{ij} = -\delta_{ji}$, the solid has undergone a rigid rotation without deformation. The tensor shear strains ϵ_{ij} are, therefore, defined as the symmetric part of δ_{ij}, so that they are insensitive to rigid rotations:

$$\epsilon_{ij} = \frac{1}{2}(\delta_{ij} + \delta_{ji}) \tag{33}$$

The tensile strains are unaffected by this transformation.

To convert the strains to matrix notation, ϵ_m is defined as ϵ_{ij} when $m = 1$, 2, or 3, but ϵ_m is twice the corresponding tensor component when $m = 4$, 5, or 6.

Strains are related to stresses through the fourth-rank elastic compliance tensor s,

$$\epsilon_{ij} = s_{ijkl}\sigma_{kl} \tag{34}$$

and stresses are related to strains, the inverse relation, through the elastic stiffness tensor c,

$$\sigma_{ij} = c_{ijkl}\epsilon_{kl} \tag{35}$$

The 81 components of either s or c may be reduced to 36 by means of the matrix notation in which the first and last pairs of subscripts are each replaced by a single subscript which equals 1–6:

$$\epsilon_m = s_{mn}\sigma_n \tag{36}$$

$$\sigma_m = c_{mn}\epsilon_n \tag{37}$$

The s_{mn} components are equal to the corresponding tensor components multiplied by the factor 1, 2, or 4, depending on whether neither, either, or both m and n are greater than 3, respectively.

A more complete description of these definitions is provided in a book by Nye [8].

With these definitions it is now possible to describe the converse piezoelectric coefficient which relates the strain induced by an electric field to the magnitude of the electric field E at constant stress:

$$\epsilon_{jk} = d_{ijk}{}^c E_i \tag{38}$$

In matrix notation this equation is

$$\epsilon_m = d_{im}{}^c E_i \tag{39}$$

where $m = 1$–16.

It is important to point out again, as was done in the case of the pyroelectric coefficient, that care must be taken in comparing the measured direct piezoelectric coefficients with the coefficients defined above. An experimental measurement usually consists of applying a stress and measuring the charge displaced between the electrodes which are shorted together. To obtain accurate results, the electrodes, typically very thin evaporated metal films, must deform with the sample without distorting the stress applied and, thus, the measured charge is

$$\frac{\partial Q}{\partial \sigma} = A \frac{\partial P}{\partial \sigma} + P \frac{\partial A}{\partial \sigma} \tag{40}$$

where Q is the measured charge and A is the area of the electrodes. The change in polarization is then

$$\frac{\partial P}{\partial \sigma} = \frac{1}{A} \frac{\partial Q}{\partial \sigma} - \frac{P}{A} \frac{\partial A}{\partial \sigma} \tag{41}$$

The necessity of correcting the experimental data with the second term on the right-hand side of Eq. (41) has long been recognized, but the correction is usually small. In pliable polymers, however, it can become dominant. Dvey-Aharon and Taylor [100] have discussed this correction in some detail.

Also, it is important to point out that the direct and converse piezoelectric coefficients are not equal [101], as is usually assumed. The difference tends to be small, ~10%, in hard and brittle ferroelectrics, but as will be demonstrated later in this chapter, the correction term can dominate in soft, pliable polymers. The relation between the direct and converse piezoelectric coefficient is typically determined by deriving a Maxwell relation from a thermodynamic potential. A thorough description of Maxwell relations and thermodynamic potentials is presented in a book by Callen [90].

The usual procedure involves using reduced variables and the relations $dW_s = \sigma_{jk}\, d\epsilon_{jk}$ for the amount of work done by stress and $dW_E = E_i\, dD_i$ for the amount of work done by the electric field. Then, with s and u here denoting the entropy and energy densities (rather than the molar quantities), the change in internal energy density is

$$du = \sigma_{jk}\, d\epsilon_{jk} + E_j\, dD_i + T\, ds. \tag{42}$$

If a thermodynamic potential ϕ is defined as

$$\phi = \sigma\epsilon + ED + Ts - u \tag{43}$$
$$d\phi = \epsilon_{jk}\, d\sigma_{jk} + D_i\, dE_i + s\, dT$$

and

$$\left.\frac{\partial\phi}{\partial\sigma_{jk}}\right|_{E,T} = \epsilon_{jk}, \quad \left.\frac{\partial\phi}{\partial E_i}\right|_{\sigma,T} = D_i \tag{44}$$

Taking the second partial derivatives with respect to the opposite variables in Eq. (44) and making use of the fact that the order of differentiation makes no difference, the relationship of the direct and converse piezoelectric coefficients is found to be

$$\left.\frac{\partial P_i}{\partial\sigma_{jk}}\right|_{E,T} = \left.\frac{\partial\epsilon_{jk}}{\partial E_i}\right|_{\sigma,T} \tag{45}$$

since $D = \kappa_0 E + P$, as is frequently stated.

The problem with this derivation is that it does not properly take into account the important effect of changes in sample dimensions [101], because $dW_E + E_i\, dD_i$ is true in general only if the sample dimensions do not change [90]. To correct this problem the Maxwell relations can be derived using F, x, V, and Q as the variables rather than the normalized quantities, where F is the total force applied to the sample, dx is the distance over which the force is applied, V is the voltage applied between the electrodes, and dQ is the change on the total free charge on the electrodes [101]. In this derivation it is assumed that the electrodes expand and contract with the sample.

Following the procedure used in Eqs. (41)–(44), a Maxwell relation similar to Eq. (45) is obtained [81,101],

$$\left.\frac{\partial Q}{\partial F}\right|_{V,T} = \left.\frac{\partial x}{\partial V}\right|_{F,T} \tag{46}$$

and this equation can be converted to reduced variables using the relations $F_{jk} = A_j\sigma_{jk}$, $x_{jk} = l_j\epsilon_{jk}$, $V_i = l_i E_i$, and $Q_i = A_i P_i$, where A and l are the areas and lengths required to give the correct total quantity. In differentials such as

$$dF_{jk} = A_j\, d\sigma_{jk} + \sigma_{jk}\, dA_j \tag{47}$$

the second term on the right-hand side vanishes at zero stress or electric field and can, therefore, be neglected; but for the charge

$$dQ_i = A_i \, dP_i + P_i \, dA_i \tag{48}$$

both terms on the right must be kept because of the remanent polarization at $V = 0$. Then the correct relation between the direct and converse piezoelectric coefficients is seen to be [81,101]

$$\left.\frac{\partial P_i}{\partial \sigma_{jk}}\right|_{E,T} + \frac{P_i}{A_i} \left.\frac{\partial A_i}{\partial \sigma_{jk}}\right|_{E,T} = \left.\frac{\partial \epsilon_{jk}}{\partial E_i}\right|_{\sigma,T} \tag{49}$$

B. Measurement Techniques

As pointed out in the previous section, there are five independent piezoelectric coefficients in a poled, oriented PVDF film, d_{31}, d_{32}, d_{33}, d_{15}, and d_{24}.

The simplest way to measure d_{31} and d_{32} is to prepare long, thin samples with electrodes on opposite sides, and to attach one end to a rigid support. Then, with an elecrtrometer connected between the electrodes, measure the charge displaced from one electrode to the other when a small weight is attached to the free end of the sample to apply some stress. The electrodes are typically thin, evaporated metal films, and it is important that these films be thin enough that they do not contribute significantly to the mechanical properties of the sample. In fact, for accurate measurements the electrodes must be this thin for all the experimental techniques discussed. The experiment described gives the direct piezoelectric coefficient, but it is equally easy to determine the converse piezoelectric coefficients $d_{31}^{\,\circ}$ and $d_{32}^{\,\circ}$ using essentially the same technique by measuring the change in length of a long, thin sample when an electric field is applied [8].

Extensive use has been made, for the measurement of d_{31} and d_{32}, of a technique originally developed by Fukada et al. [102] Thin films are mounted in an apparatus which holds the sample under tension, and the longitudinal stress on the sample is varied sinusoidally. The stress or strain and charge exchange between electrodes on opposite sides of the thin-film sample are measured simultaneously. Generally it is found that in polymers the piezoelectric charge signal is not in phase with the stress or strain when this technique is used, presumably because polymers are viscoelastic, but the phase difference may result from stress-induced reversible changes in crystallinity, which lag behind the application of stress. It has been proposed that stress-induced changes in crystallinity contribute significantly to the piezoelectric coefficients in PVDF [92]. Because of the phase difference, the piezoelectric coefficients are found to be complex quantities. Both the frequency and the temperature dependencies can be readily measured using the technique, and many studies of piezoelectricity in a wide variety of polymers have been reported.

Ohigashi [99], Bui et al. [103], and Schewe [104] have made measurements of the piezoelectric coefficients at high frequencies using piezoelectric resonances. Schewe has carried out a particularly complete set of measurements including dielectric and mechanical as well as piezoelectric experiments. His most complete set of measurements were made using the quasi-static technique, described above, at 10 Hz, but length extensional and thickness resonances at 25 kHz and 41 MHz, respectively, were also studied. The resonance technique consists of measuring the impedance of a freely vibrating piezoelectric element as a function of frequency near a resonance. The variation in the impedance with frequency is determined by the sample elastic constants, the acoustic loss factor Q, and the electromechanical coupling coefficients.

In the piezoelectric resonance technique, the electromechanical coupling coefficients are

$$k_{ij} = \frac{\kappa_{ij}{}^\sigma - \kappa_{ij}{}^\epsilon}{\kappa_{ij}{}^\sigma} \tag{50}$$

where $\kappa_{ij}{}^\epsilon$ is the clamped dielectric permittivity, that is, the permittivity observed when the dimensions of the sample are held constant, and $\kappa_{ij}{}^\sigma$ is the free dielectric permittivity. The relation between the electromechanical coupling coefficients and the piezoelectric coefficients is derived from the following relations, or some variation of them,

$$dD_i = \kappa_{ij}{}^\sigma \, dE_j + d_{ijk} \, d\sigma_{jk} \tag{51}$$

$$d\epsilon_{ij} = s_{ijkl}{}^\sigma \, d\sigma_{kl} + d_{ijk} \, dE_k \tag{52}$$

and it is assumed that the direct and converse piezoelectric coefficients are equal. As was shown theoretically earlier in this chapter and will be shown experimentally later, they are not equal [101]. In the case of low-modulus materials such as polymers, the error from the assumption of equality can become large. As we pointed out in a previous review [105], the equation that should be used instead of Eq. (51) is

$$dD_i = \kappa_{ij}{}^\sigma \, dE_j + \left(d_{ijk} - \frac{P_i}{A_i} \frac{\partial A_i}{\partial \sigma_{jk}} \right) d\sigma_{jk} \tag{53}$$

With the aid of matrix notation and a one-dimensional approximation [99], Eqs. (52) and (53) lead to

$$k_{31}{}^2 = \frac{d_{31}{}^2 c}{\kappa_{31}{}^\sigma} \left(1 - \frac{1}{d_{31}} \frac{P}{A} \frac{\partial A}{\partial \sigma} \right) \tag{54}$$

rather than

$$k_{31}{}^2 = \frac{d_{31}{}^2 c}{\kappa_{31}{}^\sigma} \tag{55}$$

as is obtained from Eqs. (51) and (52). In these equations, c is the eleastic stiffness. To show how important this correction can be, Ohigashi calculated a value of 30.7 pC/N for d_{31} in PVDF from a measured value for k_{31} of 0.151. If d_{31} is recalculated using Eq. (54) and the values given by Ohigashi for the various constants and assuming that $A^{-1}(\partial A/\partial s) = 2.43 \times 10^{-10}$ Pa^{-1}, we obtained [105] 37 pC/N. As will be shown later, the correction term can even be dominant in some cases. The error introduced by the one-dimensional assumption does not seem to have been investigated.

Rezvani and Linvill [106] have developed an all-electrical technique for the measurement of piezoelectric parameters of PVDF. A thin polymer film with thin evaporated-metal electrodes on each side is held under a slight tension between two rigid supports. The metal electrode on one side is etched away in the middle to form two separate electrodes, one on each end. A voltage applied between one of the electrodes and the electrode on the opposite side develops a stress in the film, which in turn changes the voltage between the second electrode and the electrode on the opposite side. The piezoelectric coefficient d_{31} was then calculated from the relation

$$d_{31} = \frac{2v_2}{v_1} \frac{\kappa^\sigma}{c^E} \tag{56}$$

where v_1 is the applied voltage, v_2 is the measured voltage, κ^σ is the dielectric constant under constant stress, and c^E is Young's modulus of the film under constant electric field. This technique allowed them to study the effect of high DC bias fields on the piezoelectric coefficient d_{31}.

It is very difficult to measure the direct piezoelectric coefficient d_{33} without constraining the sample dimensions, thus introducing errors. A commonly used technique to determine d_{33} is first to measure the hydrostatic piezoelectric coefficient, d_h, as well as d_{31}, and then to calculate d_{33} using the relation

$$d_h = d_{31} + d_{32} + d_{33} \tag{57}$$

Kepler and Anderson [81] measured the converse piezoelectric coefficient d_{33}^c by suspending an electroded sample between two needles, one rigidly attached to a substrate and the other free to move as the sample thickness changed. The needles held the sample in place and provided electrical contact to the electrodes. The change in thickness was measured by mounting a mirror on the needle which was free to move and then measuring its displacement with applied field using a laser interferometer technique.

Nix and Ward [107] have developed a technique for the measurement of the shear piezoelectric coefficients. They glued uniaxially oriented PVDF sheets between blocks of Perspex and attached electrodes on the edges of the sample perpendicular to the 1 and 2 directions. When the sample and blocks of Perspex were mounted in an appropriate rig, a simple shear could be applied in the 1–3 plane and the change in polarization detected by the electrodes perpendicular to the 1 direction to determine d_{15}. Similarly, shear in the 2–3 plane and detection of the change in polarization with the electrode mounted perpendicular to the 2 direction gives d_{24}.

C. Experimental Results

The results of measurements by Kepler and Anderson [81] using the static technique are presented in Table 1. All the data presented in this and other tables in this chapter are the experimentally determined values, uncorrected for changes in electrode area, and therefore are not the true piezoelectric coefficients. As we showed earlier in this chapter, it is the experimentally determined value of the direct piezoelectric coefficient that should be equal to the converse piezoelectric coefficient and, in general, it appears that it is the experimentally determined values of other coefficients that should be used for comparison and calculations.

The magnitude of this discrepancy can become very large. Using the static technique to determine the direct and converse piezoelectric coefficients in PVDF, Kepler and Anderson [81] found that the experimentally determined direct and converse coefficients, the values found before the correction for the change in electrode area was applied, were equal to within a few percent. If Eq. (45) was correct, the experimentally determined value of the direct piezoelectric coefficient $\partial Q/\partial F$ should be corrected by subtracting $(P/A)(\partial A/\partial\sigma) = P(s11 + s12)$. For the sample used the polarization P was about 5×10^{-2} C/m^2 and the elastic compliances s_{11} and s_{12} were 4×10^{-10} Pa^{-1} and -1.6×10^{-10} Pa^{-1}, respectively. Therefore the "correction" is -12×10^{-12} C/N, a value almost three times as large as the experimental value and of the opposite sign. These results clearly show experimentally, as was shown earlier theoretically, that Eq. (45) is not correct.

It is perhaps worthwhile to point out that the relationship between direct and converse piezoelectric coefficients is not the only Maxwell relation affected. Any relation which involves a differential of the polarization will contain a similar error.

For example, it is usually stated that

$$\left.\frac{\partial P_i}{\partial T}\right|_{\sigma,E} = \left.\frac{\partial s}{\partial E_i}\right|_{\partial,T} \tag{58}$$

that is, that the pyroelectric coefficient is equal to the electrocaloric coefficient. It is easy to show, using the procedure described above, that

$$\left.\frac{\partial P_i}{\partial T}\right|_{\sigma,E} + \frac{P_i}{A_i}\left.\frac{\partial A_i}{\partial T}\right|_{\sigma,E} = \left.\frac{\partial s}{\partial E_i}\right|_{\sigma,T} \tag{59}$$

Equation (59) has been tested experimentally by Brossat et al. [108]. They measured the pyroelectric and electrocaloric coefficients in PVDF and found them to be equal. They concluded that they had shown Eq. (58) to be valid, but they did not correct the pyroelectric coefficient experimental data for the change in electrode area with temperature. The correction term for this experiment would be only 25% or 50%, depending on the values of the coefficients of thermal expansion and the polarization of the sample used, but the results would be inconsistent with Eq. (58) after correction. Equation (59), on the other hand, predicts that the two experimentally observed values should be equal as observed.

Getting back to the general experimental results, the data on mechanical properties and piezoelectric coefficients of PVDF obtained by Schewe [104], using his dynamic measurement techniques at ambient temperature and at three different frequencies, are shown in Table 2.

Humphreys et al. [109,110] undertook an investigation of the effect of processing conditions on the mechanical properties and on the three piezoelectric coefficients d_{31}, d_{32}, and d_{3h}. From their studies they concluded that piezoelectricity results from a number of mechanisms and that dimennsional changes is one of them, but not necessarily the most important.

A complete set of direct piezoelectric coefficients, including the shear coefficients which were measured using the technique described in the section on experimental techniques, have been reported by Nix and Ward [107]. For d_{15} the values that they reported range from -13.1 to -27 pC/N and for d_{24} from -23 to -38.3 pC/N for differently processed films.

Wang [111] and Yagi et al. [112] have studied the effect of rolling as well as drawing. As discussed in the proof of ferroelectricity section, rolling produces increased orientation of the a and b axes of the unit cell, tending to orient the a axis in the plane of the film. Wang found that, in the drawn and rolled films, d_{31} was about 25% higher than in films that had only been drawn.

It has also been shown that annealing can have a large effect on the magnitude of the piezoelectric coefficients. Takase et al. [61] have found that annealing stretched PVDF films at temperatures between 160 and 180°C has a marked effect on the magnitude of

Table 2 Mechanical, Dielectric, and Piezoelectric Properties of PVDF Determined at Different Frequencies

Material parameter	Frequency		
	10 Hz	25 kHz	41 MHz
d_{31} (pC/N)	28	17.5	
d_{32} (pC/N)	4	3.2	
d_{33} (pC/N)	−35		
d_h (pC/N)	−3		
e_{31} (10^{-3} C/m^2)	42		
e_{32} (10^{-3} C/m^2)	−6		
e_{33} (10^{-3} C/m^2)	−59		−90.2
s_{11} (10^{-10} Pa^{-1})	3.65	2.49	
s_{22} (10^{-10} Pa^{-1})	4.24	2.54	
s_{33} (10^{-10} Pa^{-1})	4.72		
s_{12} (10^{-10} Pa^{-1})	−1.10		
s_{13} (10^{-10} Pa^{-1})	−2.09		
s_{23} (10^{-10} Pa^{-1})	−1.92		
c_{33} (10^9 Pa)	5.4		9.55
ν_{21}	0.25		
ν_{31}	0.57		
ν_{32}	0.45		
κ_{33}	15	13.6	4.9
tan δ		0.06	0.22
k_{31} (%)	13	10.2	
k_{32} (%)	1.7	1.8	
k_t (%)			14.4

Source: From Ref. 104.

the piezoelectric coefficients and remanent polarization. In the samples studied, d_{31} increased from 20 to 28 pC/N and the remanent polarization from 0.056 to 0.085 C/m^2. Ohigashi and Hattori [62] have observed lamellar crystals of PVDF about 0.2 μm thick and 10 μm wide in films that were crystallized at high pressures (2–5 kbar) and high temperatures (266–300°C). The piezoelectric coefficients were not reported but the crystal structure was β phase, and the remanent polarization of the PVDF films was found to be 0.1 Cm^2. It is well established that the magnitude of the piezoelectric coefficients depend linearly on the remanent polarization in PVDF [113].

The temperature dependence of the piezoelectric coefficients has been investigated by several groups [21,29,99,114–116]. In general it is found that the piezoelectric coefficients increase with increasing temperature, increasing more rapidly above the glass transition temperature, which is in the vicinity of −50°C.

Bauer and co-workers [64,117] have conducted extensive studies of the piezoelectric behavior of PVDF films in shock environments up to 200 kbar and have shown that they can make excellent transducers for shock-wave phenomena.

D. Theoretical Models

In contrast to the situation with respect to pyroelectricity, quite good agreement between theoretical predictions and experimental results has been obtained for piezoelectricity. Two effects are found to dominate, the change in polarization resulting simply from changes in sample dimensions in the rigid dipole approximation, and changes in the dipole moment of the monomer units resulting from changes in the local field with sample dimension changes. There is no compelling evidence that dipole libration, a mechanism first discussed [24] in connection with pyroelectricity and later extended to piezoelectricity, makes a significant contribution to either the pyroelectric or piezoelectric effect. The relatively good agreement between theory and experiment may be fortuitous, however. Theoretical investigations by Dvey-Aharon et al. [92] suggest that a contribution to piezoelectricity from reversible changes in crystallinity caused by stress might well be quite large, and such an effect has not been included in the theories.

An early attempt to develop a theory for piezoelectricity in PVDF was undertaken by Murayama [27]; he suggested that it results from trapped charge. However, it is easy to show that trapped charge will not produce a piezoelectric effect unless there are mechanical inhomogeneities within the sample.

The first attempt at a theory of piezoelectricity in PVDF which included the effect of volume changes and local field effects was undertaken by Broadhurst et al. [68] This work was an extension of a theory developed earlier for electrets by Mopsik and Broadhurst [69] and included, in addition to volume changes and local field effects, the effect of molecular libration and possible space-charge compensation effects. This is the model discussed previously in the section on pyroelectricity. They calculated the hydrostatic piezoelectric coefficient and concluded that 37% of the experimentally measured coefficient arose from the local field effect, 10% from dipole fluctuations, and 60% from dimensional changes. The model predicted a slightly larger coefficient than is observed experimentally.

Purvis and Taylor [73,98] have calculated the piezoelectric coefficients that are predicted by their point dipole and orthorhombic lattice model for the local field. They did not consider dipole libration or space-charge compensation effects but were able to obtain quite good agreement with experimental values including d_{31} and d_{32}. Al-Jishi and Taylor [75] extended the calculations of Purvis and Taylor to include a better approximation for the dipoles, one in which the dipoles are represented by separated charge [74]. This model, as well as the others, was discussed previously in the section on remanent polarization and is the one which agrees best with the remanent polarization experimental results. The predictions of these three models and experimental results are presented in Table 3.

Tashiro et al. [118] have shown that variations in the monomer dipole moment and the dielectric constants of both the crystalline and the amorphous regions with applied stress are effects that should be considered. Al-Jishi and Taylor [75] believe that if they had included these effects in their model, it would have improved the agreement between theory and experiment.

Karasawa and Goddard [119] have recently reported an interesting set of calculations. Starting from force fields for molecular dynamics simulations, they calculated the piezoelectric coefficients as well as other physical properties for PVDF crystals.

Ohigashi [99] has suggested that stress applied parallel to the long axis of the polymer molecules might reduce the steric hindrance that causes the statistical displacement

Table 3 Predictions for the Magnitude of the Piezoelectric and Pyroelectric Coefficients by Various Models Compared to Experimental Results, in Units of pC/N for the Piezoelectric Coefficients d_{ij} and 10^{-5} C/m^{2}K for the Pyroelectric Coefficient p_3

Coefficient	Theoretical model			
	Broadhurst et al.[a]	Purvis and Taylor	Al-Jishi and Taylor	Experiment[b]
d_{31} (pC/N)		18.1	15.0	21.4
d_{32} (pC/N)		4.8	4.8	2.3
d_{33} (pC/N)		−28.9	−29.3	−31.5
d_h (pC/N)	−12.9	−6.2	−9.3	−9.5
p_3 (10^{-5} C/m^{2}K)	−2.5	−2.4	−2.8	−2.74

[a]Assuming that the remanent polarization is 0.06 C/m^2.
[b]From Ref. 81.

of the fluorine atoms along the molecule, thus increasing the effective dipole moment and contributing to the piezoelectric effect. Wada and Hayakawa have considered a model involving spontaneously polarized spheres embedded in a matrix.

VI. NONLINEAR OPTICAL PROPERTIES

PVDF films are essentially transparent from 400 nm to 2 μm and absorbing over the range from 3 μm to 300 μm. They are crystalline films, however, so they do scatter light. The generation of second-harmonic light at 532 nm in thin films of PVDF has been studied by Bergman et al. [2] and McFee et al. [4] They cut wedge-shaped samples from stretch oriented films 20 μm thick and mounted them in an index-matching fluid to minimize scattering and reflection losses at the edges where the light entered and left the sample. The wedges were cut to taper from about 400 μm to 100 μm wide. The light from a neodymium-doped yttrium aluminum garnet laser ($\lambda = 1.06$ μm) with a peak power of about 100 W was incident on the edge of the film so that the beam propagated in the plane of the film and was focused so that its waist was contained in the film. Wedges were cut in two orientations, with the length parallel to the 1, or draw direction, and with the length perpendicular to the draw direction. The 3 direction, perpendicular to the plane of the film and the direction in which the dipoles were oriented, was always perpendicular to the plane of the wedge.

In the *mm*2 symmetry of stretch-oriented and poled PVDF films, there are three independent second-order nonlinear coefficients, and the components of the second-harmonic polarization are given in terms of the electric field components of the incident light by [4]

$$P_1 = 2\delta_{31}E_1E_3 \tag{60}$$

$$P_2 = 2\delta_{32}E_2E_3 \tag{61}$$

$$P_3 = \delta_{31}E_1^2 + \delta_{32}E_2^2 + \delta_{33}E_3^2 \tag{62}$$

When the incident light was propagating parallel to the molecular axis, the 1 direction, second-harmonic light polarized in the 3 direction was observed if the polarization of the incident light was in either the 2 or 3 direction, as predicted by Eqs. (60)–(62).

If the wedge was cut so that the light propagated in the plane of the film and parallel to the 2 direction, which is perpendicular to the chain axis, and the polarization of the incident beam was parallel to the 1 direction, no second-harmonic light was observed, indicating that δ_{32} is very small compared to δ_{31} and δ_{33}.

When the wedge was translated parallel to its length, the length of the light path in the sample changed and oscillatory behavior in the intensity of the second harmonic was observed. Such interference fringes result from phase-matching requirements between the fundamental and second harmonic and a coherence length of about 30 μm was determined from the fringe spacing. By comparing the intensity of the second-harmonic signal generated in the polymer with that generated in a quartz crystal wedge, the magnitude of the nonlinear optical coefficients of PVDF relative to that of quartz were found to be

$$\delta_{33}(\text{PVDF}) \approx 2\delta_{31}(\text{PVDF}) \approx \delta_{11}(\text{SiO}_2)$$

$$\delta_{32}(\text{PVDF}) \approx 0.$$

VII. SUMMARY

In spite of the fact that the mechanical properties of PVDF films are those one typically associates with polymer films—tough, flexible, and strong—and that PVDF is highly processible, it has been shown to be an excellent ferroelectric material with properties comparable to those of the more familiar, hard and brittle inorganic ferroelectrics. The remanent polarization is 0.05–0.1 C/m^2, leading to large piezoelectric and pyroelectric coefficients, and the coercive field and Curie temperature are larger than those of typical inorganic ferroelectrics.

The mechanism by which the orientation of the dipole moment of a unit cell is changed is very different from that in inorganic ferroelectrics and is quite interesting. The long polymer chains are rotated about their chain axis when a kink defect is created thermally. If a high electric field is applied to make reorientation favorable, the kink defect propagates rapidly down the chain, leaving it reoriented. The crystal structure of PVDF is such that rotations in increments of 60° are energetically possible.

Local field effects are not only important to the magnitude of the remanent polarization, they also contribute significantly to the magnitude of the pyroelectric and piezoelectric coefficients. The ferroelectric phase transition cannot be observed in PVDF because the transition temperature is above the melting point of the polymer, ~175°C. It can be observed in PVDF which has been specially synthesized to contain many defects and in copolymers with trifluoroethylene and tetrafluoroethylene. Studies of the copolymers suggest that the transition temperature is about 195°C in standard PVDF.

The pyroelectric coefficients are large and comparable to those observed in many inorganic ferroelectrics. Attempts to observe a primary pyroelectric contribution, a contribution observed when the sample dimensions are not allowed to change when the temperature is changed, have been unsuccessful. Two major contributions to the pyroelectric effect come from a change in the number of dipoles per unit volume with changes in temperature and changes in the dipole moment of the monomer units induced by changes in the local field. There is no evidence that changes in the amplitude of libration of the monomer dipoles is an important contributor to the pyroelectric effect. Changes in the amplitude of libration produces a change in the average dipole moment in the direction of polarization and was the first mechanism proposed to explain the large pyroelectric effects. These mechanisms cannot explain all the pyroelectric effects, however, and it has been shown that reversible changes in crystallinity in PVDF are an important effect and an important contributor to the pyroelectric effect. This is a new type of contribution to pyroelectric effects and one which is not observed in inorganic materials.

It appears that the magnitude of the piezoelectric effects in PVDF can be understood in terms of just two effects, the change in the number of dipoles per unit volume with stress and the change in the dipole moment of the monomer units resulting from changes in the local field with stress. A theoretical investigation into the possibility that reversible changes in crystallinity with stress might contribute has indicated that this effect would be large, but no experimental evidence to support that conclusion has been found.

Attempts to show that the direct and converse piezoelectric coefficients are equal in the pliable PVDF films have lead to the demonstration that they are not equal and that they, in fact, should not be equal. It is usually stated that the change in polarization with stress is equal to the change in strain with electric field. That definition suggests that experimental measurements in the change in polarization with stress made by changing the stress and measuring the charge displaced between the electrodes must be corrected for changes in the electrode area. It has been shown experimentally and theoretically that the uncorrected data should be used for comparison and that the uncorrected pyroelectric and piezoelectric coefficients should be used for calculations. When the experimental data are compared to theoretical predictions, it is important to keep this distinction in mind.

In view of the fact that films of PVDF are pliable, tough, and strong and, at the same time, these films exhibit ferroelectric properties comparable to those of hard and brittle inorganic ferroelectrics, there is a bright future for PVDF in many applications.

ACKNOWLEDGMENTS

The work at Sandia National Laboratories was supported by the U.S. Department of Energy under contract number DE-AC04-76DP00789.

REFERENCES

1. H. Kawai, Piezoelectricity of poly(vinylidene fluoride), *Jpn. J. Appl. Phys. 8*:975 (1969).
2. J. G. Bergman, J. H. McFee, and G. R. Crane, Pyroelectricity and optical second harmonic generation in polyvinylidene fluoride, *Appl. Phys. Lett. 18*:203 (1971).
3. A. M. Glass, J. H. McFee, and J. G. Bergman, Pyroelectric properties of polyvinylidene fluoride and its use for infrared detection, *J. Appl. Phys. 42*:5219 (1971).
4. J. H. McFee, J. G. Bergman, and G. R. Crane, Pyroelectric and nonlinear optical properties of poled polyvinylidene fluoride films, *Ferroelectrics 3*:305 (1972).
5. G. M. Sessler (ed.), *Electrets*, Vol. 33, Springer-Verlag, Berlin, 1987.
6. M. E. Lines and A. M. Glass, *Principles and Applications of Ferroelectrics and Related Materials*, Clarendon Press, Oxford, 1977.
7. O. Heaviside, *Electrical Papers*, Vol. 1, Macmillan, London, 1892.
8. J. F. Nye, *Physical Properties of Crystals*, Oxford University Press, New York, 1957.
9. T. A. Ford and W. E. Hanford, U.S. Patent 2,435,537.
10. A. J. Lovinger, Poly(vinylidene fluoride), in *Developments in Crystalline Polymers-1* (D. C. Bassett, ed.), Applied Science Publishers, London and New Jersey, 1982, p. 195.
11. R. E. Cais and J. M. Kometani, Synthesis and two-dimensional ^{19}F NMR of highly aregic poly(vinylidene fluoride), *Macromolecules 18*:1354 (1985).

12. A. J. Lovinger, D. D. Davis, R. E. Cais, and J. M. Kometani, The role of molecular defects on the structure and phase transitions of poly(vinylidene fluoride), *Polymer 28*:617 (1987).

13. B. L. Farmer, A. J. Hopfinger, and J. B. Lando, Polymorphism of poly(vinylidene fluoride): Potential energy calculations of the effects of head-to-head units of the chain conformation and packing of poly(vinylidene fluoride), *J. Appl. Phys. 43*:4293 (1972).

14. P. DeSantis, E. Giglio, A. M. Liquori, and A. Ripamonti, *J. Polymer Sci. A 1*:1383 (1963).

15. J. B. Lando and W. W. Doll, The polymorphism of poly(vinylidene fluoride). 1. The effect of head-to-head structure, *J. Macromol. Sci.-Phys. B2*:205 (1968).

16. G. T. Davis, J. E. McKinney, M. G. Broadhurst, and S. C. Roth, Electric field induced phase changes in poly(vinylidene fluoride), *J. Appl. Phys. 49*:4998 (1978).

17. A. J. Bur, J. D. Barnes, and K. J. Wahlstrand, A study of thermal depolarization of polyvinylidene fluoride using x-ray pole-figure observations, *J. Appl. Phys. 59*:2345 (1986).

18. J. E. Dohaney, A. A. Dukert, and S. S. Preston III, Vinylidene fluoride polymers, in *Encyclopedia of Polymer Science and Technology*, Vol. 14 (N. M. Bikales, ed.), John Wiley, New York, 1971, p. 600.

19. J. B. Lando, H. G. Olf, and A. Peterlin, Nuclear magnetic resonance and x ray determination of the structure of poly(vinylidene fluoride), *J. Polymer Sci. A-1 4*:941 (1966).

20. R. G. Kepler and R. A. Anderson, Ferroelectricity in poly(vinylidene fluoride), *J. Appl. Phys. 49*:1232 (1978).

21. M. Tamura, S. Hagiwara, S. Matsumoto, and N. Ono, Some aspects of piezoelectricity and pyroelectricity in uniaxially stretched poly(vinylidene fluoride), *J. Appl. Phys. 48*:513 (1977).

22. D. Naegele and D. Y. Yoon, Orientation of crystalline dipoles in poly(vinylidene fluoride) films under electric field, *Appl. Phys. Lett. 33*:132 (1978).

23. Y. Takahashi, Y. Nakagawa, H. Miyaji, and K. Asai, Direct evidence for ferroelectric switching in poly(vinylidene fluoride) and poly(vinylidene fluoride-trifluoroethylene) crystals, *J. Polymer Sci. Part C: Polymer Lett. 25*:153 (1987).

24. E. W. Aslaksen, Theory of the spontaneous polarization and the pyroelectric coefficient of linear chain polymers, *J. Chem. Phys. 57*:2358 (1972).

25. H. Dvey-Aharon, T. J. Sluckin, P. L. Taylor, and A. J. Hopfinger, Kink propagation as a model for poling in poly(vinylidene fluoride), *Phys. Rev. B 21*:3700 (1980).

26. G. Pfister, M. Abkowitz, and R. G. Crystal, Pyroelectricity in polyvinylidene fluoride, *J. Appl. Phys. 44*:2064 (1973).

27. N. Murayama, Persistent polarization in poly(vinylidene fluoride). I. Surface charges and piezoelectricity of poly(vinylidene fluoride) thermoelectrets, *J. Polymer Sci. A-2 13*:929 (1975).

28. N. Murayama, T. Oikawa, T. Katto, and K. Nakamura, Persistent polarization in poly(vinylidene fluoride). II Piezoelectricity of poly(vinylidene fluoride) thermoelectrets, *J. Polymer Sci. A-2 13*:1033 (1975).

29. M. Oshiki and E. Fukada, Piezoelectric effect in stretched and polarized polyvinylidene fluoride film, *Jpn. J. Appl. Phys. 15*:43 (1976).

30. R. J. Shuford, A. F. Wilde, J. J. Ricca, and G. R. Thomas, Characterization and piezoelectric activity of stretched and poled poly(vinylidene fluoride). Part 1 effect of draw ratio and poling conditions, *Polymer Eng. Sci 16*:25 (1976).

31. W. R. Blevin, poling rates for films of polyvinylidene fluoride, *Appl. Phys. Lett. 31*:6 (1977).

32. P. Buchman, Pyroelectric and switching properties of polyvinylidene fluoride film, *Ferroelectrics 5*:39 (1973).

33. P. D. Southgate, Room-temperature poling and morphology changes in pyroelectric polyvinylidene fluoride, *Appl. Phys. Lett 28*:250 (1976).

34. J. C. Hicks and T. E. Jones, Frequency dependence of remanent polarization and the correlation of piezoelectric coefficients with remanent polarization in polyvinylidene fluoride, *Ferroelectrics 32*:119 (1981).

35. T. Furukawa and G. E. Johnson, Measurements of ferroelectric switching characteristics in poly(vinylidene fluoride), *Appl. Phys. Lett.* 38:1027 (1981).

36. J. D. Clark and P. L. Taylor, Effect of lamellar structure on ferroelectric switching in poly(vinylidene fluoride), *Phys. Rev. Lett* 49:1532 (1982).

37. T. Furukawa, M. Date, and G. E. Johnson, Polarization reversal associated with rotation of chain molecules in β-phase poly(vinylidene fluoride), *J. Appl. Phys.* 54:1540 (1983).

38. E. Fatuzzo and W. J. Merz, *Ferroelectricity*, John Wiley, New York, 1967.

39. W. J. Merz, Switching time in ferroelectric BaTiO3 and its dependence on crystal thickness, *Phys. Rev.* 95:690 (1954).

40. H. L. Stadler, BaTiO3 switching field dependence, *J. Appl. Phys.* 29:1485 (1958).

41. A. Odajima, T. T. Wang, and Y. Takase, An explanation of switching characteristics in polymer ferroelectrics by a nucleation and growth theory, *Ferroelectrics* 62:39 (1985).

42. R. Al-Jishi and P. L. Taylor, Influence of electrostatic interactions on switching characteristics in poly(vinylidene fluoride), *Ferroelectrics* 73:343 (1987).

43. Y. Takase and A. Odajima, γ-Ray radiation-induced changes in switching current of polyvinylidene fluoride, *Jpn. J. Appl. Phys.* 22:L318 (1983).

44. Y. Takase, A. Odajima, and T. T. Wang, A modified nucleation and growth model for ferroelectric switching in form 1 poly(vinylidene fluoride), *J. Appl. Phys.* 60:2920 (1986).

45. R. G. Kepler and R. A. Anderson, unpublished results.

46. M. Tamura, K. Ogasawara, N. Ono, and S. Hagiwara, Piezoelectricity in uniaxially stretched poly(vinylidene fluoride), *J. Appl. Phys.* 45:3768 (1974).

47. R. G. Kepler, Saturation remanant polarization of poly(vinylidene fluoride), *Org. Coatings Plast. Chem.,* 38:706 (1978).

48. T. Furukawa, M. Date, and E. Fukada, Hysteresis phenomena in poly(vinylidene fluoride) under high electric field, *J. Appl. Phys.* 51:1135 (1980).

49. T. Furukawa, K. Nakajima, T. Koizumi, and M. Date, Measurements of nonlinear dielectricity in ferroelectric polymers, *Jpn. J. Appl. Phys.* 26:1039 (1987).

50. I. L. Guy and J. Unsworth, Conformational and crystallographic changes occurring in poly(vinylidene fluoride) during the production of displacement-electric field hysteresis loops, *J. Appl. Phys.* 61:5374 (1987).

51. J. P. Luongo, Far-infrared spectra of piezoelectric poly vinylidene fluoride, *J. Polymer Sci. A-2* 10:1119 (1972).

52. M. Latour, Infra-red analysis of poly(vinylidene fluoride) thermoelectrets, *Polymer* 18:278 (1977).

53. D. K. Das-Gupta and L. Doughty, Changes in x-ray diffraction patterns of polyvinylidene fluoride due to corona charging, *Appl. Phys. Lett.* 31:585 (1977).

54. D. K. Das-Gupta and L. Doughty, Piezo- and pyroelectric behavior of corona-charged polyvinyledene fluoride, *J. Phys. D* 11:2415 (1978).

55. H. Sussner, D. Naegele, R. D. Diller, and D. Y. Yoon, Polarization of α-form poly(vinylidene fluoride), *Org. Coatings Plast. Chem.* 38:266 (1978).

56. D. Naegele, D. Y. Yoon, and M. G. Broadhurst, Formation of a new crystal form (α_p) of poly(vinylidene fluoride) under electric field, *Macromolecules* 11:1297 (1978).

57. B. Servet and J. Rault, Polymorphism of poly(vinylidene fluoride) induced by poling and annealing, *J. Phys.* 40:1145 (1979).

58. G. R. Davies and H. Singh, Evidence for a new crystal phase in conventionally poled samples of poly(vinylidene fluoride) in crystal form II, *Polymer* 20:772 (1979).

59. H. Dvey-Aharon, P. L. Taylor, and A. J. Hopfinger, kink propagation α–δ transformation, *J. Appl. Phys* 51:5184 (1980).

60. A. J. Lovinger, Molecular mechanism for α → δ transformation in electrically poled poly(vinylidene fluoride), *Macromolecules* 14:225 (1981).

61. Y. Takase, J. I. Scheinbeim, and B. A. Newman, Annealing effects of phase I poly(vinylidene fluoride), *J. Polymer Sci., Part B: Polymer Phys.* 27:2347 (1989).

62. H. Ohigashi and T. Hattori, Growth of single crystals of poly(vinylidene fluoride) under high pressures, *Jpn. J. Appl. Phys. 28*:L1612 (1989).

63. Y. Takase, H. Tanaka, T. T. Wang, R. E. Cais, and J. M. Kometani, Ferroelectric properties of form I perdeuterated poly(vinylidene fluoride, *Macromolecules 20*:2318 (1987).

64. F. Bauer, PVF_2 polymers: Ferroelectric polarization and piezoelectric properties under dynamic pressure and shock wave action, *Ferroelectrics 49*:231 (1983).

65. L. M. Lee, R. A. Graham, F. Bauer, and R. P. Reed, Standardized Bauer PVDF piezoelectric polymer shock gauge, *J. Phys. Colloq. 49(C-3)*:651 (1988).

66. L. Pauling, *The Nature of the Chemical Bond*, 2nd ed., Cornell University Press, Ithaca, NY, 1948, p. 68.

67. R. Hasegawa, Y. Takahashi, Y. Chatani, and H. Tadokoro, Crystal structure of three crystalline forms of poly(vinylidene fluoride), *Polymer J. 3*:600 (1972).

68. M. G. Broadhurst, G. T. Davis, J. E. McKinney, and R. E. Collins, Piezoelectricity and pyroelectricity in polyvinylidene fluoride—A model, *J. Appl. Phys. 49*:4992 (1978).

69. F. Mopsik and M. G. Broadhurst, Molecular dipole electrets, *J. Appl. Phys. 46*:4204 (1975).

70. H. Kakutani, Dielectric absorption in oriented poly(vinylidene fluoride), *J. Polymer Sci. A-2 8*:1177 (1970).

71. M. G. Broadhurst and G. T. Davis, Piezo- and pyroelectric properties, *Topics in Modern Physics-Electrets* 2nd ed. (G. M. Sessler, ed.), Springer-Verlag, Berlin, 1987, p. 285.

72. C. K. Purvis and P. L. Taylor, Dipole-field sums and Lorentz factors for orthorhombic lattices, and implications for polarizable molecules, *Phys. Rev. B 26*:4547 (1982).

73. C. K. Purvis and P. L. Taylor, Piezoelectricity and pyroelectricity in poly(vinylidene fluoride): Influence of the lattice structure, *J. Appl. Phys. 54*:1021 (1983).

74. R. Al-Jishi and P. L. Taylor, Field sums for extended dipoles in ferroelectric polymers, *J. Appl. Phys. 57*:897 (1985).

75. R. Al-Jishi and P. L. Taylor, Equilibrium polarization and piezoelectric and pyroelectric coefficients in poly(vinylidene fluoride), *J. Appl. Phys. 57*:902 (1985).

76. H. Ogura and A. Chiba, Calculation of the equilibrium polarization of vinylidene fluoride-trifluoroethylene copolymers using the iteration method, *Ferroelectrics 74*:347 (1987).

77. A. J. Lovinger, D. D. Davis, R. E. Cais, and J. M. Kometani, Compositional variation of the structure and solid-state transformations of vinylidene fluoride/tetrafluoroethylene copolymers, *Macromolecules 21*:78 (1988).

78. T. Yagi, M. Tatemoto, and J.-i. Sako, Transition behavior and dielectric properties in trifluoroethylene and vinylidene fluoride copolymers, *Polymer J. 12*:209 (1980).

79. A. J. Lovinger, D. D. Davis, R. E. Cais, and J. M. Kometani, On the Curie temperature of poly(vinylidene fluoride), *Macromolecules 19*:1491 (1986).

80. H. Burkard and G. Pfister, Reversible pyroelectricity and inverse piezoelectricity in polyvinylidene fluoride, *J. Appl. Phys. 45*:3360 (1974).

81. R. G. Kepler and R. A. Anderson, Piezoelectricity and pyroelectricity in poly(vinylidene fluoride), *J. Appl. Phys. 49*:4490 (1978).

82. R. G. Kepler and R. A. Anderson, On the origin of pyroelectricity in poly(vinylidene fluoride), *J. Appl. Phys. 49*:4918 (1978).

83. R. G. Kepler and R. A. Anderson, The role of order in pyroelectricity of PVF_2, *Mol. Cryst. Liq. Cryst. 106*:345 (1984).

84. E. L. Nix, J. Nanayakkara, G. R. Davies, and I. M. Ward, Primary and secondary pyroelectricity in highly oriented poly(vinylidene fluoride), *J. Polymer Sci. Part B: Polymer Phys. 26*:127 (1988).

85. S. Kavesh and J. M. Schultz, Lamellar and interlamellar structure in melt-crystallized polyethylene.1. Degree of crystallinity, atomic positions, particle size, and lattice disorder of the first and second kinds, *J. Polymer Sci. A-2 8*:243 (1970).

86. J. M. Schultz, J. S. Lin, R. W. Hendricks, R. R. Lagasse, and R. G. Kepler, Temperature-dependent small-angle x-ray scattering from poly(vinylidene fluoride), *J. Appl. Phys. 51*: 5508 (1980).

87. R. A. Anderson, R. G. Kepler, and R. R. Lagasse, Time dependence of pyroelectric effects in poly(vinylidene fluoride), *Ferroelectrics 33*:91 (1981).

88. R. G. Kepler, R. A. Anderson, and R. R. Lagasse, Electric field dependence of crystallinity in poly(vinylidene fluoride), *Phys. Rev. Lett. 48*:1274 (1982).

89. R. G. Kepler, R. A. Anderson, and R. R. Lagasse, Pyroelectricity and the electric-field dependence of crystallinity in poly(vinylidene fluoride), *Ferroelectrics 33*:91 (1984).

90. H. B. Callen, *Thermodynamics*, John Wiley, New York, 1960, p. 243.

91. J. Clements, G. R. Davies, and I. M. Ward, A broad-line NMR study of oriented poly(vinylidene fluoride), *Polymer 26*:208 (1985).

92. H. Dvey-Aharon, T. J. Sluckin, and P. L. Taylor, Theory of reversible crystallization and electric effects in poly(vinylidene fluoride), *Ferroelectrics 32*:25 (1981).

93. R. Hayakawa and Y. Wada, Piezoelectricity and related properties of polymer films, *Adv. Polymer Sci. 11*:1 (1973).

94. R. Hayakawa and Y. Wada, Theory of piezoelectricity, *Rep. Prog. Polymer Phys. Jpn. 19*: 321 (1976).

95. Y. Wada, Piezoelectricity and Pyroelectricity of Polymers, *Jpn. J. Appl. Phys. 15*:2041 (1976).

96. Y. Wada, Piezoelectricity and pyroelectricity, in *Electronic Properties of Polymers* (J. Mort and G. Pfister, eds.), John Wiley, New York, 1982, p. 109.

97. M. G. Broadhurst, C. G. Malmberg, F. I. Mopsik, and W. P. Harris, Piezo- and pyroelectricity in polymer electrets, in *Electrets, Charge Storage and Transport in Dielectrics* (M. M. Perlman, ed.), Electrochemical Society, Princeton, NJ, 1973, p. 492.

98. C. K. Purvis and P. L. Taylor, Piezoelectric and pyroelectric coefficients for ferroelectric crystals with polarizable molecules, *Phys. Rev. B 26*:4564 (1982).

99. H. Ohigashi, Electromechanical properties of polarized polyvinylidene fluoride films as studied by the piezoelectric resonance method, *J. Appl. Phys. 47*:949 (1976).

100. H. Dvey-Aharon and P. L. Taylor, Thermodynamics of pyroelectricity and piezoelectricity in polymers, *Ferroelectrics 33*:103 (1981).

101. R. A. Anderson and R. G. Kepler, Inequality of direct and converse piezoelectric coefficients, *Ferroelectrics 32*:13 (1981).

102. E. Fukada, M. Date, and K. Hara, Temperature dispersion of complex piezoelectric modulus of wood, *Jpn. J. Appl. Phys. 8*:151 (1969).

103. L. N. Bui, H. J. Shaw, and L. T. Zitelli, Study of acoustic wave resonance in piezoelectric PVF_2 films, *IEEE Trans. Sonics Ultrason. SU-24*:331 (1977).

104. H. Schewe, Piezoelectricity of uniaxially oriented polyvinylidene fluoride, *1982 Ultrasonics Symposium Proceedings*, Vol. 1 (B. R. McAvoy, ed.), IEEE, New York, 1982, p. 519.

105. R. G. Kepler and R. A. Anderson, Piezoelectricity in polymers, *CRC Crit. Rev. Solid State Mater. Sci. 9*:399 (1980).

106. B. Rezvani and J. G. Linvill, Measurement of piezoelectric parameters versus bias field strength in polyvinylidene fluoride (PVF2), *Appl. Phys. Lett. 34*:828 (1979).

107. E. L. Nix and I. M. Ward, The measurement of the shear piezoelectric coefficients of polyvinylidene fluoride, *Ferroelectrics 67*:137 (1986).

108. T. Brossat, G. Bichon, C. Limonon, M. Royer, and F. Micheron, Pyroelectric and electrocaloric coefficients, *C.R. Acad. Sci. Paris Ser. B 288*:53 (1979).

109. J. Humphreys, I. M. Ward, J. C. McGrath, and E. L. Nix, The measurement of the piezoelectric coefficients d_{31} and d_{3h} for uniaxially oriented polyvinylidene fluoride, *Ferroelectrics 67*:131 (1986).

110. J. Humphreys, E. L. V. Lewis, I. M. Ward, E. L. Nix, and J. C. McGrath, A study of the mechanical anisotropy of high-draw, low-draw, and voided PVDF, *J. Polymer Sci. Part B: Polymer Phys.* 26:141 (1988).
111. T. T. Wang, Piezoelectricity in β-phase poly(vinylidene fluoride) having a "single-crystal" orientation, *J. Appl. Phys.* 50:6091 (1979).
112. T. Yagi, Y. Higashihata, K. Fukuyama, and J. Sako, Piezoelectric properties of rolled vinylidene fluoride and trifluoroethylene copolymer, *Ferroelectrics* 57:327 (1984).
113. T. Furukawa and T. T. Wang, Measurements and properties of ferroelectric polymers, *The Applications of Ferroelectric Polymers* (T. T. Wang, J. M. Herbert, and A. M. Glass, eds.), Blackie and Son, Glasgow, 1988, p. 66.
114. E. Fukada and S. Takashita, Piezoelectric effect in polarized poly(vinylidene fluoride), *Jpn. J. Appl. Phys.* 8:960 (1969).
115. E. Fukada and T. Sakurai, Piezoelectricity in polarized poly(vinylidene fluoride) films, *Polymer J.* 2:656 (1971).
116. T. Furukawa, J. Aiba, and E. Fukada, Piezoelectric relaxation in poly(vinylidene fluoride), *J. Appl. Phys.* 50:3615 (1979).
117. F. Bauer, Ferroelectric properties of PVDF polymer and vinylidene, *Ferroelectrics* 115:247 (1991).
118. K. Tashiro, M. Kobayashi, H. Tasokoro, and E. Fukada, Calculation of elastic and piezoelectric constants of polymer crystals by a point charge model: application to poly(vinylidene fluoride) form I, *Macromolecules* 13:691 (1980).
119. N. Karasawa and W. A. Goddard III, Force fields, structures, and properties of poly(vinylidene fluoride) crystals, *Macromolecules* 25:7268 (1992).

4
PVDF and Its Blends

B.-J. Jungnickel
German Plastics Institute, Darmstadt, Germany

I. INTRODUCTION

There is a large number of reasons and procedures for changing the composition of a polymeric material:

Improvement of workability
Enhancement of performance
Creation of new properties
Reduction of prizes
Support for the possibility, and enhancement of evidence, of scientific investigations

All these may be achieved by

Chemical modification of the initial polymer
Creation of composites (with ceramic powders, glass fibers, carbon fibers, etc.)
Blending with high- or low-molecular-weight materials
Sandwich arrangements, laminates, or coating with other materials

The ferroelectric and piezoelectric properties of poly(vinylidene fluoride) (PVDF), in particular, can be improved by chemical modification through copolymerization with trifluoroethylene or tetrafluoroethylene [1]. The addition of ceramic ferroelectrics such as lead zirconium titanate in order to create composites with improved electric performance is common [2,3]. Sandwich arrays with poly(tetra fluoroethylene) or poly(ethylene terephthalate) [4] have been shown to affect strongly the magnitude and temporal stability of ferroelectric polarization. Laminates with other polymers can exhibit interesting new properties [5,6]. These issues will be treated in detail in other parts of this handbook. In this particular chapter, the ferroelectric properties of PVDF to which other polymers (''polyblending'') or organic low-molecular-weight substances (''plastification'') have been added will be considered. Such modification can enhance optical [7], processing [8–11], or mechanical [10–13] properties, respectively, of the resulting materials, and can contribute to an understanding of the microscopic structural and dynamic basis of the electroactivity of PVDF [14–16].

II. THERMODYNAMIC PRELIMINARIES

Materials which consist of more than one component are called blends or alloys. Their properties depend to a large extent on the miscibility of the components, i.e., on the ability to mix stably on a molecular level. We speak of "mixtures," and of miscibility in a narrow sense, if the intimately mixed state is favored by thermodynamic forces, and is therefore thermodynamically stable. In contrast, we speak of "compatibility" if the intimate mixing can be enforced, or is only enforced, by suitable processing conditions. Although the resulting structure is then not stable from a thermodynamic point of view, such a structure can nevertheless be sufficiently stable over the lifetime of the material because of the lack of structural relaxation possibilities. The properties of such a material are then almost the same as those of a truly miscible blend.

Two substances, 1 and 2, are miscible if the Gibbs free energy of the mixture is smaller than that of the components:

$$\Delta G_m = G_m - (G_1 + G_2) < 0 \tag{1}$$

(G_m, G_1, and G_2 are the Gibbs free energies of the mixture and of components 1 and 2, respectively). ΔG_m can be separated into an enthalpic (ΔH_m) and an entropic (ΔS_m) contribution, respectively,

$$\Delta G_m = \Delta H_m - T \, \Delta S_m \tag{2}$$

which are both in turn composed of two parts. With respect to ΔS_m, we have

$$\Delta S_m = \Delta S_m^{(i)} + \Delta S_m^{(ex)} \tag{3}$$

where, according to the "lattice theory," [17]

$$\Delta S_m^{(i)} = -k \frac{V}{V_s} \left(\frac{y_1}{Q_1} \ln y_1 + \frac{y_2}{Q_2} \ln y_2 \right) \tag{4}$$

(V is the sample volume, V_s is the lattice cell volume, y_i is the volume ratio V_i/V of component i, and Q_i is the degree of polymerization) is the "ideal" entropy which results from the distribution of the monomer units over the cells of a thought lattice. $\Delta S_m^{(ex)}$ is the "excess" entropy embracing all other entropic contributions. Similarly, we can write

$$\Delta H_m = \Delta H_m^{(i)} + \Delta H_m^{(ex)} \tag{5}$$

where the ideal contribution

$$\Delta H_m^{(i)} = z \, \Delta e_{12} y_1 y_2 \frac{V}{V_s} = kT \chi_{12}^{(h)} y_1 y_2 \frac{V}{V_s} \qquad \chi_{12}^{(h)} = \frac{z \, \Delta e_{12}}{kT} \tag{6}$$

(z is the coordination number of the assumed lattice, Δe_{12} is the interaction energy between the monomer units of components 1 and 2, $\chi_{12}^{(h)}$ is the enthalpic Flory-Huggins interaction parameter) is that part which describes the pure energetic interactions between the monomer units. The excess component $\Delta H_m^{(ex)}$ considers all the contributions from volume changes with mixing and similar effects.

Miscibility is the exception rather than the rule with polymers. This is caused by the small increase in entropy, particularly of its ideal component, when blending large molecules. Thus, enthalpic interactions, in particular their excess component, become important. As a result, a number of unusual types of phase diagrams—i.e., patterns which illustrate the miscibility range in the composition–temperature plane—are possible. There are particularly such ones with demixing with rising temperature (lower critical solution temperature behavior, LCST), in contrast to the usual upper critical solution temperature (UCST) behavior, as known from blends of low-molecular-weight substances. The classical lattice theories which take into account only $\Delta H_m^{(i)}$, and which assume that

$$\Delta S_m^{(ex)} = kT\chi_{12}^{(s)}\, y_1 y_2 \frac{V}{V_s} \tag{7}$$

($\chi_{12}^{(s)}$ is the entropic interaction parameter) such that

$$\Delta G_m = \Delta S_m^{(i)} + kT\chi_{12} y_1 y_2 \frac{V}{V_s} \tag{8}$$

with

$$\chi_{12} = \chi_{12}^{(h)} + \chi_{12}^{(s)} \tag{9}$$

are not able to describe such patterns. They must therefore be replaced by more general theories by which the Gibbs free energy is considered and calculated as a whole. The several enthalpic and entropic terms can then be deduced separately from this Gibbs free energy. These approaches are called equation-of-state theories [18].

A more thorough calculation shows that miscibility is usually reduced, an UCST is increased, and a LCST is decreased by an increase in molecular weight [19].

The melt–crystal coexistence temperature $T_c^{(0)}$ of a crystallizable material is given by

$$T_c^{(0)} = \frac{\Delta H_{mc}^{(0)}}{\Delta S_{mc}^{(0)}} \tag{10}$$

($\Delta H_{mc}^{(0)}$ and $\Delta S_{mc}^{(0)}$ are the enthalpy and entropy difference between the crystal and the melt, respectively). The $T_{cm}^{(0)}$ of component 2 in a homogeneous mixture, however, is lower than that of the pure substance,

$$\frac{\Delta T_{mc}^{(0)}}{T_m^{(0)}} = \frac{T_m^{(0)} - T_{mc}^{(0)}}{T_m^{(0)}} \approx \frac{R\chi_{12}}{\Delta H_{mc}^{(0)}} \cdot (1 - y_2)^2 \tag{11}$$

(R is the general gas constant; $\Delta T_{mc}^{(0)}$ is the melting-point depression), since the entropy gap between melt and crystal is enlarged by mixing.

Miscibility is checked best by turbidity measurements (materials where the components that are mixed are transmittent, whereas phase-separated blends are opaque) and by investigation of the blend glass transition temperature T_g. Whereas phase-separated blends exhibit the T_g's of the components at their usual, possibly frequency-dependent temperatures, intimate mixing down to the monomer unit level causes the T_g's of the components to vanish. Instead, a mixture T_g occurs that is located between the component T_g's according to a certain mixture rule. The most common mixture rule is the Fox equation [20],

$$\frac{1}{T_g^{(m)}} = \frac{y_1}{T_g^{(1)}} + \frac{y_2}{T_g^{(2)}} \tag{12}$$

Measurement of the glass transition can be performed by a number of techniques, the most important of which are thermal methods, e.g., differential scanning calorimetry, and relaxation procedures such as mechanical or dielectric relaxation.

The distinct energetic interactions which, as pointed out above, promote miscibility are delivered best by dipole or van der Waals interactions, respectively, or by hydrogen bonding. PVDF has a strong electric dipole in its monomer unit, and it is therefore not surprising that it is miscible with a number of other polymers which in turn exhibit a strong molecular electric moment. A survey is given in Table 1. Among them are some polymers with commercial importance such as poly(methyl methacrylate) (PMMA), poly(ethyl acrylate) (PEA), or poly(vinyl acetate) (PVAc), all of which carry a carbonyl group. In these instances, the miscibility most probably arises from hydrogen bonding between the double-bonded oxygen of the carbonyl group and the acidic hydrogen of the CH_2-CF_2 group [42]. Miscibility with chemically closely related polymers such as poly(vinyl fluoride) (PVF) or poly(trifluoroethylene) (PTrFE) may seem to be trivial.

III. PVDF/PMMA BLENDS

A. Thermodynamics and Morphology

Thermodynamics and morphology of PMMA/PVDF blends have been investigated thoroughly in the past. The blend has been a standard for the study of mixing phenomena in polymers with crystallizable components because of its easy availability and because the interesting structural and thermodynamic phenomena—mixing, crystallization, glass transition—occur in an easily accessible temperature range.

The components are miscible at all usual processing temperatures in the whole composition range [12,24,43]. This has been proved by a single glass transition which changes continuously with composition in quenched samples as detected by thermal analysis, and can be confirmed by turbidity measurements. However, the simultaneous existence of a LCST has been reported at 350°C [11,26,27], which is far above the decomposition temperature of PMMA (remarkable decomposition starts at 230°C [44]) and an UCST at 140°C [27,28], i.e., below the crystallization temperature of PVDF. Usually, PMMA is not stereoregular and therefore cannot crystallize, but it affects strongly the crystallization kinetics of PVDF by virtue of changes in glass transition temperature and in chain mobility in general. The phase diagram, including the equilibrium melting-point depression curve, the miscibility range, and the glass transition temperatures, is given in Figure 1. The Flory-Huggins interaction parameter has been calculated from melting-point depression according to Eq. (11) [13,25,43,45,46] and neutron scattering [47] to be negative and to depend on composition and temperature. Those values from melting-point depression yield $\chi(T) \approx -110$ K/T [13,43,45] and, at ambient temperature, a composition dependence as $\chi(T) \approx -(0.10-0.26)$ for PMMA-rich compositions and $\chi(T) \approx -(0.50-1.45)$ for blends with high PVDF content [25,45]. The negative sign of χ indicates that miscibility is due to favorable energetic interactions between the two polymers rather than the mixing entropy. This energy gain with mixing is partly attributed to interactions between the PMMA carbonyl groups and the electric

Table 1 Polymers Which Are Miscible with PVDF [21–23]

Polymer	Remarks		References
Poly(methyl methacrylate), PMMA	CH_3 $\|$ $[-CH_2-C-]_x$ $\|$ $O=C-O-CH_3$	LCST = 350°C UCST = 140°C	24,25 11,26,27 27,28
Poly(ethyl methacrylate), PEMA	$CH_2-H\ CH_3$ $\|$ $[-CH_2-C-]_x$ $\|$ $O=C-O-CH_3$	LCST = 150–230°C in blends with isotactic PEMA	24,29,30 31
Poly(methyl acrylate), PMA	$[-CH_2-CH-]_x$ $\|$ $O=C-O-CH_3$		32
Poly(ethyl acrylate), PEA	$[-CH_2-CH-]_x$ $\|$ $O=C-O-CH_2-H_3$	LCST = 150°C at y_{PVDF} = 80%	26,32,33 34
Poly(vinyl methyl ketone), PVMK	$[-CH_2-CH-]_x$ $\|$ $O=C-CH_3$		35
Poly(tetramethylene adipate), PTMA	O $\|\|$ $[-(CH_2)_4-O-C$ $\|$ $_x[-O-C-(CH_2)_4$ $\|\|$ O	Complete miscibility in the melt; partial miscibility in the semicrystalline state	21
Poly(vinyl acetate), PVAc	$-[CH_2-CH-]_x$ $\|$ $O=C-O-CH_3$		36,37
Poly(vinyl fluoride), PVF	$[-CH_2-CHF-]_x$		38
Poly(trifluoroethylene), PTrFE	$[-CF_2-CHF-]_x$		38
Poly(N-vinyl-2-pyrrolidone)	$[-CH_2-CH-]_x$ $\|$ N $/\ \ \backslash$ $CH_2\ \ \ C=O$ $\|\ \ \ \ \ \|$ CH_2-CH_2		22
Poly(N-methyl ethylenimine)	$[-\ N-CH_2-CH_2-]_x$ $\|$ $CH_3-C=O$	Miscible for y_{PVDF} > 50%	39

Table 1 *Continued*

Polymer		Remarks	References
Poly (N,N-dimethyl acrylamide)	$[-CH_2-CH-]_x$ $\|$ $C=O$ $\|$ CH_3-N-CH_3		22
Poly-ϵ-caprolactone, PCL	$[-(CH_2)_5-C-O-]_x$ $\|$ O	LCST = 140°C at $y_{PVDF} = 0.55$	40
Poly-3-hydroxybutyrate, PHB	$[-CH-CH_2-C-O-]_x$ $\|$ $\quad$ $\|$ CH_3 $\quad$ O		41
Poly (neopentyl adipate)	$CH_3 \qquad\qquad O$ $\|\qquad\qquad\quad\|$ $[-CH_2-C-CH_2-O-C-(CH_2)_4-C-O-]_x$ $\|\qquad\qquad\quad\|$ $CH_3 \qquad\qquad O$		24

moments of the PVDF monomer units [21], and partly to hydrogen bonding between the PMMA carbonyl oxygens and the PVDF protons [42].

From steric reasons, the described interactions favor trans conformations in both polymers, i.e., chain extension. Depolarized light-scattering studies [48] revealed that, at noncrystallizable compositions (Figure 2) and in the melt, the two polymers exhibit local ordering with a nematiclike chain arrangement on a local level due to the same interactions which cause miscibility. Similar conclusions have been drawn from electrooptical investigations [44].

Pure PVDF crystallizes from the homogeneous isotropic melt in its α modification and to a less extent in its γ modification [49] in spherulitic supermolecular structure with a lamellar twisting period of about 2 μm [50,51]. The amount of β crystallites is usually negligible. The twisting period increases with addition of PMMA and reaches 11 μm for $y_{PMMA} = 60$ vol% and crystallization at 160°C [50,51], the amount of γ spherulites increases, and the spherulites become coarser [50,52]. Figure 3 illustrates the transition from the regular and compact α spherulites in pure PVDF to the more dendritic and open spherulites with fingery boundaries at high PMMA content of the blend [66]. It can clearly be seen that PVDF/PMMA blends can exhibit a large variety of different crystalline and supermolecular structures, depending on the composition and the time and temperature of crystallization. It has been reported that blending with PMMA occasionally induces partial crystallization in the crystalline β modification, as proved by X-ray diffraction [53] and infrared spectroscopy [54], and that its relative amount enlarges with increasing PMMA content. Blends with 20 wt% PMMA shall already exhibit equal amounts of α and β modifications, and for higher PMMA content the β modification can dominate. This may be due to the increase in extended trans conformations in the PVDF

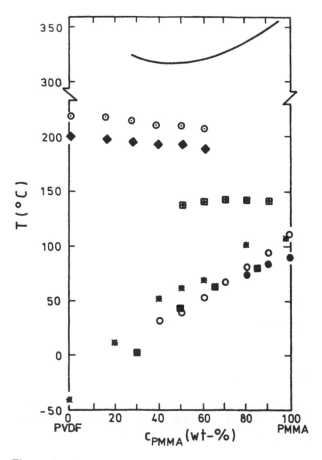

Figure 1   Phase diagram of the PVDF/PMMA blend: ●●●, glass transition temperature T_g [64]; ○○○, T_g [12]; ■■■, T_g [24]; ✳✳✳ T_g [65]; ———, LCST [27]; ▭▭▭, UCST [27]; ⊞⊞⊞, UCST [28]; ◆◆◆, melt/crystal coexistence temperature $T_c^{(0)}$ of α modification of PVDF [45,50]; ☉☉☉, $T_c^{(0)}$ of γ modification of PVDF [50], is the c weight content.

chains upon blending with PMMA, as stated above. This, in turn, promotes crystallization in the β modification, where the chain units assume all-trans conformation. It may, moreover, be due to the change in glass transition and a corresponding change in the relative growth rates of α and β spherulites. The degree of crystallinity of pure PVDF amounts about 50–60 wt%. It decreases roughly proportional to the composition in blends with PMMA and falls below detectibility when the PMMA content exceeds 50% (Fig. 2). Then, crystallization is kinetically supressed due to the rise in glass transition temperature and the long diffusion paths. The degree of crystallinity increases slightly upon drawing, which causes transformation of the α form into the β form which is necessary for electroactivity (cf. Fig. 2).

At first glance, the amorphous phase of crystallized PVDF/PMMA blends should be homogeneously mixed. The two components are completely miscible at all possible crystallization temperatures, and demixing is inhibited from kinetic reasons when the temperature is lowered below the UCST. There are, however, a number of indications that

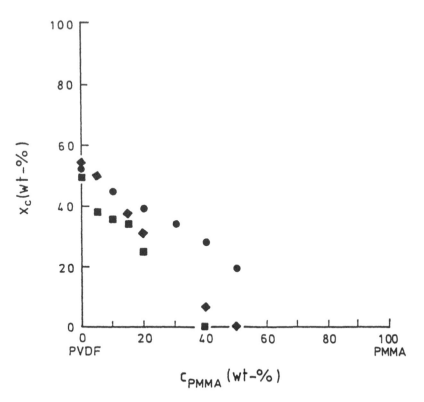

Figure 2 Degree of crystallinity x_c of PVDF/PMMA blends versus composition c_{PMMA}: ●●●, x_c by DSC, pure α modification grown at 147°C for 72 h [45]; ■■■, by WAXS, melt extruded films [15]; ◆◆◆, by WAXS, melt extruded films drawn to a draw ratio of 4 at 85°C [15].

the amorphous phase in the immediate vicinity of crystal surfaces consists of pure PVDF distinct from the amorphous bulk, and that the amorphous phase is therefore a two-phase system [11,12,55–60]. Such a demixing region has been predicted by theory, according to which a surface layer of about 2 nm thickness must exist on the crystals for entropic reasons [61]. The experimental finding is based mainly on dielectric measurements which will be presented in more detail in section C. The same conclusion can be drawn, however, from mechanical spectroscopy [11,12], and is evident also from SAXS measurements [55]. The latter reveal the crystal/amorphous interphase to be 2.0–2.5 nm thick, independent of composition. Besides, from nuclear magnetic resonance (NMR) investigations, it is speculated on local composition heterogeneities with sizes larger than 2 nm [62,63] which distinctly exceed thermodynamic concentration fluctuations which are smaller than 1 nm [25].

B. Mechanical and Rheological Properties

The mechanical properties of homogeneously mixed blends reflect to some degree the composition dependence of glass transition temperature, and the possibility of the components to attain or not to attain their individual relaxation modes in the mixture. They are also influenced by the blend-induced changes in crystalline and supermolecular struc-

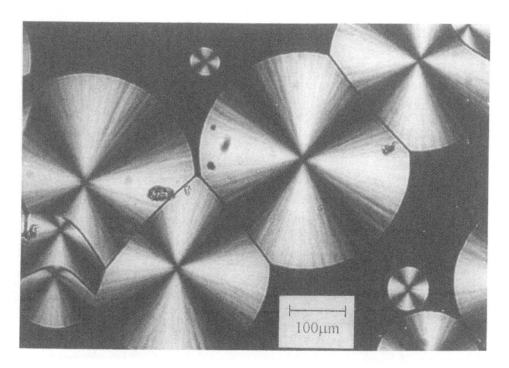

Figure 3 Spherulitic morphologies in pure PVDF and in PVDF/PMMA blends [66]. (top) Pure PVDF. Annealing temperature $T_a = 160°C$, annealing time $t_a = 115$ min. Regularly grown α spherulites with small twisting period. (bottom) PVDF/PMMA = 50/50 (wt%). $t_a = 118$ h, $T_a = 160°C$. Irregular, dendritic α spherulites.

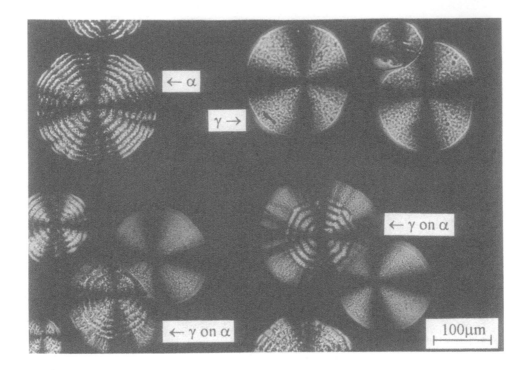

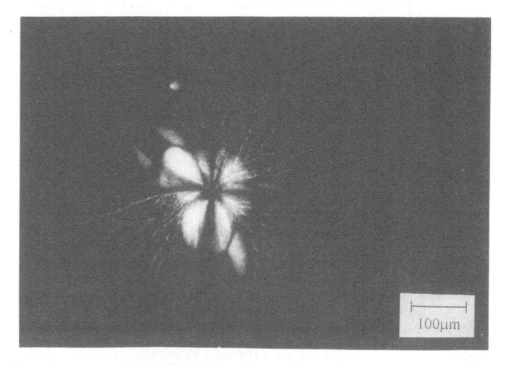

Figure 3 *Continued* (top) PVDF/PMMA = 50/50 (wt%). t_a = 118 h, T_a = 160°C. γ Spherulites nucleated on as-grown α spherulites. (bottom) PVDF/PMMA = 50/50 (wt%). t_a = 6 days, T_a = 164°C. α Spherulites, β spherulites, and γ spherulites nucleated on as-grown α spherulites.

tures. The rheological behavior is, moreover, determined by the mixing-induced conformation changes of the individual polymer chains and by the entanglements between dissimilar chains which are ruled by the thermodynamics of the blend.

The stress–strain behavior of PVDF/PMMA blends in that composition range where no crystallisation occurs exhibits a gradual change from the brittleness of the pure PMMA to the pronounced yielding of the PVDF [65]. The tensile modulus E [$E(c_{PVDF} = 0) = 2.8$ GPa; $E(c_{PVDF} = 60$ wt%$) = 0.9$ GPa] and the tensile strength σ_R [$\sigma_R(c_{PVDF} = 0) = 83$ MPa; $\sigma_R(c_{PVDF} = 60$ wt%$) = 20$ MPa] decrease accordingly proportional to the composition. The elongation at break $\epsilon_R = 4\%$ is independent of composition but increases suddenly if the PVDF content exceeds the crystallizability limit (cf. Fig. 2). The decrease of microhardness with increasing PVDF content reflects the composition dependence of the glass transition temperature [10].

The zero-shear melt viscosity of PVDF/PMMA blends is distinctly reduced in comparison to that of the components, a negative deviation from the linear mixing rule [9]. This is caused by the reduced entanglement density between the two chain types as determined by a suitable analysis of the plateau value of the mechanical loss modulus in the melts.

Both PVDF and PMMA exhibit several relaxations in dynamic mechanical measurements. In a temperature-dependent experiment, the relaxation at the highest temperature, called α relaxation, changes its location on the temperature scale gradually with composition [11,12]. This, on the one hand, indicates miscibility of the two polymers. On the other hand, this α relaxation must be linked with the glass transition of the blend. Some other relaxations do not shift with composition; they are discussed in more detail in the next section, where dielectric relaxation is treated.

The hydrostatic pressure dependence of the glass transition in PVDF/PMMA blends has been reported on in [67]. The volume compressibilities of these materials decrease rather evenly with increasing PVDF content and increasing pressure [15,16]. Typical values at 60°C are 28 GPa^{-1} for PMMA at ambient pressure and 19 GPa^{-1} for PVDF at 200 MPa.

C. Dielectric Properties

The ability of materials to form ferroelectric phases, and their possible pyroelectric and piezoelectric behavior are closely linked with their dielectric properties. Knowledge of the latter has value not only for its own sake, but is a necessary prerequisite for an understanding of the former. Besides, measurement of the complex permittivity in dependence on temperature and frequency allows insight into a number of structural and dynamic features on a microscopic level.

Pure PVDF exhibits four relaxations when temperature and frequency are changed [68,69]: an α relaxation at 1 kHz/80°C, a β' relaxation at 1 kHz/50°C, a β relaxation at 1 kHz/−20°C, and a γ relaxation at 1 kHz/−80°C. They have been attributed in older papers to

(α)   Mobilities associated with defects in the crystals (activation enthalpy, i.e., height of the energy barrier between competing sites: 24 kcal/mol [68])

(β)   ''Micro-Brownian'' motion in the amorphous phase which is linked with the glass transition since the relaxation seems to exhibit ''WLF behavior'' [70] (activation enthalpy: 120 kcal/mol [55], 50 kcal/mol [68], 100 kcal/mol [71])

(β′)   Fold motions in the amorphous phase
(γ)   Chain rotations in the amorphous phase (activation enthalpy: 12 kcal/mol [68])

Blending with PMMA almost does not affect the locations of all these relaxations on the temperature or frequency scale, respectively (Fig. 4 [56–60,72,73]). With respect to the α and γ relaxations, this is in accordance with the proposed molecular origins, since crystalline relaxations cannot be influenced by blending. The β relaxation, however, should change evenly with temperature after addition of PMMA between the T_g's of the components [T_g(PVDF) ≈ −40°C; T_g(PMMA) ≈ 110°C] if it is really due to the glass transition. This contradictory observation and its explanation, which also has consequences for ferroelectric behavior, will be discussed in more detail below.

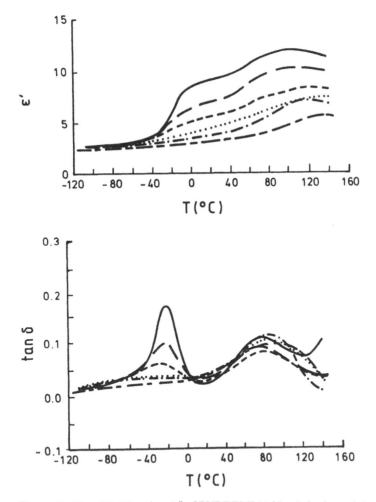

Figure 4   Permittivities $\epsilon' + i\epsilon''$ of PVDF/PMMA blends in dependence on temperature at 1 kHz [56]. PVDF content (wt%): [100] ——, [90] – –, [80] — — —, [60] ···, [40] -·-·-, [20] — -. (Above) dielectric constant ϵ', (below) loss factors tan δ = ϵ''/ϵ'. The peak in ϵ'' at about −20°C indicates the β_{PVDF} relaxation, that around 80°C is a superposition of the α_{PVDF} relaxation, of the β_{PMMA} relaxation, and of the glass transition of the homogeneously mixed amorphous phase.

The permittivities in the frequency range $\nu = 0.005-300$ kHz, in the temperature range $T = 20-60°C$, and in the pressure range $p = 0.1-260$ MPa, their real and imaginary parts as well, coincide well after application of suitable composition independent shift factors [71]. These shift factors are

$$\frac{\Delta(1/T)}{\Delta \log(\nu/\text{Hz})} = (-2 \pm 0.3) \times 10^{-4} \text{ K}^{-1} \text{ at atmospheric pressure} \tag{13}$$

and

$$\frac{\Delta p}{\Delta \log(\nu/\text{Hz})} = (-140 \pm 50) \text{ MPa at } 20°C \tag{14}$$

The Gibbs free energy of activation is

$$\Delta E_A = \Delta E_A^{(0)} + V_A \, \Delta p \tag{15}$$

where Δp is the pressure difference from atmospheric pressure. The quantity V_A has the meaning of an "activation volume." It amounts to 0.067 nm^3, independent of composition. This value should be compared with the volume of 0.055 nm^3 which a monomer unit occupies in the crystal lattice [49]. Each monomer unit, obviously, relaxes individually after a pressure jump. In contrast, the stress activation volume has been reported to diminish the activation energy [74]. It is larger by two orders of magnitude and is interpreted as cooperative inclusion of up to 50 monomer units into the mechanical load-induced thermodynamic work. From the shift factor Eq. (13), ΔE_A^0 is found to be approximately 100 kJ/(mol of activated entities), again independent of composition. By combination of the shift rules Eqs. (13) and (14), it is finally found that $\Delta T/\Delta p = 0.13$ K/MPa, this value being half that of 0.28 K/MPa derived from mechanical relaxation measurements [75]. In summary, the pressure-dependent activation energy amounts to $\Delta E_A = (100 + 0.02 \text{ MPa}^{-1} \Delta p)$ kJ/mol, independent of composition.

The experimental findings prove consequently that the β relaxation of PVDF is thermally activated in the mentioned (ν, T, p) range. The strengths of the α and β relaxations also decrease in the blended PVDF roughly in proportion to the overall degree of crystallinity as measured by X-ray scattering (cf. Figs. 2 and 4). They vanish for c_{PVDF} < 0.5. This excludes the amorphous bulk as the location, and the glass transition as the origin of the β relaxation. On the one hand, therefore, the β_{PVDF} relaxation must be linked with the crystalline phase, since it disappears with disappearing crystallinity. On the other hand, the relaxation must nevertheless be linked with the amorphous phase, since addition of PMMA can influence only the dynamics and structure of this phase. It follows that the contributing chain segments must necessarily lie in the crystal–amorphous interface ("interphase"), which must be accordingly broad. These experimental findings and conclusions are consistent with theoretical predictions on an interphase layer on polymeric crystals. These follow from consideration of the dissipation of order from the crystal to the amorphous bulk, and from the entropic effects on the miscibility connected with that [58]. This has already been outlined above in the thermodynamics and structure section.

In any case, the presence of PMMA does not significantly influence the energy barrier of the motions of the PVDF units which contribute to the β_{PVDF} relaxation, since the activation enthalpy is independent of composition. Addition of PMMA, however, shifts the relaxation to higher frequencies and pressures and thus acts as a diluent that facilitates relaxation from a basically lower energetic level. It can be concluded that addition of PMMA breaks the interactions and correlations between the PVDF permanent moments in the interface, thus enhancing the dielectric relaxation possibilities. Similar conclusions can be drawn from the composition and stress dependence of the piezoelectric constants [14,15].

If the electric field strength during the dielectric experiment is sufficiently increased, PVDF turns out to exhibit a strong nonlinear dielectric behavior [76,77]. Defining

$$D = \epsilon_0[(1 + \chi_1)E + \chi_2 E^2 + \chi_3 E^3 + \ldots] \tag{16}$$

(D is the dielectric displacement, ϵ_0 is the permittivity of free space, χ_i is the dielectric susceptibility of ith order, E is the applied electric field), one gets the values of Table 2. It should be pointed out that the component χ_2 should vanish in unpoled samples for thermodynamic reasons. Blending obviously reduces the third-order susceptibility.

D. Ferroelectric, Piezoelectric, and Pyroelectric Properties

As it has been outlined in the introduction to this chapter, blending can be an interesting tool to improve performance properties of a material. Some attempts have been made by blending with PMMA to meet this objective with respect to the electroactivity of PVDF. The background of these works is that:

Possibly immediate crystallization in the polarized and electroactive crystalline β modification can be achieved by this means [53,54], this possibly making an additional drawing step unnecessary.

The blend can have better mechanical or electrical performance properties by virtue of the changes in elasticity and glass transition temperature. With respect to pyroelectricity and piezoelectricity, the mechanical or thermal-to-electric work transversion coefficients, respectively, can possibly be changed favorably by this means.

Addition of PMMA allows preparation of a wider variety of supermolecular structures, this possibly allowing more detailed insight into the processes on the molecular and submolecular levels which enable and govern electroactivity of PVDF.

1. Polarization

Only the polar crystalline α′, β, and γ phases [49] can carry permanent polarization and contribute to an electric field-induced macroscopic remanent polarization in pure PVDF at ambient temperature, since the glass transition temperature T_g is about −40°C. Electric field-induced dipole orientations in the amorphous phase would therefore randomize after short times. This holds also for PVDF/PMMA blends with a detectable degree of crystallization (cf. Figs. 1 and 2), where T_g is accidently also below room temperature. In

Table 2 Dielectric Susceptibilities Perpendicular to a Film Plane of PVDF/PMMA Blends: Oriented Material with Crystalline β Modification ($T = 18°C$, $\nu = 25$ Hz)

Susceptibility	Pure PVDF	PVDF/PMMA = 90/10 (wt%)
χ_1	11	8
$\chi_1/10^{-9}$ m/V	1	2
$\chi_3/10^{-17}$ m²/V²	10	7

Source: After Ref. 76.

contrast, blends with PMMA contents higher than 50 wt% are not able to crystallize remarkably, and have a T_g above room temperature. They can, however, also not attain polarization for longer times since this would randomize by the γ relaxation of the PVDF and the β relaxation of the PMMA. A remnant polarization, if present at all, would then be due only to captured charges as in usual electrets. The polarization of such electrets amounts to 0.05–5 mC/m² [54], as found in the respective PVDF/PMMA blends (Fig. 5). Besides captured charges, there is, however, possibly another origin of the polarization of the noncrystallized blends. As outlined in more detail in Section III.A, the energetic interactions between the blend components enforce nematiclike chain conformations which are not crystalline entities in a crystallographic sense but can carry a permanent polarization after poling in a similar manner as true crystals.

From these considerations, it might be expected that the remanent polarization of PVDF is proportional to the degree of crystallinity x_c, which in turn can be adjusted by blending with PMMA (Fig. 2). It has been observed, however, that the polarization drops with addition of PMMA much faster than x_c (Fig. 5) until reaching the steady value characteristic for the amorphous blend both for isotropic materials with predominantly α-crystalline morphology and for drawn films with β crystallites [15,54,76,78–80]. The amorphous phase which surrounds individual crystallites therefore strongly influences the amount and stability of the crystalline polarization. It must be recalled here (cf. Secs. III.A and III.C) that the amorphous phase in the immediate vicinity of the crystallites has another composition and other chain conformation than the amorphous bulk [55,56,60]. There are a number of ways by which this amorphous–crystalline interphase can act in this respect. Strong evidence has been found by thermally stimulated current [81,82] and by polarization switching [4] experiments, and it has been concluded from theoretical considerations too [83], that captured charges as injected during the poling procedure reside in this interphase, which shield external and modify internal electric fields, respectively. Their mobility should depend strongly on the structure and dynamics of the interphase. They inhibit the reorientation of electric moments at the crystal surface by virtue of the electric fields which they cause [4]. Such a reorientation, in turn, would occasionally lead to reorientation "nuclei," which finally could grow and cause polarization reversal of the whole crystallite [84]. The charges thus stabilize a given polarization structure. This reorientation procedure is supported by the composition- and crystal shape-dependent chain conformation and dynamics, respectively, in the crystal–amorphous interphase.

Externally applied and internal electric fields can differ strongly. The differences can be linked with the structure and the dielectric properties of the material under investigation, and they can be due to the existence and mobility of charges which reside permanently in the material or which are injected by the external field. Poled crystallites in amorphous surroundings, moreover, create a permanent internal electric field E_i even when the external poling field E_e is switched off. These internal fields are stabilized by the injected charges and have been shown to exceed the initial poling field remarkably [86]: $E_i = 200$ MV/m for $E_e = 80$ MV/m has been reported for a PVDF/PMMA = 80/20 (wt%) blend. However, E_i decreases severely with annealing at elevated temperature. This strong field can cause orientation of dye molecules which reside in the amorphous phase. This, in turn, can lead to interesting nonlinear optical properties of the material. The strong internal field can, moreover, lower the photoionizing threshold of charge transfer complexes which are introduced into the blend [87]. This process, in turn, enlarges the number of active charge carriers.

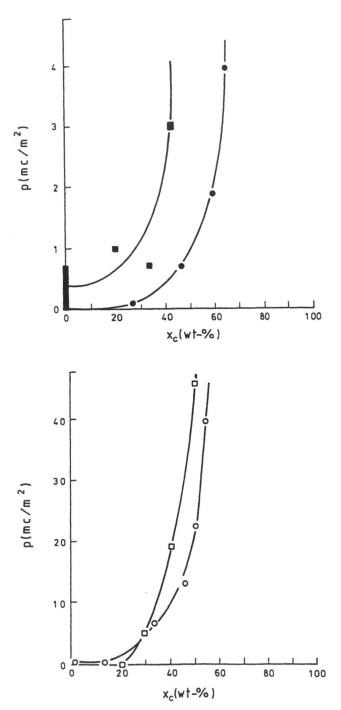

Figure 5 Permanent polarization P after equivalent poling in (above) as-crystallized and (below) drawn PVDF/PMMA blend films versus degree of crystallinity x_c as adjusted by blending with PMMA. Values taken from: [54,80] ■■■; [15,16] ●●●; [54,80] □□□; [15,16] ○○○. The long bar in the lower left corner of the upper figure indicates the scattering range of the electret polarization of amorphous blends.

It has been concluded from DC conduction measurements [76,88] that the afore-mentioned charge carriers are generated by emission from the electrodes and from trapped ionic impurities within the materials, and that they are conducted to the bulk of the material and their final residence by a hopping process along the polymeric chains.

Both remnant (P_r) and maximal (P_s) polarization, respectively, of pure PVDF in-crease with increasing poling temperature in accordance with the behavior of inorganic or ceramic ferroelectrics [76]. In contrast, a reverse behavior is found in blends (Fig. 6). The differences between blended and pure PVDF vanish with increasing poling temper-ature T_p. In all cases, moreover, the materials become ferroelectrically harder with in-creasing T_p, since the differences between P_s and P_r decrease. The polarization-reversal switching time is the same for the blends and for the pure PVDF and amounts to about 30 μs at room temperature and at a poling field strength of 200 MV/m [76]. The facts have been explained in terms of a composition- and temperature-dependent nucleation and charges injection rate, respectively, and by the temperature dependence of the injected charges.

Co-polymers of vinylidene fluoride and trifluoroethylene, P(VDF/TrFE), are miscible with PMMA similarly to the PVDF homopolymer [78,79]. In contrast to PVDF, however, the copolymer exhibits a distinct ferroelectric–paraelectric phase transition of the first order at a Curie temperature [85]. This Curie temperature does not change in blends with PMMA, although the copolymer melting temperature shows the usual depression which indicates miscibility [78]. This result indicates that the conformational changes of the crystalline chains which accompany the ferroelectric–paraelectric phase transition are not

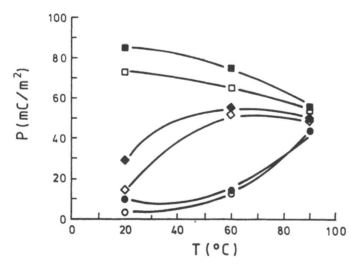

Figure 6 Saturation (P_s) and remnant (P_r) polarisation, respectively, of PVDF/PMMA blends of PVDF content c_{PVDF} versus temperature T: ■■■, P_s (c_{PVDF} = 100 wt%); ▭▭▭, P_r (c_{PVDF} = 100 wt%); ◆◆◆, P_s (c_{PVDF} = 90 wt%); ◇◇◇, P_r (c_{PVDF} = 90 wt%); ●●●, P_s (c_{PVDF} = 80 wt%); ○○○, P_r (c_{PVDF} = 80 wt%). (After Ref. 76.)

affected by the amorphous surroundings of the crystallites, in contrast to those which are linked with the polarization reversal or creation, respectively.

2. Piezoelectricity and Pyroelectricity

It is well known that the strong piezoelectricity and pyroelectricity of PVDF is linked with the existence of a remnant polarization P_r, and that theory demands the pyroelectric (p) and the piezoelectric (d) coefficients be proportional to P_r. These coefficients must consequently drop strongly after blending PVDF with PMMA to an extent which is comparable to that of the permanent polarization. This is indeed observed [15,16,53,80]. It is therefore advantageous to discuss the normalized values d/P_r and p/P_r (Figs. 7 through 9). They decrease slightly with increasing PMMA content. These changes with composition can be understood completely by the composition dependence of permittivity and of mechanical properties, i.e., of compliance (at stress piezoelectricity) and compressibility (for the hydrostatic piezoelectric coefficient). It is remarkable that these fea-

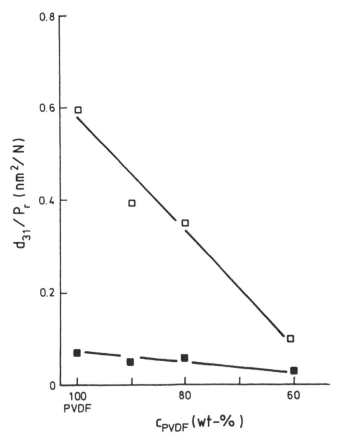

Figure 7 Composition dependence of the normalized piezoelectric coefficient d_{31}/P_r of PVDF/PMMA films with crystalline β modification (subscript 31: polarization change measured on the film surfaces after application of stress in draw direction): ■■■, $T = -100°C$; ▢▢▢, $T = 20°C$. (After Ref. 80.)

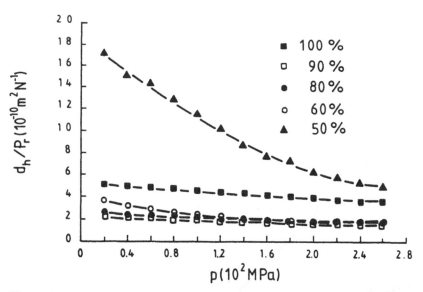

Figure 8 Normalized hydrostatic piezoelectric coefficients d_h/P_r of isotropic PVDF/PMMA blends at 60°C versus hydrostatic pressure p. Parameter: PVDF content c_{PVDF} (wt%). (After Ref. 16.)

tures hold regardless of whether the material contains the crystalline α modification of the PVDF or whether it is drawn and contains the β crystallites. They hold also for those blends with high PMMA content which do not exhibit measurable crystallinity but nevertheless have a permanent polarization. It must be concluded that the molecular origin of the polarization plays no role with respect to the magnitude of the pyroelectric and piezoelectric properties.

It can be read from Figure 8 that the hydrostatic piezoelectric coefficients of PVDF/PMMA blends depend on pressure; hydrostatic piezoelectricity is consequently a nonlinear phenomenon, since the induced polarization is not proportional to the applied pressure [15,16]. Similar observations can be made with regard to stress piezoelectricity [14,54,89]. The latter, however, is a true nonlinearity in the sense that it occurs in a mechanically distinctly linear-reversible regime, whereas the nonlinearity of the former reflects mainly the pressure dependence of compressibility.

The electromechanical coupling coefficient which describes the efficiency of the mechanical-to-electrical work transformation by piezoelectricity becomes negligibly small if the PMMA content exceeds 20% [73]. This is again due to the decrease in permittivity and the increase in glass transition temperature and, consequently, in brittleness, with blending, and proves PVDF/PMMA blends to be not a suitable material for sonic transducers.

IV. BLENDS WITH OTHER POLYMERS

There are only a few publications which treat the dielectric and ferroelectric properties and the related features of blends of PVDF with polymers other than PMMA [53,97].

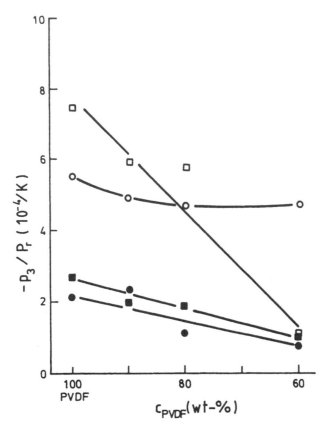

Figure 9 Dependence of normalized pyroelectric coefficients p_3/P_r of PVDF/PMMA films on PVDF content c_{PVDF} (subscript 3: polarization change as measured on the film surfaces): ▬▬, drawn materials with crystalline β modification, $T = -100°C$; ⊡⊡⊡, drawn materials with crystalline β modification, $T = 20°C$; ●●●, undrawn materials with crystalline α modification, $T = -100°C$; ○○○, undrawn materials with crystalline α modification, $T = 20°C$. (After Ref. 80.)

On the one hand, this may be because in fact blend-induced changes in ferroelectricity can be studied best with PVDF/PMMA blends, but, on the other hand, may be due to the experience that the ferroelectric properties of this blend are in general worse than those of the pure polymer. Moreover, all that can be observed with such other blends can be understood and explained by the structural and dynamic features and models developed from the findings on PVDF/PMMA blends.

The pyroelectric coefficient of isotropic blends of PVDF with polyvinylfluoride, PVF, changes between the respective values of the components [p(PVF) = 3 μC/Km²] but exhibits a maximum at a PVF content of 11.5 wt%, which exceeds the value for PVDF by 50% [53]. This seeems to reflect, and is in accordance with, a corresponding dependence on composition of the crystallinity sum of both components. Moreover, the weight ratio of the polar crystal modifications depends on composition in the same manner.

The piezoelectricity of blends from the two copolymers P(VDF/TeFE) and poly (vinylidene fluoride/hexafluoroacetone), P(VDF/HFA), has been investigated too [97].

The two components are one of the rare examples of cocrystallizing polymer pairs. The amount of polar crystal modifications increases with increasing P(VDF/TeFE) content. This in turn is imagined by the according variation of the piezoelectric coefficients.

V. ADDITION OF PLASTICIZERS

It has been stressed several times in this chapter that the dynamics and the structure of the amorphous surroundings of the polarized crystalline entities influences decidingly the strength and the stability of the electrical properties. This had been concluded particularly from the di-, ferro-, pyro-, and piezoelectric behavior, respectively, of PVDF/PMMA blends where the structure and the dynamics of the amorphous phase can be varied to a large extent. In blends, however, controlling these features of the amorphous phase separate from those of the crystalline phase is not possible. Blending always simultaneously influences the crystallization kinetics, which changes the crystallinity and structure of the crystalline phase, occasionally in a disadvantageous manner. This difficulty can be overcome by "impregnation" of already-crystallized materials with low-molecular-weight substances which are miscible with PVDF. Then, blending is performed by diffusion rather than by mixing. It is clear that such additives cause a strong drop in glass transition temperature even when present only to some percent. They are consequently called plasticizers. These plasticizers usually reside in the amorphous phase, regardless of whether they are introduced into the material by mixing or by impregnating, but in the latter case there is a large number of additional preparational routes: Impregnation can be performed in the amorphous or already-crystallized material, respectively, and prior to or after the poling procedure. Furthermore, doping of materials with different crystal modifications, and of drawn and undrawn materials, respectively, is possible.

Impregnation of PVDF films with polar liquids such as propylene carbonate which exhibit large dielectric constants and dissipation factors reduces the breakdown tendency at application of strong electric fields [90,91]. This is due to the smoothing of structural defects, and to the enhancement of dielectric strength by a mixing rule for the permittivities.

Several attempts have been made to affect the ferroelectric, pyroelectric, and piezoelectric properties of PVDF films and of P(TrFE/VDF) films by addition of tricrecyl phosphate (TCP [92–96]). A TCP content of about 3 wt% is achieved after dipping at 100°C for 10 h [95]. This specific plasticizer diffuses only into the amorphous phase and does not affect crystallinity when applied up to 5 wt% [95]. Crystallinity, however, can be enlarged during the poling procedure by growth of already-existing crystals [94]. This causes the remnant polarization to increase in turn by about 20% at both low and at high poling field strengths when inserting only 1 wt% TCP [92]. The polarization is then stable up to about 140°C, whereas the polarization of unloaded PVDF is partially destroyed upon heating above 90°C. Pyroelectric and piezoelectric stress ($\partial P/\partial\sigma$) coefficients increase accordingly [94,95]. The piezoelectric strain coefficient $\partial P/\partial\epsilon$ decreases, however, by virtue of the drop in mechanical strength by 20% with doping [95]. Ferroelectric switching by electric force reversal is enhanced by a strong decrease in coercive force. The activity of the plasticizer is explained by an increase in nucleation and growth efficiency at the crystal–amorphous interface which induces the reversal of polarization [92,94]. The glass transition temperature of PVDF decreases by about 10°C when adding 5 wt% [95]. The results are basically independent of whether the investigations are performed with drawn [92,94,95] or with undrawn materials [93,96].

VI. CONCLUDING REMARKS

It has been the objective of this chapter to show that blending of PVDF with high- and with low-molecular-weight materials can be an interesting and successful tool to enhance performance, to create new properties, and, in particular, to contribute to an understanding of the microscopic structural and dynamic basis of the electroactivity of PVDF.

Most attempts have been made so far to meet these aims by blending with PMMA. The background of these works is that, by this means, immediate crystallization in the polarized and electroactive crystalline β modification can be achieved. Addition of PMMA, moreover, allows the preparation of a wide variety of supermolecular structures. This allows more detailed insight into the processes on the molecular and submolecular levels which enable and govern electroactivity of PVDF. Whereas blending with PMMA, however, does not in general improve the ferroelectric properties of the material, it allows detailed conclusion on the deciding role which the amorphous phase plays for the magnitude and the dynamics of the polarization and for the related features in PVDF.

In contrast, plastification of PVDF is an interesting new approach to modify the structure and the dynamics of the amorphous surroundings of polarized crystals in PVDF. Most of the ferroelectric properties of PVDF can be improved with this technique. The approach may yield a number of additional interesting results in the future.

PVDF can be mixed thermodynamically stably not only with PMMA but with a large number of other polymers. However, so far only a few investigations have been reported on the electric properties of such blends. Some more publications which treat the thermodynamics and the structure of PVDF blends which have not been referred to in the present chapter are nevertheless included in the reference list at its end [97–117].

REFERENCES

1. T. T. Wang, J. M. Herbert, and A. M. Glass (eds.), *The Applications of Ferroelectric Polymers*, Chapman & Hall, New York, 1987.
2. T. Yamada, T. Ueda, and T. Kitayama, Piezoelectricity of a high-content lead zirconate titanate/polymer composite, *J. Appl. Phys. 53*:4328 (1982).
3. C. Dias and D. K. Das-Gupta, Electroactive and dielectric properties of corona and thermally poled polymer-ceramics composites, *Proc. 7th Int. Symp. on Electrets*, Berlin, Germany, 1991, pp. 495–500.
4. G. Eberle, E. Biehler, and W. Eisenmenger, Polarization dynamics of VDF-TrFE copolymers, *IEEE Trans. Elect. Insul., 26*:69 (1991).
5. Y. U. Diikova, Y. Z. Lyakhovskii, and B. I. Sazhin, Influence of the constant electrical field on the effective inductive capacity of multilayer dielectrics on the base of poly(vinylidene fluoride) and poly(ethylene terephthalate), *Vysokomol. Soed. A34*:116 (1992), in Russian.
6. K. Mazur, Piezoelectricity of PVDF/PUE, PVDF/PMMA, and PVDF/(PMMA + BaTiO₃) laminates, *IEEE Trans. Elect. Insul. 27*:782 (1992).
7. B. R. Hahn and J. H. Wendorff, Compensation method for zero birefringence in oriented polymers, *Polymer 26*:1619 (1985).
8. G. A. Gallagher, R. Jakeways, and I. M. Ward, The structure and properties of drawn blends of poly(vinylidene fluoride) and poly(methyl methacrylate), *J. Polymer. Sci. Part B: Polymer Phys. 29*:1147 (1991).
9. S. Wu, Entanglement between dissimilar chains in compatible polymer blends: PMMA and PVDF, *J. Polymer Sci. Part B: Polymer Phys. 25*:557 (1987).
10. J. Martínez-Salazar, J. C. Canalda Cámera, and F. J. Baltá Calleja, Mechanical study of poly(vinylidene fluoride)-poly(methyl methacrylate) amorphous blends, *J. Mat. Sci. 26*:2579 (1991).

11. Y. Hirata and T. Kotaka, Phase separation and viscoelastic behavior of semicompatible polymer blends: Poly (vinylidene fluoride)/poly (methyl methacrylate) system, *Polymer J. 13*:273 (1981).

12. D. R. Paul, J. O. Altamirano, Properties of compatible blends of poly (vinylidene fluoride) and poly (methyl methacrylate), *Adv. Chem. Ser. 142*:32 (1975).

13. T. J. Moon and H. G. Kim, Transition behavior of poly (vinylidene fluoride)/poly (methyl methacrylate) blends, *Polymer* (Korea) *12*:144 (1988).

14. P. Harnischfeger and B.-J. Jungnickel, Features and origin of the dynamic and the nonlinear piezoelectricity in poly(vinylidene fluoride), *Polymer Adv. Tech. 1*:171 (1990).

15. F. Schaffner, R. Wellscheid, and B.-J. Jungnickel, The hydrostatic piezoelectric coefficient of PVDF/PMMA blends, *IEEE Trans. Elect. Insul. 26*:78 (1991).

16. R. Schaffner and B.-J. Jungnickel, The electric moment contribution to the piezoelectricity of PVDF/PMMA blends, *IEEE Trans. Dielectr. Elect. Insul., 1*:553 (1994).

17. O. Olabisi, L. M. Robeson, and M. T. Shaw, *Polymer-Polymer Miscibility*, Academic Press, New York, 1979.

18. L. P. McMaster, Aspects of polymer-polymer thermodynamics, *Macromolecules 6*:760 (1973).

19. L. A. Kleintjens and R. Koningsveld, Thermodynamics of polymer solutions and blends, *Makromol. Chem. Macromol. Symp. 20/21*:203 (1989).

20. T. G. Fox, Influence of diluent and of copolymer composition on the glass temperature of a polymer system, *Bull. APS 1*:123 (1956).

21. C. Reckinger and J. Rault, Étude du mélange PVDF-PTMA, *Rev. Phys. Appl. 21*:11 (1986).

22. M. Galin, Miscibility of poly (vinylidene fluoride) with poly (N-vinyl-2-pyrrolidone) and poly (N.N-dimethylacrylamide), *Makromol. Chem. Rapid Commun. 5*:119 (1984).

23. S. Krause, Compatible polymers, *Polymer Handbook*, 3rd ed. (J. Brandrup and E. H. Immergut, eds.), John Wiley, New York, 1989, p. VI/347.

24. J. S. Noland, N. N.-C. Hsu, and J. M. Schmitt, Compatible high polymers: Poly (vinylidene fluoride) blends with homopolymers of methyl and ethyl methacrylate, *Adv. Chem. Ser. 99*:15 (1970).

25. J. H. Wendorff, Concentration fluctuations in poly (vinylidene fluoride)-poly (methyl methacrylate) mixtures, *J. Polymer Sci. Polymer Lett. Ed. 18*:439 (1980).

26. R. E. Bernstein, C. A. Cruz, D. R. Paul, and J. W. Barlow, LCST behavior in polymer blends, *Macromolecules 10*:681 (1977).

27. H. Saito, Y. Fujita, and T. Inoue, Upper critical solution temperature behaviour in poly (vinylidene fluoride)/poly (methyl methacrylate) blends, *Polymer J. 19*:405 (1987).

28. H. Tomura, H. Saito, and T. Inoue, Light scattering analysis of UCST behaviour in a PVDF/PMMA blend, *Macromolecules 25*:1611 (1992).

29. R. L. Imken, D. R. Paul, and J. W. Barlow, Transition behavior of poly (vinylidene fluoride)/poly (ethyl methacrylate) blends, *Polymer Eng. Sci. 16*:593 (1976).

30. T. K. Kwei, G. D. Patterson, and T. T. Wang, Compatibility in mixtures of poly (vinylidene fluoride) and poly (ethyl methacrylate), *Macromolecules 9*:780 (1976).

31. E. Roerdink and G. Challa, LCST behaviour in blends of poly (vinylidene fluoride) and isotactic poly (ethyl methacrylate), *Polymer 21*:1161 (1980).

32. D. C. Wahrmund, R. E. Bernstein, J. W. Barlow, and D. R. Paul, Polymer blends containing poly (vinylidene fluoride). Part I: Poly (acryl acrylates), *Polymer Eng. Sci. 18*:677 (1978).

33. R. M. Briber and F. Khoury, The phase diagram and morphology of blends of poly (ethyl acrylate) and poly (vinylidene fluoride), *ACS Div. Polymer Chem. Polymer Prep. 26*(2):310 (1985).

34. B. Endres, R. W. Garbella, and J. H. Wendorff, Studies on phase separation and coarsening in blends of poly (vinylidene fluoride) and poly (ethyl acrylate), *Coll. Polymer Sci. 263*:361 (1985).

35. R. E. Bernstein, D. C. Wahrmund, J. W. Barlow, and D. R. Paul, Polymer blends containing poly (vinylidene fluoride). Part III: Polymers containing ester, ketone, or ether groups, *Polymer Eng. Sci. 18*:1220 (1978).

36. R. E. Bernstein, D. R. Paul, and J. W. Barlow, Polymer blends containing poly (vinylidene fluoride). Part II: Poly (vinyl esters), *Polymer Eng. Sci. 18*:683 (1978).

37. R. E. Belke and I. Cabasso, Poly (vinylidene fluoride)/poly (vinyl acetate) miscible blends. I. Thermal analysis and spectroscopic (FTIR) characterization, *Polymer 29*:1831 (1988).

38. Y. K. Godovskii, Z. F. Zharikova, and Y. M. Malinskii, On the compatibility of polyfluoroethylenes, *Polymer Sci. USSR 23*:149 (1981).

39. S. Kobayashi, M. Kaku, and T. Saegusa, Miscibility of poly (2-oxazolines) with commodity polymers, *Macromolecules 21*:334 (1988).

40. W. H. Jo, S. J. Park, and I. H. Kwon, Phase behavior of poly(ε-caprolactone)/poly (vinylidene fluoride) blends, *Polymer Int. 29*:173 (1992).

41. H. Marand and M. Collins, Crystallization and morphology of poly (vinylidene fluoride)/poly (3-hydroxybutyrate) blends, *ACS Div. Polymer Chem. Polymer Prep. 31(2)*:552 (1990).

42. C. Léonard, J. L. Halary, and L. Monnerie, Hydrogen bonding in PMMA-fluorinated polymer blends: FTIR investigations using ester model molecules, *Polymer 26*:1507 (1985).

43. T. Nishi and T. T. Wang, Melting point depression and kinetic effects of cooling on crystallization in poly (vinylidene fluoride)-poly (methyl methacrylate) mixtures, *Macromolecules 8*:909 (1975).

44a. B.-J. Jungnickel, Untersuchungen zur elektrisch induzierten Doppelbrechung (Kerr-Effekt) an Mischungen thermodynamisch kompatibler Polymere, report for Deutsche Forschungsgem., Projekt-Nr. Ju 168/1-2, 1993.

44b. C. Hartig, R. Kleppinger, and B.-J. Jungnickle, Kerr effect measurements on PVDF, PMMA, and their blends, *Polymer*, (submitted, 1994).

45. B. S. Morra, R. S. Stein, Melting studies of poly (vinylidene fluoride) and its blends with poly (methyl methacrylate), *J. Polymer Sci. Polymer Phys. Ed. 20*:2243 (1982).

46. W. H. Jo, J. T. Yoon, and S. C. Lee, Miscibility of poly (vinylidene fluoride) and poly (styrene-co-methyl methacrylate) blends, *Polymer J. 23*:1243 (1991).

47. G. Hadziioannou and R. S. Stein, Neutron scattering studies of dimensions and of interactions between components in polystyrene/poly (vinyl methyl ether) and poly (vinylidene fluoride)/poly (methyl methacrylate) amorphous blends, *Macromolecules 17*:567 (1984).

48. H. Saito, M. Matsuura, and T. Inoue, Depolarized light scattering studies on single-phase mixtures of dissimilar polymers: Evidence for local ordering, *J. Polymer Sci. Part B: Polymer Phys. 29*:1541 (1991).

49. A. J. Lovinger, PVDF, *Developments in Crystalline Polymers, Vol. I* (D. C. Bassett, ed.), Applied Science Publishers, New Jersey, 1988, p. 195.

50. B. S. Morra and R. S. Stein, The crystalline morphology of poly (vinylidene fluoride)/poly (methyl methacrylate) blends, *Polymer Eng. Sci. 24*:311 (1984).

51. K. Ulrich, Kristallisationskinetik von PVDF/PMMA-Mischungen (crystallization kinetics of PVDF/PMMA blends), Master's thesis, Technische Hochschule Darmstadt, 1980.

52. B. S. Morra and R. S. Stein, Morphological studies of PVDF and its blends with PMMA, *J. Polymer Sci. Polymer Phys. Ed. 20*:2261 (1982).

53. H. Lee, R. E. Salomon, and M. M. Labes, Pyroelectricity in polymer blends of poly (vinylidene fluoride), *Macromolecules 11*:171 (1978).

54. B. Hahn, Piezo- und pyroelektrische Eigenschaften von Polyvinylidenfluorid/Polymethylmethacrylat-Mischungen (Piezo- and pyroelectric properties of PVDF/PMMA blends), Ph.D. thesis, Technische Hochschule Darmstadt, 1983.

55. B. R. Hahn, O. Herrmann-Schönherr, and J. H. Wendorff, Evidence for crystal-amorphous interphase in PVDF and PVDF/PMMA blends, *Polymer 28*:201 (1987).

56. B. Hahn, J. Wendorff, and D. Y. Yoon, Dielectric relaxation of crystal-amorphous interphase in poly(vinylidene fluoride) and its blends with poly(methyl methacrylate), *Macromolecules 18*:718 (1985).

57. J. P. Runt, C. A. Barron, X.-F. Zhang, and S. K. Kumar, Crystal-amorphous interphases in binary polymer blends, *Macromolecules 24*:3466 (1991).

58. D. Y. Yoon, S. K. Kumar, and Y. Ando, Crystal-amorphous interphase in semicrystalline polymers and binary polymer mixtures, *Polymeric Mat. Sci. Eng. 60*:715 (1989).

59. Y. Ando and D. Y. Yoon, Phase separation in quenched noncrystalline poly (vinylidene fluoride)/poly (methyl methacrylate) blends, *ACS Div. Polymer Chem. Polymer Prep., 29*: 381 (1988).

60. Y. Ando, D. Y. Yoon, Dielectric characterization of crystal-amorphous interphase in semi-crystalline PVDF and PVDF/PMMA blends, *ACS Div. Polymer Chem. Polymer Prep., 28*: 26 (1988).

61. S. K. Kumar and D. Y. Yoon, Lattice model for interphases in binary semicrystalline/amorphous polymer blends, *Macromolecules 22*:4098 (1989).

62. P. Tékély, F. Laupretre, and L. Monnerie, Local composition heterogeneities occuring during crystallization in PVF$_2$/PMMA blends as studied by high-resolution solid-state ^{13}C NMR, *Polymer 26*:1081 (1985).

63. A. P. A. M. Eijkelenboom, W. E. J. R. Maas, W. S. Veeman, G. H. W. Buning, and J. M. J. Vankan, Triple-Resonance ^{19}F, ^{13}C, CPMAS NMR study of the influence of PMMA tacticity on the miscibility in PMMA/PVF$_2$ blends, *Macromolecules 25*:4511 (1992).

64. G. K. Narula, G. K. Rashmi, and P. K. C. Pillai, Investigations of solution-mixed PVDF/PMMA polyblends by thermal, structural, and dielectric techniques, *J. Macromol. Sci.—Phys. B28*:25 (1989).

65. W. Ullmann and J. H. Wendorff, Mechanical properties of blends of poly (methyl methacrylate) and poly (vinylidene fluoride), *Comp. Sci. Tech. 23*:97 (1985).

66. D. Braun, M. Jakobs, and G. P. Hellmann, On the morphology of poly(vinylidene fluoride) crystals in blends, *Polymer 35*:706 (1994).

67. D. J. Woan, J. I. Scheinbeim, and P. R. Couchman, The pressure dependence of the glass transiton in PVDF/PMMA blends, *Bull. APS 28*:392 (1983).

68. S. Yano, Dielectric relaxation and molecular motion in poly (vinylidene fluoride), *J. Polymer Sci. A-2 8*:1057 (1970).

69. H. Kakutani, Dielectric absorption in oriented poly (vinylidene fluoride), *J. Polymer Sci. A-2 8*:1177 (1970).

70. N. G. McCrum, B. E. Read, and G. Williams, *Anelastic and Dielectric Effects in Polymeric Solids*, John Wiley, London-New York-Sydney, 1967.

71. F. Schaffner and B.-J. Jungnickel, The permittivities of PVDF/PMMA blends at hydrostatic pressure, *J. Macromol. Sci. B—Phys. 32*:343 (1993).

72. J. K. Krüger, A. Marx, R. Roberts, H.-G. Unruh, and J. H. Wendorff, Dielectric and Brouillin spectroscopic investigations of PVDF and PVDF/PMMA mixtures, *Ferroelectrics 55*:147 (1984).

73. C. Domenici, D. de Rossi, A. Nannini, and R. Verni, Piezoelectric properties and dielectric losses in PVDF-PMMA blends, *Ferroelectrics 60*:61 (1984).

74. B. Hartmann, Tensile yield in crystalline polymers, *Handbook of Polymer Science and Technology* (N. P. Cheremissinoff, ed.), Marcel Dekker, New York & Basel, 1989, p. 101.

75. E. J. Parry and D. Tabor, Effect of hydrostatic pressure and temperature on the mechanical loss properties of polymers: 2. Halogen polymers, *Polymer 14*:623 (1973).

76. M. Stein, Untersuchungen zur Polarisationsdynamik in Mischungen aus Polyvinylidenfluorid und Polymethylmethacrylat (Investigation of the polarization dynamics in PVDF/PMMA blends), Ph.D. thesis, Technische Hochschule Darmstadt, 1993.

77. T. Furukawa, K. Nakajima, T. Koizumi, and M. Date, Nonlinear dielectricity in ferroelectric polymers, *Jpn. J. Appl. Phys. 26*:1039 (1987).

78. K. Saito, S. Tasaka, S. Miyata, Y. S. Jo, and R. Chujo, Compatibility and piezoelectricity in polymer blends: Copoly (vinylidene fluoride-trifluoroethylene)/poly (methyl methacrylate) system, *Nippon Kagaku Kaishi* 1909 (1985).

79. K. Saito, S. Miyata, T. T. Wang, Y. S. Jo, and R. Chujo, Ferroelectric properties of a co-polymer of vinylidene fluoride and trifluoroethylene blended with poly(methyl methacrylate), *Macromolecules 19*:2450 (1986).

80. B. R. Hahn and J. H. Wendorff, Piezo- and pyroelectricity in polymer blends of poly (vinylidene fluoride)/poly (methyl methacrylate), *Polymer 26*:1611 (1985).

81. G. K. Narula and P. K. C. Pillai, Dielectric and TSC study in a semi-compatible solution-mixed PVDF-PMMA blend, *J. Mat. Sci. Lett. 8*:608 (1989).

82. H. Frensch and J. H. Wendorff, Open-circuit thermally stimulated current of PVDF/PMMA blends, *Polymer 27*:1332 (1986).

83. Y. Wada, R. Hayakawa, A model theory of piezo- and pyroelectricity of poly(vinylidene fluoride) electret, *Ferroelectrics 32*:115 (1981).

84. A. Odajima, T. T. Wang, and Y. Takase, An explanation of switching characteristics in polymer ferroelectrics by a nucleation and growth theory, *Ferroelectrics 62*:39 (1985).

85. Y. Tajitsu, A. Chiba, T. Furukawa, M. Date, and E. Fukada, Curie transition of copolymers, *Appl. Phys. Lett. 36*:286 (1980).

86. N. Tsutsumi, Y. Ueda, and T. Kiyotsukuri, Measurement of the internal field in a poly (vinylidene fluoride)/poly (methyl methacrylate) blend, *Polymer 33*:3305 (1992).

87. N. Tsutsumi, I. Fujii, and T. Kiyotsukuri, Photo-ionization of charge transfer complex by internal electric field in PVDF/PMMA blends, *ACS Div. Polymer Chem. Polymer Prepr. 33(2)*:387 (1992).

88. G. K. Narula and P. K. C. Pillai, Electrical conduction measurements in solution-mixed polyblends of poly(vinylidene fluoride) and poly(methyl methacrylate), *J. Mat. Sci. Mat. Elect. 2*:209 (1991).

89. B. R. Hahn, Studies on the non-linear piezoelectric response of poly (vinylidene fluoride), *J. Appl. Phys. 57*:1294 (1985).

90. T. R. Jow, The effect of liquid impregnants on solid polymeric materials, *1990 Ann. Rep. CEIDP*, 1990, p. 636.

91. T. R. Jow, Dielectric behavior of PVDF containing polar liquids, *Proc. 3rd Int. Conf. on Properties and Applications of Dielectric Materials*, Tokyo, Japan, 1991, Vol. 2, p. 1266.

92. Y. Takase, J. I. Scheinbeim, and B. A. Newman, Effect of plasticizer on the ferroelectric polarization reversal in poly (vinylidene fluoride), *Bull. APS 33*:548 (1988).

93. B. A. Newman, J. I. Scheinbeim, and A. Sen, The effect of plasticizer on the piezoelectric properties of unoriented poly(vinylidene fluoride) films, *Ferroelectrics 57*:229 (1984).

94. Y. Takase, J. I. Scheinbeim, and B. A. Newman, Effect of tricrecyl phosphate doping on the remanent polarization in uniaxially oriented poly (vinylidene fluoride), *Macromolecules 23*: 642 (1990).

95. S. Y. Kim, J. H. Ryu, and J. W. Cho, Effects of plasticizer on the piezoelectricity of PVDF and P(VDF-TrFE) Films, *J. Kor. Soc. Text. Eng. Chem. 25*:54 (1988).

96. A. Sen, J. I. Scheinbeim, and B. A. Newman, The effect of plasticizer on the polarization of poly(vinylidene fluoride) films, *J. Appl. Phys. 56*:2433 (1984).

97. J. W. Cho, S. Y. Kim, and S. Miyata, Piezoelectricity in the blends of poly (vinylidene fluoride-tetrafluoroethylene) and poly (vinylidene fluoride-hexafluoroacetone), *J. Kor. Soc. Text. Eng. Chem. 26*:54 (1989).

98. B. W. Brehmer and M. W. Urban, Mobility of ethyl acetate in poly (vinylidene fluoride) monitored by rheo-photoacoustic FT-IR-spectroscopy, *ACS Div. Polymer Chem. Polymer Prepr. 32(3)*:657 (1991).

99. T. Kyu and J.-Ch. Yang, Miscibility studies of perfluorinated nafion ionomers and poly (vinylidene fluoride) blends, *Macromolecules 23*:176 (1990).

100. N. Tsutsumi, M. Terao, and T. Kiyotsukuri, Thermal diffusivity and heat capacity of poly (vinylidene fluoride)/poly (methyl methacrylate) blends by flash radiometry, *Polymer 34*:90 (1993).

101. D. F. Siqueira, F. Galembeck, and S. P. Nunes, Adhesion and morphology of PVDF/PMMA and compatibilized PVDF/PS interfaces, *Polymer 32*:990 (1991).

102. W. H. Jo, T. Yoon, and S. Ch. Lee, Miscibility of poly (vinylidene fluoride) and poly (styrene-co-methyl methacrylate) blends, *Polymer J. 23*:1243 (1991).

103. C. R. Herrero, E. Morales, and J. L. Acosta, Influence of sepiolite on spherulite growth rate in polyvinylidene fluoride-based polyblends, *Polymer Int. 30*:351 (1993).

104. F. J. Viersen, P. Colantuoni, and I. Mamalis, Compatibilization of poly (2,6-dimethyl-1,4-phenylene ether)/poly (vinylidene difluoride) blends, *Angew. Makromol. Chem. 206*:111 (1993).

105. L. van Opstal and R. Koningsveld, Mean-field lattice equations of state: 5. Influence of pressure on liquid-liquid phase behaviour of polymer blends, *Polymer 33*:3445 (1992).

106. C. R. Herrero, E. Morales, and J. L. Acosta, The influence of sepiolite on the dynamic moduli and thermal transition of compatible and incompatible blends based on poly (vinylidene fluoride), *Angew. Makromol. Chem. 205*:97 (1993).

107. C. del Rio and J. L. Acosta, Thermal transitions, microstructure and miscibility in ternary polyblends based on poly (vinylidene fluoride), *Polymer Int. 30*:47 (1993).

108. A. Linares, C. del Rio, and J. L. Acosta, Effect of compatibility on dielectric relaxation and on the microstructure of semicrystalline polymer blends, *J. Non-Cryst. Solids 131–133*:1149 (1991).

109. W. E. J. R. Maas, W. A. C. van der Hejden, W. S. Veeman, J. M. J. Vankan, and G. H. W. Buning, A triple resonance ^{19}F, ^{1}H, ^{13}C cross polarization magic angle spinning nuclear magnetic resonance investigation of the miscibility in poly(ethyl ethacrylate)/poly(vinylidene fluoride) blends, *J. Chem. Phys. 95*:4698 (1991).

110. C. H. Wang, Q.-L. Liu, and B. Y. Li, Brouillin scattering of oriented films of PVDF and PVDF/PMMA blends, *J. Polymer Sci. Part B: Polymer Phys. 25*:485 (1987).

111. Rashmi, G. K. Narula, and P. K. C. Pillai, Dielectric and structural properties of the solution-mixed polyblends of poly(vinylidene fluoride) and poly(methyl methacrylate), *J. Macromol. Sci.—Phys. B26*:185 (1987).

112. W. Medycki, B. Hilczer, J. K. Krüger, and A. Marx, Dielectric and TSC study of semicompatible PVDF/PMMA blends, *Polymer Bull. 11*:429 (1984).

113. D. Yang and E. L. Thomas, Effect of PMMA on the morphology and $\alpha \rightarrow \beta$ phase transition of oriented PVDF/PMMA blends, *J. Mat. Sci. Lett. 6*:593 (1987).

114. D. Song, D. Yang, and Z. Feng, Formation of β-phase microcrystals from the melt of PVDF-PMMA blends induced by quenching, *J. Mat. Sci. 25*:57 (1990).

115. G. Natta, G. Allegra, I. W. Bassi, D. Sianesi, G. Caporiccio, and E. Torti, Isomorphism phenomena in systems containing fluorinated polymers and in new fluorinated copolymers, *J. Polymer Sci. Part A 3*:4263 (1965).

116. E. M. Genies, A. A. Syed, and M. Salmon, Electrochemical study of some chemically prepared *para*-substituted poly-*N*-phenylpyrroles, *Synth. Met. 11*:353 (1985).

117. J. I. Scheinbeim, B. A. Newman, Z. Y. Ma, and J. W. Lee, Electrostrictive response of elastomeric polymers, *ACS Div. Polymer Chem. Polymer Prepr. 33(2)*:385 (1992).

5
Poly(trifluoroethylene)

Naokazu Koizumi
Kyoto University, Kyoto, Japan

I. SOME PHYSICAL PROPERTIES

The chain conformation of poly(trifluoroethylene) (PTrFE) crystal was first reported by Lando et al. to be a 3/1 helix [1]. Tashiro et al. have studied the structure and ferroelectric transitions of vinylidene fluoride (VDF) and trifluoroethylene (TrFE) co-polymers and concluded that the crystal structure of PTrFE is the same as the cooled phase which has the all-trans conformation with skew bonds [2]. The latter structural model permits the formation of the spontaneous polarization in TrFE. Lovinger et al. carried out X-ray measurements on PTrFE and proposed the conformation of an irregular succession TG, $T\bar{G}$, and TT from the lattice spacing parallel to the carbon chain [3].

The glass transition of PTrFE is found at 32°C from the variation of the heat capacity with temperature [4]. The linear thermal expansion-versus-temperature curve indicates two break points at 31 and −40°C [5]. The break point at 31°C is in good agreement with 32°C as found by the thermal property, giving the glass transition temperature T_g.

The relaxation behavior in PTrFE has been studied by mechanical, ultrasonic, and dielectric measurements [5–7]. The dynamic mechanical properties of PTrFE were investigated over a wide range of temperature by Choy et al. [7] and Murata et al. [5], and two relaxation processes α and β were found in decreasing order of temperature. The dispersion of tensile modulus and the peak size of mechanical loss tangent are much larger for the α relaxation than for the β relaxation. The temperatures for the loss peaks in the α and β processes are listed in Table 1, where the difference in temperature for the loss peak is due to the frequency employed for measurements. Considering that the α relaxation takes place above the glass transition of 31°C, the α process is attributable to the large-scale rearrangement or the micro-Brownian motion of the carbon-chain backbone in amorphous regions, and the β process is assigned to the molecular motions of short segments of molecular chains in the glassy state. These two relaxation processes

Table 1 Temperature for Peak Mechanical Loss Tangent in α and β Relaxations

Frequency (Hz)	Temperature (°C)		Reference
	α	β	
10	45	−35	5
110	60	−20	7

are also found in the dielectric properties [5,7]. Unlike the mechanical relaxation, the magnitude of the β relaxation is much greater in the dielectric behavior. The loss peak of the α relaxation is overlapped by the loss process due to DC conductance, so that it hardly appears as a separated loss peak. These features in the dielectric behavior are well demonstrated in a contour diagram of the dielectric loss ϵ'' as a function of frequency and reciprocal of temperature in Figure 1, where the β relaxation appears as a mountain range descending toward low frequency and temperature. The dotted broken line means the relaxation frequency for the peak of dielectric loss ϵ'' as a function of frequency at constant temperature, while the broken line denotes the frequency for the peak in the temperature dependence of ϵ'' at constant frequency. It is seen in Figure 1 that the loss peak for the α relaxation is mostly concealed by the conductive loss at low frequency and high temperature. The relaxation processes in PTrFE are similar to those in poly(chlorotrifluoroethylene) [8], although no relaxation in crystalline regions at high temperature for the latter is found for the former.

Thermally stimulated current (TSC) spectra of PTrFE for two samples I and II with crystallinities of 36% and 23%, respectively, are depicted in Figure 2, where three current peaks, P_1, P_2, and P_3, are seen in ascending order of temperature [9]. The peak temper-

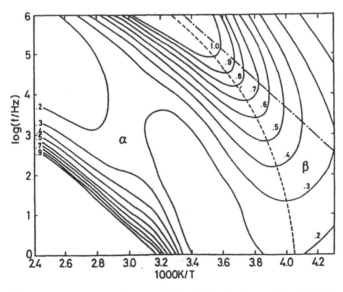

Figure 1 Contour diagram of the dielectric loss ϵ'' for PTrFE as a function of frequency and reciprocal of temperature [5].

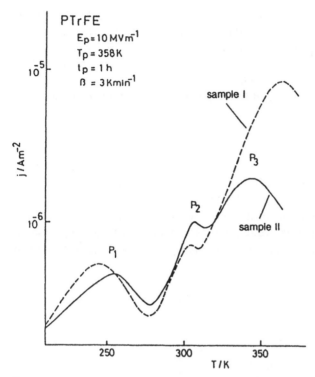

Figure 2 TSC spectra of PTrFE taken at a heating rate of 3 K/min for samples I and II poled under 10 MV/m at 85°C for 1 h [9].

atures for sample II are 255, 305, and 345 K for P_1, P_2, and P_3, respectively. P_1 and P_2 are of dipole origin, and P_3 is due to space-charge motion. As compared with the results of dielectric relaxation studies [5,7], P_1 is attributable to the β relaxation and P_2 to the α relaxation or the glass transition. The magnitude of the current peak P_2 in sample II is greater than that in sample I, being consistent with the fact that sample II has lower crystallinity than sample I. The magnitude and temperature of P_1 is considerably dependent on crystallinity.

II. PYROELECTRICITY

Ferroelectricity in poly(vinylidene fluoride) (PVDF) of β form and co-polymers of vinylidene fluoride (VDF) with trifluoroethylene (TrFE) and tetrafluoroethylene (TeFE) have been extensively studied in recent decades [10,11]. In particular, VDF/TrFE co-polymers with higher VDF content exhibit typical ferroelectric behavior such as piezoelectric and pyroelectric properties and dielectric hysteresis or polarization reversal [12]. In VDF/TrFE co-polymers with TrFE content higher than 50 mol%, however, ferroelectric nature becomes weak with increasing TrFE content. A co-polymer with 86.8 mol% TrFE still exhibits polarization reversal, although the D-E hysteresis loop is very thin; the remanent polarization is very small [13]. Regarding poly(trifluoroethylene), the polarization reversal is marginally observed, but ferroelectric properties have not been known in detail [14], although no dielectric anomaly associated with the Curie transition is found for PTrFE [15]. Unlike the pyroelectric behavior in PVDF and VDF/TrFE co-polymers with high VDF content, PTrFE exhibits anomalous pyroelectric activity as noted by Oka and Koizumi [9]: The polarity or the sign of pyroelectric coefficient changes twice over a temperature range of 100 to 330 K; the pyroelectric activity is strongly enhanced in rapidly quenched samples.

The pyroelectric coefficient p is related to the polarization P by the expression

$$p = -\frac{dP}{dT}$$

Temperature dependence of the pyroelectric coefficient p is shown for sample II poled under an electric field of 10 MV/m at 85°C for 1 h in Figure 3, where the temperature dependence of p in α-form PVDF poled under the same condition is also shown for comparison. The most noticeable feature of the pyroelectricity in PTrFE is that p changes sign twice in heating and cooling runs, while p increases with increasing temperature in PVDF [16]. It is noted in Figure 3 that p shows a broad peak with a negative sign near

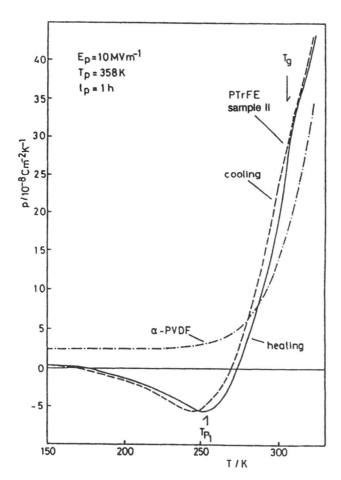

Figure 3 Variation of p with temperature T for sample II and α-form PVDF poled under 10 MV/m at 85°C for 1 h [9].

the temperature T_{P_1} for the TSC current peak P_1 and a shoulder above the temperature P_2 or the glass transition T_g as seen in Figure 2. The $p-T$ curve in a cooling run shifts slightly to lower temperature. Influence of poling time t_p on the maximum value of p is shown for sample II under the condition of fixed poling temperature T_p and poling field E_p in Figure 4a. The longer the poling time, the larger is the pyroelectricity in both directions of positive and negative signs. Effect of the poling field E_p on the $p-T$ curve is depicted for sample II in Figure 4b, where increasing E_p induces a larger pyroelectric activity. Thus the pyroelectricity in PTrFE is enhanced by increasing t_p or E_p. The increase in p tends to level off with increasing t_p, while leveling off of p is not as obvious in the case of increasing E_p. It is noted that the temperature at which p changes signs is unchanged, regardless of the magnitude of p. The temperature for the broad peak of negative p and the shoulder of positive p also remains the same [9]. The maximum value of p at the negative broad peak p_m can be taken as a measure of the magnitude of pyroelectricity.

A. Effect of Crystallinity on Pyroelectricity

PTrFE samples with different densities or crystallinities are listed in Table 2. Temperature dependence of the pyroelectric coefficient p is shown for samples A, B, and C in Figure 5 as poled under the fixed condition of poling field E_p = 20 MV/m, poling temperature T_p = 133°C, and poling time t_p = 3 h [17]. The $p-T$ curves for these samples have an essentially similar profile peculiar to the pyroelectricity in PTrFE. Starting with a positive sign from below 180 K in a heating run, p changes sign at 180–185 K, has a negative maximum at 250–255 K, changes sign again at 270–275 K, and shows a steep increase with a broad peak or a shoulder around 315 K. In the temperature dependence of p, the pyroelectricity in PTrFE is clearly distinguishable from that in ordinary poling-induced pyroelectric polymers such as PVDF [16], PVF [18], and some co-polymers of VDF. In most pyroelectric polymers, p increases with increasing temperature without change in sign. The tendency for the $p-T$ curve to show a minimum is noted for VDF/TrFE co-polymers with high TrFE content, as illustrated in Figure 6 [14].

Another striking feature of the pyroelectricity of PTrFE is that greater pyroelectric activity is induced in samples with lower crystallinity; that is, the magnitude of p decreases with increasing order of crystallinity from sample A to sample C as seen in Figure 5. Thus the value of p at the negative maximum at 250–255 K is taken as a measure of the magnitude of the pyroelectricity p_m. The dependence of p_m on T_p is shown in Figure 7 for samples A, B, and C poled under the conditions E_p = 10 MV/m and t_p = 1.5 h. The greatest pyroelectricity is observed for sample A, and the next greatest for sample B, in agreement with the results in Figure 5. Thus the pyroelectric activity induced by poling is closely related to the crystallinity of the sample before poling. It should be also noted in Figure 7 that poling below 30°C results in no pyroelectricity for all samples, and p_m increases drastically with increasing T_p. Considering that the glass transition T_g of PTrFE is 31°C, it is inferred that the relation $T_p > T_g$ is required for the formation of pyroelectricity by poling.

The effect of heat treatment before poling on the formation of pyroelectricity is shown in Figure 8, where sample A is heat treated at 123°C for preheating or aging time t_h followed by poling under 20 MV/m for 1 h. The magnitude of the pyroelectricity p_m is plotted against the aging time t_h. It is clearly seen in Figure 8 that heat treatment before poling suppresses the formation of pyroelectricity to a great extent. Heat treatment of semicrystalline polymers usually promotes crystallization, but this is considered to have

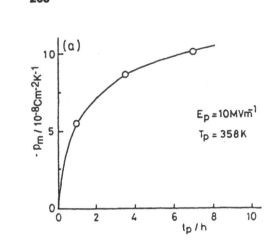

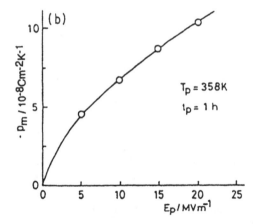

Figure 4   Variation of the maximum value p_m of p near T_{p1} with (a) poling time t_p and (b) poling field E_p [9].

Table 2   PTrFE Samples

Unoriented sample	Density (g/cm^3 at 25°C)	Crystallinity (%)	
A	1.953	18	
B	1.972	23	
C	1.992	31	
Oriented sample	Density (g/cm^3 at 25°C)	Draw ratio	
s-A	1.949	3.5	Sample A drawn at 50°C
s-C	1.980	5.5	Sample C drawn at 50°C

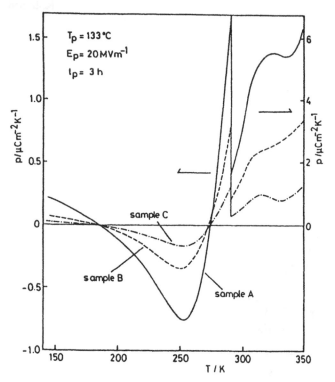

Figure 5 Variation of p with temperature for samples A, B, and C poled under 20 MV/m at 133°C for 3 h [17].

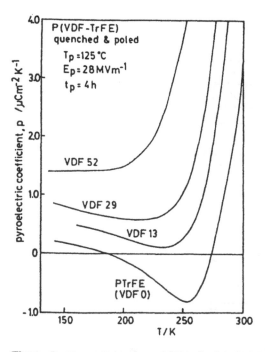

Figure 6 Temperature dependence of p in a low-temperature range for VDF/TrFE polymers with high PTrFE content [14].

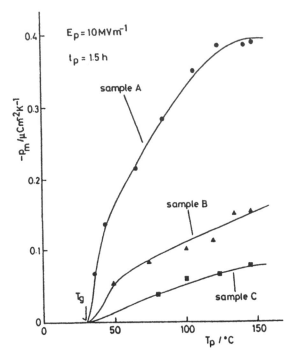

Figure 7   Variation of p_m with T_p for samples A, B, and C poled under 10 MV/m for 1.5 h [17].

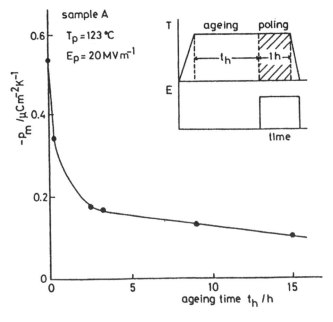

Figure 8   Effect of heat treatment before poling on p_m for sample A. The sample is heat treated for aging time t_h at 123°C, followed by poling under 20 MV/m for 1 h, as depicted in the inset [17].

an adverse effect on the formation of pyroelectricity in PTrFE. The rate of crystallization in PTrFE for samples A, B, and C is represented by plots of the increase in density Δd at 25°C against the annealing time at 120°C in Figure 9. In sample A, with the most pyroelectricity, Δd increases drastically with increasing annealing time, while in sample B and C the increase in Δd becomes smaller in an order corresponding to decreasing order of pyroelectricity. A close relation between the increase in the density and the pyroelectricity is shown in Figure 10, where the pyroelectricity p_m is plotted against the increase in the density and where a linear relation holds between p_m and Δd. Thus the crystalline region developed in the course of a poling procedure carries a spontaneous polarization, whereas the crystalline region already present before poling does not.

The decay of pyroelectricity by aging after poling is shown for sample A, poled under 20 MV/m at 123°C for 6 h, in Figure 11. Although an aging temperature 130°C higher than the poling temperature 123°C was employed, the decay of pyroelectricity is of the order of 35% after 300 h. This implies that it is difficult for the carbon-chain backbone in the crystalline region with dipoles oriented in the direction of an electric field to be rerotated by thermal activation leading to a random arrangement of dipoles. From another aspect, the carbon-chain backbone in the crystalline region can hardly rotate to align the dipoles in the direction of an electric field by poling. So the alignment of dipoles can be brought about only in the crystallization process under an electric field in the developing crystalline region.

In order to identify the characteristic of the temperature dependence of p, the effect of poling with a reversed polarity on the temperature dependence of p for a poled sample is examined as shown in Figure 12. Sample A was initially poled under 20 MV/m at 100°C for 4 h and then an electric field of the same magnitude but with reversed polarity was applied at 33°C for time t_r. The change in the p–T curve is presented for $t_r = 10$

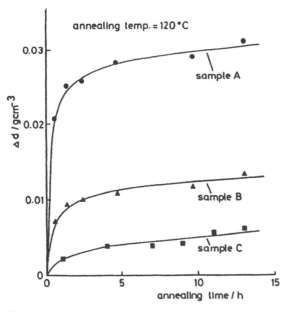

Figure 9 Increases in density at 25°C as a function of annealing time for samples A, B, and C [17].

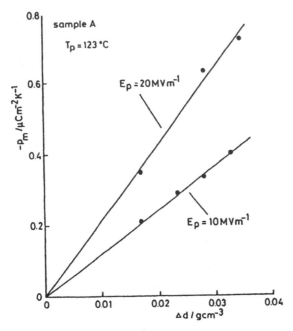

Figure 10 Relationship between p_m and Δd as induced by poling for sample A poled under 10 and 20 MV/m at 123°C [17].

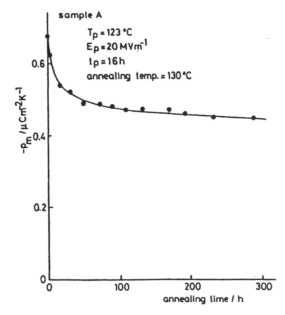

Figure 11 Decay of pyroelectricity by heat treatment for sample A. The sample poled under 20 MV/m at 123°C for 16 h is annealed at 130°C [17].

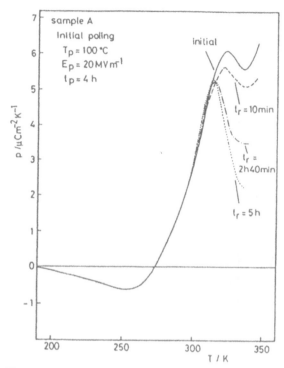

Figure 12 Effect of poling with reversed polarity on the p–T curve for sample A. The sample initially poled under 20 MV/m at 100°C for 4 h is subsequently poled with reversed polarity under 20 MV/m at 33°C for indicated time t_r.

min, 2 h 40 min, and 5 h in Figure 12, where, in the higher-temperature range above 315 K, the value of p is considerably reduced by increasing t_r, leaving a clear peak around 315 K; while in the lower-temperature range the p–T curve remains unchanged. This indicates that the spontaneous polarization originating from the dipole orientation in the crystal is not affected by the poling with a reversed polarity at 33°C close to T_g. Therefore the increase in p above 315 K in the initially poled sample should be due to a mechanism other than the temperature change of the intrinsic polarization of the dipoles. The space-charge effect would be responsible for the unstable pyroelectricity above 315 K. Consequently, the p–T curve above 300 K inherent to the polarization dipoles is considered to have a peak around 315 K and then to descend rapidly to zero. Pyroelectricity is defined as a reversible change in spontaneous polarization with temperature, and the pyroelectric coefficient p is given by $p = -dP/dT$, where P is polarization. In Figure 13 is shown variation of P with temperature obtained by charge measurements during heating of the sample, which is cleaned by poling with a reversed polarity. The p–T curve obtained by the relation $p = -dP/dT$ is compared with observed values of p. The calculated p–T curve reproduces the observed values, indicating the validity of the temperature dependence of P.

The temperature dependence of P shown in Figure 13 is seen to be quite unusual, especially in the temperature range 185–273 K, where P increases with increasing temperature, resulting in anomalous temperature dependence and the double sign change in

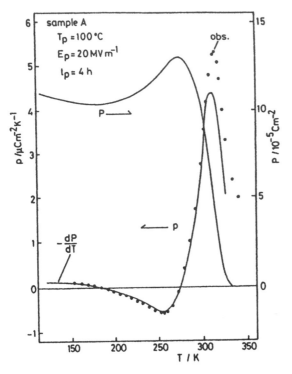

Figure 13 Variation of P and p with temperature for sample A poled under 20 MV/m at 100°C for 4 h. The solid curve of p is obtained by $p = - (dP/dT)$, and the closed circles are the observed values of p [17].

the p–T curve. In ordinary pyroelectric polymers such as PVDF [16], PVF [18], and co-polymers of VDF [12], P exhibits typical temperature dependence, decreasing monotonically with increasing temperature, and thus p never changes signs. In this respect PTrFE is regarded as an unusual pyroelectric polymer. The reason for the anomalous temperature dependence of P in PTrFE is still open to question. However, some ferroelectric materials have been reported to exhibit an unusual temperature dependence of P similar to the case of PTrFE; that is, P increases with increasing temperature, for example, in ammonium sulfate, etc. [19]. In particular, the polarization mechanism in ammonium sulfate is interpreted in terms of ferrielectricity [20] or the two-sublattice model [21,22].

B. Oriented PTrFE

The p–T curve of the oriented sample is found to be essentially identical to that for the unoriented sample in Figure 5. The value of p changes signs twice, at around 185 and 273 K, with a negative broad maximum at 255 K, and then increases sharply with increasing temperature, reaching a peak above 300 K [23]. Figure 14 shows the dependence of the pyroelectric activity p_m on the poling temperature T_p for samples s-A and s-C, as listed in Table 2, with other poling conditions fixed as E_p = 10 MV/m and t_p = 1.5 h, giving the results for samples A and C for comparison. It is seen that poling above the glass transition T_g ensures the pyroelectricity, implying that the mechanism by which the pyroelectricity is formed in the oriented sample is identical to that by which it is formed

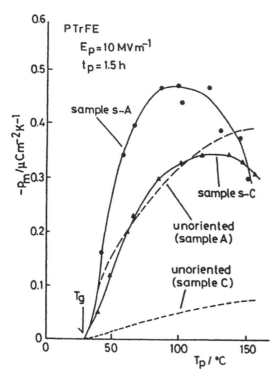

Figure 14 Variation of p_m with T_p for samples s-A and s-C (solid lines) poled under 10 MV/m for 1.5 h. Broken lines are for samples A and C [23].

in the unoriented one. That is, the spontaneous polarization is induced in the new crystalline region during the isothermal crystallization process under an electric field and not in the crystalline region originally present before poling. In the range of lower T_p the value of p_m increases sharply with T_p for the oriented sample. Also, the formation of pyroelectricity is greatly enhanced in the oriented sample as compared with that in the corresponding predrawn ones, and the effect is outstanding—especially in sample s-C. As T_p is raised above 100 and 130°C for samples s-A and s-C, respectively, it is noticed that the pyroelectricity is reduced by increasing T_p.

The enhancement of pyroelectric activity in the oriented samples, especially in the range of lower T_p, may be attributable to the greater increase in the amount of the new crystalline region after poling and also to the effect of the orientation of the molecular chain for the following reasons: The densities of the oriented samples become smaller after drawing, as given in Table 2, which means that the initial crystallinity before poling is smaller for the oriented samples than for the predrawn ones. Therefore the increase in density Δd by heat treatment, which is proportional to the increase in crystallinity, is expected to be larger for the oriented samples than for the predrawn ones. In Figure 15, Δd is plotted against annealing time at 120°C for the oriented samples, together with the results for the predrawn ones for comparison. It is seen in Figure 15 that Δd for the oriented sample is greater than that for the predrawn ones. Thus the amount of the new crystalline region in which the spontaneous polarization is induced is larger in the oriented samples, resulting in a higher pyroelectric activity in the oriented samples. Com-

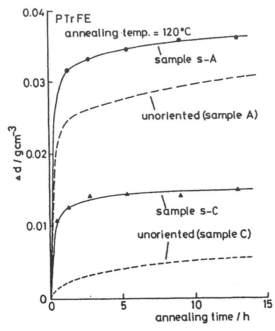

Figure 15   Increases in density at 25°C as a function of annealing time at 120°C for samples s-A and s-C (solid lines). Broken lines are for samples A and C [23].

paring the results in Figure 14 with those in Figure 15, one can see that the increase in pyroelectric activity in the oriented samples is larger than that expected from only the increase in density or crystallinity. Thus it is considered that the chain orientation which aligns the dipoles perpendicular to the film surface plays an additional role in enhancing the pyroelectric activity. This effect is prominent in sample s-C, with a higher draw ratio.

Although the value of Δd increases with increasing T_p, the pyroelectric activity decreases at higher T_p. To investigate the mechanism of this anomalous behavior in the oriented samples, the irreversible depolarization immediately after poling was observed as TSC spectra for various values of T_p and the results are presented in Figures 16a and 16b for samples s-A and s-C, respectively. It is noticed in Figure 16a that a peak around 305 K is missing in TSC curves a and b at lower T_p, and as T_p increases, a shoulder appears in curve c in the temperature range of interest. Finally, a peak takes place in curve d at the highest T_p employed. Another peak in the higher-temperature range is possibly due to the depolarization of space-charge polarization and hence the reversible polarization of crystalline dipoles being retained after heating. A similar result is obtained for sample s-C, as shown in Figure 16b.

The peak around 305 K is attributable to the primary or α relaxation near the glass transition in PTrFE [5]. Since the carbon-chain backbones, i.e., the main chains in the amorphous region of an as-drawn sample, are in tension as a result of cold drawing, they are not liable to exhibit the α relaxation in dielectric measurements. At lower T_p the annealing effect is not strong enough for the main chains to become movable, so the TSC peak is still unobservable. However, at high T_p the main chain is released from tension due to the annealing effect, and thus the peak becomes clear. By making a

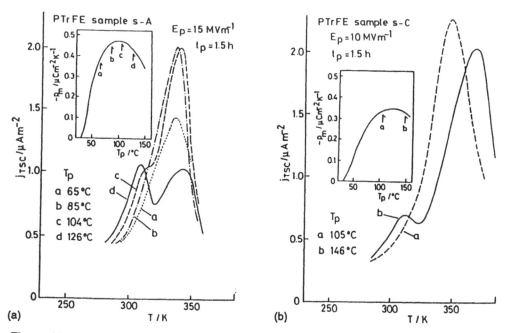

Figure 16 TSC spectra of depolarization after poling for 1.5 h at various T_p for samples s-A (a) and s-C (b). T_p is indicated by arrows on the p_m-versus-T_p curves in the inset [23].

comparison between the location of T_p in the p_m-versus-T_p curve indicated by arrows in the inset of Figure 16 and the profiles of the corresponding TSC curves, it is seen that the appearance of the peak is closely related to the temperature range of decreasing pyroelectric activity. To obtain further information about the effect of annealing on the drawn sample relating to the decrease in pyroelectric activity, results of X-ray and infrared measurements before and after heat treatment are shown in Figure 17 for sample s-A annealed at 140°C for 1.5 h. The annealing conditions are well within the temperature range of decreasing pyroelectric activity for sample s-A in Figure 14. It is seen from the X-ray pattern as shown in Figure 17 that the arc of unresolved (110) and (200) reflections become sharper after heat treatment, indicating that a more perfect crystalline state is obtained by heat treatment.

Figure 17 also shows IR dichroic spectra before and after heat treatment for the CH stretching mode at 2995 cm^{-1}, which has a transition moment perpendicular to the main chain. Apparently, the perpendicular band becomes stronger than the parallel band after heat treatment, implying that the degree of chain orientation becomes higher. The relation between T_p and the dichroic ratio $D_\perp/D_{//}$ of the 2995 cm^{-1} band is depicted for sample s-A in Figure 18, together with the T_p-versus-p_m relation of Figure 14, where $D_\perp$ and $D_{//}$ are the absorbances for perpendicular and parallel bands, respectively. In Figure 18, $D_\perp/D_{//}$ remains constant after heat treatment in the T_p range of increasing $-p_m$ and increases above the T_p of 100°C, where p_m starts to decrease. Since $D_\perp/D_{//}$ for the CH stretching mode is an indication of the chain orientation, the increase in $D_\perp/D_{//}$ corresponds to an increase in the degree of chain orientation parallel to the film surface, which is closely connected with the decrease in the pyroelectric activity in the range of high T_p.

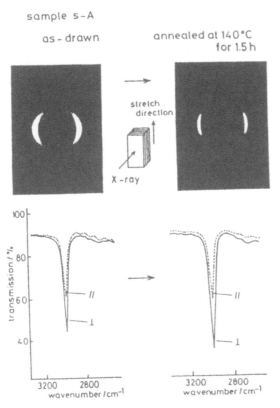

Figure 17 X-ray diffraction patterns (up) and IR dichroic spectra at 2995 cm^{-1} (down) for sample s-A (left) as drawn and (right) after annealing at 140°C for 1.5 h [23].

On the basis of the results of TSC, X-ray, and IR measurements, the following mechanism is inferred regarding the reduction of the pyroelectric activity with increasing T_p in oriented PTrFE: The effect of simultaneous heat treatment and poling on the orientation of the crystallites and the resulting spontaneous polarization at low and high T_p are illustrated schematically in Figure 19. As depicted in Figure 19, the initial as-drawn sample before poling has a lower degree of both orientation and crystallinity, and the b axis or polar axis indicated by arrow tends to lie more or less parallel to the film surface as a result of strip biaxial drawing. At low T_p the amorphous region is still in tension and the degree of orientation remains the same after poling; thus the new crystalline region can align the polar axis along an electric field regardless of the orientation of the mother crystal, resulting in a higher spontaneous polarization. As T_p is increased, the amorphous region becomes free from the tension, so the main chain forming the new crystalline region can move over a wide range to match the orientation with that of the mother crystal, as in epitaxial crystal growth. At the same time the annealing effect increases the orientation of the crystalline region, so the polar axis tends to lie more along the film surface. Therefore the spontaneous polarization perpendicular to the film surface induced in the new crystalline region decreases with increasing T_p. The tendency of an electric field to align the polar axis perpendicular to the film surface overcomes

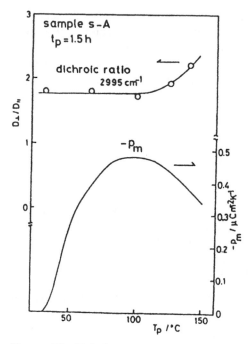

Figure 18 Variation of dichroic ratio $D_\perp/D_{//}$ of 2995 cm^{-1} band with T_p and the p_m-versus-T_p curve of Figure 14 for sample s-A [23].

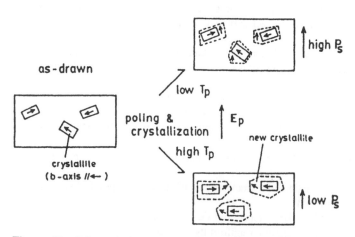

Figure 19 Schematic illustration of isothermal crystallization and formation of spontaneous polarization by poling at low and high T_p for oriented PTrFE in a cross-sectional view perpendicular to the draw direction. The polar axis (b axis) is indicated by the arrow [23].

that of the chain orientation parallel to the film surface when T_p is low, but the situation is reversed when T_p is high, depending on the state of the amorphous region. This results in the anomalous decrease in the pyroelectric activity with increasing T_p in oriented PTrFE. This unusual behavior in the oriented sample originates from the mechanism of formation of spontaneous polarization peculiar to PTrFE, namely, the isothermal crystallization under an electric field.

REFERENCES

1. R. R. Kolda and J. B. Lando, The effect of hydrogen-fluorine defects on the conformational energy of polytrifluoroethylene, *J. Macromol. Sci. Phys. B11*:21 (1975).
2. K. Tashiro, K. Takano, M. Kobayashi, Y. Chatani, and H. Tadokoro, Structural study of ferroelectric phase transition of vinylidene fluoride-trifluoroethylene copolymers (III). Dependence of transitional behavior on VDF molar content, *Ferroelectrics 57*:297 (1984).
3. A. J. Lovinger and R. E. Cais, Structure and morphology of poly(trifluoroethylene), *Macromolecule 17*:1939 (1984).
4. W. K. Lee and C. L. Choy, Heat capacity of fluoropolymers, *J. Polymer Sci. Polymer Phys. Ed. 13*:619 (1975).
5. Y. Murata, J. Hagino, and N. Koizumi, Dielectric relaxation in polytrifluoroethylene, *Kobunshi Ronbunshu 36*:697 (1979), in Japanese.
6. S. F. Kwan, F. C. Chen, and C. L. Choy, Ultrasonic studies of three fluoropolymers, *Polymer 16*:481 (1975).
7. C. L. Choy, Y. K. Tse, S. M. Tsui, and B. S. Hsu, Mechanical and dielectric relaxations in polytrifluoroethylene, *Polymer 16*:501 (1975).
8. J. D. Hoffman, G. Williams, and E. Passaglia, Analysis of the α, β, and γ relaxations in polychlorotrifluoroethylene and polyethylene: Dielectric and mechanical properties, *J. Polymer Sci. Part C 14*:173 (1966).
9. Y. Oka and N. Koizumi, Pyroelectricity in polytrifluoroethylene, *Jpn. J. Appl. Phys. 22*:L281 (1983).
10. T. Furukawa, Ferroelectric properties of vinylidene fluoride copolymers, *Phase Transitions 18*:143 (1989).
11. K. Tashiro and M. Kobayashi, Structural phase transition in ferroelectric fluorine polymers: X-ray diffraction and infrared/Raman spectroscopic study, *Phase Transitions 18*:213 (1989).
12. T. Furukawa, Piezoelectricity and pyroelectricity in polymers, *IEEE Trans. Elect. Insul. 24*: 375 (1989).
13. N. Koizumi, Y. Murata, and H. Tsunashima, Polarization reversal and double hysteresis loop in copolymer of vinylidene fluoride and trifluoroethylene, *IEEE Trans. Elect. Insul. 21*:543 (1986).
14. Y. Oka, N. Koizumi, and Y. Murata, Ferroelectric order and phase transition in polytrifluoroethylene, *J. Polymer Sci. Part B: Polymer Phys. 24*:2095 (1986).
15. Y. Higashihata, J. Sako, and T. Yagi, Piezoelectricity of vinylidene fluoride copolymers, *Ferroelectrics 32*:85 (1981).
16. D. K. Das-Gupta, On the nature of pyroelectricity in polyvinylidene fluoride, *Ferroelectrics 33*:75 (1981).
17. Y. Oka and N. Koizumi, Formation of pyroelectricity in unoriented polytrifluoroethylene, *Jpn. J. Appl. Phys. 23*:748 (1984).
18. S. B. Lang, A. S. DeReggi, M. G. Broadhurst, and G. T. Davis, Effects of poling field time on pyroelectric coefficient and polarization uniformity in polyvinyl fluoride, *Ferroelectrics 33*:119 (1981).
19. H. G. Unruh, The spontaneous polarization of $(NH_4)_2SO_4$, *Solid State Commun. 8*:1951 (1970).

20. A. Sawada, S. Ohya, Y. Ishibashi, and Y. Takagi, Ferroelectric phase transition in $(NH_4)_2SO_4$-K_2SO_4, *J. Phys. Soc. Jpn. 38*:1408 (1975).

21. V. Dvorák and Y. Ishibashi, Two-sublattice model of ferroelectric phase transitions, *J. Phys. Soc. Jpn. 41*:548 (1976).

22. K. Ohi, J. Osaka, and H. Uno, Ferroelectric phase transition in Rb_2SO_4-$(NH_4)_2SO_4$ and Cs_2SO_4-$(NH_4)_2SO_4$ mixed crystals, *J. Phys. Soc. Jpn. 44*:529 (1978).

23. Y. Oka and N. Koizumi, Pyroelectricity in oriented polytrifluoroethylene, *Jpn. J. Appl. Phys. 24*:669 (1985).

6
Ferroelectric Nylons

Hari Singh Nalwa
Hitachi Research Laboratory, Hitachi Ltd., Hitachi City, Ibaraki, Japan

I. INTRODUCTION

Ferroelectric properties of organic polymers have been investigated since the 1960s, when piezoelectricity was first discovered, in polypeptides, poly(methyl methacrylate) (PMMA), poly(vinyl chloride) (PVC), and poly(vinylidene fluoride) (PVDF). Ferroelectric polymers began to attract much attention when a significant enhancement of piezoelectricity was observed for PVDF by electret formation. Over the years, a number of applications of PVDF were investigated that include audiofrequency transducers, ultrasonic transducers, electromechanical transducers, pyroelectric detectors, robotics, and biosensors [1]. To date, PVDF has been the most successful material because of its wide range of applications in electronics and biomedical engineering. With the discovery of PVDF, the field was confronted with challenges to develop novel organic polymers with large pyroelectric, piezoelectric, and ferroelectric responses. With this quest, new ferroelectric polymeric systems have been sought for potential use in a variety of electronic and photonic devices. Virtually together, co-polymerization of VDF with trifluoroethylene and tetrafluoroethylene were found to be useful, since co-polymers crystallize spontaneously into a ferroelectric phase and exhibit a high degree of crystallinity as well as large piezoelectric response. The most promising polymers found so far include PVDF and its co-polymers, co-polymers of poly(vinylidene cyanide), odd-numbered nylons, polyureas, and polymer composites of piezoelectric ceramics [2,3]. Odd-numbered nylons have emerged as a new class of ferroelectric polymers similar to PVDF. The dielectric, pyroelectric, piezoelectric, and ferroelectric properties of various odd-numbered nylons are discussed in this chapter.

Polyamides, commonly known as nylons, are polymeric materials that have the recurring amide (—CO—NH—) linkage in the repeating hydrocarbon units of the polymer chain. Generally, nylons are named after the number of carbon atoms present in a repeating unit of the polymer backbone. Nylon chains have a number of methylene groups between amide groups. The chemical structure of the repeating unit in the nylon chain can be written as

$$\left[HN-(CH_2)_x-CO \right]_n$$

Here, if $x = 2$, then the corresponding nylon will be identified as nylon-3 because it has three carbon atoms in a repeating unit. Similarly, if the number of x is 3, 4, 5, 6, 7, 8, 9, 10, or 11, then the corresponding polymers are nylon-4, nylon-5, nylon-6, nylon-7, nylon-8, nylon-9, nylon-10, nylon-11, and nylon-12, respectively. Nylons with an even number of carbon atoms are called even nylons, whereas nylons having an odd number of carbon atoms in the repeating unit are known as odd nylons. The typical synthetic route for a nylon can also be expressed as follows:

$$H_2N-(CH_2)_x-NH_2 + HOOC-(CH_2)_y-COOH$$
$$\downarrow$$
$$\left[HN-(CH_2)_x-NH-CO-(CH_2)_y-CO \right]_n$$

Here x and y represent the number of carbon atoms in the repeating units of the diamine and dicarboxylic acid. For example, the polyamide prepared from hexamethylene diamine $[H_2N-(CH_2)_6-NH_2]$ and adipic acid $[HOOC-(CH_2)_4-COOH]$ is termed nylon-6,6, which is an even-even nylon. Similarly, polymerization of hexamethylene diamine with sebacic acid or dodecanoic acid yields nylon-6,10 or nylon-6,12, respectively. The odd-odd nylons have an odd number of carbon atoms in the repeating units of the diamines and dicarboxylic acid: nylon-5,7, nylon-7,7, nylon-7,5, nylon-9,5, etc. The physical properties of nylons are governed by the building blocks in the chain, such as methylene groups. Aromatic nylons can also be synthesized from aromatic diamines and aromatic diacids. Nylons are prepared by melt polymerization, solution and interfacial polymerization, ring-opening polymerization, and anionic polymerization. Nylons having linear, branched, and cross-linked structures can be prepared, and the nature of linkage significantly affects the molecular structure. The morphology and degree of crystallinity of nylons depend on the basic structure of the chemical linkage. The unique physical properties of nylons, such as toughness; stiffness; resistance to chemicals, temperature, and wear, low coefficient of friction, higher melting points, crystallinity, and size of spherullites depend mainly on the chemical species integrated into the polymer chains and on the resulting strong interactions between amide groups on neighboring chains. Highly crystalline nylons are grown by crystallizing near their melting points. More details about nylons can be found in Refs. 4–8. Nylons have been widely used in molding, wire straps, electrical connections, fasteners, gears, sportswear, and equipment.

Odd nylons and odd-odd nylons constitute an important class of ferroelectric polymers. Because the dipoles in odd nylons are aligned in the same direction and the all-trans conformation give rises to a large dipole moment and spontaneous polarization in the unit cell of the crystalline phase, they show very interesting pyroelectric, piezoelectric, and ferroelectric properties. Ferroelectric nylons were initially investigated by a research team at Rutgers University, who made excellent contributions in this field by studying pyroelectric, piezoelectric, and ferroelectric properties of various odd-numbered nylons. Recently, a variety of alicyclic, aromatic, and fluorinated nylons has been reported by research groups in Japan. This review mainly describes the results on nylons reported by these research groups. Chemical aspects, physical properties, molecular and crystal structure, dielectric properties, pyroelectric, piezoelectric, and ferroelectric properties of odd and odd-odd nylons are discussed in this chapter.

II. ODD-NUMBERED NYLONS

A. Spectroscopy

General physical properties and spectroscopic analysis of various types of nylons can be found in Ref. 8. Details of infrared spectroscopy and X-ray diffraction dealing with odd nylons are briefly mentioned here. Table 1 lists the chemical structures and thermal properties of odd nylons, odd-odd nylons, and fluorinated odd-odd nylons, which will be discussed in this chapter. Kinoshita [9] measured the melting points, density, infrared spectrum, and X-ray diffraction of nylons with varying chain lengths. The melting point of nylons increases with the concentration of amide groups because hydrogen bonds cause strong intermolecular interactions. The density of nylons follows the same trend. Infrared spectral data indicated that perfect hydrogen bonds are formed between N—H and C=O groups in nylons having an odd number of methylene groups. The glass transition temperature of odd nylons decreases as the number of methylene groups increases in the polymer chain. The quenched nylon-5,7 shows a melting point of 215°C, while a melting point of 228°C was measured for slow-cooled samples by Lin et al. [10] Fluorinated odd-odd nylons have much lower T_g than the corresponding odd-odd nylons having the same number of carbon atoms, indicating that methylene groups lead to higher T_g than fluoromethylene groups [12].

The important infrared bands in nylon backbones are the N—H stretching region from 3100 to 3500 cm^{-1} and the amide I region from 1600 to 1700 cm^{-1}. Infrared studies of nylons have been reported in the literature [8,9,13,14]. For nylon-11, the N—H stretching at 3298 cm^{-1}, amide I mode dominated by the C=O stretching band at 1637 cm^{-1} and amide II mode having contributions from the N—H in-plane bend, and slight contributions from the C$_\alpha$—C, N—C$_\alpha$, and C—N stretches at 1551 cm^{-1} were reported by Roberts and Jenekhe [13]. The symmetric and assymetric CH$_2$ stretching vibrations appeared at 2851 and 2921 cm^{-1}, respectively. Skrovanek et al. [14] reported Fourier transform infrared temperature studies of nylon-11. Nylon-11 shows the hydrogen bonded amide I at 1638 cm^{-1} and 1645 cm^{-1} for ordered and disordered conformations, respec-

Table 1 Chemical Structure and Thermal Properties of Odd-Numbered Nylons [10–12]

Nylon	Chemical structure	T_g (°C)	T_m (°C)
Nylon-7	—[—NH—(CH$_2$)$_6$—CO—]$_n$—	83	235
Nylon-9	—[—NH—(CH$_2$)$_8$—CO—]$_n$—	70.5	—
Nylon-11	—[—NH—(CH$_2$)$_{10}$—CO—]$_n$—	68	195
Nylon-5,7	—[—NH—(CH$_2$)$_5$—NH—CO—(CH$_2$)$_5$—CO—]$_n$—	—	215
Nylon-7,5	—[—NH—(CH$_2$)$_7$—NH—CO—(CH$_2$)$_3$—CO—]$_n$—	190	228
Nylon-9,5	—[—NH—(CH$_2$)$_9$—NH—CO—(CH$_2$)$_3$—CO—]$_n$—	170	209
Nylon-3,5F	—[—NH—(CH$_2$)$_3$—NH—CO—(CF$_2$)$_3$—CO—]$_n$—	—	215
Nylon-5,5F	—[—NH—(CH$_2$)$_5$—NH—CO—(CF$_2$)$_3$—CO—]$_n$—	38	201
Nylon-7,5F	—[—NH—(CH$_2$)$_7$—NH—CO—(CF$_2$)$_3$—CO—]$_n$—	25	158
Nylon-9,5F	—[—NH—(CH$_2$)$_9$—NH—CO—(CF$_2$)$_3$—CO—]$_n$—	25	156

tively, and hydrogen-bonded N—H stretching mode at 3299 cm^{-1} at 30°C. The infrared spectra of the N—H stretching region of nylon-11 recorded as a function of increasing temperature from 30 to 210°C show steady increase in the breadth of the band, decrease in total area, and shift in frequency from 3299 to 3332 cm^{-1}, which is associated with the degree of order. The absorptivity coefficient of the hydrogen-bonded N—H stretching mode shows strong dependence as a function of hydrogen-bond strength. The hydrogen-bonded N—H groups in ordered and disordered conformations showed no difference, instead reflecting the overall distributions of hydrogen-bonded strength in nylons. In particular, the amide I mode was sensitive to the conformation through dipole–dipole interactions, and infrared bands associated with ordered and disordered hydrogen-bonded amide groups were easily distinguishable. The infrared spectra of different polymorphs of odd-numbered nylons is discussed in other sections.

B. Polymorphism and Crystal Structure

Polymorphism and crystal structure of different forms of odd-numbered nylons have been reported by several research groups. Slichter [15] reported crystal structure of odd-numbered and even-numbered nylons. The X-ray diffraction patterns of nylon-7 and nylon-11 were found to be similar to that of nylon-6,6. Table 2 lists the unit-cell crystal parameters and density of odd-numbered nylons and odd-odd numbered nylons. Newman et al. [16] investigated X-ray diffraction of nylon-11 at high pressure and high temperatures and pointed out that experimental data do not completely fit to the model reported by Slichter [15]. The temperature-dependent X-ray diffraction scans revealed a crystal transition in nylon-11 from the triclinic α phase to a pseudo-hexagonal γ phase at 95°C. Though this transition was reported for the first time, it was supported by relaxation studies from mechanical, dielectric, and thermal measurements. The temperature-dependent X-ray diffraction study showed that peaks from the (010) and (100) planes approach each other as

Table 2 Unit-Cell Dimensions in Odd Nylons

| Parameters | Nylon-7 [15] | Nylon-11 | | | Nylon-5,7 [10] |
		Form I (α) [15]	Form II [17]	Form III (γ) [17]	
Crystal system	Triclinic	Triclinic	Monoclinic	Monoclinic	Monoclinic (Pb)
a (Å)	4.9	4.9	9.75	9.48	4.83
b (Å)	5.4	5.4	15.0	29.4	9.35
c (Å)	9.85	14.9	8.02	4.51	16.61
α (o)	49	49	—	—	—
β (o)	77	77	65	118.5	—
γ (o)	63	63	—	—	58.9
Z	1	1	4	4	2
Density (g/cm^3)	1.21	1.15	1.145	1.10	1.168

the temperature increases and merge into a single peak at 95°C. Upon cooling, a reverse process takes place because both peaks reappear in the spectrum. The *d* spacing of the (010) planes increased with temperature while the spacing of the (100) planes decreased with temperature up to the transition temperature. Kawaguchi et al. [17] reported three crystal modifications of nylon-11. Form I (α phase) was obtained from a 0.005 wt% solution in glycerine at 160°C. Form II was prepared from a solutions in water containing 5% formic acid at 160°C. Form III (γ phase) was obtained from solution in water of triethyleneglycol. Cell dimensions for different polymorphs of nylon-11 are shown in Table 2. Nylon-11 has a triclinic α form, monoclinic form II, and a pseudo-hexagonal γ form. Nylon-11 exhibits a structural phase transformation with an increase of temperature from a triclinic α phase to a pseudo-hexagonal γ phase.

Slicher [15] reported crystal structures for nylon-11 and nylon-7 and suggested that X-ray diffraction patterns of both nylons closely resemble those of nylon-66. Newman et al. [16] carried out X-ray diffraction studies of nylon-11 at elevated temperatures and pressures and reported a crystal phase transition from the α phase to the γ phase at 95°C, supported by mechanical and dielectric relaxation studies. Two types of crystal structures of nylon-11 were observed during differential scanning calorimetric measurements. Rapid crystallization leads to a γ-phase structure, whereas slow crystallization yields an α-phase structure at room temperature. The α-phase nylon-11 is polar in nature, and the dipoles are hydrogen-bonded and aligned in the same direction in a particular crystallite. Non-polar γ-phase nylon-11 has a pseudo-hexagonal structure in which the amide group lies in the plane perpendicular to the chain axis. Jacobs and Hicks [18] prepared α-phase nylon-11 by annealing the film at 170°C for 14 h under vacuum and then slowly cooling to room temperature, whereas the γ-phase nylon-11 was obtained by annealing the film at 75°C for 14 h under vacuum. Polymorphs of nylon-9 have also been prepared; α-phase nylon-9 was obtained from cooling of the melt at room temperature, whereas the γ phase was obtained from trifluoroacetic acid treatment. Figure 1 shows the formation of different polymorphs of nylon-11. Nylon-11 shows five polymorphs: α, γ, δ, α′, and δ′, at least three of which have stable crystalline phases and the others a metastable phase. Metastable hexagonal δ′ phase is obtained from the melt or from the δ phase.

Balizer et al. [19,20] prepared different polymorphs of nylon-11 by exposing quenched films of the δ′ phase to plasticizers and to acid vapors. A mixture of α and δ′ phases was obtained by exposing the quenched films to both D_2O and a solution of 3% phenol in D_2O. The γ phase was obtained by vapor-phase doping of unoriented and oriented quenched films with DCl or HCl resulting from the crystal–crystal transition from the δ′ to the γ phase. The isotope effect was studied when the monomer aminoundecaneoic acid was exposed to either D_2O or water at 120°C for 24 h and then polymerized for 5 h at 216°C. The crystalline phases and their transitions were investigated by both FTIR spectroscopy and X-ray diffraction. The deuteration as a function of time in D_2O and its phenol solution depend on orientation and texture. FTIR spectra of nylon-11 were recorded before and after deuteration. The N—H vibration bands occurred at 3300 cm^{-1}, whereas the N—D vibration doublet appeared at 2416 and 2470 cm^{-1}. The Fermi resonance between the N—D vibration and a deuterated amide (II+III) combination band leads to the doublet. The N—D vibration shifted to a lower frequency compared with combined band for strong hydrogen bonds, and 2416 cm^{-1} absorption dominates as for the γ phase. On the other hand, as the strength of the hydrogen bond decreases, the N—D vibration band reaches near that of the combination band, and further decrease in the strength of hydrogen bond shifts the N—D vibration band above that of the combined

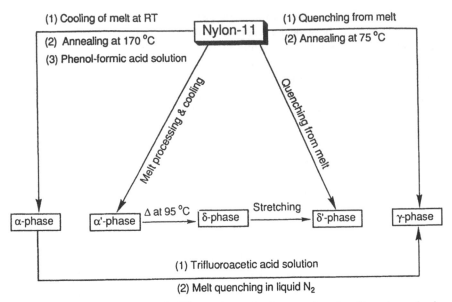

Figure 1 Interconversion of different crystal phases of nylon-11 using physicochemical techniques.

band and increases the 2470 cm^{-1} band as for the α phase. The crystalline change as a function of exposure time was studied by wide-angle X-ray diffraction. The unoriented film showed three peaks which were superimposed due to two crystalline phases. The middle peak occurred from the hexagonal δ' phase, while another peak occurred from the triclinic α phase. Exposure to both phenol solution and D_2O did not produce α phase in the oriented samples. A X-ray diffraction peak similar to the δ' phase was observed when unoriented films with α phase were cold-drawn 2.6:1. Longer solution exposure broadened the peak width toward higher scattering angles, which indicates a decrease in hexagonal lattice dimension of some of the δ' phase due to solution annealing effects. Similar sensitivity was not observed in FTIR deuterated spectra. The X-ray diffraction of DCl-treated unoriented films showed hexagonal lattice deformation and a broad scattering at 21° and a new peak at 12.5°. After exposure to D_2O, a major peak due to a hexagonal lattice of smaller dimension than in the δ' phase and a minor peak indicating the γ phase were formed. After orientation, a broadened peak shifted toward the position of the δ' phase. Exposure of oriented films to DCl vapor and then to D_2O vapor resulted in the γ phase. The vapor-treated γ-phase films showed different features.

Lin et al. [10] reported the crystal structure of nylon-5,7, which is monoclinic and belongs to noncentrosymmetric space group P_b. The crystal structure of nylon-7,7 was reported by Kinoshita [21], who pointed out that the amide is tilted approximately 30° with respect to the fiber axis. The methylene chain orients nearly parallel to the fiber axis and brings favorable orientation around C—C' and C—N single bonds of the C—C'=O—N—H group. X-ray diffraction studies supported γ-phase crystal structures for nylon-5,7 and nylon-7,7. The X-ray diffraction patterns of nylons with odd and even numbers of methylene groups show somewhat different features arising from the alignment of chains from the orientation of amide and methylene groups.

III. DIELECTRIC PROPERTIES

Nylons exhibit interesting dielectric properties, which change significantly with temperature and frequency. The dielectric constant of oriented nylon-11 films was measured as a function of temperature by Litt et al. [22]. Nylon-11 has a dielectric constant of 3 at 25°C, which increases with increasing temperature. In general, both dielectric constant and loss spectra showed no apparent peak, but on an expanded scale, shoulders appeared at 88, 81, and 73°C for 10 kHz, 1 kHz, and 100 Hz, respectively. The peak at 73°C for 100 Hz was assumed to be an α peak in nylon-11. The dielectric loss was found to decrease with increasing frequency above 90°C. Figure 2 shows the frequency dependence of the dielectric constant at different temperatures for poled and annealed nylon-7 and nylon-11 films, respectively [23]. Both ϵ' and ϵ'' increase as the frequency decreases for temperatures above the glass transition of these nylons. The ϵ'' increases more rapidly than the ϵ' above the glass transition temperature. Nylon-11 shows an ϵ' of about 80 and an ϵ'' of 100 at 100 Hz. Nylon-7 shows similar ϵ' and ϵ'' characteristics. For nylon-7, the ϵ' reaches higher than 100 and the ϵ'' is around 100 at 100 Hz. A significant increase in ϵ'' of nylon-7 occurred for temperatures of 114, 134, 155, and 175°C. Both nylon-7 and nylon-11 follow the Cole-Cole curve of dielectric data except at the very high and the very low frequency regions. The shape of the ϵ'-versus-ϵ'' curve changes with the Cole-Cole plot to the opposite sign when ionic contribution becomes larger. Both the static value and the instantaneous value of the dielectric constant were determined by extrapolation of the data. Instantaneous values of 3.15 and 3.35 versus static values of 18 and 16 were obtained for nylon-11 and nylon-7, respectively. Takase et al. [23] reported the dielectric constant of nylon-7 and nylon-11 as a function of temperature at 1, 10, and 100 KHz. The dielectric constant showed α, β, and γ relaxations.

Wu et al. [24] reported X-ray diffraction and infrared spectra of polymorphs of nylon-9 and nylon-11. Figure 3 shows X-ray diffraction diagrams of polymorphs of nylon-9 and nylon-11. The α phases of nylon-9 and nylon-11 show patterns of triclinic form with three rings. The lattice spacings were 3.72, 4.33, and 9.77 Å for α-phase nylon-9 and 3.77, 4.38, and 11.9 Å for α-phase nylon-11. The γ forms of nylon-9 and nylon-11 show two rings. The spacings of the two rings were 4.07 and 11.8 Å for the α-phase nylon-9 and 4.08 and 14.4 Å for the α-phase nylon-11. The infrared spectra of α and γ forms of nylon-9 and nylon-11 show different features. For α-phase nylon-9 and nylon-11, the amide V and VI bands were observed at 687 and 584 cm^{-1}, respectively; these bands disappeared for the γ phase, though a new band appeared at 627 cm^{-1}. The dielectric constants for these polymorphs were studied as a function of temperature. Temperature dependence of the real and imaginary parts of complex dielectric constants at 10 Hz, 100 Hz, 1 kHz, and 10 kHz for α-nylon-9 has been reported by Wu et al. [24]. Dielectric constant (ϵ') was about 2.5–3 at various frequencies below 0°C and increased with increasing temperature, more rapidly from 20°C to 100°C. Around 100°C at 10 Hz, the ϵ' increased to above 30. Three distinct peaks were observed for the dielectric loss (ϵ''). At 10 Hz, the γ peak appeared at about -110°C due to the local mode motion, while the β peak appeared at about -30°C due to the absorbed water and motions of the dipolar amide groups. The α relaxation appeared at about 90°C. These three peaks shifted to higher temperature regions as the measurement frequency was increased. Temperature dependence of dielectric constant for α-nylon-11 and γ-nylon-11 samples at 10 Hz yielded higher values for γ-nylon-11 that for α-nylon-11 at higher temperature. Similar features were also observed for α-nylon-9 and γ-nylon-9 samples.

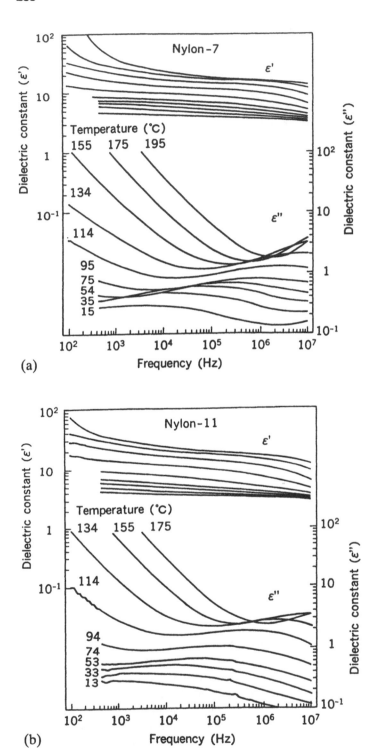

Figure 2 Frequency dependence of the dielectric constant at different temperatures for poled and annealed nylon-7 (a) and nylon-11 (b) films. (Reprinted with permission from Ref. 23, the American Chemical Society.)

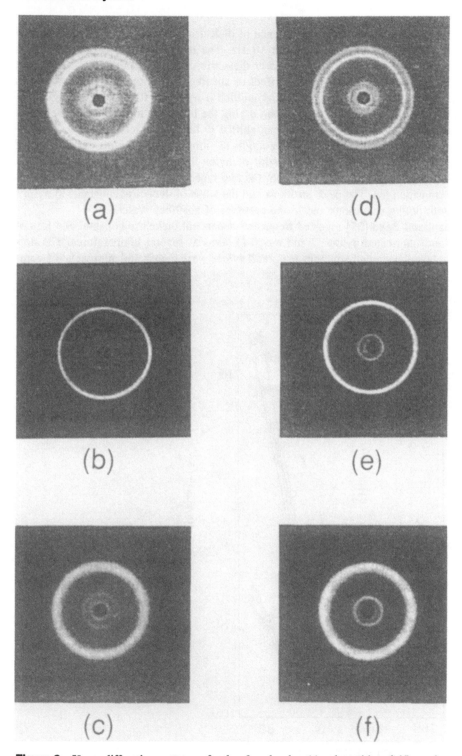

Figure 3 X-ray diffraction patterns of nylon-9 and nylon-11, where (a) and (d) are for α phase, (b) and (e) are for γ phase, and (c) and (f) are for mixtures of α phase and γ phase, respectively. (Adapted from Ref. 24.)

Figure 4 shows the temperature dependence of dielectric constant for α-nylon-9 and α-nylon-11 poled and unpoled samples at 10 Hz. The α-nylon-9 and α-nylon-11 films poled at 300 kV/cm at 80°C showed lower dielectric constant compared with unpoled samples at the same temperature. The effect of absorbed water on dielectric properties of α-nylon-11 films was also studied. The undried α-nylon-11 films showed higher dielectric constant than the dried films. Upon drying the films, the magnitude of the relaxation peak of ϵ'' decreased and the α peak shifted to the higher temperatures. The dielectric constant of both α-nylon-9 and α-nylon-11 films decreased with evaporation of the absorbed water. The dielectric behavior of nylon-9 and nylon-11 was similar, and three dielectric loss peaks originated from the two types of local motions (γ and β) and the glass transition (α). The peak positions and the value of dielectric constants in nylons change with poling conditions and in the presence of absorbed water.

Yemani and Boyd [25] reported frequency-dependent dielectric constant and loss of oriented and unoriented nylon-77 and nylon-11 films at several temperatures. The normalized relaxation strengths in odd-numbered nylons were higher and approached a com-

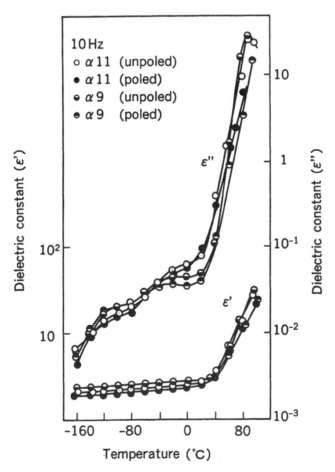

Figure 4 Temperature dependence of dielectric constant for α-nylon-9 and α-nylon-11 poled and unpoled samples at 10 Hz. (Adapted from Ref. 24.)

mon level as the chain length between dipoles was increased. The dielectric constant and loss of oriented nylon-77 was measured in the directions parallel and perpendicular to the extrusion direction. Overall orientation induced anisotropy in the α loss process. The extrusion-induced enhanced intramolecular dipolar orientation was not apparent as intensity in both parallel and perpendicular directions was reduced compared to the unoriented sample. This reduction in intensity could be associated with a decrease in the amount of amorphous material and not from the change in dipole correlation induced by orientation. The dielectric constant and relaxation strength related to the amorphous-phase glass–rubber relaxation was higher in odd nylons than in even nylons.

IV. PYROELECTRICITY

Pyroelectric properties of odd-numbered nylons have not been addressed very much; only limited results have been reported on nylon-11. Litt et al. [22] reported pyroelectricity in nylon-11. Uniaxially oriented films of nylon-11 were prepared by stretching 300% at about 150–160°C, and pyroelectricity was measured by cycling the poled sample between 20 and 75°C. The contribution of space charge and other effects was evaluated by measuring the pyroelectricity of vacuum-dried samples that were first immersed in distilled water at 25°C for 30 h. The activation energy increased to 2.03 eV and the spontaneous current decreased one-third at 60°C. Pyroelectric coefficient was measured as a function of the temperature, which decreased by a factor of 2 though similar shape was obtained. Water-soaking treatment neutralized some of the volume polarization and caused significant reduction in pyroelectric response. Resoaking the same sample in water at 60°C for 30 h and vacuum drying as before led to further decrease in spontaneous current. Here again, the activation energy increased to 2.33 eV and the spontaneous current decreased by a factor of 5.5 at 60°C, though pyroelectric coefficient remained unchanged. Therefore, water-soaking treatment was found to be useful in eliminating the volume polarization effect on the pyroelectric response. Before water-soaking treatment, the pyroelectric coefficient increased in the glass transition temperature region, which disappeared after water soaking. The pyroelectricity in nylon-11 originated due to orientation of dipoles in the crystalline regions rather than from the volume polarization.

Nylon-5,7 has only γ form, which can be produced by thermal or mechanical treatments. Nylon-5,7 has a net dipole moment, and both poled as well as unpoled samples exhibit pyroelectricity. Litt and Lin [26] reported a pyroelectric constant of 100 pC/cm^{2}K. The effect of water on piezoelectricity of odd nylons has been studied by Kim [27]. The piezoelectric constant increases with increasing water content, probably due to the crystalline change. Lin et al. [10] reported a pyroelectric coefficient of 10–100 C/cm^{2}K for nylon-5,7.

V. PIEZOELECTRICITY

Figure 5 shows the polar hydrogen-bonded sheet structures for odd-numbered nylons and even-numbered nylons. As can be seen from the molecular dipole alignment, in odd-numbered nylons there exists a net polarization in hydrogen-bonded sheets. In nylon-7, all the N—H···O=C hydrogen bonds point in the same direction, and this arrangement leads to a polar structure (Fig. 5a). In contrast, the N—H···O=C hydrogen bonds point in the opposite direction to each other in nylon-6, which causes a cancellation of the

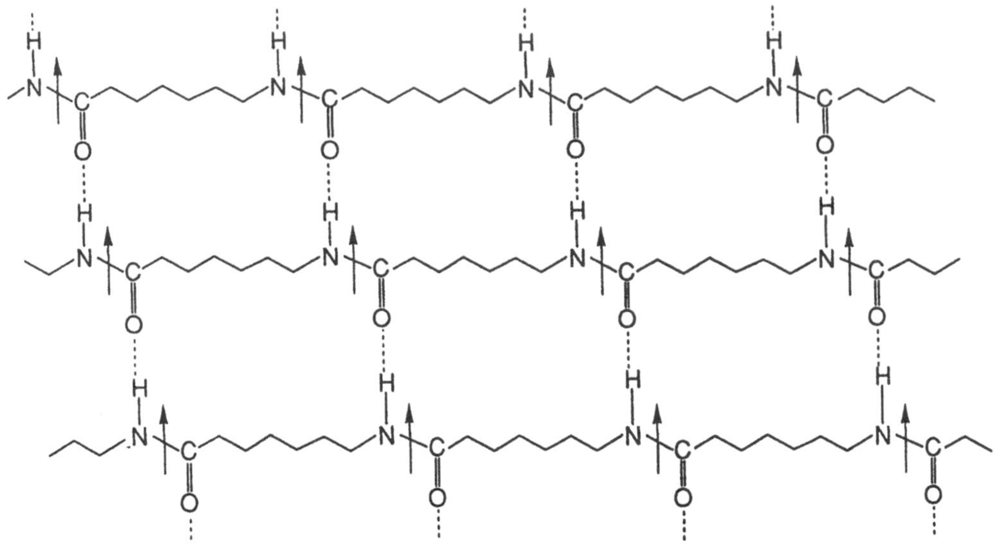

(a)

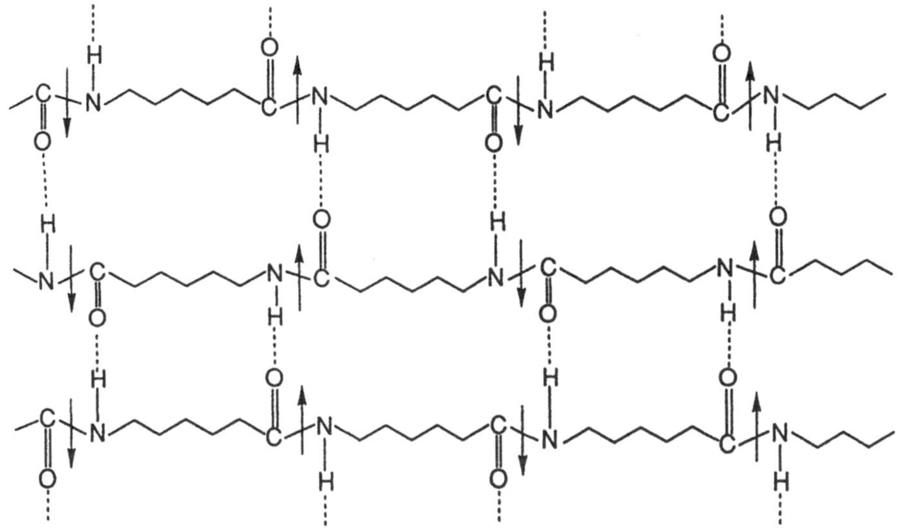

(b)

Figure 5 (a) Polar hydrogen bonding sheet structures for odd-numbered nylon-7. (b) Polar hydrogen bonding sheet structures for even-numbered nylon-6.

polarization (Fig. 5b). A similar type of hydrogen bonding occurs in other odd-numbered nylons and even-numbered nylons. The all-trans conformations in odd-numbered and even-numbered nylons are shown in Figure 5c. Odd-numbered nylons having all dipoles oriented in the same direction possess a net polarization due to hydrogen-bond formation between N—H and C=O groups. The molecular chain arrangement in odd-numbered nylons, through close packing of paraffinic groups and hydrogen bonds, holds the key to the appearance of pyroelectric, piezoelectric, and ferroelectric properties.

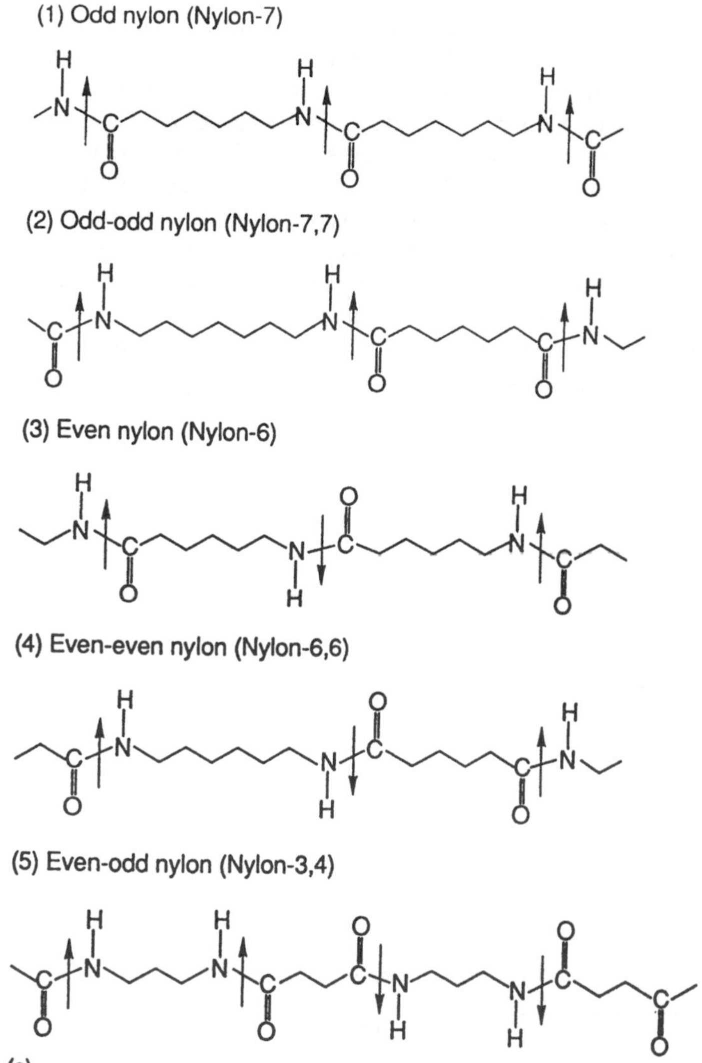

(1) Odd nylon (Nylon-7)

(2) Odd-odd nylon (Nylon-7,7)

(3) Even nylon (Nylon-6)

(4) Even-even nylon (Nylon-6,6)

(5) Even-odd nylon (Nylon-3,4)

(c)

Figure 5 (Continued) (c) All trans-conformations for odd-numbered and even-numbered nylons. Arrows indicates the dipole direction. (Adapted after Ref. 12.)

Newman et al. [28] reported the piezoelectric strain constant d_{31} of α-phase and γ-phase nylon-11 films with variation of poling conditions. Both poling field and temperature significantly affect the piezoelectric strain coefficient d_{31} and piezoelectric stress coefficient e_{31} of nylon-11. At the highest poling field and temperature, the largest d_{31} = 3.2 pC/N was measured. Both d_{31} and e_{31} increased sharply at 500 kV/cm when the poling temperature exceeded 45°C; however, at low poling field such enhancement was less. Figure 6 shows the poling time dependence of d_{31} for γ-phase nylon-11 films poled at 330 kV/cm at 90°C. The d_{31} increased sharply in the first 5 min of poling and showed a saturation tendency as the poling time increases. A comparison of the temperature dependence of d_{31} and e_{31} of nylon-11 films containing α phase and γ phase showed much larger piezoelectric response of γ-phase than of α-phase films under identical poling conditions. The d_{31} and e_{31} for the γ-phase structure were about three and two times larger than the α-phase structure at 330 kV/cm, respectively. Scheinbeim [29] measured the temperature dependence of d_{31} for γ-form nylon-11 film poled at 85°C for 30 min at 300 kV/cm. The d_{31} increased slowly as temperature approached the glass transition (T_g = 60°C), and beyond that it increased sharply with increasing temperature. The largest d_{31} of 12 pC/N was obtained at the maximum temperature of 107°C. The temperature dependence d_{31} showed no decrease in polarization at 95°C, where a transition from α phase to γ phase occurs. This characteristic difference in piezoelectric activities of α phase and γ phase was explained in terms of the breaking and re-forming of hydrogen bonds under a strong electric field. The γ phase has a more regular arrangement of dipoles and stronger hydrogen bonding. The dipole moment of the γ phase can assume up to six possible orientations by means of rotation about the chain axis. It is believed that alignment of the dipoles occurs under the influence of an applied electric field. Piezoelectric

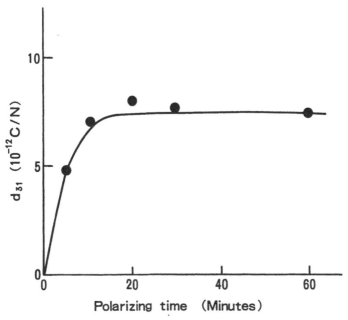

Figure 6 Poling time dependence of d_{31} for γ-phase nylon-11 films poled at 330 kV/cm at 90°C. (Adapted from Ref. 28.)

properties of nylon-11 were reported by Litt et al. [22], who pointed out that poling conditions and relaxation procedures should have a significant effect on piezoelectric coefficients. Gelfandbein et al. [30] reported a piezoelectric constant d_{31} of 0.26 pC/N for nylon-11.

To improve the piezoelectricity in nylon-11, Scheinbeim et al. [31] studied piezo-electricity for heavily plasticized films. Nylon-11 films were prepared by dissolving the nylon-11 in the plasticizer 2-ethyl-1,3-hexanediol at 150°C. The cast films were heated in a vacuum oven to remove excessive plasticizer. The nylon-11 films contained about 30% by weight of the plasticizer. The films were melted at 200°C, pressed and then quenched into ice water. The effect of poling fields and poling time on piezoelectricity was also investigated. The plasticizer content varied with poling time, with simultaneous increase in crystallinity. In the first 30 min under high vacuum, plasticizer was released which leveled off at about 11% by weight. Simultaneously, crystallinity also increased, which then remained at the same level as the evaporation of plasticizer decreased. The variation of d_{31} and e_{31} was studied as a function of poling time, which increased with increasing poling time. The saturation occurred at 30 min, where d_{31} and e_{31} values were 2 pC/N and 0.7 mC/m^2, respectively. On the other hand, the d_{31} and e_{31} also increased with increasing poling field but did not show any saturation. Figure 7 shows poling field dependence of d_{31}. The d_{31} increased rapidly with poling field, and no dielectric break-down was noticed up to 350 kV/cm. Similarly, e_{31} increased rapidly with poling field up to 350 kV/cm. The magnitude of d_{31} and e_{31} increased as ramping time was lowered. A ramping time of 7 min was the lowest that could be used, and dielectric breakdown occurred at about 10 min. The d_{31} increased from 4.0 to 7.1 pC/N and e_{31} from 1.2 to 2.2 mC/m^2 as the poling-field ramping time decreased from 30 min to 7 min. Most important, the piezoelectric response in plasticized nylon-11 was found to be more than double compared to unplasticized films. The piezoelectric response was influenced re-markably by applied poling field when the contents of plasticizer was high, and so was

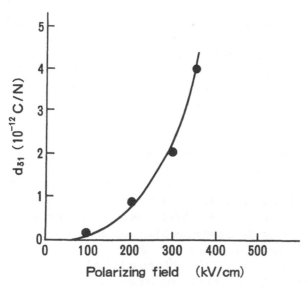

Figure 7 Poling field dependence of d_{31} for plasticized nylon-11 films. (Adapted from Ref. 31.)

the mobility of the amorphous material. The crystallinity was recorded in the presence of the poling field as the content of plasticizer was removed. The crystallinity increased from 15.2% to 43.6% for films poled at 350 kV/cm with a ramping time of 7 min. Significant development in crystallinity was noticed without any indication of orientation of the crystallites. The poling-field ramping time studies demonstrated that piezoelectricity in nylon-11 probably originates from the oriented hydrogen bonds in the amorphous regions rather than from the crystalline regions.

A. Factors Affecting Piezoelectricity

1. Anisotropy

Wu et al. [24] studied the effects of anisotropy and absorbed water on piezoelectric properties of nylon-9 and nylon-11. The piezoelectric measurements on polymorphs of nylon-11 and nylon-9 were carried out at various poling fields and temperatures. Piezoelectric constant for uniaxially stretched and poled α-nylon-9 was measured as a function of angle between the orientation axis and applied tensile stress. The maximum piezoelectric constant d_{31} was recorded when the tensile stress was applied in the extended direction, whereas the minimum d_{32} was obtained when the stress was applied to the film perpendicular to the extended direction. The d_{31}/d_{32} have a ratio of 10. Nylon-11 showed similar results. The angle dependence of the piezoelectric constants and the ratio of d_{31}/d_{32} were similar to the oriented form I PVDF film.

Takase et al. [23] reported temperature dependence of the elastic modulus for nylon-11 at 10 Hz along the draw direction and perpendicular to the draw direction. The value of the elastic modulus perpendicular to the draw direction was about 35–40% that along the draw direction from −196 to 150°C, indicating a high anisotropy. Nylon-11 films easily break when stress is applied perpendicular to the draw direction.

2. Absorbed Water

Temperature dependence studies of e_{31}', e_{31}'', and d_{31}'' of poled α-phase nylon-9 showed no significant change from −70°C to room temperature, though the values increased rapidly as temperature was raised above room temperature, and a peak was observed from 50 to 60°C which disappeared when the samples were cooled from 100°C to 0°C and their piezoelectric constants were remeasured as a function of temperature [24]. This peak reappeared when the samples were left at room temperature for a few days. The appearance of the peak was associated with absorbed water. Figure 8 shows the temperature dependence of elastic modulus and piezoelectric strain constant for α-phase nylon-9. The elastic modulus decreased while the d_{31}' increased from 20 to 60°C. From 60 to 80°C, the d_{31}' decreased, whereas the elastic modulus did not change. In the second heating cycle, the elastic modulus showed rapid decrease from 20 to 40°C and a gradual decrease from 20 to 80°C. Both elastic modulus and piezoelectric strain constant were found to increase as the content of absorbed water increased. The variation in elastic modulus may be associated with the absorbed water. Water acts as a plasticizer which causes a decrease in elastic modulus of the amorphous phase and an increase in the dielectric constant at the same time. As the water evaporates above 50°C, the d_{31}' starts decreasing and gives rise to a peak near 60°C. The difference in both the dielectric constant and elastic modulus between the amorphous and crystalline phases becomes larger compared with dried sample, and the piezoelectric response also becomes larger.

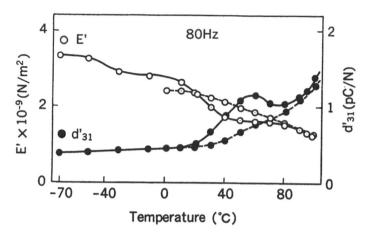

Figure 8 Temperature dependence of elastic modulus and piezoelectric strain constant for α-phase nylon-9. (Adapted from Ref. 24.)

Figure 9 compares the temperature dependence of elastic modulus and piezoelectric strain constant for α-phase nylon-9 and α-phase nylon-11. Nylon-9 has larger piezoelectric strain constant than that of nylon-11, related to the larger dipole density of in nylon-9 in the crystalline phase.

3. Orientation

Figure 10 shows piezoelectric constants measured as a function of temperature for stretched and unstretched nylon-9 poled films. A comparison of the temperature dependence of the piezoelectric coefficient of α-phase nylon-9 and nylon-11 is shown in Figure 11. There was no apparent change in the d_{31}' of stretched and unstretched films, though e_{31}' increased upon stretching. The e_{31}' of stretched film was 1.6 times larger than that

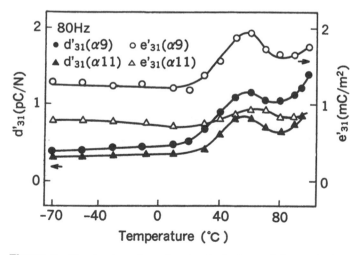

Figure 9 Temperature dependence of elastic modulus and piezoelectric strain constant for α-phase nylon-9 and α-phase nylon-11. (Adapted from Ref. 24.)

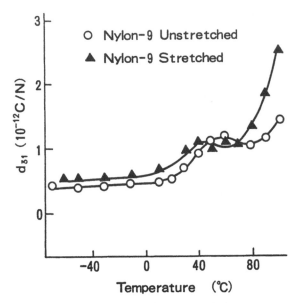

Figure 10   The piezoelectric constants measured as a function of temperature for stretched nylon-9 poled films. (Adapted from Ref. 24.)

of unstretched film (1.3 mC/m^2) at 25°C. This increase may be associated with the increase in elastic modulus along the draw direction by stretching. Similar effect of orientation was observed in nylon-11. Table 3 summarizes the piezoelectric strain constant d_{31}' and piezoelectric stress constant e_{31}' of different specimens of nylon-9 and nylon-11. Nylon-9 containing a mixture of α and γ phases showed the largest d_{31}', while the uniaxially oriented α-phase nylon-9 showed the largest e_{31}'.

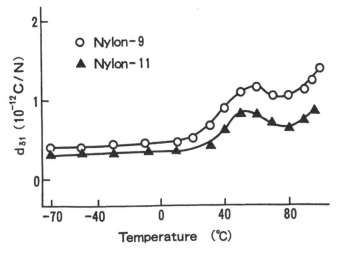

Figure 11   The piezoelectric constants measured as a function of temperature for α-phase nylon-9 and nylon-11 poled films. (Adapted from Ref. 24.)

Table 3 Piezoelectric Strain Constant d'_{31} and Piezoelectric Stress Constant e'_{31} of Nylon-9 and Nylon-11

Nylons	d'_{31} (pC/N)	e'_{31} (mC/m^2)
α-Phase nylon-9	0.6	1.3
α-Phase nylon-11	0.4	0.7
γ-Phase nylon-11	0.4	0.7
Nylon-9 containing a mixture of α and γ phases	1.1	1.8
Nylon-11 containing a mixture of α and γ phases	0.6	1.0
Uniaxially oriented α-phase nylon-9	0.8	2.1
Uniaxially stretched α-phase nylon-11	0.5	1.3
Uniaxially stretched γ-phase nylon-11	0.5	1.5

Source: After Ref. 24.

Infrared spectroscopy and X-ray diffraction studies demonstrated a triclinic α-phase structure for nylon-9. The poled α-phase nylon-9 showed a piezoelectric constant d_{31} of 0.6 pC/N, whereas the largest d_{31} value of 1.1 pC/N was obtained for a mixture of α and γ forms at room temperature. The higher piezoelectric constant of a mixture rather than the pure α and γ forms are associated with the orientational behavior of the dipole moments in the crystallites under an external electric field. Furthermore, uniaxially oriented thin films of α form showed larger piezoelectric constant d_{31} of 0.8 pC/N than did those of γ forms. Moreover, the piezoelectric constant d_{31} of 0.6 pC/N of nylon-9 is larger than for nylon-11 ($d_{31} = 0.4$ pC/N). It has been suggested that odd nylons possessing smaller numbers of methylene groups should show stronger piezoelectricity because of the higher dipole density in the crystallite.

Jacobs and Hicks [18] reported electric field-induced morphological changes by studying X-ray diffraction patterns of unpoled nylon-11 in α, γ, and mixed or poorly formed α phase. By annealing α-phase films at 95°C, a mixture of both α and γ phases or poorly formed α phase was obtained. A complete transition from the γ phase to the α phase was observed by annealing close to the melting point and slow-cooling to room temperature. The X-ray diffraction reflection patterns changed for mixed or poorly formed α-phase nylon-11 before and after poling, and the (010) peak was reduced in the poled sample. The transmission X-ray scans showed an increase of the (010) peak in size relative to the (100) peak. The changes in X-ray diffraction indicate an orientation of crystal planes containing dipoles in the direction of the poling field and a correlation of the poling field with dipoles affecting the crystal morphology. The increase in the piezoelectricity of mixed phases or poorly formed α phase over the pure α phase was explained by alignment of dipoles in the α phase and with existing γ phase. Wu et al. [24] also reported the largest d_{31} value for nylon-9 containing a mixture of α and γ phases at room temperature. A definite correlation between the crystal orientation and piezoelectricity was observed from the change in the X-ray scans after poling.

Takase et al. [23] reported high-temperature piezoelectric strain constant d_{31}, stress constant e_{31}, electromechanical coupling coefficients κ_{31} of nylon-11 and nylon-7. The films of nylons were prepared by melt-press process and then quenching molten films into ice and uniaxially stretching to a draw ratio of 3:1 at room temperature. The temperature dependence of the elastic modulus (real part c' and imaginary part c'') of nylon-11 and nylon-7 films was measured at 104 Hz. Three relaxation processes were observed:

the high temperature α relaxation at 80–100°C, β relaxation at −55°C for nylon-11 and at −33°C for nylon-7 and the γ relaxation at −156°C for nylon-11 and at −143°C for nylon-7. The α and γ relaxations in modulus were the strongest. The dielectric constant ϵ was measured at frequencies of 1, 10 and 100 kHz as a function of temperature. Likely, α, β and γ relaxation processes were observed where the α relaxation was stronger while the α relaxation was weaker. With the increase of temperature, the micro-Brownian motion of the hydrocarbon segments of the nylons is excited in the noncrystalline regions whereas the amide group are not due to the hydrogen bonding. Moreover, the hydrocarbon segments in both nylons are larger than the amide segment and only the amide group contributes to a strong dipole moment. As a consequence, only modulus measurements showed a strong and glass-transition-like γ relaxation process. With the increase of temperature, breaking of some hydrogen bonds will occur and hydrocarbon as well as amide groups begin micro-Brownian motion in the noncrystalline regions. The α relaxation can be designated as the glass transition process.

Figure 12a shows the static value of the electromechanical coupling coefficient κ_{31} of nylon-7, nylon-11, and PVDF. The κ_{31} was obtained from the static dielectric constant by using the following equation:

$$\kappa_{31} = d_{31} \left(\frac{c}{\epsilon_0 \epsilon} \right)^{1/2} \tag{1}$$

where ϵ_0 is the permittivity of free space. The κ_{31} of nylon-7 and nylon-11 is higher than that of PVDF. The maximum κ_{31} values of 0.054, 0.049, and 0.019 were estimated for nylon-7, nylon-11, and PVDF, respectively. The κ_{31} of nylon-7 does not change noticeably from 120 to 200°C, since the plot shows flatness. In the case of nylon-11, the κ_{31} decreases in the temperature region above the glass transition temperature. The decrease in κ_{31} with increasing temperature is related to the increase of dielectric constant. Electromechanical coupling as a function of temperature for annealed nylon-7, nylon-11, and PVDF is shown in Figure 12b. The values of κ_{31} of nylon-7 and nylon-11 are larger than that of PVDF in the low-temperature region below −50°C after annealing of PVDF at 175°C. Both nylon-7 and nylon-11 have κ_{31} of 0.02, while that of PVDF is 0.008. At higher temperatures, above T_g, the κ_{31} of nylon-7 and nylon-11 are still larger than that of PVDF because of the large increment in dielectric constant of nylons. The κ_{31} rapidly decreases as the temperature increases.

Figure 13 shows variation of piezoelectric strain constant d_{31} measured at 104 kHz from 50°C to temperature near the melting points of nylon-11, nylon-7, and PVDF. A large increase in d_{31} of nylons was observed around their glass transition temperatures. Similar features as a function of temperature appeared in e_{31} data. Both d_{31} and e_{31} exhibited a shoulder between 120 and 140°C. The maximum d_{31} and e_{31} values were 14 pC/N and 21 mC/m^2 for nylon-11 and 17 pC/N and 27 mC/m^2 for nylon-7, respectively. The d_{31} of PVDF remained constant while the value of e_{31} decreased linearly with increasing temperature. PVDF showed a d_{31} of 5 pC/N and e_{31} of 7 mC/m^2 at 100°C. The high-temperature d_{31} and e_{31} values of nylon-11 and nylon-7 are larger than that of PVDF.

4. Annealing Temperature

Takase et al. [23] reported the effect of annealing on the piezoelectric coefficient d_{31} as a function of temperature for nylons. Figure 14 shows the temperature dependence of d_{31}' and d_{31}'' for unannealed and annealed nylon-11 films. The unannealed sample refers to the first measurement after poling, and annealed to the measurements done after an-

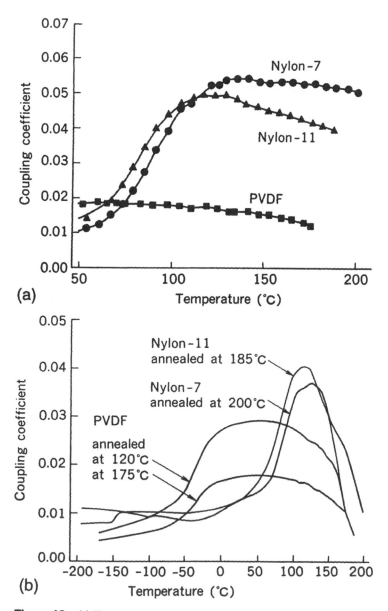

Figure 12 (a) Temperature-dependent static values of the electromechanical coupling coefficient κ_{31} of nylon-7, nylon-11, and PVDF. (b) Temperature dependent electromechanical coupling coefficients κ_{31} of annealed nylon-7, nylon-11, and PVDF films. (Reprinted with permission from Ref. 23, the American Chemical Society.)

nealing of the poled samples. Nylon-11 was annealed at 185°C for 2 h, and measurement frequency was 104 Hz in both cases. There is a significant change in d_{31} in the temperature region above the T_g. The unannealed nylon-11 showed a maximum d_{31}' of 21 pC/N at 163°C, while a constant d_{31}' of 14 pC/N was observed after annealing. Nylon-7 showed similar temperature-dependent d_{31}' characteristics. The unannealed nylon-7 had

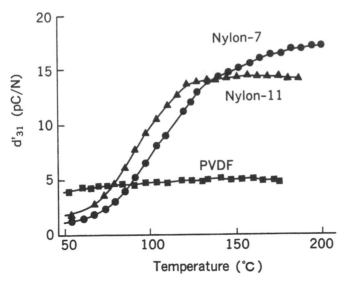

Figure 13 Variation of piezoelectric strain constant d_{31} measured at 104 kHz from 50°C to temperatures near the melting points of nylon-11, nylon-7, and PVDF. (Reprinted with permission from Ref. 23, the American Chemical Society.)

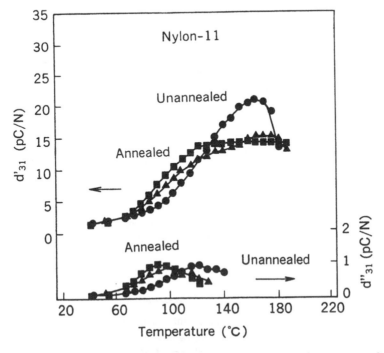

Figure 14 The effect of annealing on d_{31} as a function of temperature for nylon-11. (Reprinted with permission from Ref. 23, the American Chemical Society.)

a maximum d_{31}' of 30 pC/N at 180°C and a constant d_{31}' of 17 pC/N after annealing. The maximum e_{31}' values were 26 mC/m² at 150°C for nylon-11 and 34 mC/m² at 168°C for nylon-7. After annealing, the e_{31}'-versus-temperature curves showed a plateau with constant e_{31}' values of 21 mC/m² for nylon-11 and 26 mC/m² for nylon-7. The values of d_{31}' and e_{31}' decreased to about 65–75% of the initial value after annealing. The decreases in piezoelectric coefficients of nylons in the temperature region above T_g may be associated with Poisson's ratio caused by crystallographic and morphological changes induced by annealing [23]. The annealing process leads to a significant decrease in spacing between hydrogen-bonded sheets, which tends to decrease the reorientation of amide groups with increasing thermal motion.

VI. FERROELECTRICITY

A. Nylon-7 and Nylon-11

Ferroelectric properties of nylons have been investigated by several research groups. In preliminary studies, the D-versus-E hysteresis characteristics showed no hysteresis loop in nylon-11 [18,32], and further attempts were made to study the ferroelectric hysteresis in nylon-11 in the temperature range from room temperature to about 60°C and frequencies ranging from 0.1 to 40 Hz. No indication of the $D–E$ hysteresis was observed in any of the samples tested. The authors provided an explanation that very slow switching times coupled with large conduction effects make ferroelectric response difficult to detect. Mathur et al. [33] reported ferroelectric hysteresis effects in uniaxially stretched nylon-11 films melt-quenched (δ' form) or slow-cooled (α' form) from the melt. The polarity of the piezoelectric constant was found to reverse and exhibit a hysteresis-like behavior for static positive and negative poling fields. Lee et al. [34] investigated the $D–E$ hysteresis characteristics in nylon-11 by preparing samples under different conditions. The nylon-11 films were prepared by melt-pressing, moltened at 205–210°C and quenched in water at room temperature. The quenched film was uniaxially oriented by stretching to a draw ratio of 2.8:1 at room temperature. Cold-drawn nylon-11 film showed the current density-versus-electric field (J-versus-E) characteristics in the temperature range from 20 to −60°C. The $J–E$ plots of nylon-11 were similar to the DC conduction curve. Nylon-11 showed a sharp current peak which indicated a polarization-reversal phenomenon. The coercive fields were 65, 98, 125, 160, and 215 MV/m for 20, 0, −20, −40, and −60°C, respectively. For comparison, the temperature dependence of the coercive field of nylon-11 was much larger than that of PVDF. Figure 15 shows the $D–E$ characteristics of nylon-11 which were obtained by integrating current density (J) in the temperature range from 20 to −60°C. The remanent polarizations estimated by the intercept of each loop with the D axis were 68, 64, 56, 46, and 31 mC/m² for 20, 0, −20, −40, and −60°C, respectively. Ferroelectricity was observed only in the cold-drawn or samples annealed at ~100°C or below. The time dependence of the switching current densities J_f (forward) and J_r (reverse) were measured at 20°C under a pulsed electric field. At the pulsed field of 100 MV/m or more, the J_r increased more sharply than the J_f. The time dependence of the charge densities D_f (forward) and D_r (reverse) were measured under the same conditions. The polarization reversal was found to complete in a few tens of milliseconds under 140 MV/m at 20°C. The fast polarization reversal and the rectangular $D–E$ hysteresis loops evidenced the dipolar origin. Cold-drawn nylon-11 showed ferroelectricity at or below room temperature. The spontaneous polarization of

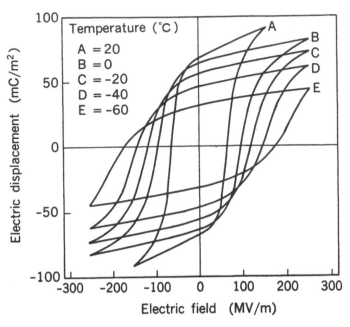

Figure 15 D–E characteristics of nylon-11 which were obtained by integrating current density (J) in the temperature range from 20 to −60°C. (Adapted from Ref. 34.)

a nylon-11 crystal was estimated as 48 mC/m². From the remanent polarization, 68–46 mC/m² at temperatures from 20 to −40°C, approximately 96% of the dipoles were estimated to be oriented by the external electric field.

Lee et al. [35] investigated the effect of annealing on the ferroelectric behavior of nylon-11 and nylon-7. Nylon-11 and nylon-7 films were prepared by melting in a hot press at 210 and 240°C, respectively. The molten films were quenched in ice and the stretched to a draw ratio of 3:1 at room temperature. Annealing of the films was carried out under vacuum at 55, 70, 115, 145, and 185°C for 2 h. The J–E hysteresis characteristics were measured after annealing nylon-11 and nylon-7 samples. The coercive fields of 62 and 79 MV/m at 25°C were obtained for the as-stretched nylon-11 and nylon-7, respectively. The amide groups are responsible for the large difference in coercive fields because the amide groups in nylon-7 are more difficult to rotate under electric field compared with nylon-11. The typical D–E hysteresis characteristics of nylon-7 and nylon-11 are shown in Figures 16 and 17, respectively. The estimated remanent polarization was 51 mC/m² for nylon-11 and 86 mC/m² for nylon-7. The remanent polarization decreases with increasing annealing temperature, and almost no remanent polarization was observed at the annealing temperature of 185°C. Nylon-11 films annealed at 185°C before poling showed no hysteresis loop because the coercive field required exceeded the dielectric strength. On the other hand, the coercive field increases as the annealing temperature increases. The increase in coercive fields (62 to 115 MV/m for nylon-11 and 79 to 97 MV/m for nylon-7) due to annealing treatment and decrease in the ferroelectric behavior indicate a rearrangement of the hydrogen-bonded sheet structure in both nylons.

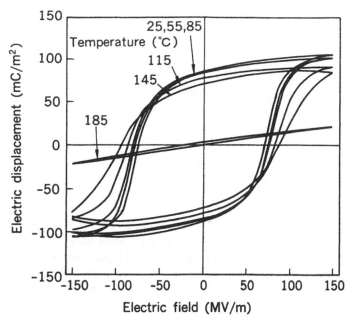

Figure 16 *D–E* hysteresis characteristics of nylon-7. (Adapted from Ref. 35.)

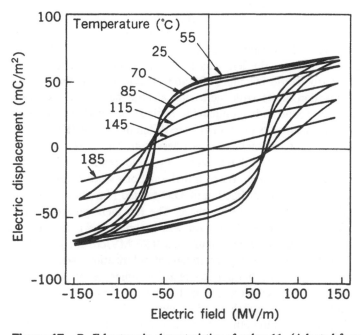

Figure 17 *D–E* hysteresis characteristics of nylon-11. (Adapted from Ref. 35.)

The remanent polarization decreased from 51 to 17.3 mC/m^2 for nylon-11 films and 86 to 70.5 mC/m^2 for nylon-7 films with increasing annealing temperature from 25 to 145°C. Nylon-7 has remanent polarization as high as 86 mC/m^2, which is significantly larger than that of PVDF films. Figure 18 shows the current density-versus-electric field plots and D–E hysteresis curves of nylon-7, nylon-11, and PVDF [23]. Nylons have higher current density than PVDF. Remanent polarizations of 93, 58, and 55 mC/m^2 and coercive fields of 92, 56, and 46 MV/m were determined for nylon-7, nylon-11, and PVDF, respectively.

Wide-angle X-ray diffraction patterns of nylon-11 and nylon-7 at different annealing temperatures are shown in Figures 19 and 20, respectively. For both nylons, the crystallinity increased with increasing annealing temperature and the position of the diffraction peak shifted to higher 2θ angles. The 2θ angle of the diffraction peak of as-stretched nylon-7 was larger than that of as-stretched nylon-11, and the position of the diffraction peak in both nylons comes closer as the annealing temperature reaches 185°C. The different changes were also noticed in the peak position with increasing annealing temperature between the reflection and transmission modes, which indicates the existence of some orientation. Only reflection modes showed the significant shift in 2θ angle and the large increase in diffraction intensity with increasing annealing temperature while not in the transmission modes. The preferential orientation was developed during the stretching process, as it was not apparent in melt-quenched, unstretched, and annealed samples. Clear ferroelectric behavior was observed for quenched, cold-drawn, and annealed nylon-11 and nylon-7. Since nylon-12 showed no ferroelectricity, the origin of ferroelectric behavior in nylon-11 and nylon-7 was considered to be from the crystalline regions.

The D–E hysteresis characteristics and ferroelectric switching in nylon-11 have recently been reported by Furukawa et al. [36]. The nylon-11 films for measurements were prepared by hot-pressing and melting sample at 230°C followed by quenching in cold water. These films were stretched to three times their original length. The D–E hysteresis loop of nylon-11 measured at 0.1 Hz at room temperature is shown in Figure 21. At low fields, a simple line is observed whose shape changes as the field increases. At room temperature, the rising time of the electric field was 10 µs. The coercive field was 90 mV/m. Figure 22 shows the ferroelectric switching of nylon-11. Polarization reversal was observed as well as peak position shifts with increasing electric field. The switching characteristics of nylon-11 were found to be comparable to those of vinylidene fluoride co-polymers. The switching characteristics of nylon-11 are different from PVDF. Balizer et al. [19,20] reported the possibility of metastable γ phase with disordered hydrogen bonds and γ′-to-α crystalline transformation during stretching. Acid treatment of oriented δ′ phase also leads to an oriented γ′ phase. These transformations were confirmed by an unpolarized infrared beam. The X-ray diffraction of D$_2$O-treated film showed a peak at 21.5° assigned to the (100) reflection of the γ′ phase. The remanent polarization of nylon-11 films soaked in D$_2$O and in phenol solution was measured as a function of exposure time. The coercive fields E_c were 0.60, 0.62, and 0.86 MV/cm for quenched, D$_2$O-, and phenol-treated samples, respectively. The remanent polarizations were 55, 42, and 13.6 mC/m^2 for the quenched, D$_2$O-, and phenol-treated samples, respectively. The polarization of the oriented films in D$_2$O decreased with exposure time, about 10% of that of the δ′ phase (55 mC/m^2). The switching of γ phase indicates that structural hindrance rather than hydrogen-bond strength restricts the more stable α phase from exhibiting switching behavior. The oriented γ-phase films showed switching behavior, as

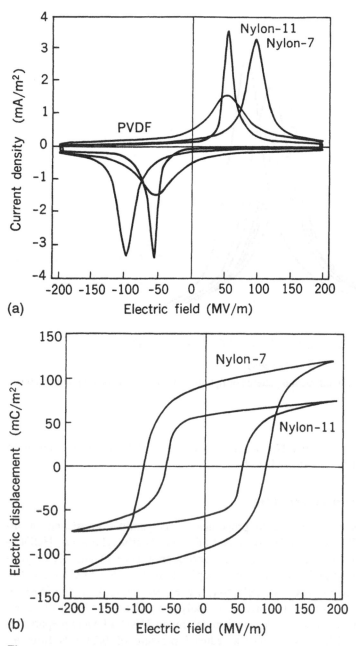

Figure 18 Current density-versus-electric field plots and *D–E* hysteresis curves of nylon-7, nylon-11, and PVDF. (Adapted from Ref. 35.)

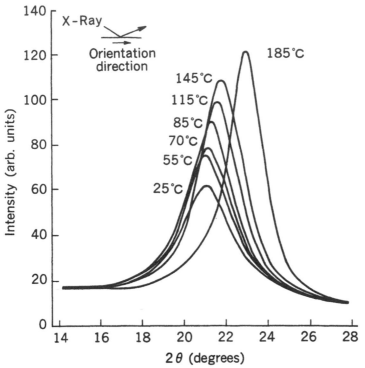

Figure 19 Wide-angle X-ray diffraction patterns of nylon-11 at different annealing temperatures. (Adapted from Ref. 35.)

it has the strongest hydrogen bonds. No switching was observed for the phase transition obtained from stretching the unoriented γ phase.

B. Alicyclic and Aromatic Nylons

Murata et al. [37] reported ferroelectric properties of polyamides consisting of m-xylylenediamine, $H_2N-CH_2-C_6H_4-CH_2-NH_2$, and aliphatic dicarboxylic acids, $HOOC-(CH_2)_x-COOH$ with number of carbon atoms (x) varying from 6 to 11 and 13. Poly(m-xylylene adipamide), nylon-MXD-6, poly(m-xylylene pimelamide), nylon-MXD-7, poly(m-xylylene suberamide), nylon-MXD-8, poly(m-xylylene azelaamide), nylon-MXD-9, poly(m-xylylene sebacaamide), nylon-MXD-10, poly(m-xylylene undecanediamide), nylon-MXD-11, and poly(m-xylylene tridecanediamide), nylon-MXD-13, were prepared by melt-polycondensation at 220–264°C under a reduced pressure of 0.13 kPa from m-xylylenediamine and adipic, pimelic, suberic, azelaic, sebacic, undecanedioic, and tridecanedioic acids, respectively. Nylon-MXDs have the following chemical structure:

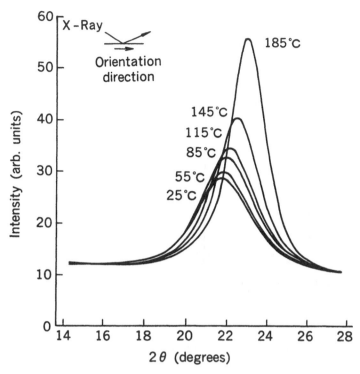

Figure 20 Wide-angle X-ray diffraction patterns of nylon-7 at different annealing temperatures. (Adapted from Ref. 35.)

Thin films were obtained by melt-pressing at temperatures of 30 K above their melting points and then slowly cooled or quenched in cold water. Table 5 lists the thermal properties and remanent polarization for nylon-MXDs. Nylon-MXD-13 showed no glass transition temperature, but peaks for both crystallization and melting. The quenched nylon-MXD-9 showed no exothermic crystallization, but T_g and a dispersed melting peak.

Table 4 Comparison of Various Parameters for Nylon-7 and Nylon-11

Parameter	Nylon-7	Nylon-11
1. Piezoelectric strain coefficient, d_{31} (pC/N)	17	14
2. Piezoelectric stress coefficient, e_{31} (mC/m^2)	27	21
3. Electromechanical coupling coefficient, k_{31} (pC/N)	0.054	0.049
4. T_m (°C)	235	195
5. Coercive field (MV/m)	92	56
6. Remanent polarization (mC/m^2)	93	58
7. Dielectric constant		
ϵ'	401 (at 200°C)	1014 (at 187°C)
ϵ''	1566 (at 200°C)	1894 (at 187°C)

Source: After Ref. 23.

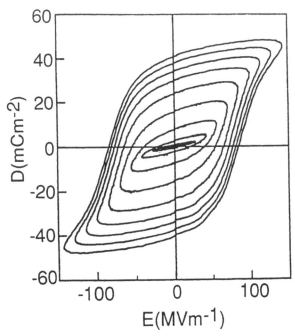

Figure 21 *D–E* hysteresis characteristics of nylon-11. (Adapted from Ref. 36.)

The dispersed melting peaks in nylon-MXD-7 and nylon MXD-9 indicated very low crystallinity. The quenched samples of nylon-MXD-6, nylon-MXD-8, nylon-MXD-10, nylon-MXD-11, and nylon-MXD-13 were found to be either amorphous or with very low crystallinity. As slowly cooled samples of these nylons showed no crystallization peak, they may have higher crystallinity.

Figures 23 and 24 show the *D–E* hysteresis loop of quenched nylon-MXD-6 and nylon-MXD-9, respectively, indicating their ferroelectric nature. At a maximum field of 194 MV/m, the coercive field E_c and remanent polarization P_r were 64 MV/m and 67 mC/m^2, respectively. The remanent polarization for nylon-MXD-6 was the largest and decreased for nylons having higher number of carbon atoms in dicarboxylic acids. The density of amide polar groups decreases with the increasing carbon atoms in dicarboxylic acids. Here nylon-MXD-7, nylon-MXD-9, nylon-MXD-11, and nylon-MXD-13 are odd-numbered nylons, since the *m*-xylylene group in the polyamide chain corresponds to pentamethylene groups. Nylon-MXD-6, nylon-MXD-8, and nylon-MXD-10 are not odd-numbered nylons, but they belong to the same type. The remanent polarization of nylon-MXD-6 was larger than that of nylon-11 (51 mC/m^2) but lower than that of nylon-7 (86 mC/m^2). The slowly cooled nylon-MXD-6 samples and the annealed nylon-MXD-11 exhibited slender hysteresis curves with very small remanent polarizations. The quenched nylons showed a *D–E* hysteresis loop with remanent polarization of 23–67 mC/m^2. The ferroelectricity in these nylons occurs due to the amide group orientation by the electric field, as amides are the only polar groups in the polyamide chains. Two possibilities for the origin of ferroelectricity were discussed, one related to the crystalline region and another to the amorphous regions. Ferroelectricity may originate from the amorphous region in nylons, since quenched samples have very low crystallinity. If this is the case,

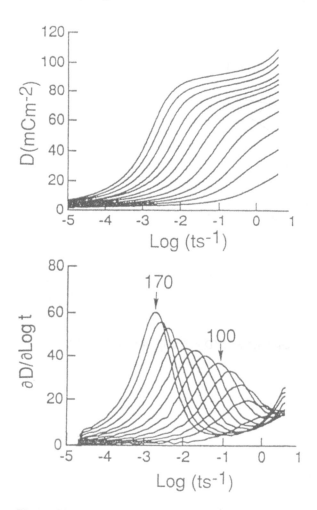

Figure 22 Ferroelectric switching characteristics of nylon-11. (Adapted from Ref. 36.)

Table 5 Thermal Properties, Crystallization Temperature (T_c), and Remanent Polarization (P_r) for Quenched Nylon-MXDs

Polyamide	T_m (K)	T_c (K)	T_g (K)	P_r (mC/m²)
Nylon-MXD-6	510.7	423.0	344.5	67
Nylon-MXD-7	451.4	—	326.4	53
Nylon-MXD-8	475.7	390.3	318.9	44
Nylon-MXD-9	425.8	—	315.5	43
Nylon-MXD-10	465.9	375.5	324.2	25
Nylon-MXD-11	442.2	348.3	313.3	41
Nylon-MXD-13	443.8	318.0	—	23

Source: After Ref. 37.

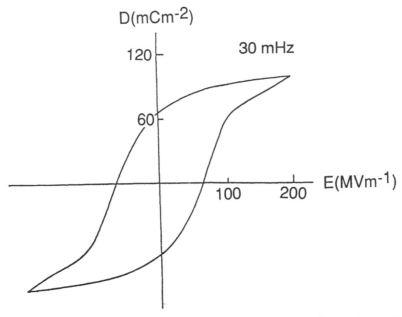

Figure 23   *D–E* hysteresis loop of quenched nylon-MXD-6. (Adapted from Ref. 37.)

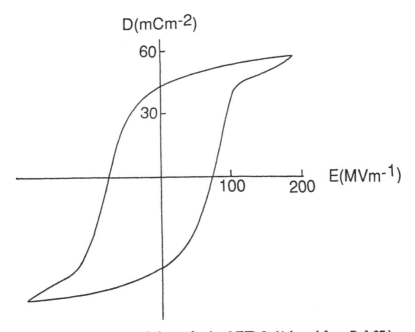

Figure 24   *D–E* hysteresis loop of nylon-MXD-9. (Adapted from Ref. 37.)

then the smaller P_r for slowly cooled nylon-MXD-6 and annealed nylon-MXD-9 were considered due to the higher crystallinity of these polyamides.

Murata et al. [38] reported the relationship between electric displacement D and electric field E for alicyclic and aromatic polyamides. Alicyclic polyamides was prepared from 1,3-bis(aminomethyl)cyclohexane (BAC) and aliphatic dicarboxylic acids. These polymers are referred as BAC_n, where n is the number of carbon atoms in the dicarboxylic acid. Nylons BAC6, BAC7, and BAC10 were obtained from dicarboxylic acids HOOC–$(CH_2)_x$––COOH with adipic ($n = 4$), pimelic ($n = 6$), and sebacic ($n = 8$) acids, respectively. Nylon prepared from hexamethylenediamine (HMD) and isophthalic acid (I) and from HMD and 70 mol% I and 30 mol% terephthalic acid (T) were designed as 6I and 6I/6T, respectively. Differential thermal analysis (DTA) yielded glass transition temperatures of 103, 89, 73, 121, and 122°C for nylons BAC6, BAC7, BAC10, 6I and 6I/6T, respectively. The low melting-point peak of nylon-BAC10 showed very low crystallinity. Nylon-6I and 6I/6T were found to be amorphous from X-ray diffraction studies. Thin films of nylons-BAC_n were prepared from melt-pressing at temperatures of 30 K above their melting points, and for nylon-6I and 6I/6T above their T_g, and were quenched in cold water. Nylon-BAC_n and nylon-PI have the following chemical structures:

Nylon-BAC_n

Nylon-PI

The D–E hysteresis curve from thin films of nylon-BAC7 was obtained at room temperature (Fig. 25). Nylon-BAC7 showed a coercive field E_c of 82 MV/m and a remanent polarization P_r of 33 mC/m^2 at a maximum field of 215 MV/m. The D–E hysteresis curve of nylon-BAC10 was similar to that of nylon-BAC7. The remanent polarization P_r of nylon-BAC6 was 26 mC/m^2 at 53°C, which reduced to half at room temperature (Fig. 26). The depolarization currents increased with increasing temperatures, showing peaks near their T_g and then decreasing abruptly. On the other hand, the remanent polarization decreased with increasing temperature and vanished above T_g. Remanent polarizations of 28, 33, and 39 mC/m^2 were obtained for nylon-BAC6, -BAC7, and -BAC10, respectively. The pyroelectric coefficient of nylon-BAC7 was 6 μC/m^{2}K with a remanent polarization of 22 mC/m^2. Nylon-6I showed D–E hysteresis curves at 104°C, though significant D–E hysteresis was observed at 37°C, perhaps because the coercive field was too high (Fig. 27). Nylon-6I/6T also showed a D–E hysteresis loop at the same temperatures as nylon-6I. The depolarization current of nylon-6I increased with increasing temperature and showed a peak at 118°C. As nylon-6I and nylon-6I/6T are amorphous and quenched nylon-BAC_n's have low crystallinity, the origin of the D–E hysteresis is attributed to amide groups in the amorphous region. Amorphous nylons showed D–E hysteresis curves at temperatures below their T_g. The dipoles in the amorphous region

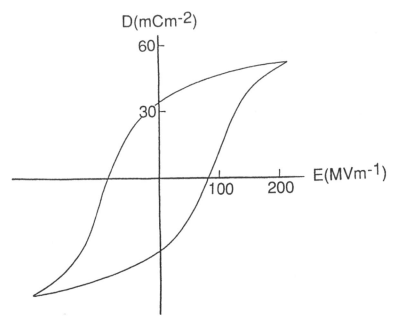

Figure 25 *D–E* hysteresis loop of quenched nylon-BAC7 at room temperature. (Adapted from Ref. 38.)

align in the parallel direction but can be reversed via spontaneous polarization by the applied electric field and still remain when the field is switched off.

C. Fluorinated Odd-Odd Nylons

Tasaka et al. [12,39–42] reported ferroelectric behavior in fluorinated nylons having an odd number of carbon atoms. Fluorinated nylons where $x = 3$, 5, 7, and 9, namely, nylon-

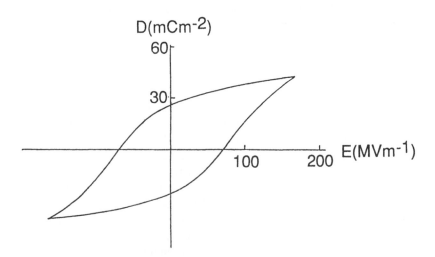

Figure 26 *D–E* hysteresis loop of quenched nylon-BAC6 at 326 K. (Adapted from Ref. 38.)

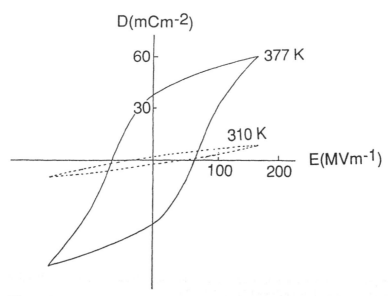

Figure 27   *D–E* hysteresis loop of nylon-PI. (Adapted from Ref. 38.)

35F, nylon-55F, nylon-75F, and nylon-95F, respectively, were synthesized by interfacial condensation of corresponding alkyldiamines with perfluoroglutalochlorides, as shown below.

$$H_2N-(CH_2)_x-NH_2 \ + \ ClCO-(CF_2)_3-COCl$$
$$\downarrow$$
$$(HN-(CH_2)_x-NH-CO-(CF_2)_3-CO)_n$$

The melting and glass transition temperatures of the fluorinated nylons were found to be relatively lower than those of corresponding nylons (Table 1). Thin films from white powders were made by melting them above their melting temperatures and then pressing. These fluorinated nylons are soluble in polar solvents and show resistance to moisture. They have electrical conductivities of the order of 10^{-12} S/cm up to 120°C.

The temperature dependence of dielectric constant and dielectric loss as a function of frequency for quenched nylon-75F is shown in Figure 28. Three relaxation processes, (a) a low-temperature relaxation around 30°C due to micro-Brownian motion, (b) the largest relaxation associated with crystallization or phase transition, and (c) a higher-temperature relaxation due to crystal defect or impurity. The peak for second relaxation increases as the frequency decreases, which indicates lower cooperative motion of molecular dipoles. The sharp decrease in relaxation with increasing temperature may be associated with ferroelectric-to-paraelectric transition or vice versa. Figure 29 shows hydrogen-bonded sheets of odd-numbered nylons and physical changes occurring due to the annealing that results in a ferroelectric-to-paraelectric transition. The annealing process causes a significant decrease in spacing between hydrogen-bonded sheets, which lessens the reorientation of amide groups with increasing thermal motion. The effect of annealing that leads to almost constant values of piezoelectric coefficients of nylon-7 and nylon-11 may be associated with the changes in the dimensions through ferroelectric-to-paraelectric transition.

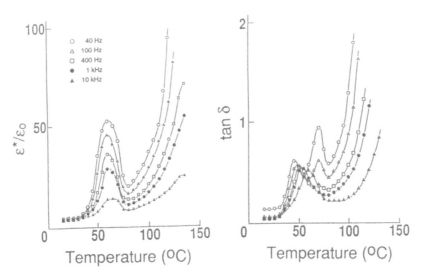

Figure 28 The temperature dependence of dielectric constant and dielectric loss as a function of frequency for quenched nylon-75F. (Adapted from Ref. 12.)

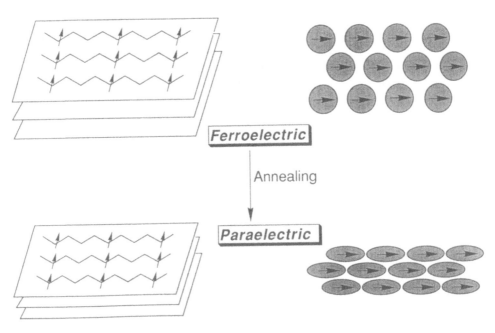

Figure 29 Ferroelectric-to-paraelectric transition in odd nylons after annealing. (Adapted from Ref. 12.)

The D–E hysteresis loop for quenched and annealed nylon-75F at 25°C is shown in Figure 30. The remanent polarization was 30 mC/m² lower than for nylon-7, due to the dipole moment cancellation between fluorine and amides. Figure 31 shows the D–E hysteresis loop of nylon-75 at 90°C and at 0.01 Hz, which shows a remanent polarization of 80 mC/m² and a coercive field E_c of 34 MV/m. For comparison, the DSC curves for quenched nylon-75F and nylon-75, shown in Figure 32, which are characteristically different from each other. Nylon-75F has two peaks, an exothermic and an endothermic one, while nylon-75 shows only an endothermic peak at relatively higher temperature. Similarly, the X-ray diffraction patterns of quenched nylon-75F and nylon-75 are somewhat different (Fig. 33). Figure 34 shows the D–E hysteresis loop of nylon-77 recorded at 70°C and 25°C at a frequency of 0.017 Hz. The hysteresis loop expands at 25°C.

D. Mechanism of Ferroelectricity

The mechanism of ferroelectric polarization in odd-numbered nylons has not been addressed so far except in the case of nylon-11. Scheinbeim et al. [43] explained the ferroelectric polarization mechanism in poled nylon-11 films with infrared spectroscopy and X-ray diffraction studies. Infrared spectra of as-stretched, as-poled, and poled-annealed nylon-11 films were recorded. The C=O stretching vibrations at 1640 cm^{-1} and the N—H stretching vibrations at 3300 cm^{-1} of the as-stretched and the as-poled films showed noticeable differences in absorption intensities, indicating that the hydrogen bonds break and dipolar rotation toward the electric field direction to re-form bonds in a new direction is retained even after removing the electric field. The absorption intensities of the C=O stretching vibrations and the N—H stretching vibrations from the as-poled and the poled-annealed films showed very little difference, supporting the idea that significant change in orientation of C=O and N—H dipoles is not caused by the annealing treatment (under vacuum at 185°C for 2 h). This high-temperature stability of dipole

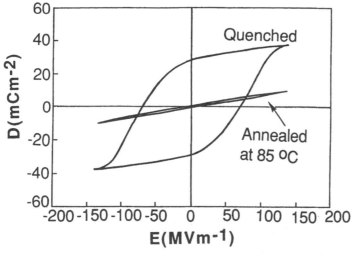

Figure 30 D–E hysteresis loop for quenched and annealed nylon-75F at 25°C. (Adapted from Ref. 12.)

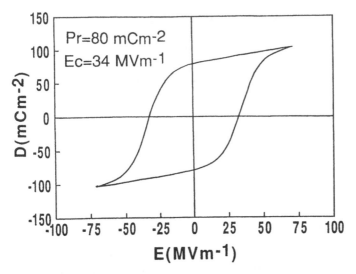

Figure 31 *D–E* hysteresis loop of nylon-75 at 90°C and at 0.01 Hz. (Adapted from Ref. 12.)

orientation in the crystalline region sheds light on the polarization mechanism for poled nylon-11 films.

The X-ray diffraction patterns of as-annealed and poled-annealed nylon-11 films were obtained. Different diffraction patterns were obtained from the geometric arrangements of nylon-11 films which differ only by a rotation angle of 90° about the draw direction (*c* axis). Nylon-11 showed three-dimensional texture rotated by 90° about the *c* axis with respect to initial orientation. Following electrical poling and then annealing,

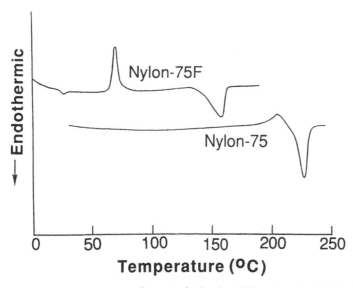

Figure 32 DSC curves for quenched nylon-75F and nylon-75. (Adapted from Ref. 12.)

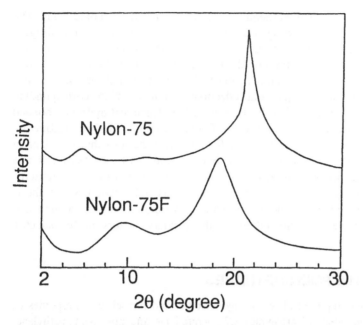

Figure 33 X-ray diffraction patterns of quenched nylon-75F and nylon-75. (Adapted from Ref. 12.)

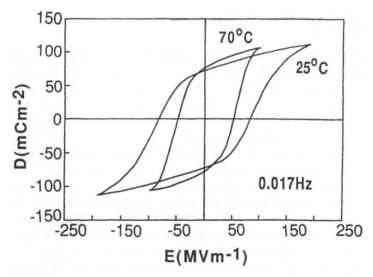

Figure 34 *D–E* hysteresis loop of nylon-77 recorded at 70 and 25°C at a frequency of 0.017 Hz. (Adapted from Ref. 12.)

the hydrogen-bonded sheets become perpendicular to the plane of the poled films. The strong dipoles associated with the hydrogen bonds remain in the plane of the hydrogen-bonded sheets. The dipole switching of 90° takes place with applied electric field for switching dipoles in the plane of the film into the field direction. A 180° dipole switching (antiparallel packing) mechanism of chains in the hydrogen-bonded sheets occurs with electric field reversals. Very strong equatorial reflections with $d = 3.89$ Å, corresponding to the spacing between hydrogen-bonded sheets, provided evidence that hydrogen-bonded sheets are parallel to the plane of the films. Both IR spectroscopy and X-ray diffraction studies of three-dimensionally oriented nylon-11 films support the idea that polarization eventually takes place in the crystalline regions. Balizer et al. [20] reported FTIR and X-ray diffraction study of polymorphs of nylon-11 and their relation to ferroelectricity. The δ' phase, having stronger hydrogen bonds, switches its polarization, while the α phase, with weaker hydrogen bonds, does not switch. Hydrogen-bonding strength and the crystalline structures were considered to be the primary factors in ferroelectric switching.

VII. FERROELECTRIC POLYURETHANES

Recently, Tasaka et al. [44] reported dielectric, pyroelectric, and ferroelectric properties of poly(trimethylene-heptamethylene dicarbamate) 3,7-polyurethane and poly(pentamethylene-hexamethylene dicarbamate) 5,6-polyurethane (structures shown below). Thermal analysis showed a glass transition temperature (T_g) of 31°C for both polyurethanes and melting temperatures of 142°C for 3,7-polyurethane and 160°C for 5,6-polyurethane. Temperature dependence of dielectric constant and tanδ at 100 Hz for 3,7-polyurethane has been measured. After annealing, the dielectric loss peak shifted to low temperature and dielectric relaxation strength decreased in the glass transition region. The annealed samples also showed a steep rise in dielectric behavior above 125°C.

$$-(CH_2)_3-O-CO-NH-(CH_2)_7-NH-CO-O-)_n$$
3,7-Polyurethane
$$-(CH_2)_5-O-CO-NH-(CH_2)_6-NH-CO-O-)_n$$
5,6-Polyurethane

Both 3,7-polyurethane and 5,6-polyurethane show polarization reversal phenomena. The current density-versus-electric field features of 3,7-polyurethane yielded coercive fields of 90, 75, and 40 MV/m at 75, 85, and 95°C, respectively. Figure 35 shows the displacement-field (D–E) hysteresis loop for 3,7-polyurethane recorded at 75°C. A remanent polarization of 50–60 mC/m^2 was measured for both polyurethanes. With annealing, the remanent polarization was found to increase for 3,7-polyurethane and decrease for 5,6-polyurethane, which indicates that the polarization is related to the crystal polymorphism. The remanent polarization was found to be stable over T_g up to 115°C for annealed 3,7-polyurethane. The pyroelectric coefficient of 5 C/m^{2}K was determined, and pyroelectricity was considered to be attributed to the urethane dipole polarization in the crystal or related phase. The polarization reversal was observed after crystallization upon annealing, and pyroelectricity was improved. This phenomenon is quite different from those of nylon-7 and nylon-11, which show no polarization reversal after annealing. The spontaneous polarization of about 50–55 mC/m^2 was calculated from the 2.8-Debye dipole moment of the urethane group. The smaller value of spontaneous polarization of about 55 mC/m^2 at 75°C may be related to the lower crystallinity or insufficient dipole

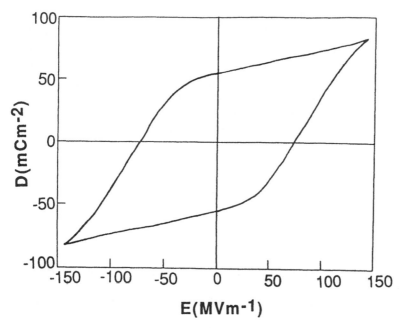

Figure 35 *D–E* hysteresis loop for 3,7-polyurethane recorded at 75°C. (Adapted from Ref. 44.)

orientation in the crystallites. The X-ray diffraction showed a main peak around 20° in the annealed 3,7-polyurethane due to the increased crystallinity. In 5,6-polyurethane, a broad single peak was seen around 20° in quenched sample, which splitted to 20° and 23.5° after annealing. These results indicate that polyurethanes with odd-odd as well as even-even numbers of methylene groups are likely to form ferroelectric crystals. Poly-urethanes are an entirely new class of ferroelectric polymers, and more needs to be done to realize their industrial uses.

VIII. CONCLUDING REMARKS

The spectroscopic analysis and various physical, dielectric, and electrical properties of odd-numbered nylons have been summarized. Odd-numbered nylons show polymor-phism. Nylon-11 shows at least five different crystal forms. The γ form seems to be the most common polymorph in many odd-numbered nylons. The piezoelectric and ferro-electric behaviors of odd-numbered nylons such as nylon-7 and nylon-11 are comparable or better than for PVDF and depend on the crystal form. The desired stability of the piezoelectric response in odd-numbered nylons is related to the ferroelectric reorientation of the amide group dipoles followed by densely packed hydrogen-bonded sheets in the crystalline regions induced by poling and annealing.

Ferroelectricity has been observed in nylon-7, nylon-11, nylon-75, nylon-77, fluor-inated nylon-35, nylon-55, nylon-75 and nylon-95, as well as several alicyclic and aro-matic nylons. Ferroelectricity has been observed in nylons prepared from *m*-xylylenedi-amine and aliphatic dicarboxylic acids, referred as nylon-MXDs. Alicyclic and aromatic nylons prepared from 1,3-bis(aminomethyl)cyclohexane (BAC) and aliphatic dicarboxylic acids were also found to be ferroelectric. The newly developed fluoro-nylons having odd numbers of carbon atoms also exhibit very interesting ferroelectric behavior. Interestingly, ferroelectricity has also been observed for odd-odd and even-even polyurethanes. It is surprising that after more than 15 years of study of piezoelectricity in odd-numbered nylons, this class of ferroelectric polymers somehow remained unexplored. Much of the work on ferroelectric properties of nylons summarized here was in fact reported in the last few years. Odd-numbered nylons seem to be promising ferroelectric polymers, so their potential warrants further extensive studies focused both on novel high-performance materials and on their use in electronic and photonic devices.

ACKNOWLEDGMENTS

The author is grateful to Professor Shigeru Tasaka of Shizuoka University for generously providing information in advance of publication on fluorinated odd-numbered nylons.

REFERENCES

1. T. T. Wang, J. M. Herbert and A. M. Glass (eds.), *The Applications of Ferroelectric Polymers*, Blackie, Glasgow, 1988.
2. H. S. Nalwa, Recent developments in ferroelectric polymers, *J. Macromol. Sci. Rev. Macromol. Chem. Phys. 29*:341 (1991).
3. R. G. Kepler and R. A. Anderson, Ferroelectric polymers, *Adv. Phys. 41*:1 (1992).
4. W. E. Nelson, *Nylon Plastic Technology*, Butterworths, London, 1976.
5. R. Vieweg and A. Muller, *Kunststoff Handbuch, Vol. VI, Polyamides*, Carl Hanser Verlag, Munich, 1966.
6. K. H. Inderfurth, *Nylon Technology*, McGraw-Hill, New York, 1953.
7. M. I. Kohan, *Nylon Plastics*, Society of Plastics Engineers, Greenwich, CT, USA, 1974.
8. W. Sweeny and J. Zimmerman, Polyamides, *Encyclopedia of Polymer Science and Technology*, John Wiley, New York, 1969, vol. 10, p. 483.
9. Y. Kinoshita, An investigation of the structure of polyamide series, *Makromol. Chem. 33*:1 (1959).
10. J.-C. Lin, M. H. Litt, and G. Froyer, X-ray and thermal studies of nylon-5,7, *J. Polymer Sci. Polymer Chem. Ed. 19*:165 (1981).
11. D. C. Prevorsek, R. H. Butler, and H. K. Reimschuessel, Mechanical relaxation in polyamides, *J. Polymer Sci. Part A-2, 9*:867 (1971).
12. S. Tasaka, M. Ohtani, A. C. Jayasuriya, and N. Inagaki, Ferroelectric behavior in fluoro-nylons with odd number of carbons, *New Polym. Mat.*, (1995) in press.
13. M. F. Roberts and S. A. Jenekhe, Site-specific reversible scission of hydrogen bonds in polymers. An investigation of polyamides and their Lewis acid-base complexes by infrared spectroscopy, *Macromolecules, 24*:3142 (1991).
14. D. J. Skrovanek, P. C. Painter, and M. M. Coleman, Hydrogen bonding in polymers. 2. Infrared temperature studies of nylon-11, *Macromolecules 19*:699 (1986).
15. W. P. Slichter, Crystal structures in polyamides made from ω-amino acids, *J. Polymer Sci. 36*:259 (1959).
16. B. A. Newman, T. P. Sham, and K. D. Pae, A high-pressure X-ray study of nylon-11, *J. Appl. Phys. 48*:4092 (1977).
17. A. Kawaguchi, T. Ikawa, Y. Fujiwara, M. Tabuchi, and K. Monobe, Polymorphism in lamellar single crystals of nylon-11, *J. Macromol. Sci. Phys. B20*:1 (1981).
18. E. W. Jacobs and J. C. Hicks, Electric field induced morphological changes in nylon-11, *Appl. Phys. Lett. 44*:402 (1984).
19. E. Balizer, J. Fedderly, D. Haught, B. Dickens, and A. S. DeReggi, Deuterated ferroelectric nylon-11, Annual report on Conference on Electrical Insulation and Dielectric Phenomena, IEEE Publication No. 91, CH3055-1, Piscataway, NJ, 1991, p. 193.

20. E. Balizer, J. Fedderly, D. Haught, B. Dickens, and A. S. DeReggi, FTIR and X-ray study of polymorphs of nylon-11 and relation to ferroelectricity, *J. Polymer Sci. Polymer Phys. Ed.* 32:365 (1994).

21. Y. Kinoshita, The crystal structure of polyheptamethylene pimelamide (nylon 77), *Makromol. Chem.* 33:21 (1959).

22. M. H. Litt, C. H. Hsu, and P. Basu, Pyroelectricity and piezoelectricity in nylon-11, *J. Appl. Phys.* 48:2208 (1977).

23. Y. Takase, J. W. Lee, J. I. Scheinbeim, and B. A. Newman, High-temperature characteristics of nylon-11 and nylon-7 piezoelectrics, *Macromolecules* 24:6644 (1991).

24. G. Wu, O. Yano, and T. Soen, Dielectric and piezoelectric properties of nylon-9 and nylon-11, *Polymer J.* 18:51 (1986).

25. T. Yemani and R. H. Boyd, Dielectric relaxation in the odd-numbered polyamides: nylon 7-7 and nylon 11, *J. Polymer Sci. Polymer Phys. ed.* 17:741 (1979).

26. M. H. Litt and J. C. Lin, Dielectric and pyroelectric properties of nylon 5,7 as a function of molecular orientation, *Ferroelectrics* 57:171 (1984).

27. K. G. Kim, Ph.D. thesis, Rutgers University, New Brunswick, NJ, 1985.

28. B. A. Newman, P. Chen, K. D. Pae, and J. I. Scheinbeim, Piezoelectricity in nylon-11, *J. Appl. Phys.* 51:5161 (1980).

29. J. I. Scheinbeim, Piezoelectricity in γ-form nylon-11, *J. Appl. Phys.* 52:5939 (1981).

30. V. Gelfandbeim and D. Katz, Pyroelectric response and crystal structure of nylon-11, *Ferroelectrics* 33:111 (1981).

31. J. I. Scheinbeim, S. C. Mathur, and B. A. Newman, Field-induced dipole reorientation and piezoelectricity in heavily plasticized nylon-11 films, *J. Polymer Sci. Polymer Phys. Ed.* 24: 1791 (1986).

32. D. Katz and V. Gelfandbeim, Ferroelectric behavior of α-nylon-11, *J. Phys. D: Appl. Phys.* 15:L115, 1982.

33. S. C. Mathur, J. I. Scheinbeim, and B. A. Newman, Piezoelectric properties and ferroelectric hysteresis effects in uniaxially stretched nylon-11 films, *J. Appl. Phys.* 56:2419 (1984).

34. J. W. Lee, Y. Takase, B. A. Newman, and J. I. Scheinbeim, Ferroelectric polarization switching in nylon-11, *J. Polymer Sci. Polymer Phys.* 29:273 (1991).

35. J. W. Lee, Y. Takase, B. A. Newman, and J. I. Scheinbeim, Effect of annealing on the ferroelectric behavior of nylon-11 and nylon-7, *J. Polymer Sci. Polymer Phys.* 29:279 (1991).

36. T. Furukawa, Y. Takahashi, D. Suzuki, and M. Kutani, Ferroelectric switching in odd nylons, *Polymer Preprints Japan* 41:4562, 1992 (in Japanese).

37. Y. Murata, K. Tsunashima, N. Koizumi, K. Ogami, F. Hosokawa, and K. Yokoyama, Ferroelectric properties in polyamides of *m*-xylylenediamine and dicarboxylic acids, *Jpn. J. Appl. Phys.* 32:L849 (1993).

38. Y. Murata, K. Tsunashima, and N. Koizumi, Dielectric hysteresis loop in alicyclic and aromatic polyamides, *Jpn. J. Appl. Phys.* 33:L354 (1994).

39. S. Tasaka, M. Ohtani, K. Maeda, and N. Inagaki, Electrical properties of odd number nylons, *Polymer Preprints Japan* 40:1227 (1991) (in Japanese).

40. M. Ohtani, T. Shouko, S. Tasaka, and N. Inagaki, Ferroelectric behavior of odd polyamides and odd polyureas, *Polymer Preprints Japan* 41:4559 (1992) (in Japanese).

41. M. Ohtani, H. Tohuhisa, S. Tasaka, and N. Inagaki, Hydrogen bonding in ferroelectric odd nylons, *Polymer Preprints Japan* 42:1432 (1993) (in Japanese).

42. T. Syoko, M. Ohtani, S. Tasaka, and M. Imagaki, Ferroelectric behavior on polyamides and polyureas with odd number of CH2 groups, *Sen-i-Gakkai Symp. Preprints*, 1992, p. B97.

43. J. I. Scheinbeim, J. W. Lee, and B. A. Newman, Ferroelectric polarization mechanism in nylon-11, *Macromolecules* 25:3729 (1992).

44. S. Tasaka, T. Shouko, K. Asami, and N. Inagaki, Ferroelectric behavior in aliphatic polyurethanes, *Jpn. J. Appl. Phys.* 33:1376 (1994).

7
Cyanopolymers

Shigeru Tasaka
Shizuoka University, Shizuoka, Japan

I. INTRODUCTION

Ferroelectrics are electrically sensitive polar crystals. Organic polymers, in either glass or crystal, pack by very weak cohesive forces and are easily deformed by an external field. Therefore, a polymer with large dipoles such as a polar polymer can be a ferroelectric of the order–disorder type if its structure is controlled appropriately.

Cohesive force of the polymer chains is usually governed by dipole interactions and dispersion force, which depend on the magnitude of electric dipole moment and electronic polarizability, respectively. In order to obtain polar packing between polymer chains, therefore, it is important to have polarity in the whole chain—namely, polar chain conformation. Adjacent dipoles in the chain of a polar polymer show strong dipole–dipole interactions and orient to give an energy minimum, resulting in a chain conformation which is polar. Further, to form a polar crystal, such polar chains must pack together in such a manner that they overcome the spatial and electrostatic repulsion problems. Consequently, a necessary condition of molecular design for a ferroelectric polymer is to realize a polar chain with a relatively weak cohesive force. From the this point of view, one can see poly(vinylidene fluoride) (PVDF) as a typical example of ferroelectric polymer design; of course, PVDF is actually ferroelectrics.

The cyano group, especially C–CN, has as a special feature a large dipole moment and complex forming ability with transition metals. Cyanopolymers discussed here mean polyacrylonitriles, poly(vinylidene cyanide)s, and the polymers with a CN group in a side chain. All the polymers have been described as strong cohesive polymers. The van der Waals volume of the C–CN group is not large, but it is rigid and bulky, and has rotational freedom along the chain. For a well-packed polymeric crystal containing aligned cyano groups, therefore, it may be difficult to change the polarity by dipole rotation under applied electrical force, although a loose-packed state—for example, an amorphous state at high temperature—might make it possible. In this chapter, cyano-polymers showing interesting ferroelectric behavior—hopefully, real ferroelectrics, al-though the issue is still to be resolved—are described. In particular, dielectric, piezo-electric, and ferroelectric properties of various cyanopolymers are discussed.

II. PHYSICAL CHEMISTRY OF THE CN GROUP [1]

A. Electronic Structure

In the cyano group, the nitrogen and carbon atoms are approximately diagonally (*sp*) hybridized. The bonding, therefore, consists of a σ bond and further of two π bonds at right angles to each other, giving a linear arrangement R—CN. The bond moment (3.5 D) is caused mainly by the lone-pair orbital centered on the nitrogen atom and directed along the CN axis. Furthermore, the π orbitals are displaced toward the nitrogen, and the charge distribution may therefore be represent as

$$\overset{\delta+}{R-C}\equiv\overset{\delta-}{N}--- \tag{1}$$

in which the dashed line represents the direction of the lone-pair orbital.

The lone-pair electrons are mainly responsible for the coordination of the cyano group. Thus, protonation occurs at the nitrogen atom, and hydrogen bonding and complex formation to Lewis acids generally take place through the lone pair. However, weak com-plexes formation involving the CN π electrons can occur. The large dipole moment of the cyano group can lead to dipole–dipole interactions in pure nitriles (self-association), or form nitriles to other molecules with polar groups. Further, association may take place as a result of interaction between the partial positive charge on the carbon atom and lone-pair electrons on other molecules.

Therefore, many chemical reaction proceed through complexes as intermediates, and the solvent properties of the nitriles are certainly connected to their coordination ability.

B. Size and Shape of the CN Group

The size and shape of the CN group make it unique in relation to other substituents. The cyano group is effectively a rod surrounded by a cylindrical cloud of π electrons which can interact with an adjacent π system regardless of rotatory orientation (Fig. 1). It has a strong dipole, oriented with the negative and toward the nitrogen.

In substitution on an organic residue, the width of the group is an important factor; the cyano group is much larger than the C—F group and about the same size as chloro group.

An obvious consequence of the rodlike shape of the cyano group is that most organic molecules can be polysubstituted by cyano groups without steric interference, so that interaction with an adjacent π-system is not disturbed.

C. Hydrogen Bonding and Dipole Association

Due to the strongly directional character of the lone-pair electrons on the *sp*-hybridized nitrogen atom, a nitrile should be a good base for hydrogen bonding. However, the nitriles are not included among the well-recognized hydrogen-bonding systems. The cyano group is certainly a poor π-electron donor, resulting in weak intermolecular hydrogen bonding.

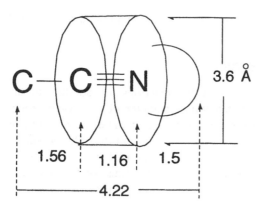

Figure 1 Molecular size of C—CN group [1].

Because of the high bond moments of the cyano group, the nitriles are self-associated and interact with other molecules that have polar groups by dipole–dipole forces. The nitriles have considerably higher boiling points, viscosities, and heats of vaporization than the corresponding hydrocarbons, strongly suggesting association. However, in cyano compounds with acidic hydrogens, such as succinonitrile and glutaronitrile, hydrogen bonding may contribute to the association. Two models for dipole pairs have been proposed from spectroscopic data in acetonitrile (Fig. 2).

Intermolecular hydrogen bonding between CN nitrogens and tertiary hydrogens had previously been considered as the main interaction responsible for the fiber-forming capacity of polyacrylonitrile (PAN) [2]. Although such hydrogen bonds may be present to a certain extent, their importance is now considered small compared with the intermolecular dipole–dipole interaction of CN groups, for the following reasons. Whereas the energy gained by a dipole–dipole interaction is in the range of 20–30 kJ/mol, the energy of a hydrogen bond of the type N · · · HC may be assumed to be in the range of 4–5 kJ/mol [3].

There has been some discussion as to whether the water–nitrile interaction involves the lone-pair orbital of nitrogen, or the π orbitals of the CN bond [4]. The interaction between the H$_2$O and the nitrile group has a profound influence on a various properties of PAN. Under normal pressure, PAN is strongly plasticized by water. However, unlike dimethylformanide (DMF) or dimethylsulfoxide (DMSO), water is able to dissolve PAN. The high degree of aggregation of the water molecules seriously reduces their diffusion within the polymer. The migration of water into the film is sufficient to produce a plasticizing effect. The chain mobility is increased, as indicated by a decrease of the glass

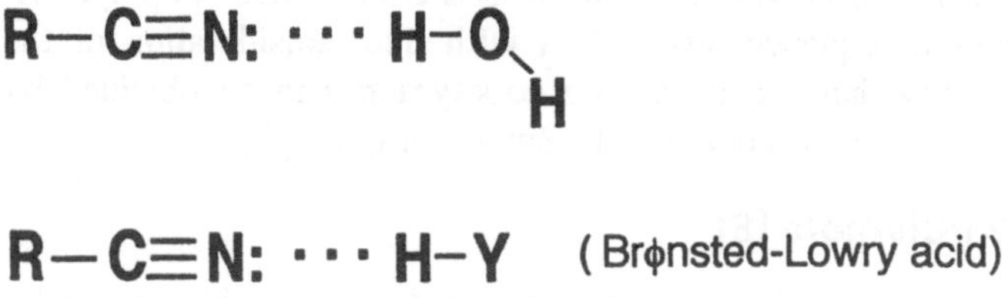

$$\text{R}-\text{C}{\equiv}\text{N:} \cdots \text{H}-\underset{\text{H}}{\text{O}}$$

$$\text{R}-\text{C}{\equiv}\text{N:} \cdots \text{H}-\text{Y} \quad (\text{Brønsted-Lowry acid})$$

Figure 2 Schematic model of hydrogen bonding of CN groups [4].

transition temperature by 35–50°C. This must also affect the residual polarization formed by electrical poling.

III. SYNTHESIS OF CYANOPOLYMERS

In monomers that contain cyano groups, many of the reactions are addition to the carbon–carbon double bond. The electron-withdrawing character of the cyano group, however, enables the monomer to participate readily in a large number of additions of the Michael type, resulting in cyanoethylated products for acrylonitrile. The polymerization, which is undoubtedly the most important reaction of the carbon–carbon double bond, can occur through free-radical and ionic mechanisms. Its discussion below is limited to the general features in acrylonitrile and vinylidene cyanide. These monomers are very lacrimators and highly toxic.

A. Acrylonitrile [5]

Acrylonitrile has been manufactured by a variety of routes, the most important being those starting from propylene, and is produced as a high-grade chemical with virtually no interfering impurities. It is usually stored in the presence of a small concentration of polymerization inhibitor, such as ammonia or the monomethyl ether of hydroquinone. This should be usually removed for use in polymerization processes by flash distillation.

The homopolymer of acrylonitrile is quite insoluble in the monomer; consequently, during bulk polymerization the polymer is precipitated. The free-radical polymerization under these conditions shows a number of interesting features such as heterogeneous polymerization. Acrylonitrile does not appear to polymerize in the absence of initiators. The conventional initiators, azobisisobutyronitrile and benzoyl peroxide, are convenient to use at temperatures below 100°C. Ultraviolet light irradiation of wavelength less than 2900 Å, of course, readily forms the polymer by a free-radical mechanism. Oxygen, which is a very powerful inhibitor of the polymerization, sometime makes peroxides and induces an explosive reaction.

Acrylonitrile is one of the most reactive monomers toward anionic catalysts. A wide range of initiators of this type has been used and include the alfin catalysts, alkoxides, butyllithium, metal ketyls, and solutions of alkali metals in ethers. In a number of anionic polymerizations, there is no termination reaction if pure reagents are used, and so-called living polymers are formed. Such "living" systems are more difficult to observe in the case of acrylonitrile owing to the insolubility of the polymer in most of the usual solvents. It is possible to produce block co-polymers with acrylonitrile from other "living" polymeric anions.

The structure of PAN obtained by anionic or radical polymerization is usually non-stereoregular, that is, atactic. It also well known that highly stereoregular (isotactic) PAN can be prepared by radiation in a urea canal complex of acrylonitrile [6].

Most useful polymers containing acrylonitrile are indeed copolymers. Copolymerization in radical process gives a sequence distribution with due consideration of the reactivity ratio. It is worth noting that the alternating copolymers can be obtained by radical polymerization using a 1:1 molar acrylonitrile-ZnCl$_2$ complex [7].

B. Vinylidene Cyanide Polymers [8]

Vinylidene cyanide is an extremely reactive monomer that undergoes rapid ionic polymerization in the presence of almost any weak base to form a hydrolytically unstable homopolymer. The monomer polymerizes readily with a wide variety of comonomers such as vinylacetate, stylene, and dienes to form alternating rather than random copolymers.

Vinylidene cyanide can be prepared by several methods, for example, pyrolysis of 1,1,3,3,-tetracyanopropane as follows:

$$2H_2C(CN)_2 + H_2CO \rightarrow HC(CN)_2CH_2(CN)_2CH \xrightarrow{150-200°C} H_2C=C(CN)_2 + H_2C(CN)$$

The pyrolysis product has to be purified by distillation, crystallization, or a combination of both. The boiling point of this monomer is 154°C at 1 atm (the melting point is −9.7°C).

Vinylidene cyanide polymerizes in the cold upon contact with water to form a hard, white, infusible polymer. This type of initiation occurs with alcohols, amines, and ketones. Anionic polymerization appears to proceed owing to the polarization of π electrons of the double bond and the presence of unshared electrons of the nitrile groups giving the polarized structure. Solvents for the homopolymer include DMF, tetramethyl urea, and diethylcyanamide. Depolymerization begins at 160°C. This polymer is sensitive to moisture and degrades in contact with water or bases.

The copolymerization of vinylidene cyanide is very easy, and three mechanisms have been observed: (a) autocatalytic, (b) anionic, and (c) free-radical polymerization. The strong tendency of vinylidene cyanide to alternate during copolymerization is immediately apparent from reactivity ratio data ($r_1 \cdot r_2 = 0$).

IV. SPECTROSCOPIC CHARACTERIZATION

A. NMR and IR

The relative tacticity of the samples can be measured from the NMR signal. In the ^{13}C NMR spectrum of PAN, the methine resonances show a distinct triplet, with very little splitting exhibited by the methylene resonances. The greatest sensitivity toward configuration occurs for the nitrile resonances, where an almost ideal Bernoullian distribution is observed. The triad chemical shift sequence with respect to an increasing magnetic field strength for the CN-carbon resonance (around 120 ppm) of PAN take in order as mm (isotactic), mr (heterotactic), and rr (syndiotactic), as shown in Figure 3a [9,10]. The tacticity can be determined by IR absorption bands with a relative absorption ratio of 1250 cm^{-1} (wagging mode of methine group) and 1230 cm^{-1} (twisting mode of methylene group coupled with the methine group). When the isotacticity of PAN increases, the intensity of the 1230-cm^{-1} band increases proportionately. In the case of the copolymers of PAN, it is necessary to analyze the tacticity distribution with sophistication, because an inhomogeneity is often involved even in one chain of the copolymer.

Detailed characterization of VDCN copolymers by NMR was performed by Jo et al. [11]. The VDCN copolymers with electron-donating-type monomers are almost perfect alternating copolymers. Only the tacticity of alternate comonomers, ϵ tacticity, is important for the configurational analysis, because VDCN monomer has mirror symmetries. If we assume a VDCN-VAc repeating unit in the copolymer of VDCN and VAc(vinyl

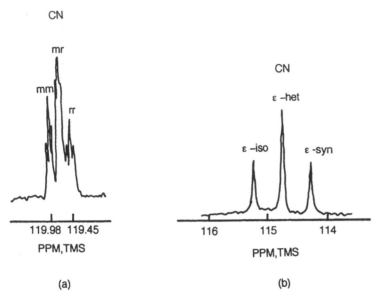

Figure 3 ^{13}C-NMR spectrum of polyacrylonitrile and poly(vinylidene cyanide/vinyl acetate) [10,11].

acetate), two configurationally different structures in the dyad sense are possible:

Here configuration I is defined as ε isotactic, and the configuration II as ε syndiotactic. The splitting of the cyanide carbon resonant of the VAc-VDCN-VAc triad into three peaks is caused by the ε-tactic arrangement (Fig. 3b). The three peaks correspond to ε isotactic, ε heterotactic, and ε syndiotactic, respectively. The intensity ratio of the peaks is 1:2:1, which indicates an "ε-atactic sequence." This is a reason that this copolymer is amorphous in X-ray diffraction.

B. Thermal Analysis

No cyanopolymers have clear melting points on the DSC curve because of high coersive force and low thermal stability. PAN shows several endothermic transitions in the tem-

perature range 20–380°C, including glasslike transition, melting point, autocatalytic cy-
clization, and decomposition leading to gaseous products. Extremely high heating rates
are necessary to determine the melting point of PAN. A small endothermic peak around
320°C, superimposed on the large exothermic maximum of decomposition, can be observed and ascribed to the melting point at a heating rate of 40°C/min.

No clear picture can be gained from the literature concerning the number and exact
temperature range of glass transitions of PAN. Most authors appear to agree on two
major transitions, one at about 80–100°C, and the other in the vicinity of 140°C. Interestingly, the lattice spacing of the ordered region increases with temperature in a discontinuous way in the range 80–120°C [12]. The motions related to the lower transition are
localized in the more ordered regions, like a paracrystal. The higher transition may be
ascribed to a loosening of intermolecular dipole–dipole interaction, which might be related to T_{ll} (liquid-state transition) [13] originating from an intermolecular segment–
segment melting.

Thermal properties of VDCN copolymers are typical of amorphous materials. However, a considerable enthalpy relaxation is observed at the glass transition temperature
by annealing below its temperature (Fig. 4) [14]. Interestingly, a new ordering in X-ray
diffraction appears at the lower angle after the annealing. The large dipole interaction of
the dicyano group may make aggregation between chain molecules possible. The ordering
is destroyed by micro-Brownian motion above the glass temperature.

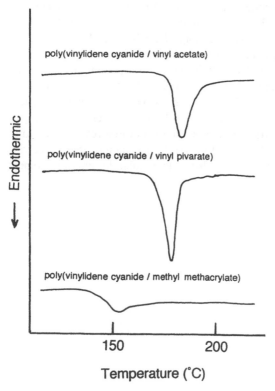

Figure 4 DSC curves for VDCN polymers annealed for 8 h at $T_g - 10°C$ [14].

V. FERROELECTRICITY AND RELATED PROPERTIES

A. Polyacrylonitriles

Polyacrylonitrile (PAN), shown in Figure 5, is a conventional organic polymer used in commercial fabrics and a component of composite materials for high-impact-strength purposes. Usually, PAN has been synthesized by radical polymerization, so the polymer tacticity is atactic. The dominant characteristic of the PAN molecule is the presence of strongly polar nitrile groups at an intramolecular distance of only a few tenths of a nanometer. The CN groups have a variety of possibilities for interacting with the surroundings. The dipole moment (3.5 D) causes strong attraction or repulsion with other molecules.

1. Intra- and Intermolecular Interaction and Structure of PAN [16]

As a result of the strong dipole moment of CN groups, adjacent nitrile groups of the same molecule repel each other. It has been accepted that this intramolecular repulsion makes a helical conformation in an essentially isotactic PAN. However, a zigzag conformation is more stable, according to recent energy calculations [17,18]. Since, technically, PAN is an atactic polymer with a certain extent of syndiotactic component, molecular conformation and packing are expected to be imperfect. Therefore, the concept of ordered and disordered domains existing in PAN seems to supported by the appearance of a meridional maximum in small-angle X-ray scattering, as has been recorded during annealing. In the wide-angle X-ray diffraction pattern of PAN, all of the equatorial reflections are in a pseudo-hexagonal packing and diffuse. Furthermore, no meridional reflection takes place [19]. The density, however, calculated with a hexagonal packing of X-ray data and the normal length of extended zigzag chain is 1.10–1.13 g/cm^3; the experimentally determined value ranges from 1.17 to 1.22 g/cm^3 [20].

These results suggest that the kinking of the chain results in a contraction along the chain axis. If the repulsion forces depending on the tacticity of PAN are turned on, a polymer molecule would probably assume the extended planar zigzag configuration. When the repulsion forces are turned on again, there would occur a certain kinking or twisting of the molecule. The resulting structure is a kinked chain that resembles a kind of symmetrical rigid rod, which packs laterally into a hexagonal pattern that gives rise to the 5.2 Å. Such randomly kinked chains may have no regular longitudinal repeat

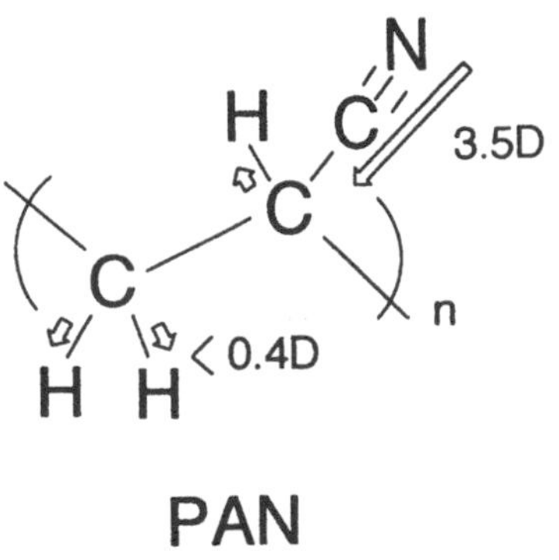

Figure 5 Chemical structure and dipole moment of polyacrylonitrile.

distance, accounting for the absent of any meridional diffraction. Figure 6 shows the kinked chain structure of PAN [17].

In a sufficiently stretched PAN film, individual kinked chains may be considered to be packed side by side, and bonded laterally by intermolecular dipolar interactions; these ordered zones are interrupted by amorphous regions where the intermolecular forces have not been effective. A periodic arrangement of more or less ordered regions has been found by Hinrichsen and Orth [21], as orthohombic, with dimensions $a = 1.06$, $b = 1.16$, and $c = 5.04$ Å. The periodicity in the c direction is exactly twice in a planar zigzag chain length per monomer. It was suggested that only the syndiotactic chain sequence would contribute in building up the crystal regions. Acrylonitrile may be expected to have a certain preponderance of syndiotactic growth introduced during the polymerization process, due to the electrostatic repulsion of the polar side groups. Recent results of deuteron NMR [22], however, indicate no regions showing different kinds of mobility

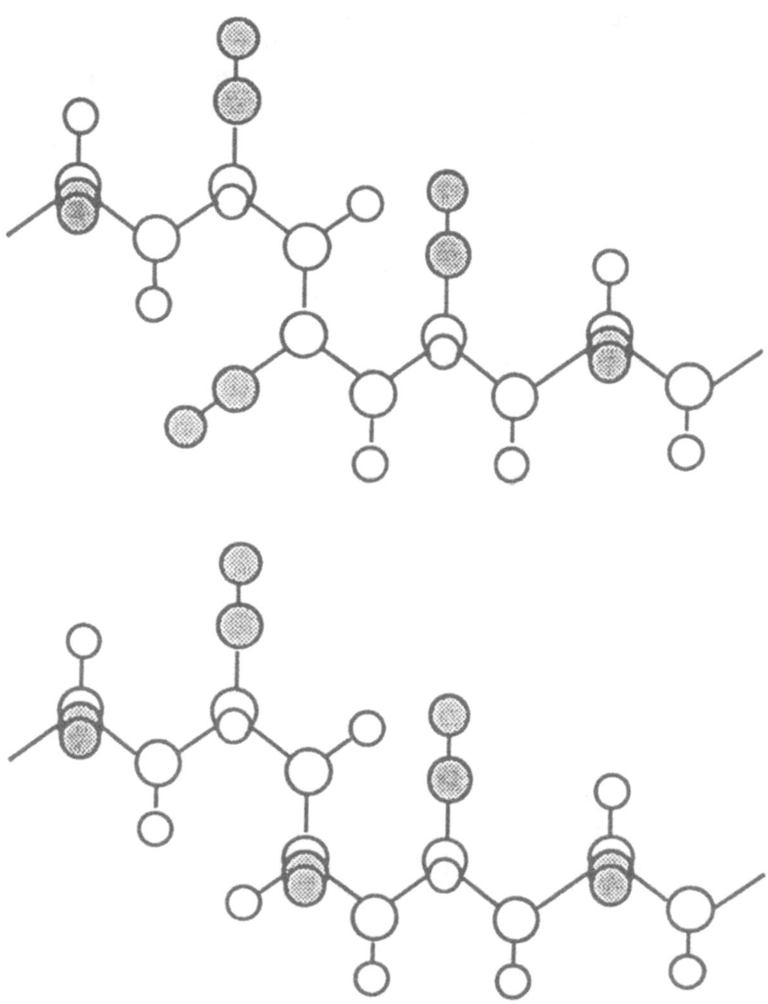

Figure 6 Zigzag chain of polyacrylonitrile with kinks at isotactic sequence [17].

like the crystal and amorphous regions. We should imagine a more uniform phase consisting of only conformationally disordered crystals, like that in liquid crystalline polymers.

The presence of a longer sequence of syndiotactic placements, interrupted by a single isotactic position or short sequences of isotactic or atactic configuration, has actually been confirmed by spectroscopic techniques. The high perturbation in the crystalline region would then be caused primarily by natural distribution of lengths of the syndiotactic sequences, which forces sequences of other tacticity into the crystal region, disturbing the crystal lattice. In the following discussion, ferroelectric behavior arising in PAN and its copolymers might be related to the microscopic syndiotactic component forming a polar crystalline packing.

Most commercial fibers from acrylonitrile are actually made from copolymers, containing 3–7% of vinyl monomer, mostly methyl acrylate, methyl methacrylate, or vinyl acetate. The introduction of such a small amount of comonomer greatly enhances the internal mobility of the polymer segments, reducing the sequences of acrylonitrile molecules capable of interacting with neighboring sequences.

The mobile additives have a greater chance to approach PAN-CN dipoles in the most favorable direction. Therefore, the intermolecular repulsion of adjacent —CN groups within the polymer molecules will be partly neutralized. At the same time, part of the intermolecular dipole–dipole interactions between polymer molecules will be replaced by interactions of polymer CN groups with dipoles of the additives.

2. Ferroelectric Behavior of PAN

Stupp and Carr [23] reported a structural rearrangement of poled PAN by infrared (IR) spectroscopy in 1978. This was mainly related to CN dipole orientation. However, they also showed field-induced structural change of PAN by X-ray and IR spectroscopy [24]. Figure 7 shows the annealing and poling effect of PAN. It seems that the different characteristics between the two treatments are related to dipole orientation and kinking

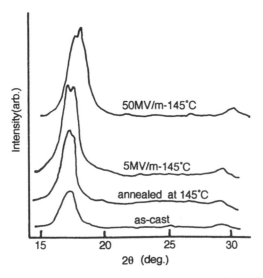

Figure 7   X-ray diffraction patterns of polyacrylonitrile annealed and poled at 145°C [24].

of the chain. Further, IR data suggested the high orientation of CN dipoles normal to the film surface above the glass transition temperature. We cannot determine whether the CN dipole in PAN is in an amorphous or a paracrystal state. At this stage, we might be able to call it "ferroelectric behavior."

The first detailed study of the piezo- and pyroelectricity of PAN was reported by Ueda and Carr in 1984 [25]. The effect of molecular orientation on piezoelectricity in PAN was very large, like that in poly(vinylidene fluoride). Though stretching of PAN before poling makes a large remanent polarization, the piezoelectric d_{31} constant obtained was low (only a few pC/N), 10 times smaller than that of PVDF (Fig. 8). The piezoelectric constant is generally proportional to the dielectric relaxation strength in amorphous polymers. The dielectric relaxation strength of PAN in the glass transition region, which is sometimes called "paracrystal relaxation," is about 40 [26,27], this value is large enough to cause piezoelectricity by electrical poling.

The piezoelectric constant of PAN was improved by von Berlepsh et al. [28–30], by using the copolymer with methylacrylate. The optimum value of the piezoelectric constant, which is about 3 pC/N in the copolymers of acrylonitrile (93%) and methylacrylate (6–7%), is independent of poling temperature above 75°C (Fig. 9). A more remarkable feature is a D-E hysteresis loop in a stretched film (Fig. 10). The measurement temperature was 68°C, below the glass transition temperature. However, thermally stimulated current (TSC) behavior, shown in Figure 11 [31], suggests that 68°C is in the depolarizing region, and especially TSC spectra of stretched and unstretched samples were found to be quite different. In addition, more detailed polarization reversal data were given. [32] The copolymers at lower temperatures have lower residual polarization and higher coersive field (Fig. 12). The polarization reversal should exist even in low

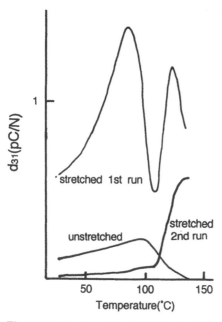

Figure 8 Temperature dependence of piezoelectric constant d_{31} for stretched and/or poled polyacrylonitriles [25].

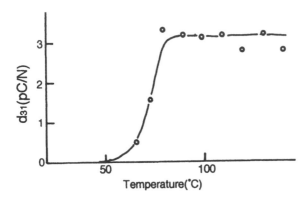

Figure 9 Poling temperature dependence of piezoelectric d_{31} constant for stretched acrylonitrile/methylacrylate (93:7) copolymer at room temperature. Poling conditions: E_p = 80 MV/m, t_p = 10 s [28].

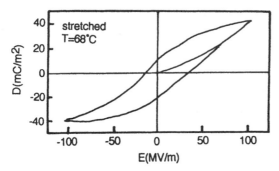

Figure 10 D-E hysteresis loop of stretched acrylonitrile/methylacrylate (93:7) copolymer at 0.2 Hz [29].

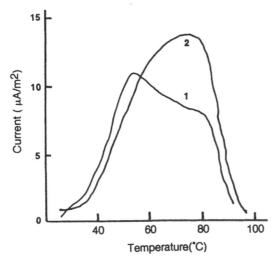

Figure 11 Thermally stimulated current curves of acrylonitrile/methylacrylate (93:7) copolymer. Poling conditions: E_p = 41 MV/m, T_p = 10 s, heating rate = 2°C/min [32].

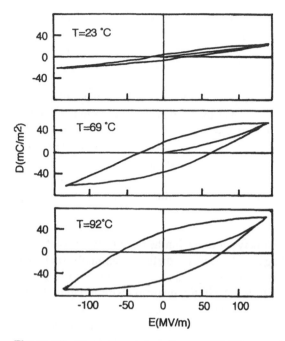

Figure 12 Temperature dependence of D-E hysteresis loop for acrylonitrile/methylacrylate (93 : 7) copolymer (0.2 Hz) [32].

temperature, but electrical breakdown at higher field seems to suppress their measurement. This depends strongly on the sample's synthesis method and its impurities. In addition, PAN homopolymer is likely to have a fatal weakness against water and ionic impurities.

Tasaka et al. [33] selected other copolymers of PAN having mainly alternating sequence, namely, poly(allylcyanide [35%]/acrylonitrile [65%]) prepared by radical polymerization. This copolymer has unusual properties: high crystallinity (paracrystallinity), high optical transparency, and large dielectric relaxation strength in glass transition region, similar to that of PAN. Figure 13 shows the chemical structure of this copolymer. According to the kinetics of radical polymerization, the monomer reactivity ratios (r_1, r_2) for acrylonitrile (M_1) and allylcyanide (M_2) are $r_1 = 1$ and $r_2 = 0$. This suggests little allylcyanide sequence in the copolymer. For example, the copolymer with 35% allyl-

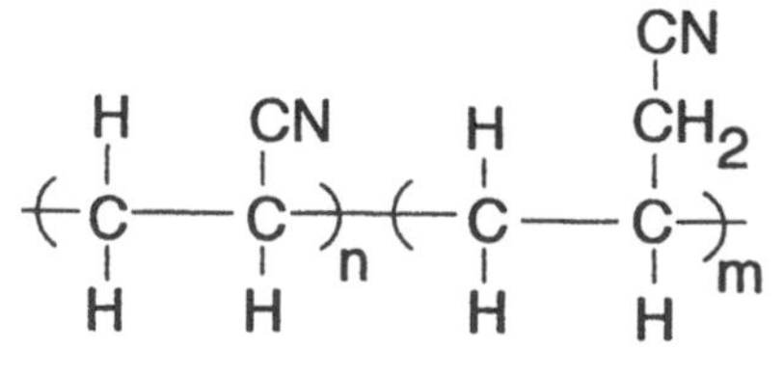

P(AN-ALCN)

Figure 13 Chemical structure of acrylonitrile/allylcyanide copolymer.

cyanide is expected to contain a large number of alternating sequence. The structure of the copolymer may be a hexagonal-packing paracrystal, similar to PAN. Although the atactic tacticity—that is, statistical sequence—may be realized by radical polymerization, the X-ray diffraction pattern showed sharp peaks whose intensities increased remarkably after annealing (Fig. 14).

The copolymer shows a D-E hysteresis loop. The value of the ramanent polarization is quite large, i.e., 200–700 mC/m^2 at 105°C, due to the dipole polarization. At higher temperatures, DC conduction of ionic impurities would superimpose on the dipole orientation current. Therefore, we attempted to conduct measurements at low temperatures, and recorded a D-E hysteresis loop that has a remanent polarization of 30 mC/m^2 (Fig. 15) [34]. This value of remanent polarization coincides with the integration of thermally depolarization current, which may be attributed to the CN dipole orientation.

3. Phase Transitions in PAN

Normal dielectric relaxation due to the paracrystal dipersion has been observed at about 100°C, its relaxation strength was as large as 40, and no dielectric anomaly such peak behavior existed, in which remanent polarization disappear. In VDF copolymer, there is a local conformational change from trans to gauche in the phase-transition region. Cyano groups have large cohesive energy, so only a macroscopic conformational change occurs. Polar paracrystal would change to nonpolar phase, and the polarity would be held in a microcluster by the strong dipole–dipole interaction. Kinking, if could exist, might be a part of domain walls. Apparent glass transition temperature expresses a starting point of segment motion in the paracrystalline phase. In this temperature range, new packing by annealing occurs. The thermal expansion coefficient of the lattice spacing changes in this temperature region [33]. Consequently, it might be possible to propose a new type of gradual phase transition with a relaxation process from ferroelectric to antiferroelectric. At least, no ferroelectric–paraelectric phase transition has been found yet.

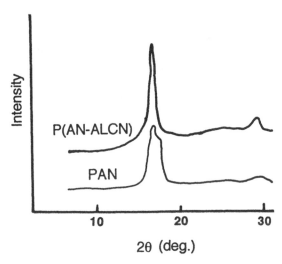

Figure 14 X-ray diffraction patterns for polyacrylonitrile and acrylonitrile/allycyanide (65:35) copolymer [33].

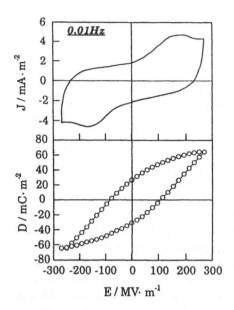

Figure 15   J-E and D-E hysteresis loops in stretched acrylonitrile/allylcyanide (65:35) copolymer at room temperature [33].

B. Vinylidenecyanide Polymers

Vinylidene cyanide is an extremely reactive monomer that undergoes rapid ionic polymerization in the presence of almost any base to form hydrolytically unstable homopolymer. The monomer polymerizes readily with a wide variety of comonomers, such as vinyl acetate or stylene, to form alternating copolymers. Many of these copolymers have high melting or softing points. A fiber from 1:1 copolymer of vinylidene cyanide with vinylacetate was developed by B. F. Goodrich Company and was produced commercially under the trade name Darvan. Since a large piezoelectric constant have been found in vinylidene cyanide/vinyl acetate copolymer after poling, this polymer received much attention as an amorphous piezoelectric material. The dipole moment of repeat unit is as large as 4.5 D in trans conformation.

1. Poly(vinylidene Cyanide) Homopolymer [8]

Poly(vinylidene cyanide) is a hard, white, infusible polymer with a density of 1.31 g/cm^3. It can be dissolved in solvents such as dimethylformamide, diethylcyamide, etc. The polymer does not show any sharp melting point, and the X-ray diffraction pattern shows little crystallinity. Depolymerization take place at 160°C or even at lower temperatures, the rate of depolymerization increasing as the temperature is raised. The polymer is sensitive to moisture, turning dark upon standing in moist air, and degrades in contact with water or, especially, with bases. So far, no piezo- or ferroelectric measurements have been achieved for this polymer.

2. Poly(vinylidene Cyanide) Copolymers

Vinylidene cyanide copolymerizes readily with a wide variety of common monomers. The strong tendency of vinylidene cyanide to alternate during copolymerization results from the monomer reactivity ratio product r_1r_2 of less than 10^{-3} for many monomers. The ^{13}C NMR spectrum gives microstructural information that this copolymer is an alternating copolymer in a head-to-tail arrangement and has a nonstereoregular structure [11]. Therefore, the copolymers are considered to be amorphous, and this has also been confirmed by X-ray diffraction [35,36]. However, copolymers with vinyl esters show a large energy absorption during annealing below the glass transition temperature by DSC measurement [14,37]. This phenomenon is called "enthalpy relaxation," and involves not only densification in the liquidlike amorphous packing but also a change in conformational energy. Furthermore, the X-ray diffraction of these copolymers shows a special equatorial peak at a lower angle in addition to typical amorphous peaks [14]. In P(VDCN-VPiv) or P(VDCN-VPr) (see Fig. 16), this lower-angle diffraction disappears above the glass transition temperature, and with decreasing temperature, the diffraction appears again. Interestingly, VDCN alternating copolymers with longer side chains show a very sharp diffraction corresponding to double-layer structure with extended side chains of alkyl group now involving polar or nonpolar aggregation. Owing to the nonstereoregularity of vinyl esters, these copolymers have been considered to be amorphous, but the large dipole interaction of the cyanide group makes aggregation between chain molecules possible and the weak ordering is destroyed by micro-Brownian motion above the glass transition temperature. The ordering may be useful to form a remanent polarization or to create a cooperative motion of dipoles.

3. Dielectricity and Piezoelectricity

Miyata et al. [38] first reported high piezoelectricity in vinylidene cyanide-vinyl acetate (VDCV-VAc) copolymers after poling. The piezoelectricity (d_{31}) of drawn films becomes larger than that of prestine films. The copolymer shows piezoelectric activity quite similar to PVDF in the temperature range 20–100°C (Fig. 17). If the d_{31} of these two polymers is compared below the glass transition temperature, then the value for the copolymer is 10 times larger than that for PVDF. This piezoelectricity is stable up to the T_g, around 180°C.

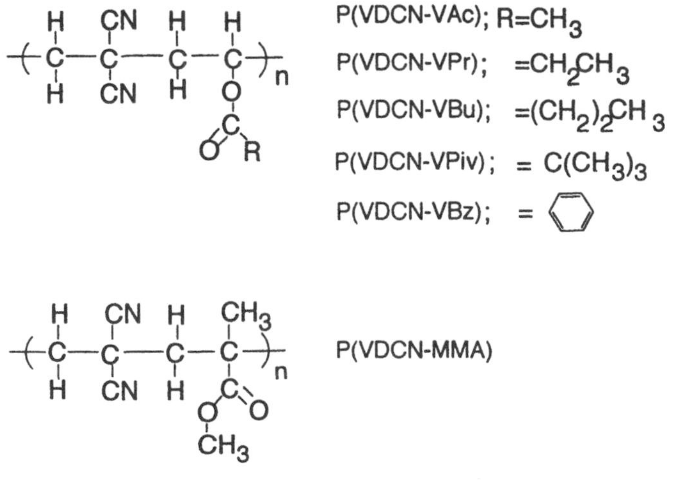

Figure 16 Chemical structures of vinylidene cyanide copolymers.

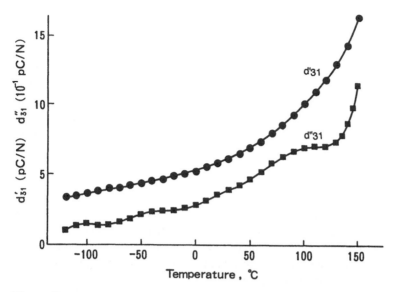

Figure 17 Temperature dependence of piezoelectric d_{31} constant in drawn poly(vinylidene cyanide/vinyl acetate). Poling conditions: $E_p = 20$ MV, $t_p = 0.5$ h, $T_p = 150°$C [38].

Piezo- and pyroelectric constants and dielectric relaxation strengths of VDCN copolymers are listed in Table 1 [14,38]. Piezoelectric constant seems to be directly proportional to residual polarization (P_r), and further related to dielectric relaxation strength. Generally, piezoelectric constant in polymer electrets is proportional to the remanent polarization as related by following equation:

$$e = K \cdot P_r$$

where K is Poisson's ratio of the samples, and $e = 0.48$ (for e_{31}) in P(VDCN-VAc). [39] If the residual polarization was formed by a poling field (E) above the glass transition temperature and was frozen by cooling below the glass temperature, the relation between

Table 1 Physical Properties of VDCN Copolymers[a]

Copolymer	T_g (°C)	ϵ	$\Delta\epsilon$	d_{31} (pC/N)	e_{31} (mC/m^2)	p (μC/Km^2)	P_r (mC/m^2)
P(VDCN-VAc)	178	5.6	120	7.0	14.5	10	35
P(VDCN-VBz)	184	5.6	115	5.2	8.6	10	21
P(VDCN-VPr)	176	5.8	85	9.8	16.6	30	28
P(VDCN-VPiv)	172	5.8	100	7.0	14.0	12	33
P(VDCN-MMA)	146	5.4	30	2.2	5.4	7	12
P(VDCN-IB)	75	5.0	16	1.0	1.8	3.5	6

[a]VAc = vinyl acetate; VBz = vinyl benzoate; VPr = vinyl propyonate; VPiv = vinyl pivarate; MMA = methyl methacrylate; IB = isobutylene; T_g = glass transition temperature (from DSC); ϵ = dielectric constant (20°C, 10 Hz); $\Delta\epsilon$ = dielectric relaxation strength at T_g; d, e = piezoelectric constants; p = pyroelectric constant; P_r = remanent polarization.
Source: From Ref. 14.

them can be written as

$$P_r = \Delta\epsilon \cdot E$$

Therefore, a simple relationship is introduced in these polymers. The associated problem is the origin of large dielectric relaxation strength over 100, one of the largest values among polymers. Such a relaxation strength is considered to be due to micro-Brownian motion in amorphous dipoles, with a large cooperativity. The dielectric behavior in P(VDCN-VAc) has been discussed in detail by Furukawa et al. [40,41]. They concluded that the large dielectric relaxation strength originates from the cooperative motion of 10 or more CN dipoles.

Table 2 shows a comparison of dielectric behavior in P(VDCN-VAc) with that in PAN and PVAc. P(VDCN-VAc) behaves as a typical noncrystalline dielectric with a normal relaxation process except for the large linear and nonlinear permittivity. This indicates the existence of local order in the VDCN copolymer, which might be related to the structure shown by the X-ray diffraction results. Moreover, the observed dielectric behavior was discussed by assuming a freely rotating dipole system [42], whose polarization follows the Langevin function as given by

$$P = N\mu \cdot L\left(\frac{\mu F}{kT}\right)$$

where L is the Langevin function, N is the number of dipoles, μ is the effective moment, F is the internal field, k is Boltzmann's constant, and T is the absolute temperature. This function can be further extended as

$$\frac{F}{E} = \frac{3E(0)}{2E(0) + n^2}$$

$$\mu = \frac{\mu_v(n^2 + 2)}{3}$$

where $E(0)$ is the equivalent permittivity, n is the refractive index, and μ_v is the vacuum moment. A combination of linear and nonlinear dielectric increments allowed evaluation of the number and effective moment of equivalent dipoles. The values obtained were $N = 3.4 \times 10^{26}$ m^{-3} and $\mu = 2.1 \times 10^{-28}$ Cm. The exceptionally large permittivity of the copolymer could be interpreted in terms of the large equivalent dipoles, where more than 10 monomer units are involved. Several samples of P(VDCN-VAc) show remarkable dielectric peaks around their glass transition temperature, similar to a ferroelectric transition [43].

Table 2 Dielectric Parameters of Various Cyanopolymers[a]

Polymer	Dipole moment, μ $(10^{-30}$ Cm)	Dipole density $(10^{28}$ m$^{-3})$	Dielectric relaxation strength, $\Delta\epsilon$
PVAc	6.0	0.83	6.5
PAN	11.3	1.34	38
P(VDCN-VAc)	13/6	0.44	125

[a]PVAc = polyvinylacetate.
Source: From Ref. 40.

Zuo et al. [44] reported dielectric properties of some VDCN copolymers having long aliphatic side chains, which may have a layered structure, and which showed three unusual phenomena: (a) extremely large dielectric relaxation strength; (b) dielectric transition phenomena whose strengths change suddenly at a critical temperature; and (c) unusual temperature dependence of the strength above the transition temperature, as shown in Figure 18. These may form "local ordering," which depends on the balance between the mobility and the cohesive force of the side chain and the main chain with CN dipoles around the critical point.

Piezoelectricity is proportional to remanent polarization, as discussed above, which depends directly on polar conformation of the polymer chain. Jo et al. [45] pointed out that the amorphous glass structure of P(VDCN-VAc) and its piezoelectricity depends on the casting solvent species as supported by NMR technique. It is well known that crystal modifications of PVDF depend on the film casting solvents: α crystal (TGTḠ) comes from acetone and γ crystal (TTTGTTTḠ) from DMA (dimethylacetamide). The crystallization of polar polymer usually reflects the conformation of chain when using a recrystallization solvent. In P(VDCN-VAc), polar molecules with all-trans-rich conformation are obtained from DMSO (dimethlsulfoxide), which show a large piezoelectric activity.

P(VDCN-VAc) has been used in ultrasonic transducers as a thickness extensional (TE) mode resonator. The transducer of piezoelectric fluoropolymers, PVDF and/or P(VDF-TrFE), is well known in medical applications. Table 3 lists data on the electromechanical properties of P(VDCN-VAc), PVDF, and PZT. Electromechanical coupling factor K_t of these polymers is 0.2–0.3, which is larger than that of quartz but substantially

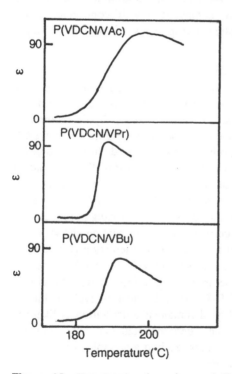

Figure 18 Temperature dependence of dielectric constant for P(VDCN-VAc), P(VDCN-VPr), and P(VDCN-VBz) [44].

Table 3 Comparison of Electromechanical Properties of Various Ferroelectric Polymers

Property	P(VDCN-VAc) [14,57]	β-Phase PVDF [58]	PZT [60]
Density (10^3 kg/m^3)	1.20	1.78	7.5
Sound velocity (km/s)	2.62	2.26	4.63
Acoustic impedance (10^6 kg/m^{2}s)	3.14	4.02	34.4
Stiffness constant (10^9 N/m^2)	8.2	9.1	159
Coupling constant, K_t	0.22	0.20	0.51
K_{31}	0.05	0.16	0.33
Piezo. constant, e_{33} (C/m^2)	−0.18	−0.14	15.1
d_{31} (pC/N)	7	27	2.7
Dielectric constant	5.8	6.2	635
Dielectric loss tangent	0.1	0.25	0.004
Mechanical loss tangent	0.05	0.1	0.004
Coersive field at 20°C (MV/m).	200	45	0.7

smaller than those of conventional ferroelectric ceramics such as PZT. The acoustic impedance of P(VDCN-VAc) is smaller than that of the others, and close to those of water and the human body. This makes P(VDCN-VAc) exceptionally useful as a transducer material for medical ultrasonics and nondestructive testing with organic liquid couplers. This is because the acoustic energy generated is damped effectively by transmission media. The piezoelectric behavior of P(VDCN-VAc) is quite similar to that of VDF copolymers. [46] The most distinguishable feature compared to VDF copolymers is the low density, low mechanical anisotropy, and optical transparency of P(VDCN-VAc), a amorphous polymer.

4. Ferroelectric Properties of P(VDCN-VAc)

According to X-ray diffraction results, P(VDCN-VAc) should be more noncrystalline than PAN copolymers. Dielectric dispersion behavior gives some information about dipole rotation above the glass transition temperature. One can easily imagine dipole rotation by poling above T_g. However, CN dipole rotation occurs even below T_g. Figure 19 shows CN dipole orientation in P(VDCN-VAc) while using poling below T_g. The dichroic ratio of the CN stretching mode (2200 cm^{-1}) changes with the change of field direction because of alignment of the dipoles. The piezoelectric constant increases with time, and its polarity is inverted with the change of the electric field. The polarization formation rate is very slow compared with that in VDF copolymers, though it increases with increasing poling electric field or temperature. Furthermore, depolarization current released on heating shows a very sharp peak at around T_g, which may be attributed to the dipole reorientation (Fig. 20).

Recently, Tasaka et al. [47] recorded a D-E hysteresis curve in P(VDCN-Pr) with a sharp dielectric peak and a special X-ray peak, as mentioned before. It was difficult to observe a J-E curve (Fig. 21) above T_g or good D-E curves because of DC conduction and dielectric breakdown of samples. The stability of remanent polarization is likely to be related to the enthalpy relaxation behavior. This demonstrates that annealing and poling below T_g force the samples to crystallize or stabilize into a suitable conformation and packing corresponding to the chemical structure.

The remanent polarization formed during poling disappears for P(VDCN-VAc) below or near T_g, while for P(VDCN-VPr) it remains even above T_g. Ikeda et al. [48] have

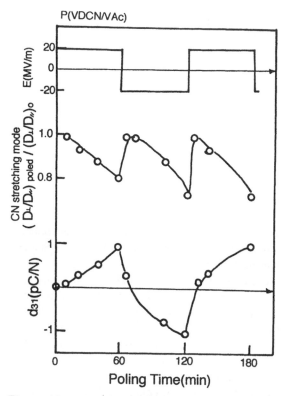

Figure 19 Infrared dichroic ratio change (CN stretching band) and piezoelectric constant for stretched P(VDCN-VAc) at room temperature, as a function of poling field at 140°C [38].

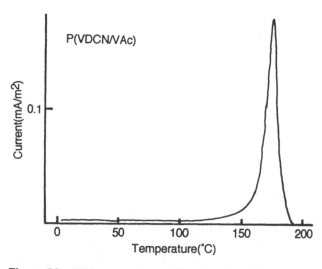

Figure 20 TSC curve of stretched and poled P(VDCN-VAc). Poling condition: E_p = 40 MV/m, t_p = 30 min, T_p = 160°C, heating rate = 6°C/min [38].

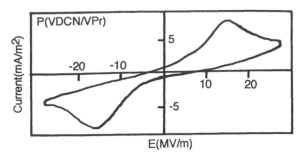

Figure 21   J-E curve of P(VDCN-VPr) at 170°C. Remanent polarization = 40 mC/m² [47].

observed ferroelectric behavior by bias field dependence of second-order dielectric constant for P(VDCN-VAc) below T_g, as shown in Figure 22. The dielectric constant vanishes unless spontaneous polarization breaks the sample symmetry. Therefore, this hysteresis behavior indicates the formation of remanent polarization by the bias field.

We may consider T_g as a kind of Curie point. There is a dielectric anomaly with peak maximum at T_g; this behavior, however, should be observed in all polar polymers around their T_g. The VDCN copolymers are emphasized here because they have a large dielectric constant. If extremely low frequency and high electric field were realized, a typical dielectric anomaly would appear around T_g. Since the freezing of dipole rotation in glass transition gives a substantial cooperativity in a small region of VDCN polymers, a ferroelectric transition may exist around T_g. Especially, some relaxation processes may further complicate the situation of the phase transition. Consequently, the ferroelectricity in VDCN copolymers has not been recognized yet, although further studies could make clear the ferroelectric behavior in VDCN copolymers.

C. Miscellaneous Cyanopolymers

There are many specialty polymers containing the cyano group. In this section, the polymers showing large piezo- and pyroelectric constant will be discussed briefly. Piezoelectric activity in poly(1-bicyclobutanecarbonnitrile) has been recently reported by Hall et al. [49] The polymer, prepared by radical polymerization, has a rigid structure which might restrict helical conformation, unlike PAN. The polymer has a piezoelectric coefficient, d_{31}, of 0.30 pC/N. The polar cyanide group may contribute to the piezoelectric activity. The structure of this polymer seems to be amorphous and to make polar chain conformation difficult (Fig. 23). Padias and Hall [50,51] attempted to enhance the dipole concentration by using multicyano-substituted small rings, such as cyclopropane or cyclobutanes. In larger rings, the substituents can be either axial or equatorial, and they may interchange. In small rings the substituents are pointed roughly in the same direction, and the conformations are much more restricted. Tetracyanocyclopropane-containing monomer was prepared and polymerized, though it was difficult to obtain polymers.

Cyanopolymers containing methacrylonitrile and cyanoacrylates as target monomers were also synthesized by the same research group [51]. These monomers can be copolymerized with vinyl esters by a free-radical polymerization to obtain polymers with good mechanical properties. Among these polymers, cyanofumaronitrile and dicyanoacrylate copolymers and acrylonitrile copolymer have an alternating sequence. These polymers have large pyroelectric constant (Table 4). Their structure are shown in Figure 24. These

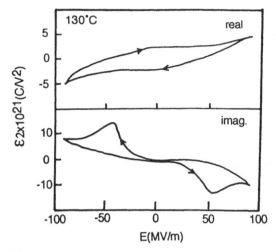

Figure 22 Bias field dependence of second-order dielectric constant for P(VDCN-VAc) at 130°C [48].

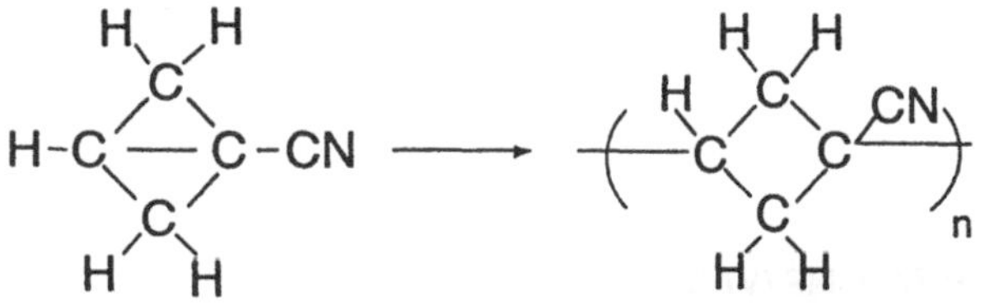

H-C—C-CN ⟶ Poly(1-bicyclobutanecarbonitrile)

Poly(1-bicyclobutanecarbonitrile)

Figure 23 Chemical structure of poly(1-bicyclebutanecarbonitrile).

Table 4 Pyroelectric Constants for Various Nitrile Copolymers[a]

Copolymer	Poling condition		Pyroelectric constant (10^{-6} C/m²K)
	Field (MV/m)	Temp. (°C)	
AN/VAc	16	70	1.9
AN/VFor	75	65	2.0
MAN/VAc	75	95	1.6
MAN/VFor	100	88	2.8
MAN/IPOAc	100	98	3.9
CNF/VAc	75	137	2.8
DCNA/VAc	69	148	5.9
MeCNA/VAc	50	134	4.1
EtCNA/VAc	80	103	6.7
EtCNA/MMA	50	118	1.9

[a]AN = acrylonitrile; VAc = vinyl acetate; VFor = vinyl formate; MAN = methacrylonitrile; IPOAc = isopropenyl acetate; CNF = dimethyl cyanofumarate; DCNA = Methyl-β-β-dicyanoacrylate; MeCNA = Methyl-α-cyanoacrylate; EtCNA = Ethyl-α-cyanoacrylate; MMA = methyl methacrylate.
Source: From Ref. 51.

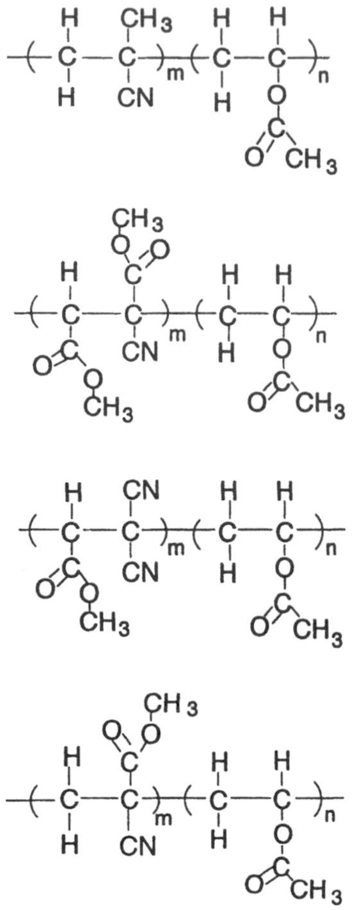

Figure 24 Chemical structure of cyanopolymers.

polymers are likely to be amorphous, therefore CN dipoles in amorphous regions contribute to the pyroelectric activity, which may depend on the chain conformation. The pyroelectricity of amorphous copolymers of acrylonitrile and vinyl chloride has also been investigated [59]. The molecular volume of vinyl chloride is similar to that of acrylonitrile, while the pyroelectric activity is likely to be relatively small because of irregular conformations.

Polycyanoaryl ether, one of the aromatic super-engineering plastics, is a crystalline polymer (Fig. 25). From crystal structure analysis, it is known that it belongs to a noncentrosymmetrical space group, namely, nonpolar. The uniaxially drawn amorphous film shows a piezoelectric activity (d_{31}) of 1–3 pC/N [52]. Table 5 lists the physical properties of this polymer. This amorphous sample shows a D-E hysteresis loop below the glass transition temperature [53], as shown in Figure 26. Remanent polarization is as small as 10 mC/m², and coercive field is very high, which may be due to the rigidity of the main chain in the glassy state.

Polymers containing cyanoethyl groups as side chains have been investigated as materials for large dielectric constants, which are useful as binders to ZnS powder in

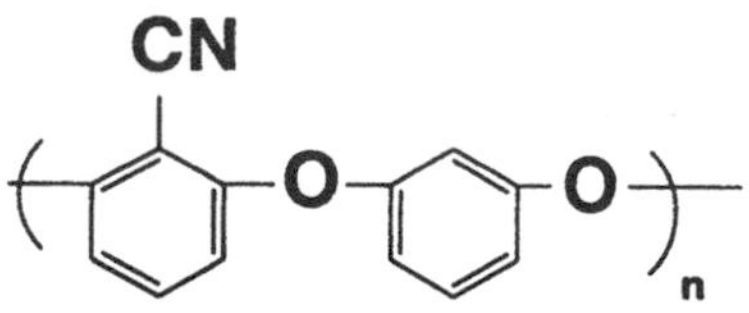

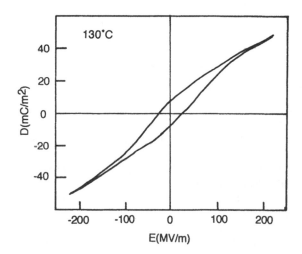

Figure 25 Chemical structure of polycyanoaryl ether.

Table 5 Physical Properties of Polycyanoallylether

Density (g/cm³)	1.32
Elastic constant (GPa)	3.6
T_g (°C)	145
T_m (°C)	340
Dielectric constant (25°C, 1 kHz)	3.55
Conductivity (S/cm)	1.8×10^{-16}
Piezoelectric constant, d_{31} (pC/N)	1–3

Figure 26 D-E hysteresis loop of amorphous polycyanoaryl ether film at 120°C [53].

electroluminescence devices. The dielectric constants of these kinds of materials usually are 10–30 at room temperature (Table 6). In particular, cyanoethylated polysaccharides show large dielectric constants even below the glass transition temperature (110°C) because a free side-chain motion along rigid chains occurs beginning from −30°C [54,55]. Poled cyanoethylated polymers form a small remanent polarization near T_g, compared with other cyanopolymers. It appears that the dipoles apart from the main chain are difficult to freeze near T_g.

Table 6 Dielectric Constants of Various Side-Chain-Type Cyanopolymers

Polymer	T_g (°C)	ϵ (20°C, 1 kHz)
Cyanoethyl amylose	110	17
Cyanoethyl cellulose	120	16
Cyanoethyl pullulan	110	18 (20; drawn)
Cyanoethyl PVA	20	20
Cyanoethyl polyhydroxymethylene	35	10
Cyanoethyl hydroxypropyl cellulose	−40	14
Cyanoethyl dihydroxypropyl cellulose	−40	23

VI. CONCLUSIONS

The strongly polar nitrile groups are responsible for most noncovalent interactions of cyanopolymers, and essentially ferroelectric behavior can be traced back to such interactions. Various results on cyanopolymers can be summarized as follows.

1. A wide variety of cyanohomopolymers and copolymers have been synthesized.
2. The large dipole moment of the CN group assists in forming a polar structure that display spontaneous polarization after poling.
3. The cyanopolymers which can form polar conformations in the main chain have a high possibility of being ferroelectric aggregations either in crystal or in amorphous phase.

REFERENCES

1. Z. Rappoport (ed.), *The Chemistry of the Cyano Group*, Interscience, New York, 1970.
2. A. M. Saum, Intermolecular association in organic nitriles; the CN dipole-pair bond, *J. Polymer Sci. 42*:57 (1960).
3. H. Umeyama and K. Morokuma, The origin of hydrogen bonding. An energy decomposition study, *J. Am. Chem. Soc. 99*:1316 (1977).
4. J. E. Del Bere, Molecular orbital theory of the hydrogen bond. Pi electron as proton acceptor, *Chem. Phys. Lett. 24*:203 (1973).
5. C. H. Bamford and G. C. Eastmond, Acrylonitrile polymers, *Encyclopedia of Polymer Science and Technology*, John Wiley, New York, 1971, vol. 1, p. 374.
6. M. Minagawa, K. Miyano, M. Takahashi, and F. Yoshii, Infrared characterization absorption band of highly isotactic polyacrylonitrile, *Macromolecules 21*:2387 (1988).
7. M. Hrooka, Complex copolymerization, *Copolymerization*, Baifukan, Tokyo, 1976, chap. 6, p. 272.
8. (a) A. B. Crociatori, L. F. Trapasso, and R. W. Stockman, Vinylidene cyanide polymers, *Encyclopedia of Polymer Science and Technology*, John Wiley, New York, 1971, vol. 14, p. 580; (b) H. Gillbert, F. F. Miller, S. J. Averill, R. S. Schmidt, F. D. Stewart, and H. L. Turnbull, Vinylidene cyanide. III, *J. Am. Chem. Soc. 76*:1074 (1954); (c) J. C. Westfahl, Vinylidene cyanide. IX. Reaction of polyvinyliden cyanide with compounds containing a single active hydrogen atom, *J. Am. Chem. Soc. 80*:871 (1958); (d) H. Gillbert, F. F. Miller, S. J. Averill, E. J. Carlson, V. L. Fort, H. J. Heller, F. D. Stewart, R. S. Schmidt, F. D. Stewart, and H. L. Turnbull, Vinyliden cyanide. VII. Copolymerization, *J. Am. Chem. Soc. 78*:1669 (1956).
9. Y. Inoue and A. Nishioka, ^{13}C-[^{1}H] nuclear magnetic resonance spectroscopy of polyacrylonitrile, *Polymer J. 3*:149 (1972).
10. J. C. Randall, *Polymer Sequence Determination: Carbon-13 NMR Method*, Academic Press, New York, 1977, chap. 4.

11. Y. S. Jo, Y. Inoue, R. Chujo, K. Saito, and S. Miyata, ^{13}C NMR analysis of microstructure in the highly piezoelectric copolymer vinylidene cyanide-vinyl acetate, *Macromolecules, 18*: 1850 (1985).

12. G. Hinrichsen, Structural change of drawn polyacrylonitrile during annealing, *J. Polymer Sci. C38*:303 (1972).

13. R. F. Boyer, T_{11} and related liquid state transition-relaxation, *Polymer Year Book* (R. A. Pethrick, ed.), vol. 2, p. 234.

14. S. Tasaka, N. Inagaki, T. Okutani, and S. Miyata, Structure and properties of amorphous piezoelectric vinylidene cyanide copolymers, *Polymer 30*:1639 (1989).

16. G. H. Olive and S. Olive, Molecular interactions and macroscopic properties of polyacrylonitrile and model substances, *Adv. Polymer Sci. 32*:124 (1979).

17. X. D. Lin and W. Ruland, X-ray studies on the structure of polyacrylonitrile fibers, *Macromolecules 26*:3030 (1993).

18. G. Hennico, J. Delhalle, C. Boiziau, and G. Lecayon, Theoretical study of the dependence of the valence electronic levels on the tacticity and conformation of acrylonitrile model origomers, *J. Chem. Soc. Faraday. Trans. 86*:1025 (1990).

19. P. H. Lindenmayer and R. Hosemann, Application of theory of paracrystals to the crystal structure analysis of polyacrylonitrile, *J. Appl. Phys. 34*:42 (1963).

20. C. R. Bohn, J. R. Schaefgen, and W. O. Statton, Laterally ordered polymers: Polyacrylonitrile and poyl(vinyl trifluoroacetate), *J. Polymer Sci. 55*:531 (1961).

21. G. Hinrichsen and H. Orth, Zur Struktur verstreckter Folien und Fäden sowie aus verdünnten Lösungen Hergestellter Einkristalle aus Polyacrylnitril, *Kolloid Z. 247*:847 (1971).

22. T. Thomson, H. G. Zachmann, and S. Korte, Molecular motion in poly(acrylonitrile) as determined by deuteron NMR, *Macromolecules 25*:6934 (1992).

23. S. I. Stupp and S. H. Carr, Spectroscopic analysis of electrically polarized polyacrylonitrile, *J. Polymer Sci. Polymer Phys. Ed. 16*:13 (1978).

24. S. I. Stupp and S. H. Carr, Dielectric field-induced structure in poly(acrylonitrile), *Colloid Polymer Sci. 257*:913 (1979).

25. H. Ueda and S. H. Carr, Piezoelectricity in polyacrylonitrile, *Polymer J. 16*:661 (1984).

26. R. Hayakawa, T. Nishi, K. Arisawa, and Y. Wada, Dielectric relaxation in paracrystalline phase in polyacrylonitrile, *J. Polymer Sci. A-2(5)*:165 (1967).

27. A. K. Gupta and N. Chand, Glass transition in polyacrylonitrile: Analysis of dielectric relaxation data, *J. Polymer Sci. Polymer Phys. Ed. 18*:1125 (1980).

28. H. Von Berlepsch, W. Kunstler, and R. Danz, Piezoelectricity in acrylonitrile/methylacrylate copolymer, *Ferroelectrics 81*:353 (1988).

29. H. Von Berlepsch, W. Kunstler, A. Wedel, R. Danz, and D. Geiss, Piezoelectric activity in a copolymer of acrylonitrile and methylacrylate, *IEEE Trans. Elect. Insul. 24*:357 (1989).

30. H. Von Berlepsch and W. Kunster, Piezoelectricity in acrylonitrile/methylacrylate copolymer, *Polymer Bull. 19*:305 (1988).

31. H. Von Berlepsch, M. Pinnow, and W. Stark, Electrical conduction in acrylonitrile/methylacrylate copolymer films, *J. Phys. D. Appl. Phys. 22*:1143 (1989).

32. A. Wedel, H. Von Berlepsch, and R. Danz, Remanent polarization and ferroelectric behavior in acrylonitrile/methylacrylate copolymer films, *Ferroelectrics 120*:255 (1991).

33. S. Tasaka, T. Nakamura, and N. Inagaki, Ferroelectric behavior in copolymers of acrylonitrile and allycyanide, *Jpn. J. Appl. Phys. 31*:2492 (1992).

34. T. Nakamura, S. Tasaka, and N. Inagaki, Ferroelectric behavior in acrylonitrile-copolymers, *Polymer Prep. Jpn. 42*:4456 (1993).

35. A. J. Yanko, A. Hawthorne, and J. W. Born, X-ray diffraction pattern of a vinylidene cyanide/vinyl acetate copolymer, *J. Polymer Sci. 27*:145 (1958).

36. T. Yurugi, A. Yamaguchi, and T. Ogihara, X-ray study on the fiber prepared from vinylidene cyanide/vinyl acetate copolymer, *Sen-i Gakkaishi 14*:1197 (1961).

37. Y. S. Jo, S. Tasaka, and S. Miyata, Piezoelectricity and enthalpy relaxation in the copolymer of vinylidene cyanide and vinyl acetate, *Sen-i Gakkaishi 39*:451 (1983).

38. S. Miyata, M. Yoshikawa, S. Tasaka, and M. Ko, Piezoelectricity in the copolymer of vinylidene cyanide and vinylacetate, *Polymer J. 12*:857 (1980).

39. S. Tasaka, K. Miyasato, M. Yoshikawa, S. Miyata, and M. Ko, Piezoelectricity and remanent polarization in vinylidene cyanide/vinylacetate copolymer, *Ferroelectrics 57*:267 (1988).

40. T. Furukawa, M. Date, K. Nakajima, T. Kosaka, and I. Seo, Large dielectric relaxations in an alternate copolymer of vinylidene cyanide and vinyl acetate, *Jpn. J. Appl. Phys. 25*:1178 (1986).

41. T. Furukawa, M. Date, K. Nakajima, T. Kosaka, and I. Seo, Nonlinear dielectric relaxation in a vinylidene cyanide/vinyl acetate copolymer, *Jpn. J. Appl. Phys. 27*:200 (1988).

42. T. Furukawa, Non-linear dielectric relaxation of polymers, *J. Noncrystalline Solids 131*:1154 (1991).

43. T. T. Wang and Y. Takase, Ferroelectriclike dielectric behavior in the piezoelectric amorphous copolymer of vinylidene cyanide and vinyl acetate, *J. Appl. Phys. 62*:3466 (1987).

44. D. Zou, S. Iwasaki, T. Tsutusi, S. Saito, M. Kishimoto, and I. Seo, Anomaly in dielectric relaxation in alternating copolymers of vinylidene cyanide and fatty acid vinyl ester, *Polymer 31*:1888 (1990).

45. Y. S. Jo, M. Sakurai, Y. Inoue, R. Chujo, S. Tasaka, and S. Miyata, Solvent-dependent coformations and piezoelectricity of the copolymer of vinylidene cyanide and vinyl acetate, *Polymer 28*:1583 (1987).

46. H. Ohigashi, Ultrasonic transducers in the megahertz range, *Application of Ferroelectric Polymers* T. T. Wang et al., eds.), Blackie, Glasgow, 1988, chap. II.

47. S. Tasaka, T. Nakamura, and N. Inagaki, Dielectric hysteresis near glass transition in vinyldene cyanide copolymers, *Polymer J.*, in press.

48. S. Ikeda, H. Kiba, M. Kutani, and Y. Wada, Polarization reversal in copolymer of vinylidene cyanide and vinyl acetate studied by measurement of dielectric constants, *Rep. Prog. Polymer Phys. Jpn. 31*:389 (1988).

49. K. H. Hall, Jr., R. J. Chou, J. Oku, O. R. Hughes, J. Scheinbeim, and B. Newman, Piezoelectric activity in films of poly(1-bicyclobutanecarbonitrile), *Polymer Bull. 17*:135 (1987).

50. A. B. Padias and H. K. Hall, Jr., Synthesis of potentially piezoelectric polymers containing tetracyanocyclobutyl side groups, *Polymer Bull. 24*:195 (1990).

51. A. B. Padias, Novel cyano-polymers as piezoelectric materials, *Polymer Prep. Jpn. 40*:54 (1991).

52. K. Koga, S. Ueta, and M. Takayanagi, Piezoelectricity of polycyanoarylether, *Polymer Prep. Jpn. 41*:1216 (1992).

53. S. Tasaka, T. Nakamura, and N. Inagaki, Ferro- and pyroelectricity in amorphous polycyanoarylether, *Jpn. J. Appl. Phys. 33*:5838 (1994).

54. S. Tasaka, T. Chiba, N. Inagaki, and S. Miyata, Electric properties of cyanoethylated polysaccharides, *Sen-i Gakkaishi 44*:70 (1989).

55. S. Kimpara, S. Tasaka, and N. Inagaki, Molecular motion and dielectricity in polymers with cyanoethyl group, *Sen-i Gakkaishi 49*:48 (1992).

56. T. Sato, Y. Tsujii, Y. Kita, T. Fukuda, and T. Miyamoto, Dielectric relaxation of liquid crystalline cyanoethylated O-(2,3-dihydrooxypropyl)cellulose, *Macromolecules 24*:4691 (1991).

57. I. Seo, Supersonic resonator using vinylidene cyanide copolymers, *1st Symp. on Organic Material for Optics and Electronics, Abstracts*, Tokyo, 1984, p. 25.

58. H. Ohigashi, Electromechanical properties of polarized polyvinylidene fluoride films as studied by the piezoelectric resonance method, *J. Appl. Phys. 47*:949 (1976).

59. H. Lee, R. E. Salomon, and M. M. Labes, Pyroelectricity due to a space charge mechanism in a copolymer of acrylonitrile and vinylidene chloride, *J. Appl. Phys. 50*:3773 (1979).

60. T. T. Wang et al. (eds.), *Application of Ferroelectric Polymers*, Blackie, Glasgow, 1988, p. 244.

8
Polyureas and Polythioureas

Eiichi Fukada
Kobayashi Institute of Physical Research, Tokyo, Japan, and Institute for Super Materials, ULVAC Japan, Ltd., Tsukuba, Japan

Shigeru Tasaka
Shizuoka University, Shizuoka, Japan

Hari Singh Nalwa
Hitachi Research Laboratory, Hitachi, Ltd., Hitachi City, Ibaraki, Japan

I. INTRODUCTION

Polyureas and polythioureas fall in the class of amino resins, which have been known from over 100 years ago. Amino resins, which are thermosetting polymers, are the products of polycondensation of aldehydes, commonly formaldehyde, and diamino compounds such as urea, thiourea, melamine, guanidine, and other related diamino compounds. Amino resins have been used commercially in molding, adhesives, laminating, textile finishing, protective coatings, paper manufacture, stoving, lacquers, enamels, fillers, varnishes, binders for inorganic fibers and foundry core sand, graphite resistors, foam structures, ion-exchange resins, leather tanning, plywood bonding, and electrical devices [1–4]. In the past two decades, very significant progress in the chemistry and technology of polyureas has been made by synthesizing a wide variety of aliphatic as well as aromatic polyureas from different diamines and diisocyantes [5]. The polar structures of polyureas lead to higher melting points than those of the corresponding polyimides, due to the greater molecule cohesion, though aliphatic polyureas have lower thermal stability than their aromatic counterparts. Thermal stability of polyureas is higher than polyurethanes but lower than polyamides. Polyurea fibers have tensile strengths between those of polyamides and polyester fibers. Polyureas have been used for cable insulation, foams, moldings, lubricants, and biomedical applications. Thermosetting polymers of polyurea and polythiourea were long considered suitable only as insulating materials, and the study of electrical properties of polyureas and polythioureas is of recent origin. The first attempt in this direction was made by Nalwa et al. in 1978 [6], who proposed to study the electrical behavior of thiourea when incorporated as a part of a polymer strand, because thiourea is known for its very interesting dielectric, pyroelectric, piezoelectric, and ferroelectric properties. The influence of thiourea structure on thermally stimulated depolarization effects, as well as dielectric and pyroelectric properties of a variety of polythioureas, was investigated [6,7]. Pyroelectricity in a polythiourea was also first reported in 1979 by the same research team [8]. There was no report on the electrical properties of polythioureas and polyureas until that time.

Pyroelectricity, piezoelectricity, and ferroelectricity in polyurea and polythiourea have not attracted much interest until recently. Urea resins are highly cross-linked amorphous polymers and have been used as insulating plastics for many years. Polyurea and polythiourea are usually synthesized by condensation or addition polymerization, and the products are mostly in the form of powders. Owing to their insolubility, preparation of films was not possible.

A technique of vapor deposition polymerization has recently been developed. This method was first applied successfully to synthesize thin layers of polyimide [9] and aromatic polyamide and aromatic poly(amide-imide) [10].

To prepare thin films of polyimide, pyromellitic dianhydride (PMDA) and 4,4'-diaminodiphenyl ether (ODA) are evaporated inside a vacuum chamber and deposited on a glass substrate. Two monomers deposited on the substrate react to form a thin layer of polyamic acid, as shown in Figure 1. By heating the film near 200°C, imidization of the product is completed.

The evaporation temperatures were 180°C for PMDA and 160°C for ODA. The temperature of the substrate was room temperature. The pressure in the chamber was 2×10^{-4} Pa. The deposition rate was 2.4 μm/h [9].

To prepare polyurea, a diamine monomer and a diisocyanate monomer were evaporated simultaneously in a vacuum chamber and deposited on a substrate. A variety of polyureas, both aromatic and aliphatic, have been prepared in the range of thicknesses between 200 nm and 10 μm. After being poled under a high electric field at elevated temperatures, these polyurea layers exhibited pyroelectric and piezoelectric activities.

The vapor deposition polymerization of polythiourea has not been undertaken yet, although there is ample possibility of preparation. Polythiourea was prepared by the ordinary solution method and pressed into pellets [8].

This chapter describes the most recent developments in pyroelectricity, piezoelectricity, and ferroelectricity of polyurea and polythiourea.

II. POLYUREAS

A. Aromatic Polyureas

1. Synthesis and Processing

The preparation of thin films of aromatic polyurea has been made possible by using the method of vapor deposition polymerization. By the usual solution polymerization method, the product is in the form of solid powders. Because of strong insolubility, it has not been possible to prepare very thin films of polyurea.

The urea bond is formed by reaction between an amino (NH_2) group and an isocyanate (CNO) group. A number of combinations of diamine monomer and diisocyanate monomer can be utilized to synthesize the different kinds of polyurea. The most typical example is shown in Figure 2.

A monomer, 4,4'-diphenylmethane diisocyanate (MDI), and another monomer, 4,4'-diamino diphenylmethane (MDA), were placed in vessels in the bottom of a vacuum chamber and heated to evaporation. A substrate, typically a slide glass, was placed about 30 cm above the monomer vessels. A commercial polyimide film (Kapton) with a thick-

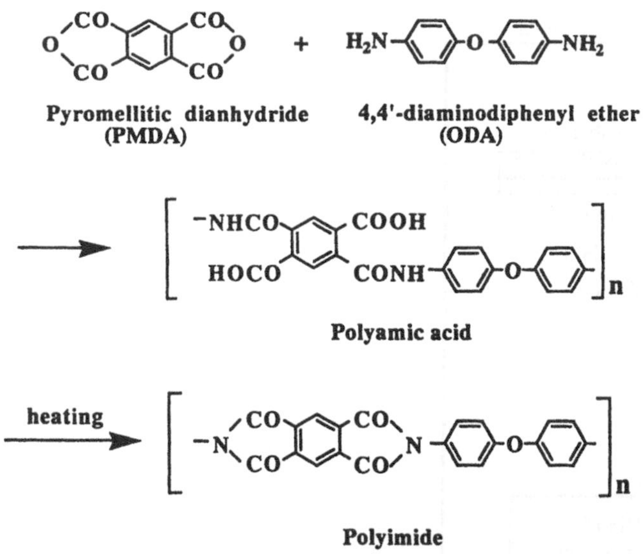

Pyromellitic dianhydride   4,4'-diaminodiphenyl ether
(PMDA)   (ODA)

Polyamic acid

Polyimide

Figure 1   A reaction to synthesize aromatic polyimide.

ness of 25 μm was used as the substrate when the samples for piezoelectric measurements were prepared. The temperature of the substrate was usually kept at room temperature.

The vapors of monomers deposit onto the surface of the substrate. The monomers with thermal energies diffuse on the surface and react with each other to form urea bonds between them. When the deposited quantities of monomers are balanced stoichiometrically, the addition polymerization proceeds successively. The thickness of the deposited layer is monitored by a quartz resonator. A schematic picture of the polymerization chamber is shown in Figure 3.

In a typical example, to prepare a stoichiometrically balanced sample of P (MDA/MDI), the evaporation temperatures of MDA and MDI were 100°C and 70°C, respectively. When the temperature of the substrate was room temperature, the deposition rate was about 1.8 μm/h and the pressure inside the chamber was about 3×10^{-3} Pa.

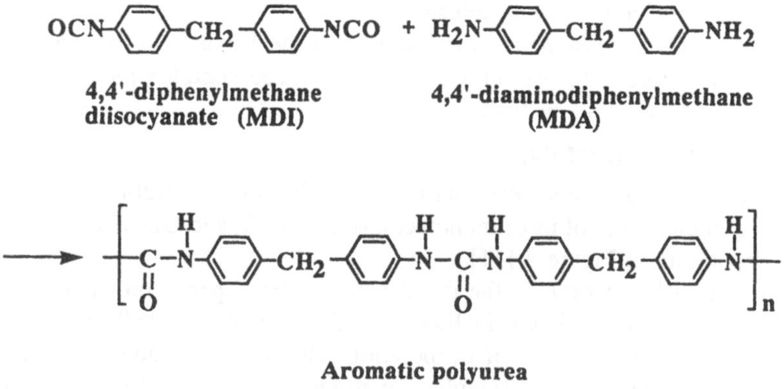

4,4'-diphenylmethane   4,4'-diaminodiphenylmethane
diisocyanate (MDI)   (MDA)

Aromatic polyurea

Figure 2   A reaction to synthesize aromatic polyurea.

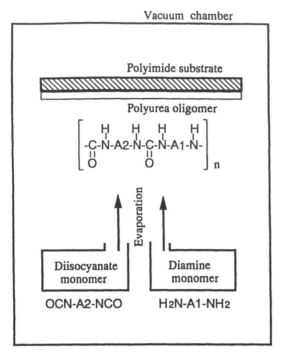

Figure 3 Schematic diagram of the apparatus for the vapor deposition polymerization of polyurea from diamine and diisocyanate monomers.

For poling treatment, aluminum as an electrode was deposited on the substrate before the deposition of monomer vapors and also on the surface of the as-deposited layer, both with thicknesses of approximately 100 nm. The typical thickness of the polyurea layer used for measurements was 500 nm.

A typical poling procedure was to apply an electric field of about 120 MV/m, increase the temperature to about 200°C, keep the sample at that temperature for 10 min, and follow by decreasing the temperature under the field.

To measure the piezoelectric constant, a sinusoidal strain of 10 Hz was given the substrate polyimide film on which the polyurea was deposited, and the electric charge produced on the aluminum electrodes was detected.

Rectangular coordinates are assigned to the film so that the 1 axis is in the plane of the film and the 3 axis is normal to the surface of the film. The complex piezoelectric stress constant is given by $e_{31} = e_{31}' + ie_{31}''$.

2. Spectroscopic Characterization

The rates of evaporation of monomers were controlled by adjusting the heating currents. For example, the evaporation rates of three monomers, MDI, MDA, and ODA, are shown as a function of temperature in Figure 4 [11].

The composition ratio MDA/MDI in the deposited samples depends on the evaporation rates of both monomers, as shown in Figure 5. The stoichiometrically balanced products are obtained in a limited range of evaporation rate of both monomers. In the regions with oblique lines, the deposited samples are slightly opaque due to the precipitated excess monomers. The films with balanced MDA/MDI composition are transparent.

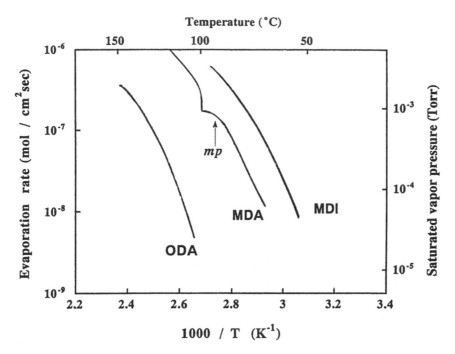

Figure 4 Evaporation rate and saturated vapor pressure of monomers, MDI, MDA, and ODA as a function of temperature.

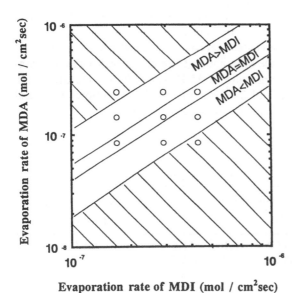

Figure 5 Relation between the composition ratio MDA/MDI and the evaporation rates of MDA and MDI.

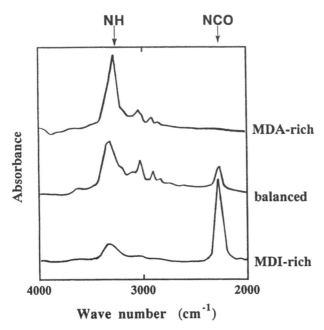

Figure 6 Comparison of the IR spectrum for the MDA-rich, balanced, and MDI-rich films of P (MDA/MDI).

Even for transparent films, the excess of either monomer is detected by infrared spectroscopy as shown in Figure 6. The film (MDA-rich) with excess MDA shows a strong peak at 3300 cm^{-1} due to NH stretching vibration. The film (MDI-rich) with excess MDI shows a strong peak at 2270 cm^{-1} due to N=C=O stretching vibration. The stoichiometrically balanced film (Balanced) shows both peaks with medium strengths.

Table 1 illustrates the results of elemental analysis of three as-deposited films of P (MDA/MDI) [12]. For these samples, the evaporation temperature of MDA was maintained at 100°C and the evaporation temperature of MDI was 65°C for the MDA-rich sample, 70°C for the balanced sample, and 80°C for the MDI-rich sample. The ideal molecular structures supposed for these samples are given in Figure 7.

Table 1 Elemental Analysis of As-Deposited Samples of Polyurea.

As-deposited samples	Weight percent				Molar ratio MDI/MDA
	O	N	H	C	
MDA-rich	5.8	13.0	6.1	75.6	0.67
($l = 5$)	(5.8)	(12.8)	(5.7)	(75.7)	(0.66)
Balanced	6.7	12.9	5.7	75.6	0.87
(m = 4 or 6)	(7.1)	(12.5)	(5.4)	(75.0)	(1.0)
MDI-rich	8.7	12.3	5.5	73.8	1.68
($n = 5$)	(8.3)	(12.2)	(5.1)	(74.3)	(1.5)

MDA-rich $NH_2-(R - NHCONH)_{l-1} R-NH_2$ $l=3, 5, 7 - - -$

balance $NH_2-(R - NHCONH)_{m-1}R-NCO$ $m=2, 4, 6 - - -$

MDI-rich $OCN-(R - NHCONH)_{n-1} R-NCO$ $n=3, 5, 7 - - -$

$$R= C_6H_4-CH_2-C_6H_4$$

Figure 7   Ideal molecular structures for as-deposited oligomers.

The best agreement is obtained between measured and calculated values for the weight percents of oxygen, nitrogen, hydrogen, and carbon atoms and for the molar ratio MDI/MDA, if the values of l, m, and n are assumed to be 5. This suggests that as-deposited samples consist not of polymers but of oligomers of polyurea.

Figure 8 depicts the results of differential scanning calorimetry (DSC) for the three samples [12]. For the balanced sample, an exothermic peak is observed over 100°C for the first run of heating curve. The peak appears neither in the second run for the balanced sample nor in the first and second runs for the unbalanced samples. This exothermic peak for the balanced sample should represent the heat of polymerization of oligomers.

Both ends of the unbalanced molecules are occupied by the same functional group, NH_2 or NCO. The polymerization of oligomers, therefore, is not possible. For the MDA-rich samples, it is supposed that during heating the cross links between NH_2 groups and NHCONH bonds take place and yield a highly cross-linked network structure. For the MDI-rich samples, dimerization may occur and the cross links between molecules would also yield a highly cross-linked network structure. Good orientation of urea bond dipoles is not expected after poling for the unbalanced samples.

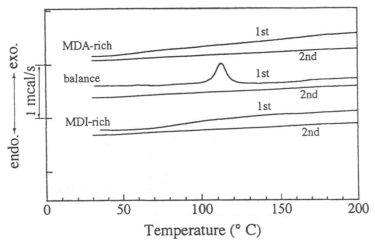

Figure 8   Differential scanning calorimetry for the MDA-rich, balanced, and MDI-rich films. (From Ref. 12.)

When the balanced samples are subjected to an electric field at elevated temperatures, a residual polarization due to the dipolar orientation is produced, which leads to piezoelectric and pyroelectric activities.

Figure 9 shows a comparison of infrared spectroscopy for an as-deposited sample, a poled sample, and a heat-treated sample of the balanced polyurea P (MDA/MDI) [13]. The peak at 2270 cm^{-1} observed for the as-deposited film, which is due to NCO stretching vibration at the chain end, disappears for the poled and heat-annealed samples. This also suggests the occurrence of polymerization of oligomers during heating. The comparison of the heat-treated (unpoled) sample and the poled sample indicates the decrease by poling of the peaks at 3300 cm^{-1} and 1650 cm^{-1} due to NH and CO stretching vibrations, respectively. Since the incident IR ray is perpendicular to the film surface, these observations indicate that the NH and CO groups are oriented by the poling procedure perpendicular to the surface of the film. The urea bond, NHCONH, with a dipole moment of 4.9 D, rotates under a high electric field.

Figure 10 shows the change of X-ray diffraction pattern by heat treatment. The as-deposited sample is almost amorphous. With increasing temperature, crystallization takes place at about 120°C. A rough estimation of the degree of crystallization deduced from the comparison of the area of peaks and the diffuse part of the diffraction pattern is given in Figure 11. The annealed sample has a degree of crystallinity of about 30%.

Summarizing the experimental results described above, we may reach an interesting conclusion about the process of producing a large residual polarization in aromatic polyureas. The as-deposited balanced films are oligomers consisting of about five monomers. With increasing temperature over about 120°C, the oligomers polymerize to high-molecular polymers. Under the poling process, the as-formed urea bond dipoles are ori-

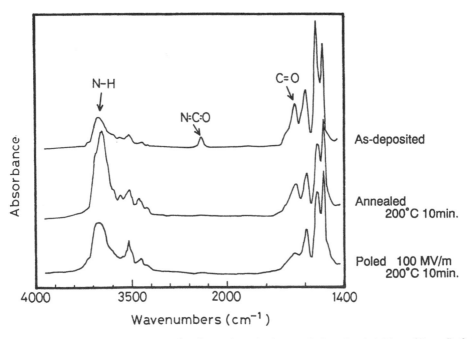

Figure 9 Infrared spectroscopy for the as-deposited, annealed, and poled films. (From Ref. 13.)

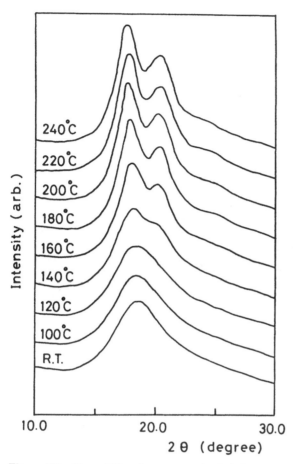

Figure 10   X-ray diffraction patterns for P (MDA/MDI) during heat treatment.

ented in the field direction. Parallel packing of polymers induces crystallization at about
120°C. In other words, the polymerization of oligomers, the formation of dipolar urea
bonds, and crystallization all take place at the same time under a high poling field. Thus
a stabilized large residual polarization is produced, and high piezoelectric and pyroelectric
properties are provided to aromatic polyurea. Owing to the high decomposition temperatures of aromatic polymers, good thermal stability up to about 200°C is also provided.

3.   Dielectric Properties

The dielectric constant $\epsilon_3 = \epsilon_3' - i\epsilon_3''$ and loss tan $\delta_\epsilon = \epsilon_3''/\epsilon_3'$ were determined at 10 Hz
as a function of temperature for the MDA-rich, balanced, and MDI-rich samples of
P (MDA/MDI). Figure 12 shows the results for poled films and Figure 13 those for
annealed samples. The dielectric constant for the MDA-rich film of P (MDA/MDI) is
about 10 for poled films and about 14 for annealed films; these figures are constant up
to 200°C. The dielectric loss for both poled and annealed films is less than 0.01 in the
whole temperature range. The MDA-rich films of P (MDA/MDI) should be an excellent
material for a capacitor, usable up to 200°C.

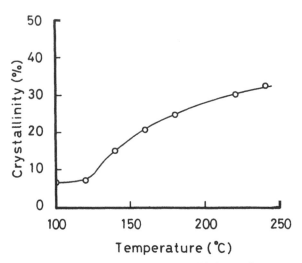

Figure 11   Degree of crystallinity with increasing temperature.

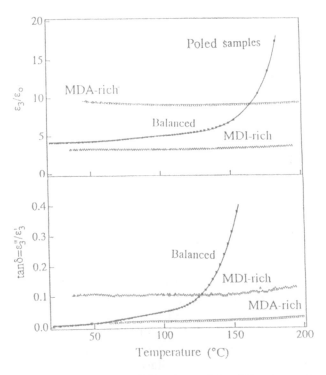

Figure 12   Temperature dependence of the dielectric constant $\epsilon_3 = \epsilon_3' - i\epsilon_3''$ and the dielectric loss tan $\delta_\epsilon = \epsilon_3''/\epsilon_3'$ for the MDA-rich, balanced, and MDI-rich films of poled P (MDA/MDI).

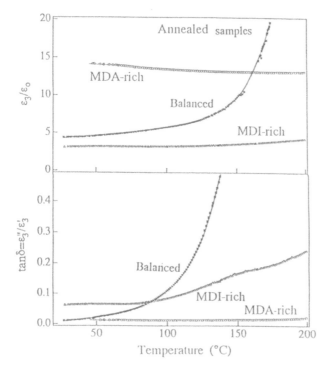

Figure 13 Temperature dependence of the dielectric constant $\epsilon_3 = \epsilon_3' - i\epsilon_3''$ and the dielectric loss tan $\delta_\epsilon = \epsilon_3''/\epsilon_3'$ for the MDA-rich, balanced, and MDI-rich films of annealed P (MDA/MDI).

The MDI-rich films of P (MDA/MDI) also show good thermal stability until 200°C, though the value of the dielectric constant is as small as 3 and the dielectric loss is as large as about 0.1. These unbalanced films of P (MDA/MDI) are so highly cross-linked that the thermal motions of molecules are suppressed until the decomposition temperature. The higher values of the dielectric constant for the MDA-rich films compared to the MDI-rich films may be due to an abundance of NH groups in the structure of the MDA-rich films.

The dielectric constant of balanced films is about 4 at room temperature and increases with rising temperature above about 100°C. The dielectric loss also increases sharply with rising temperature.

The piezoelectric constant e_{31} is highest for the balanced films. However, to avoid the increase of dielectric constant and loss at higher temperatures, the films with the monomer composition shifted slightly to the MDA-rich seem to be better suited for practical applications.

Figure 14 shows the temperature dependence of the dielectric constant and loss for the balanced films of P (MeMDA/MDI), P (ODA/MDI), and P (SDA/MDI). Similar to the balanced film of P (MDA/MDI), increases of ϵ_3 and tan δ_ϵ with increasing temperature are observed. In the balanced films, high-molecular molecules of polyurea do not form a cross-linked network structure as in the unbalanced films, but are oriented or crystallized in a local region and form a semicrystalline structure as a whole. The strong residual polarization produced at oriented molecular regions and crystallites gives rise to pyro-

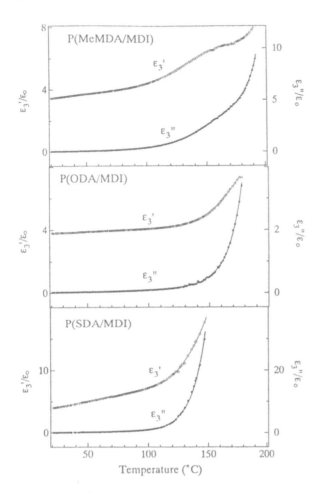

Figure 14 Temperature dependence of the dielectric constant and loss for the balanced films of P (MeMDA/MDI), P (ODA/MDI), and P (SDA/MDI).

electric and piezoelectric activities. With increasing temperature, molecular motions can be thermally excited in noncrystalline regions, which causes the increase of the dielectric constant and loss.

4. Pyroelectric and Piezoelectric Properties

Figure 15 shows the dependence of pyroelectric and piezoelectric constants on the poling temperature when the poling field is 100 MV/m and the poling time is 10 min. Both the pyroelectric and piezoelectric constants start to increase at about 120°C and level off at about 200°C.

Pyroelectric and piezoelectric constants are largest for the films with balanced composition. Figure 16 shows the dependence of these constants for P (MDA/MDI) on the evaporation temperature of MDI when the evaporation temperature of MDA is maintained

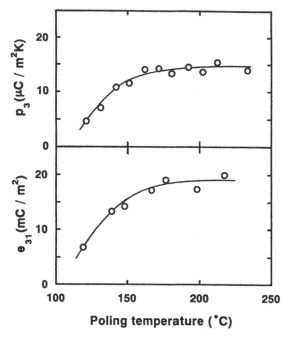

Figure 15 Dependence of the pyroelectric constant p_3 and piezoelectric constant e_{31} on the poling temperature, when the poling field is fixed at 100 MV/m.

at 100°C. The maximum pyroelectric constant, p_3, is 18 $\mu C/m^2 K$, and the maximum piezoelectric constant, e_{31}, is 22 mC/m^2 for the balanced films.

Figure 17 shows that the two kinds of piezoelectric constants are almost unchanged in the temperature range from room temperature to about 200°C. The dynamic piezoelectric stress constant $e_{31} = e_{31}' + i e_{31}''$ expresses a coefficient for electric displacement divided by strain. The dynamic piezoelectric strain constant expresses a coefficient for electric displacement divided by stress. The excellent thermal stability of aromatic polyurea is noticeable.

Besides P (MDA/MDI), different aromatic polyureas were synthesized by combining different diamine monomers with MDI monomer. The diamine monomers used were 4,4′-diamino diphenylmethane (MeMDA), 4,4′-diamino diphenyl ether (ODA), and 4,4′-diamino diphenyl thioether (SDA).

Figure 18 shows the dependence of piezoelectric constant e_{31} on the evaporation temperature of MDI when the evaporation temperature of the diamine monomers is maintained at fixed temperature: 126°C for MeMDA, 128°C for ODA, and 116°C for SDA. The poling condition is the same as that for (MDA/MDI), i.e., a field of 100 MV/m at 210°C for 10 min.

In comparison with Figure 16, the curve in Figure 18 is more flat, indicating that the balanced oligomers can be obtained in a broader range of evaporation temperature of monomers. The temperature dependence of the piezoelectric constant, defined as $e_{31} = e_{31}' + i e_{31}''$, for P (MeDA/MDI), P (ODA/MDI), and P (SDA/MDI) is shown in Figure 19. P (ODA/MDI) shows the highest piezoelectric constant as 26 mC/m^2, which remains unchanged until 200°C. P (SDA/MDI) decomposes above about 120°C.

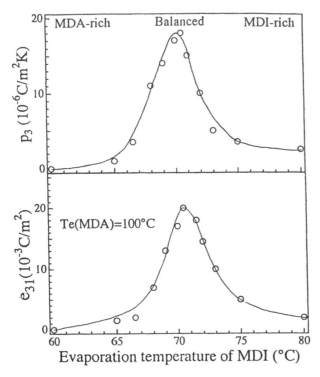

Figure 16 Dependence of the pyroelectric and piezoelectric constants p_3 and e_{31} of P (MDA/MDI) on the evaporation temperature T_e of MDI monomer, when the T_e of MDA monomer is kept at 100°C. (From Ref. 12.)

5. Ferroelectric Properties

N-phenylated polyurea was prepared by reacting an equimolar mixture of α,α′-dianilino-p-xylene with 4,4′-diphenylmethane diisocyanate in *sym*-tetrachloroethane as shown in Figure 20. The N-phenylated polyurea shows excellent solubility and can be easily processed in colorless, transparent, flexible thin films that exhibit a large second-order optical nonlinearity and transparency in the UV region [16,17]. Interesting properties seem to arise from the chemical structure of polyurea, where the lower density of hydrogen bonding introduced by N-substituted structure and bulky main chain are responsible for an amorphous structure.

Figure 21 shows the temperature dependence of the dielectric constant for N-phenylated polyurea, abbreviated NPUR. Two relaxation processes were observed: first with a small relaxation strength in the temperature range 100–120°C, probably due to processes such as a local motion, and second in the range 160–170°C, where the dielectric constant increases to about 12 due to the glass transition. The DC conduction above 160°C increased drastically. The NPUR has a dielectric constant of 3.4 at room temperature. The X-ray diffraction pattern shows a typical amorphous structure for NPUR.

Figure 22 shows the electric field dependence of polarization current of a NPUR sample at 120°C. The measurement using triangular waves should give a rectangular shape in the lower field, and the height of the rectangle is proportional to a linear dielectric constant. Inclination of the rectangle indicates the magnitude of DC conduction.

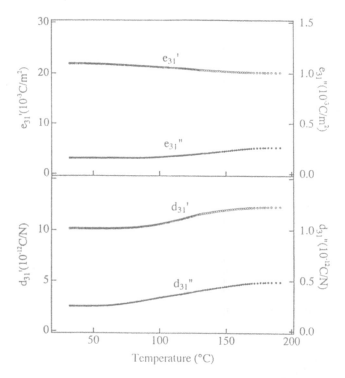

Figure 17 Temperature dependence of the piezoelectric stress constant $e_{31} = e_{31}' + ie_{31}''$ and the piezoelectric strain constant $d_{31} = d_{31}' + id_{31}''$ for a balanced film of P (MDA/MDI). (From Ref. 12.)

With increasing electric field, DC conduction decreases and a peak current appears. At high electric fields, the orientation polarization may reach saturation and simultaneously form hydrogen bonds between C=O and N—H in the polyurea. Appling a higher electric field induces conformational change to a metastable structure with dipole orientation. Finally, hydrogen bonding may form a remanent polarization and a D-E hysteresis loop was observed as shown in Figure 23. The remanent polarization obtained was found to be about 3 mC/m², which is smaller than the estimated value of 10–20 mC/m² assuming a reasonable conformation of the polyurea chain. The sample with Pr gives a large pyroelectric constant (3 mC/m²), which seems stable at least up to the glass transition temperature (170°C). IR spectra showed strong hydrogen-bonded N—H stretching mode, which indicates the aggregation of dipoles even in the amorphous phase. This suggests that a polar amorphous phase is partially possible and shows the ferroelectric behavior.

B. Aliphatic Polyureas

1. Synthesis and Processing

Since the vapor pressures and evaporation rates of aliphatic diamine and diisocyanate monomers are much higher than those of aromatic monomers, vapor deposition polymerization of aliphatic polyurea is not possible when the substrate is kept at room temperature. The times of residence of aliphatic monomers on the substrate are too short to

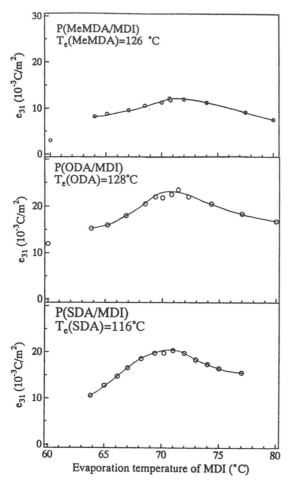

Figure 18 Dependence of the piezoelectric constant e_{31} for P (MeMDA/MDI), P (ODA/MDI), and P (SDA/MDI) on the evaporation temperature of MDI.

allow the addition polymerization. The preparation of thin films of aliphatic polyureas is possible only when the substrate is cooled down to temperatures below 0°C.

The reaction between 1,9-diaminononane and 1,9-diisocyanatononane yields poly-urea-9 as shown in Figure 24. Here 9 indicates the number of carbon atoms between urea bonds. If this number is even, the pyroelectric and piezoelectric activities do not appear, because the alternative urea bond dipoles orient in the antiparallel direction and the dipole moment as a whole is cancelled out. The situation is similar to that for even nylons.

Figure 25 depicts the deposition rate of polyurea-9 plotted against the temperature of substrate. The synthesis of polymer was possible only as the temperature of the sub-strate was lowered below −20°C. The evaporation temperatures of 1,9-diaminononane and 1,9-diisocyanatononane kept in silicone oil baths were 25°C and 50°C, respectively. The pressure inside the vacuum chamber was 2×10^{-3} Pa. A glass substrate was fixed at the bottom of a liquid nitrogen bath surrounded by an electric heater. A deposition

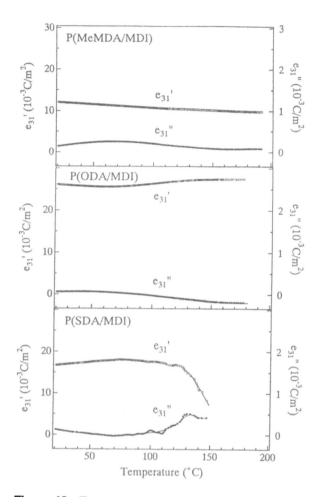

Figure 19 Temperature dependence of the piezoelectric constants $e_{31} = e_{31}' + ie_{31}''$ for P (MeMDA/MDI), P (ODA/MDI), and P (SDA/MDI).

rate of about 4 μm/h was obtained when the temperature of the substrate was maintained from $-80°C$ to $-20°C$ [14].

The deposited films of aromatic polyurea stick strongly to the surface of the substrate and cannot be separated from the substrate. On the contrary, the deposited films of aliphatic polyurea are easily separated from the glass substrate. Therefore, the films for piezoelectric and dielectric measurements were prepared using a deposited polyurea-9 film with a thickness of 10 μm as a flexible substrate. First, aluminum was deposited to the thickness of about 0.2 μm, then polyurea-9 about 0.3 μm, and then again aluminum of about 0.2 μm. Poling of the films was carried out with electric fields of 50–350 MV/m and at room temperatures of 20–150°C.

2. Spectroscopic Characterization

Figure 26 shows an X-ray diffraction pattern measured at room temperature for polyurea-9 prepared at different substrate temperatures T_s. It is seen that the diffraction peak becomes sharper with increasing substrate temperature, suggesting the increase of crys-

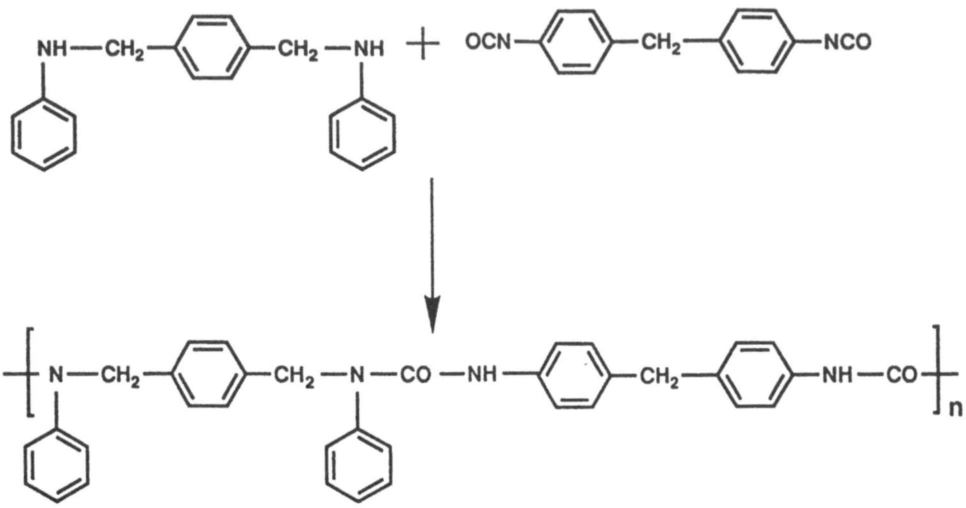

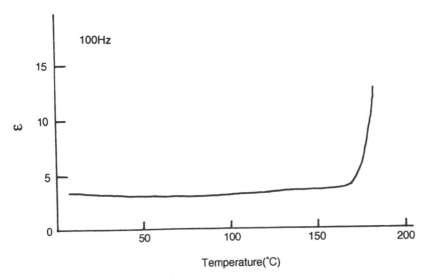

Figure 20   Polymerization of N-phenylated aromatic polyurea.

tallinity with T_s. The diffraction angle also changes between $-70°C$ and $-60°C$ of T_s. In analogy with nylon-11, [15] it has been assumed that the crystals grown at $T_s \geq -60°C$ (type I) has a triclinic structure and that the crystals grown at $T_s \leq -70°C$ (type II) has a pseudo-hexagonal structure. It is very probable that a large number of hydrogen bonds are formed between adjacent molecules. The parallel orientation of the hydrogen-bonded sheets is supposed to form the triclinic structure.

Figure 27 shows the IR spectrum for polyurea-9 prepared at $T_s = -20°C$. The two absorption peaks around 2950 cm^{-1} are due to CH$_2$ symmetric and antisymmetric vibra-

Figure 21   Temperature dependence of dielectric constant for N-phenylated polyurea.

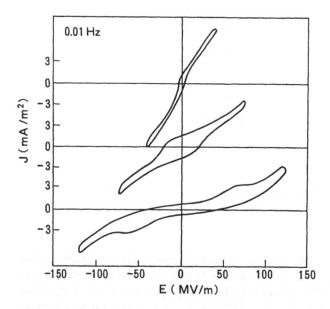

Figure 22 Relationship between polarization current density and external field for N-phenylated aromatic polyurea at 120°C. (From Ref. 18.)

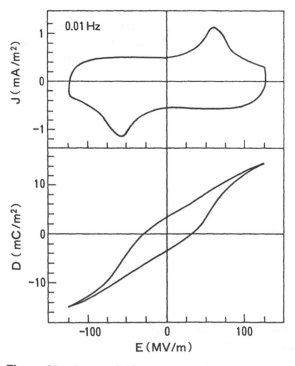

Figure 23 Current–field curve excluding DC conduction current and D-E hysteresis loop of N-phenylated aromatic polyurea at 120°C. (From Ref. 18.)

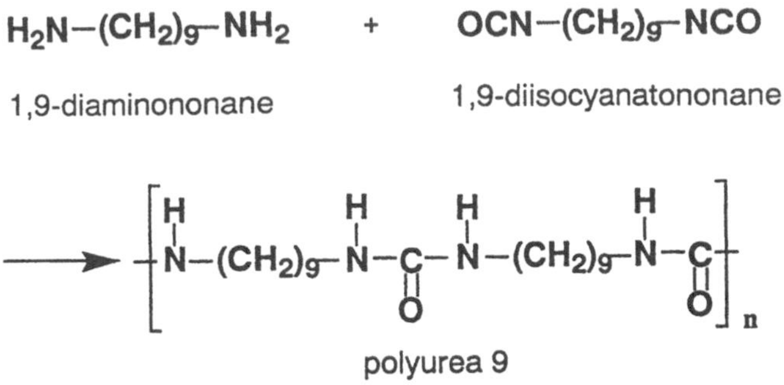

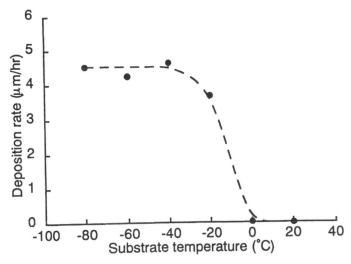

$H_2N-(CH_2)_9-NH_2$ + $OCN-(CH_2)_9-NCO$

1,9-diaminononane   1,9-diisocyanatononane

polyurea 9

Figure 24   Reaction between 1,9-diaminononane and 1,9-diisocyanatononane to produce polyurea-9.

tions in the hydrocarbon chains. The peak at 2270 cm^{-1}, due to NCO vibration, disappears after poling or annealing treatment. Peaks of CO and NH stretching vibrations are seen at 1620 cm^{-1} and 1570 cm^{-1}, respectively, for film prepared at $T_s = -20°C$. However, these peaks are seen at 1628 cm^{-1} and 1574 cm^{-1}, respectively, for film prepared at $T_s = -80°C$. These shifts of wavenumber for CO and NH peaks suggests that stronger hydrogen bonds are formed between urea bonds in type I films compared to type II films.

3.   Elastic and Dielectric Properties

The dynamic elastic constant $c_1 = c_1' + ic_1''$ and the dielectric constant $\epsilon_3 = \epsilon_3' - i\epsilon_3''$, both at 10 Hz, were measured as a function of temperature for a film of polyurea-9 with a thickness of 20 μm. The results are shown in Figures 28 and 29. It is seen that the substrate temperature T_s has a marked influence on the physical properties of deposited polyurea-9, probably due to the change of its crystalline structure. As described before,

Figure 25   Dependence of the deposition rate of aliphatic polyurea-9 on the substrate temperature. (From Ref. 14.)

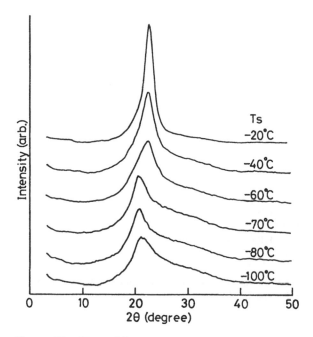

Figure 26 X-ray diffraction pattern for aliphatic polyurea-9 prepared at different substrate temperatures T_s.

we call the sample prepared at $T_s = -20°C$ type I, and the sample prepared at $T_s = -80°C$ type II.

The larger value of c_1' and smaller value of ϵ_3' for type I than for type II indicates that type I has more crystallized structure than type II. Three relaxations are observed at about $-150°C$, $-50°C$, and $100°C$ (for type II) and $140°C$ (for type I), which are designated as γ, β, and α relaxations, respectively.

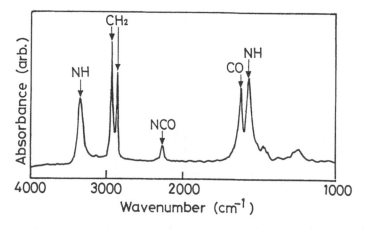

Figure 27 IR spectrum for aliphatic polyurea-9 prepared at $T_s = -20°C$.

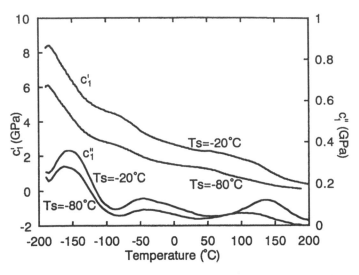

Figure 28 Temperature dependence of elastic constant $c_1 = c_1' + ic_1''$ for polyurea-9, prepared at $T_s = -20°C$ (type I) and at $T_s = -80°C$ (type II). (From Ref. 14.)

It is supposed that the γ relaxation is caused by the micro-Brownian motion of hydrocarbon sections, that the β relaxation is caused by the local thermal motion of long-chain segments including hydrocarbon sections and urea bonds, and that the α relaxation is caused by thermal motion on a large scale of long chains including hydrocarbon segments and urea bonds. The α relaxation may take place in both crystalline and noncrystalline regions because polyurea molecules are highly hydrogen bonded in both crystalline and noncrystalline regions. It is assumed that the temperatures for glass transition and crystalline relaxation overlap and cannot be separated.

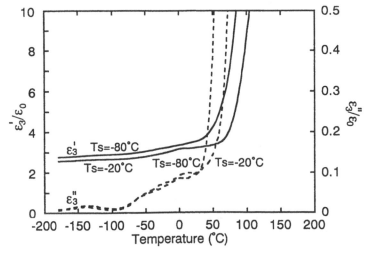

Figure 29 Temperature dependence of dielectric constant $\epsilon_3 = \epsilon_3' - i\epsilon_3''$ for polyurea-9, prepared at $T_s = -20°C$ (type I) and at $T_s = -80°C$ (type II). (From Ref. 14.)

It is noted that if the substrate temperature T_s is below the β relaxation, the less ordered structure type II is obtained, while if T_s is above the β relaxation, the more ordered structure type I is obtained. The thermal motion of molecules on the substrate facilitates the crystallization during deposition.

4. Pyroelectric and Piezoelectric Properties

The piezoelectric constant depends on the poling field and the poling temperature as well as the crystalline structure. Figure 30 shows the dependence of the piezoelectric constant e_{31} on the poling electric field when the poling temperature is kept at 25°C. The piezoelectric constant is proportional to the residual polarization. It is seen that the orientation of dipoles occurs easily at low fields and soon saturates for type II, but hardly occurs at low fields and increases sharply at higher fields for type I.

Figure 31 shows the dependence of e_{31} on the poling temperature when the poling field is kept at 150 MV/m. The piezoelectric constant increases with increasing poling temperature, but decreases above 100°C for type II and above 130°C for type I, corresponding to the α relaxation. The increase of DC conduction due to the α relaxation appears to shield the internal electric field to induce polarization. The highest value of the piezoelectric constant e_{31} so far obtained for aliphatic polyurea-9 is about 5 mC/m², which has been obtained with a poling field of 350 MV/m and a poling temperature of 25°C.

Figure 32 shows the temperature dependence of the piezoelectric constant e_{31} for type I and type II. It is believed that the piezoelectric constant e_{31} is proportional to the product of the Poisson ratio times residual polarization. The increase of e_{31} at about −150°C should be due to the increase of Poisson ratio caused by the decrease of the elastic constant at the γ relaxation. The influence of the β relaxation is hardly seen. The e_{31} increases again above 50°C by the increase of Poisson ratio due to the α relaxation and decreases finally by the increase of DC conduction above 100°C for type II and above 130°C for type I.

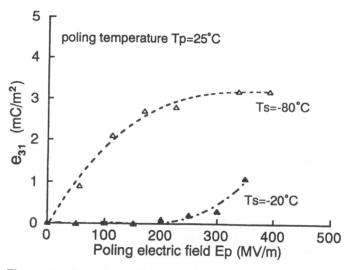

Figure 30 Dependence of the piezoelectric constant e_{31} for polyurea-9 on the poling electric field when the poling temperature is 25°C.

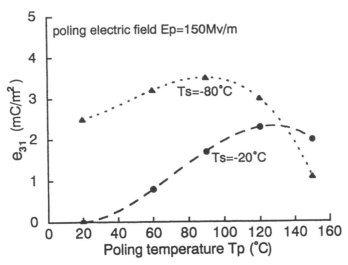

Figure 31 Dependence of the piezoelectric constant e_{31} for polyurea-9 on the poling temperature when the poling field is kept at 150 MV/m.

5. Ferroelectric Properties

Aliphatic polyureas have been used commercially as fabric materials, though commercial production was stopped when nylon and polyester fibers appeared in the market due to their better mechanical performance. Aliphatic polyurea fibers show good mechanical properties, and that is now transferred to a part of hard segment formation of elastic urethane fibers. Aliphatic polyureas are generally synthesized by the condensation of diamine and urea as shown in scheme (1) or by the addition of diisocyanate and diamine

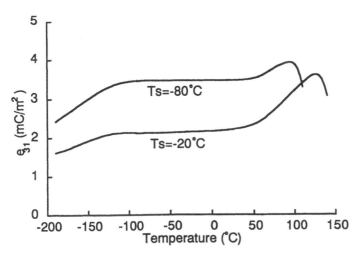

Figure 32 Temperature dependence of the piezoelectric constant e_{31} for type I and type II films of polyurea-9. (From Ref. 14.)

as shown in Scheme (2).

$$H_2N-R-NH_2 + CO(NH_2)_2 \xrightarrow{-NH_3} (HNCONH-R-)_n \qquad (1)$$

$$H_2N-R-NH_2 + OCN-R'-NCO \rightarrow (HNCONH-R-)_n \qquad (2)$$

The physical properties of the aliphatic polyureas are similar to those of aliphatic polyamides, namely, nylons. The melting temperatures and glass transition temperatures of nylons are a little bit lower than those of the polyureas with the same number of carbon atoms. This indicates a certain strength of hydrogen bonding in the polyureas. The density of hydrogen bonding in polyureas can be controlled by the copolymerization technique and by the combination of the number of carbons (even/even or odd/odd type). Especially, an odd number of carbon atoms in a polyurea chain can make a polar structure in a planar zigzag chain. The chemical structures of polyureas with an odd number of methylene groups are shown in Figure 33. These polyureas were synthesized by condensing diamine and urea. For example, approximately equal amounts of hexamethylenediamine or nonamethylenediamine and urea were heated up stepwise at 260°C and kept for 1 h under an aspirator to remove the ammonia as a by-product.

Figure 34 shows the temperature dependence of the dielectric constant at various frequencies for poly(heptametylene/nonamethyleneurea), abbreviated PU79, in a heating run [19]. This polymer has a melting point at 180°C and a glass transition temperature at 51°C from DSC measurements. Since the measurement sample is quenched to room temperature, below the glass transition temperature, its crystallinity is probably very low and the quality of the crystal is also low. However, IR measurements suggested strong hydrogen bonding even in quenched samples, and X-ray diffraction showed a defected-crystal pattern. The PU79 changes the crystal form and quality during the heating process, as confirmed by X-ray diffraction. The dielectric peaks above 125°C observed in only one heating run increase with decreasing frequency, and their magnitude is quite large. Therefore, this dielectric anomaly may be related to the crystal transition from a mobile dipole state to a rigid dipole state corresponding to different hydrogen-bonding states.

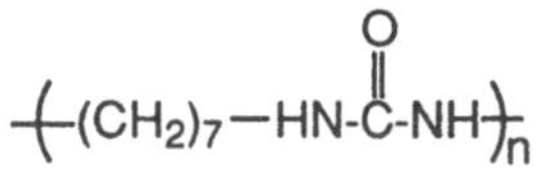

$$\left.\left(-(CH_2)_7-HN-\overset{\overset{\displaystyle O}{\|}}{C}-NH\right.\right)_n$$

Polyurea 7

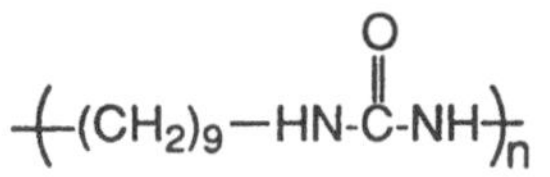

$$\left.\left(-(CH_2)_9-HN-\overset{\overset{\displaystyle O}{\|}}{C}-NH\right.\right)_n$$

Polyurea 9

$$\left.\left(-(CH_2)_7-HN-\overset{\overset{\displaystyle O}{\|}}{C}-NH\right)_m\right.\left(-(CH_2)_9-HN-\overset{\overset{\displaystyle O}{\|}}{C}-NH\right)_n$$

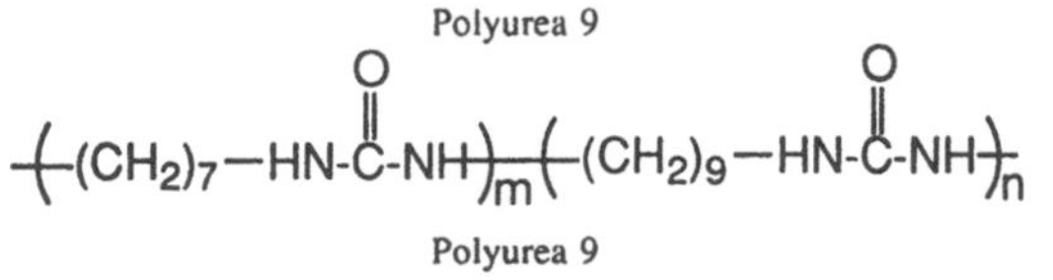

Polyurea 9

Figure 33 Chemical structures of polyureas.

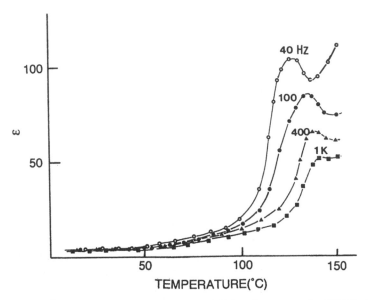

Figure 34 Temperature dependence of dielectric constant for PU-79. (From Ref. 19.)

Figure 35 shows the D-E hysteresis characteristics of PU79 and poly (nonamethyleneurea) PU9 obtained from the current density–electric field curve, which excludes the DC conduction current of the samples [19]. The value of remanent polarization (P_r) is 200–440 mC/m^2 at 90°C, which may be attributed to an apparent effect from ionic currents of impurities. The reproducibility of P_r depends on the crystallinity and molecular orientation, but the same D-E hysteresis loop can be obtained for more than 100 cycles

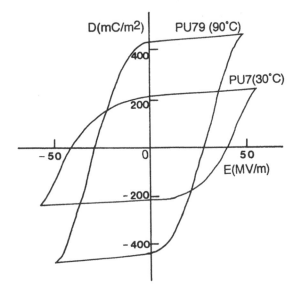

Figure 35 D-E hysteresis loop of PU-79 and PU-9 at 0.017 Hz. (From Ref. 19.)

for the same conditions. The existence of a large reversible pyroelectric constant of 20 mC/m^{2}K indicates the polarization reversal by the external electric field. The remanent polarization disappears above 110°C. These results suggest that the rearrangement of hydrogen bonding around 110°C induces crystal transition and depolarization. The polarization reversal observed at 30°C in polyurea-9 below its glass transition temperature indicates that this behavior is not attributed to the mobile ions but to the urea dipole in the crystal. However, no polarization reversal or higher coercive field was observed after crystallization by annealing. The origin is probably the cooperative motion of urea dipoles in the metastable phase. The dipole moment of the urea group is 4.9 D, and the spontaneous polarization P_s of polyurea-79 calculated by means of a simple sum of the vaccum moment over a unit would be 90–100 mC/m^2. A very large P_r value of 440 mC/m^2 at 90°C may be explained by the local field effect, namely, the increase of the instantaneous dielectric constant in the ferroelectric phase.

X-ray diffraction studies on polyurea-79 show that the melt-cast film has a disordered crystal with hexagonal packing and simple lateral order. This crystal may consist of polar chains to make a ferroelectric crystal. On the other hand, in the case of annealed polyurea-79, the rearrangement of hydrogen bonding during recrystallization can give rise to irreversible ferroelectric-to-paraelectric transition. The residual polarization is about 450 mC/m^2 for unannealed sample, which reduces drastically for annealed sample [20]. Polyurea-79 has two crystal phases: a metastable and a stable crystal phase. The difference between these two phases may be attributed to an intermolecular hydrogen-bonding condition. Such a phenomenon is also observed for aliphatic polyurethanes [21].

III. POLYTHIOUREAS

A. Synthesis and Spectroscopic Characterization

Thiourea is usually used as a comonomer in polycondensation or addition condensation, like urea. Polythioureas are not so popular commercially because of their low thermal stability. The interest in design and synthesis of polymers of thiourea originates from the fact that thiourea is a well-known pyroelectric, piezoelectric, and ferroelectric material. Thiourea has a chemical structure $H_2N\!-\!CS\!-\!NH_2$, where the amino group is an electron-donor group and CS is an electron-acceptor group. Thiourea single crystals which are orthrohombic can be grown from a saturated methanol solution. Ferroelectricity in thiourea crystals was first reported by Solomon [22]. Later, Goldsmith and White [23] reported ferroelectric properties in detail. Thiourea exhibits five different phases: a ferroelectric phase I below −104°C; an antiferroelectric phase II between −104 and −97°C; a ferroelectric phase III between −97 and −93°C; a paraelectric or antiferroelectric phase IV between −93 and −71°C, and an antiferroelectric phase V above −71°C. Studies on various electrical properties such as temperature- and frequency-dependent dielectric constant, spontaneous polarization, pyroelectric, piezoelectric, and ferroelectric effects have shown thiourea to be an interesting organic molecular material. The terminal amino groups are versatile chemical species suitable for preparing thiourea polymers via condensation polymerization. Therefore, it was considered of great interest to study the electrical properties of such ferroelectric materials when incorporated as a part of a polymer matrix.

Polythioureas should be regarded as main-chain polymers in which thiourea moieties are bonded covalently into a single polymer strand. Thiourea polymers can be prepared

by condensing thiourea with formaldehyde under different conditions. A simple polymerization reaction is shown below, where water is eliminated during condensation.

$$H_2N — CS — NH_2 + HCHO$$

$$\downarrow -H_2O$$

$$\text{─}(NH — CS — NH — CH_2 — NH\text{─})_n$$

The reaction between the initial materials and polymerization of the intermediate condensates is complex and is influenced by reaction conditions such as (a) mole ratio and
nature of the reactants, (b) reaction time and temperature, (c) reaction medium and pH,
(d) catalyst, and (e) concentration of the reactants. Synthesis of polythioureas is similar
to that of polyureas. For example, the reaction of dithiocyanates with water, diamines
with carbon disulfide, or thiourea with formaldehyde makes linear polythioureas under
mild conditions, but otherwise may result in cross-linked polymers or degraded structures.
The chemical structure of polythioureas selected for ferroelectric properties may have a
polar chain in zigzag conformation, like odd nylons. Though the chemical structures of
polythioureas depend on the reaction conditions, it seems rather difficult for polythioureas
to have a highly crystalline state because side reactions takes place during the polymerization process, leading to a structural disorder that forms an amorphous part.

To study electrical properties, two types of thiourea-formaldehyde polymers were
synthesized: one in basic medium using NaOH at elevated temperatures [7,8], and another
in acidic medium using glacial acetic acid at room temperature [24], which will be
referred as PTUFB and PTUFA, respectively. The PTUFB was a pale yellow material,
while the PTUFA was a white material. These PTUFs were characterized spectroscopically by IR, TGA, DTA, and DSC techniques, which show features of polythiourea
condensates. Both PTUFA and PTUFB are rather low-molecular-weight condensates. The
IR spectra indicated the presence of methylol and amino end groups in PTUFA samples
which have a linear chain structure. The PTUFB which was prepared under basic conditions has some cyclic structure, as indicated by the IR spectrum. This may be possible
because polymerization for PTUFB was carried out at higher temperature.

B. Dielectric Properties

The first work on polythioureas in the field of ferroelectrics was published in 1978 using
a thiourea-formaldehyde condensation polymer as a target material. Figure 36 shows the
variation of the dielectric constant of PTUFB at 10 kHz as a function of temperature [8].
Even at room temperature, the PTUFB condensate has a relatively high dielectric constant
of 20–30 as compared to a value of 1–10 shown by most of the conventional polymers.
There is a slight increase in the dielectric constant upon heating from 100 to 130°C,
though a sharp increase was noted at higher temperature and a maximum was obtained
at 145°C. Dielectric constant as high as 320 was observed at 145°C in PTUFB samples.
This transition at 145°C was found to be reproducible, but the magnitude of the dielectric
constant peak varied depending on the samples. The temperature dependence of dielectric
loss also showed a maximum at 145°C. The change in dielectric constant was also recorded both under continuous heating as well as after allowing the samples to equilibrate
for about 15 min. PTUFB samples preheated at 145°C for an hour exhibited similar
characteristics. It is seen that once again a sharp maximum occurs at the same temperature

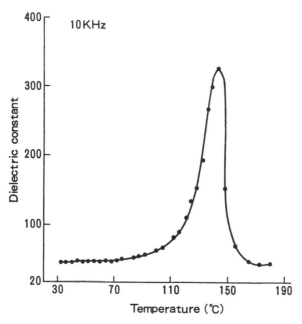

Figure 36 Temperature dependence of dielectric constant for thiourea-formaldehyde polymer at 1 kHz.

but the net effect varies depending on the sample conditions. TSD spectra of poled PTUFB samples showed two peaks: one at 156°C and the other between 178 and 188°C [7], but sometimes a TSD peak also occurred at around 145°C, similar to that of dielectric constant. PTUFB shows dielectric anomaly and spontaneous pyroelectric current around 140°C. This behavior may be related to the ferroelectric transition of thiourea as well as to the cooperative orientational change of thiourea segments. The main disadvantage of PTUFB is that it is difficult to process into thin film, hence the dielectric measurements were done by preparing a powder pellet form.

The dielectric constant (ϵ) measured as a function of temperature on pellets of PTUFA showed no peak, unlike PTUFB. The ϵ of PTUFA was found to be between 10 and 15 at room temperature and slightly increased with the rise of temperature. Thermally stimulated depolarization (TSD) effects have studied in detail for PTUFB by varying polarization and depolarization conditions such as poling temperature, poling field, electrode material, and heating rate. Figure 37 shows typical TSD current spectra recorded for PTUFA at a heating rate of 2 and 4°C/min when the sample was poled at 60°C applying an electric field of 4 kV/cm for 2 h [24]. A TSD peak appearing at 122°C splits into two peaks of lower magnitude at a low heating rate. Figure 38 shows the TSD spectra of PTUFA at various heating rates from 4 to 10°C/min [25]. When the heating rate increases, the initial polarization is released faster while the dielectric responds less quickly, therefore the TSD peak increases and shifts to a higher temperature. On the other hand, a decrease in heating rate will simultaneously decrease the intensity of the electric current and resolved TSD peaks which overlap. Figure 39 shows the effect of electrode material on the TSD peak, which is a reliable method for distinguishing between dipolar and space-charge mechanism due to the difference in metal work function. The

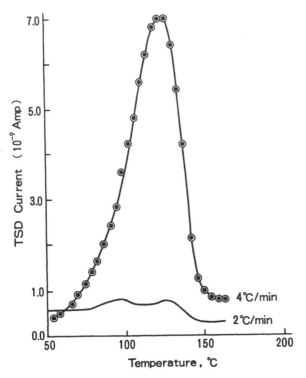

Figure 37 TSD current spectra of PTUFA at heating rates of 2 and 4°C/min, showing the splitting of TSD peak at low heating rate. (From Ref. 24.)

electrode material does not affect the dipolar peak unless it is accompanied by a space-charge peak. The TSD studies show three peaks in PTUFA at 122, 130, and 138°C. The resistivity and TSD measurements along with the variations in IR spectra at TSD peak temperatures, gas chromatograph, and thermal analysis indicates the presence of two distinct transition regions in the temperature ranges 94–100°C and 122–126°C. The various physical properties of PTUFA and PTUFB are listed in Table 2, which indicates that thiourea-formaldehyde polymers prepared under different media are entirely different materials.

Recently, Ohishi et al. [26] measured dielectric properties of aliphatic polythioureas which were synthesized by thermal condensation of diamine with carbon disulfide as shown below. Polymerization of carbon disulfide and aliphatic diamine polymer is called polythiourea-N, where N is the number of CH_2 groups.

$$H_2N — (CH_2)_n—NH_2 + CS_2$$

$$\downarrow \; -H_2S$$

$$\left[(CH_2)_n—NH—\overset{\overset{\displaystyle S}{\|}}{C}—NH \right]_n$$

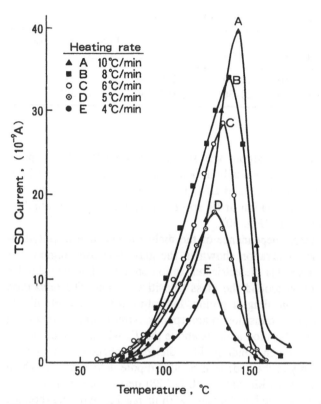

Figure 38 Effect of heating rate on TSD current spectra of PTUFA. (From Ref. 25.)

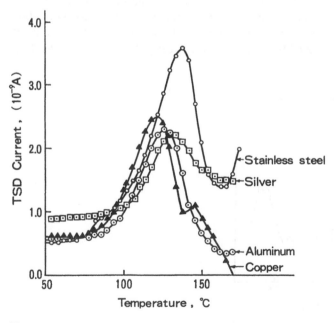

Figure 39 Effect of electrode material on TSD current spectra of PTUFA. (From Ref. 25.)

Table 2   Physical Properties of Thiourea-Formaldehyde Polymers (PTUF)

Polymer	T_d (°C)	T_d (°C)	ρ (Ω cm)	ϵ (1 kHz)	TSD peaks (°C)	Pyroelectric constant (10^{-9} C/cm²)
PTUFA	126	200	2.5×10^{11}	10–15	122, 130, 138	—
PTUFB	140	198	8.9×10^{12}	20–30	140, 156, 166	3.0

A polythiourea with an odd number of carbons was selected in order to obtain a polar chain and polar packing. The polymer is typically an amorphous material having a T_g at 50°C as evidenced by X-ray measurements and DSC technique. According to the NH stretching band (3240 cm^{-1}) with association of IR spectrum, strong hydrogen bonding exists in the polymer even above its T_g (Fig. 40).

Figure 41 shows the temperature dependence of the dielectric constant of polythiourea-9 (PTU-9). A large dielectric relaxation shown in the glass transition region was observed for all the samples. From this dielectric relaxation strength, thiourea dipoles probably have a cooperative motion around the glass transition region. The relaxation strength of PTU-9 decreases with annealing or poling, indicating the changeable dipole mobility due to crystallization or phase transition in amorphous. After poling or annealing around T_g, a new, extremely large relaxation appears above the T_g, which may be due to a new ordering formed by the hydrogen bonding. This dielectric phenomenon may be similar to the relaxation, due mainly to the hydrogen-bonded dipole defect motion which has been observed in aliphatic alcohols. Meakins et al. [27,28] derived theoretically that the dielectric relaxation and the dielectric loss increase with increasing hydrogen-bonded chain length. Consequently, hydrogen-bonded thiourea dipoles in the intermolecular chains are able to rotate easily by external electric field above T_g. Thioura [SC(NH$_2$)$_2$], which has a large dipole moment (5.4 D) is well known as one of the order–disorder type ferroelectrics [23]. The ferroelectric transitions in thiourea is essentially due to both crystal and "incommensurate-phase" natures, which are attributed not to hydrogen bonding, but to dipole interactions [29,30]. It is anticipated that polymers possessing large

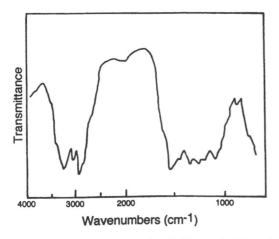

Figure 40   IR spectrum of polythiourea-9. (From Ref. 26.)

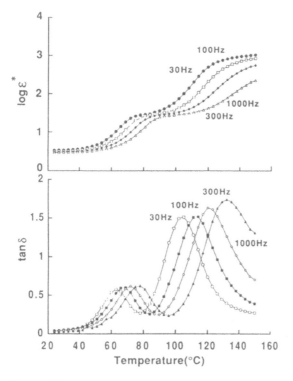

Figure 41 Temperature dependence of dielectric constant for polythiourea-9 annealed and poled at 100°C (50 MV/m). (From Ref. 26.)

thiourea dipoles would be able to have a polar structure with hydrogen bonding by control of chemical structure. The structure of such thiourea polymers may be called "incommensurate phase."

C. Pyroelectricity and Ferroelectricity

Interestingly, the unpoled PTUFB samples released current on heating and both dielectric constant as well as dielectric loss showed a peak around 145°C, therefore the possibility of pyroelectric effect was examined [8,31]. Pyroelectricity was studied by heating the TUF samples with and without poling at a uniform heating rate of 4°C/min. Figure 42 shows the pyroelectric coefficient of unpoled TUF samples as a function of temperature up to 140°C. At room temperature, the pyroelectric coefficient was in the range of 3–4 $\times 10^{-9}$ C/cm^2, and it increased sharply with increase of temperature. The value of 60 $\times$ 10^{-9} C/cm^2 was estimated around 140°C. Though the value of pyroelectric coefficient decreased after repeated heating cycles, still after the fourth heating cycle, the pyroelectric coefficient as high as 32 $\times$ 10^{-9} C/cm^2 at 140°C was obtained. Similar pyroelectric behavior was also observed for poled PTUFB samples. The TUF sample was poled at 50°C under an electric field of 2 kV/cm for 2 h (Fig. 42b). The poled sample was heated as a function of temperature, and the magnitude of the pyroelectric coefficient was found to be higher than that of the unpoled PTUFB samples and increased sharply with rising temperature. The pyroelectric coefficient of poled sample around 140°C was found to be

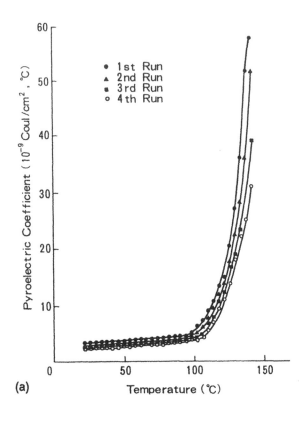

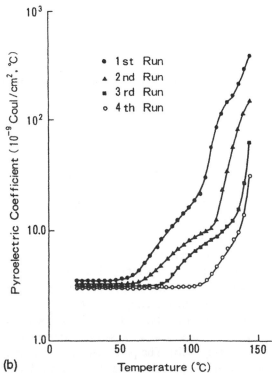

Figure 42 Temperature dependence of pyroelectric constant of (a) unpoled and (b) poled PTUFB. (From Ref. 31.)

several times larger than that of unpoled sample. Like unpoled samples, the pyroelectric coefficient decreased after repeated heating cycles. Therefore the poling significantly enhances the pyroelectric effect in PTUFB, as has been reported for other pyroelectric materials. It has been seen that PTUFB samples compare favorably with polyvinylidene fluoride (PVDF). Lang also reported similar pyroelectric behavior as a function of temperature in the case of PVDF. The absolute value of the static pyroelectric coefficient for PVDF was reported to be $0.5-1.0 \times 10^{-9}$ C/cm^2 in the temperature region 100–250°C, but by heating up to 350°C, the pyroelectric coefficient increased up to 200×10^{-9} C/cm^2. In this case also a decrease in pyroelectric coefficient values after repeated heating cycles was noted, similar to the present study.

Polythioureas discussed above show large reversible pyroelectric constant (higher than 50 mC/m^{2}K) after poling [32]. The pyroelectric constant in drawn sample is stable up to 120°C (Fig. 43). The structure of both drawn and undrawn PUT-9 samples by X-ray diffraction is almost amorphous, both having no difference. Further, D-E hysteresis is observed in drawn sample above T_g, as shown in Figure 44 [32]. Coersive field and remanent polarization of PUT-9 were 18 MV/m and 16 mC/m at 80°C, respectively. This is the first report on the ferroelectric behavior of amorphous polythioureas above the T_g. If amorphous materials having some mobility were present below the glass transition temperature, the dipoles which are oriented under high electric field would be frozen by low molecular motion. However, above T_g, dipoles are easy to rotate in a chain by thermal

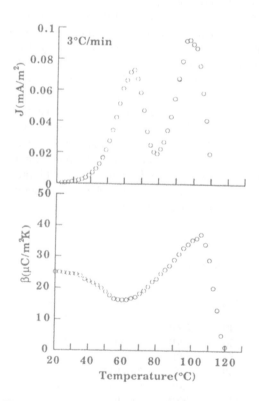

Figure 43 Temperature dependence of depolarization current and pyroelectric constant (10 Hz) for drawn and poled (55°C, 50 MV/m) polythiourea-9. (From Ref. 32.)

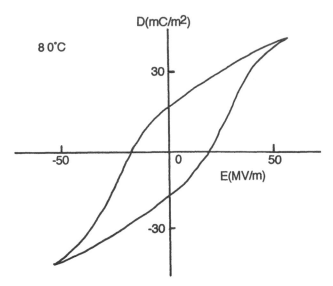

Figure 44 D-E hysteresis loop for polythiourea-9 at 80°C (0.03 Hz). (From Ref. 32.)

vibration, and reorient instantly during the depolarization process. Hydrogen bonding, generally, is stronger than van der Waals force. The possibility of forming a hydrogen-bonded part in an amorphous material is lower than for a crystalline material, though dipole orientation by electrical poling may significantly enhance such possibility. Consequently, the remanent polarization can be stabilized by the hydrogen bonding, and this stabilization should contribute to the ferroelectric state and large dielectric relaxation.

Though only a small amount of chains can form sufficient hydrogen bonding, while there remains a large part of non-hydrogen-bonded chains, that leads to a low value of remanent polarization. The structural control induced by hydrogen bonding in aromatic polyureas gives rise to ferroelectric behavior [18,19]. According to these results, a large amount of hydrogen bonding may contribute to structural stability, so that dipole rotation energy is higher. The molecular design can assist in generating ferroelectric polythioureas if the quality and quantity of hydrogen bonding are controlled by the chemical structure. The electrical and thermal properties of polythioureas are summarized in Table 3.

Table 3 Physical Properties of Polyureas and Polythioureas

Polymer	Dielectric constant 20°C (1 kHz)	T_g (°C)	T_m (°C)	P_r (mC/m²)	p (μC/m² K)
PTUFB	20–35	140	200*	—	—
PTU9	3.5	50	170*	10–80	20–100
PUR 7	3.5	51	242	—	3–10
PUR9	3.5	50	220	30–100	5–20
PU79	3.5	51	180	50–100	5–20
NPUR	3.4	175	300*	4–10	1~3

T_m: melting temperature, *: decomposition temperature.

IV. MELAMINE-THIOUREA POLYMERS

A. Synthesis and Spectroscopic Characterization

Like urea and thiourea, melamine forms polycondensation products with formaldehyde. The basic reaction between an aldehyde and melamine is very similar to other amino compounds such as urea, thiourea, and guanidine. Under proper conditions, the primary reaction results in the attachment of the aldehyde to the nitro of an amino group accompanied by the splitting of water. As discussed earlier, thiourea polymers exhibit very interesting electrical properties, so it was of considerable interest to examine the electrical properties of thiourea, which is a well-known ferroelectric compound when incorporated in a polymer matrix of melamine-thiourea. The condensation polymerization of melamine-thiourea polymer is shown below. A white product can be obtained after stirring melamine, thiourea, and formaldehyde for up to 5–6 h [33].

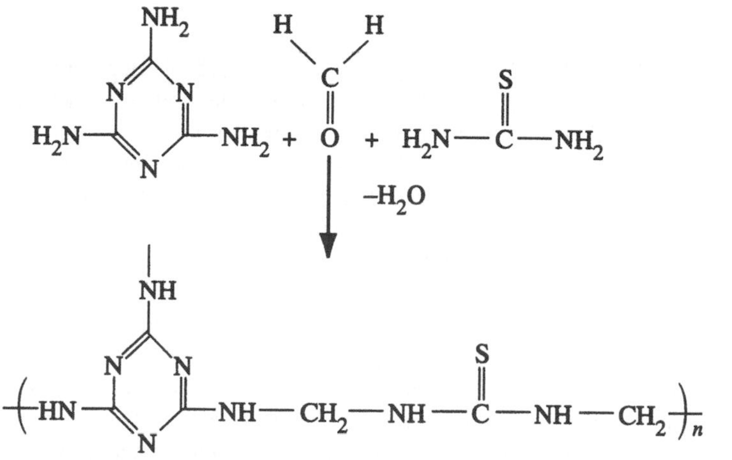

The infrared spectrum of melamine-thiourea-formaldehyde (MTUF) polymer showed a strong band at 813 cm^{-1} and a number of strong bands in the 1250–1670 cm^{-1} region which are characteristics of the triazine ring. The C=S vibrations due to thiourea appeared at 710 and 1540 cm^{-1}. The NH band at 3340 cm^{-1} and methylene bridge bands at 1260 and 1460 cm^{-1} supported the presence of melamine and thiourea units links. The IR spectrum evidenced the existence of the triazine ring, —NH, —N—CH$_2$—N—, and C=S groups [33]. Both thermogravimetry analysis (TGA) and derivative thermogravimetry (DTG) showed thermal transitions at 160 and 370°C, where 160°C should be related to the initial decomposition temperature. Differential thermal analysis (DTA) also showed an endothermic peak at 160–170°C.

B. Electrical Properties

MTUF polymer is an insulator having an electrical resistance (ρ) of 4.78 × 10^{11} Ω cm at room temperature. Temperature dependence study showed a break at 80°C and at 250°C; the ρ was 1.97 × 10^{10} Ω cm. The activation energies calculated from the Arrhenius equation were 0.17 and 0.53 eV for the temperature regions 30–80°C and 80–250°C.

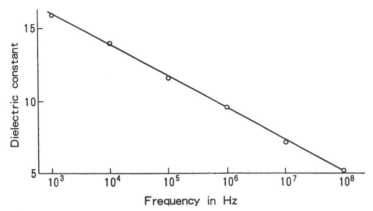

Figure 45 Frequency dependence of dielectric constant for melamine-thiourea-formaldehyde polymer. (From Ref. 33.)

The dielectric constant of the MTUF polymer was between 5 and 16, depending on the measurement frequency (10^3–10^8 Hz) [33]. Figure 45 shows the frequency dependence of dielectric constant for MTUF polymer recorded at room temperature. Upon heating the sample, the dielectric constant increased from 15 to 19 in the temperature region 140–250°C. The dielectric constant of mineral-filled melamine-formaldehyde resin has been reported as 14, 9, and 7.5 at 50 Hz, 1 kHz, and 1 MHz, respectively [34]. The dielectric constant of MTUF polymer is rather higher compared with conventional organic polymers. The TUF polymers show dielectric constant in the range of 10 to 15 for samples prepared in acidic medium, and between 20 and 30 for samples synthesized in basic medium. The slightly higher dielectric constant of MTUF polymer may result from the presence of thiourea moieties in the polymer strand. It can be seen that incorporation of the ferroelectric thiourea moiety in a melamine-formaldehyde resin slightly influence the electrical properties compared with a melamine-formaldehyde polymer. The incorporation of thiourea may be useful in generating novel polymers with unusual electrical properties.

REFERENCES

1. C. P. Vale, *Aminoplastics*, Cleaver-Hume, London, 1950.
2. J. F. Blais, *Amino Resins*, Reinhold, New York, 1959.
3. A. Bachmann and T. Bertz, *Aminoplatics*, VEB Deutsch Verlag, Leipzig, 1970.
4. J. Urbanski, Amino resins, *Handbook of Analysis of Synthetic Polymers and Plastics*, Ellis Horwood, Chichester, 1977.
5. C. I. Chiriac, Polyureas, *Encyclopedia of Polymer Science and Engineering*, John Wiley, New York, vol. 13, p. 212.
6. H. S. Nalwa, K. L. Taneja, U. S. Tewari, and P. Vasudevan, Transitions in thiourea-formaldehyde resins: A study of electrical and thermal properties, *Proc. Nuclear Physics and Solid State Physics Symp.*, 1978, 21C, 712.
7. H. S. Nalwa, P. S. Viswanathan, and P. Vasudevan, Thermally stimulated depolarization currents in thioura formaldehyde polymers, *Angew. Makromol. Chem.* 82:39 (1979).
8. P. Vasudevan, H. S. Nalwa, K. L. Taneje and U. S. Tewari, Pyroelectricity in thiourea formaldehyde polymer, *J. Appl. Phys.* 50:4324 (1979).

9. Y. Takahashi, M. Iijima, K. Inagawa, and A. Itoh, Synthesis of aromatic polyimide film by vacuum deposition polymerization, *J. Vacuum Sci. Technol. A5*:2253 (1987).

10. Y. Takahashi, M. Iijima, Y. Oishi, M. Kakimoto, and Y. Imai, Preparation of ultrathin films of aromatic polyamides and aromatic poly(amide-imides) by vapor deposition polymerization, *Macromolecules 24*:354 (1991).

11. Y. Takahashi, K. Matsuzaki, M. Iijima, E. Fukada, S. Tsukahara, Y. Murakami, and A. Maesono, Determination of evaporation rate and vapor pressure of organic monomers used for vapor deposition polymerization, *Jpn. J. Appl. Phys. 32*:L875 (1993).

12. X. S. Wang, M. Iijima, Y. Takahashi, and E. Fukada, Dependence of piezoelectric activities of aromatic polyurea thin films on monomer composition ratio, *Jpn. J. Appl. Phys. 32*:2768 (1993).

13. Y. Takahashi, S. Ukishima, M. Iijima, and E. Fukada, Piezoelectric properties of thin films of aromatic polyurea prepared by vapor deposition polymerization, *J. Appl. Phys. 70*:6983 (1991).

14. T. Hattori, M. Iijima, Y. Takahashi, E. Fukada, Y. Suzuki, M. Kakimoto, and Y. Imai, Synthesis of aliphatic polyurea film by vapor deposition polymerization and their piezoelectric properties, *Jpn. J. Appl. Phys. 33*:4647 (1994).

15. A. Kawaguchi, T. Ikawa, Y. Fujiwara, M. Tabuchi, and K. Monobe, Polymorphism in lamellar single crystals of nylon 11, *J. Macromol. Sci. Phys.* B20:1 (1981).

16. H. S. Nalwa, T. Watanabe, A. Kakuta, A. Mukoh, and S. Miyata, N-phenylated aromatic polyurea: A new nonlinear optical material with large second harmonic generation and UV transparency, *Polymer 34*:657 (1993).

17. H. S. Nalwa, T. Watanabe, A. Kakuta, A. Mukoh, and S. Miyata, Aromatic polyurea exhibiting large second harmonic generation and optical transparency down to 300 nm, *Appl. Phys. Lett. 62*:3223 (1993).

18. S. Tasaka, K. Ohishi, H. S. Nalwa, T. Watanabe, and S. Miyata, Ferroelectric polarization reversal stabilized by hydrogen bonding in N-phenylated aromatic polyurea, *Polymer J.*, 26: 505 (1994).

19. S. Tasaka, T. Shouko, and N. Inagaki, Ferroelectric polarization reversal in polyurea with odd number of CH_2 groups, *Jpn. J. Appl. Phys. 31*:L1086 (1992).

20. T. Syoko, M. Ohtani, S. Tasaka, and N. Inagaki, Ferroelectric behavior in polyamises and polyureas with odd number of CH_2 groups, *Sen-i Gakkai Symp. Prepr.*, p. B97 (1992).

21. S. Tasaka, T. Shouka, and N. Inagaki, Ferroelectric behavior in aliphatic polyureas, *Jpn. J. Appl. Phys. 33*:1376 (1994).

22. A. L. Solomon, Thiourea, a new ferroelectric, *Phys. Rev. 104*:1191 (1956).

23. G. J. Goldsmith and J. G. White, Ferroelectric behavior in thiourea, *J. Chem. Phys. 31*:1175 (1959).

24. H. S. Nalwa and P. Vasudevan, Electrical properties of thiourea-formaldehyde condensates, *Eur. Polymer J.* 17:145 (1981).

25. H. S. Nalwa and P. Vasudevan, Thermally stimulated depolarization effect in thiourea-formaldehyde condensate, *Polymer 24*:1197 (1983).

26. K. Ohishi, S. Tasaka, and N. Inagaki, Electrical properties of polythioureas, *Polymer Prep. Jpn. 42*:1430 (1993).

27. R. J. Meakins and R. A. Sack, The dielectric properties of symmetrical long-chain secondary alcohols in the solid state, *Aust. J. Sci. Res. A4*:213 (1951).

28. R. A. Sack, The dielectric properties of systems containing straight polar chains, *Aust. J. Sci. Res. A5*:135 (1952).

29. K. Hamano, T. Sugiyama, and H. Sakata, Nature of dielectric anomaly at 161K in Thiourea, *J. Phys. Soc. Jpn. 59*:4476 (1990).

30. H. Futama, Y. Shiozaki, A. Chiba, E. Tanaka, T. Mitsui, and J. Furuichi, Satellite X-ray scattering by thiourea, *Phys. Lett. 25A*:8 (1967).

31. H. S. Nalwa and P. Vasudevan, Pyroelectricity in polymers, 28th Macromolecular Symp. (IUPAC Macro-82), University of Massachusetts, July 12–16, 1982, Preprint no. 435.

32. S. Tasaka, K. Ohishi, and N. Inagaki, Ferroelectric behavior in aliphatic polythioureas, *Ferroelectrics*, (1995) in press.

33. H. S. Nalwa, Electrical properties of melamine-thiourea formaldehyde condensates, *Makromol. Chem. Rapid Commun. 4*:45 (1983).

34. W. J. Roff and J. R. Scott, *Fibers, Films, Plastics, and Rubbers: A Handbook of Common Polymers*, Butterworths, London, 1971.

9
Piezoelectricity and Pyroelectricity of Biopolymers

Eiichi Fukada
Kobayashi Institute of Physical Research, Tokyo, Japan, and Institute for Super Materials,
ULVAC Japan, Ltd., Tsukuba, Japan

I. INTRODUCTION

Pyroelectricity and piezoelectricity of biopolymers were first reported for keratin in 1941 by Martin [1]. When a bundle of wool or hair was immersed in liquid air, an electric voltage of a few volts was generated between the tip and the root. When pressure was applied on the cross section of the bundle, an electric voltage was also generated.

Later, in 1973, Menefee [2] observed a depolarization voltage in the axial direction for a section of porcupine quill at the temperature where the helical conformation of keratin molecules undergoes thermal denaturation.

These observations are explained by the structure of keratin, which is a typical kind of protein. Keratin molecules assume an α-helical form in which the dipoles of CONH peptide bond directly in the same axial direction. A large dipole moment exists in a helical molecule. In many natural structures such as hair and quill, these polar molecules are aligned in parallel with a preferred direction of the polar axis to form the crystalline structure. Therefore, such structures can be regarded as natural electrets, having a polarization in the direction from the root to the tip. Because of this intrinsic polarization, pyroelectricity and piezoelectricity in the axial direction can be observed.

Another popular crystalline structure for protein is the β structure formed by extended protein molecules. A pleated sheet structure is formed by hydrogen bonds between adjacent extended molecules aligned in parallel. Piezoelectricity in β-form protein was first observed by Fukada in 1956 for silk fibroin [3]. Since there is no intrinsic polarization in the β structure, pyroelectricity and tensile piezoelectricity are not observed. However, the piezoelectricity due to shear is observed. If the shear is applied such as to cause slip between oriented molecules, polarization is induced in the direction perpendicular to the plane of the shear. Shear piezoelectricity is observed for most natural biopolymers, not only keratin but also many other proteins.

Shear piezoelectricity for polysaccharides was first investigated in detail for wood cellulose by Bazhenov in 1951 [4]. Shear piezoelectricity for bone and tendon collagen was first reported by Fukada and Yasuda in 1957 [5] and 1964 [6].

Starting from these initial studies, a large number of works have been published on the piezoelectric properties of biopolymers of both natural and synthetic origin. This chapter outlines the work carried out for the shear piezoelectricity of biopolymers as well as optically active polymers.

The bending piezoelectricity of bone was first discovered by Yasuda in 1953 [7]. In connection with the piezoelectric phenomena in bone, it has been found that the growth of bone can be accelerated by applying an electric current. The study of the electric stimulation of bone has developed so extensively that an international society, the Bioelectrical Repair and Growth Society, was founded in 1980 [8].

II. PIEZOELECTRIC MATRIX FOR UNIAXIAL SYMMETRY

Piezoelectricity may exist for certain symmetries of crystalline structures. Most piezoelectric polymers possess a uniaxial symmetry in their semicrystalline textures. Three kinds of symmetry, D_∞, C_∞, and $C_{\infty v}$, were originally proposed by Schubnikov [9] for uniaxially oriented crystalline textures.

Using rectangular coordinates where the 3 axis is the axis of orientation, the three components of polarization, P_1, P_2, P_3, are related to the six components of stress, T_1, T_2, T_3, T_4, T_5, T_6, by the 18 coefficients, d_{ij} ($i = 1, 2, 3, j = 1, 2, 3, 4, 5, 6$), where d_{ij} are called as the piezoelectric strain constants. Some of them are null, according to the symmetry of the material [10]. Similarly, the coefficients relating the polarization P_i to the strain S_{ij} are called the piezoelectric stress constants e_{ij}.

In the past four decades, four kinds of piezoelectric matrix have been found in different polymers as shown below.

1.

D_∞ (∞ 2):

$$\begin{vmatrix} 0 & 0 & 0 & d_{14} & 0 & 0 \\ 0 & 0 & 0 & 0 & -d_{14} & 0 \\ 0 & 0 & 0 & 0 & 0 & 0 \end{vmatrix}$$

2.

C_∞ (∞):

$$\begin{vmatrix} 0 & 0 & 0 & d_{14} & d_{15} & 0 \\ 0 & 0 & 0 & d_{15} & -d_{14} & 0 \\ d_{31} & d_{31} & d_{33} & 0 & 0 & 0 \end{vmatrix}$$

3.

$C_{\infty v}$ (∞ m):

$$\begin{vmatrix} 0 & 0 & 0 & 0 & d_{15} & 0 \\ 0 & 0 & 0 & d_{15} & 0 & 0 \\ d_{31} & d_{31} & d_{33} & 0 & 0 & 0 \end{vmatrix}$$

4.

C_{2v} (**2mm**):

$$\begin{vmatrix} 0 & 0 & 0 & 0 & d_{15} & 0 \\ 0 & 0 & 0 & d_{24} & 0 & 0 \\ d_{31} & d_{32} & d_{33} & 0 & 0 & 0 \end{vmatrix}$$

The symmetry of D_∞ is observed for most natural biopolymers, which show only shear piezoelectricity. The 3 axis is taken as the axis of orientation. In this symmetry the relation $d_{25} = -d_{14}$ holds. The symmetry of C_∞ is found in certain structures of living systems such as bone and tendon. The preferred alignment of the polar axis of polymer crystallites in the 3 axis provides the piezoelectric constants d_{31} and d_{33} as well as the pyroelectric effect in the 3 axis, although they are not so large.

The symmetry of $C_{\infty v}$ holds for the poled composite films consisting of piezoelectric ceramic powders and polymers. When elongated films of polar polymer are poled, the symmetry of C_{2v} appears. The 3 axes for $C_{\infty v}$ and C_{2v} are in the direction of poling. In 1969, Kawai first discovered large tensile piezoelectricity in elongated and poled films of poly(vinylidene fluoride) [11].

III. SHEAR PIEZOELECTRICITY

Shear piezoelectricity is observed almost universally for the oriented textures of biopolymers. It is also observed for the oriented films of a variety of synthetic polymers with optical activity. Rectangular coordinates are assigned to the film sample as shown in Figure 1. The 3 axis is taken in the direction of orientation, and the 1 axis is perpendicular to the plane of the film. The film is cut at a slant, at an angle θ with the 3 axis. The tension applied at $\theta = 45°$ gives a shear stress T_4, which is positive. Then the polarization P_1 or the electric displacement D_1, which is negative for L polymers and positive for D polymers, is induced in the direction of the 1 axis.

Now the apparent piezoelectric constant is defined as $d_0 = D_1/T$, where T is the stress applied at the angle θ with the 3 axis. Figure 2 shows the dependence of d_0 on the angle

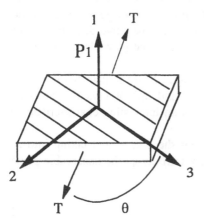

Figure 1 A rectangular coordinate assigned to the sample for the measurement of piezoelectric constant. θ is the angle between the orientation axis and the direction of applied stress. The tensile stress T at the angle θ produces the polarization P_1.

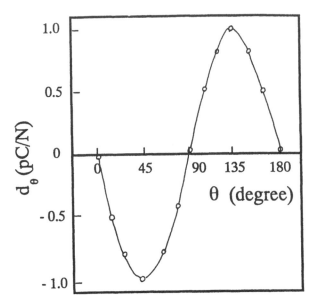

Figure 2 Anisotropy of the piezoelectric effect for an oriented film of collagen. θ is the angle between the orientation axis (the 3 axis) and the direction of applied stress.

θ measured for an elongated film of collagen. The relation of d_0 to θ is derived by the coordinate transformation for the symmetry of D_∞ or C_∞ as follows:

$$d_0 = \frac{d_{14}}{2} \sin 2\theta \tag{1}$$

By fitting the experimental data with the equation, one gets $d_{14} = -2.5$ pC/N for oriented collagen.

Shear piezoelectricity is observed in oriented films of many synthetic polymers. During casting from solution, the polar axes of polymer crystals are oriented at random in the film. Therefore, after elongation of the film, the polar axes of polymer crystallites are distributed with equal probability in both the positive and negative directions of the 3 axis. Thus the constants d_{31} and d_{33} become null, although the piezoelectric constants d_{31}^c and d_{33}^c of crystals may have finite values.

Nevertheless, the constants d_{14} and d_{25} for the elongated film remain finite because the contributions from the piezoelectric constants d_{14}^c and d_{25}^c of crystals oppositely directed in the 3 axis do not cancel each other. This is an interesting characteristics of shear piezoelectricity [12].

Assuming uniform distribution of crystallites about the 3 axis, and taking into account the orientation function F_c and the degree of crystallinity ϕ, the observable piezoelectric constant d_{14} is given by

$$d_{14} = \frac{\phi F_c (d_{14}^c - d_{25}^c)}{2} \tag{2}$$

where $F_c = (3\langle\cos^2\chi\rangle - 1)/2$ and χ is the angle between the 3 axis of a crystallite and the 3 axis of the film [13].

In the following figures showing experimental results, the piezoelectric strain constant is expressed by either d_{14} or d_{25} and the piezoelectric stress constant is expressed by either e_{14} or e_{25}. These have the same magnitude but opposite sign. In the explanations, for the sake of simplicity, d and e will be often used instead of $-d_{14} = d_{25}$ and $-e_{14} = e_{25}$. The unit generally used for d is pC/N, where 1 pC/N = 3 × 10^{-8} cgs esu and the unit generally used for e is mC/m², where 1 mC/m² = 3 × 10^2 cgs esu.

IV. PIEZOELECTRIC RELAXATION

A. Complex Piezoelectric Constant

There has been a great deal of work on elastic and dielectric properties of polymers. An important characteristic of elastic and dielectric properties of polymers is that relaxational phenomena take place with change of temperature and frequency (or time scale) of measurement. Therefore, the elastic and dielectric constants determined by dynamic measurements are represented by complex quantities as $c = c' + ic''$ and $\epsilon = \epsilon' - i\epsilon''$.

It was first found in 1968 that the piezoelectric constant also exhibits relaxation and should be represented as a complex quantity such as $d = d' - id''$ [14]. Other piezoelectric constants are also represented by complex quantities as $e = e' - ie''$, $g = g' - ig''$, and $h = h' - ih''$, where relations $e = dc$, $g = d/\epsilon$, and $h = e/\epsilon$ hold.

For a system with a single relaxation time, the elastic, dielectric, and piezoelectric relaxations are represented by the following equations:

$$c = c_\infty + \frac{c_0 - c_\infty}{1 + i\omega\tau} \qquad \Delta c = c_0 - c_\infty < 0 \tag{3}$$

$$\epsilon = \epsilon_\infty + \frac{\epsilon_0 - \epsilon_\infty}{1 + i\omega\tau} \qquad \Delta\epsilon = \epsilon_0 - \epsilon_\infty > 0 \tag{4}$$

$$d = d_\infty + \frac{d_0 - d_\infty}{1 + i\omega\tau} \qquad \Delta d = d_0 - d_\infty \lesseqgtr 0 \tag{5}$$

where $\tau = \tau_0 \exp(\Delta E/kT)$ is the relaxation time, τ_0 is the frequency factor, ΔE is the activation energy, k is Boltzmann's constant, and T is the temperature.

For elastic and dielectric relaxations, the relations $\Delta c < 0$ and $\Delta\epsilon > 0$ always hold. However, for piezoelectric relaxations, experimental observations have shown that Δd can be either positive or negative. In one case, d' increases with decreasing frequency or increasing temperature ($\Delta d > 0$) and the phase of polarization lags behind the applied elastic stress (tan $\delta = d''/d' > 0$). In the other case, d' decreases with decreasing frequency and increasing temperature ($\Delta d < 0$) and the phase of the polarization leads beyond the applied elastic stress (tan $\delta = d''/d' < 0$). This is a unique feature of the piezoelectric relaxation. One example will be shown for wood containing moisture.

The piezoelectric constant may consist of a nonrelaxing component and a relaxing component. The signs of these two are not necessarily the same. As a result, the reversal of the sign of the observable piezoelectric constant takes place while changing the measuring frequency or temperature. Examples will be shown for cellulose triacetate and amylose.

B. Two-Phase Model

Most piezoelectric polymers are semicrystalline materials and can be regarded as a texture consisting of crystalline regions and noncrystalline (amorphous) regions. As the thermal motions of molecules are activated at a certain temperature such as the glass temperature, the relaxations of both elastic and dielectric properties may take place in amorphous regions but not in crystalline regions, while the piezoelectric effects can be present only in crystalline regions and not in completely amorphous regions. It should be added that the well-oriented amorphous regions are also able to display shear piezoelectricity, as will be shown later.

To understand the piezoelectric relaxations, one must consider the composite structure of polymers consisting of piezoelectric crystalline regions and nonpiezoelectric noncrystalline regions. The simplest model for such composite structures is a spherical dispersion system as shown in Figure 3.

We assume that spheres with elastic constant c_2 and dielectric constant ϵ_2 and piezoelectric constants e_2 and d_2 are dispersed uniformly in a medium with elastic constant c_1 and dielectric constant ϵ_1.

The elastic, dielectric, and piezoelectric constants for the system, c, ϵ, d, and e, are theoretically derived as a first approximation as follows [15]:

$$c = c_1 \frac{3c_1 + 2c_2 - 3\phi(c_1 - c_2)}{3c_1 + 2c_2 + 2\phi(c_1 - c_2)} \tag{6}$$

$$\epsilon = \epsilon_1 \frac{2\epsilon_1 + \epsilon_2 - 2\phi(\epsilon_1 - \epsilon_2)}{2\epsilon_1 + \epsilon_2 + \phi(\epsilon_1 - \epsilon_2)} \tag{7}$$

$$e = \phi e_2 \frac{5c_1}{3c_1 + 2c_2 + 2\phi(c_1 - c_2)} \frac{3\epsilon_1}{2\epsilon_1 + \epsilon_2 - \phi(\epsilon_1 - \epsilon_2)} \tag{8}$$

$$d = \phi d_2 \frac{5c_2}{3c_1 + 2c_2 - 3\phi(c_1 - c_2)} \frac{3\epsilon_1}{2\epsilon_1 + \epsilon_2 - \phi(\epsilon_1 - \epsilon_2)} \tag{9}$$

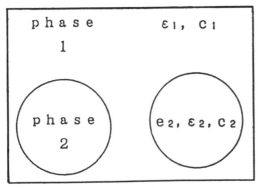

Figure 3 A spherical dispersion model. The spherical crystallites (phase 2) with piezoelectric constant e_2, elastic constant c_2, and dielectric constant ϵ_2 are dispersed uniformly in the amorphous medium (phase 1) with elastic constant c_1 and dielectric constant ϵ_1.

where ϕ is the volume fraction of spheres or the degree of crystallinity. If ϕ is small, Eqs. (8) and (9) are reduced to

$$e = \phi e_2 \frac{5c_1}{3c_1 + 2c_2} \frac{3\epsilon_1}{2\epsilon_1 + \epsilon_2} \qquad (10)$$

$$d = \phi d_2 \frac{5c_2}{3c_1 + 2c_2} \frac{3\epsilon_1}{2\epsilon_1 + \epsilon_2} \qquad (11)$$

We may assume that the crystalline properties c_2 and ϵ_2 remain constant with increasing temperature. At the glass transition temperature, the amorphous properties c_1 and ϵ_1 change with increasing temperature. The increase of ϵ_1 results in the increase of both e and d, but the increase of c_1 results in the increase of e and the decrease of d. Various examples of piezoelectric relaxation will be described later.

V. PIEZOELECTRICITY OF BIOPOLYMERS

A. Wood Cellulose

In the following description, the piezoelectric constant d indicates the shear piezoelectric constant $-d_{14}$ or d_{25}. For symmetry C_∞ and D_∞, the relation $-d_{14} = d_{25} = d$ holds. In most cases, a sinusoidal stress at 10 Hz is given to the sample and both in-phase component and $\pi/2$ out-of-phase component of the resulting sinusoidal polarization are detected. The ratio of polarization to stress is the complex piezoelectric strain constant $d = d' - id''$, and the ratio of polarization to strain is the complex stress constant $e = e' - ie''$.

Figure 4 shows the absolute value of the complex piezoelectric constant $d = d' - id''$ and the loss factor $\tan \delta = d''/d'$, where $d = -d_{14} = d_{25}$, for Japanese cypress (Hinoki) treated with liquid ammonia against increasing temperature. For the vacuum-dried sample, the absolute value of d gradually increases with rising temperature and $\tan \delta$ takes almost the same positive values against temperature. For the air-dried sample containing about 8% moisture, the absolute value of d decreases above 0°C because of the increased conductivity due to the unfrozen water. It increases again with the evaporation of water. The rise of ionic conductivity neutralizes the polarization produced in crystallites of cellulose. When the d constant decreases, $\tan \delta$ takes negative values. Obviously, the principle of time–temperature equivalence holds for the phenomenon [16].

B. Cellulose Derivatives

Chemical treatment of cellulose produces a change in the piezoelectric properties of cellulose. When wood samples were treated with ethylenediamine, sodium hydroxide, and liquid ammonia, the piezoelectric constant attained values two to four times that of untreated sample. It is believed that the crystalline structure is changed by such hydrogen bond-breaking chemicals [17].

Acetylation of cellulose produces significant effects on piezoelectric properties. Cellulose diacetate is an almost amorphous polymer. As shown in Figure 5, the real part of the piezoelectric constant gradually increases with increasing temperature. Two relaxation peaks are seen for the imaginary part of the piezoelectric constant, at −80°C and 80°C, suggesting thermal molecular motions in amorphous regions [14].

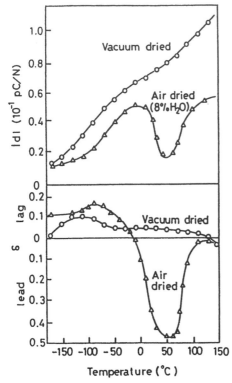

Figure 4   Temperature dependence of piezoelectric constant d and phase angle δ_d for Japanese cypress (Hinoki) treated by liquid ammonia. (From Ref. 16.)

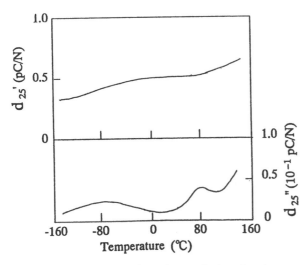

Figure 5   Temperature dependence of piezoelectric constant for cellulose diacetate. (From Ref. 14.)

Cellulose triacetate is a semicrystalline polymer. As shown in Figure 6, d' for the dry state increases with increasing temperature and changes its sign at 38°C [18]. The temperature for the sign inversion is lowered to −60°C for a wet sample with a water content of 2.4%. X-ray diffraction studies showed that water molecules are included inside the crystalline phase and expand the lattice spacing. It is interesting to note that d' of cellulose triacetate crystals is negative at low temperatures such as −150°C, but positive at high temperatures such as 150°C. The reason is not yet clear. It is suspected that the piezoelectric contributions from main chains and from side chains are different in sign.

C. Chitin

Chitin is the major biopolymer of exoskeletons of invertebrates such as crabs and lobsters and a β(1→4)-linked polymer of N-acetyl-D-glucosamine. Figure 7 shows the temperature dependence of d' and d'' for two chitin samples, dried and with 19% moisture content [19]. For the dried state, d' increases gradually with increasing temperature. The value of d'' is a maximum at −100°C, reflecting local thermal motions of chitin molecules in the amorphous phase. The presence of moisture causes the decrease of d' and a negative maximum of d'' above −50°C. The behavior is similar to that of wood cellulose.

D. Amylose

The temperature dependence of the piezoelectric constants e and d, the elastic constant c, and the dielectric constant ϵ at 10 Hz for an elongated film of amylose in the dry state is shown in Figure 8 [20]. Here all imaginary parts show a maximum at about −80°C, indicating a relaxation phenomenon common for piezoelectric, elastic, and dielectric properties. The origin of these relaxations is interpreted as being due to the rotation of a methylol (CH_2OH) group attached to a glucose ring.

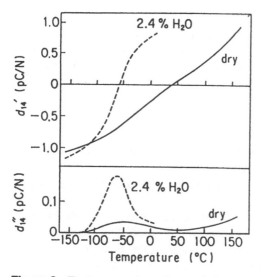

Figure 6 Temperature dependence of piezoelectric constant for cellulose triacetate. (From Ref. 18.)

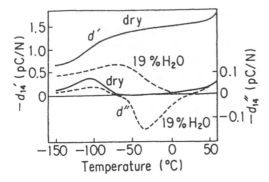

Figure 7 Temperature dependence of piezoelectric constant for demineralized lobster apodeme (chitin) with different hydrations. (From Ref. 19.)

Apparently, the piezoelectric constant d or e consists of two components, one negative and nonrelaxing and the other positive and relaxing at about $-80°C$. It is suggested that the nonrelaxing polarization is caused by dipolar displacement accompanied by an instantaneous shift of the potential curve acting on the dipoles, and that the relaxing polarization is caused by the dipolar rotation thermally activated over a potential barrier.

It is of interest that the nonrelaxing component of the piezoelectric constant is negative at low temperatures and positive at high temperatures, similar to cellulose triacetate.

E. Bone and Collagen

Extensive work on piezoelectric properties of bone have been carried out [21]. Figure 9 shows the temperature dependence of d' and d'' for bone with different water contents [22]. Bone consists of oriented collagen fibers embedded in a matrix of hydroxyapatite. Figure 10 shows the temperature dependence of d' and d'' for decalcified bone, which consists only of oriented collagen fibers.

The value of d' for dry bone is almost constant with increasing temperature. Hydration increases d' as well as the elastic constant c' at low temperature such as $-150°C$. It is presumed that absorbed water forms interchain hydrogen bonds inside the triple helixes of collagen and increases the crystallinity. However, absorbed water higher than 8% in collagen appears to expand the distance between triple helixes and decrease the value of d' for decalcified bone as seen in Figure 10.

Relaxation of the d constant around $-100°C$ is clearly observed for wet bone but obscure for dehydrated bone. The elastic and dielectric relaxations are also obviously observed at the same temperature range both for bone and dehydrated bone [23]. It is presumed that this relaxation is caused by the local mode oscillation of the collagen main chains.

Since bone consists of piezoelectric collagen fibers and nonpiezoelectric hydroxyapatite, Eq. (9) for a two-phase model can be qualitatively applied. The increment of the dielectric constant for dehydrated bone (collagen) is much greater than that for bone. It is seen in Eq. (9) that the increase of ϵ_2 brings about the decrease of d. It is presumed that the piezoelectric relaxation around $-100°C$ for bone is caused mainly by the dielectric relaxation in collagen.

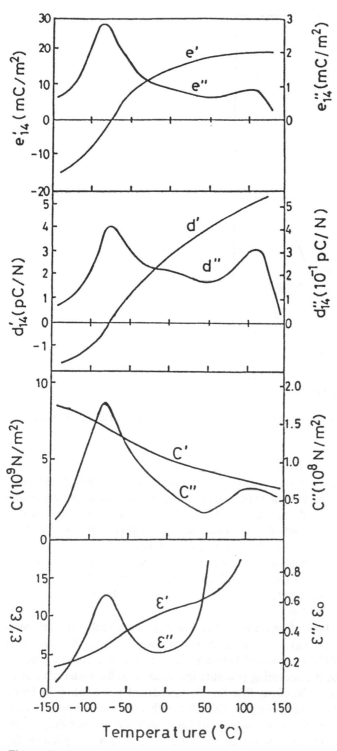

Figure 8 Temperature dependence of piezoelectric, elastic, and dielectric constants for amylose. (From Ref. 20.)

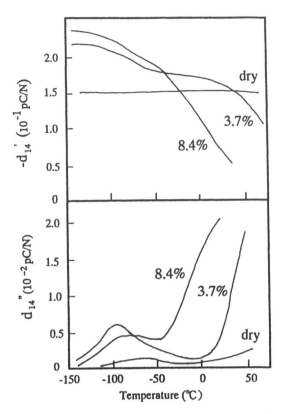

Figure 9 Temperature dependence of piezoelectric constant for bone with different water contents. (From Ref. 22.)

The decrease of constant d above about 0°C for both bone and dehydrated bone appears to be caused by the increase of ionic conductivity, which neutralizes the piezoelectric polarization in collagen fibers. Figure 11 shows the dependence of constant d at room temperature on the water content. The water content of bone saturates at about 10%, owing to the dense structure embedded in a matrix of hydroxyapatite.

Piezoelectricity in bone by bending stress was also investigated quantitatively [24]. The bending piezoelectric constant f is the coefficient of polarization P_1 related to the stress derivative dT_3/dx_1. It was reported that $f = 2 \times 10^{-18}$ Cm/N for dried bone and $f = 2 \times 10^{-17}$ Cm/N for dried tendon.

If wet bone is bent, the flow of liquid takes place through canaliculi from the compressed region to the expanded region. Since the flow is accompanied by the movement of ions, the streaming potential is observed between the concave and convex parts of bent bone. The magnitude of the streaming potential increases with the water content of bone. It is generally accepted, therefore, that the stress-induced polarization in living bone is generated mostly by the streaming potential effect rather than by the piezoelectric effect in collagen fibers. Since the water content of bone may vary depending on the location, and the distribution of stress and strain in living bone is very complicated, the separation of the piezoelectric effect and streaming potential effect in living bone appears to be difficult.

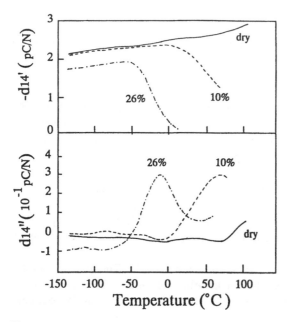

Figure 10 Temperature dependence of piezoelectric constant for decalcified bone (collagen) with different water contents. (From Ref. 22.)

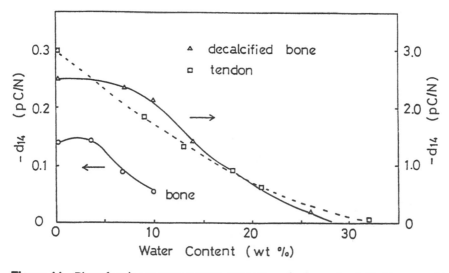

Figure 11 Piezoelectric constant at room temperature for bone, decalcified bone, and tendon as a function of water content.

F. Keratin

Molecules of keratin form an α-helical conformation with a large dipole moment. In keratinous structures, α-helical molecules are assembled in parallel. Axial polarity remains to a certain degree because the direction of the polar axis of helixes is not statistically uniform. Meneffee [2] observed the thermally stimulated current in the axial direction for porcupine quill at the denaturation temperatures and estimated that about 40% of keratin molecules are oriented with the same polarity. Since the current flows from the tip to the root in an external circuit, the direction of dipole of peptide groups (CONH) in helix is such that oxygen atoms direct toward the tip and hydrogen atoms toward the root.

Figure 12 shows the thermally stimulated current and the piezoelectric d constant observed for bovine horn during heating at a rate of 2°C/min [25]. The depolarization current reverses sign at about 130°C. The sense of current below 130°C is axially toward the root and radially outward, and above 130°C both components reverse direction, showing a two-step denaturation process.

The piezoelectric constant d shows a maximum at about 175°C. The gradual increase of d' is ascribed to the increasing flexibility of keratin molecules. The decrease of d' above 175°C is due to the denaturation of keratin. Once the denaturation is complete, piezoelectricity is no longer observed.

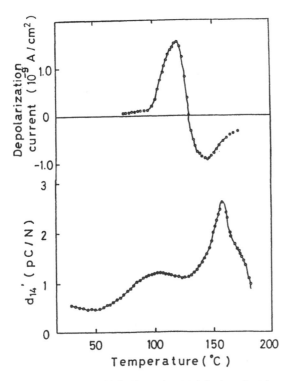

Figure 12 Depolarization current and piezoelectric constant measured for bovine horn (keratin) during heating at a rate of 2°C/min. (From Ref. 25.)

Figure 13 shows the temperature dependence of the d constant for rhinoceros horn keratin with water contents of about 2 and 12 wt%. The value of d' increases gradually with increasing temperature and decreases above about 0°C due to the conductivity of unfrozen water.

G. Various Proteins

Animal organs are composed of a variety of proteins. It appears that most fibrous proteins in the oriented state display piezoelectric effects. Figure 14 shows the anisotropy of piezoelectric effect observed for swine aorta [26]. The blood vessels were cut longitudinally along the tubular axis and elongated approximately 50% of the natural length either longitudinally (sample L) or transversally (sample T). The vessels were then dehydrated under tension by soaking in alcohol for several weeks.

The horizontal coordinate θ in Figure 14 indicates the angle between the tubular axis and the direction of applied stress. The results for sample L accord with Eq. (1), and the anisotropy of d is similar to that of previously described proteins such as collagen and keratin. The results for sample T show the opposite sign of d, because the orientation of molecules lies in the transverse direction.

Similar anisotropy of piezoelectric effect depending on the direction of elongation was observed for vena cava. Ligament and intestine showed the same anisotropy as aorta L, and trachea showed the same anisotropy as aorta T, in accordance with the direction of orientation of protein molecules. The proteins constituting blood vessel walls are collagen, elastin, and muscle proteins.

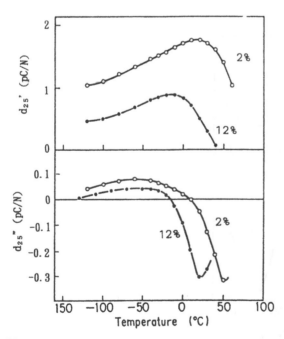

Figure 13 Temperature dependence of piezoelectric constant for rhinoceros horn (keratin) with different water contents.

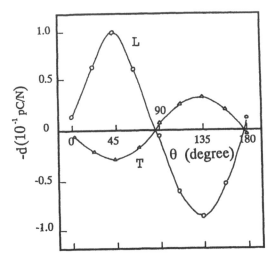

Figure 14 Piezoelectric anisotropy for swine aorta elongated longitudinally (*L*) or transversally (*T*). (From Ref. 26.)

Figure 15 shows the temperature dependence of the piezoelectric constant *d* for Psoas muscle with a few percent water content [27]. The decrease of *d′* above 0°C should be due to the conductivity of adsorbed water. The oriented films of myosin and of actin extracted from muscle showed similar behavior [28].

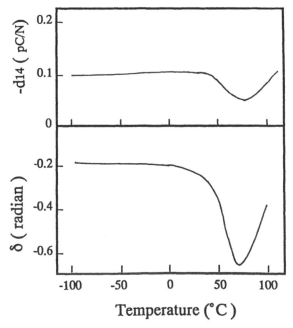

Figure 15 Temperature dependence of piezoelectric constant for Psoas muscle. (From Ref. 27.)

Silk fibroin, which is the main constituent of silk fibers, has a β-form molecular conformation. A sheet structure is formed by hydrogen bonds between adjoining molecules. Oriented films of silk fibroin consist of β-form crystals. The temperature dependence of the d constant is similar to that for dry collagen film.

Fibrin clot was prepared by adding thrombin to a buffer solution of fibrinogen. By stretching the dehydrated clot, oriented films of fibrin were formed. Fibrin contains both α-helix and β-form structures. The temperature dependence of the d constant obtained for oriented fibrin films is similar to that for collagen [29].

H. Deoxyribonucleic Acid

Piezoelectricity of DNA was reported by Duchesne [30]. It is known that the molecular conformation of sodium salt of deoxyribonuclic acid (DNA) is double-helix with a regular stacking of purine–pyrimidine base pairs above 75% relative humidity, but it becomes disordered state below 55% relative humidity [31].

Figure 16 shows the temperature dependence of the d constant for oriented films of NaDNA equilibrated with different water contents [32]. When the water content is less than about 16%, the negative value of d' increases and passes through a maximum with increasing temperature, similar to the proteins. In the low-hydration state, DNA molecules assume a denatured conformation. It is presumed that the piezoelectric polarization is caused by the internal rotation of dipoles in the backbone chains.

When the water content is higher than about 50%, the sign of d' is positive and decreases markedly at about −75°C. In the high-hydration state, water molecules are

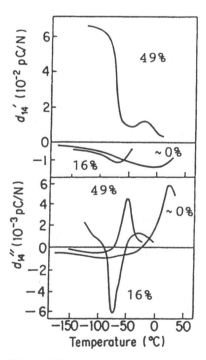

Figure 16 Temperature dependence of piezoelectric constant for NaDNA with different water contents. (From Ref. 32.)

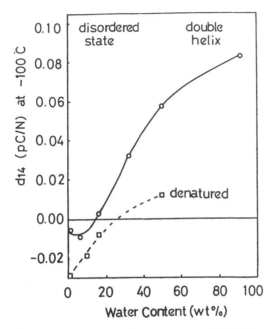

Figure 17 Piezoelectric constant at $-100°C$ for oriented films of DNA as a function of water content.

adsorbed to bases and stabilize the double-stranded helix structure of DNA molecules. In this crystalline state of DNA, it is presumed that the internal rotation of dipoles in purine and pyrimidine bases also contributes to the piezoelectric polarization. It is assumed that the stress-induced polarization in the main chains and that in the side chains possess opposite sign.

A marked relaxation of dielectric constant is observed at $-75°C$, owing to the thermal motion of bases in the double-helix molecules. The decrease of d' at $-75°C$ should be associated with the relaxation of the dielectric constant. Presumably the piezoelectric polarizations in double-helix molecules are shielded by the increase of the dielectric constant of the molecules themselves.

Figure 17 shows the piezoelectric constant d at $-100°C$ as a function of water content. The results for an oriented film of denatured NaDNA are also shown. The double-helix film with low water content shows properties similar to those of the denatured film. The ordered structure of base pairs in double-helix film with high water content appears to contribute to the positive value of the constant d. The sign of the piezoelectric constant for oriented DNA molecules without the ordered double-helix structure is negative, similar to proteins.

VI. PIEZOELECTRICITY OF POLYPEPTIDES

A. Poly-γ-methylglutamate

Protein molecules are poly(amino acids) consisting of a variety of amino acid residues. The studies on piezoelectricity of proteins naturally lead to work on synthetic poly(amino acids), particularly homopolymers of a single amino acid residue. Poly-γ-methyl-L-glu-tamate (PMLG) is one of the popular synthetic polypeptides.

Figure 18 shows the temperature dependence of the piezoelectric constants e and d for an oriented film of PMLG determined at 20 Hz by a dynamic method [33]. The signs of e and d are negative, as in proteins. Both e' and d' show a maximum at about 0°C. When e' and d' increase with rising temperature, e'' and d'' are positive, indicating that the phase of polarization lags behind those of applied strain and stress. When e' and d' decrease with rising temperature, e'' and d'' are negative, indicating that the phase of polarization leads those of applied strain and stress. The temperature–time equivalence principle clearly holds. Similar changes are observed at about 100°C.

The elastic and dielectric relaxations are also observed at about 0°C and 100°C. The former is ascribed to the thermal activation of molecular motions of side chains of methyl glutamate, and the latter is ascribed to the thermal motion of helical main chains.

The piezoelectric polarization in helical molecules of polypeptides is produced by the internal rotation of peptide bonds (NHCO) under the action of shearing force. The side chains of methyl glutamate are attached to an asymmetric carbon atom connected to peptide bonds. In the crystalline structure of pseudo-hexagonal symmetry, the helical main chains are aligned in parallel in a continuous medium of methyl glutamate side

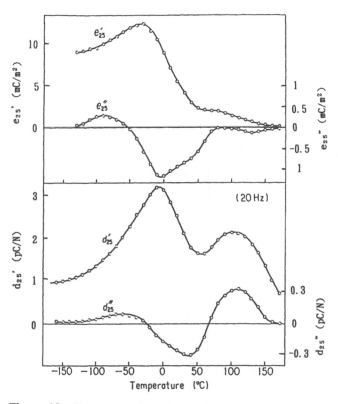

Figure 18 Temperature dependence of piezoelectric constants e and d for poly-γ-methyl-L-glu-tamate. (From Ref. 33.)

chains. The thermal motion of side chains can be activated even in such crystallites as well as in noncrystalline regions.

A qualitative interpretation for the peculier retardation followed by relaxation behavior shown in Figure 18 is as follows. The piezoelectricity is caused by the internal rotation of peptide dipoles attached to the asymmetric carbon atoms, to which are connected the side chains. If the thermal motion of the side chains is activated, elastic relaxation takes place and the constraints to the CONH dipoles decrease. As a result, the stress-induced rotation of dipoles is increased and the piezoelectric constant is increased. The thermal motion of the side chains also induces dielectric relaxation for CONH dipoles. As a result, the stress-induced polarization is shielded and the piezoelectric constant is decreased. More detailed explanation of the mechanism taking into account the higher-order structure of polypeptides will be given later.

The sign of the piezoelectric constant d_{14} is negative for the L polymer and positive for the D polymer. Figure 19 depicts the temperature dependence of d_{14} for poly-γ-methyl-D-glutamate (PMDG) with α-helix and β-form conformations [34]. The side-chain relaxation around 0°C is not observed for the β-form sample. The temperature dependence of d_{14} for the D polymer is similar to that for the L polymer except for the reversal of sign.

B. Poly-γ-benzyl-L-glutamate

Oriented films of poly-γ-benzyl-L-glutamate (PBLG) display piezoelectric properties similar to those of PMLG. Highly oriented films are prepared by means of magnetic ori-

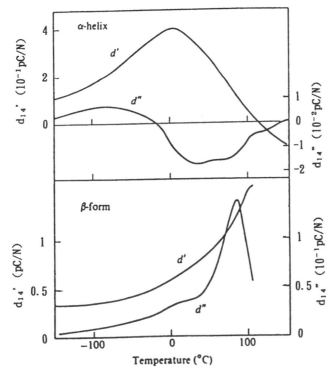

Figure 19 Temperature dependence of piezoelectric constant for oriented films of poly-γ-methyl-D-glutamate with α-helix and β-form conformation, respectively. (From Ref. 34.)

entation [35]. PBLG is dissolved in methylene bromide (CH_2Br_2) at concentrations from 6 to 30 wt% and left in a glass vessel for a week until PBLG molecules form a liquid crystalline state. Then the glass vessel is placed between the two poles of a permanent magnet. Under a magnetic field of 7000 gauss, the solvent is evaporated very slowly for a period of 1 or 2 days. Casted films are finally evacuated for 12 h at 120°C.

The temperature dependence of the piezoelectric constant for magnetically oriented films of PBLG is shown in Figure 20 [36]. When the initial concentration of PBLG in CH_2Br_2 is 15%, the highest piezoelectric constant as well as the highest crystallinity and orientation are obtained. Relaxation due to side-chain motions at about 25°C and the relaxation due to main-chain motions at about 100°C are seen. Contrary to the mechanically oriented films, the latter is more significant than the former. The degree of crystallinity is about 55%, and the degree of orientation is about 85%. The peak value of d_{25}' at 100°C is about 7 pC/N. It is supposed that, due to the interaction of benzene rings in the side chain with the strong magnetic field, α-helical main chains are well oriented, with their axes oriented along the field.

It is also reported that PBLG dissolved in CH_2Br_2 with a concentration above 6% is well oriented by applying an electric field of about 8 kV/m [37]. It is presumed that bundles of PBLG molecules forming liquid crystal possess a polar axis. The senses of the dipole moment of α-helical molecules are not statistically random, but preferentially oriented in one direction.

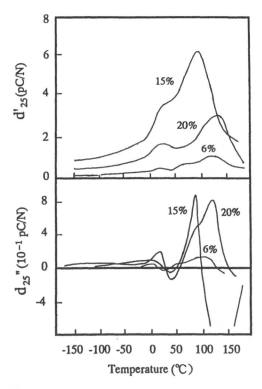

Figure 20 Temperature dependence of piezoelectric constant for magnetically oriented films of poly-γ-benzyl-L-glutamate. (From Ref. 35.)

If the sense of the dipole moment of α helixes could be oriented in the same direction, there would be a possibility to obtain a dipolar electret with a huge remanent polarization, which should exhibit very large pyroelectricity as well as both tensile and shear piezoelectricity.

Figure 21 shows the temperature dependence of the piezoelectric constants e and d, the elastic constant c, and the dielectric constant ϵ, measured at 10 Hz, for a mechanically oriented film of PBLG [13]. The influence of side-chain relaxation is clearly seen in all the constants.

Most significant is the temperature dependence of the d constant. With increasing temperature, d' increases, passes through a peak at about 25°C, and decreases. When d' increases, d'' shows a negative peak, indicating that polarization lags in phase behind stress. When d' decreases, d'' shows a positive peak, indicating that polarization leads in phase beyond stress. The time–temperature equivalence holds similar to the elastic retardation and the dielectric relaxation. The occurrence of retardation followed by relaxation is characteristic of the temperature dependence of the d constant of PBLG.

Figure 22 shows the frequency dependence of the piezoelectric constants d and e measured at different temperatures for oriented film of PBLG. The frequency was varied from 0.1 to 30 Hz at temperatures between 0 and 50°C. The curves were shifted horizontally in the frequency axis. Adding small vertical shifts, master curves of d and e

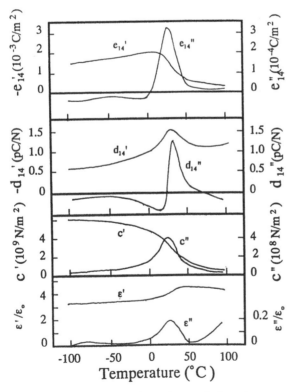

Figure 21   Temperature dependence of piezoelectric constants d and e, elastic constant c, and dielectric constant ϵ for poly-γ-benzyl-L-glutamate. (From Ref. 13.)

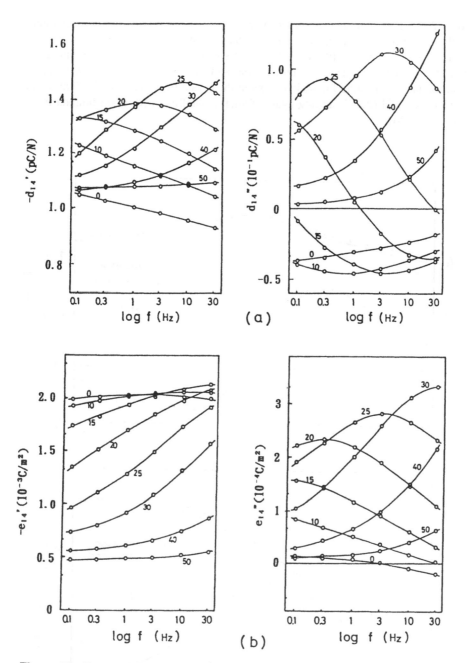

Figure 22 Frequency dependence of (a) $d_{14} = d'_{14} - id''_{14}$ and (b) $e_{14} = e'_{14} - ie''_{14}$ for poly-γ-benzyl-L-glutamate. (From Ref. 13.)

reduced to 20°C were obtained as shown in Figure 23. Both retardation and relaxation exist in the master curve of d in the frequency range from 10^{-4} to 10^4 Hz.

The chemical structure of PBLG molecules is shown in Figure 24a. The main chains of PBLG assume a rodlike helical conformation, and the side chains attached to asymmetric carbon atoms surround the main chains in the crystallites as shown in Figure 24b. The oriented film of PBLG is an aggregate of such crystallites with grain boundaries as shown in Figure 24c.

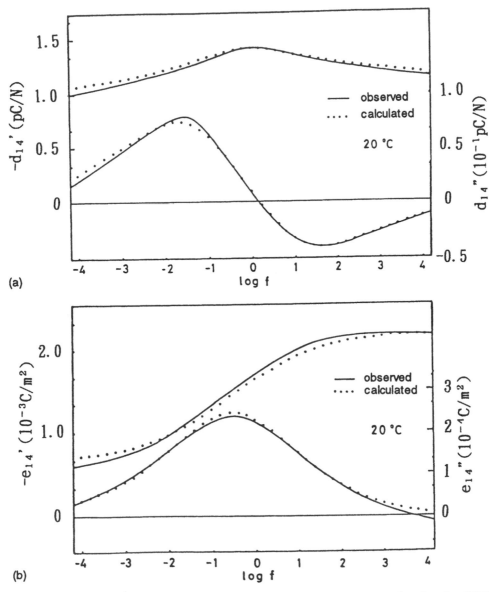

Figure 23 Frequency dependence of (a) $d_{14} = d'_{14} - id''_{14}$ and (b) $e_{14} = e'_{14} - ie''_{14}$ reduced to 20°C for poly-γ-benzyl-L-glutamate. (From Ref. 13.)

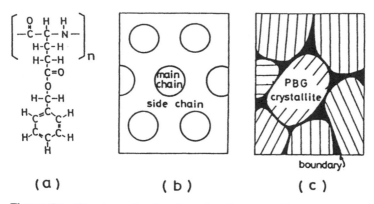

Figure 24 Structure of poly-γ-benzyl-L-glutamate: (a) molecular structure; (b) structure of a crystal; (c) structure of an oriented film. (From Ref. 13.)

To interprete the frequency dependence of d and e in terms of the side-chain relaxation, Furukawa [13] proposed an extended three-element model as shown in Figure 25. Springs a and c, with spring constants G_a and G_c, represent the boundary regions of crystallites. Spring b, with a spring constant G_b, represents a crystallite of PBLG. Spring b is comprised of springs 1 and 3 with spring constants G_1 and G_3, representing the side chain, and spring 2, with a spring constant G_2, representing the main chain. It is assumed

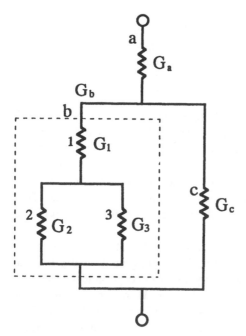

Figure 25 Extended three-element model to express the higher-order structure of an oriented film of polypeptide. (From Ref. 13.)

that spring 2 is piezoelectric and nonrelaxing and springs 1 and 3 are nonpiezoelectric and relaxing. It is also assumed that springs a and c are nonpiezoelectric and nonrelaxing.

The spring constant G of the model as a whole is

$$G = \frac{G_a(G_b + G_c)}{G_a + G_b + G_c} \tag{12}$$

where

$$G_b = \frac{G_1(G_2 + G_3)}{G_1 + G_2 + G_3} \tag{13}$$

The ratio of the local force F_2 in spring 2 to the applied force F is

$$\frac{F_2}{F} = \frac{G_2}{G_2 + G_3} \frac{G_b}{G_b + G_c} \tag{14}$$

The ratio of the local deformation Δl_2 in spring 2 to the applied deformation Δl is

$$\frac{\Delta l_2}{\Delta l} = \frac{G_1}{G_1 + G_2 + G_3} \frac{G_a}{G_a + G_b + G_c} \tag{15}$$

Assuming a relaxation function with a distribution of relaxation time for the side-chain relaxation and assuming appropriate parameters, these equations were calculated as a function of frequency. The dotted curves in Figure 23 are the results calculated, which show a good fit to the experimental results [13].

When temperature or time of measurement increases, the decrease of G_3 causes an increase of the first term of Eq. (14). The decrease of G_1 and G_3 causes the decrease of G_b, which in turn causes the decrease of the second term of Eq. (14). Thus the increase followed by the decrease of the d constant with increasing temperature or time of measurement is realized. In this interpretation it is assumed that no relaxation takes place in G_2, G_a, and G_c. The influences of the dielectric relaxation are ignored because its relaxation strength is one order smaller than that of the elastic relaxation.

C. Miscellaneous Polypeptides

A few kinds of synthetic polypeptides other than PMLG and PBLG have been investigated for their piezoelectric properties. Table 1 shows the results of typical polypeptides [38]. These polymers were cast from solution and elongated by rolling or stretching. Polypeptides may assume three different conformations; α helix, β form, and random coil. Piezoelectricity is not observed for the films with random coil conformations. The piezoelectric effect was observed for both α and β conformations in PMLG. The effect was also observed for poly-β-benzyl-L-aspartate with an ω-helix conformation, which is slightly different from the α-helix one.

Polypeptides are assigned the L or D prefix according to the atomic configuration about an asymmetric carbon atom in the main chain. The coiling of the helix is right-handed for the L polymer and left-handed for the D polymer. An exception is poly-benzyl-L-aspartate, where the helix winds in the left-handed direction. However, the sign of the piezoelectric constant for this polymer is the same as those of other L polymers. Comparison of L and D polymers, both with α-helix conformation, such as those of PMLG, shows a reversal of the sign of the piezoelectric constant. These results indicate that the sign of the piezoelectric constant is determined not by the sense of winding of helix, but

Table 1 Piezoelectric Strain Constant for Synthetic Polypeptides at Room Temperature

Polymer	Molecular conformation	Oriented by	Elongation ratio	d_{14} (pC/N)
Poly-L-alanine	α	Roll	1.5	−1
Poly-γ-methyl-L-glutamate	α	Stretch	2	−2
Poly-γ-methyl-L-glutamate	β	Roll	2	−0.5
Poly-γ-methyl-D-glutamate	α	Stretch	2	1.4
Poly-γ-benzyl-L-glutamate	α	Magnetic field		−4
Poly-β-benzyl-L-aspartate	α	Roll	2	−0.3
Poly-β-benzyl-L-aspartate	ω	Roll	2	−3
Poly-γ-ethyl-D-glutamate	α	Roll	2	0.6

Source: Ref. 38.

by the atomic configuration about the asymmetric carbon atoms. The piezoelectric effect is closely associated with the optical activity. These observations support the concept that the piezoelectricity in polypeptides is caused by a stress-induced rotation of dipoles around asymmetric carbon atoms such as NHCO groups.

Since proteins are copolymers of a number of different amino acids, the study of piezoelectricity in synthetic copolypeptides seems to be very interesting. Published so far is work on copolymers of benzyl-L-glutamate [Glu(OBz)] and L-leucine (Leu) [39].

Figure 26 shows the temperature dependence of the real parts of the piezoelectric constants e_{25}' and d_{25}' for homopolymers of Glu(OBz) and Leu and their copolymers. The orientation function F_c and the diffraction angle 2θ for the (110) reflection are indicated at the top of the figures. Peaks of e' and d' are seen at about 20°C for PBLG and at about −100°C for poly-Leu, owing to the side-chain relaxation. The side chain of PBLG includes a polar group CO—O, but the side chain of poly-Leu, —CH$_2$—CH(CH$_3$)$_2$, includes no polar group. The elastic relaxation due to side-chain motion is observed for both PBLG and poly-Leu, but the dielectric relaxation due to the side-chain relaxation is observed only for PBLG, not for poly-Leu.

With increasing amounts of Leu, the piezoelectric constant in the low-temperature range increases and the relaxational behavior due to the side-chain motions becomes broader. Figure 27 shows the piezoelectric constant divided by the orientation function, e_{25}'/F_c, measured at −50°C and 100°C, as a function of glu(OBz) content. This quantity indicates the piezoelectric constant for a perfectly oriented sample. A maximum occurs at about 20% Glu(OBz).

It is believed that the piezoelectricity in α-helical polypeptides originates from stress-induced rotation of dipoles of peptide bonds in the main chains. Since peptide bonds are connected to side chains, the compliance of side chains affects the magnitude of the piezoelectric effect. If the compliance of side chains increases, then the piezoelectric constants increase. For copolymers of Glu(OBz) and Leu, the introduction of Leu weakens the interaction between Glu(OBz) in side chains and increases the compliance of the side chains, and thus the piezoelectric constant. On the other hand, the hydrophobic interaction in Leu decreases the compliance of the side chains and thus the piezoelectric constant.

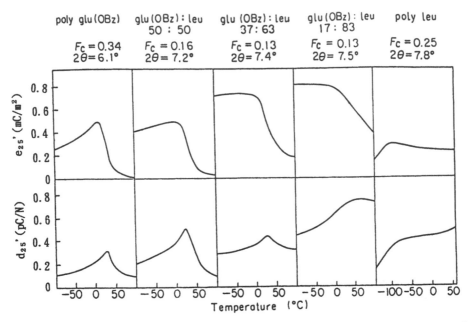

Figure 26 Temperature dependence of the real parts of the piezoelectric constants e and d for poly-γ-benzyl-L-glutamate, poly-L-leucine, and copolymers with molar ratios 50:50, 37:63, and 17:83. (From Ref. 39.)

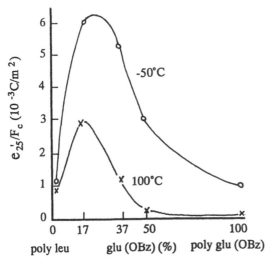

Figure 27 Piezoelectric constant, e_{25}', divided by orientation function, F_c, measured at -50 and 100°C, plotted against molar ratio of Leu:Glu (OBz). (From Ref. 39.)

VII. PIEZOELECTRICITY OF OPTICALLY ACTIVE POLYMERS

A. Polypropylene Oxide

As a synthesized, optically active polymer besides polypeptides, polypropylene oxide was first investigated for piezoelectric properties [40]. A film of poly-D-propylene oxide (PDPO) was cast from a benzene solution and elongated by about 50% of its original length. The degree of crystallinity was about 40%.

Figure 28 shows the temperature dependence of the piezoelectric constant d, the elastic constant c, and the dielectric constant ϵ for an oriented film of PDPO. All quantities show a relaxation at the glass temperature of $-60°C$.

B. Poly-β-hydroxybutyrate

Poly-β-hydroxybutyrate (PHB) and its copolymers are produced by microorganisms and thus are biodegradable. The chemical structure of the monomer is [—O—CH(CH₃)—CH₂—CO—] for hydroxybutyrate (HB) and [—O—CH(C₂H₅)—CH₂—CO—] for hydroxyvalerate (HV). Cast films were stretched about four times the original length.

Figure 29 shows the temperature dependence of the piezoelectric constants e_{14} and d_{14}, the elastic constant c, and the dielectric constant ϵ at 10 Hz for an oriented film of PHB [41]. Since PHB is a D polymer, the signs of e_{14} and d_{14} are positive. All quantities show the relaxation at the glass transition temperature of about 15°C. The

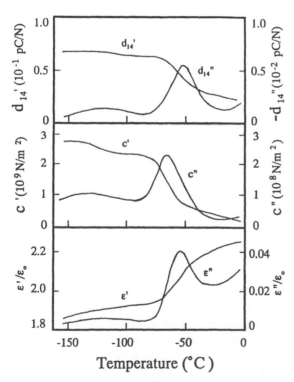

Figure 28 Temperature dependence of piezoelectric constant, elastic constant, and dielectric constant for oriented polypropylene oxide. (From Ref. 40.)

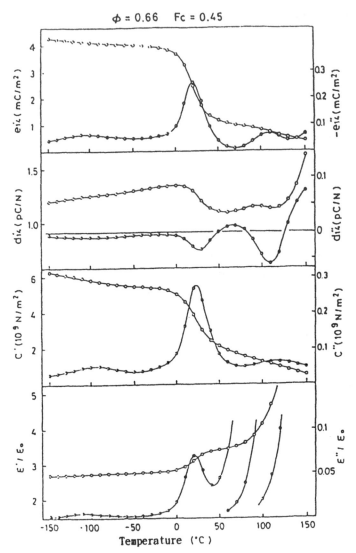

Figure 29 Temperature dependence of piezoelectric constants *e* and *d*, elastic constant *c*, and dielectric constant ϵ at 10 Hz for an oriented film of poly-β-hydroxybutyrate (PHB). (From Ref. 41.)

degree of crystallinity $\phi = 0.66$, and the degree of orientation $F_c = 0.45$ was obtained by the X-ray diffraction studies.

Figure 30 shows experimental results for a copolymer [P(25HV/75HB)] which consists of 25 mol% HV and 75 mol% HB that was elongated about 4.6 times the original length [42]. Relaxations are observed for all quantities at the glass transition temperature, similar to PHB. The difference is that d_{14}' for PHB decreases with rising temperature, but d_{14}' for P(25HV/75HB) increases with rising temperature.

To interpret these observations, the spherical dispersion model was utilized. Intuitively, the piezoelectric crystalline phase should not undergo relaxation at the glass tran-

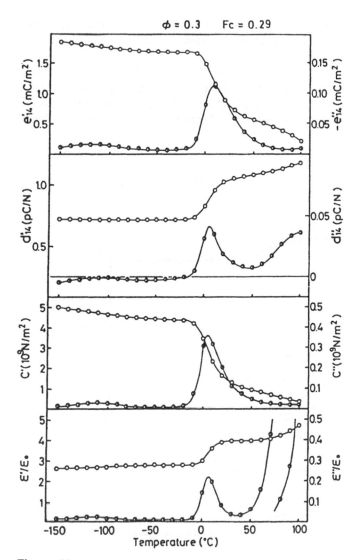

Figure 30 Temperature dependence of piezoelectric constants e and d, elastic constant c, and dielectric constant ϵ at 10 Hz for an oriented film of a copolymer of (25HV/75HB). (From Ref. 42.)

sition temperature, but nonpiezoelectric amorphous phase should undergo relaxation at the glass temperature. We tried to find out the piezoelectric constant for the crystalline phase by means of the spherical dispersion theory.

We assume that the both elastic and dielectric constants c_2 and ϵ_2 for the crystallites (phase 2) are independent of temperature and possess the values observed at the low temperature $-150°C$. Then, using Eqs. (6) and (7), we can calculate the elastic and dielectric constants c_1 and ϵ_1 for the amorphous regions (phase 1) as a function of temperature from the observed data for the elastic and dielectric constants c and ϵ.

Now, using Eqs. (8) and (9), we can calculate the piezoelectric constants e_2 and d_2 for the crystalline phase (phase 2) as a function of temperature from the observed data of e and d as well as the calculated values of c_1 and ϵ_1.

The results are shown in Figure 31 for PHB and in Figure 32 for P(25HV/75HB). Figure 32 shows that both e_2' and d_2' remain unchanged in the glass transition temperature range and both e_2'' and d_2'' are very small. The results agree reasonably with the anticipation that the crystalline phase undergoes no relaxation. However, Figure 31 shows

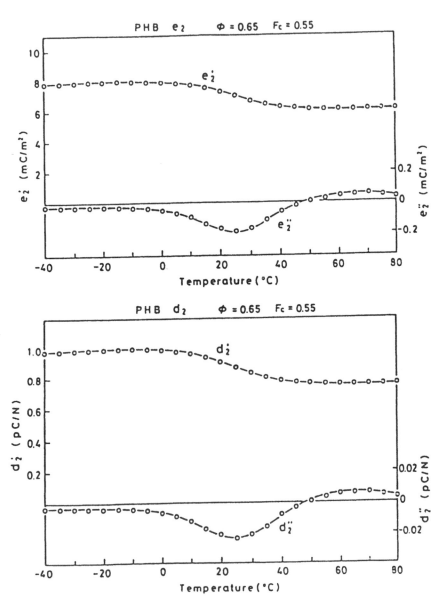

Figure 31 Calculated values of piezoelectric constants e_2 and d_2 for piezoelectric phase of PHB corresponding to Figure 29. (From Ref. 42.)

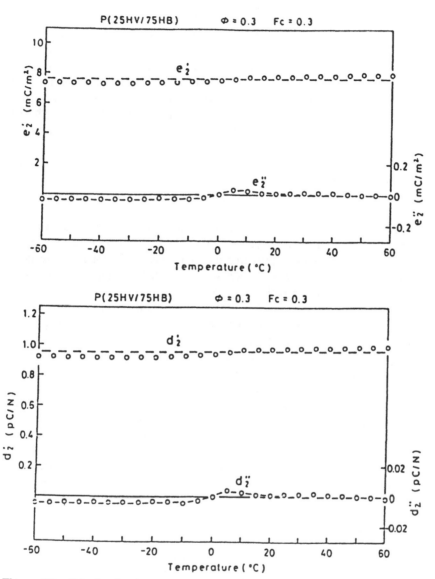

Figure 32 Calculated values of piezoelectric constants e_2 and d_2 for a copolymer of (25HV/75HB) corresponding to Figure 30. (From Ref. 42.)

that both e_2' and d_2' decrease and both e_2'' and d_2'' show a maximum in the glass transition temperature region, suggesting a relaxation taking place in phase 2.

Piezoelectric effect may take place not only in the crystalline phase but also in the interface between the crystalline phase and the amorphous phase, where molecules are well oriented. The oriented regions where molecules are aligned parallel, but with non-uniform intermolecular spacing may have a symmetry required for piezoelectric effect. However, such oriented amorphous phase may show relaxational behavior, as indicated in the temperature variation of e_2 and d_2 in Figure 31. We assume that piezoelectric phase

2 includes both crystalline phase and oriented amorphous phase and that the crystalline phase displays no relaxation but the oriented amorphous phase displays relaxation.

A copolymer of 17 mol% HV and 83 mol% HB gave experimental results similar to those for PHB [42]. The calculated values of e_2 and d_2 showed a relaxation similar to PHB. To remove the internal stress of the oriented film of copolymer, the film was exposed to the vapor of chloroform in a closed vessel for 30 min and then kept in vacuum at 40°C for 24 h. At room temperature the film contracted about 40% in length. The orientation function F_c decreased to 0.2 from the original value of 0.53, though the degree of crystallinity $\phi = 0.3$ was unchanged.

After chloroform treatment, the copolymer P(17HV/83HB) gave experimental results similar to those for P(25HV/75HB). The calculated values of e_2 and d_2 showed no relaxation similar to P(25HV/75HB). The treatment by chloroform removed the relaxations of e_2 and d_2 for P(17HV/83HB). This suggests that the residual stress in the highly stretched structure is removed and the oriented amorphous phase almost disappears.

Chloroform treatment of PHB and P(25HV/75HB) had almost no effect on experimental results. These observations suggest that the molecular chains of PHB are rather stiff and the oriented amorphous phase is persistent, and that the copolymerization of HV introduces greater flexibility in molecular chains.

It should be noted that e_2' and d_2' for PHB decrease with increasing temperature. The sign of the relaxing piezoelectric constant should be negative and opposite to that in the crystalline phase. Figure 33 shows a plot of d_{14}'/F_c against ϕ for PHB films with the different orientation function F_c and the different degree of crystallinity ϕ. The intercept of the linear line at $\phi = 1$ gives the piezoelectric constant d_c for the crystalline phase, which is positive. The intercept at $\phi = 0$ gives the piezoelectric constant d_a for the amorphous phase, which is negative [41]. This observation accords with the results that the relaxing piezoelectric constant in phase 2 has a negative sign.

The reason why the sign of the relaxing piezoelectric constant in the oriented amorphous phase is opposite to the sign of the nonrelaxing piezoelectric constant in the crystalline phase is not yet clearly understood. We assume that the observed piezoelectric polarization consists of an instantaneous component and a relaxational component. The

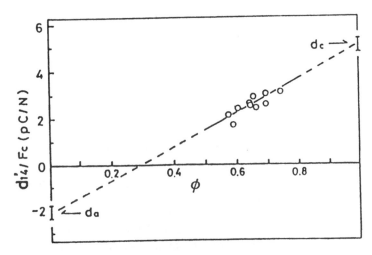

Figure 33   Relation between d_{14}'/F_c and ϕ at 0°C for PHB. (From Ref. 41.)

former is caused by the shift of the potential curve acting on a dipole in crystalline molecules. The latter is caused by the rotation of dipoles due to the stress-induced change of the potential curve. The direction of the dipolar rotation for instantaneous and relaxational polarizations do not need to be the same but may be opposite. The relaxational polarization may take place in the defects in the crystalline phase and mostly in the oriented amorphous phase. The local field acting on dipoles in such phases may be different from the perfectly crystalline phase. The reversal of the piezoelectric constant with increasing temperature is also known for the oriented films of cellulose triacetate [18] and amylose [20].

Frequency dependence from 1 to 100 Hz of the piezoelectric, elastic, and dielectric constants for P(25HV/75HB) was measured at a temperature range from -40 to $30°C$ and master curves reduced to $0°C$ were obtained [43]. Using these master curves, the calculation of e_2' and e_2'' was undertaken. It was shown that over the frequency range from 10^{-5} to 10^7 Hz, $e_2' = 13$ mC/m^2 and e_2'' is approximately zero. The spherical dispersion theory was applied successfully to analyze the master curves for P(25HV/75HB), in which the crystallites seem to be well separated from the amorphous regions.

C. Poly-lactic Acid

Poly-lactic acid, $[-O=CH(CH_3)-CO-]_n$, is a biocompatible and biodegradable polymer. A thorough study of the crystalline structure and optical activity of this polymer was recently published [44].

The temperature dependence of the piezoelectric constants e and d, the elastic constant c, and the dielectric constant ϵ for a film of poly-L-lactic acid (PLLA) elongated 4 times the original length is depicted in Figure 34 [45]. The results are similar to those for PHB. Using the same method as for PHB and assuming that the degree of crystallinity $\phi = 0.3$, the piezoelectric constants e_2 and d_2 for the crystalline phase were calculated. Figure 35 shows the results. The values of e_2' and d_2' are unvaried in the glass transition temperature range, though there is some scattering of data in the higher temperature range. The values of e_2'' and d_2'' are very small. These results accord with the prediction that the piezoelectric constant in the crystalline phase does not change with varying temperature.

The estimated piezoelectric constants in the crystalline phase for PLLA are as large as $e_{14}' = -60$ mC/m^2 and $d_{14}' = -30$ pC/N, respectively. If the degree of crystallinity is assumed to be $\phi = 0.5$, these becomes approximately $e_{14}' = -40$ mC/m^2 and $d_{14}' = -20$ pC/N. These are the highest values so far obtained for the shear piezoelectric constant in polymer crystals [46].

Table 2 shows a comparison of physical quantities for three optically active polymers described above. With an increase of polarity in chemical structure, the magnitude of the piezoelectric constant increases remarkably, although the degree of crystallinity and the degree of orientation are not exactly the same for the three polymers. The chemical structure of poly-lactic acid is the simplest form to couple an asymmetric carbon atom and a polar group CO—O and is most suitable for displaying the piezoelectric effect in this series.

VIII. PYROELECTRICITY OF BIOPOLYMERS

Pyroelectric effect of bone and tendon was first reported by Lang in 1966 [47]. The magnitude of the pyroelectric constant is as small as 3×10^{-9} C/m^2. It is mostly the

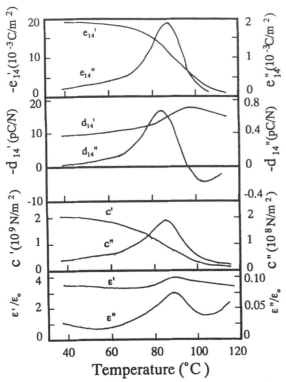

Figure 34 Temperature dependence of piezoelectric constants e and d, elastic constant c, and dielectric constant ϵ at 10 Hz for an oriented film of poly-L-lactic acid. (From Ref. 45.)

secondary effect due to the piezoelectric effect in the axial direction. Nevertheless, the effect should come from the unidirectional orientation of the polar axis of collagen molecules. In artificially prepared film of biopolymers, the sense of polar axis of molecules and crystallites is directed at random. In living tissues, however, a preferred alignment of the sense of polar axis exists, which results in tensile piezoelectricity in the polar axis and also pyroelectricity. The magnitude of the piezoelectric constant d_{33} is about 0.08 pC/N for dry tendon and less than that for dry bone [6].

A number of reports have been published by Athenstaedt on the pyroelectric effect in various living tissues of animal and plant, particularly in exoskeletons consisting of collagen or chitin [48]. Pyroelectric and piezoelectric effects are observed in most living systems. The effects are better observed in the dry state. In the wet state the conductivity due to water diminishes the observable polarization.

IX. CONCLUDING REMARKS

Starting with the early observations of piezoelectricity and pyroelectricity in bone and tendon and piezoelectricity in wood, a large number of biological polymers have been investigated for their piezoelectric and pyroelectric activities [49,50]. The piezoelectric effect has been also detected in synthetic, optically active polymers. The magnitude of the piezoelectric constant depends on the chemical structure and physical conditions of

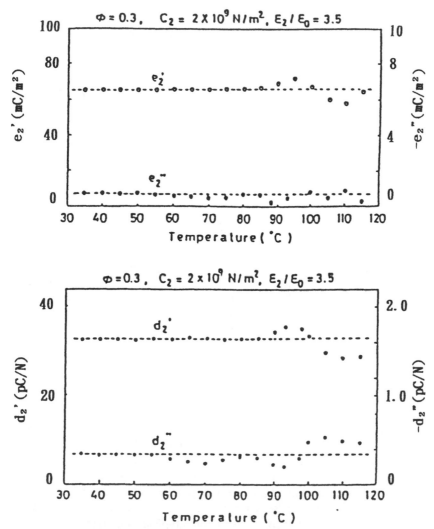

Figure 35 Calculated values of piezoelectric constants e_2 and d_2 for poly-L-lactic acid. (From Ref. 46.)

polymers. The magnitude of the piezoelectric constant observable on average for dried proteins is comparable with the piezoelectric constant of quartz crystal, $d_{11} = 2.2$ pC/N.

The measurement of pyroelectricity and piezoelectricity is essentially the observation of internal rotation or displacement of dipolar groups in the molecule under the action of a mechanical or electrical field. In the direct effect, one observes macroscopic polarization produced by internal rotation of dipoles under the macroscopic mechanical stress. In the converse effect, one observes macroscopic mechanical strain produced by the internal rotation of dipoles under the macroscopic electrical field. The internal strain and the internal polarization always appear together.

Theoretical calculation based on lattice dynamics of the piezoelectric constant of α-helical polypeptides was performed by Namiki et al. [51]. They estimated the piezoelec-

Table 2 Comparison of the Piezoelectric Constant for Three Optically Active Polymers

Polypropylene oxide		$T_g = -60°C$	At $-100°C$

$$\left(- O - \overset{\overset{\textstyle CH_3}{|}}{\underset{\underset{\textstyle H}{|}}{C}} - \overset{\overset{\textstyle H}{|}}{\underset{\underset{\textstyle H}{|}}{C}} - \right)_n$$

$T_g = -60°C$

At $-100°C$
$d = 0.1$ pC/N
$e = 0.2$ mC/m^2
$\epsilon = 1.9\ \epsilon_0$
$c = 2$ GN/m^2

Poly-β-hydroxyburyrate

$$\left(- O - \overset{\overset{\textstyle CH_3}{|}}{\underset{\underset{\textstyle H}{|}}{C}} - \overset{\overset{\textstyle H}{|}}{\underset{\underset{\textstyle H}{|}}{C}} - \overset{\overset{\textstyle}{}}{\underset{\underset{\textstyle O}{\|}}{C}} - \right)_n$$

$T_g = 20°C$

At $0°C$
$d = 1.3$ pC/N
$e = 4$ mC/m^2
$\epsilon = 3\ \epsilon_0$
$c = 4.5$ GN/m^2

Poly-lactic acid

$$\left(- O - \overset{\overset{\textstyle CH_3}{|}}{\underset{\underset{\textstyle H}{|}}{C}} - \overset{\overset{\textstyle}{}}{\underset{\underset{\textstyle O}{\|}}{C}} - \right)_n$$

$T_g = 85°C$

At $50°C$
$d = 10$ pC/N
$e = 18$ mC/m^2
$\epsilon = 3.5\ \epsilon_0$
$c = 2$ GN/m^2

tric constants $d_b = 6.2$ pC/N, $e_b = 67$ mC/m^2, and the dielectric constant $\epsilon_b/\epsilon_0 = 6.09$, and the elastic constant $c_b = 1.08 \times 10^{10}$ N/m^2, where the subscript b refers to the backbone of an α-helical polypeptide molecule. Using a dispersion model of rod particles, the piezoelectric constants $d = 0.5$ pC/N and $e = 2.4$ mC/m^2, the elastic constant $c = 4.9 \times 10^9$ N/m^2, and the dielectric constant $\epsilon/\epsilon_0 = 3.0$ were obtained for oriented film of polypeptides, which were approximately the same as the observed values. Table 3 lists the chemical structures and piezoelectric constants of some typical biopolymers.

Elastic and dielectric relaxations are commonly observed for most polymers. The onset of thermal molecular motions influences the elastic and dielectric properties. Piezoelectricity is a cross effect of the elastic and dielectric effects. Since the piezoelectricity expresses the internal strain of the polymer, the piezoelectric relaxation reflects the change of the internal strain. This is the most interesting characteristic of the piezoelectric relaxation.

The theory of spherical dispersion has been applied successfully to some oriented polymers, and the piezoelectric constants for the crystalline phase have been estimated. However, there still remain problems such as the interface between the crystalline phase and the amorphous phase. Actually, for oriented films of poly-γ-benzylglutamate, the simple spherical dispersion model is not applied satisfactorily, and the extended three-element model, taking account of grain boundaries, has to be introduced to explain the experimental results. Further elaboration of this model would be useful to understand the texture and properties of a composite such as a mixture of piezoelectric ceramic particles and polymers.

Table 3 The Chemical Structure and Piezoelectric Constant for Typical Biopolymers

Polymer	Chemical Structure	$-d_{14} = d_{25}$ (pC/N)
Cellulose		0.15
Chitin		1.5
Amylose		3.5
Proteins	Collagen / Keratin R: amino acid residue	2.5 1.5
Poly-γ-methyl glutamate		3
Poly-γ-benzyl glutamate		3.5
DNA	B: base	0.01

Because of the relatively small magnitudes of piezoelectric constants, practical applications have not progressed so far. A trial microphone using an elongated film of poly-γ-methylglutamate was reported [52].

Physiological significance of the pyroelectric and piezoelectric effects in biopolymers has been discussed [49]. The relations with the sense of heat and the sense of pressure were speculated. However, no quantitative relation has been found so far. The piezoelectric and pyroelectric effects are originated from the dipolar effects and not from ionic effects, which seem to play most important roles in electrophysiological phenomena.

Electrical stimulation of cell activities is an independent problem of the piezoelectric effect in polymers, although the initial work was closely related to the stress-induced electric potential in bone. The enhancement of cell activities in bone, cartilage, and other tissues by the action of electric current has been investigated in great detail. The stress-induced potential in bone is believed to be caused mostly by the ionic streaming potential, which is proportional to the ζ potential in canaliculi.

The interactions between the dipoles and ions in biopolymers are important in many respects. The physiological significance of piezoelectric and pyroelectric effects, if any, could be the influence of the electrical polarization of biopolymers on the behavior of ions. One possible example would be the interaction of the ion channel proteins with ions. The applied stress, mechanical or electrical, may change the molecular conformation, which is associated with the dipolar rotation. The adsorption and desorption of ions to the protein molecule could be affected by the conformational change of molecule. An enzymatic reaction accompanied by a change of molecular conformation could be another example.

Piezoelectric polymer films have been tested as implants to stimulate bone growth. Elongated films of poly-γ-methylglutamate and collagen and poled films of poly(vinylidene fluoride) implanted near periosteum of femur or tibia of animals induced callus around them in a few weeks [53].

In experiments on fixation of osteotomies, screws made of oriented poly-L-lactic acid generated about 50% more callus than stainless steel screws [54]. The amount of callus formed by PLLA rod or film implanted inside or around femur increased in proportion to the draw ratio of PLLA, which was linearly related to the piezoelectric constant. It is presumed that the piezoelectric polarization generated in the implanted polymer by the motion of the animal induces the ionic current to neutralize the polarization, and that the ionic current influences some biochemical reactions in cell membrane and leads to proliferation of bone cells [8]. Since the hydrolysis of PLLA takes place gradually in living tissue, PLLA is a biodegradable piezoelectric material, which is promising for orthopedic applications.

REFERENCES

1. A. J. P. Martin, Tribo-electricity in wool and hair, *Proc. Phys. Soc.* 53:186 (1941).
2. E. Menefee, Thermocurrent from alpha-helix disordering in keratin, *Electrets* (M. M. Perlman, ed.), The Electrochemical Society, Princeton, NJ, 1973, p. 661.
3. E. Fukada, Piezoelectric effect of silk, *J. Phys. Soc. Jpn.* 12:1301 (1956).
4. V. A. Bazhenov, *Piezoelectric Properties of Wood*, Consultants Bureau, New York, 1961.
5. E. Fukada and I. Yasuda, On the piezoelectric effect of bone, *J. Phys. Soc. Jpn.* 12:1158 (1957).
6. E. Fukada and I. Yasuda, Piezoelectric effects in collagen, *Jpn. J. Appl. Phys.* 3:117 (1964).

7. I. Yasuda, Piezoelectricity of bone and electric callus, *Clin. Ortho. Rel. Res. No. 124*:5 (1977).
8. C. T. Brighton and S. R. Pollack (eds.), *Electromagnetics in Medicine and Biology*, San Francisco Press, San Francisco, 1991.
9. A. V. Shubnikov, *Piezoelectric Textures*, Izd-vo AN SSSR, 1946.
10. J. F. Nye, *Physical Properties of Crystals*, Oxford University Press, London, 1960.
11. H. Kawai, The piezoelectricity of poly(vinylidene fluoride), *Jpn. J. Appl. Phys. 8*:975 (1969).
12. E. Fukada, Piezoelectricity of natural biomaterials, *Ferroelectrics 60*:285 (1984).
13. T. Furukawa and E. Fukada, Piezoelectric relaxation in poly(γ-benzyl-L-glutamate), *J. Polymer Sci. Polymer Phys. 14*:1979 (1976).
14. E. Fukada, M. Date, and T. Emura, Temperature variation of complex piezoelectric modulus in cellulose acetate, *J. Soc. Mat. Sci. Jpn. 17*:335 (1968).
15. M. Date, The piezoelectric constant for dispersed systems, *Polymer J. 8*:60 (1976).
16. E. Fukada, M. Date, and K. Hara, Temperature dispersion of complex piezoelectric modulus of wood, *Jpn. J. Appl. Phys. 8*:151 (1969).
17. E. Fukada, M. Date, and N. Hirai, Effect of temperature on piezoelectricity in wood, *J. Polymer Sci. C 23*:509 (1968).
18. S. Sasaki and E. Fukada, Deformation of the crystal lattice by water absorption and piezoelectricity of cellulose triacetate, *J. Polymer Sci. Polymer Phys. 14*:565 (1976).
19. Y. Ando, E. Fukada, and M. J. Glimcher, Piezoelectricity of chitin in lobster shell and apodeme, *Biorheology 14*:175 (1977).
20. K. Nishinari and E. Fukada, Viscoelastic, dielectric, and piezoelectric behavior of solid amylose, *J. Polymer Sci. Polymer Phys. 18*:1609 (1980).
21. C. T. Brighton, J. Black, and S. R. Pollack (eds.), *Electrical Properties of Bone and Cartilage*, Grune & Stratton, New York, 1979.
22. H. Maeda and E. Fukada, Effect of water on piezoelectric, dielectric, and elastic properties of bone, *Biopolymers 21*:2055 (1982).
23. E. Fukada, H. Ueda, and R. Rinaldi, Piezoelectric and related properties of hydrated collagen, *Biophys. J. 16*:911 (1976).
24. W. S. Williams and L. Breger, Analysis of stress distribution and piezoelectric response in cantilever bending of bone and tendon, *Ann. N.Y. Acad. Sci. 238*:121 (1974).
25. E. Fukada, R. L. Zimmerman, and S. Mascarenhas, Denaturation of horn keratin observed by piezoelectric measurements, *Biochem. Biophys. Res. Commun. 62*:415 (1975).
26. E. Fukada and K. Hara, Piezoelectric effect in blood vessel walls, *J. Phys. Soc. Jpn. 26*:777 (1969).
27. E. Fukada and H. Ueda, Piezoelectric effect in muscle, *Jpn. J. Appl. Phys. 9*:844 (1970).
28. H. Ueda and E. Fukada, Piezoelectricity in myosin and actin, *Jpn. J. Appl. Phys. 10*:1650 (1971).
29. H. Ueda and E. Fukada, Piezoelectricity, dielectricity, and viscoelasticity of elongated fibrin films, *J. Soc. Mat. Sci. Jpn. 21*:397 (1972).
30. J. Duchesne, J. Depireux, A. Bertinchamps, N. Cornet, and J. M. van der Kaa, Thermal and electrical properties of nucleic acids and proteins, *Nature 188*:405 (1960).
31. M. Falk, K. A. Hartman, Jr., and R. C. Lord, Hydration of deoxyribonucleic acid II. An infrared study, *J. Am. Chem. Soc. 85*:387, 391 (1963).
32. Y. Ando and E. Fukada, Piezoelectric properties of oriented deoxyribonucleate films, *J. Polymer Sci. Polymer Phys. 14*:63 (1976).
33. M. Date, S. Takashita, and E. Fukada, Temperature variation of piezoelectric moduli in oriented poly(γ-methyl-L-glutamate), *J. Polymer Sci. A-2 8*:61 (1970).
34. E. Fukada and S. Takashita, Piezoelectric constant in oriented β-form polypeptides, *Jpn. J. Appl. Phys. 10*:722 (1971).
35. Y. Go, S. Ejiri, and E. Fukada, Magnetic orientation of poly-γ-benzyl-L-glutamate, *Biochim. Biophys. Acta 175*:454 (1969).

36. T. Konaga and E. Fukada, Piezoelectricity in oriented films of poly(γ-benzyl-L-glutamate), *J. Polymer Sci. A-2* 9:2023 (1971).
37. E. Iizuka, Orientation of poly-γ-benzyl-L-glutamate in a very low electric field, *Biochim. Biophys. Acta* 175:457 (1969).
38. E. Fukada, Piezoelectric properties of organic polymers, *Ann. N.Y. Acad. Sci.* 238:7 (1974).
39. E. Fukada, T. Furukawa, E. Baer, A. Hiltner, and J. M. Anderson, Piezoelectric relaxation in homopolymers and copolymers of γ-benzyl-L-glutamate and L-leucine, *J. Macromol. Sci. Phys.* B8:475 (1973).
40. T. Furukawa and E. Fukada, Piezoelectric effect and its temperature variation in optically active polypropylene oxide, *Nature* 221:1235 (1969).
41. Y. Ando and E. Fukada, Piezoelectric properties and molecular motion of Poly(β-hydroxy-butyrate) films, *J. Polymer Sci. Polymer Phys.* 22:1821 (1984).
42. E. Fukada and Y. Ando, Piezoelectric properties of poly-β-hydroxybutyrate and copolymers of β-hydroxy butyrate and β-hydroxyvalerate, *Int. J. Biol. Macromol.* 8:361 (1986).
43. Y. Ando, M. Minato, K. Nishida, and E. Fukada, Primary piezoelectric relaxation in a co-polymer of β-hydroxy butyrate and β-hydroxyvalerate, *IEEE Trans. Elec. Insul.* EI-21:505 (1986).
44. M. Ichiki, T. Asahi, T. Watanabe, J. Kobayashi, and E. Fukada, Optical and structural studies on poly-lactic acid, *Rep. Prog. Polymer Phys. Jpn.* 35:441,445 (1992).
45. E. Fukada, Piezoelectric properties of poly-L-lactic acid, *Rep. Prog. Polymer Phys. Jpn.* 34: 269 (1991).
46. E. Fukada, Bioelectret and biopiezoelectricity, *IEEE Trans. Elec. Insul.* 27:813 (1992).
47. S. B. Lang, Pyroelectric effect in bone and tendon, *Nature* 212:704 (1966).
48. H. Athenstaedt, Permanent electric polarization and pyroelectric behaviour of the vertebrate skeleton, I, II, III, IV, *Z. Zellf.* 91:135 (1968); 92:428 (1968); 93:484 (1969); 97:537 (1969).
49. E. Fukada, Piezoelectric properties of biological polymers, *Quart. Rev. Biophys.* 16:59 (1983).
50. E. Fukada, Piezoelectricity of biological materials, *Electronic Conduction and Mechano-electrical Transduction in Biological Materials* (B. Lipinski, ed.), Marcel Dekker, New York, 1982, p. 125.
51. K. Namiki, R. Hayakawa, and Y. Wada, Molecular theory of piezoelectricity of α-helical polypeptide, *J. Polymer Sci. B* 18:993 (1980).
52. E. Fukada, I. Yamamuro, and M. Tamura, Polypeptides piezoelectric transducer, *6th Int. Cong. Acoustics D* 69 (1968).
53. E. Fukada, Piezoelectricity of bone and osteogenesis by piezoelectric films, *Mechanisms of Growth Control* (R. O. Becker, ed.), Charles C Thomas, Springfield, IL, 1981, p. 192.
54. Y. Matsusue, T. Yamamuro, S. Yoshii, M. Oka, Y. Ikada, S. H. Hyon, and Y. Shikinami, Biodegradable screw fixation of rabbit tibia proximal osteotomies, *J. Appl. Biomaterials* 2:1 (1991).

10
Ferroelectric Liquid Crystal (FLC) Polymers

G. Scherowsky
Technical University of Berlin, Berlin, Germany

I. INTRODUCTION

A. Ferroelectricity in Liquid Crystals

Whereas ferroelectricity in polymers of the poly(vinylidene fluoride) (PVDF) type is induced by stretching and poling of polymers films and is not characteristic for undisturbed samples, spontaneous polarization appears in liquid crystalline polymers with intrinsic lower symmetry C_2.

The occurrence of ferroelectricity in liquid crystals was predicted 1974 by Meyer [1], based on symmetry arguments. He suggested the appearance of spontaneous polarization in any lamellar system consisting of inclined and chiral molecules having a nonzero dipole moment normal to the direction of the long molecular axis. One year later this brillant prediction was verified experimentally [2].

Liquid crystals mostly consist of rodlike or disklike molecules exhibiting a fourth state of matter. At their melting point they partially lose crystalline order, generating a fluid but ordered state. They can form *smectic phases* (layered structures) with different kinds of ordering of the molecules in the layers, and *nematic phases*, with an approximately parallel orientation of the molecular long axis and no correlation of their centers of mass. Disklike molecules mostly form *columnar phases*.

Reviews of ferroelectric liquid crystal polymers can be found in the following:

1. P. Le Barny and J. C. Dubois, The chiral smectic liquid crystal side chain polymers, *Side Chain Liquid Crystal Polymers* (C. B. McArdle, ed.), Blackie, Chapman & Hall, New York, 1989, p. 130.

2. M. V. Kozlowsky and L. A. Beresnew, *Phase Transition 40*:129–169 (1992).
3. Ferroelectric Liquid Crystal Polymers, Special issue, *Polymers for Advanced Technologies 3*(5) (1992) F. Kremer, guest editor).

Approximately 20 different smectic phases have been identified up to now [3]. Eight among them consist of so-called tilted phases, i.e., the long axes of the molecules are tilted with respect to the layer normal (S_C, S_I, S_F, S_G, S_H, S_J, S_K, and S_M). If these latter mesophases consist of chiral molecules, they in principle match the requirement for intrinsic ferroelectric polarization. In six of these tilted chiral phases (denoted herein by an asterisk), spontaneous polarization has been measured (S_C^*, S_F^*, S_I^*, S_G^* S_J^*, S_M^*). For technological application in electrooptical devices, the chiral smectic C phase S_C^* is prominent due to its lowest ordering and hence highest fluidity, making reorientation processes caused by electric fields very fast.

There are three main requirements for the appearance of spontaneous polarization in a liquid crystal: (a) a center of chirality, preferably located near to the mesogenic (aromatic) molecular unit; (b) a dipole moment positioned at the chiral center and acting transverse to the molecular long axis; (c) the existence of a tilted smectic phase.

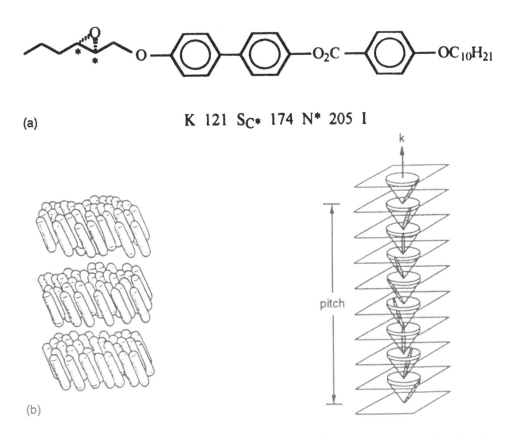

(a)   K 121 S_{C*} 174 N* 205 I

(b)

Figure 1   (a) Molecular arrangement for the S_C^* phase. (b) Rotation of the tilt direction shown for one molecule in each layer.

An example of a chiral molecule that forms a chiral smectic C phase and the molecular arrangement for the S_C^* phase is shown in Figure 1. The rodshaped molecules stand upright in the layers with their long axes tilted with respect to the normal of the layer planes. Because of the chirality of the molecues, their tilt direction alters from layer to layer, creating a helical arrangement (Fig. 1b).

In every layer a spontaneous polarization exists parallel to the layer planes, caused by a preferred syn orientation of the transverse dipoles of the molecules due to librational motion around the molecular axis (Fig. 2). As the direction of the polarization (● coming out, ○ going into) is also moving from layer to layer around the helical axes, due to the coupling with the tilt direction, the spontaneous polarization cancels out in the bulk S_C^* phase.

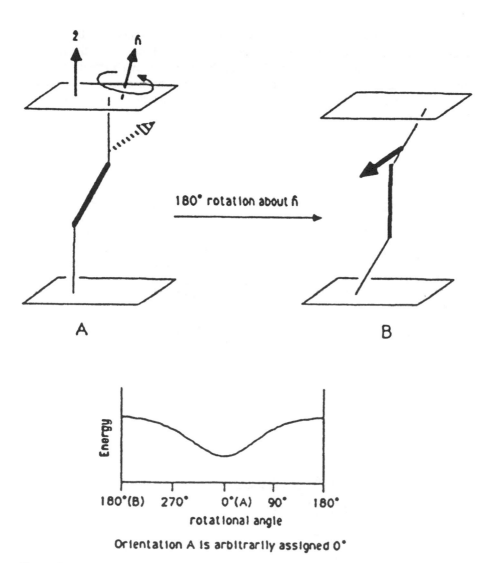

Figure 2 Energetically preferred direction of the lateral dipolemoment in a smectic layer.

B. Application in Electrooptical Devices

Unwinding or suppressing the helix leads to macroscopic polarization. This has been successfully realized in surface-stabilized thin cells with a 2-μm gap. This invention by Clark and Lagerwall in 1980 [4] was the starting point of the development of fast-switching ferroelectric liquid crystal displays. The principle of their operation is shown in Figure 3.

There are two energetically equivalent orientational states, depending on whether the molecules are inclined to the right or the left. The direction of the polarization is opposite for the two states (Fig. 3). A weak electric field allows switching from one state to the other, because the spontaneous polarization always tends to align parallel to an applied electric field. Here the main axes of the molecules are precessing on the surface of a cone whose opening angle equals twice the tilt angle of the molecules in the smectic phase.

Between two polarizers set crosswise to each other, the arrangement can be switched from dark to bright. Because of the short switching time, which can be in the range of microseconds, this new generation of liquid crystals has attracted much attention in fundamental research and electrooptical applications. There has been considerable activity aimed at synthesizing new materials with large values of spontaneous polarization (P_S) and short electrooptical response times.

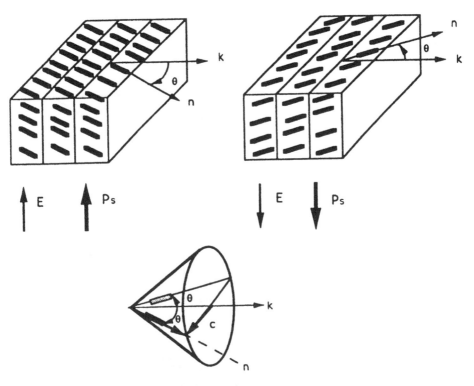

Figure 3   Switching in a surface-stabilized FLC cell.

II. FERROELECTRICITY IN LIQUID CRYSTALLINE POLYMERS

A. Introduction

The fascinating features of low-molar-mass ferroelectric crystals (FLCs) posed the question as to whether liquid crystalline polymers (FLCPs) containing ferroelectric low-molar-mass building blocks may also show electrooptical switching properties and what the response times achievable in such polymers might be. In nematic polymers, response times in the range of seconds have been measured [5]. Low-molar-mass nematic LCs, used in commercial twisted nematic displays, are switching in milliseconds. This reflects the high viscosity which is characteristic of polymeric systems. This situation could, however, be very different in the smectic C* phase of a ferroelectric LC side-chain polymer, where the ferroelectric building blocks are pending at the backbone and connected via a flexible spacer (Fig. 4).

The change in molecular orientation constituting the ferroelectric switching is the precession of the ferroelectric side group about the lateral area of a cone. This rotation is essentially a "free rotation" if the spacer effectively decouples the side group from the polymer backbone. Hence short electrooptic response times were thinkable in those systems.

Some ferroelectric LC polymers developed in the last few years indeed are switching in the range of milliseconds, in favorable cases down to a few hundred microseconds, depending on their molecular weight and the temperature range of the S_C^* phase.

B. First Representatives

The first report on an electrooptical switching *ferroelectric LC polymer* (polymer 1, Fig. 5) appeared in 1988 [6]. Ushida et al. measured a small spontaneous polarization of 3

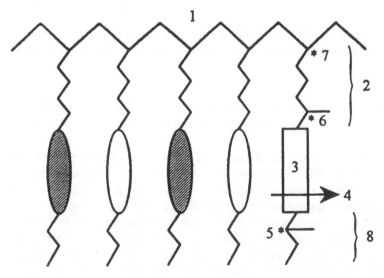

Figure 4   Sketch of S_C^* side-group polymer: (1) backbone; (2) spacer; (3) mesogen; (4) lateral dipolemoment; (5,6,7) possible chiral centers; (8) terminal alkyl chain.

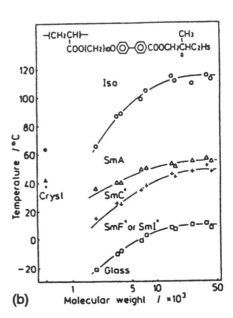

1

Figure 5 First electrooptical switching FLCP: Temperature dependence of (a) the spontaneous polarization; (b) the response time in samples of different molecular weight, M_n. (From Ref. 6.)

nC/cm² in maximum and response times between 3 and 50 ms (Fig. 5a) (depending on the molecular weight Mn), when applying an electric field of 200 V across a 10-μm cell (Fig. 5b).

Spontaneous polarization in the chiral smectic C phase of a LC polymer was described for the first time in 1984 by Shibaev et al. [7]. The detailed structure of this polymer (2a) was disclosed in a later paper [8]. The spontaneous polarization was calculated from the temperature dependence of the pyroelectric coefficient (Fig. 6).

By increasing the lateral dipolemoment at the chiral center (polymer 2b in Table 1), a slight increase of the spontaneous polarization from 1 to 3 nC/cm² was observed. In comparison with their monomers, the polymers exhibit a three to five times lower spontaneous polarization [9]. More details on the physical investigation of those two polymers, especially concerning the relaxation times for helix untwisting (Goldstone mode) and the soft mode, are presented in Ref. 10 (see also Sec. II.I).

Table 1 Structures and Properties of Polymers 2a and 2b

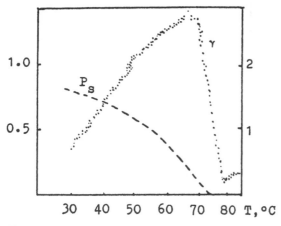

Polymer	$\overline{M}$	$\dfrac{\overline{M}_w}{\overline{M}_n}$	Transition temperatures (°C)	Tilt angle (deg)
2a	48,000	4.1	g 45 S_C^* 73 S_A 85 I	24
2b	31,000	3.1	g 50 S_C^* 71 (S_A) 79 I	34

C. Efforts to Synthesize S_C^* Polymers

1. S_C^* Polyacrylates

Following the detection of spontaneous polarization in FLC polymers, a large effort has been launched directed at the synthesis of chiral polymers exhibiting smectic C* phases. Decobert et al. [11] synthesized structural modifications of the Shibaev polymer (2a,2b).

Figure 6 Temperature dependence of pyrocoefficient γ and spontaneous polarization P_S for polymer 2a. (From Ref. 7.)

Table 2 Structures and Thermal Behavior of Polymers 3a–3d

3a–d

3	X	Y	n	T_g	T_m		$T_{S1,S5}$		T_C
a	H	—	2	65	—	S_C^*	110	S_{A2}	146
b	CH$_3$	—	2			S_C^*	110	S_{A2}	155
c	CH$_3$	—	11	10–40	58	S_C^*	90	S_{A1}	106
d	CH$_3$	—CH=CH—	11	35	65	S_C^*	115	S_{A1}	142

In four examples (3a–3d, Table 2) of 13, an S_C^* phase was found. In the whole series the glass transition temperature (T_g) decreases as the spacer length increases. An increase of T_g occurs when passing from the polyacrylate to the polymethacrylates or the polychloroacrylates ($X = Cl$) due to the increasing stiffness of the polymer backbone. The effects of $X = CH_3$ and $X = Cl$ are nearly the same [12,13].

Poly-α-chloroacrylates do not exhibit S_C^* phases, only S_{A2} phases. "Bilayer" structures (S_{A2}) are observed at short spacers ($n = 2$). Despite their S_C^* phases, the polymers 3a–3d could not be proved to be ferroelectric. No spontaneous polarization could be measured because of misalignment in the samples [14].

2. S_C^* Polysiloxanes

Side-chain polymers with polysiloxane backbones are known for their low viscosity and low glass transition temperatures. Those properties are favorable for applications of switchable ferroelectric LC polymers. The first examples exhibiting a S_C^* phase were reported by Hahn and Percec [15] (polymers 4a and 4b, Fig. 7). The S_C^* mesophases were identified only by optical microscopy. Investigation of ferroelectric properties was not reported.

The same holds for the polysiloxanes (5) presented by Keller [16] (Table 3). A comparison of polysiloxanes 5a–5c shows:

1. They have crystalline order instead of a glassy state.
2. S_C^* phases occur at good decoupling of the backbone and mesogenic part.
3. Phase type and transition temperature are not affected by the degree of polymerization (in the range of 25–80 units).

D. Identification of the S_C^* Phase in LC Polymers

Mesophase structures are deduced from a combination of differential scanning calorimetry (DSC), polarization microscopy, and X-ray measurements. Identification of LCP mesophases is often more difficult than in the case of low-molar-mass LCs, due to fine

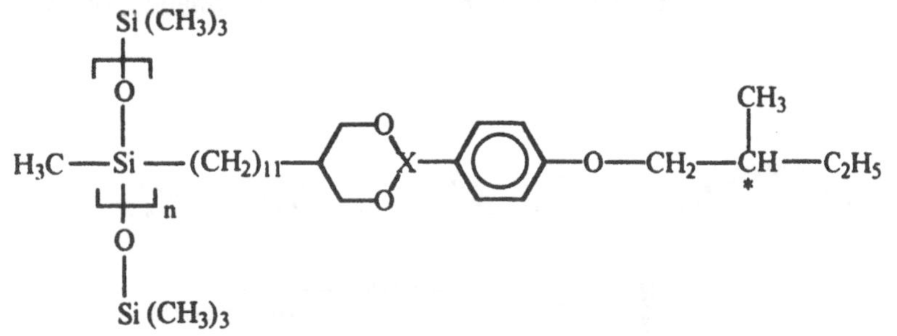

4a-b

4	X	T_g		T_1		T_C
a	CH	4		50	S_C^*	97
b	B	-7	S_2	19	S_C^*	85

Figure 7 LC polysiloxanes 4a–4e.

Table 3 Phase Transitions of Polysiloxanes 5a–5e

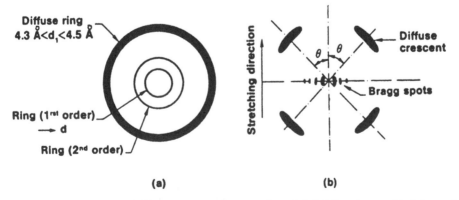

5a-e

n	x	Phase transitions
4	80	K 50 S_A 143 I
6	80	K 88 S_A 146 I
10	80	K 78 S_3 90 S_C^* 106 S_A 138 I
10	25	K 75 S_3 90 S_C^* 105 S_A 133 I
10	36	K 75 S_3 90 S_C^* 106 S_A 134 I

Source: From Ref. 16.

grain textures, which are often not characteristic. Sometimes, after long annealing times, characteristic, well-defined textures develop. Even then it is often impossible to distinguish the S_C^* texture from the texture of the higher-ordered, also tilted S_I^* or S_F^* phase.

Mixing experiments of unknown phases with known ones, applied successfully to identify low-molar-mass LCs, appeared mostly to fail in polymeric systems. One obtains more information from X-ray investigations. An X-ray diffraction pattern from an unoriented sample of a S_C^* LC polymer is displayed schematically in Figure 8a. The broad, diffuse outer ring reflects the absence of ordering within the layer planes. One or several well-defined inner rings are related to the smectic layer distance. If this distance is smaller than the calculated length of the side chain in its most extended conformation, one can

Figure 8 (a) Schematic X-ray pattern of an unoriented S_C^* LC polymer. (b) Schematic X-ray pattern of polymer 3a obtained from an oriented fiber drawn out of the S_C^* phase. (From Ref. 14.)

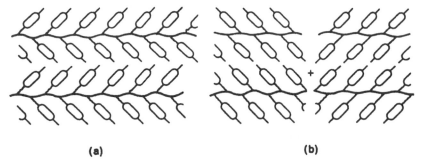

(a)   (b)

Figure 9 Possible models for side-chain ordering in the S_C* phase of polymer 3a. (From Ref. 11.)

conclude that the phase is a tilted one. An oriented sample for a S_C* phase, obtained either in a magnetic field or by drawing a fiber, will give a diffraction pattern as shown schematically in Figure 8b. The small-angle Bragg spots on the equator correspond to the three first orders of reflection on the layer plane, indicating that the smectic layers and hence the polymer backbone are parallel to the fiber axis. The four diffuse crescents at large angles indicate a liquidlike order within the layers. The two distinct orientations of the mesogens in the side chain, which are tilted by the angle θ with respect to the fiber axis, are in accord with two models shown in Figure 9 [11].

Distinction between S_C* and the hexagonal ordered S_I* and S_F* phases affords X-ray diffraction from a monodomain, the X-rays incoming perpendicular to the layer planes, in order to identify the hexagonal ordering.

E. Switchable Ferroelectric LC Polymers

Ferroelectrics are defined as materials in which the unit cell of the crystal is polar (in LCs the smectic layer), and the direction of polarization can be changed by the application of an electric field [17].

1. Evaluation of Ferroelectric Properties

To prove ferroelectricity in LC polymers, the magnitude of polarization has to be determined. Among the methods available to measure the spontaneous polarization, the reversal-current technique is most widely used. For a review on methods for polarization measurement, see Martinot-Lagarde [18]. In the repolarization technique, a triangular or rectangular field is applied to a cell with an aligned FLC polymer sample. The integrated area of the repolarization current reduced to the sample surface unit yields the spontaneous polarization (Fig. 10).

Because of the slowness of response in polymeric materials, the frequencies used are rather low (0.5–0.05 Hz). Generally, the magnitude of observed P_S depends on the quality of orientation, the cell thickness, and the frequency. FLC polymers exhibit larger values of P_S than their monomers. It is supposed that the molecular rotation around the long axis is slowed down by bonding to the main chain, thus increasing the magnitude of P_S [19].

2. Electrooptical Switching

A homogeneous oriented sample with suppressed helical structure can be switched between two stable states of molecular orientation by changing the sign of the applied field

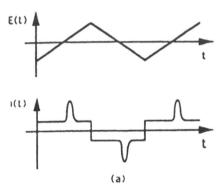

(a)

Figure 10 Determination of the spontaneous polarization. Current response $i(t)$ on a triangular waveform $E(t)$.

(recall Fig. 3). The switching time is normally defined as the time of the change in transparency from 10% to 90% in an oriented sample of a surface-stabilized LC cell (SSFLC [4]) between crossed polarizers. The switching time of a ferroelectric device is given approximately by the formula

$$\tau = \frac{\eta}{P_s \cdot E}$$

where τ is the reorientation time, η is the rotational viscosity, P_s is the spontaneous polarization, and E is the applied electric field [20].

The rotational viscosity can be estimated by measuring the spontaneous polarization and the optical response time τ. The optical response depends on the tilt angle, and is not precisely equal to τ defined by the equation given above. Taking this dependence into account, Endo et al. [21] evaluated rotational viscosities of their polymers by using the optical response time (t_{10-90}), the tilt angle, and the spontaneous polarization. As a result, they found the rotational viscosity of the FLC polymers to be two or three orders of magnitude larger than that of low-molar-weight FLC. Furthermore, they found the rotational viscosity to be dependent to the second power of the weight-average molecular weight (M_W).

Kawasaki et al. studied the dynamic molecular motion of a polysiloxane side-chain FLC polymer in an applied electric field by using time-resolved FT-IR technique [22]. They observed a synchronous change of absorbance for the side groups and the main chain as well. The movement of the side chain caused by an applied field is influenced and restricted by the coupling via the main chain. This reveals why the electrooptical response of FLC polymers is 100 to 1000 times as long as that of low-molar-weight FLCs.

It was proposed that the spontaneous polarization has to be increased to realize short switching times. However, it was found that high spontaneous polarization created relatively strong internal fields, which have to be overcome before the molecules will latch. Strong lateral dipole moments in the molecules, leading to high spontaneous polarization, also increase the viscosity. Therefore a compromise has to be found between those counteracting properties. In low-molar-mass systems, short switching times have been achieved by making mixtures of achiral materials that exhibit smectic C phases of low viscosity and chiral dopants that have large polarization but are not necessarily liquid crystals. As mixtures of polymers and low-molar-mass materials often suffer from phase separation, only few examples of induced ferroelectric mixtures are known. They are described in Section II.G.1.

3. Structures and Properties of the First Switchable FLC Polymers

In 1988, when the first paper on switchable FLC polymers appeared [6], two other groups also reported on this topic: Scherowsky et al. [23], and Suzuki et al. [24]. Scherowsky et al. pursued two different approaches to succeed in getting FLC polymers:

1. Introduction of a center of chirality in a LC polymer that has a broad smectic C phase
2. Synthesis of a monomeric ferroelectric LC with broad S_C^* phase and high spontaneous polarization, hoping to preserve this quality when polymerizing this monomer.

The first approach lead to a S_C^* polymer, but no switching properties emerged [25]. The second approach was successful.

Figure 11 shows a comparison of properties of a typical monomer and the corresponding polymer. The monomer has six LC phases. After polymerization, four of them are lost (S_A, N^*, BP_I, BP_{II}). In polymer 6 there are two subphases in the S_C^* phase. They exhibit an identical X-ray pattern but different electrooptical behavior. They are marked by the subscript x for the low-temperature phase, and y for the high-temperature phase.

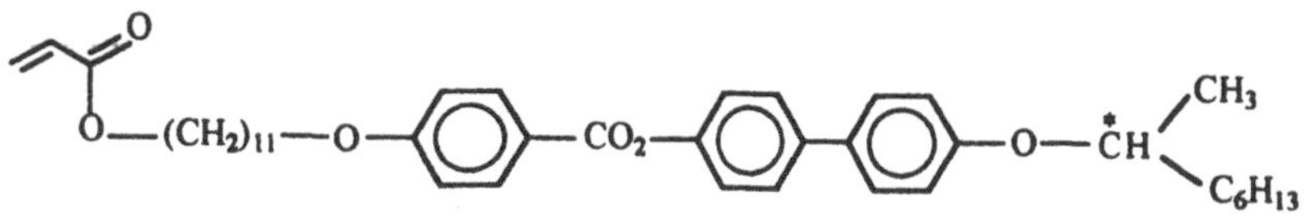

Cr 57.1 S_B 62.2 S_C^* 75.4 S_A 92.3 N* 93.3 BP_I 93.3 BP_{II} 95.4 I

6

T_g 50 S_B 92 $S_{C_X}^*$ 125 $S_{C_Y}^*$ 142 I   (M_w = 15 000)

Figure 11 Structure and phase transitions of a FLC monomer and the corresponding polymer 6. S_A, smectic A phase (molecules oriented perpendicular to the smectic layer planes); S_B, smectic B phase, (molecules directed as in S_A phase, but additionally hexagonal ordered); N*, chiral nematic (cholesteric) phase; BP, blue phase.

A plot of the switching time versus temperature (Fig. 12) shows a range of response between 1 and 2.7 ms. Apart from the rapidity of this switching in a polymer, another exiting phenomenon was the shorter response time in the lower-temperature phase (S_{Cx}^*). There is a distinct difference between the switching processes in the S_{Cx}^* and S_{Cy}^* phases. Whereas true ferroelectric switching between two stable states takes place in the S_{Cy}^* phase, caused by the reorientation of the director over the conical surface, the switching in the S_{Cx}^* phase has no bistability: It is an electroclinic process, similar to that found for the S_A phase in low-molar-mass LCs by Garoff and Meyer 1977 [26]. The electroclinic effect consists of a linear increase of the tilt angle with applied field strength. The switching time is independent of the field strength. In Figure 13 the switching time for polymer 6 versus voltage is displayed for the S_{Cx}^* and S_{Cy}^* phases. The independence of the response time on the field strength in the S_{Cx}^* phase is typical for the electroclinic effect. Its decrease in the S_{Cy}^* phase, i.e., the linear increase to the reciprocal voltage (shown in the inset), is in accordance with a ferroelectric switching process. We measured response times between 200 and 400 µs for a sample of polymer 6 with $M_w = 15,000$ (Fig. 14). This is an extremely fast switching in a polymer.

The interesting ferroelectric properties of polyacrylate 6 inspired us to change the backbone to a polysiloxane, to use different spacer lengths, and to introduce the α-chlorocarbonic ester group as chiral unit, which is known for its ability to create high spontaneous polarization. Due to their low glass transition temperatures, polysiloxane side group polymers seemed to be favorable candidates for fast-switching FLC polymers operating at room temperature.

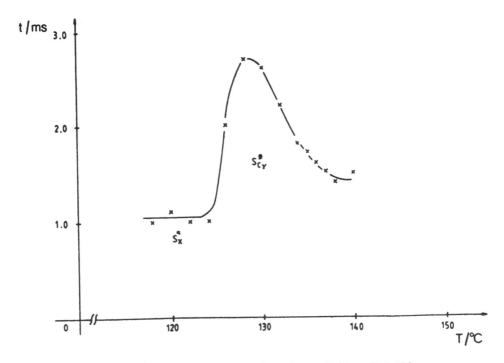

Figure 12 Switching time versus temperature for polymer 6. (From Ref. 23.)

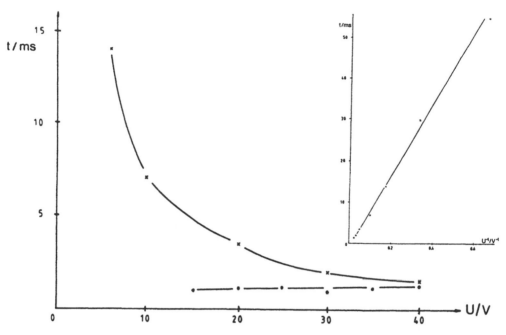

Figure 13 Switching time versus applied DC voltage for polymer 6 in the $S_{C\alpha}^*$ phase (125°C) and S_{Cy}^* phase ($\times$ 137°C). (From Ref. 23.) Inset: switching time versus reciprocal voltage (S_{Cy}^* phase at 137°C).

In Figure 15 the structures of the olefinic precursors 7a–7e and the derived polysi-loxanes 8a–8e and in Table 4 their phase transitions are displayed. The smectic phases of the polymers 8a–8c could not be identified by microscopy, due to their uncharacteristic fine grain textures. Because of their high viscosity, those polymers could not be drawn into thin cells (4 μm) by capillary attraction. In 25-μm cells after annealing in electric

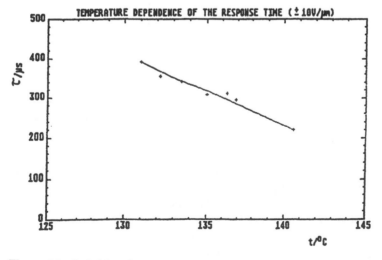

Figure 14 Switching time versus temperature for polymer 6 ($M_w = 15,000$). (From Ref. 23.)

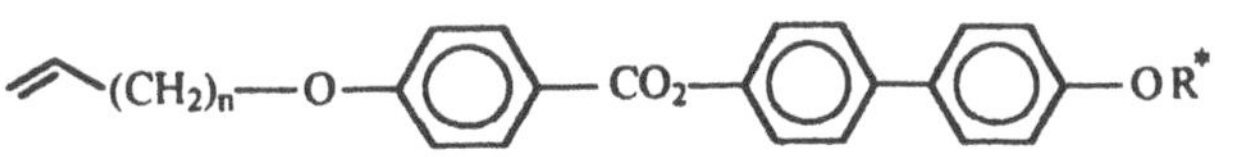

Figure 15 First switchable polysiloxanes. (From Ref. 23.)

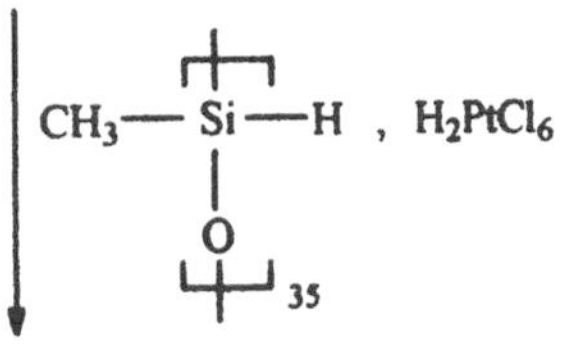

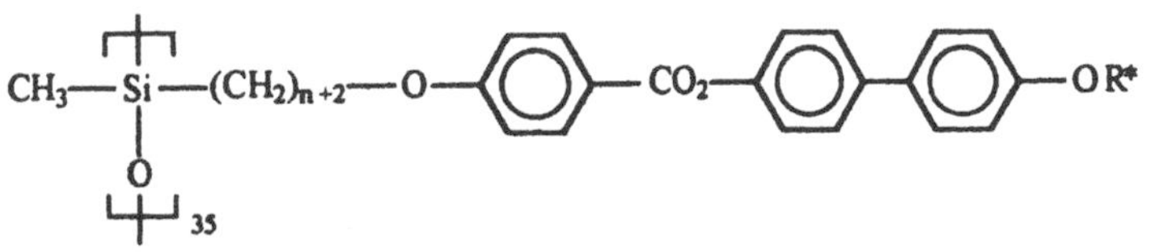

8	n	R*
a	8	(R)-2-(octyl)
b	6	(R)-2-(octyl)
c	4	(R)-2-(octyl)
d	8	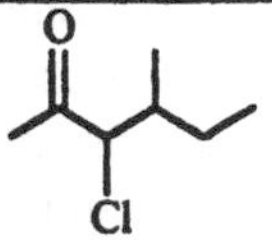 (2R,3S)
e	4	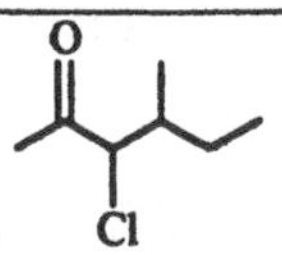 (2R,3S)

fields, domains develop, which can be switched between dark and bright states in a DC field of 30 V. In polymer 8b, the switching process is observable in two different smectic phases between 80 and 150°C. The response time is 200 ms at 130°C. Due to poor alignment, no spontaneous polarization was measurable.

The first report about measurements of spontaneous polarization in chiral polysiloxane FLCs came from Suzuki et al. [24]. The polymer structure and the phase transitions are displayed in Table 5. Polymers 9a–9c exhibit S_C^* phases at room temperature. Incorporation of an oxygen atom in the spacer (10a–10b) resulted in loss of the S_C^* phase in favor of a S_A phase. The spontaneous polarization for 9b and 9c is shown in Figure 16. Elongation of the spacer by only one CH_2 group cause a distinct increase of the polarization. The electrooptical response decreases about one order of magnitude when

Table 4 Polysiloxanes 8a–8e: Transition Temperature (in °C) of the Precursor Alkenes (7a–7e) and the Polymers

No.								Switchable
7a	C 61.7		S_C^* 101.7	S_A 105.6	N^* 114.6	BP_I 116.0	BP_{II} 116.1 I	+
8a		S_B 66.7	S_1 66 S_2 172 I					+
7b		S_x 53.8	S_C^* 75.6	S_A 79.1	N^* 101.8	BP_I 104.5	BP_{II} 104.7 I	+
8b		S_1 72 S_2 155 I						+
7c	C 44.9		S_C^* 76.2		N^* 104.9	BP_I 107.6 I		+
8c		S_1 111 S_2 170 I						+
7d	C 84.0		S_C^* 130.0	S_A 139.8	N^* 153.9	BP_I 154.4	BP_{II} 154.7	+
8d	T_G 250 I							−
7e	C 64.3	S_x 92.1	S_C^* 112.8	S_A 116.1	N^* 164.4	BP_I 165.0	BP_{II} 165.1 I	+
8e	T_G 248 I							−

Source: From Ref. 23.

Table 5 Polysiloxanes 9a–9c and 10a–10b and Their Phase Transition Temperatures

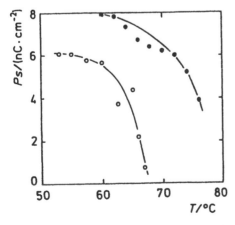

9a-c

10a-b

9: -R- = -(CH$_2$)$_{\overline{n}}$

10: -R- = -(CH$_2$)$_3$-O-(CH$_2$)$_{\overline{n}}$

	9a	b	c	10a	b
n	6	10	11	6	8

9a	SmC* $\underset{66\text{–}60}{\overset{66\text{–}72}{\rightleftarrows}}$ I
b	SmC* $\xleftarrow{78}$ SmA $\underset{112\text{–}101}{\overset{107\text{–}118}{\rightleftarrows}}$ I
c	SmC* $\xleftarrow{70}$ SmA $\xleftarrow{122\text{–}128}$ I
10a	SmA $\underset{73}{\overset{72\text{–}79}{\rightleftarrows}}$ I
b	SmA $\underset{83\text{–}75}{\overset{73\text{–}91}{\rightleftarrows}}$ I

Source: From Ref. 24.

Figure 16 Spontaneous polarization (P_s) for polymers 9b and 9c measured by the triangular wave method. (Ref. 24.)

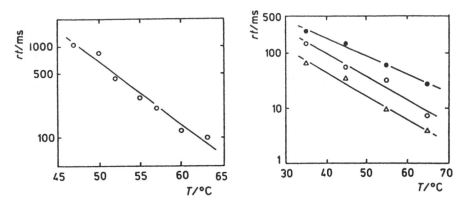

Figure 17 Electrooptical response time of 9c (left, applied voltage 20 Vpp/μm) and 9b (right, appl. voltage (Δ) 20 Vpp/μm, (○) 10 Vpp/μm, (●) 4 Vpp/μm. (From Ref. 24.)

elongating the spacer from 6 to 11 CH_2 groups (Fig. 17). This demonstrates the importance of a good decoupling of the ferroelectric unit from the backbone to achieve short switching times.

4. Variation of the Chiral Moiety

D. Walba at al. were the first to introduce the chiral oxirane ring in a low-molar-mass ferroelectric LC. The fixed combination of two chiral centers and a strong lateral dipole-moment in the three-membered ring proved to be advantageous to obtain high values of spontaneous polarization [27]. By appending the low-molar-mass FLC described in [27] via a terminal double bond to a H-siloxane polymer, they obtained polymer 11 [28].

11

$$K \xrightarrow{95} S_C^* \xrightarrow{147} I$$

$$T_g \xleftarrow{85} S_C^* \xleftarrow{147} I$$

By using a novel combination of interaction with a rubbed polymer alignment layer on the surface of the electrodes and shear, an excellent alignment was achieved in a 3-μm cell, affording a monodomain of ordered polymerfilm. An electrooptical switching time of 3 ms at 85°C (15 V/μm driving field) was measured. At the same temperature a spontaneous polarization of +65 nC/cm² was determined. The low-molar-mass FLC itself showed only +45 nC/cm².

Very short switching has been achieved by Dumon et al. [29] with "diluted" polysiloxane FLC polymers (12) containing the powerful α-chloro-β-methylpentanoate chiral

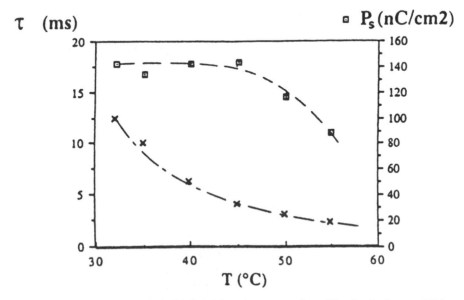

12

$$n = 8 \text{ or } 11 \qquad x = \frac{a}{a+b} \cdot 100 = 9, 17, 30 \text{ and } 100\%$$

$$a+b = 30$$

group. In these polymers only one-third of the silicon atoms in the backbone is connected with side groups. This lowers the transition temperature and the viscosity of these compounds. In a sample with a 30% fixation rate of the mesogens at the backbone, spontaneous polarization of about 140 nC/cm^2 has been measured (Fig. 18). The response time is in the range 3–12 ms. Two degrees below the S_C^*/S_A phase transition (73°C), a switching time of 300 μs under 2 V/μm has been measured. One has to point out that this

Figure 18 Spontaneous polarization (□) and response time (X) of polysiloxane 12 ($n = 11$, $x = 30$). (From Ref. 29.)

diluted polymer is rather an oligomer, because it consists of 30 silicon units with only 10 mesogenic side groups.

A further distinct improvement in the value of spontaneous polarization was achieved by Zentel et al. [30] when introducing a nitro group in ortho position to the chiral substituent at the terminal aromatic core of the mesogen (13). The nitro group causes a

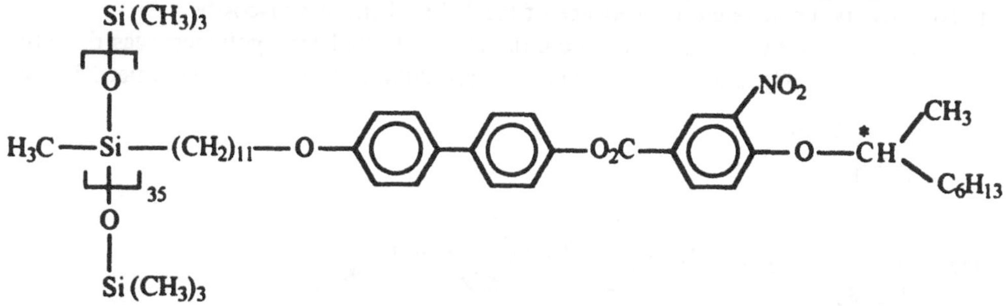

13

$$g \ 30 \ S_C^* \ 156 \ S_A \ 187 \ I$$

strong dipole moment perpendicular to the long axis of the mesogen in close proximity to the chiral center. In the broad S_C^* phase (30°C S_C^* 156°C), a spontaneous polarization between 250 and 420 nC/cm^2 was measured in the temperature region 110–155°C. Its unusual decrease with decreasing temperature was attributed to the increase of the viscosity. The strong dipoles create strong internal fields. This also causes a moderate switching speed between 2 ms at the transition to the S_A phase at 155°C and 90 ms at 106°C.

By replacing 6% of the chiral nitrosubstituted part of the mesogen in polymer 13 with the nonlinear optical chromophore 2-nitro-5-dimethylaminobenzoic acid, a frequency doubling of a Nd:YAG laser beam has been demonstrated.

By replacing the chiral nitrosubstituted benzoate moiety of the mesogen with the α-chloro-β-methylpentanoate group (derived from isoleucine), a polarization of $P_S = 130$ nC/cm^2 was found. It should be mentioned that the introduction of those chiral moieties was accomplished by a polymer analogous esterification leaving 10–20% of the biphenylol hydroxy groups unreacted (21% in the case of α-chloro-β-methylpentanoate). Therefore a higher value of spontaneous polarization is to be expected when accomplishing a quantitative esterification.

The highest spontaneous polarization with more than 1.5 mC/cm^2 has been measured by H. Coles at al. [31] on polyacrylate (14). The response time was 10 ms at 20°C below

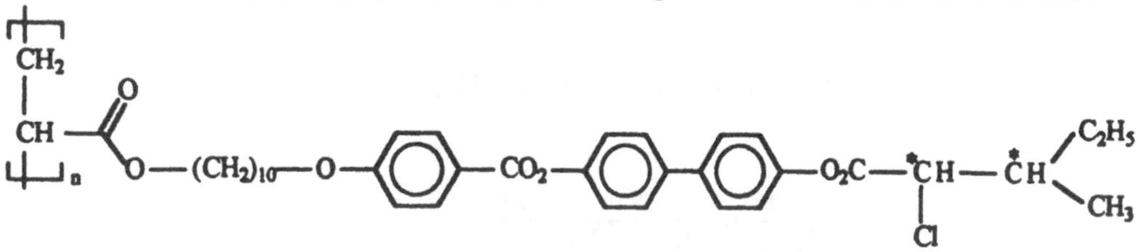

14

DSC (°C)   g -20 S_1 60 S_C^* 105 S_A 164 I
microscopy (°C) S_1 65 S_C^* 105 S_A 135-164 I

the $S_C^* \rightarrow S_A$ transition, decreasing to submilliseconds at 5°C below this transition. Anomalous behavior was observed insofar as the polarization decreases with decreasing temperature. This has also been observed in some other low-molar-mass and polymeric materials [31]. Apparently anomalous behavior also was observed for the tilt angle and the response time in the S_C^* phase below 80°C, and was discussed with respect to polydispersity, temperature dependence of the helix pitch, and viscosity.

An interesting variation in the chiral moiety for ferroelectric polymers was done by Zentel et al. [32,33] in using a molecule of axial chirality as shown in polymer 15. The

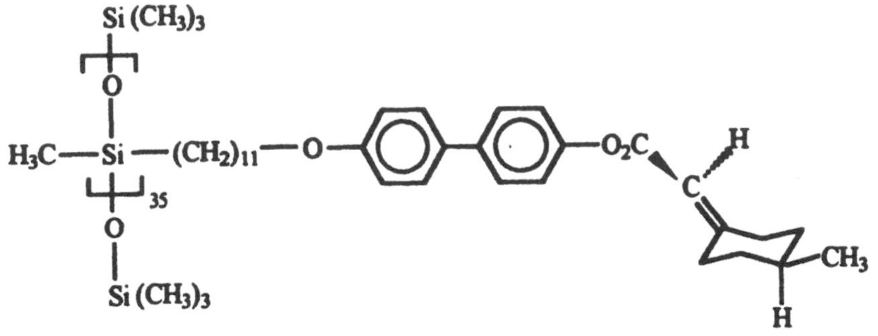

15

g 51 S_X 118 S_C^* 151 I

low polarization of $P_s = 18$ nC/cm^2 was attributed to the low formanisotropy of the chiral cyclohexene system. The use of a trifluoromethyl group as polar substituent at the chiral center of liquid crystals was found to be advantageous to achieve high spontaneous polarizations. Kitazume et al. [34] introduced trifluoromethyl-2-alkanoate moieties in polyacrylates and -methacrylates such as 16. They used no spacer to decouple the motion

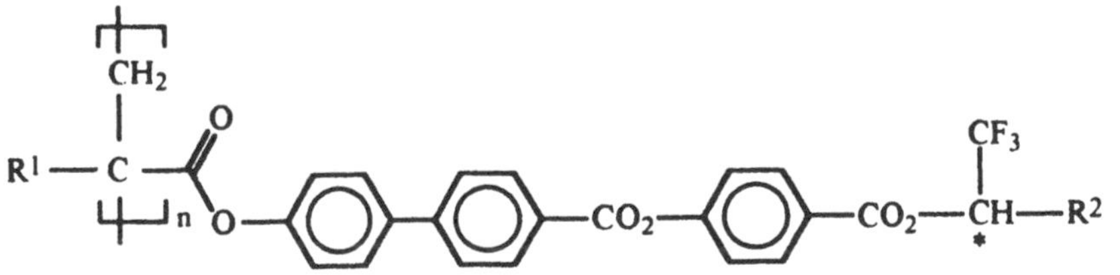

16

$R^1 =$ H, CH$_3$, $R^2 =$ C$_6$H$_{13}$, C$_8$H$_{17}$

$$g \xrightarrow[16]{20} S_{C^*} \xrightarrow[47]{58} S_A \xrightarrow[116]{127} I$$

$$(R^1 = H,\ R^2 = C_6H_{13},\ M_W = 51\ 500,\ \frac{M_w}{M_n} = 1.30)$$

of the ferroelectric side group from the main chain. The monomers did not show any mesophases. Nevertheless, the polymers exhibit S_C^* phases with the onset at room temperature. Response times are from 7.5 to 9.5 ms. Unfortunately, no data for polarization are reported.

When using a spacer [35], the response time is drastically reduced to about 1 ms for a short spacer of two methylene groups. With elongation of the spacer the response time increases [~4 ms for $(CH_2)_8$ and ~9 ms for $(CH_2)_{12}$]. It also increases when the CF_3 polar group is replaced by a $-CHF_2$ or a $-CF_2CF_3$ group. The increase in response time is related to a decrease in spontaneous polarization. The highest polarization ($P_S = 17$ nC/cm^2) was found for the short spacer polyacrylate. The respective polymethacrylates are distinctly slower in switching. A variety of 20 polymers, each of them in two or three different molecular weights ranging from $M_W = 37,000$ to 64,000, having a polydispersity of about 1.3, are presented in the above paper. The S_C^* phases start at ambient temperatures and are around 40°C broad.

A S_C^* phase ranging from -11°C to 53°C was found for the polysiloxane 17, which

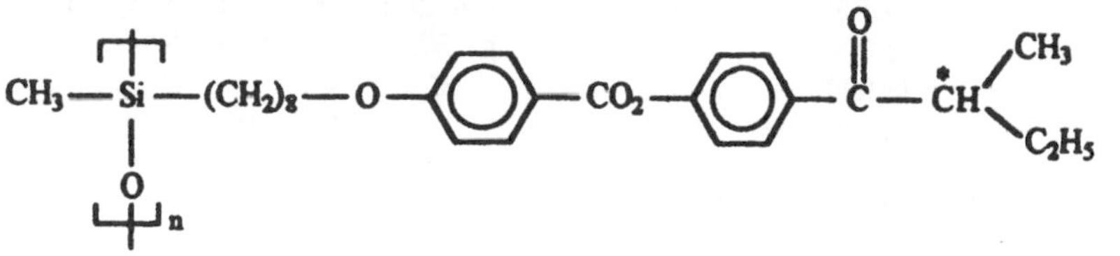

17

$$\text{g} \ -11 \ S_C^* \ 53 \ S_A \ 110 \ I \quad (\overline{M}_n = 7300)$$

contains a branched ketone as chiral moiety. Despite the low average molecular weight number ($M_n = 7300$), the response is rather slow (300 ms at 30°C and 33 ms at 43°C, that is, 10°C below the $S_A \rightarrow S_C^*$ transition). At this latter temperature the spontaneous polarization is 30 nC/cm^2 and the tilt angle is 16°. The authors, Takahashi et al. [36], stress the excellent memory effect above 99% even in a thick cell (18 μm), resulting from the smaler relaxation due to the viscosity of the polymer.

A favorable chiral moiety, which is derived from cheap lactic acid, has been used by Shashidar et al. [37] to produce polysiloxane polymers (e.g., 18) and copolymers with

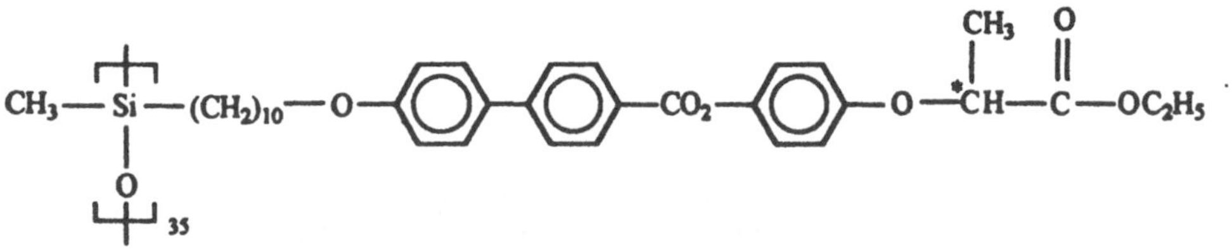

18

$$\text{g} \ 52 \ S_I^* \ \text{or} \ S_F^* \ 58 \ S_C^* \ 178 \ S_A \ 215\text{-}222 \ I$$

polarizations up to $P_S = 180$ nC/cm^2 and extremely fast switching down to 150 μs.

The temperature range of the S_C^* phase for the monomeric side group is about 40 K. This range is considerably enhanced to 120 K in the polymer. The advantages of the "diluted" copolymers are described in Section II.F.2.

Table 6 Phase Transitions of Polymers 19 with Different Spacer Lengths

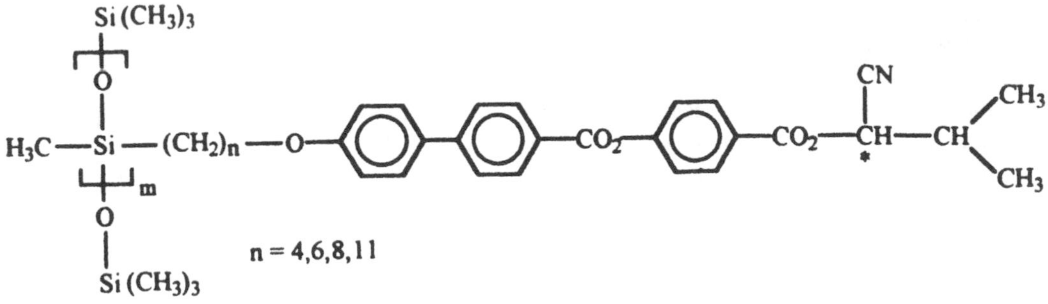

n = 4,6,8,11

19

Polymer	n	M_n	γ	Transition temperatures
19a	2	15,300	1.9	g 47.5 T₁ 106 S_C^* 174.9 S_A 230.3 I
19b	4	13,000	2.8	g 37.1 S_C^* 185.1 N* 257.9 I
19c	6	28,000	1.6	C 48.9 S_C^* 177.3 N* 275.6 I
19d	9	16,800	2.1	g 39.0 N* 271.7 I

M_n = number-average molecular weight; γ = polydispersity.
Source: From Ref. 38.

Chien et al. [38] introduced a cyanohydrinester as chiral building block with the strong polar nitril group located at the chiral center. The polysiloxanes 19a–19c exhibit broad S_C^* phases, as shown in Table 6. With a spacer length of 11 methylene units, only a cholesteric phase occurs. No ferroelectric properties have been reported. For the monomer of 19b, spontaneous polarization of more than 500 nC/cm^2 was measured.

5.   Chirality in the Spacer

According to one of the concepts outlined above, it was attempted to obtain FLC polymers by introducing a center of chirality into the spacer of a polymer that has a broad smectic C phase. Polymer 20, from Shibaev et al. [39], was chosen, which exhibits a S_C phase from 45 to 145°C.

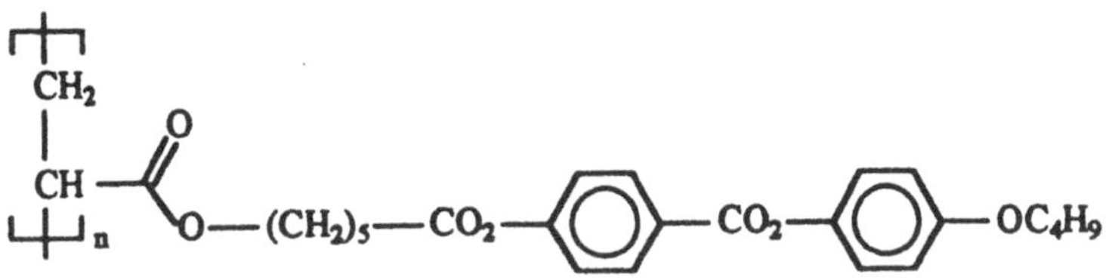

20

S_F 45 S_C 145 I

When incorporating a chiral branching in the α position to the ester carbonyl group near the mesogen, the tilted smectic phase was lost [25]. However, the introduction of a chiral branching in the β position resulted in polymer 21, which exhibited a S_C^* phase

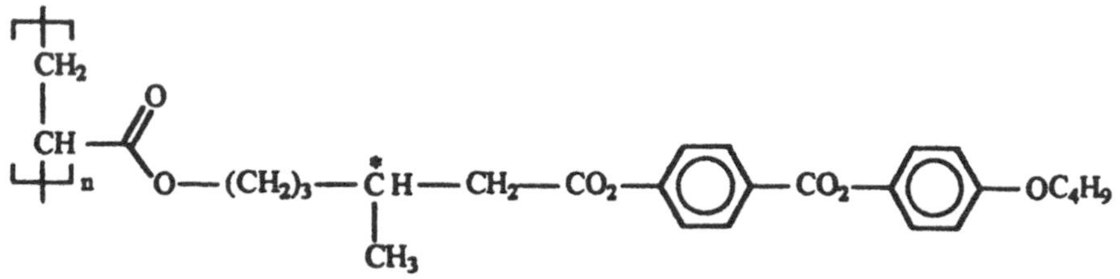

21

$$g \; 33 \; S_C^* \; 78 \; I$$

between 33 and 78°C. A distinct drop of the phase-transition temperatures was caused by the branching.

Unfortunately, no ferroelectric switching could be observed in this polymer [25]. Therefore we decided to elongate the spacer and to use the same mesogen as in polymer 6. As a result we obtained polymer 22, which exhibits a broad S_C^* phase that can be supercooled to room temperature [40].

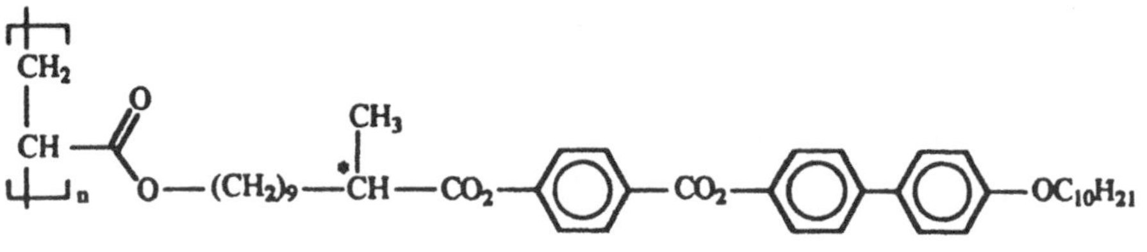

22

This first FLC polymer with a center of chirality in the spacer exhibits an interesting switching behavior. Between 70 and 95°C the switching times were independent of the applied field strength up to 30 V/μm, implying an electrolinic switching mechanism. An applied triangular waveform produced a "triangular" optical response. Above 95°C the switching times and optical responses became field dependent. As shown in Figure 19, at low voltages the electrooptic response behaved like a typical electroclinic effect ($E = 7$ V/μm), while at high voltages ($E = 25$ V/μm) ferroelectric switching is obvious. This seems to be the first case in which both electroclinic and ferroelectric switching could be observed at the same temperature by variation of the applied voltage [41]. Values of P_S were measured from 40 to 120 nC/cm^2 between 105°C and 123°C.

6. Two Remote Centers of Chirality in the Side Group

In polymer 23, one chiral center is located in the spacer, a second in the terminal alkyoxy

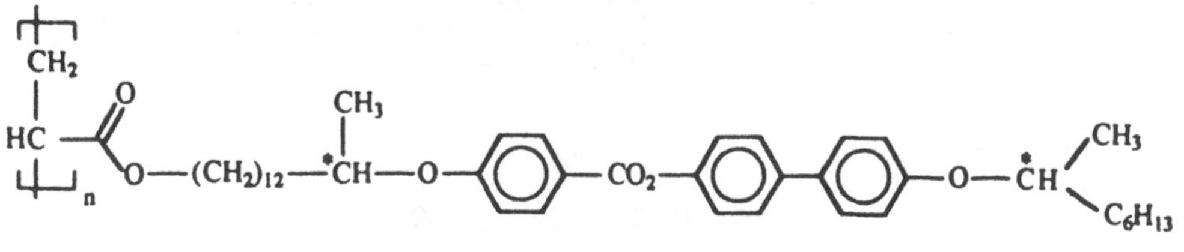

23

$$S_I \; 43 \; S_C^* \; 56 \; I$$

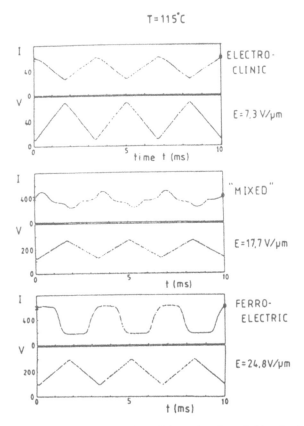

T = 115°C

Figure 19 Electroclinic and ferroelectric switching in the S_C^* phase of polymer 22 at different voltages and constant temperature (115°C). (From Ref. 40.)

group [40,41]. Both have the same absolute configuration R. Compared to polymer 6, which has only one chiral center terminally, polymer 23 has only two smectic phases. The additional branching causes a dramatic shift of the S_C^* region of about 75 K toward lower temperatures. The spontaneous polarization of about $P_s = 15$ nC/cm² is lowered by half.

A strong increase in the polarization by introducing a second chiral center was observed at polymer 24 by Takahasi et al. [42]. The value of $P_s = 330$ nC/cm² is higher

24

$$S_X \ 83 \ S_C^* \ 117 \ S_A \ 159 \ I$$

than the sum of the polarization for the two corresponding polymers, which have only one of those chiral centers in their side group ($P_s = 180 + 39 = 219$ nC/cm²). The corrected spontaneous polarization with regard to the tilt angle for 24 is $P_0 = 640$ nC/cm^2. One disadvantage is the shrinking of the temperature range of the S_C^* phase from about 100 K to 35 K, caused by the inclusion of the second chiral branching.

7. Two Adjacent Centers of Chirality in the Side Group

A powerful chiral building block to achieve high values of polarization—as already discussed above—consists of the α-chloro-β-methyl-pentanoate moiety (derived from isoleucin, using natures chiral pool). The second chiral center in the β position has a distinct impact on the size of the polarization. In polymer 25 (Dumon et al.), the chiral

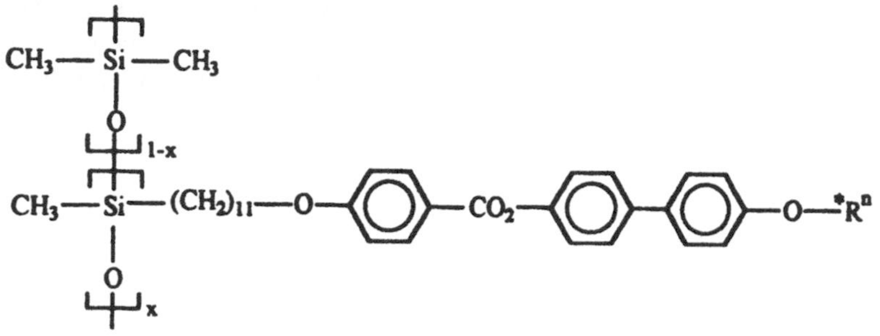

25

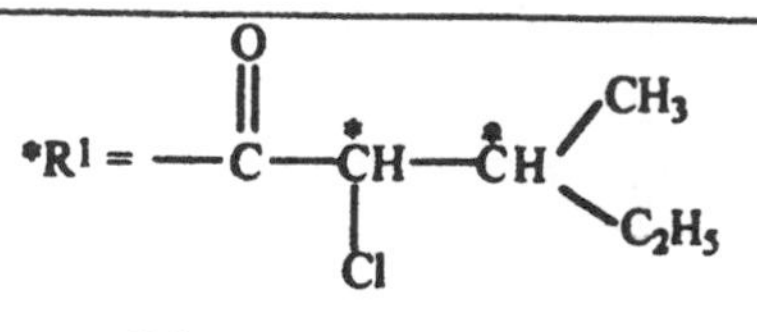

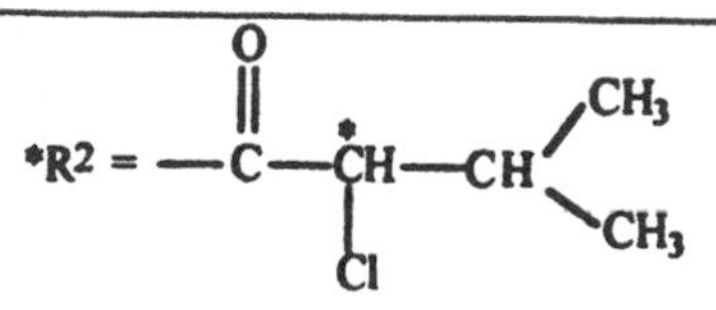

end group ;$^*R^1$ is responsible for a 30% higher polarization ($P_S = 95$ nC/cm^2 at $T_C - T = 45°C$) compared with the polymer having the chiral end group $^2R^*$ with only one chiral center [43]. Another favorable chiral moiety consists of an oxirane ring that contains two adjacent asymmetric-substituted carbon atoms, which are fixed to the laterally acting dipole of the three-membered ring. In polymer 26 a cis-disubstituted oxirane carboxylate

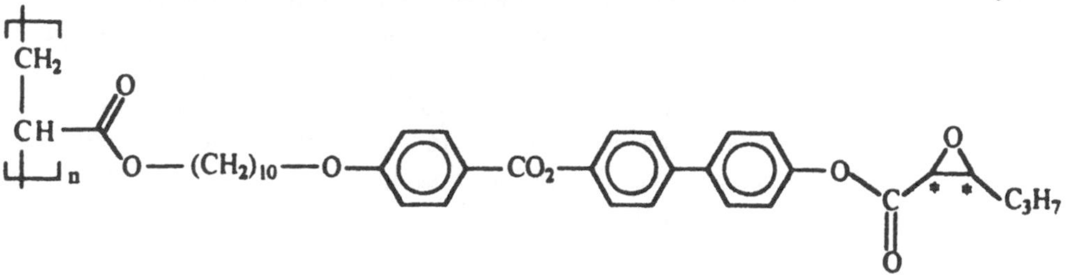

26

$$S_1 \ 93 \ S_2 \ 145 \ S_C^* \ 161 \ S_A \ 194 \ I$$

created a spontaneous polarization of $P_S = 90$ nC/cm^2 [40,41]. This polymer showed electrooptical switching in three different smectic phases. From 88°C to 143°C the switching is electroclinic-like, from 143.4°C to 160°C it is ferroelectric, and from 160°C to 169°C it is again electroclinic. The response time varies only from 500 μs to 4.5 ms (5 V/μm) over the whole temperature range (88–170°C).

By shifting the oxirane ring from the terminal position into the spacer, the properties of the polymer change dramatically. One observes a sign reversal of the spontaneous polarization within the $S_C{}^*$ phase. Details will be discussed in Section II.H.

8. Chirality in the Backbone

So far only one example has been reported of a FLC polymer which contains centers of chirality in the backbone [44]. It consists of an R,R-tartaric-acid-butanediol-polyester main chain and nitroazobenzene mesogenic side groups pending via ether linkages and spacers at the chiral centers (27). Despite the fact that the centers of chirality are located

27

Cr 48.5 $S_C{}^*$ 95.5 N* 114.5 I   ($\overline{M}_n = 6000$)

Cr 42.0 $S_C{}^*$ 108 N* 121.0 I   ($\overline{M}_n = 9000$)

far apart from the mesogens, a large spontaneous polarization between 300 and 500 nC/cm^2 was measured. This may be due to a restriction in internal rotation around the C2*–C3* bond in the tartrate unit, in part caused by the smectic ordering of the mesogenic side chains, leading to stereopolar coupling. The response time required for the realignment of P_S was 50 ms under an electric field of 100 V_{rms}/2 Hz at 80°C. From X-ray analysis a tilt angle of the mesogenic groups of 16° was derived. From a conoscopic figure an optically biaxial state was deduced for a sample aligned on application of an AC electric field of 6 MV_{rms}//5 kHz.

9. Influence of the Type of Backbone

The most popular polymeric main chains are of the polyacrylic and polysiloxane types. The former are prepared by radical polymerization, thus having sometimes wide molar mass distribution. It is not easy to repeat a synthesis to obtain approximately the same number-average molecular weight, M_n. With polysiloxanes it is possible to obtain ma-

Table 7 Polyacrylates of Low to Moderate Degree of
Polymerization: Moleculer Structure and Phase Transitions (in °C)

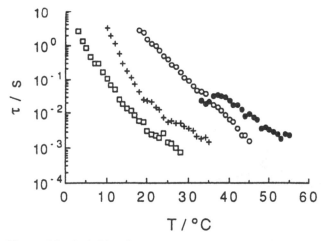

28

$\overline{M}_n$	g	$S_C{}^*$	S_A	I
3,100	−5	10	52	
4,800	−1	14	61	
5,900	9	29	90	
19,000	16	42	107	

Source: From Ref. 45.

terials with known molar mass. Their low glass transition temperatures and the possibility of having $S_C{}^*$ phases at ambient temperatures favor them for applications in displays.

$S_C{}^*$ phases at temperatures from below 0°C to 10–40°C have also been achieved with polyacrylates having a low to moderate degree of polymerization (M_n = 3100–19,000) [45]. The molecular structure and phase-transition temperatures of the corresponding polymer (28) are given in Table 7.

Figure 20 displays the switching times for polymer 28. With increasing M_n, the curves are shifted nearly parallel upwards. The response time is in the range from seconds to milliseconds. Because of polydispersity, the phase-transition temperatures from $S_C{}^*$ to S_A are imprecise. In the broad coexistence region of both phases to electrooptical switch-

Figure 20 Switching time versus temperature for different M_n of polymer 28. (From Ref. 40.)

ing times for the sample with the highest $M_n = 19,000$ is mainly electrocliniclike, as can be seen from the small tilt angle (Fig. 21), hence the contrast achievable is low.

A new type of backbone has been introduced in FLC polymers by the Idemitsu research group [45,46], leading to polyoxyethylenes like 29 (Table 8). They exhibit 100-K-broad S_C^* phases, with an onset at room temperature. Unfortunately, no data on the degree of polymerization were presented. The tilt angle especially for polymer P_3 of 29, is nearly constant over a broad temperature range (Fig. 22).

In Figure 23 the response times of the polymers are shown in comparison with their monomeric side groups. The value of τ for the polymers were 10^2–10^3 times larger. The steep decrease of τ at higher temperature (for example, $T > 125°C$ for P_1) is caused by a change of the type of switching which became electroclinic.

Further interesting results emerged when oxygen atoms were included in the spacer [46]. The inclusion of one oxygenation near to the main chain (30) resulted in an exten-

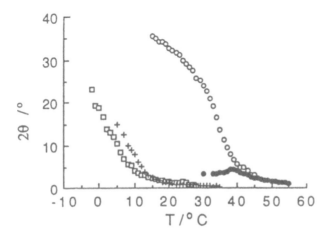

$X = O, CH_2$

30

sion of the S_C^* region and a decrease of the response time (Fig. 24). Further inclusion of oxyethylene units in the spacer caused the contrary, an increase in the response time. Polymers with a low degree of polymerization were obtained by ring-opening polymerization of terminally epoxidized alkenes in the presence of tin(IV) chloride. Only low degrees of polymerization have been achieved ($M_n = 1000$–5000).

Figure 21 Tilt angle versus temperature for different M_n of polymer 28. (From Ref. 40.)

Table 8 Structures and Phase Transitions of Polymers 29 (P₁–P₃) Determined on Cooling

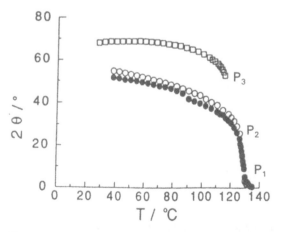

29

	P₁	P₂	P₃
R*:	$-\overset{*}{CH}\overset{C_3H_7}{\underset{CH_3}{<}}$	$-\overset{*}{CH}\overset{CO_2-C_2H_5}{\underset{CH_3}{<}}$	$-\overset{*}{CH}\overset{C_6H_{13}}{\underset{CH_3}{<}}$

g		S_C^*	S_A	I
P₁		30	130	148
P₂		23	132	152
P₃		22	122	?

Source: From Ref. 45.

A first example of a polyvinyl ether exhibiting a small S_C^* phase has been reported by Chiellini et al. [47]. No investigations on the ferroelectric properties have been published yet.

A comparison of polymers having the same side group but different main chains has been done by C.-S. Hsu et al. [48]. From the phase transitions of the polymethacrylate

Figure 22 Dependence of 2θ on temperature for polymers 29 (P₁–P₃). (From Ref. 45.)

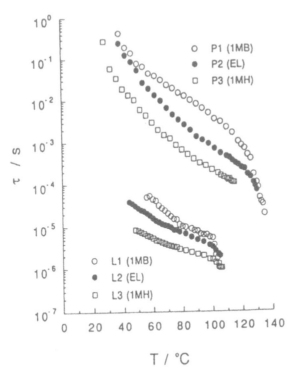

Figure 23 Comparison of the response times of the polymers 29 (P₁–P₃) with their corresponding monomeric side groups (L₁–L₃). (From Ref. 45.)

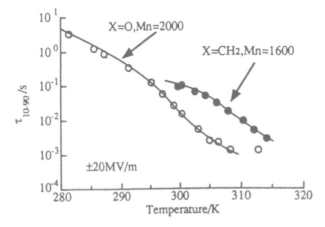

Figure 24 Influence of an oxygen atom in the spacer on phase behavior and response time of polymer 30. (From Ref. 46.)

31a versus polysiloxane 31b, one can derive that the more flexible polysiloxane backbone causes a lower glass transition and wider mesomorphic ranges.

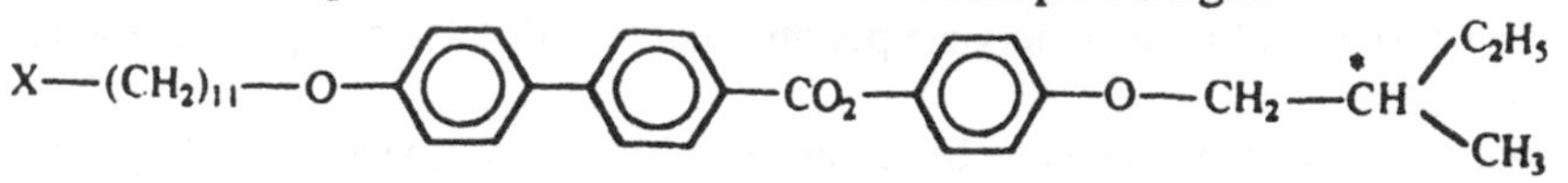

$$X-(CH_2)_{11}-O-\text{(structure)}-CO_2-\text{(ring)}-O-CH_2-\overset{*}{C}H\overset{C_2H_5}{\underset{CH_3}{<}}$$

31a-b

$$X = \quad CH_3-\overset{\overset{\displaystyle CH_2}{|}}{C}H-CO_2- \qquad g\ 30.5\ Cr\ 101\ S_C*\ 132\ S_A\ 197\ I$$

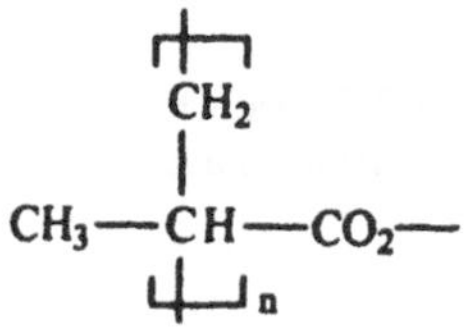

$$X = \quad CH_3-\overset{\overset{\displaystyle O}{|}}{\underset{m}{Si}}- \qquad g\ 20\ \ S_B\ 109\ S_C*\ 143\ S_A\ 218\ I$$

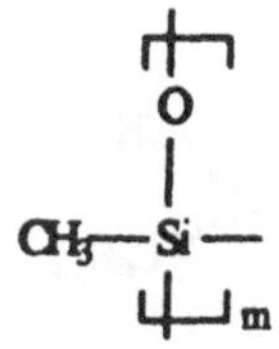

Variation of the spacer length revealed a chiral smectic C phase starting to exist from six methylene units in the case of the siloxane, whereas in the polymethacrylate series 11 methylene units are necessary to obtain an S_C* phase. No investigation of the ferroelectric properties has been reported.

Very recently, Tsai et al. [49] published a polyacrylate (32) in which the side chain

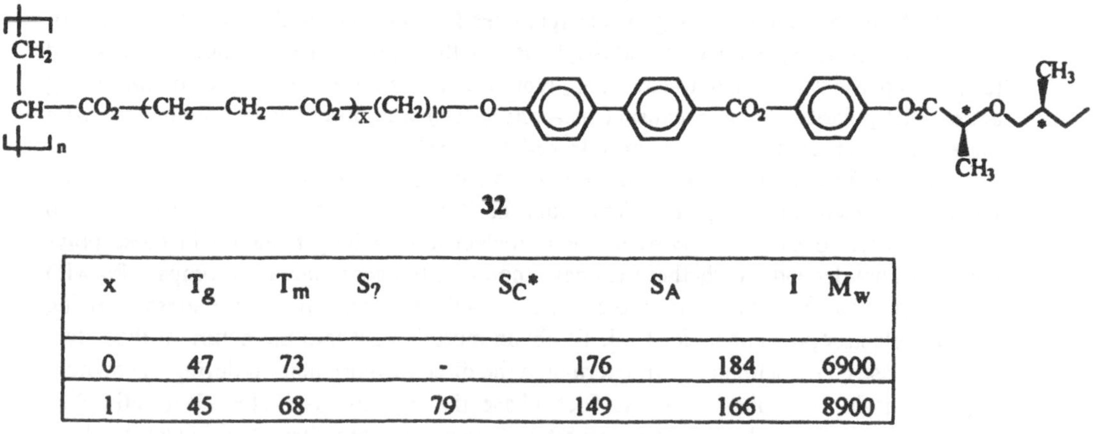

$$\overset{\overset{\displaystyle CH_2}{|}}{\underset{n}{CH}}-CO_2-(CH_2-CH_2-CO_2)_{\overline{x}}(CH_2)_{10}-O-\text{(structure)}-CO_2-\text{(ring)}-O_2C-\overset{*}{C}-O-\text{(chain)}$$

32

x	T_g	T_m	$S_?$	S_C*	S_A	I $\overline{M}_w$
0	47	73	-	176	184	6900
1	45	68	79	149	166	8900

was attached via an additional acrylate unit to the backbone. As a result, an additional smectic phase occurs below the S_C* phase and the phase temperature ranges shrink. The S_C* phase narrows from 103 K to 60 K. The spontaneous polarization ($P_s = 62$ nC/cm^2) and the response time ($\tau = 3$ ms) are not affected by the additional acrylate unit.

10. Spacer Length Variations

It was the general opinion that a spacer—consisting mainly of 6–12 methylene groups— is necessary to decouple the motions of the main chain and the smectogenic mesogenic part in the side group, thus allowing the latter to form the tilted smectic C phase. Very few systematic investigations are known concerning the influence of spacer length on the formation of S_C^* phases and their ferroelectric properties. Hsu et al. [48] showed for a polysiloxane that at least six methylene groups are necessary for the occurrence of the S_C^* phase. The corresponding methacrylate polymer needed 11 CH_2 groups to allow the S_C^* phase formation.

For polyacrylates, S_C^* phases have been observed for a short spacer (2 CH_2 groups) and long spacers (10–11 CH_2 groups) [13], but not in the medium range (4–8 CH_2 groups). With a chiral branching in the spacer a S_C^* phase was found with a spacer length of six carbon atoms (structure 20) [25].

Kitazume et al. [34] described a polyacrylate (16) with a broad S_C^* phase, using no spacer at all. A systematic investigation of polymer 33 concerning phase behavior and

33

ferroelectric properties depending on the spacer length (n = 2, 6, 8, 11) and comparison with the monomers has been done by Kühnpast et al. [50]. As displayed in Table 9, the monomers only show smectic phases with a monolayer structure (subscript 1), whereas the smectic phases of the polymer samples exhibit a rich polymorphism mainly of the bilayer type. Only in the case of a long spacer (polymer $P/_{11}$) do two modifications of a mono-layered S_{C1}^* phase appear (S_{C1x}^* and S_{C1y}^*). Below these phases a S_C^* phase was observed [51]. This phase is presumed to have a S_{C2}-type structure, with bilayer regions of alternating antiparallel polarity. Another unusual mesophase (UI) was observed in polymer $P/_2$ with the short spacer, exhibiting a chevronlike ordering [51].

Table 9 further shows that some polymers have S_C^* subphases which do not show ferroelectric switching properties. This puts the "polymeric" S_C^* phases in contrast to the monomeric ones. The occurrence of ferroelectric switching behavior in those poly-mers does not depend on whether they have mono- or bilayer structures (compare $P/_6$ with $P/_{11}$). In polymer $P/_{11}$, there are two S_C^* phases with an identical X-ray pattern showing different electrooptical behavior [52]. To obtain more insight into the nature of those two S_C^* subphases, Giesselmann [50] determined the director reorientation during the switch-ing process in both phases. In the S_{Cx}^* phase the director oscillates electrocliniclike around a position which is located accurately between the two ferroelectric states, which are adopted in the S_{Cy}^* phase, thus representing a third state. As reason for the fact that the director in the S_{Cx}^* phase cannot be forced by the field to reach the ferroelectric states, an eclipsed conformation of the side groups with respect to the polymer backbone was discussed.

In Figure 25a the spontaneous polarization for three polymers ($P/_6$, $P/_8$, and $P/_{11}$) is plotted versus temperature. In case of polymer $P/_8$, both S_{C2}^* phases (x and y) exhibit ferroelectric properties. The values of the spontaneous polarization for the polymers tend to be higher compared with the corresponding monomers (Figure 25b). This may be due to a slight increase in the rotational hindrance of the mesogens around their molecular long axes as a result of the link to the polymer chain.

Table 9 Transition Temperature (in °C) of the Monomers $M/_n$ and the Corresponding Polyacrylates $P/_n^{[a]}$

33

	n	$M_w^{[b]}$ (g/mol)	Transition temperatures (°C)
$M/_2$	2		C U1$^{[c]}$ 192 S$_{C2}^*$ 76.8 S$_{A1}$ 82.1 N* 91.6 BP 92.3 I
$P/_2$	2	117,000	g 74 214 S$_{A2}$ I
$M/_6$	6		C 77.3 S$_{A1}$ 94.2 N* 99.3 BP 99.6 I
$P/_6$	6	83,000	g 56 S$_{F2}^*$ 80 S$_{C2x}^*$ 148 S$_{C2y}^*$ 197 S$_{A2}$ 216 I
$M/_8$	8		C 79.9 S$_{A1}$ 94.8 N* 97.1 BP 97.5 I
			↑
$P/_8$	8	68,000	g 45 S$_{B1}$ 46.7 S$_{C1}^*$ 59 S$_{C2x}^*$ 131 S$_{C2y}^*$ 62.0 I
$M/_{11}$	11		C 53.4 S$_{B1}$ 58.5 S$_{C1}^*$ 123 74.6 S$_{A1}$ 91.5 N* 93.5 BP 167 I
$P/_{11}$	11	38,000	g 40 S$_{F2}^*$ 84 S$_{C}^*$ 123 S$_{C1x}^*$ 141 **S$_{C1y}^*$ 158** I

[a] Phases which exhibit ferroelectric properties are shown in boldface type. The subscript 1 denotes monolayer, and the subscript 2 denotes bilayer.
[b] Measured by GPC (polystyrene standard).
[c] Phase of the S$_C^*$ type [1].
Source: From Ref. 50.

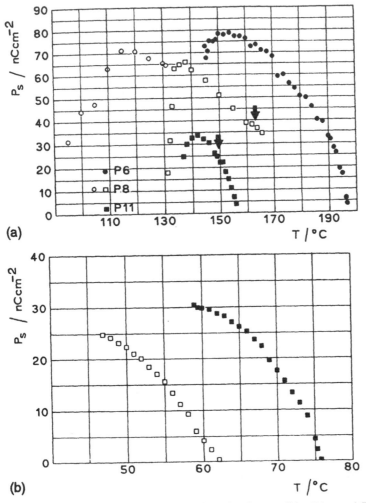

(a)

(b)

Figure 25 (a) Spontaneous polarization of polymers $P/_6$, $P/_8$, and $P/_{11}$ as a function of temperature (the arrows indicate the temperatures at which the polymers began to clear). (b) Spontaneous polarization of monomers M/8 and M/11 versus temperature. (From Ref. 52.)

11. Molecular-Weight Dependencies

During variation of the molecular weight, Kühnpast et al. [52] observed a change in the type of mesophase. In Figure 26 the DSC traces for three samples of polymer 33 are shown. The sample with a M_w = 15,000 g/mol (M_w/M_n = 1.5) exhibits a S_B phase at low temperatures. At higher molecular weights the low-temperature phase changes to a double-layer smectic F phase (Sm_{F2}^*). The sample with M_w = 240,000 exhibits three subphases in the Sm_C^* region, a double-layer (Sm_{C2}^*) phase and two monolayer smectic C phases (Sm_{1x}^* and Sm_{1y}^*), which show different electrooptical switching behavior (as discussed above). Going down to lower molecular weights, the double-layered smectic phase changes

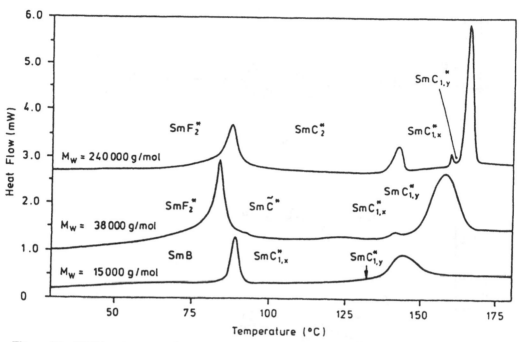

Figure 26 DSC heating traces for samples of polymer 33 with different molecular weights (heating rate 10 K/min). (From Ref. 52.)

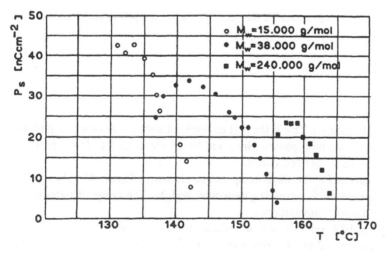

Figure 27 Temperature dependence of the spontaneous polarization for samples of different molecular weights (polymer 33). (From Ref. 52.)

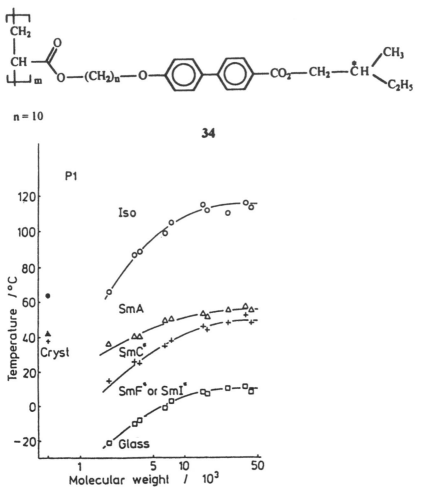

34

Figure 28 Molecular-weight dependence of the phase-transition temperatures for polymer 34. (From Ref. 53.)

to a smectic C tilde phase ($S\tilde{m}_C^*$). This is the first example of such phase type in a polymer [52]. The two differently switching phases (Sm_{1x} and Sm_{1y}) are shifted to lower temperatures and their transition peaks are broadened.

The polarization curves for the three samples are shown in Figure 27. The spontaneous polarization decreases with increasing molecular weight. The range of ferroelectric behavior is shifted to higher temperatures.

For another type of FLC polyacrylate the phase transitions for different molecular weights have been determined by K. Yuasa et al. [53]. As shown in Figure 28, the phase-transition temperatures increase with increasing molecular weight, leveling off around $M_n = 25,000$ g/mol. The S_A and S_F^* or S_I^* regions are expanding with the degree of polymerization, the S_C^* phase on the contrary becomes narrower and levels off.

In Figure 29 the optical response is plotted for different molecular weights at ±30 V applied across a 2-μm-thick cell. The response time reached submilliseconds at higher temperatures. It was one or two orders of magnitude below those of nematic low-molar-mass liquid crystals, but still two to three orders above those of low-molar-weight FLC. The response time became longer with increasing molecular weight but saturated around $M_n = 2 \times 10^4$.

For a polysiloxane FLC with two remote centers of chirality in the side group (polymer 24, page 461), K. Takahashi et al. [42] determined the dependence of phase transitions, spontaneous polarization, and response time on the molecular weight (Figs. 30–32). As the molecular weight increases, S_A and S_X phase ranges become broader, however, no change is observed above molecular weights of 10,000 (Fig. 30).

The spontaneous polarization (Fig. 31) decreased with increasing molecular weight, becoming almost constant above $M_n = 7,000$.

The response time for two samples is displayed in Figure 32. It decreases nearly linear with increasing temperature for both samples, although the slopes are different in rise. Since the spontaneous polarization is almost the same, the viscosity has to be the main reason for the different response times. The shortest response time is around 100 μs at 115°C for $M_n = 7,700$.

An extension of the idea to achieve fast electrooptical switching by using oligomers resulted in the synthesis of chiral twin dimers 35, composed of ferroelectric mesogens

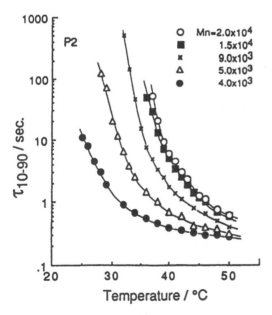

Figure 29 Temperature dependence of optical response time for various molecular weights of polymer 34. (From Ref. 53.)

linked by a dimethylsiloxane spacer ([54]). Compound 35 exhibits both ferroelectric and

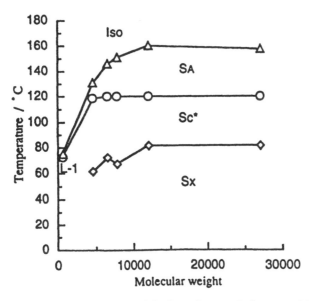

35

Sc* 37.5 S$_A$ 50 biphasic 60 I

electroclinic switching. The S$_C$* phase extends from subambient temperatures to 37.5°C
followed by an S$_A$ phase up to 50°C. At room temperature a tilt angle of 23°C was
measured, close to the optimum value for birefringance displays of 22.5°C. Within the
S$_C$* phase the response is fast, i.e., <100 μs over a range of 7 K below the S$_C$*–S$_A$*
transition (for an applied field of 12.5 V$_{rms}$/μm). A pronounced electroclinic effect with
fast response times (< 50 μs) and large induced tilt angles (17°) has been observed.

As the value of the helical pitch in the S$_C$* phase is of importance in connection
with the application in electrooptical displays, as discussed earlier, its dependence on the

Figure 30 Molecular-weight dependences of phase-transition temperature for polymer 24. (From
Ref. 42.)

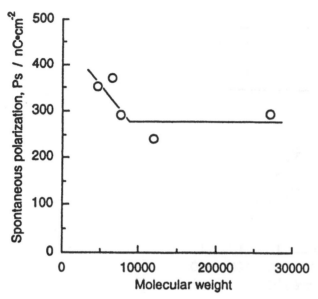

Figure 31 Spontaneous polarization versus molecular weight. (From Ref. 42.)

molecular weight was investigated by Endo et al. [55]. As shown in Figure 33 for M_n = 4,000–15,000, the pitch increases slightly with increasing molecular weight.

12. Influence of Polydispersity

There is only one report discussing the effect of the polydispersity of a FLC polymer on the polarization. Endo et al. [56] synthesized the polysiloxane 37, having a degree of

37

$$T_g \text{ -18 } S_C\text{* 65 } S_A \text{ 75 I}$$

$$M_W = 12\,000, \quad M_n = 8\,800; \quad M_W/M_n = 1.36$$

polymerization $P = 35$ and a broad S_C* phase from −18°C up to 65°C. At 40°C the spontaneous polarization was $P_s = 120$ nC/cm^2. From the shape of the measured reversal current signal in comparison with calculated signal shapes for different molecular weights was presumed that the components having different molecular weights respond cooper-

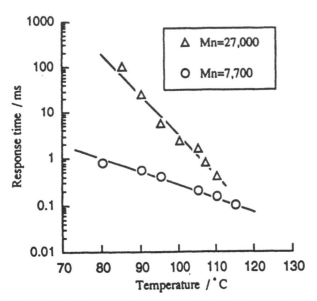

Figure 32   Response time versus molecular weight. (From Ref. 42.)

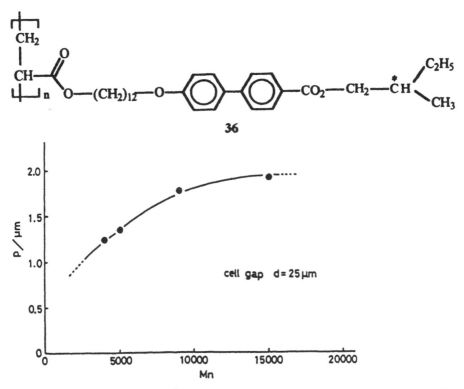

Figure 33   Dependence of the helical pitch on molecular weight (M_n), measured in a 25-μm cell at $T_{SA/SC}{}^* - T = 5$ K for polymer 36. (From Ref. 55.)

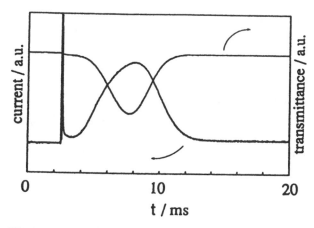

Figure 34 Simultaneous measurement of the change in polarization reversal current and the transmitted light intensity. (From Ref. 56.)

atively with each other to the applied voltage. Thus the shape of the current signal is independent of the polydispersity of the polymer. The authors also measured simultaneously the change of the polarization reversal current and the transmitted light intensity with the applied square wave voltage. As shown in Figure 34, the positions of the peaks are mostly the same. They also were independent of temperature. From this result it was deduced that the core moiety of the side group (whose optical anisotropy is responsible for the light transmittance) moves as one unit with the dipole on applying the electric field.

The evaluation of the rotational viscosity versus temperature using the data of polarization and response time resulted in the Arrhenius plot shown in Figure 35. The reason for the remarkably larger rotational viscosity of the polymer compared with low-molar-mass FLSs was discussed in terms of resistance forces caused by the backbone. The motion of the backbone was observed by using time-resolved FT-IR. It was found that

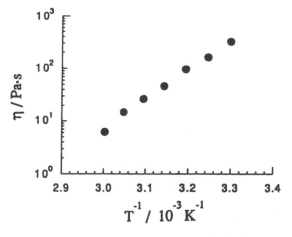

Figure 35 Arrhenius plot for the rotational viscosity of polysiloxane 36. (From Ref. 56.)

the Si—O bond in the main chain moves at the same time as the side chain moves. This coupling of motions between main chain and side group is supposed to be the origin of a resistance that causes the apparent large rotational viscosity of FLCPs.

F. Ferroelectric LC Copolymers

1. Copolymers with Different Chiral Side Groups

As the natural helical structure of the S_C* phase causes the compensation of the spontaneous polarization in a bulk sample, one has to unwind or suppress the helix to achieve a net polarization in a display configuration. The helical pitch has to be larger than the cell gap. In low-molar-mass FLCs an extension of the helical pitch has been achieved by mixing right- and left-handed chiral components (preferably with the same sign of polarization). In FLC polymers this goal was realized by copolymerization of right- and left-handed monomers of different molecular structure [56].

In Figure 36 the structure of the copolymer 38 is shown together with the dependence of the helical pitch on the composition m of the copolymer. The sign of p^{-1} represents the sense direction of the helicoidal twist. The value of p^{-1} almost exhibits a linear dependence on the composition of the copolymer (m). The pitch becomes infinite when the composition m is nearly 0.5. In all compositions, the pitch value for a 25-μm cell was larger than that of a 100-μm cell, presumably because of wall anchoring effects.

2. "Diluted" FLC Polymers

To obtain short response times, a decoupling of the mesogenic moiety from the backbone by using an appropriate spacer length—as discussed earlier—proved to be advantageous. A further improvement in this respect was achieved by introducing spacers in the backbone, thus keeping the mesogenic pending groups apart from each other. Those polymers we call "diluted" ones. The striking effect of this procedure has been demonstrated by Dumon et al. [29] and by Shashidhar and Naciri [37].

Figure 37 shows the chemical structures of the ferroelectric side group precursor 39 with a terminal double bond and the copolymer 40, obtained by the hydrosilylation reaction between 39 and the copoly (methyl-hydrogen siloxane) together with the phase-transition temperatures for 39, 40, and the homopolymer 18. The temperature range of the S_C* phase for both polymers is 120 K, thus considerable enhanced compared with the side group precursor 39, which has only a 40-K-wide S_C* range. The S_C* phase of the copolymer is extended to room temperature.

The effect of dilution is clearly demonstrated in Figures 38 and 39. The spontaneous polarization for the homopolymer reached 105 nC/cm^2 at 110°C, whereas for the copolymer the highest P_S was 180 nC/cm^2 at 40°C. A comparison of the switching times for both polymers is shown in Figure 39.

For both polymers the switching is very fast, less than a millisecond close to the S_C*–S_A transition. Also, copolymer 40 exhibits a clear bistable ferroelectric switching even at room temperature (albeit slow, $t \approx 600$ ms). For both polymers a pronounced electrolinic effect in the S_A phase was observed. The shortest electronic switching time for the copolymer was less than 30 μs at $T–T_{S_C*}–S_A = 10$ K. Close to the S_C*–S_A transition, a tilt angle of nearly 18° is induced for an applied field of 100 V across a 10-μm cell.

An investigation of the influence of different spacer lengths in copolymer 40 on the switching time [57] revealed the shortest switching for the longest spacer [(—CH2)$_{10}$—] duo to the better decoupling of mesogen and backbone.

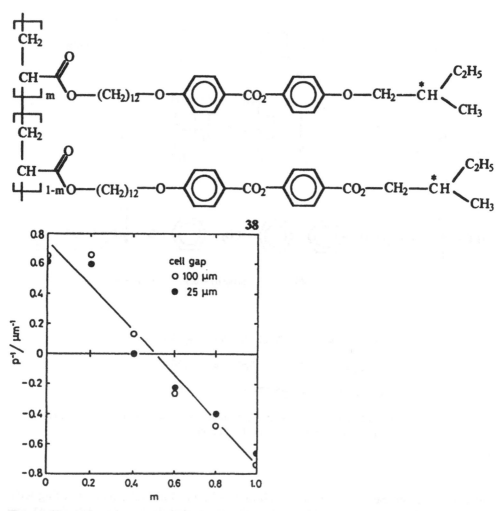

Figure 36 Dependence of the helical pitch on composition m of copolymer 38. The pitch was measured at $T_{SA-SC}^* - T = 5$ K. (From Ref. 56.)

In addition to the dilution, another principle to tune the S_C^* phase temperature range has been developed by Shashidar et al. [58]. They incorporated two different chiral mesogenes as side groups attached to the siloxane backbone to yield the twin copolymer 41, which exhibits a stable S_C^* phase down to 20°C. It shows very unusual P_s behavior. Instead of the typical power-law dependence, a linear variation with temperature exists throughout the S_C^* phase (see Fig. 40). As a result, the pyroelectric coefficient ($p = P_s/T$) of this material is temperature dependent. This is a very important feature for a pyroelectric device designed to operate as an infrared detector over a wide range of ambient temperatures. If a pyroelectric detector is used as a voltage source, its figure of merit is determined by the ratio p/ϵ (p = pyroelectric coefficient, ϵ = dielectric constant). In this respect the new ferroelectric polymer 41 compares well with solid ferroelectric materials that are currently used for infrared detection and imaging. Advantages of the

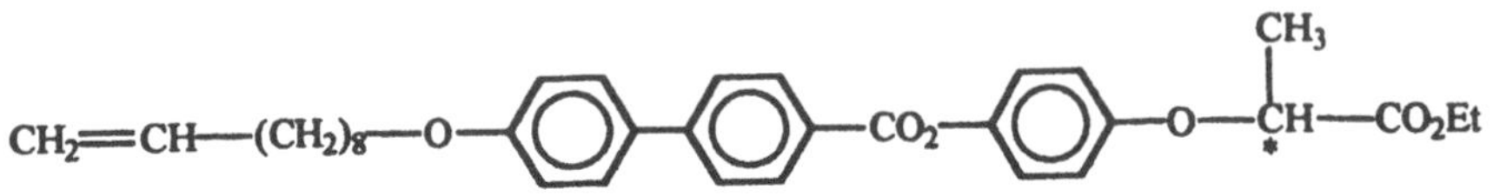

$$CH_2{=}CH{-}(CH_2)_8{-}O{-}\bigcirc{-}\bigcirc{-}CO_2{-}\bigcirc{-}O{-}\overset{\underset{|}{CH_3}}{\underset{*}{CH}}{-}CO_2Et$$

39

$$\left[\underset{\underset{H_3C}{|}}{\overset{\overset{CH_3}{|}}{Si}}{-}O\right]_a\left[\underset{\underset{H_3C}{|}}{\overset{\overset{H}{|}}{Si}}{-}O\right]_b$$

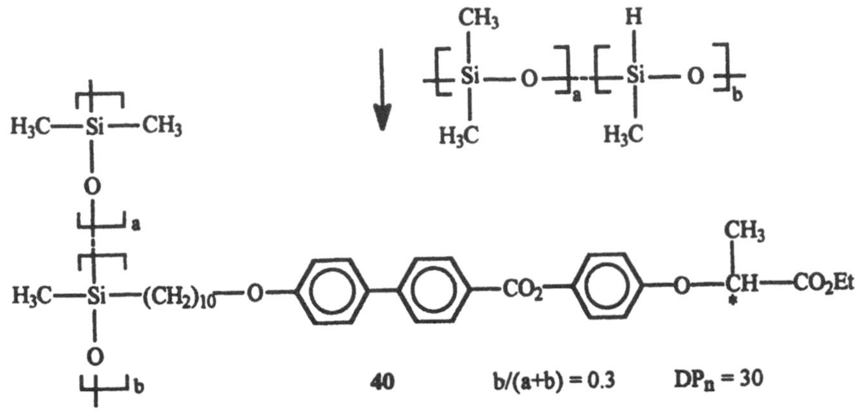

40 b/(a+b) = 0.3 DP$_n$ = 30

39 (S$_I$ or S$_F$ 55.5) Cr 67 S$_C$* 95.4 S$_A$ 125.4 I

40 T$_g$ (not found) Cr 15 S$_C$* 136 S$_A$ 150–162 I

18 T$_g$ 25 Cr 58 S$_C$* 178 S$_A$ 215–222 I

Figure 37 Structures of side-group precursor 39, diluted polymer 40, and phase transitions of 39, the homopolymer 18, and copolymer 40 (in °C). (From Ref. 37.)

FLC material are easily processable, well-oriented thin films and already existing technologies such as direct interfacing with a charge-coupled device. The data for 41 are: p = 6.8 nC/cm^{2}K; ϵ = 8; p/ϵ = 0.85 nC/cm^{2}K.

The effect of dilution in the area of the side groups has been investigated by K. Grüneberg [59]. A copolymerization of a ferroelectric acrylate monomer with hexylacrylate yielded the copolymers 42a–42c. Table 10 shows a comparison of the homopolymer and its diluted samples, 42a–42c. A 1:1 dilution shifts the S$_C$* phase more than 50 K downward, approaching ambient temperatures. Further dilution shrinks the S$_C$* phase, and at a dilution of 1:5 the mesophases disappear. The interesting tristable switching observed for the homopolymer (discussed in Sec. II.A), changes to a bistable ferroelectric switching in the diluted samples.

The spontaneous polarization for 42a is 30 nC/cm^2 at 40°C. The switching time decreases from 18 ms at 40°C to 2 ms at 90°C.

Chien et al. [38] investigated the effect of dilution in both the backbone and the side groups. The polysiloxane 43 contains, in addition to the chiral side group, the nonchiral strong polar cyanobiphenyl side group. As shown in Table 11, the glass transition temperature of the copolymer decreases with increasing content of the cyanobiphenyl moiety. The dilution effect of the cyanobiphenyl side group favors the formation of the S$_C$* phase,

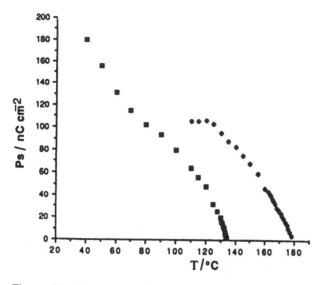

Figure 38 Temperature dependence of P_s for homopolymer 18 (■) and copolymer 40 (♦). (From Ref. 37.)

which was not present in the homopolymer with the same spacer length (see earlier). The same results from the dilution in the backbone (see 44). The very broad $S_C{}^*$ phase regions are impressive. Unfortunately, no investigation of the ferroelectric properties of those polymers are described in the paper.

3. Colored and Fluorescent FLC Polymers and Copolymers

Colored LC displays applying the guest host effect of dichroic dyes dissolved in a ferroelectric liquid crystal can afford considerable advantages over the birefringence device. Specifically, the stringent constraints on cell thickness, being 2 μm, inherent in the birefringence device, are relaxed in the guest host device, as dye absorption is far less susceptible to variations in the device thickness.

Coles et al. were the first to investigate the dye guest host effect in a FLC polymer [60]. The high-polarization polyacrylate 14 [31] (see above), containing 4% of a blue azo dye (M 483 from Mitsui Toatsu Chemical, Inc.), gave the absorption spectra displayed in Figure 41. From the absorption parallel and perpendicular to the direction of molecular orientation (director), an order parameter of 0.57,

$$S_{dye} = \frac{A_{\parallel} - A_{\perp}}{A_{\parallel} - 2A_{\perp}}$$

was calculated. This was almost identical to the value of 0.56 obtained for a 4% w/w solution of the same blue dye in BDH low-molar-mass ferroelectric mixture, indicating that the dichroic dye was cooperatively ordered by the polymer side groups in the same way as in the low-molar-mass system. The tilt angle of the guest host mixture was reduced approximately 10° compared with the homopolymer at equivalent reduced temperatures. The optical response time was enhanced, and submillisecond switching was achieved over a 10°C range below the $S_C{}^* \rightarrow S_A$ phase transition. One drawback to this

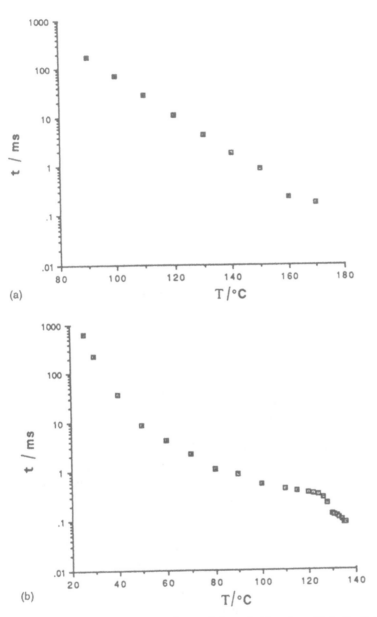

Figure 39 Temperature dependence of the switching times (t) in the S_C^* phase of homopolymer 18 (a) (10-µm cell, $U = \pm 34$ V) and the S_C^* and S_A phases of copolymer 40 (b). (From Ref. 37.)

type of guest host system is the poor solubility of dyes, especially in polymeric liquid crystals. By copolymerization of dye monomer with a ferroelectric monomer, this problem has been overcome. Colored FLC polymers containing 5% and 15% of an azo dye were synthesized by Scherowsky et al. and investigated by H. Coles [61]. The structure of the colored copolymer 45 and its properties are displayed in Table 12.

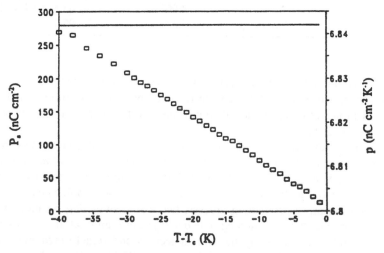

41   a + b = 0.3 DP_n = 30   a ~ 0.21, b ~ 0.09

Figure 40 Temperature variation of P_s (□) and the pyroelectric coefficient (solid line, $P = P_s/T$) for polymer 41. (From Ref. 58.)

The homopolymer and the copolymer containing 5% azo dye monomer show nearly identical phase-transition temperatures. With 15% dye content the transition temperatures decrease. The electrooptical switching starts at 60°C, in comparison with 80°C for the homopolymer, and is faster at the same temperatures. This copolymer exhibits switching in three different smectic phases, as shown in Figure 42.

In the hitherto unidentified S_x phase and the S_A phase, the electrooptical response was linear in the applied field (i.e., a triangular optical response was recorded for an applied triangular wave). In the S_C^* phase one needed fields of 25 V/μm to achieve true ferroelectric switching. For the spontaneous polarization an anomalous increase with temperature was observed, reaching ≥350 nC/cm². The orientational order parameter of

Table 10 Molecular Weights, Polydispersity, and Phase Transitions for a Homopolymer and Its Diluted Samples, 42a–42c

n = 1
m = 1, 3, 5

42 a-c

	M_w/M_n	M_w (g/mol)	Mesophases
Homopolymer	2.08	17,500	S_X 95 S_C* 150 S_A 175 I
Copolymer			
42a	2	4,000	S_F or S_I 40 S_C* 100 S_A 135 I
42b	2	4,000	T_G 85 S_C* 97 S_A 120 I
42c	2.5	3,500	T_G 95 I

the azodye component of the copolymer was measured at room temperature in the frozen smectic phase. Values of $S \geq 0.82$ were recorded near the absorption maximum at $\lambda = 360$ nm, demonstrating the high ordering of the guest molecules cooperatively achieved by the mesogenic host.

A further adaptation of the dye guest host ferroelectric device is the inclusion of an anisotropic fluorescent dye in a ferroelectric material. The operation principle of such device is displayed in Figure 43. If the dichroic dye molecules are oriented with their optical axes parallel to the director of the ferroelectric host molecules and follow their reorientation during the switching process, two stable switching positions for the fluorescent dye emerge. In the switching state displayed in Figure 43, the incoming polarized ultraviolet light is absorbed by the dye molecules and transformed into visible light, which is emitted to all directions and passes the UV filter. One achieves a bright appearance and a large viewing angle. In the second switching state the incoming UV light cannot be absorbed by the dye molecules; it passes the guest host cell and is absorbed in the UV filter. Thus one achieves a completely dark state.

Such display has been realized by using low-molar-mass liquid crystals and anthraquinone dyes [62]. One limitation is the poor solubility of such dyes in liquid crystals. A 0.5 wt% solution was investigated.

Scherowsky and Beer [63] included a fluorescent dye in a FLC polymer by copolymerization. As shown in formula 46, the fluorescent dye consists of a donor-acceptor substituted naphthalicimide. 46b, with the hexyloxy group as donor, exhibits the same phase-transition temperatures as the homopolymer, when containing 5 wt% dye comonomer. The spontaneous polarization of 46b is reduced of about 20% compared with the homopolymer (in maximum from 155 to 115 nC/cm^2). The tilt angle is diminished less than 10% (e.g., from 36° to 33° at 110°C). The response time of the fluorescent copolymer 46b is distinctly shorter, as shown in Figure 44. It reaches values below 1 ms

Table 11 Transition Temperatures, Molecular Weights, and Polydispersity of Copolymers 43a–43c and Polymer 44

Polymer	x	y	z	M_n	M_w/M_n	Transition temperatures
43a	5	95	—	13,000	2.5	g 4.2 S_C^* 154.9 I
43b	50	50	—	15,000	3.0	g 17.2 T, 43.2 S_C^* 215 I
43c	75	25	—	16,400	2.1	C 78.1 S_C^* 249.7 I
44	33.3	—	66.7	8,400	1.6	C 56.9 S_C^* 196.7 I

Source: From Ref. 38.

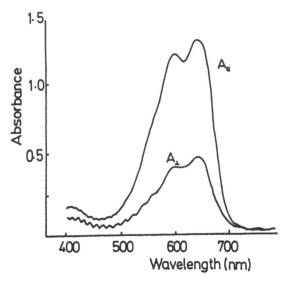

Figure 41   Absorption spectra of the blue dye (M 483) in the FLC polymer 14, showing parallel and perpendicular components relative to the director. (From Ref. 60.)

above 125°C. When using the principle of dilution with this type of copolymer, fluorescent devices with fast response at ambient temperature are foreseeable.

The dichroic order parameter of 46b measured at room temperature was $S = 0.53$. For the dye monomer, dissolved in the commercial FLC mixture FELIX 007 (Hoechst AG), an order parameter of $S = 0.67$ was measured.

When fixing a second methoxydonor group at the dye (see copolymer 46c), the order parameter is improved to $S = 0.63$, but now the S_C^* phase disappears, leaving a phase sequence S_B 75 S_A 160 I ($M_w = 11{,}800$ g/mol). Nevertheless, a fast electrooptical response was found. An electroclinic switching process in the S_A^* phase takes place with response times between 2.6 ms (80°C) and 400 μs (125°C). A tilt angle up to 12° was induced at voltages of 10 V/μm.

G.   Induction of Ferroelectricity in Polymers

In the case of low-molar-mass LCs, the phenomenon of induced spontaneous polarization is well known and established in commercial FLC mixtures. By doping smectic C materials with chiral dopants, spontaneous polarization can be induced. Advantages over pure S_C^* liquid crystals are tunable ferroelectric properties and high switching speed achievable by using a nonchiral, low-viscous S_C basic material and doping with small amounts of high-polarization S_C^* substances.

1.   Induction of Ferroelectricity by Chiral Dopants

The transformation of an S_C phase in a main-chain polymer into an S_C^* phase by adding a chiral mesogenic dopant has been described by Noel et al. [64]. Induced ferroelec-

Table 12 Structure of Colored Copolymer 45 and Its Properties

45

	Homopolymer	Copolymer 5% dye	Copolymer 15% dye
M_w	10,900	14,600	19,300
$E = \dfrac{M_w}{M_n}$	1.24	1.39	2.33
Mesophases and transition temperatures	g 75°C S_x 100°C S_C^* 140°C S_A 200°C I	g 75°C S_x 100°C S_C^* 140°C S_A 205°C I	g 60°C S_x 94°C S_C^* 110°C S_A 190°C I
Optical switching	20 ms (80°C)–1 ms (138°C)	10 ms (76°C)–400 μs (120°C)	30 ms (60°C)–below 100 μs (140°C)

Source: From Ref. 61.

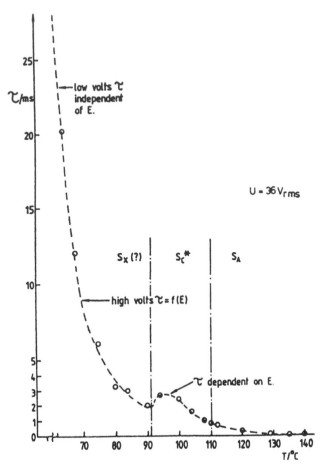

Figure 42   Response time τ versus temperature for copolymers 45 (15%, sample thickness 4 μm, $U = 36$ V$_{rms}$). (From Ref. 60.)

tricity was not reported. The first investigation on induced ferroelectricity in a polymer was done with the racemic polyacrylate 47 and the chiral dopant 48

47   (M_w = 12 500 g/mol, Mw/M$_n$ = 1.43)

$$S_B \xrightarrow{98.1} S_X \xrightarrow{109.6} S_C \xrightarrow{140.6} I$$

$$S_B \xleftarrow{87.0} S_X \xleftarrow{101.9} S_C \xleftarrow{136.6} S_A \xleftarrow{138.1} I$$

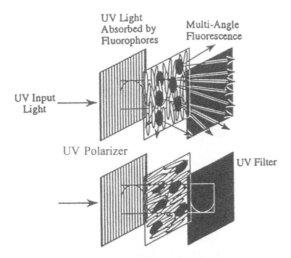

Figure 43 Dichroic absorption mode of a fluorescent dye-containing FLC device.

(Scherowsky et al. [65]). At concentrations of up to 15 mol% dopant, no change in the phase sequence of the doped polymer was observed. The phase-transition temperatures decreased about 15 K in maximum. The induced spontaneous polarization, depending on temperature and dopant concentration, is shown in Figure 45. P_S exhibits a maximum around 104°C, coinciding with the phase transition from S_X to S_C^* phase. Coming down from the S_i to a higher-ordered phase (S_X presumably is a S_i phase), one would expect a further increase in the spontaneous polarization. A similar decrease in P_S has also been observed in other systems (Ushida et al. [6]). The dependence of P_{Smax} on the dopant concentration is displayed in Figure 46. It shows a nonlinear increase.

With the goal of improving the switching speed of FLC polymers, Ido et al. [66] synthesized S_C oligomers with a polyethylenoxid backbone and induced spontaneous polarization by different chiral dopants. The best result was obtained with oligomer 49

$$\left[\begin{array}{c} O \\ | \\ CH_2 \\ | \\ CH \\ \end{array} \right]_n - (CH_2)_8 - O - \bigcirc - CO_2 - \bigcirc - CO_2 - C_4H_9$$

49

T_g -13 S_C 55 S_A 91 I

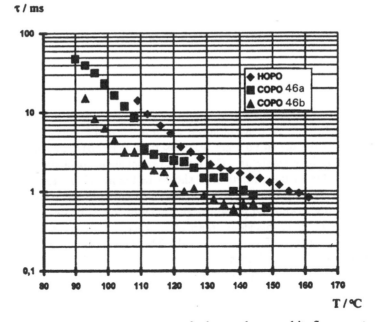

46 b: S_B 91 S_C^* (158 S_A) ~161 I   (M_w = 12 800)

Figure 44 Response time τ versus temperature for homopolymer and its fluorescent copolymers (5 wt% dye) 46b.

and the dopant 50. The dependence of the tilt angle (2θ is the switching angle) and

$$H_{21}C_{10}-\bigcirc-CO_2-\bigcirc-CO_2-\overset{*}{C}H-CO_2Et$$
$$\underset{50}{\phantom{H_{21}}}\qquad\qquad\qquad\underset{CH_3}{}$$

10 wt % mixture : T_g 5 S_C^* 45 S_A 87 I

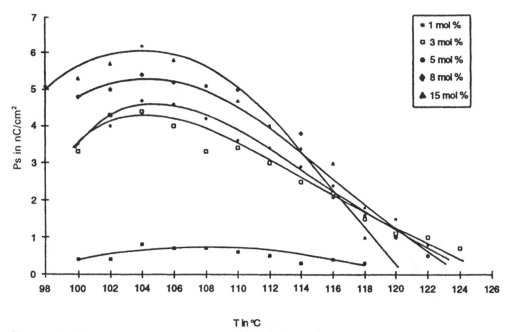

Figure 45 Dependence of induced spontaneous polarization for polymer 47 on temperature and concentration of dopant 48. (From Ref. 65.)

the spontaneous polarization ($10 \ \mu C/m^2 = 1 \ nC/cm^2$) on the temperature for a 10 wt% mixture of the dopant in oligomer 49 is plotted in Figure 47. Due to the small polarization, the switching time is—as shown in Figure 48—between 300 ms at 14°C and below 10 ms at 40°C near the S_C^*–S_A transition.

2. Induction of Ferroelectricity by Chiral Comonomers

Another approach to inducing ferroelectricity in a macromolecule consists of the introduction of chiral units as comonomers. Kozlovsky et al. [67] realized this approach by synthesizing a copolymer from a chiral monomer (which alone gives a homopolymer without a tilted smectic phase) and a monomer that alone gives a smectic F homopolymer.

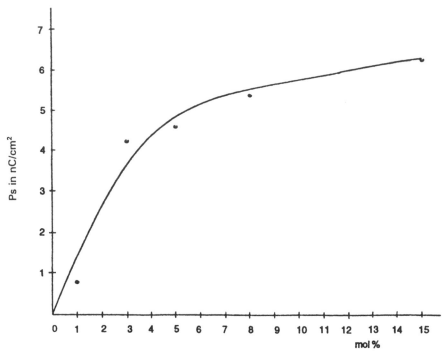

Figure 46 Dependence of the maximum of spontaneous polarization on dopant concentration. (From Ref. 65.)

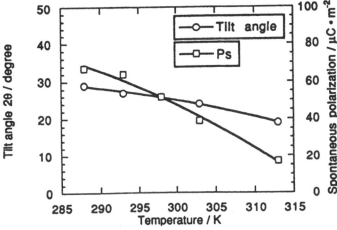

Figure 47 Tilt angle and spontaneous polarization versus temperature for 49 mixed with 10 wt% dopant 50. (From Ref. 66.)

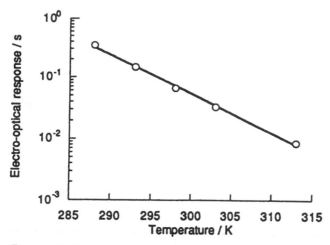

Figure 48 Electrooptical response versus temperature for 49 mixed with 10 wt% dopant 50. (From Ref. 66.)

Introduction of up to 37 mol% of the chiral monomer into the macromolecule of the smectic F polymer does not destroy the tilted smectic structure. At lower (8%) and higher (37%) percentages of chiral monomer, a monotropic S_C^* phase occurs. The spontaneous polarization which was calculated from the temperature dependence of the pyroelectric coefficient reached a maximum at 25 wt% chiral monomer ($P_S = 3$ nC/cm^2) in the S_F^* phase.

3. Induction of S_C^* Phases by Hydrogen Bonding

A novel approach to ferroelectric side-chain polymers has been developed by Kumar et al. [68]. Their strategy involves self-association of free carboxylic acid groups in side groups of polysiloxanes with pyridines containing mesogenic moieties as shown in formula 52. Intensive infrared absorption bonds at 2500 and 1930 cm^{-1} are proof of strong hydrogen bonding between the carboxylic acid and the pyridyl-nitrogen atom. The nonionic character of the complexes was confirmed by NMR measurements. The clearing points of the complexes are higher than those of the uncomplexed polymers and higher than the melting point of the alkoxy stilbazol component (m.p. = 111°C).

Another proof of the nonionic character of the hydrogen-bonded complexes is the fact that polarization measurement could be performed with the Sawyer-Tower method, generating a hysteresis loop. The temperature dependence of the spontaneous polarization for the complexes 52a (plot a) and 52b (plot b) is shown in Figure 49. The higher P_S values for 52b are caused by the shorter spacer length (5 CH$_2$ groups). The unusual steep increase above 125°C is not easy to interpret, despite the fact that this phenomenon has also been observed with other S_C^* polymeric systems.

H. Polymer-Monomer Binary FLC Mixtures

The combination of different LC materials in a mixture to produce mesophases with wide temperature ranges, enhanced birefringence, and low viscosity is an important practice in the development of materials for device applications. Investigations of mixtures of polymer 11 and the monomeric FLC 53, shown in Figure 50, revealed some interesting

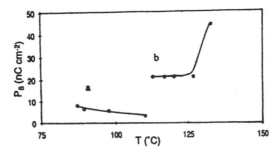

52 a-d

$(M_n = 7\ 000 - 12\ 000\ g/mol)$

52	n	x	K		S_C^*		I
a	5	0.29	•	85	•	154	
b	5	0.43	•	90	•	162	
c	8	0.29	•	88	⌀	151	
d	8	0.43	•	91	⌀	170	

Figure 49 Spontaneous polarization versus temperature for (a) 52a and (b) 52b. (From Ref. 68.)

results [69]. Both components were completely miscible. The mixture exhibits an alignment transition: At polymer concentrations greater than 50%, the polymer backbones line up along the rubbing direction (of the rubbed nylon surfaces used), resulting in smectic layer orientation in the same direction. At lower polymer concentration, the usual alignment of smectic layers *normal* to the rubbing direction is obtained.

A phase diagram of the mixture is shown in Figure 51. Up to ~80% concentration of the polymer 11, a smectic A phase exists in the mixture. The glass transition is decreased by about 25 K in the 1:1 mixture. The aligned mixture exhibits high-contrast electrooptic switching. The absence of zigzag walls at high polymer concentrations indicates that the smectic layers are more or less normal to the plates and do not form chevron textures. Although the optical response is slow owing to the high rotational viscosity of the polymer, the switching mechanism seems to be the same as for low-

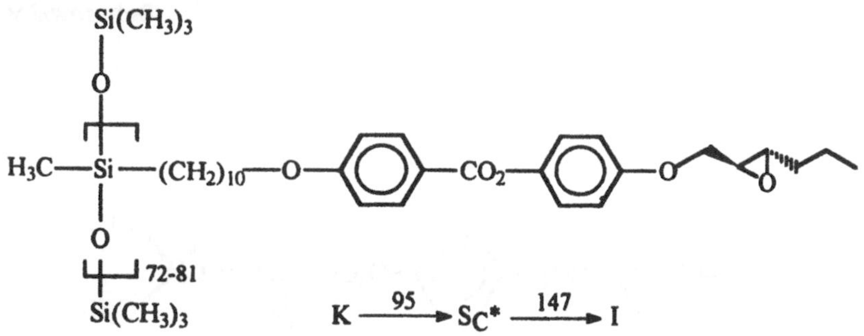

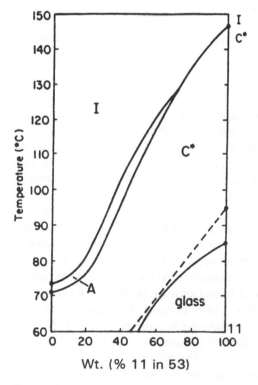

11

$$K \xrightarrow{\ 95\ } S_{C^*} \xrightarrow{\ 147\ } I$$

$$T_g \xleftarrow{\ 85\ } S_{C^*} \xleftarrow{\ 147\ } I$$

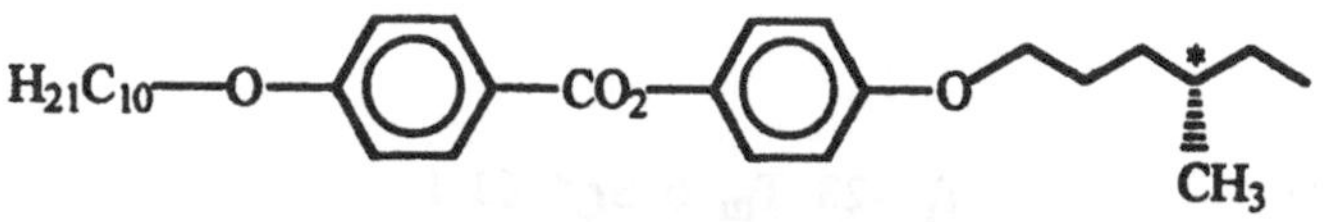

53

$$K \xrightarrow{\ 35\ } S_{C^*} \xrightarrow{\ 70.5\ } S_A \xrightarrow{\ 73.7\ } I$$

$$K \xleftarrow{\ 24\ } S_B \xleftarrow{\ 30.5\ } S_{C^*} \xleftarrow{\ 70.5\ } S_A \xleftarrow{\ 73.3\ } I$$

Figure 50 Structure and transition temperatures for 11 and 53. (From Ref. 69.)

Figure 51 Phase diagrams for 11 and 53. (From Ref. 69.)

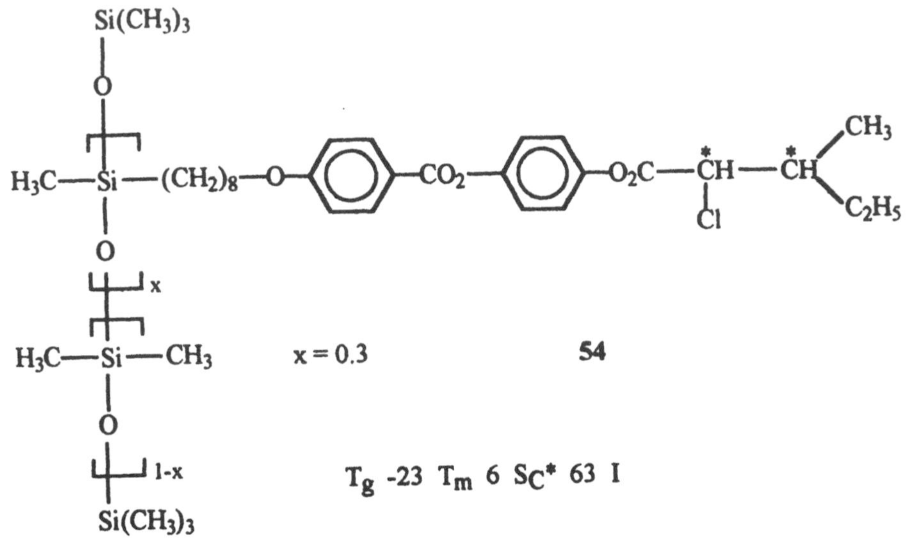

Si(CH$_3$)$_3$

H$_3$C—Si—(CH$_2$)$_8$—O—⟨◯⟩—CO$_2$—⟨◯⟩—O$_2$C—$\overset{*}{C}$H—$\overset{*}{C}$H—CH$_3$ / C$_2$H$_5$ (Cl)

H$_3$C—Si—CH$_3$ x = 0.3   **54**

Si(CH$_3$)$_3$

T_g -23 T_m 6 S_C* 63 I

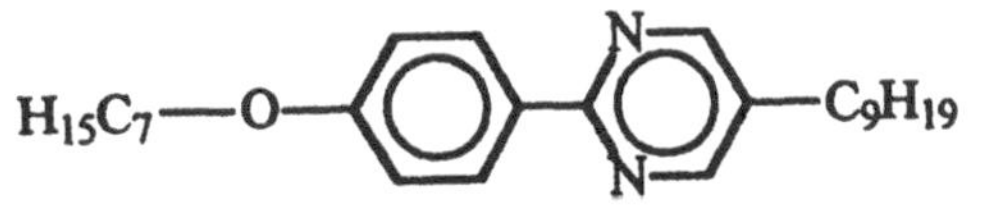

H$_{15}$C$_7$—O—⟨◯⟩—⟨N◯⟩—C$_9$H$_{19}$ **55**

K 51 S_A 55 N 69 I

Figure 52 Structures, mesophases, and transition temperatures (°C) for 54 and 55. (From Ref. 70.)

molar-mass FLCs. The mixtures reveal the total polarization as a sum of the two components, giving linearity to the P_s-versus-concentration curves.

Dumon et al. [70] investigated binary mixture of "diluted" FLC polymers, e.g., 54, and different chiral and nonchiral monomers, e.g., the low-viscous 55 (see Fig. 52). They observed fairly large stabilization of the S_A and S_C* phases and richer mesomorphic sequences: for example, a N*–S_A–S_C* sequence which may prove to be very useful for the alignment processes of the S_C* phases. Moreover, mixtures with more than 50 wt% of FLC polymer do not crystallize and exhibit an S_C* phase down to room temperature. Figure 53 shows the polarization and response time dependencies for different mixtures of 54 and 55. The response times are distinctly reduced by the addition of the low-viscous 55 and becomes almost independent of concentration for small polymer concentrations. The first systematic measurements of ferroelectric properties for polymer/monomer binary mixtures throughout the entire phase diagram between a mesogenic side-chain polymer 56 and its respective low-molar-weight compound 57 (Fig. 54) were performed by Heppke et al. [71]. The spontaneous polarization was found to decrease continuously with increasing amount of the polymer. The tilt angle showed only a slight influence of the composition. The activation energy for the ferroelectric switching, surprisingly, exhibits the largest increase at the addition of only small amounts of the pol-

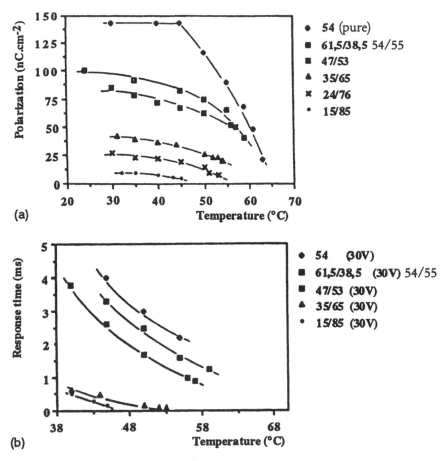

Figure 53 (a) Polarization and (b) Response time versus temperature for mixtures of 54 and 55. (From Ref. 70.)

ymer. The viscosity, calculated from the values of response time, spontaneous polarization, and tilt angle, showed a strong increase with polymer fraction and with decreasing temperature. The value for the polymer being three orders of magnitude higher than for the low-molar-mass component is remarkable when taking into account the relatively low degree of polymerization of about 10 for the polymer.

Shashidhar et al. [72] studied mixtures of the high-polarization "diluted" FLC polymer 58 and its side-group antecedent monomer 59 (Fig. 55). Figure 56 shows the phase diagram for mixtures of 58 and 59. The $S_C{}^*$ phase, which is about 115 K broad in the polymer, shrinks with increasing monomer concentration to a minimum of about 37 K at 70% monomer content.

The spontaneous polarization (Fig. 57) is an increasing monotonic function of the monomer weight percentage, saturating at about 70% monomer content. The ferroelectric response times (Fig. 58) are in a broad temperature range below 1 ms. They are monotonic, decreasing with increasing concentration of monomer. The large tilt angles of the polymer (~40°) and the monomer (~33°) show small variations in the mixtures.

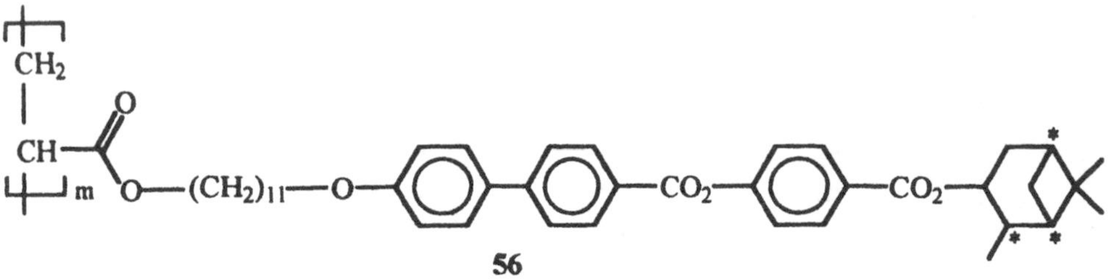

56

$$S_X \ 68 \ S_C^* \ 131 \ S_A \ 174 \ I$$

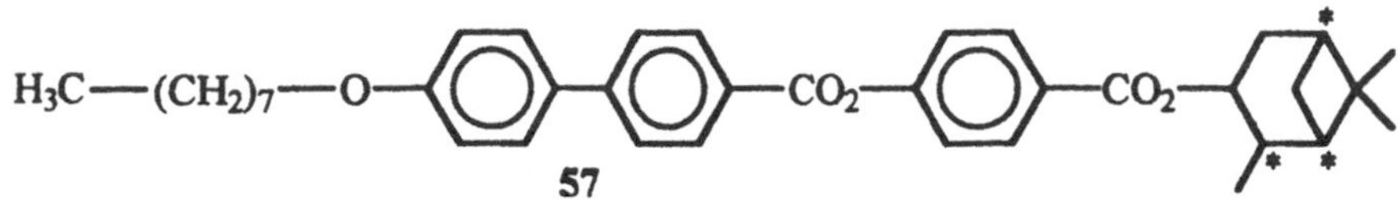

57

$$Cr \ 122 \ S_C^* \ 132 \ S_A \ 175 \ N^* \ 183 \ I$$

Figure 54 Structures and phase-transition temperatures (°C) for polymer 56 and monomer 57. (From Ref. 70.)

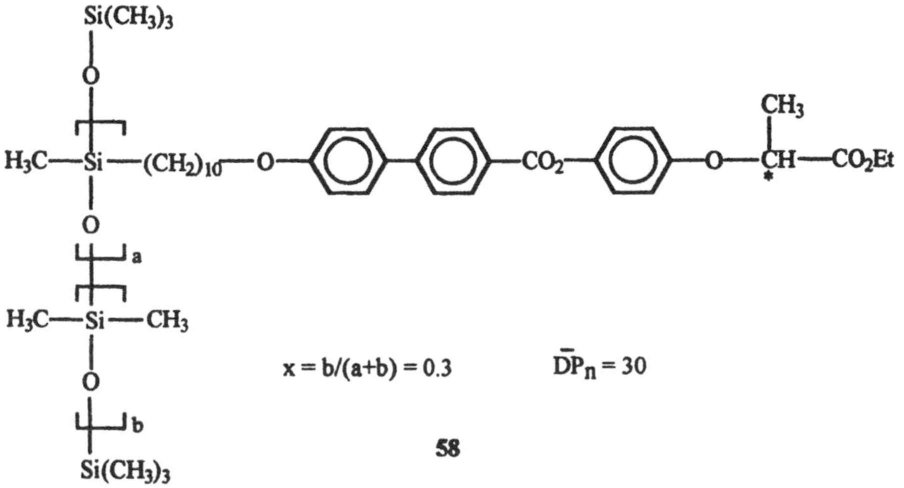

$$x = b/(a+b) = 0.3 \qquad \bar{D}P_n = 30$$

58

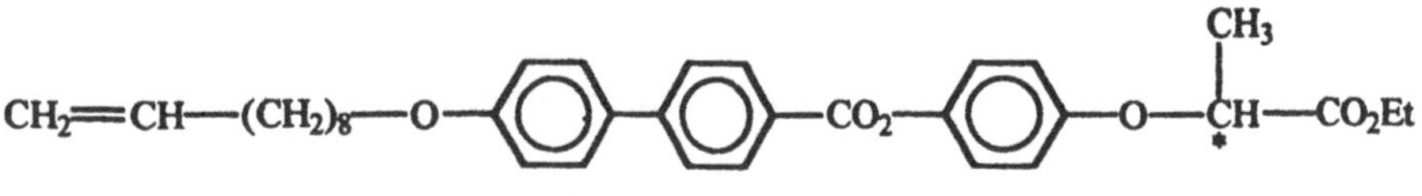

59

Figure 55 Structures of polymer 58 and monomer 59. The mean degree of polymerization is P = 30. The side groups are bonded to an average of 30% of the siloxane in its backbone. (From Ref. 72.)

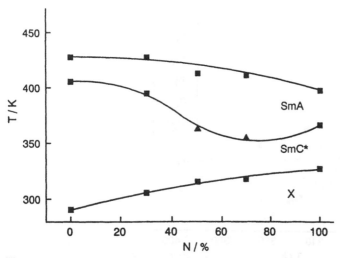

Figure 56   Phase diagrams for the binary mixtures of polymer 58 and its side-group antecedent 59. N is the monomer weight percentage in a mixture. The squares denote data from DSC, triangles denote data determined by the onset of spontaneous polarization, X denotes an unidentified phase. (From Ref. 72.)

I. Dielectric and Special Electrooptical Properties

1. Dielectric Properties

A comprehensive review of dielectric properties has very recently been published by Blinov and Haase [73]. Therefore we will restrict ourselves mainly to some basic and some new results.

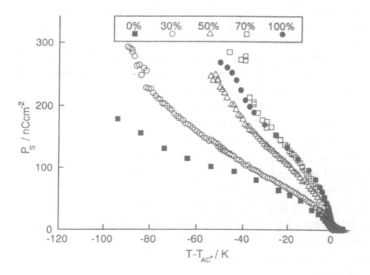

Figure 57   Spontaneous polarization as a function of $T - T_{SA/SC}*$ for mixtures of 58 and 59 (0% = polymer, 100% = monomer). (From Ref. 72.)

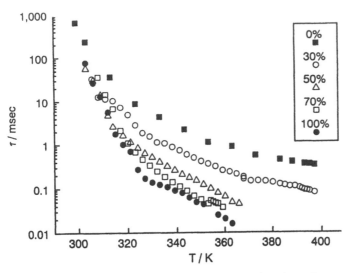

Figure 58 Electrooptical response time τ as a function of temperature for 58, 59, and their mixtures. (From Ref. 72.)

Ferroelectric liquid crysals have two specific modes of motion of their molecular director, the soft mode and the Goldstone mode. Those motions are monitored by the dielectric spectroscopy as responses with different characteristic frequencies and intensities. The soft mode is connected with the change of the molecular tilt angle θ. It usually has a characteristic frequency of 10^5–10^6 Hz for low-molar-mass liquid crystals. The Goldstone mode describes the motion of the molecular director along a cone surface with the angle 2θ. The molecular assignment of the Goldstone mode is the fluctuation of the phase of the helical superstructure which is connected to the different polarization vectors of the smectic layers. This mode is usually observed at 10^2–10^3 Hz for low-molar-mass LCs. The Goldstone mode can be suppressed by a DC field.

The first description of Goldstone mode and soft mode in a polymeric liquid crystal was presented by Vallerien et al. [74]. Dielectric measurements on the combined chiral polymer 60 (Fig. 59) resulted in the three-dimensional plot shown in Figure 60, where the dielectric loss is plotted versus temperature and the logarithm of frequency. The Goldstone mode is present only in the S_C^* phase at a frequency of about 1 kHz. In comparison to low-molar-mass FLCs, the Goldsteone mode is only slightly shifted (10 times) to lower frequencies. The soft mode was detected in all three smectic phases. The existence of the Goldstone and soft modes in the S_C^* phase of the oriented polymer is a proof of ferroelectricity in this nonswitchable system.

In the side-chain polyacrylate 61 (Fig. 61), Pfeiffer et al. [75] measured a value of the soft mode relaxation frequency of about $\sim10^2$ Hz, which is distinctly lower than in low-molar-mass FLCs (10^4–10^5 Hz). No Goldstone mode was observed. These results were discussed in terms of a sharp rise of the rotational viscosity with increasing molecular weight of FLC molecules.

Besides the two collective dynamics, a third relaxation process can be observed at higher frequencies (10^3–10^7 Hz). It has a rather small intensity and is associated with the rotation of the mesogens around their long axis (β relaxation). The common explanation of ferroelectricity in FLCs is based on the conjecture that the mesogens, which rotate "freely" in the $S_A{}^*$ phase, are facing a hindrance or slowing down of this rotation in the $S_C{}^*$ phase, giving rise to spontaneous polarization. Careful dielectric investigations of the β relaxation at the phase transition S_A–$S_C{}^*$ revealed no decrease of its relaxation rate and dielectric strength (Kremer et al. [76] and references cited therein). Therefore it is suggested that β relaxation has to be comprehended as a librational motion, which has an isotropic distribution of lateral dipole orientations in the S_A phase, which becomes anisotropic in the tilted $S_C{}^*$ phase, thus giving rise to local ferroelectricity in the smectic layers.

2. Field-Induced Ferroelectricity

In orthogonal smectic phases (S_A, S_B, S_E) consisting of nonracemic chiral molecules, an applied electric field induces a tilt of the molecular axis when the electric field is applied along the smectic layers. By this means a polarization is induced. This so-called electroclinic effect [23] has already been discussed in preceding sections of this chapter, in which electrooptical investigations of ferroelectric FLC polymers revealed, in addition to the bistable switching in the $S_C{}^*$ phase, a linear electrooptical effect (electroclinic switching) in adjacent orthogonal smectic A and B phases. A main feature of the electroclinic effect is the fast switching time, below 1 ms in LC polymers. In this section LC polymers are discussed that are aimed to have broad $S_A{}^*$ phases for studying their electroclinic behavior.

Chiellini et al. synthesized S_A side-group polymers with a biphenyl-mesogenic group and different chiral moieties [77,78]. With a polyacrylate backbone an induced tilt angle of ~0.6° in maximum was measured [79,80]. The switching time varied from 760 μs down to 17 μs at the S_A–isotropic transition. Approaching the $S_A \rightarrow S_C{}^*$ transition the soft-mode viscosity diverges [80].

Upon changing the polyacrylate to the polysiloxane backbone, a remarkable increase in the induced tilt angle (up to 15°) was detected [79]. In earlier papers induced tilt angles of 0.5° (50 V_{rms}/μm) [81] and up to 18° (10 V_{rms}/μm) [37] have been reported. In the $S_C{}^*$ phase of some polymers, discussed in preceding sections, the electroclinic switching

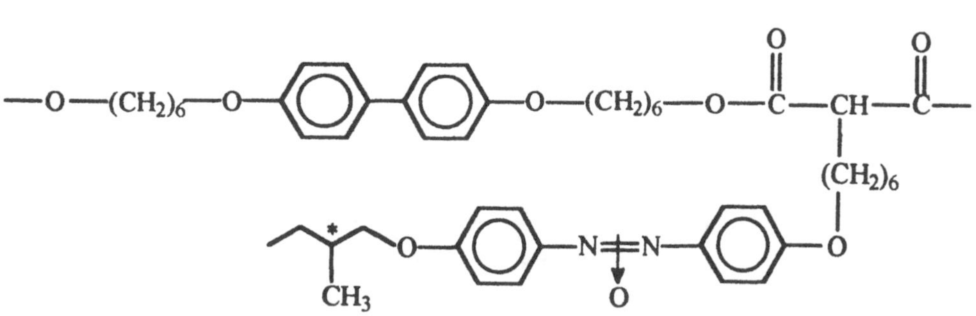

60

$S_J{}^*$ 77 $S_C{}^*$ 118 S_A 131 I   ($M = 23\ 000$ g/mol)

Figure 59 Structure and phase-transition temperatures (°C) for polymer 60. (From Ref. 74.)

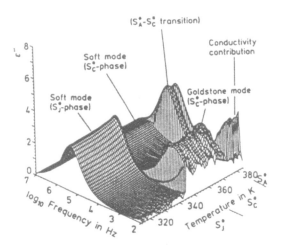

Figure 60 Dielectric loss ϵ'' versus temperature and log of frequency for polymer 60. (From Ref. 74.)

was detected at applied electric fields below the threshold voltage at which the ferro-electric bistable switching starts [50,61]. Therefore it may be anticipated that optimized polymer structures will eventually lead to significant improvements of both the induced tilt angle and speed of the electroclinic switching in LC polymers.

3. Sign Reversal of the Spontaneous Polarization

The sign of the spontaneous polarization is related to the molecular geometry, especially to the position of the lateral dipolemoment and its relation to the center of chirality. The relatively simple "Boulder model" [82] proved to be very useful to predict the sign of polarization for many FLC molecules. The Boulder concept states (Fig. 62) states that the molecules in the smectic C phase adopt a preferred orientation with respect to con-formation and rotational state by a time-average interaction with neighboring molecules

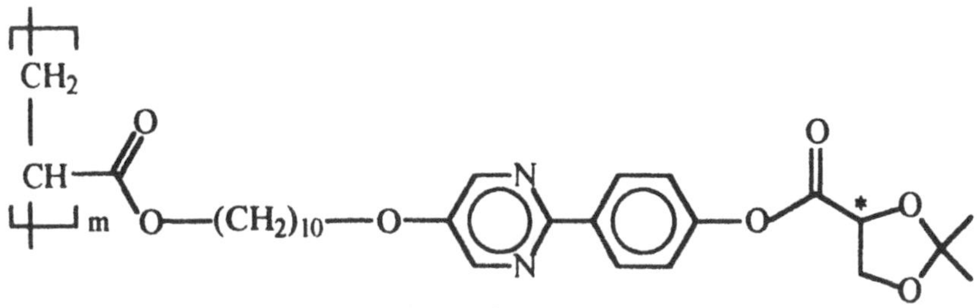

61

T_g 29 S_C* 85-88 S_A 110-115 I

(M_w = 7600 g/mol)

Figure 61 Structure and phase transitions (°C) for polymer 61. (From Ref. 75.)

(molecular recognition). The direction of a lateral dipole located at a chiral center can be derived from the all-anti conformation of the side chain, assuming that there is a well in the energy surface for rigid rotation of a conformation about the director. Thus a prediction of the sign of the spontaneous polarization is possible.

When measuring the optical response time in the $S_C{}^*$ phase of polymer 62 (Fig. 63), we observed an exciting phenomenon [83]. At 130°C the switching became very weak (Fig. 64), vanished, and returned at around 133°C but with the opposite optical contrast. Such anomalous behavior was first observed in the $S_C{}^*$ phase of a low-molar-mass FLC by Goodby et al. [84] and attributed to a change of the sign of the polarization, caused by a temperature-dependent equilibrium of different molecular conformers, which have opposite signs of polarization. This model of interpretation is still under discussion. Two theoretical approaches [85,86] interpret this phenomenon of sign reversal as being the result of a competition between polar and quadrupolar ordering depending on temperature.

To prove whether in the case of 62 a special polymer effect might exist, a comparison with the monomer was done. In Figure 65 the optical responses of polymer 62 and its monomeric unit are displayed, demonstrating the inversion of the optical answer in both cases. Measurement of the apparent tilt angle also shows a sign reversal (Fig. 66). At the inversion temperature the apparent tilt angle is zero. The inversion phenomenon is very sensitive to changes in the molecular structure. For example, a connection of the spacer at the other end of the three-aromatic mesogenic moiety leads to a disappearance of the polarization inversion. The same thing occurs when, instead of the oxirane ether, an oxirane ester was included in the spacer [59].

4. Nonlinear Optical Effects

Due to its noncentrosymmetric structure, the $S_C{}^*$ phase should exhibit second-order non-linear (NLO) responses. This includes effects such as second harmonic generation (SHG) and the Pockels effect. First reports in this field revealed NLO efficiencies so small that these materials seemed to be irrelevant for any NLO application [87,88]. The FLC compounds used for these investigations were optimized for display applications rather than for NLO performance. Kapitza et al. [89] and Walba et al. [90,91] designed FLC compounds devoted specifically to NLO application and improved the SHG efficiency by orders of magnitude. But still the highest reported values of the NLO coefficients ($d_{ij} = 0.1$–0.6 pm/V) were not competitive with state-of-the-art NLO materials such in $LiNbO_3$ or poled NLO polymers. A major breakthrough was achieved by Schmitt et al. [92a] by incorporating the NLO chromophor 4-nitroaniline in such a fashion in the mesogenic part of a FLC close to the chiral center that the direction of maximum second-order polarizability is oriented almost along the polar axis of the FLC compound, i.e., normal to the molecular longitudinal axis. Their most efficient NLO-active FLC, 63a (Fig. 67), also exhibits a high ferroelectric polarization of $P_S = 306$ nC/cm^2 in the $S_C{}^*$ phase and an even higher value of $P_S = 750$ nC/cm^2 at 50°C in the monotropic S_X phase, which can be cooled down to room temperature. Moreover, 63a can easily be supercooled, giving rise to a strong glass transition, which allows one to freeze the NLO-active orientation. Due to its high viscosity, 63a has poor homogeneous aligning properties. This problem was solved by mixing 63a with compound 63b. With this mixture the wave guiding experiments, phase matching, and the determination of the NLO coefficients were performed. As a result, the strongest second-order NLO responses reported so far were found (for example, $d_{22} = 5$ pm/V). Those values compare favorably with state-of-the-

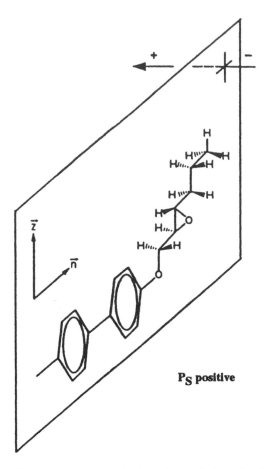

Figure 62 Preferred orientation and sign of polarization of a chiral liquid crystal in the S_C^* phase (Boulder model).

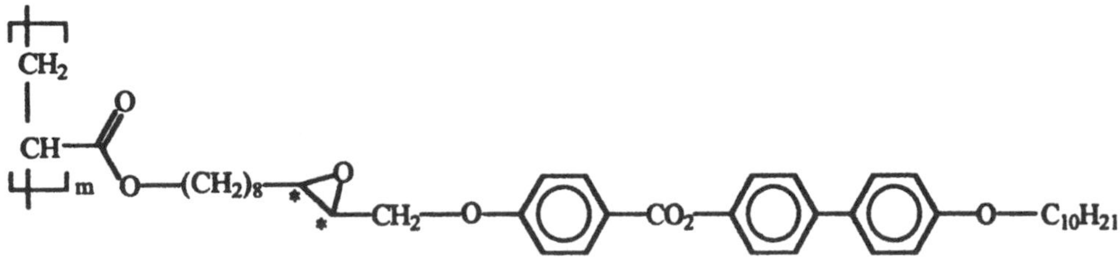

$$S_X \ 57 \ S_F \text{ or } S_I \ 122 \ S_C^* \ 177 \ I$$

Figure 63 Structure and phase transitions (°C) for polymer 62 with an [S,S]-oxirane ring in the spacer. (From Ref. 83.)

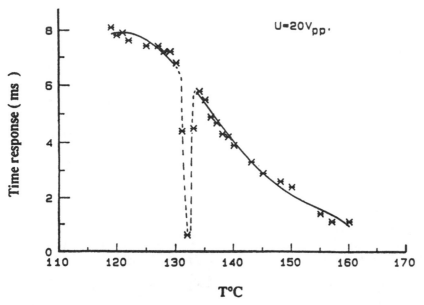

Figure 64   Temperature dependence of the electrooptical switching in polymer 62. (From Ref. 83.)

art inorganic NLO crystals such as LiNbO₃. Incorporating this type of excellent NLO active FLC as side group in a polyacrylate resulted in polymer 64 [92b], which exhibits a S$_C$* phase on cooling from 118°C down below room temperature. The results of the wave guiding experiments, phase matching, and the determination of the NLO coefficients are expected to be published in 1995. Second-harmonic generation in mechanically orientated S$_C$* elastomers is discussed in Section II.L.

J.   Electromechanical Effects

The influence of an electric field on liquid crystals is manifested primarily in changing the orientation of the director and hence the optical properties of a cell. However, the reorientation of the director is coupled to a viscous flow (e.g., the "backflow" effect in nematics) or to a layer compression in smectics (piezoelectricity). In certain cell geometries, either the flow or the layer compression may result in mechanical displacement or vibration of the cell substrates. These phenomena are called *electromechanical effects*. In the majority of liquid crystalline phases these effects are expected to be mainly nonlinear, due to the quadratic interaction between the electric field and the induced polarization. However, in ferroelectric LCs the interaction between the spontaneous polarization and the electric field is linear, so that they are sensitive to the polarity of the applied field. AC field-induced mechanical vibrations in planar aligned sandwich cells—in a geometry where one of the glass plates could move freely with respect to the other one—were measured using two different side-chain FLC polymers, 65 [93] and 6 [94,95] (Fig. 68). The measurements revealed that both FLC polymers exhibit linear and quadratic electromechanical effects. Field-induced vibrations were detected in the horizontal as well as in the vertical direction of the cells. In the horizontal direction (parallel to the

U(pp) = 60 V   U(pp) = 110 V

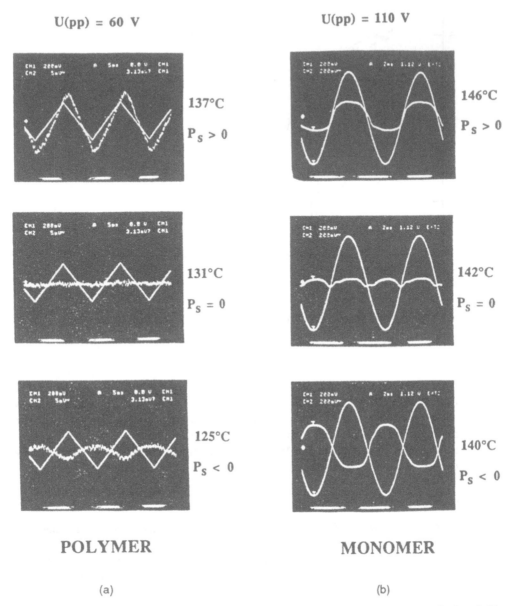

Figure 65 Electrooptical response signal (a) for the polymer (triangular voltage applied) and (b) for the monomer (sinusoidal voltage applied). Sign reversal for P_S at 131°C and 142°C, respectively. (From Ref. 83.)

layers), the linear electromechanical effect had a maximum in the frequency range of ~3.5–4 kHz. For compound 65 the temperature dependence of the vibration amplitudes followed the temperature dependence of the helical pitch, while for polymer 6 the maximum response was detected in the middle of the S_X^* phase.

The strength of the linear electromechanical effect of a FLC was defined as the maximum obtainable displacement induced by a unit electric field. For polymer 65 the

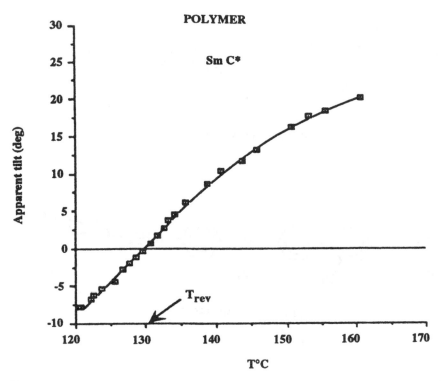

Figure 66 Apparent (optical measured) tilt angle, taken on heating for polymer 62 (T_{rev} = temperature of sign reversal). (From Ref. 83.)

value of this strength was $S = 10^{-17}$ m²/V, and for polymer 6 it was $S = 3 \times 10^{-16}$ m²/V. In low-molar-mass FLCs, values of $S = 10^{-16}$–10^{-14} m²/V were found. Taking into account the relatively poor alignment of the polymers studied, one can expect the electromechanical effects of FLC polymers to be almost as strong as in low-molar-weight FLCs.

The vertical motions (vibrations involving changes of the sample thickness) in the kilohertz range were found to be of the same order of magnitude as in the horizontal direction. The frequency dependence of the vertical vibrations for polymer 65 is shown in Figure 69. The spectrum contains a series of relatively broad resonancelike maxima. The frequency ratios of the subsequent overtones are approximately 1:3:5:8. Due to the coupling between the magnitude of the spontaneous polarization and the director tilt θ, an AC field induced an increase of θ (the electroclinic effect), which in turn leads to layer compression. This mechanical consequence of the electroclinic effect can be regarded as a piezoelectric response. A homogeneous excitation, such as the application of an AC voltage to the cell electrodes, can transfer energy only to standing waves with the fundamental frequency f, and its odd overtones ($3f$, $5f$, etc.), which was approximately measured as shown in Figure 69. The horizontal vibrations in the low-frequency range are presumed to be related to coupling between the director rotation and the viscosity flow. The electromechanical effects can be of interest for practical applications in sensors, electromechanical transducers, speakers, etc.

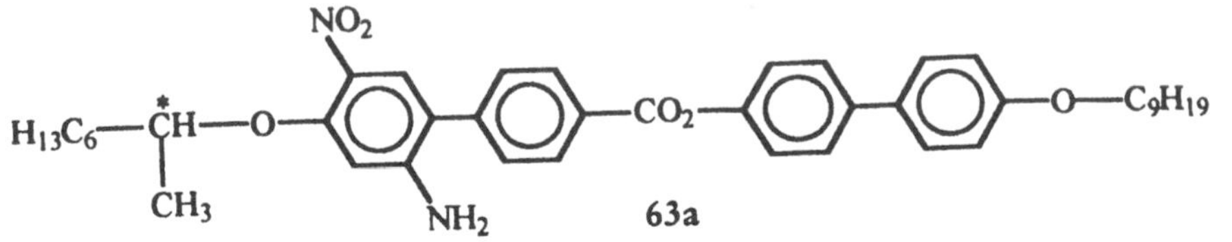

Cr 80 S$_C$* 132 S$_A$ 186 I

S$_X$ 96

Cr 110 S$_A$ 132 N* 139 I

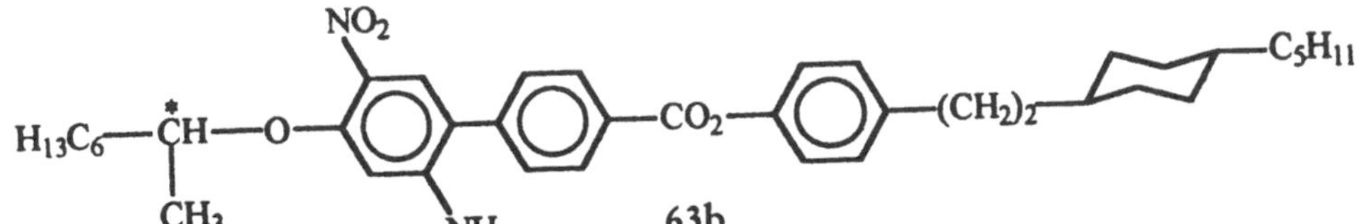

S$_C$* 62

$G \rightleftarrows S_C^* \xrightarrow{LC} \xrightarrow{132} S_A \xrightarrow{} I$
118 127

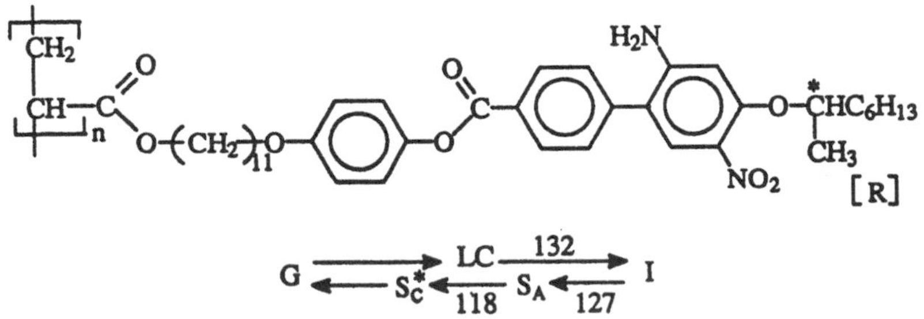

64

Figure 67 Structures, phases, and phase-transition temperatures (°C) of NLO FLC 63a and 63b (S$_X$ = unidentified ferroelectric smectic phase) and NLO-polymer 64. (From Ref. 92a,b.)

K. Langmuir Blodgett FLC Films

It is known that the ordering in the S$_C$* phase is highly influenced by an interface with a dissimilar material. In fact, the molecular ordering in the vicinity of the interface may differ significantly from the bulk where the ordering is established due to self-organization. The concept of anchoring FLC molecules at the surface of conducting glass substrates was successfully realized by Clark and Lagerwall in their surface-stabilized FLC devices, in which the macroscopic helix of the LC structure is untwisted, leading to a bistability and bulk ferroelectric properties. An alternative potential method to achieve bistable FLC materials is the transfer of monolayers of these materials by the

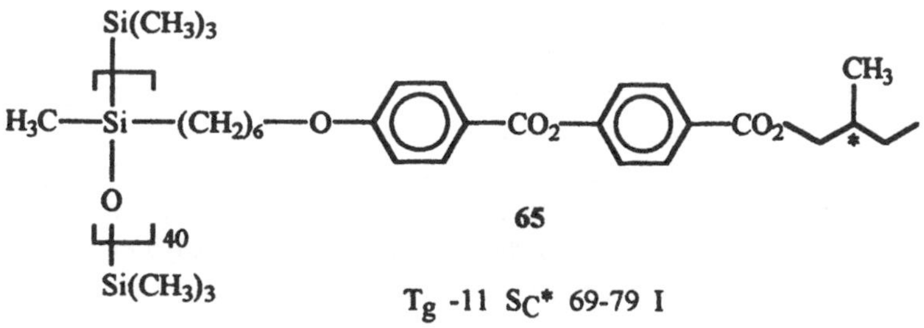

Si(CH$_3$)$_3$

H$_3$C—Si—(CH$_2$)$_6$—O—⟨benzene⟩—CO$_2$—⟨benzene⟩—CO$_2$—CH(CH$_3$)—*

O

[ ]$_{40}$

Si(CH$_3$)$_3$

65

T$_g$ -11 S$_C$* 69-79 I

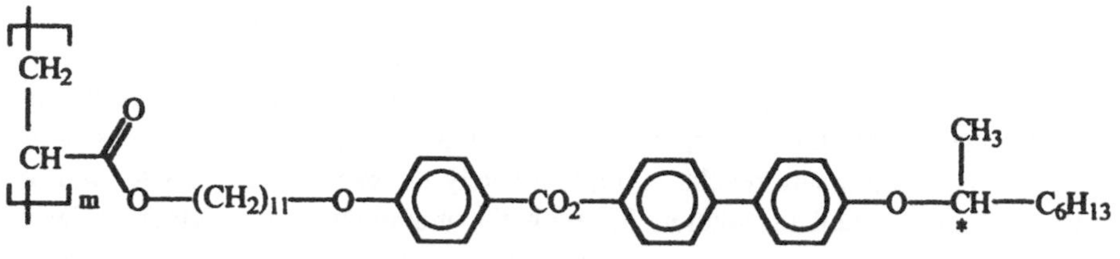

CH$_2$

CH—C(=O)—O—(CH$_2$)$_{11}$—O—⟨benzene⟩—CO$_2$—⟨biphenyl⟩—O—CH(CH$_3$)—C$_6$H$_{13}$

[ ]$_m$

6

T$_g$ 50 S$_B$ 92 S$_X$ 125 S$_C$* 142 S$_A$ 151 I

Figure 68 Structures, phases, and phase-transition temperatures (°C) for electromechanically investigated polymers 65 and 6. (From Refs. 93, 94, and 95.)

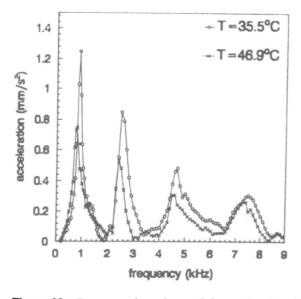

Figure 69 Frequency dependence of the acceleration of the cover plate of polymer 65 sample in the direction normal to the surfaces (U = 10 V). (From Ref. 95.)

Langmuir-Blodgett (LB) technique to a solid substrate and to produce in this way multilayers with well-known numbers of layers and molecular orientation.

To get insight and detailed knowledge of ordering in S_C^* materials near interfaces, R. Shashidhar, W. Rettig, et al., started experiments to produce monolayers at a planar air/liquid interface and to prepare Langmuir-Blodgett multilayer films [96]. They used the polysiloxane FLC polymer 66 and for comparison its side-group precursor 67 (for simplicity called the monomer) (Fig. 70). Both the monomer and the polymer formed stable monomolecular films when a chloroform solution was spread on water in a Langmuir trough. The compression curves displayed in Figure 71, measured with a floating barrier device, indicate that the monomer forms a condensed monolayer only in the region above 20 mN/m. However, the polymer organizes in the whole range as a condensed film. The collapse pressure of the polymer is substantially higher than that of the monomer. The presence of the siloxane backbone thus enhances the film stability with respect to its collapse pressure. Extrapolation of the linear high-pressure regions of both isothermes to zero pressure results in approximately the same area of 30 Å^2, indicating the same mesogenic unit orientation for both the monomer and polymer at intermediate applied surface pressures. The molecular areas are dominated by the packing of the mesogens rather than the polymer backbone. Langmuir-Blodgett film depositions of the polymer were performed successfully (despite its high viscosity) on a variety of hydrophobic substrates at pressures of 50 mN/m. Figure 72 shows that layers could be deposited on both the upstroke and downstroke (y-type deposition). However, at pressures of only 35 mN/m there was only transfer during the upstroke (z-type deposition). Up to more than 100 layers were transferred and studied by UV spectroscopy. A linear relationship of peak intensities to layer number was observed, indicating homogeneous trans-

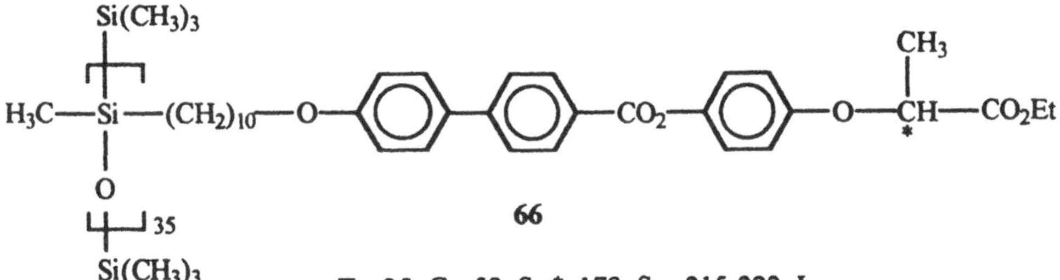

T_g 25 Cr 58 S_C^* 178 S_A 215-222 I

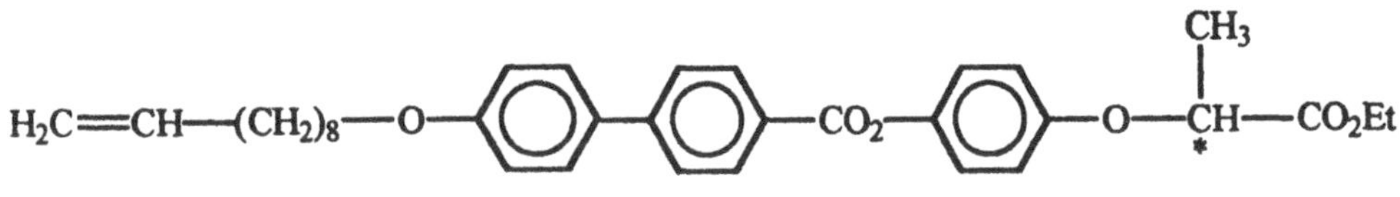

$(S_I/S_F$ 55.5) Cr 67 S_C^* 95.4 S_A 125.4 I

Figure 70 Structures, mesophases, and phase-transition temperatures (°C) for polysiloxane 66 and precursor 67. (From Ref. 96.)

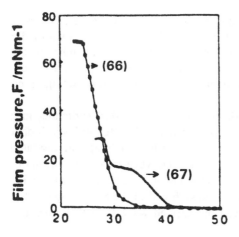

Figure 71 Compression curves of polymer 66 and its precursor 67 at $T = 27°C$. (From Ref. 96.)

fer. X-ray diffraction studies showed that the Langmuir-Blodgett (LB) layers had a tilted layered structure similar to that usually observed in S_C* phases.

Optical investigations using polarized UV or IR light did not show measurable dichroism, indicative of a random orientation of the functional groups in the plane [97]. Annealing at 110°C for 3 h increases the intensities of X-ray peaks, indicating an enhancement of ordering. To reduce the viscosity of the homopolymer films, an average 2.3-dimethylsiloxane unit was incorporated between each mesogenic monomer unit, thus diluting the polymer [98,99]. As a result, the transfer properties and the monolayer stability were improved despite a distinctly higher collapse mean molecular area and a lower collapse temperature as compared with the homopolymer.

With the aim of producing defect-free films, an alternative method of deposition was developed. A sequence of sending the substrate down to the surface film and up through the clean phase produced pure x-type films of more than 30 layers that were homogeneous in color and texture, in which previously visible defects were absent [99].

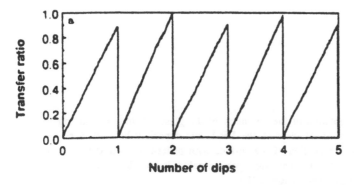

Figure 72 Transfer data versus number of dipping for polymer 66 at 50 mN/m. (From Ref. 96.)

To get insight into the ordering and the domain size of LB monolayers, Brewster angle microscopy was successfully applied [100]. Whereas the monomer 67 exhibited a liquid crystalline phase with large domains in which the molecular tilt is uniform, the homopolymer 66 was in a solid amorphous state without any uniformly tilted region. The "diluted" copolymer, in contrast to the homopolymer, is highly fluid, but again no domains with uniform tilt were observed.

The first observation of ferroelectric polarization and electrooptical switching in a LB film of a FLC copolymer (40, Fig. 37) has been reported by Pfeiffer and Shashidhar [101]. A 30-layer LB film was produced onto a glass substrate containing interdigitated electrode arrays (Fig. 73). Clear current peaks were observed when measuring the spontaneous polarization using the triangular wave method. The magnitude of P_S of the LB film was found to be similar to that of a 10-μm sandwich cell in the SSFLC geometry. The polarization current in the LB film followed the applied wave field to higher fre-

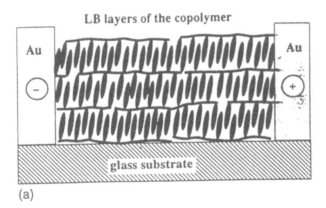

(a)

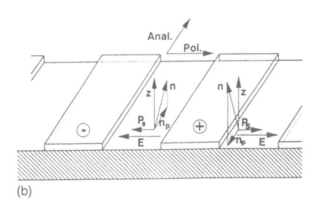

(b)

Figure 73 Schematic representation of the geometry of the glass substrate containing the gold electrode patterns. (a) Cross-sectional view of the LB layers deposited between a pair of gold electrodes. (b) Oblique view of a section of the surface with patterned electrodes. E is the field direction parallel to the layers, Z is the smectic layer normal, n is the tilt director, and n_p is the projection of n onto the substrate. (From Ref. 101.)

quencies as compared to the surface-stabilized cell. The characteristic switching time was found to be 150 ms for 40 V across the LB film at 30°C, which was shorter than in the SSFLC cell, perhaps due to the fact that the LB film is anchored at only one surface. It decreases to 65 ms for 90 V. These properties are promising features for possible applications in the areas of electronic/optical devices and thermal imaging.

With the 30-layer LB film discussed, a first detailed X-ray diffuse scattering study was performed [102]. It was shown that the observed layer undulations are induced by the roughness of the film–substrate interface. This contrasts with the case of free-standing films, wherein thermal fluctuations play the major role.

L. $S_C{}^*$ Elastomers, Piezoelectric and NLO Properties

1. Piezoelectric Elastomers

Liquid crystalline elastomers are produced by the introduction of cross linking into liquid crystal polymer systems. This cross linking results in materials with a number of unusual properties, for example, stress-induced phase transitions and spontaneous elongation of samples in the liquid crystalline phase. When they contain chiral units, $S_C{}^*$ elastomers are formed. For such materials piezoelectricity was predicted by Brand [103]. The helical structure of those systems can be untwisted by application of a mechanical stress, generating an electrical signal. This possibility provides the basis for the development of materials with piezoelectric properties. Such materials are of considerable interest, since the basis for their piezoelectric properties is rather different from that in the best-known piezoelectric polymer, poly(vinylidenefluoride) (PVDF). There exists rather more scope for the modification of their properties; for example, the nature of the chiral unit may be varied to alter the helical superstructure, or differences in cross link density can change the mechanical properties of the sample.

The experimental proof of piezoelectricity in $S_C{}^*$ elastomers was demonstrated by Vallerien and Zentel et al. [104]. They selected the combined main-chain/side-group polymer 68 (Fig. 74), assuming that after its cross linking a deformation of the network would be most effective on mesogens, which are incorporated into the polymer main chain. In the combined polymer, one-half of the mesogens are included in the backbone, but all mesogens orient parallel to each other and to the polymer chain. As chiral group, a chiral ether of 4-hydroxy-3-nitrobenzoic acid was selected, which was known to give rise to a very high spontaneous polarization in LC side-group polymers [30]. The combined polymer (68) consists of a malonate polyester backbone containing two different diole precursors, a stiff azoxybenzene moiety (C), and a flexible siloxane bridge (D). One side group (A) contains the chiral unit, the other an acrylate terminal group (B). Cross linking by thermally activated polymerization of the acrylate function creates the elastomer 69, with an 80-K-broad $S_C{}^*$ phase. A proof for the $S_C{}^*$ mesophase was established via dielectric spectroscopy, which shows a huge contribution of the Goldstone mode to the dielectric loss. In Figure 75 the results of the piezoelectric measurements are displayed. The piezoresponse was registered by a frequency of 7 Hz and an amplitude of 10 μm (corresponding to 10% change in sample thickness). Entering the $S_C{}^*$ phase at 81°C leads to a continuous increase of the piezo signal until nearly a 1000-fold value compared to the isotropic signal is reached.

Figure 75b shows a linear amplitude dependence of the piezo signal at different temperatures. As these results are qualitative, a quantitative comparison with other piezoelectric polymers [e.g., poly(vinylidene difluoride)] at this stage was not possible.

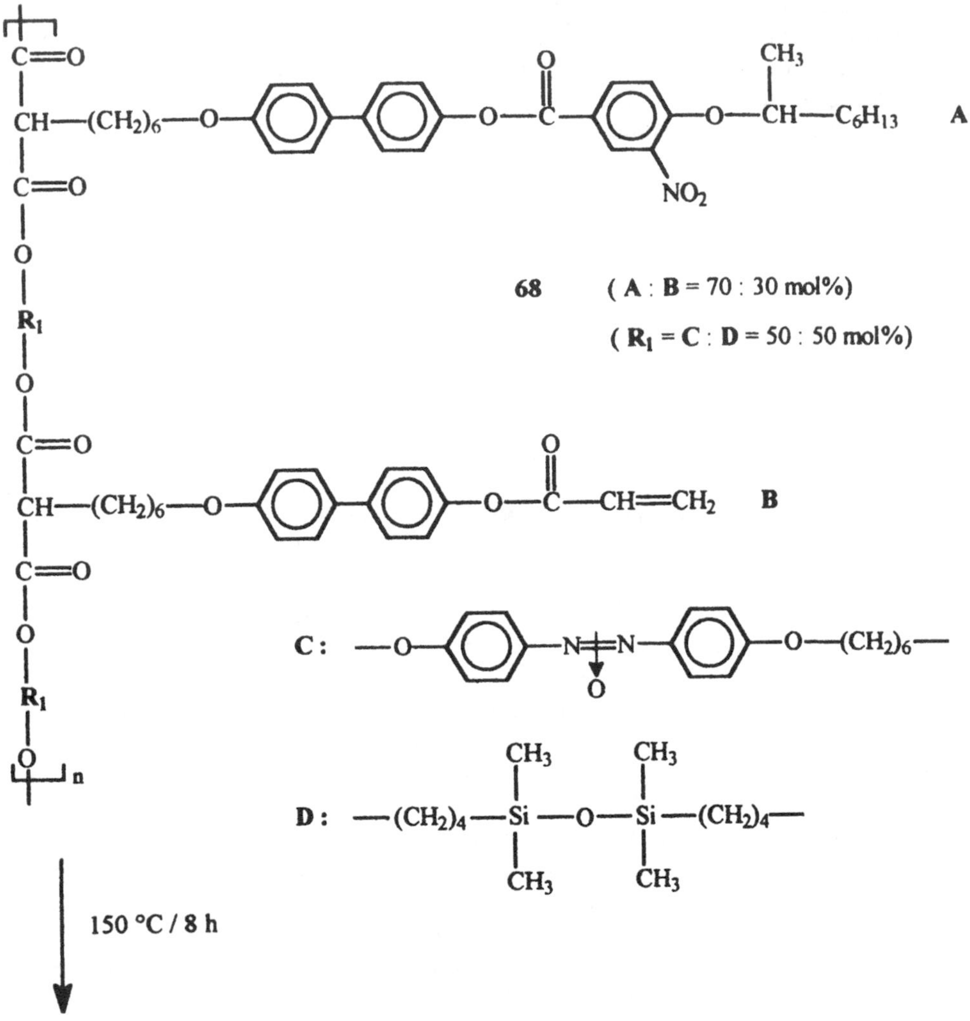

69 : T_g 1 S_C^* 81 I   (Mw = 28 000 g/mol)

Figure 74 Structure and composition of combined polymer 68 and phase behavior (°C) of the derived S_C^* elastomer 69. (From Ref. 104.)

Another approach to obtaining a piezoelectric network was used by Hikmet [105]. He mixed the diacrylate 70 with the chiral dopant 71, thus inducing a S_C^* phase (Fig. 76). Upon photo-induced polymerization of the diacrylate in the ferroelectric phase which was oriented under a DC field of 20 V/60 μm, a highly transparent, unaxially oriented network was obtained containing chiral molecules which are not chemically attached to the network. From the measured open-field voltage across the uniaxially oriented film (thickness 60 μm) during an applied sinusoidal strain in the direction parallel to the

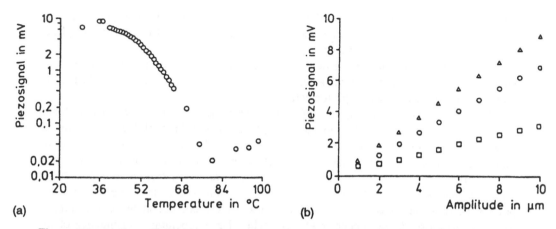

Figure 75 (a) Decadic logarithm of piezo signal versus temperature for S_C^* elastomer 69 with an amplitude of 10 μm. (b) Amplitude dependence of the piezosignals at 35 (Δ), 39 (○), and 52°C (□). (From Ref. 104.)

molecular orientation, the piezoelectric coefficient was estimated. A value of 3.1 pC/N was obtained. In the direction perpendicular to the molecular orientation, the piezoelectric coefficient was 1.4 pC/N.

Using X-ray diffraction, the origin of the piezoelectricity was related to the presence of chevrons and the tilted orientation of the molecules. It is interesting to compare the values of the piezoelectric coefficients and spontaneous polarization ($P_s = 8$ nC/cm²) of this plasticized ferroelectric network with the respective values for the β phase of PVDF,

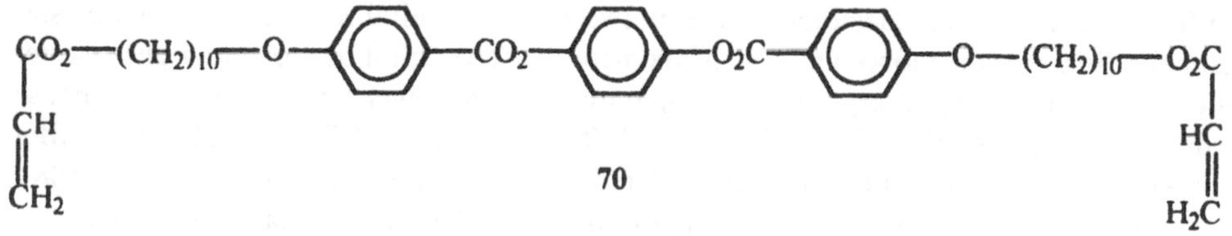

70

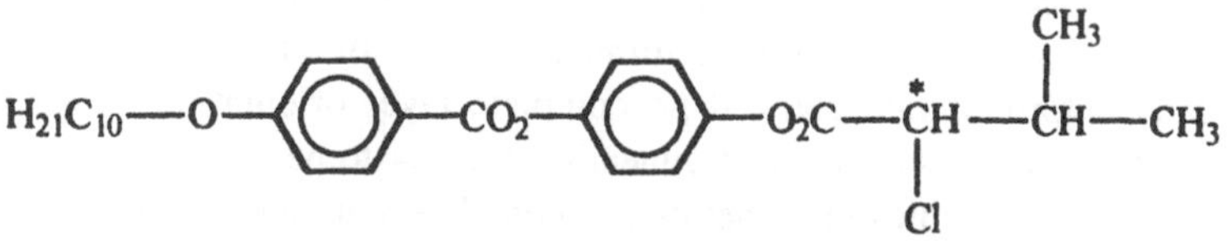

71

70: 100% K 82 S_C 108 N 149 I
85% K 80 S_{C^*} 96 N* 126 I

Figure 76 Smectic C diacrylate 70, chiral dopant 71, and phase behavior of pure 70 and an 85:15 wt% mixture with dopant 71 (transition temperatures in °C). (From Ref. 105.)

which are $P = 13$ μC/cm^2 and $d = 28$ pC/N. It can be seen that even though the network has a much lower value for the spontaneous polarization, the difference in the piezo-electric coefficients is much smaller.

The ease with which the plasticized networks are producable and the prospect of using materials with higher spontaneous polarization make them serious candidates for highly birefringent transparent piezoelectrics.

Piezoelectric coefficients up to 7 pC/N have been measured by Brehmer and Zentel [106] on a soft elastomer (73), obtained by photo-cross linking of the copolymer 72 (Fig. 77). Due to the long dimethylsiloxane segments in the backbone and the low cross linking density, the elastomer possesses enough flexibility to allow some reorientation of the liquid crystalline director and of the polar axis. Ferroelectric switching is possible at high voltages. At lower voltages an elastic stress prohibits complete reorientation of the polar axis. It is thus possible to change the piezoelectric coefficient by temperature or by reorientation of the mesogens in electric fields. The temperature dependence of the piezocoefficient is shown in Figure 78. The polarity of the smectic A phase was inter-preted by an internal "mechanoclinic" effect caused by a superposition of the potentials for the tilt in the smectic A or C phase with a mechanical field, which tries to keep the tilt angle in a position which was adopted during the cross-linking reaction.

2. NLO-Active Elastomers

As described in Section II.I.4, a polar order of donor-acceptor-substituted π systems is a precondition for second-order NLO effects such as frequency doubling (second-harmonic generation, SHG). Compared to poled polymers, which are often used for this purpose, chiral smectic C* polymers have the advantage of a stable polar structure.

To obtain strong NLO activity, well-oriented samples with an untwisted helicoidal superstructure are necessary. In S$_C$* elastomers this goal can be achieved by mechanical fields. Finkelmann et al. [107] demonstrated that by applying a shear field consistent with the phase symmetry, S$_C$* elastomers such as 74 (Fig. 79) can be macroscopically, uniformly aligned. This elastomer contains two different chiral mesogenic moieties sta-tistically linked to the monomer units of the network. The chiral benzoic acid phenylester is used to shift the liquid crystal to isotropic transition to lower temperatures. The phase sequence for 74 is: glass-7 S$_C$* 52 S$_A$ 75 I (temperatures in °C). The elastomer film (thickness 300 μm) was mechanically ordered at room temperature in two steps [108]. In a first step, a uniform director orientation is obtained by a uniaxial load. Under this condition the smectic layers are ordered with an angle $\pm\Phi$ of the layer normal to the mesogens long axis, where Φ is the tilt angle of the mesogenic units in the smectic layers (see Fig. 80). In the second step a uniaxial deformation of the sample under the angle Φ with respect to the first deformation is performed, to order the layers. X-ray patterns of the mechanically deformed film showed no complete layer orientation as drawn sche-matically in Figure 80. However, after annealing the elastomer film (50°C) for 24 h, excellent orientation was achieved, generating a completely translucent film.

The optical experiments were performed using a Q-switched Nd-YAG laser (λ = 1064 nm, 10 pulses per second, pulse width 8 ns). The geometry of the SHG experiment for measuring the Maker fringes is outlined in Figure 80. For the free-standing film the intensity of the second harmonic beam is analyzed as a function of angle, which describes the rotation of the sample around the X axis. Figure 81 shows the typical Maker fringes, consistent with the C$_2$ symmetry of the untwisted sample, also demonstrating the sharp increase of the SH intensities after the annealing process. Despite the fact that the ob-

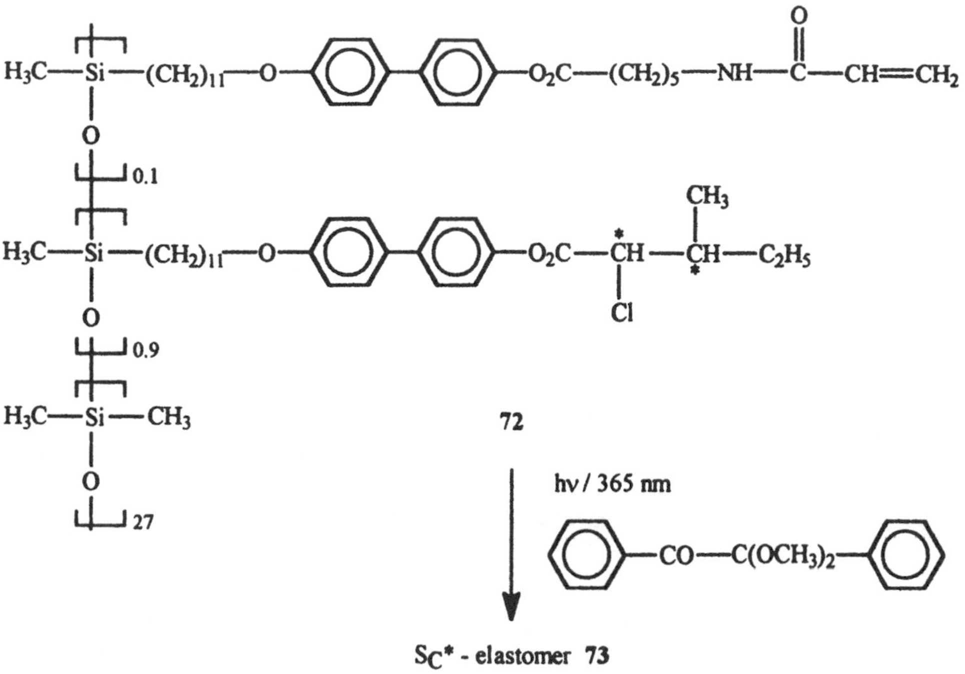

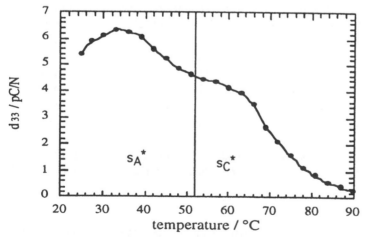

72 : S_X 29 S_C^* 53 S_A 88 I

73 : S_C^* 49 S_A

Figure 77 Copolymer 72 and phase behavior of 72 and elastomer 73. (From Ref. 106.)

Figure 78 Temperature dependence of the piezoelectric coefficient. (From Ref. 106.)

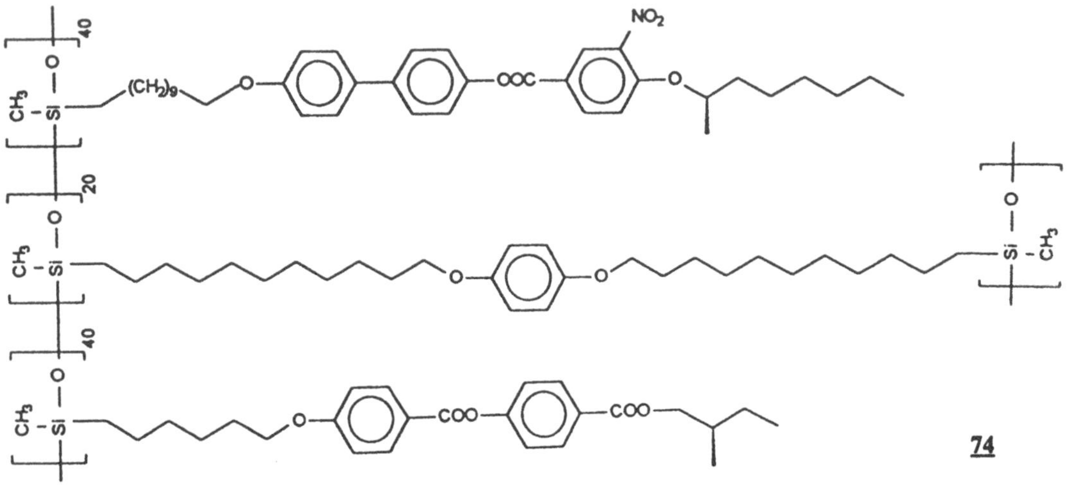

Figure 79   NLO-active S_C^* elastomer 74. (From Ref. 107.)

served SH signal was relatively low in these first experiments and a quantitative analysis of the nonlinear susceptibilities has to be done (announced for a forthcoming paper), the results reveal that S_C^* elastomers can be macroscopically uniform oriented by mechanical fields, making them suitable for SHG without any external electric poling field. Compared to organic and inorganic noncentrosymmetric single crystals, the processing and handling of S_C^* elastomers is more practicable. Furthermore, they can be optimized by choosing appropriate high-polarization ferroelectric units and high-efficiency NLO chromophores. The mechanical properties can be modified by simple variation of the cross-linking density.

M.   Polymer Dispersed Liquid Crystals (PDLC)

1.   Conventional PDLCs

Polymer dispersed liquid crystals (PDLCs) are composite materials consisting of micrometer-sized liquid crystal droplets embedded in an optically transparent polymer matrix. By using *nematic* liquid crystals as droplets in these PDLCs, large, flexible polymer films can be produced, which are switchable between a translucent "off" state and a transparent "on" state. They are used for switchable privacy windows. Recent new developments are reflective displays and colored projection displays. In comparison with conventional liquid crystal displays, PDLC films are very easy to prepare, since no orienting glass plates are required. Because no polarizers are needed, high transittance is possible.

Excellent reviews on PDLCs have been given by Doane et al. (nematic PDLCs [109]) and Kitzerow (nematic, cholesteric, ferroelectric, and antiferroelectric PDLCs [110]).

2.   Ferroelectric PDLCs

New types of PDLC devices are using smectic A and *ferroelectric* S_C^* liquid crystals [112]. In contrast to the nematic PDLCs, prealignment of the liquid crystals is essential.

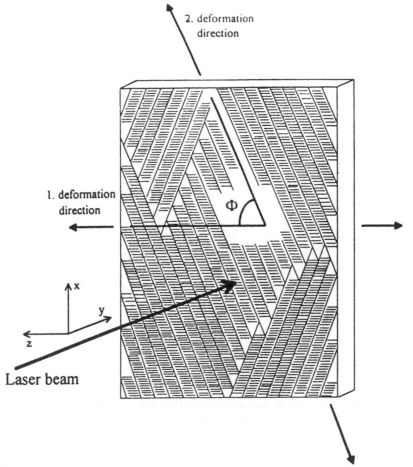

Figure 80 Sketch of S_C^* elastomer 74 after two deformations and the experimental arrangement for the Maker fringe experiment. (From Ref. 107.)

This is realized by a controlled mechanical shear during the polymerization-induced phase separation, leading to a strongly elongated form of the embedded droplets (Fig. 82). Instead of the field-induced reorientation of nemactic or cholesteric liquid crystals, the electroclinic effect in the S_A^* phase or the helical unwinding (DHF effect [111]) in a S_C^* liquid crystal is used.

The two switching effects are displayed in Figure 83. The electroclinic effect (a) has been discussed in Section II. The field-induced helical unwinding (b) is caused by the coupling of the spontaneous polarization P_s to an external field. Both effects lead to a rotation of the optical axis in the film plane. The sample is placed between crossed polarizers and, due to the rotation of the optical axis, the intensity of the transmitted light can be modulated by the applied field. The induced tilt of the optical axis has been found to be a linear function of the applied field. In the case of short-pitch ferroelectric PDLC, the tilt can be about 13° at an applied field of 10 V/μm (Fig. 84).

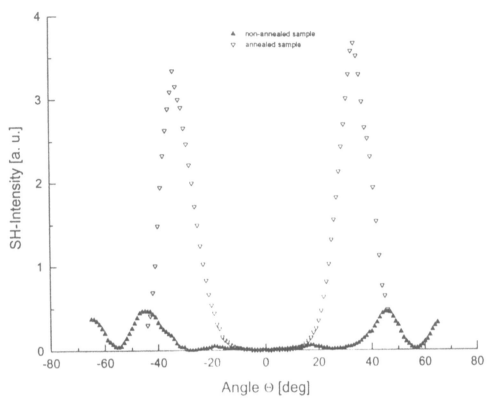

Figure 81 Maker fringe diagram for the annealed sample (including data for the nonannealed sample); θ indicates the rotation of the sample around the *X* axis. (From Ref. 107.)

The response time at ambient temperature is about 30 μs, being independent on the field strength (Fig. 85). At about 43°C the response decreases at high voltages due to a change to a ferroelectric type of switching [113].

3. Antiferroelectric PDLCs

Antiferroelectric liquid crystals form tilted smectic phases with a herringbone structure, where the tilt direction alternates on passing from layer to layer (Fig. 86). As a consequence, the sign of the spontaneous polarization alternates too, hence compensating for the bulk sample. This provides the possibility of tristable switching due to a discontinuous antiferroelectric-to-ferroelectric transition. Molsen et al. [114] were the first to observe the tristable switching in PDLC films. Although the current double hysteresis loop was not perfect and the contrast ratios (about 3:1) were rather low, the tristability indicates that the helical structure can be at least partially unwound by shearing PDLC films. Further improvements are to be expected.

III. ANTIFERROELECTRIC BEHAVIOR IN CHIRAL POLYMERS

Characteristic for antiferroelectric liquid crystals are three states of switching emerging between the zero field state with zigzag structure layer orientation and the two known

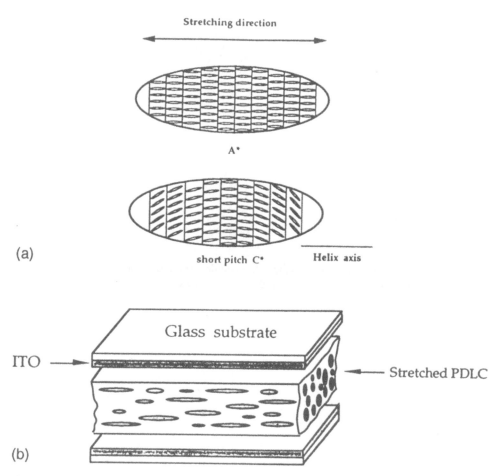

Figure 82 (a) Schematic presentation of the shear-induced orientation in the smectic A* and the short-pitch smectic C* phase in elongated droplets. (b) PDLC cell containing the stretched film. (From Ref. 113.)

ferroelectric states. As a consequence, two current response peaks occur, as shown in the schematic presentation of Figure 87, when applying a triangular field.

Antiferroelectric LCs exhibit some advantages in display applications compared with bistable switching FLCs. Accumulation of surface charges is avoided by the switching to the zero field state. A first antiferroelectric FLC display has been demonstrated, which shows 125 colors and video frame speed [115].

A. Three Switching States in FLC Polymers

The first observation of tristable switching was described for polymer 75 (Fig. 88 [116]). The structure of the chiral mesogenic unit is quite different from the structures of known low-molar-mass antiferroelectric LCs. The steps in the optical response are not so distinct as in low-molar-mass systems, but the two current response peaks are characteristic. Figure 89 shows the change of the apparent tilt angle with applied voltage, demonstrating

Figure 83 Molecular orientations for (a) the electroclinic switching in the S_A^* phase and (b) the helical unwinding in the S_C^* phase (γ = optical axis). (From Ref. 113.)

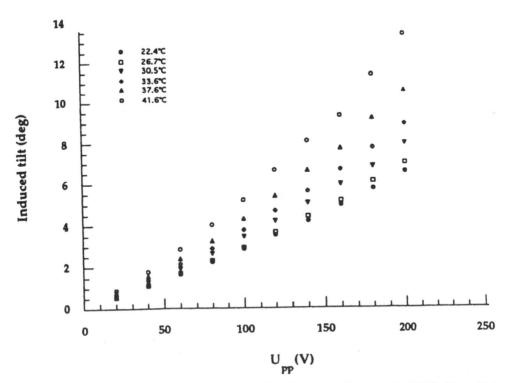

Figure 84 Field-induced tilt of the optical axis in a short-pitch ferroelectric PDLC. (From Ref. 113.)

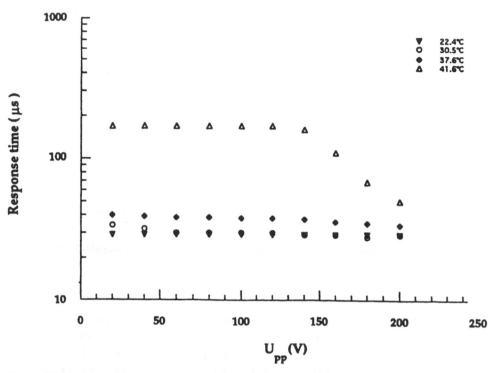

Figure 85 Voltage dependence of the response time for a short-pitch PDLC. (From Ref. 113.)

three states in the molecular orientation. In the zero field state the optical axis is normal to the layer planes of the smectic layers with their zigzag orientation of the mesogens. As is typical for antiferroelectric LCs, the increase in the optical tilt angle at low voltages (up to ~5 V/μm) is linear, increasing steeply above 10 V/μm at the antiferroelectric-to-ferroelectric transition and finally saturating. An alternative interpretation for the three-state switching in polymer 75 was given by Giesselmann et al. [117]. They suggest that conformational interactions between the side group and the main chain may cause the same behavior.

In polymer 76 (Fig. 90), which contains a chiral oxirane ring in the spacer, a three-state switching was observed in the S_C^* phase [59,118]. In SSFLC cells this polymer shows a characteristic stripe texture. The electrooptical response exhibits typical antiferroelectric properties, i.e., two current response peaks, an optical and electrical double

Figure 86 Molecular orientations in the antiferroelectric tristable switching process.

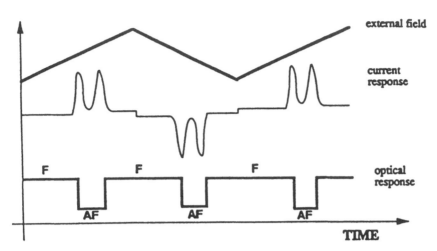

Figure 87 The schematic current and optical responses to an external triangular field in antiferroelectric LCs.

hysteresis. However, if one cools down polymer 76 under an applied electric field, a monodomain instead of the stripe texture occurs, which behaves typically ferroelectric, exhibiting bistable switching. The antiferroelectric behavior in the case of the stripe texture can be interpreted as being caused by a cell surface-induced macroscopic antiferroelectric ordering in domaines as shown schematically in Figure 91. The liquid crystal molecules are aligned along the rubbing direction of the cell surface. The stripe domains have different direction of the smectic layer normal. A DC field, depending on its polarity, is switching one or the other type of domains, thus indicating that in adjacent domains the spontaneous polarization has ''up'' or ''down'' state, respectively. At field-free condition the layer texture relaxes quickly to the initial, quasi-periodic one. This ''macroscopic antiferroelectric ordering'' is the reason for the observed three switching states, the double hysteresis and double current response peaks, which are typically for antiferroelectric phases. The occurrence of the stripe domains in the $S_C{}^*$ phase is caused by strong monostable surface anchoring. As a conclusion one has to state that the elec-

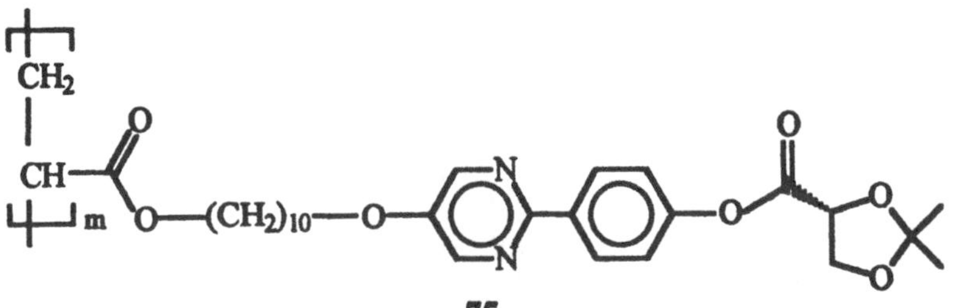

75

g/S_X 50 $S_C{}^*$ 85 S_A 110-115 I

Figure 88 Structure and phase transitions (°C) for polymer 75. (From Ref. 116.)

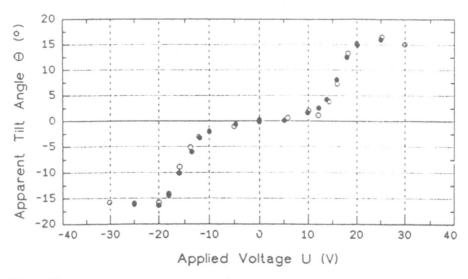

Figure 89 Apparent tilt angle as a function of the applied voltage (temperature 82°C) for polymer 75. (From Ref. 116.)

trooptical properties typically observed for antiferroelectric substances are not sufficient to distinguish between ferroelectric and antiferroelectric phases.

Another example where the possibility of an antiferroelectric structure was discussed is polymer 77 (Fig. 92 [119]). Skarp et al. [119] also found three optical switching states, the occurrence of two peaks in the polarization reversal current, and a double hysteresis loop in the current response in a shear-aligned cell of this polymer at a temperature within the S_C^* phase. In their conclusions the authors state that it would be desirable to investigate the polymer in a homeotropically aligned sample, to observe the transitions conoscopically. Free-standing films may also be a tool to distinguish antiferroelectric from ferroelectric ordering.

B. Antiferroelectric LC Polymers

First evidence for an antiferroelectric smectic phase in a chiral side-chain polymer was established by Bömelburg et al. [120] (Fig. 93). The smectic X denoted phase below the S_C^* phase exhibits the properties to be expected from an antiferroelectric phase. It is a tilted phase, with tilt angles above 20° as derived from X-ray data, whereas optical measurements establish zero tilt angle. This suggests an antiferroelectric structure, which is also strongly supported by the temperature dependence of the birefringence Δn. At the phase transition S_C^*–S_X, a strong drop in Δn indicates a partial averaging of the optical anisotropy (Fig. 94), as is to be expected for an antiferroelectric zigzag ordering in adjacent smectic layers.

The optical transmittance using crossed polarizers and applying a triangular wave voltage with a frequency of 0.01 Hz shows a typical double hysteresis. The induced apparent tilt angle (Fig. 95) measured for some temperatures within the S_X^* phase shows the transition from the antiferroelectric to the ferroelectric state.

A proof of the antiferroelectric nature of the chiral smectic phase of a polymer was offered by Nishiyama and Goodby [121]. They performed a miscibility test between the

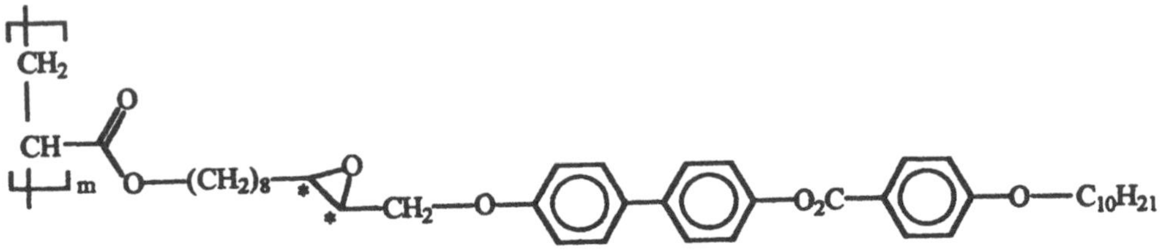

76

S$_F$ or S$_I$ 130 S$_C$* 184 I

Figure 90 Structure and phase transition (°C) for polymer 76, exhibiting three-state switching in the S$_C$* phase. (From Ref. 59, 118.)

polymer 79 and a similar monomeric liquid crystal, 78 (Fig. 96). The phase diagram shows continuous miscibility across the full composition range for the antiferroelectric phase of polymer 79 and the antiferroelectric phase of FLC 78. The same holds for the S$_A$ phases of both materials. However, the S$_C$* and ferrielectric phases were found to disappear in the binary mixture as the percentage of the polymer rose above 15 wt%. The monomer of polymer 79 exhibits the following phase sequence: isotropic–S$_A$–S$_C$*– ferrielectric–antiferroelectric–crystal. After polymerization only S$_A$ and antiferroelectric phase are left. This result shows that the antiferroelectric ordering is stabilized by polymerization. Two possible explanations for this stabilization are under discussion. As a polymer liquid crystal can be considered as a "cluster" of nomomers, where the dipole-dipole interaction between the clusters is stronger than that which occurs between the monomers, the zigzag ordering of the mesogenic groups may be strongly favored. Another reason might be that the conformational shape of the backbone determines the orientation of the appended liquid crystalline groups.

Investigation of the field dependence of the apparent tilt angle in the antiferroelectric phase of polymer 79 revealed that the tilt angle gradually increases with the field before the transition from the antiferroelectric state to the electrically induced ferroelectric state occurs (Fig. 97). The hysteresis behavior observed for low-molar-mass antiferroelectric LCs was not observed for the polymer.

IV. CONCLUSION

To combine the properties of liquid crystals with that of polymers has a long tradition, so it was a natural consequence to synthesize ferroelectric liquid crystalline polymers as well. This was achieved for the first time in 1984 by Shibaev et al. Since that time, fast development has taken place in this field. Most of the mesophases known from low-molar-mass FLCs have been also detected in FLC polymers. A vast variety of chemical structures have been synthesized. Our knowledge of structure/properties/relationships has grown considerably. Many modifications in the type of backbone, mesogenic side group, chiral moiety, and lateral-acting dipole have been performed. The variation of further structural parameters, including, for example, spacer length, molecular weight, polydispersity, degree of copolymerization, etc., allows one to tune the properties in a wide range.

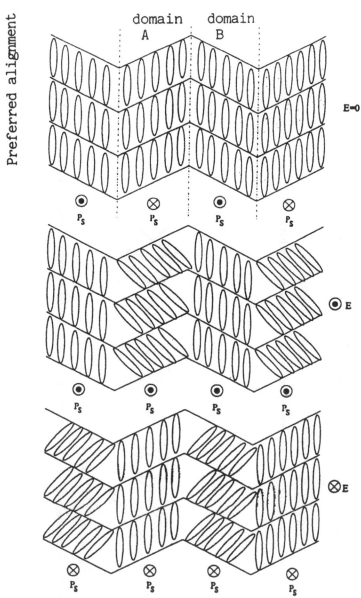

Figure 91 Schematic presentation of stripe domains and their alternating switching. (From Ref. 118.)

By this fast electrooptical switching, FLC polymers were synthesized with response times below 1 ms. This was the basis for the development of large, flexible displays. The detection of a spontaneous polarization in Langmuir-Blodgett films, of piezoelectricity in S_C^* elastomers, and of NLO properties in different S_C^* materials allows us to envisage further new applications.

After 10 years of strongly expanding research, the field of ferroelectric liquid crystal polymers is still in its infancy. The rapid progress has opened up a lot of questions for

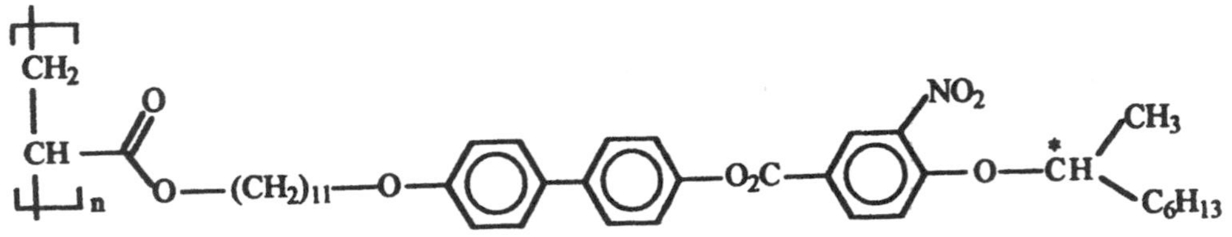

Figure 92 Structure and phase transition (°C) for polymer 76, exhibiting antiferroelectric behavior (S_X = unidentified smectic phase). (From Ref. 119.)

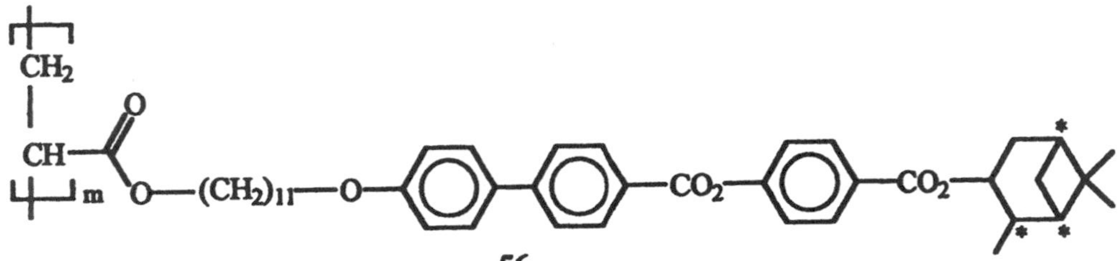

77

S_X 56 S_C^* 158 I   (M_{GPC} = 25 000)

Figure 93 Structure and phase transitions (°C) for polymer 56, exhibiting an antiferroelectric smectic phase. (From Ref. 120.)

56

S_X 68 S_C^* 131 S_A 174 I

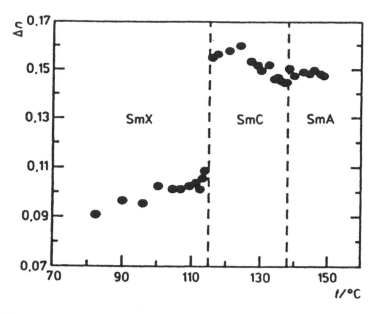

Figure 94 Temperature dependence of birefringence Δn for 56, measured using a tilting compensation. (From Ref. 120.)

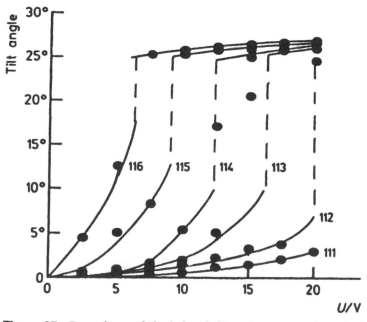

Figure 95 Dependence of the induced tilt angle versus voltage in the S_x phase at different temperatures. (From Ref. 120.)

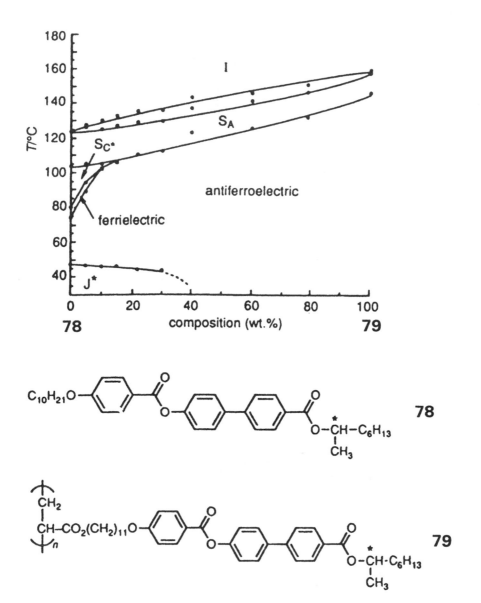

Figure 96 Miscibility phase diagram between FLC 78 and polymer 79. (From Ref. 121.)

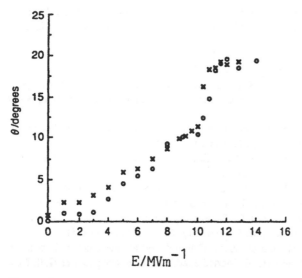

Figure 97 Field dependence of the apparent tilt angle in the antiferroelectric phase of polymer 79, measured at 144.5°C in 2.5-μm-thick cells (buffed, but uncoated): ○, increasing field; X, decreasing field. (From Ref. 121.)

basic research and opportunities for further material development and additional applications.

REFERENCES

1. R. B. Meyer, L. Liebert, and L. Strzelecki, A material with ferroelectric chiral smectic C and H phases. Contribution to the 5th Int. Liquid Crystal Conf., Stockholm, 1974.
2. R. B. Meyer, L. Strzelecki, L. Liebert, and P. Keller, Ferroelectric liquid crystals, *J. Phys. Lett.* (Paris) *36*:69 (1975).
3. (a) G. W. Gray and J. W. Goodby, *Smectic Liquid Crystals—Textures and Structures*, Leonard-Hill, Glasgow-London, 1984. (b) Y. Galerne and L. Liebert, Smectic-O films, *Phys. Rev. Lett. 64*:906 (1990). (c) A. D. L. Chandani, E. Gorecka, Y. Ouchi, H. Takazoe, and A. Fukuda, Antiferroelectric chiral smectic phases responsible for the tristable switching in MHPOBC, *Jpn. J. Appl. Phys. 28L*:1265 (1989). (d) G. Heppke, D. Lötzsch, D. Demus, S. Diele, and H. Zaschke, The S_M phase: Evidence for a new type of tilted smectic phase, *Mol. Cryst. Liq. Cryst. 208*:9 (1991).
4. N. A. Clark and S. T. Lagerwall, Submicrosecond bistable electro-optic switching in liquid crystals, *Appl. Phys. Lett. 36*:899 (1980).
5. C. M. Haws, M. G. Clark, and G. S. Attard, Dielectric relaxation spectroscopy of liquid crystalline side chain polymers, *Side Chain Liquid Crystal Polymers* (G. B. Ardle, ed.), Blackie, Chapman & Hall, New York, 1989, p. 196.
6. S. Ushida, K. Morita, K. Miyoshi, K. Hashimoto, and K. Kawasaki, Synthesis of some smectic liquid crystalline polymers and their ferroelectricity, *Mol. Cryst. Liq. Cryst. 155*: 93 (1988).

7. V. P. Shibaev, M. V. Koslovsky, L. A. Beresnev, L. A. Blinov, and N. A. Platé, Thermotropic liquid crystalline polymers. 16. Chiral smectics "C" with spontaneous polarization, *Polymer Bull.* *12*:299 (1984).

8. V. P. Shibaev, M. V. Koslovsky, L. A. Beresnev, L. A. Blinov, and N. A. Platé, Ferroelectric liquid crystalline polymethacrylates, *Vysokomolek. Soed.* *29*:1470 (1987).

9. M. V. Koslovsky, L. A. Bresnev, S. G. Kononov, V. P. Shibaev, and L. M. Blinov, Spontaneous polarization in a liquid crystalline polymer and the monomer based on it, *Solid State Phys.* *29*:98 (1987).

10. V. P. Shibaev, M. V. Kozlowsky, N. A. Platé, L. A. Beresnev, and L. M. Blinov, Ferroelectric liquid-crystalline polymethacrylates, *Liq. Crystals* *8*:545 (1990).

11. G. Decobert, S. Dubois, S. Esselin, and C. Noel, Some novel smectic C* liquid-crystalline side-chain polymers, *Liq. Crystals* *1*:307 (1986).

12. S. Esselin, L. Bosio, C. Noel, G. Decobert, and J. C. Dubois, Some novel smectic C* side-chain polymers, polymethacrylates and poly-α-chloroacrylates, *Liq. Crystals* *2*:505 (1987).

13. S. Esselin, C. Noel, G. Decobert, and J. C. Dubois, Synthesis and properties of smectic liquid crystalline side chain polymers, *Mol. Cryst. Liq. Cryst.* *155*:371 (1988).

14. P. L. Barny and J. C. Dubois, The chiral smetic C liquid crystal side chain polymers. in *Side Chain Liquid Crystal Polymers* (C. B. McArdle, ed.), Blackie, Chapman & Hall, New York, 1989, p. 150.

15. B. Hahn and V. Percec, LC polymers containing heterocycloalkane mesogenic groups. 5. Synthesis of biphasic chiral smectic polysiloxanes containing 2.5-disubstituted-1,3-dioxane- and 2.5-disubstituted-1,3,2-dioxaborinane-based mesogenic groups, *Macromolecules* *20*:2961 (1987).

16. P. Keller, FLC side-chain polysiloxanes, *Ferroelectrics* *85*:425 (1988).

17. M. E. Lines and A. M. Glass, *Principles and Application of Ferroelectrics and Related Materials*, Claredon Press, Oxford, 1977.

18. P. Martinot-Lagarde, The ferroelectric polarization of liquid crystals and its measurements, *Ferroelectrics* *84*:53 (1988).

19. K. Yuasa, S. Ushida, T. Sekya, K. Hashimoto, and K. Kawasaki, Electrooptical properties of FLC polymers, *Ferroelectrics* *122*:53 (1991).

20. N. A. Clark and S. T. Lagerwall, A microsecond-speed, bistable, threshold sensitive liquid crystal device, *Liquid Crystals of One- and Two-Dimensional Order* (W. Helfrich and G. Heppke, eds.), Springer Verlag, Berlin, 1980.

21. H. Endo, S. Hachiya, T. Sekiya, and K. Kawasaki, Rotational viscosity of FLC-polysiloxanes, *Liq. Crystals* *12*:147 (1992).

22. K. Kawasaki, H. Kidera, T. Sekya, and S. Hachiya, Molecular motion of FLC polymers, *Ferroelectrics* *148*:233 (1993).

23. G. Scherowsky, A. Schliwa, J. Springer, K. Kühnpast, and W. Trapp, Fast switching ferroelectric liquid-crystalline polymers, X. Int. Liquid Crystal Conf. Freiburg, 1988; *Liq. Crystals* *5*:1281 (1989).

24. T. Suzuki and T. Okawa, Preparation of ferroelectric liquid-crystalline polysiloxanes and electrooptical measurements, *Macromol. Chem. Rapid Commun.* *9*:755 (1988).

25. G. Scherowsky, U. Müller, J. Springer, W. Trapp, A. M. Levelut, and P. Davidson, Liquid-crystalline side chain polymers containing a chiral spacer unit exhibiting chiral smectic phases, *Liq. Crystals* *5*:1297 (1989).

26. S. Garoff and R. B. Meyer, Electroclinic effect at the A-C phase change in a chiral smectic liquid crystal, *Phys. Rev. Lett.* *38*:848 (1977).

27. D. M. Walba, R. T. Vohra, N. A. Clark, M. A. Handschy, J. Xue, D. S. Parmar, S. T. Lagerwall, and K. Skarp, Design and synthesis of new FLCs. 2. Liquid crystals containing a nonracemic 2,3-epoxyalcohol unit, *J. Am. Chem. Soc.* *108*:7424 (1986).

28. D. M. Walba, P. Keller, N. A. Clark, M. D. Wand, Design and synthesis of new ferroelectric liquid crystals. 9. An approach to creation of organic polymer thin films with controlled, stable polar orientation of functional groups, *J. Am. Chem. Soc.* *111*:8273 (1989).

29. M. Dumon, H. T. Nguyen, M. Mauzac, C. Destrade, M. F. Achard, and H. Gasparoux, New ferroelectric liquid crystal polysiloxanes, *Macromolecules 23*:355 (1990).

30. (a) H. Kapitza, R. Zentel, R. J. Twieg, C. Nguyen, S. O. Valerien, F. Kremer, and C. G. Willson, Ferroelectric liquid crystalline polysiloxanes with high spontaneous polarization and possible application in non linear optics, *Adv. Mater. 2*:539 (1990). (b) R. Zentel, H. Poths, F. Kremer, A. Schönfeld, D. Jungbauer, R. Twieg, C. G. Willson, and D. Yoon, Polymeric liquid crystals: Structural basis for ferroelectric and nonlinear optical properties, *Polymer Adv. Tech. 3*:211 (1992).

31. H. J. Coles, H. F. Gleeson, G. Scherowsky, and A. Schliwa, Ferroelectric side chain polymer liquid crystals: I. Static and dynamic properties, *Mol. Cryst. Liq. Cryst. Lett. 7*:117 (1990).

32. H. Poths, R. Zentel, S. U. Valerien, and F. Kremer, LC-polymers with axial chirality, *Mol. Cryst. Liq. Cryst. 203*:101 (1991).

33. R. Zentel, H. Poths, F. Kremer, A. Schönfeld, D. Jungbauer, R. Twieg, C. R. Willson, and D. Yoon, Polymeric liquid crystals: Structural basis for ferroelectric and nonlinear optical properties, *Polymer Adv. Tech. 3*:211 (1992).

34. T. Kitazume and T. Ohnogi, Synthesis of ferroelectric liquid crystal polymers possessing the trifluoromethyl group, *J. Fluorine Chem. 47*:459 (1990).

35. T. Kitazume, T. Ohnogi, and K. Ito, Design and synthesis of new fluorinated ferroelectric liquid crystalline polymers, *J. Am. Chem. Soc. 112*:6608 (1990).

36. K. Takahashi, S. Matsumoto, T. Tsuru, and F. Yamamoto, Electrooptical effects in new polymeric liquid crystal, *Mol. Cryst. Liq. Cryst. Lett. 8*:33 (1991).

37. J. Naciri, S. Pfeiffer, and R. Shashidhar, Fast switching ferroelectric side-chain liquid-crystalline polymer and copolymer, *Liq. Crystals 10*:585 (1991).

38. L. C. Chien, I. G. Shenouda, A. Saupe, and A. Jakli, Side-chain liquid crystalline polysiloxanes containing a cyanohydrin chiral center, *Liq. Crystals 15*:497 (1993).

39. V. P. Shibaev and N. A. Platé, Synthesis and structure of LC side chain polymers, *Pure Appl. Chem. 57*:1589 (1985).

40. H. Coles, R. Simon, H. Gleeson, J. Bone, G. Scherowsky, A. Schliwa, and U. Müller, Electrooptic and non linear optic effects in side chain LC polymers, *Polymer Preprints, Jpn.* (English ed.) *39*:5 (1990).

41. G. Scherowsky, U. Müller, A. Schliwa, P. Schreiber, K. Kühnpast, J. Springer, H. J. Coles, and P. Harnischfeger, Novel ferroelectric liquid crystal side chain polymers, *Proc. 19. Freiburger Arbeitstagung Flüssigkristalle*, 1990, p. 25.

42. K. Takahashi, S. Ishibashi, and F. Yamamoto, Properties of ferroelectric liquid crystals with two chiral groups, *Ferroelectrics*, Proc. IVth Int. Conf. on Ferroelectric Liquid Crystals, Tokyo, *148*:255 (1993).

43. M. Dumon and H. T. Nguyen, Mesomorphic and ferroelectric properties of FLCP/FLC binary mixtures, *Polymer Adv. Tech. 3*:197 (1992).

44. S. Ujiie and K. Imura, Liquid crystalline and ferroelectric properties of tartrate with two chiral centers, *Chem. Lett.*, 2217 (1989); Thermal and ferroelectric properties of chiral liquid crystalline polytartrate, *Chem. Lett.*, 1031 (1990); Ferroelectric liquid-crystalline polytartrate, *Polymer J. 23*:1483 (1991).

45. T. Sekiya, K. Yuasa, S. Ushida, S. Hachiya, K. Hashimoto, and K. Kawasaki, FLC-polymers and related model compounds with a low-moderate degree of polymerization, *Liq. Cryst. 14*:1255 (1993).

46. K. Kawasaki, F. Moriwaki, S. Hachiya, H. Endo, T. Sekyia, and K. Hashimoto, FLC-polyexyethylenes with etheric-oxygen-containing spacers, *Ferroelectrics*, Proc. IVth Int. FLC Conf., Tokyo, *148*:245 (1993).

47. E. Chiellini, G. Galli, and F. Cioni, Chiral liquid crystal polymers with potential ferroelectric properties: Synthesis and characterization, *Ferroelectrics 114*:223 (1991).

48. C.-S. Hsu, J.-H. Lin, and L.-R. Chou, Synthesis and characterization of FLC polysiloxanes and polymethacrylates containing [(S)-2-methyl-1-butoxy]phenyl-4-(alkoxy)biphenyl-4'-carboxylate side groups, *Macromolecules 25*:7126 (1992).

49. W.-L. Tsai, H.-L. Kuo, and S.-H. Yang, Liquid crystalline side chain polymers consisting of acrylate and 2(S)-[2(S)-methylbutoxy]proprionate, *Liq. Cryst. 16*:143 (1994).

50. K. Kühnpast, J. Springer, G. Scherowsky, F. Giesselmann, and P. Zugenmaier, Ferroelectric liquid crystalline side group polymers. Spacer lengths variation and comparison with the monomer, *Liq. Cryst. 14*:861 (1993).

51. P. Davidson, K. Kühnpast, J. Springer, and G. Scherowsky, Two unusual mesophases in chiral side chain polymers, *Liq. Cryst. 14*:901 (1993).

52. K. Kühnpast, J. Springer, P. Davidson, and G. Scherowsky, Spacer length and molecular weight variation on FLC-polymers, *Macromol. Chem. 193*:3097 (1992).

53. K. Yuasa, S. Ushida, T. Sekiya, K. Hashimoto, and K. Kawasaki, Ferroelectric liquid crystalline polymers for display devices, *Proc. SPIE-INT. Soc. Opt. Eng.*, 1665 (Liq. Cryst. Mater., Devices, Appl.), 1992, p. 154; Electrooptical properties of FLC-polymers, *Polymer Adv. Tech. 3*:205 (1992).

54. M. Redmond, H. Coles, E. Wischerhoff, and R. Zentel, Ferroelectric and electroclinic characterization of a new organic siloxane bimesogen, *Ferroelectrics*, Proc. IVth Int. Conf. on Ferroelectric Liquid Crystals, Tokyo, *148*:323 (1993).

55. H. Endo, S. Hachiya, and K. Kawasaki, Measurements of polarization reversal current of a ferroelectric liquid crystalline polymer, *Liq. Cryst. 13*:721 (1993).

56. H. Endo, S. Hachiya, S. Ushida, K. Hashimoto, and K. Kawasaki, Helical pitch of ferroelectric liquid-crystalline polymers and copolymers, *Liq. Cryst. 9*:635 (1991).

57. S. Pfeiffer, R. Shashidhar, J. Naciri, and S. Mery, Electrooptic properties of FLC-polymers, *SPIE 1665*:166 (1992).

58. J. Ruth, B. R. Ratna, J. Naciri, and R. Shashidhar, Ferroelectric liquid crystalline polymers with large pyroelectric coefficients for infrared detectors, *SPIE 1911*:104 (1993).

59. K. Grüneberg, Ferroelectric and antiferroelectric properties of chiral acrylates and polyacrylates, Dissertation, Technical University, Berlin, 1993.

60. H. J. Coles, H. F. Gleeson, G. Scherowsky, and A. Schliwa, Ferroelectric polymer liquid crystals: II The dye guest host effect, *Mol. Cryst. Liq. Cryst. Lett. 7*(4):125 (1990).

61. G. Scherowsky, A. Beer, and H. J. Coles, A coloured ferroelectric side chain polymer, *Liq. Cryst. 10*:809 (1991).

62. H. J. Coles, H. F. Gleeson, and J. S. Kang, Dye guest host effects in ferroelectric liquid crystals, *Liq. Cryst. 5*:1243 (1989).

63a. A. Beer, Coloured and fluorescent ferroelectric copolymers. Synthesis and electrooptical properties, Dissertation, Technical University Berlin, 1994.

63b. A. Beer, G. Scherowsky, H. Coles, and H. Owen, Fluorescent ferroelectric copolymers, *Liq. Cryst.* submitted (1995).

64. B. Fayolle, C. Noel, and G. Billard, Investigation of polymer mesophases by optical microscopy, *J. Phys.* (Paris) *40*, Colloq. C3:485 (1979).

65. G. Scherowsky, K. Grüneberg, and K. Kühnpast, Induced spontaneous polarization in a side chain polymer, *Ferroelectrics 122*:159 (1991).

66. M. Ido, K. Tanaka, S. Hachiya, and K. Kawasaki, Ferroelectricity of compositions using SmC polymers, *Ferroelectrics 1994*, FLC Conf., Tokyo, 1993, p. 109.

67. M. V. Kozlovsky, S. G. Kononov, L. M. Blinov, K. Fodor-Csorba, and L. Bata, Chiral smectic side-chain copolymers 2. Pyroelectric effect and spontaneous polarization, *Eur. Polymer. J. 28*:907 (1992).

68. (a)U. Kumar, J. M. J. Frechet, T. Kato, and S. Ujiie, Induction of mesogenity in the side chain of polysiloxanes via hydrogen bonding: From ferroelectric to non planar LC assemblies, *Polymer Mater. Eng. 67*:439 (1991). (b) U. Kumar, J. M. J. Frechet, T. Kato, S. Ujiie, and K. Jimura, Induktion von Ferroelektrizität in Polymersystemen durch

Wasserstoffbrückenbindungen, *Angew. Chem. 104*:1545 (1992); *Int. Ed. 31*(11):1531 (1992).

69. D. S. Parmar, N. A. Clark, P. Keller, D. M. Walba, and M. D. Wand, Physical properties and alignment of polymer-monomer FLC mixtures, *J. Phys. France 51*:355 (1990).

70. N. Dumon and H. T. Nguyen, Mesomorphic and ferroelectric properties of FLP/FLC binary mixtures, *Polymer Adv. Tech. 3*:197 (1992).

71. J. Bömelburg, G. Heppke, and J. Hollidt, Polarization and viscosity measurements of FLC-monomer/polymer mixtures, *Polymer Adv. Tech. 3*:237 (1992).

72. J. Ruth, J. Naciri, and R. Shashidhar, Ferroelectric properties of mixtures of a liquid crystalline polymer and its side group mesogen, *Liq. Cryst. 16*:883 (1994).

73. L. M. Blinow and W. Haase, Electro-optical properties of polymer liquid crystals, *Mol. Mater. 2*:145 (1993).

74. S. U. Vallerien, R. Zentel, F. Kremer, H. Kapitza, and E. H. Fischer, Ferroelectric modes in combined side group main chain liquid crystalline polymers, *Makromol. Chem. Rapid Commun. 10*:333 (1989).

75. M. Pfeiffer, L. A. Beresnev, W. Haase, G. Scherowsky, K. Kühnpast, and D. Jungbauer, Dielectric and electrooptic properties of a switchable FLC-side chain polymer, *Mol. Cryst. Liq. Cryst. 214*:125 (1992).

76. F. Kremer, A. Schönfeld, A. Hofmann, R. Zentel, and H. Poths, Collective and molecular dynamics in FLC: From low molar to polymeric and elastomeric systems, *Polymer Adv. Tech. 3*:249 (1992).

77. E. Chiellini, G. Galli, E. Dossi, and F. Cioni, New chiral smectic polysiloxanes from mesogenic olefines or vinylether monomers, *Macromolecules 26*:849 (1993).

78. E. Chiellini, G. Galli, F. Cioni, E. Dossi, and B. Gallot, Chiral smectic liquid-crystalline polyacrylates with variously spaced and substituted biphenylene units, *J. Mater. Chem. 3*: 1065 (1993).

79. E. Chiellini, G. Galli, F. Cioni, and E. Dossi, Chiral liquid crystalline polymers: Recent issues and perspectives, *Makromol. Chem. Macromol. Symp. 69*:51 (1993).

80. L. Komitov, S. T. Lagerwall, B. Stebler, E. Chiellini, G. Galli, and A. Strigazzi, in *Modern Topics in Liquid Crystals* (A. Buka, ed.), World Science, Singapore, 1994.

81. K. Flatischler, L. Komitov, K. Skarp, and P. Keller, Electroclinic effect in some side-chain polysiloxane liquid crystals, *Mol. Cryst. Liq. Cryst. 209*:109 (1991).

82. D. H. Walba and N. A. Clark, Molecular design of ferroelectric liquid crystals, *Ferroelectrics 84*:65 (1988).

83. G. Scherowsky, B. Brauer, K. Grüneberg, U. Müller, L. Komitov, S. T. Lagerwall, K. Sharp, and B. Stebler, Sign reversal of the spontaneous polarization in the C* phase of a side-chain polyacrylate and its monomer, *Mol. Cryst. Liq. Cryst. 215*:257 (1992).

84. J. Goodby, E. Chin, J. Geary, J. Patel, and P. L. Finn, The ferroelectric and liquid-crystalline properties of some chiral alkyl 4-*n*-alkanoyloxybiphenyl-4′-carboxylates, *J. Chem. Soc., Faraday Trans. 1, 83*:3429 (1987).

85. R. Meister and H. Stegemeyer, An extended microscopic model of the spontaneous polarization in FLC, *Ber. Bunsenges. Phys. Chem.*, *97*:1242 (1993).

86. S. Saito, K. Murashiro, M. Kikuchi, T. Inukai, D. Demus, M. Neundorf, and S. Diele, P_S Inversion in three homologous series of ferroelectric 5-(2-fluoro-alkoxy)-2(4*n*-alkylphenyl)-pyrimidines, *Ferroelectrics, 147*:367 (1993).

87. J. Y. Liu, M. G. Robinson, K. M. Johnson, and D. Doroski, Second harmonic generation in ferroelectric liquid crystals, *Opt. Lett. 15*:267 (1990).

88. A. Tagushi, Y. Ouchi, H. Takazoe, and A. Fukuda, Angle phase matching in second harmonic generation from a ferroelectric liquid crystal, *Jpn. J. Appl. Phys. 28L*:997 (1989).

89. H. Kapitza, R. Zentel, R. J. Twieg, C. Nguyen, S. Vallerien, F. Kremer, and C. G. Wilson, Ferroelectric liquid crystalline polysiloxanes with high spontaneous polarization and possible application in nonlinear optics, *Adv. Mater. 2*:539 (1990).

90. D. M. Walba and M. B. Ros, An approach to the design of FLC's with large second order electronic nonlinear optical succeptibility, *Mol. Cryst. Liq. Cryst.* 198:51 (1991).

91. J. Y. Liu, M. G. Robinson, K. M. Johnson, D. M. Walba, M. B. Ros, N. A. Clark, R. Shao, and D. Doroski, The measurement of second-harmonic generation in novel ferroelectric liquid crystal materials, *J. Appl. Phys.* 70:3426 (1991).

92a. K. Schmitt, R. P. Herr, M. Schadt, J. Fünfschilling, P. Buchecker, X. H. Chen, and C. Benecke, Strongly nonlinear optical FLC for frequency doubling, *Liq. Cryst.* 14:1735 (1993).

92b. X. H. Chen, R. Buchecker, and R. P. Herr, The synthesis of nitroaniline monomers and polymers as non-linear optical ferroelectric liquid crystals, *Liq. Cryst.* submitted (1995).

93. A. Jakli and A. Saupe, Linear electromechanical effect in a S_C^* polymer liquid crystal, *Liq. Cryst.* 9:519 (1991).

94. N. Eber, L. Bata, G. Scherowsky, and A. Schliwa, Linear electromechanical effect in a polymeric FLC, *Ferroelectrics* 122:139 (1991).

95. A. Jakli, N. Eber, and L. Bata, Electromechanical effects inf FLC polymers, *Polymer Adv. Tech.* 3:269 (1992).

96. W. Rettig, J. Naciri, R. Shashidhar, and R. S. Duran, Monolayers and Langmuir-Blodgett films of a ferroelectric side-chain polymer and its constituent mesogen, *Macromolecules* 24:6539 (1991).

97. W. Rettig, J. Naciri, R. Shashidhar, and R. S. Duran, The behaviour of FLC compounds at the air-water interface, *Thin Solid Films* 210:114 (1992).

98. J. Adams, A. Thibodeaux, J. Naciri, R. Shashidhar, and R. S. Duran, Ferroelectric l.c. monolayers at the air/water interface: Comparison between different chemical constituents, *Polymer Preprints* 33:1211 (1992).

99. A. F. Thibodeaus, R. Geer, S. Quadri, J. Adams, J. Naciri, R. Shashidhar, and R. S. Duran, Langmuir-Blodgett multilayer films of a FLC copolymer using an alternate deposition method, *Polymer Preprints* 33:1220 (1992).

100. J. Adams, W. Rettig, R. S. Duran, J. Naciri, and R. Shashidhar, Langmuir Films of LC materials: The influence of molecular architecture on morphology and properties, *J. Phys. Chem.* 97:2021 (1993).

101. S. Pfeiffer, R. Shashidhar, T. L. Fare, J. Naciri, J. Adams, and R. S. Duran, Ferroelectricity in a Langmuir-Blodgett multilayer film of a LC side chain polymer, *Appl. Phys. Lett.* 63(9): 1285 (1993).

102. R. E. Geer, R. Shashidhar, A. F. Thibodeaux, and R. S. Duran, X-ray diffuse scattering study of static undulations in multilayer films of a LC polymer, *Phys. Rev. Lett.* 71:1391 (1993).

103. H. R. Brand, Electromechanical effects in cholesteric and chiral smectic liquid-crystalline elastomers, *Makromol. Chem. Rapid Commun.* 10:441 (1989).

104. S. U. Valerien, F. Kremer, E. W. Fischer, H. Kapitza, R. Zentel, and H. Poths, Experimental proof of piezoelectricity in cholesteric and chiral smectic C* phases of LC-elastomers, *Makromol. Chem. Rapid Commun.* 11:593 (1990).

105. R. A. M. Hikmet, Piezoelectric networks obtained by photopolymerization of liquid crystal molecules, *Macromolecules* 25:5759 (1992).

106. M. Brehmer, R. Zentel, G. Wagenblast, and K. Siemensmeyer, Ferro- and piezoelectric LC-elastomers, *Macromol. Chem. Phys.*

107. K. Semmler and H. Finkelmann, Orientation of chiral smectic C elastomer by mechanical fields, *Polymer Adv. Tech.* 5:231 (1994).

108. J. Benne, K. Semmler, and H. Finkelmann, Second harmonic generation on mechanically oriented S_C^*-elastomers, *Makromol. Chem. Rapid Commun.* 15:295 (1994).

109. J. W. Doane, A. Golemme, J. L. West, J. B. Whitehead, and B. G. Wu, Polymer dispersed liquid crystals for display application, *Mol. Cryst. Liq. Cryst.* 165:511 (1988).

110. H. S. Kitzerow, Polymer-dispersed liquid crystals from the nematic curvilinear aligned phase to ferroelectric films, *Liq. Cryst.* 16:1 (1994).

111. L. A. Beresnev, V. G. Chigrinov, D. I. Dergachev, E. P. Poshidaev, J. Fünfschilling, and M. Schadt, Deformed helix ferroelectric liquid crystal display: A new electrooptic mode in ferroelectric smectic C liquid crystals, *Liq. Cryst.* 5:1171 (1989).

112. (a) H. S. Kitzerow, H. Molsen, and G. Heppke, Linear electro-optic effects in polymer-dispersed ferroelectric liquid crystals, *Appl. Phys. Lett.* 60:3093 (1992). (b) H. S. Kitzerow, H. Molsen, and G. Heppke, Helical unwinding in polymer dispersed FLC's, *Polymer Adv. Tech.* 3:231 (1992).

113. L. Komitov, S. T. Lagerwall, and G. Chidichimo, Linear light modulation by polymer dispersed chiral liquid crystals, *SPIE 2175*:160 (1994).

114. H. Molsen, H. S. Kitzerow, and G. Heppke, Antiferroelectric switching in polymer dispersed liquid crystals, *Jpn. J. Appl. Phys.* 31:L-1083 (1992).

115. Y. Yamada, N. Yamamoto, M. Yamawaki, I. Kawamura, and Y. Suzuki, *Proc. Jpn. Display 92*:57 (1992).

116. G. Scherowsky, K. Kühnpast, and J. Springer, Three switching states in a chiral smectic side-chain polymer, *Makromol. Chem. Rapid Commun.* 12:381 (1991).

117. F. Giesselmann, P. Zugenmaier, G. Scherowsky, K. Kühnpast, and J. Springer, Electrooptical investigation on the three switching states of a chiral smectic side group polymer, *Makromol. Chem. Rapid Commun.* 13:489 (1992).

118. L. Komitov, S. T. Lagerwall, B. Stebler, K. Grüneberg, and G. Scherowsky, on the seemingly antiferoelectric behavior of ferroelectric liquid crystals, *Ferroelectrics 1995*, in press.

119. (a) K. Skarp, G. Andersson, S. T. Lagerwall, H. Kapitza, H. Poths, and R. Zentel, Anomalous current and electrooptical response in a polyacrylate FLC with large spontaneous polarization, *Ferroelectrics 122*:127 (1991). (b) K. Skarp, G. Andersson, F. Gouda, T. S. Lagerwall, H. Poths, and R. Zentel, Antiferroelectric behaviour in a LC polymer, *Polymer Adv. Tech.* 3:241 (1992).

120. J. Bömelburg, G. Heppke, and J. Hollidt, Evidence for an antiferroelectric smectic phase in a chiral side chain polymer, *Makromol. Chem. Rapid. Commun.* 12:483 (1991).

121. I. Nishiyama and J. W. Goodby, Effect of polymerization on the stability of antiferroelectric LC-phases: LC-properties of some chiral acrylates and their corresponding polyacrylates, *J. Mater. Chem.* 3:169 (1993).

11
Polymer-Ferroelectric Ceramic Composites

Karol Mazur
Technical University, Zielona Góra, Poland

I. INTRODUCTION

Polymer-ferroelectric ceramic composites, after poling, control as much as they can for both theoretical and practical reasons. On the one hand, they are becoming increasingly more important as materials for the study of charge transport and its storage phenomena in multicomponent systems; what is more, there has been a rapid development in their applications in a wide range of devices. The most well-known applications of such composites include electromechanical transducers, above all microphones and hydrophones, as well as pyroelectric detectors in infrared sensors and microcalorimeters.

Ceramic materials such as barium titanate and lead zirconate titanate (PZT) have very good pyro- and piezoelectric properties and as such are used in a variety of applications [1,2]. However, their brittle nature and inflexibility limit their use in some applications. Moreover, in most electronic devices there are several phases involved and a number of materials to be optimized. The electromechanical transducer, for example, might require a combination of properties such as a large d or g piezoelectric coefficient. On the contrary, a pyroelectric detector might require a large pyroelectric coefficient, low thermal capacity, and high mechanical flexibility.

Also, ferroelectric polymers such as poly(vinylidene fluoride) (PVDF) in particular have wide applications. (Recent developments in ferroelectric polymers have been described by Kawai [3a] and Nalwa [3b]). Sometimes, ferroelectric properties are advantageous for some reasons, e.g., low electric permittivity and small thickness, which limit its application in other devices [4,5].

Therefore a composite consisting of highly piezo- and pyroelectric ceramic material combined with a polymer would be the ideal replacement to obtain the properties of both these classes. As noted by Muralidhar and Pillai [6] such a composite would exhibit the piezo- and pyroelectric coefficients of ceramics and the flexibility, strength, and lightness of the polymer.

A composite in general is a heterostructural material whose properties are determined by the number of different phases of the material, the volume fractions of the phases, the properties of individual phases, and the ways in which different phases are interconnected [7,8]. The latter is the most important feature of composites, since the mixing rules of a given property are controlled by the self-connectiveness of individual phases [8].

Two- or multiphase materials are called composites if particular components are selected and connected in such a way that the composite materials have the required properties, which rarely occur in one-phase materials [9–16].

Theoretical approaches to the linear response of two-phase systems, even in the simplest uncoupled case, are based on various model rules. Recently, it was pointed out that the state of the interface between the filler and the matrix [17–20] as well as the microgeometry [21] affect the macroscopic behavior of composites.

It should be stated, however, that the above-mentioned aspects have not been satisfactorily modeled theoretically, and a new theoretical approach together with systematic experimental studies is needed.

In this chapter, composite materials most frequently used in electromechanical transducer and the ways of their production are presented. When choosing composite materials that will be optimal for some applications, so-called figures of merit are taken into consideration, which consider the most sensitive parameters characterizing the use of properties and enable one in a simple way to choose the materials to be included in the composite. One useful figure of merit for the electromechanical transducer is the product of the hydrostatic piezoelectric coefficient d_h and the voltage hydrostatic piezoelectric coefficient $g_h = d_h/\epsilon\epsilon_0$, where ϵ_0 is the electric permittivity of free space when ϵ is the electric permittivity of the material. The way of joining phases included will of course determine the composite properties. Below we shall discuss the ways of obtaining various types of composites used in electromechanical transducters.

At present, piezoelectric and pyroelectric materials are of great interest. It is enough to mention that in his "Guide to the Literature of Piezoelectricity and Pyroelectricity" Lang [2a] cited 834 references on piezoelectric and pyroelectric properties and their applications published just during the period 1991–1992. Over 60 papers of this number refer to composite materials. Some of these papers will be cited below.

The ways of preparing and general properties of the composites of various types will first be described. Then their dielectric and electret properties will be discussed, since they determine the piezo- and pyroelectricity of polymer-ceramic composites. The last-mentioned properties will be described in the last part of this chapter.

We will try to suggest the answer to the question, "What causes the piezo- and pyroelectricity in electrets—dipoles (or spontaneous polarization), charges, or both?" The exact answer to this question was not found during the Special Discussion Session at ISE 7 in Berlin, 1991. The investigation of relaxation processes by means of the TSD method seem to lead to the answer to the above question concerning polymer-ceramic composites. It will be shown by the example of the electret and piezoelectric properties of some multilayer systems [17–19].

II. COMPONENTS, THEIR CONNECTIVITY, AND GENERAL CLASSIFICATION OF PROPERTIES IN COMPOSITES

As pointed out by van Suchtelen and other authors, the physical properties of polymer-ceramic composite materials can be divided into the sum, combined, and product properties [4,22–24]. It is interesting to note [16] that in some composites there are not only the properties of the separate modified phases (sum properties), but the composite may exhibit completely new couplings (product properties) not found in the separate phases. These properties depend not only on the particular phases but also on the way of their connectivity and some processings of the composite [4,7,8,15,16].

A. Connectivity

The ways of connectivity of the separate ceramic and polymer phases of composite were arranged by Newnham et al. [15]. In accordance with their idea, each phase in a composite may be self-connected in zero, one, two, or three dimensions. In the case of diphasic composites, there are 10 connectivities, designated as 0-0, 1-0, 2-0, 3-0, 1-1, 2-1, 3-1, 2-2, 3-2, and 3-3 (The 10 different connectivities are illustrated after Newnham [15,16] in Figure 1 using a cube as the basic building block). In general, for n phases the number of connectivity patterns is $(n + 3)!/3!n!$.

B. General Classification of Properties

A physical property relates an input physical quantity X to an output physical quantity Y. The $X–Y$ effect may be a linear relationship specified by a property coefficient $C = \partial Y/\partial X$ or may be a more complicated effect [15,16].

Sum properties are those [4,16,22] in which the $X–Y$ effect of the composite is determined by the $X–Y$ effects in phases 1 and 2. As an example, the stiffness of a composite is determined by the elastic properties of the component phases and by the mixing rule, which depend on the geometry of the phases. Similarly, permittivity ϵ of the composite consisting of two phases with ϵ_1 and ϵ_2 is determined by the sum

$$\epsilon^n = v_1\epsilon_1{}^n + v_2\epsilon^n \tag{1}$$

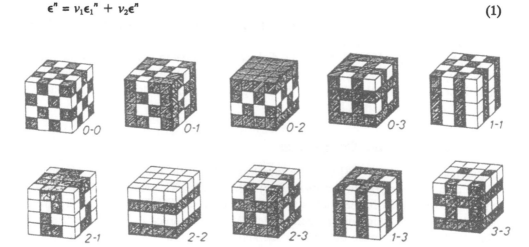

Figure 1 Ten connectivity patterns for a diphasic solid. (After Refs. 4 and 16.)

where v_1 and v_2 are the volume ratios of phases 1 and 2 and the exponent n equals $+1$ or -1 for parallel and series connections, respectively. As is seen from relation (1), the permittivity for the parallel model changes linearly from ϵ_1 to ϵ_2 with the volume fraction v_2 of the ceramic phase. For series connection the values of permittivity ϵ are lower but also change from ϵ_1 to ϵ_2 if v_2 changes from 0 to 1. Also, other dielectric characteristics plotted as a function of $v_2 = 1 - v_1$ should begin with the properties of the polymeric phase (phase 1) and finish with the properties of the ceramic phase (phase 2). However, the shape of the $f(v_2)$ characteristics between these boundary points ($v_2 = 0$ and $v_2 = 1$) has been a subject of study for many years and still is an open problem [25].

In the last years a few interesting theoretical papers concerning a powerful and versatile principle, allowing the estimation of the effective properties of nonlinear heterogeneous systems, have been introduced by Ponte Castañeda [26,27]. Some results of Castañeda theory application for two-phase composite are described in other papers of this author and his co-workers [28].

If some physical property is described by two or three coefficients, the application of the above cited rules is impossible. In this case we have to deal with the combined properties [4]. An example of a combined property in a 1–3 type composite is the electromechanical transducer consisting of nonpiezoelectric polymer matrix and polarized ferroelectric thin rods.

As is known from Suchtelen, Newnham, and other authors' papers [4,15,16], so-called product properties of composites are less expected and somewhat more complicated.

As a consequence of the interaction between phase 1 and phase 2 in a composite, two different properties of each phase can lead to the new third property. As an example of a product property, Newnham et al. [16] have considered a magnetoelectric composite made from ferroelectric $BaTiO_3$ (phase 1) and ferromagnetic $CoFe_2O_4$ (phase 2). If the ferroelectric grains are poled near the ferroelectric Curie temperature in a strong electric field, this leads to production of a piezoelectric composite. In a similar way, the magnetic poling of the ferromagnetic phase is accomplished by annealing the composite in a magnetic field. When an electric field is applied to a magnetoelectric composite of this type ($BaTiO_3$-$CoFe_2O_4$), the ferroelectric grains elongate parallel to the electric field. The change in shape of the ferroelectric grains causes the ferromagnetic grains to deform, resulting in a change in magnetization. Magnetoelectric measurements on $BaTiO_3$-$CoFe_2O_4$ composites show magnetoelectric coefficients two orders larger than the best single-phase material [29,30]. Similarly, diphase composites in which one of the phases shows high electrical conductivity and the second phase high thermal expansion can have thermistor properties [4], whereas if the first phase is piezoelectric, the composite can have pyroelectric properties [4,16].

C. Materials for 0-3 Composites

In previous papers [9,31–33] we have investigated the dielectric and electret properties of the composite poly(methyl methacrylate) (PMMA)/$BaTiO_3$ ceramic system. The 0-3 type composites used in the studies were prepared by polymerization of methyl methacrylate (MMA) in which fine particles of ferroelectric ceramics $BaTiO_3$ were dispersed. The $BaTiO_3$ ceramics were obtained by the known method of triple sintering of $BaCO_3$ and TiO_2 in stoichiometric ratio. The final process was conducted at about 1620 K for 3 h, resulting in a ceramic density of 5.57×10^3 kg/m^3. This was pulverized and sifted through an appropriate sieve. The powder was mixed in various volume ratios with methyl methacrylate in which the polymerization process had been previously initiated. After the polymerization had been concluded at the higher temperature, disk-shaped samples with 0 to about 60 vol% of BaTiO$_3$ were formed from this composite.

Furukawa et al. [12] have investigated the piezoelectric properties in composite systems of polymers and PZT ceramics. In previous papers [13,14], these authors have described the piezoelectric properties of the composite epoxy resin/PZT ceramic system on the basis of the theoretical expressions of the piezoelectric constants for a two-phase system where piezoelectric spherical particles are dispersed in nonpiezoelectric continuous media (0-3 type composite). Also, piezoelectric properties have been studied [12] for composite systems of PZT ceramics and polymers including poly(vinylidene fluoride) (PVDF), polyethylene (PE), and polyvinyl alcohol (PVA). The composites were prepared in the following ways. PVDF pellets and PZT ceramic powders were mixed at 200°C using a hot roller and then pressed into about a 200-μm-thick film at 180°C. The composite of PE and PZT was made in a similar way. The composite of PVA and PZT was cast from water to obtain a thin film.

Most often, the following materials are used as the ceramic phases in polymer-ceramic composites:

Barium titanate (BaTiO$_3$) and BaTiO$_3$ doped by Nb$_2$O$_5$ and Sb$_2$O$_5$ [6,9,31–39]
Lead titanate (PbTiO$_3$) [40–49]
Solid solutions of PbTiO$_3$ and PbZrO$_3$ (PZT): Pb(Ti$_{1-x}$, Zr$_x$)O$_3$ [12,50–64]
PZT doped by niobium [65], lanthanum [66], and other modified PZT ceramics [8,38,67]

In the different piezocomposites more complicated solutions are also used, such as (Pb,Bi) (Ti,Fe,Mn)O$_3$ [24,68], Pb(Zr,Nb)O$_3$-PbTiO$_3$ [24], (Pb,X)(Ti,Mn)O$_3$, where $X =$ La, Nd, Sm, Gd [69], Pb(Zr,Ti)O$_3$–Pb (Mn,Nb)O$_3$ [70], and (Pb,Ca)(Co,W,Ti)O$_3$ [71].

The composites have the best piezoelectric properties (hydrostatic) if the ceramic phase in the composite shows the largest piezoelectric anisotropy, for example, such ceramics as PbTiO$_3$ and (Pb,Bi) (Ti,Fe,Mn)O$_3$ [4].

We note that the effect on dielectric, electret, ferro-, piezo-, and pyrelectric properties can result from not only the above-mentioned materials but also depend the grain size of these ceramics [39,42,72–74]. This effect is very important, since electronic devices using ferroelectric ceramics, such as multilayer capacitors, have been miniaturized. There have already been many reports on grain size dependence in BaTiO$_3$ ceramic properties [73].

Yamamoto et al. [39], in recent work, have studied the particle-size dependence of hydrothermally produced BaTiO$_3$ powder, by focusing on the crystal phase, and discussed the piezoelectric properties in BaTiO$_3$/polymer composites as a function of the particle size of BaTiO$_3$ powder. For example, Figure 2 shows the BaTiO$_3$ particle size dependence of the piezoelectric coefficient (d_{33}) in BaTiO$_3$/polymer composites. Also in this work the hydrostatic piezoelectric constant (d_h), piezoelectric strain constant (d_{31}), and piezoelectric voltage coefficient (g_{33}) have been investigated in terms of their dependence on particle size. It is shown in this paper that the critical particle size is 0.1–0.2 μm for tetragonal to pseudo-cubic phase transformation at room temperature. The piezoelectric properties in pseudo-cubic BaTiO$_3$ powder were obtained in high electric field, although their values were small.

Most often, polymer materials for the performance of different type of polymer-ceramic composites are applied:

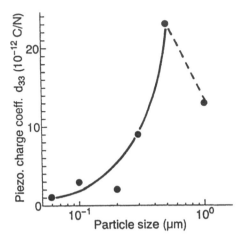

Figure 2 BaTiO₃ particle size dependence of piezoelectric charge coefficient (d_{33}) in BaTiO₃/polymer composites. (After Ref. 39.)

Poly(methyl methacrylate) (PMMA) [2,3,9,17–19]
Poly(vinylidene fluoride) (PVDF), and the random copolymers P(VDF-TrFE) of vinylidene fluoride (VDF) and trifluoroethylene (TrFE) [2–4,6,12,34,55–57,59,75–79].
Polypropylene (PP) [38]

Also applied are polyethylene (PE) [12,76,77], copolymer vinyl acetate and vinyl chloride [66], and various rubbers [35,36,40,42,65,80].

Last of all, we should mention that Daikin Industries, Inc., makes thin composite films (Piezel N25 and L25), which consist of PZT ceramics and the copolymer of VDF and TrFE. Physical properties of Piezel films, such as dielectric, piezoelectric, and pyroelectric properties, are described in papers [4,55,75].

We should mention [4] that nonferroelectric fillers, such as tartaric acid [81], have also been applied. As another example, in Mazur [82] suggests the preparation of piezoelectric PTFE/PUE systems without the ferroelectric materials.

D. Piezoelectric 1-3 and 2-2 Composites

During the years 1978–1980, Newnham et al. [16] developed processing techniques for making diphasic ceramic composites with different connectivities. Extrusion, tape casting, and replamine methods have been expecially successful.

As mentioned above, a variety of piezoelectric composite materials can be formed by combining piezoelectric ceramics with a polymer phase. Among them, the piezoelectric 1-3 composite has attracted a great deal of attention and been widely used [15,16,24,83]. A typical configuration of 1-3 and 2-2 composites are illustrated in Figures 3 and 4. According to Zhang et al. [83,84], in the 1-3 composites, piezoelectric ceramics (PZT or BaTiO₃, for example) play the active role of energy conversion between mechanical energy and electric energy, while the polymer phase acts as a passive medium, which transfers the mechanical energy between the piezoelectric ceramics and the surroundings with which the composite interacts. 1-3 composites can be realized by placing

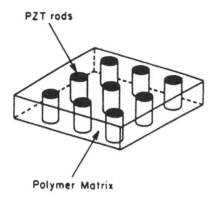

Figure 3 Schematic drawing of a 1-3 PZT-polymer composite, where PZT rods are embedded in a polymer matrix. (After Ref. 83.)

large ceramic grains, rods, or fibers in polymer matrices [4]. In these composites ceramic grains placed in the polymer matrix are so big that they reach from one electrode to another, which makes poling easier. However, this type of composite shows very large heterogeneity [65,85].

Safari [86] has made 1-3 piezocomposites by having different kinds of polymers flooding a layer of ceramic balls of PZT. Other 1-3 composites are polymer matrices with ceramic rods placed inside [87].

Usually, piezoelectric ceramics are rods of a solid solution of PZT with a constitution close to the morphotropic phase boundary; matrices are mostly different types of polymers, such as, for example, silicone rubber, epoxide resin, polysterene, polymethyl methacrylate, polyamide, and other dipolar polymers. Different methods of rod preparation are described in detail in Refs. 88–91.

Zhang et al. [83,84] have shown theoretically that the basic elastic coupling mechanism between the two components in the 2-2 piezoelectric composite is similar to that of the 1-3 composite, although the 2-2 structure is not widely used. There are two major areas where 1-3 composites have been widely used: underwater hydrophone applications and ultrasonic actuators and sensors for medical diagnostic devices [15,16,24,83,84].

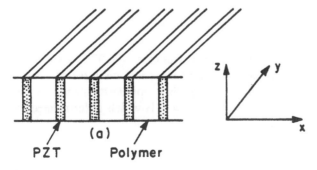

Figure 4 A schematical drawing of a 2-2 composite, where hatched plates are PZT. (After Ref. 83.)

E. 3-3, 3-1, and 3-2 Composites

In 3-3 composites, both phases, piezoelectric ceramics and a polymer, are connected together in three directions, making two interlacing skeletons. This is a structure similar to wood or coral structures. The first composites of this type were made on the basis of natural coral skeleton [15,16]. {"The most complicated and in many ways the most interesting pattern is 3-3 connectivity (Fig. 1), in which two phases form interpenetrating three-dimensional networks. Patterns of this type often occur in living systems such as coral where organic tissue and an inorganic skeleton interpenetrate one other. These structures can be replicated in other materials using the lost-wax method [15,16]. The replamine process, as it is called, can also be used to duplicate the connectivity patterns found in foam, wood, and other porous materials."} Now composites of this type are made by mixing plastic balls and PZT powder in an organic binding agent and burning out the plastic [92]. The result is porous skeleton that next must be filled with the polymer and polarized. This process is called BURPS, which is an acronym for "Burn-out Plastic Spheres." A few methods of making three-dimensional skeletons from piezoelectric ceramics have been developed in Japan by Mitsubishi Mining and Cement [93]. They used a reactive synthesis, organic additives, and grain size, and burning-out control.

The piezoelectric ceramics phase is connected in three dimensions in 3-1 and 3-2 composites but the polymer phase is one- or two-dimensional. To make such composites, skeletons of ceramic PZT made by Corning Glass Works with the structure of a honeycomb [94,95] are mostly used.

III. DIELECTRIC PROPERTIES OF COMPOSITES

A. Mixing Rules for Permittivity

In practice we frequently have to deal with the problem of determining the permittivity ϵ of a composite which is a mixture of two (or more) components. In the simplest case, it is easy to calculate ϵ for a model of a capacitor whose dielectric system consists of two different homogeneous dielectrics connected in parallel or in series [96]. In this case the ϵ of the composite, consisting of two phases with ϵ_1 and ϵ_2 permittivities, is expressed by Eq. (1).

For a more general case of m various dielectrics, Eq. (1) is transformed into the formula

$$\epsilon^n = \sum_{i=1}^{m} (v_i \epsilon_i^n) \tag{2}$$

Equations (1) and (2) can be useful in a number of practical cases [96,97]. However, in most general cases, composite dielectrics are chaotic or statistical mixtures of several components. Then the true value of permittivity of a statistic composite should lie between the values determined by Eq. (2) for $n = 1$ and $n = -1$, which is formulated by the Wiener inequalities [96],

$$\frac{1}{\sum_{i=1}^{m} v_i/\epsilon_i} \leq \epsilon \leq \sum v_i \epsilon_i \tag{3}$$

and is clearly illustrated in Figure 5.

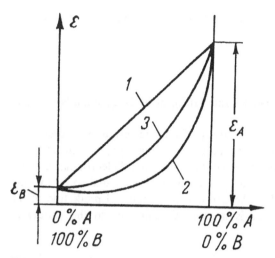

Figure 5 Permittivity ϵ of a mixture of two components A and B versus their volume content in the mixture: 1, parallel connection; 2, series connection; 3, statistic distribution of mixture components (schematic). (After Ref. 96.)

A great variety of formulas have been suggested for the calculation of permittivity of statistic composites. These formulas are derived on the basis of various theoretical presumptions and experimental data [4,7–16,25–28,57,75,96–106]. Each mixing rule, obviously, must fulfill the basic requirement that

$$\frac{\epsilon}{\epsilon_2} = f\left(\frac{\epsilon_1}{\epsilon_2}\right) \tag{4}$$

which arises from the differential equation and boundary conditions for the potential [106].

Wide recognition has been gained by the Lichtenecker logarithmic law of mixing, having for a composite of two components the form

$$\log \epsilon = v_1 \log \epsilon_1 + v_2 \log \epsilon_2 \tag{5}$$

or, in a general form for a mixture of m components,

$$\log \epsilon = \sum_{i=1}^{m} v_i \log \epsilon_i \tag{6}$$

However, the mixing rule (5) can be applied only if ϵ_1 differs slightly from ϵ_2 [97].

The effective permittivity ϵ for a dielectric with permittivity ϵ_1 incorporating uniformly distributed spherical inclusions of ceramic material with permittivity ϵ_2 can be also calculated from the equation

$$\epsilon = \epsilon_1 \frac{2\epsilon_1 + \epsilon_2 + 2v_2(\epsilon_2 - \epsilon_1)}{2\epsilon_1 + \epsilon_2 - v_2(\epsilon_2 - \epsilon_1)} \tag{7}$$

which is known as the Maxwell-Wagner mixing rule [99,100].

For symmetric microgeometry there are two structural units (spheres of two components). The resulting equation is

$$v_1 \frac{\epsilon - \epsilon_1}{2\epsilon + \epsilon_1} + v_2 \frac{\epsilon - \epsilon_2}{2\epsilon + \epsilon_2} = 0 \tag{8}$$

which is known in the literature as Landauer's rule [106], or as Bruggeman's effective medium theory [107].

Some other formulas for calculating permittivity of chaotic mixtures of dielectrics can be found in many books [4,96,97,106,108] and in the original papers [26–28].

Though the mixing rules for permittivity of nonhomogeneous dielectric has aroused the interest of scientists for a long time [99–101], it has been investigated so far by only a few [2–4,7–9,12,15,25–28,31–33]. Now the question of theoretical prediction of the dielectric properties of multiphase systems is interesting again due to the variety of applications of the composites.

B. Dielectric Behavior of 0-3 Polymer-Ceramic Composites

As mentioned above, a composite in general is a heterostructural material whose properties are determined by the contents, the number of different phases of which the material is composed, their properties, and the ways in which different phases are interconnected [7,8,67]. The latter is the most important feature of composites, since the mixing rules of a given property are controlled by the self-connectiveness of individual phases. Piezoelectric polymer-ceramic composites, for example, have a number of applications, since their properties can be tailored to the requirements of various devices by combining the superior properties of a polymer and those of ceramics [67].

The most interesting properties of polymers are their high mechanical and electrical strength and low electrical conductivity and acoustic impedance, whereas the ferroelectric ceramics exhibit good dielectric, pyroelectric, and piezoelectric properties [4,8,67].

The simplest type of polymer-ferroelectric ceramic composite consists of fine ceramic particles dispersed in a polymer matrix; the polymer and ceramic particles have three-dimensional and zero-dimensional connectivity, respectively.

We present in this section the results of a few studies of dielectric properties of 0-3 polymer-ceramic composites and compare some of those with the mixing rules described in Section II.

1. PMMA/BaTiO₃ Composites (Polar-Polymer/Ferroelectric Ceramic Systems)

Composite samples of PMMA/$BaTiO_3$ were prepared by polymerization of MMA in which fine particles of $BaTiO_3$ were dispersed [9,31–33]. Samples with 0 to about 60 vol% $BaTiO_3$ were successfully obtained without any considerable gradient of ceramic powder concentration. Further enrichment of the composite with barium titanate led to a sharp rise in porosity of the material. This is a known property of mixtures of solid dielectrics [109].

a. Relative Permittivity and tan δ as a Function of Volume Percent BaTiO₃ For investigating the dielectric properties of PMMA/$BaTiO_3$, plane electrodes were used, 2 cm in diameter and deposited on the sample surfaces from a colloidal solution of silver. The dependence of permittivity ϵ on the volume fraction of $BaTiO_3$ in composites of PMMA/$BaTiO_3$ is shown in Figure 6 (experimental data are denoted as circles, curve 1).

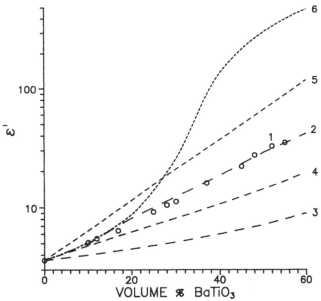

Figure 6 Dependence of permittivity ϵ' on BaTiO$_3$ fraction in PMMA/BaTiO$_3$ composite [111].

In this figure different mixing rules are compared with experimental data. Curves 3, 4, 5, and 6 represent Eqs. (1), (7), (5), and (8) rules, whereas curve 2 represents a modified Lichtenecker rule which has been introduced previously by the author of this chapter in his Ph.D. thesis and other papers [33]. This mixing rule can be written as

$$\log \epsilon = \log \epsilon_1 + (1 - k)v_2 \log\left(\frac{\epsilon_2}{\epsilon_1}\right) \tag{9}$$

where $k \simeq 0.70$ for PMMA/BaTiO$_3$ composite. As is seen from Figure 6, there is very good conformity of this modified Lichtenecker mixing rule with respect to experimental data as far as about 60 vol% BaTiO$_3$. However, the Landaueer rule agrees only for ~20 vol% of this ceramic. The values of ϵ calculated from the Landauer rule drastically exceed experimental data, whereas ones from the Maxwell-Wagner mixing rule are considerably lower than experimental values if the volume fraction of the ceramic phase increases above 20%. The predictions of the Maxwell-Wagner (7) and Bruggeman-Landauer (8) equations are very different, which was to be expected, since for DC conductivity of metal-insulator composite, Eq. (8) predicts a percolation threshold at one-third volume fraction of metal (i.e., conductivity = 0 for metal fraction less than one-third), whereas Eq. (7) tells us that for insulating inclusions, the DC conductivity vanishes only when the metal volume fraction approaches zero [25]. In no case can BaTiO$_3$ inclusions be treated exactly like the above-mentioned inclusions. Change of the real DC conductivity as a function of volume fraction of BaTiO$_3$ in PMMA/BaTiO$_3$ composite (Fig. 7) on the one hand, and foaming and porosity of the material, on the other hand, cause deviation of experimental data from the Bruggeman (8), Maxwell-Wagner (7), and classical Lichtenecker mixing rules. Thus the semiempirical rule, Eq. (9), can be useful in preparation of this type of composite in order to obtain desirable dielectric properties.

In Figure 8 we present tan δ in dependence on volume percent BaTiO$_3$ in composite PMMA/BaTiO$_3$.

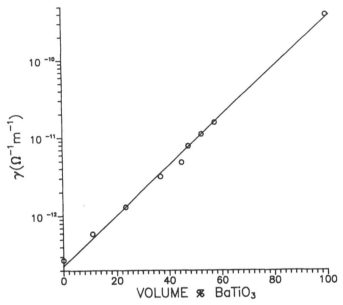

Figure 7 Dependence of electrical conductivity γ on BaTiO$_3$ fraction in PMMA/BaTiO$_3$ composite [9].

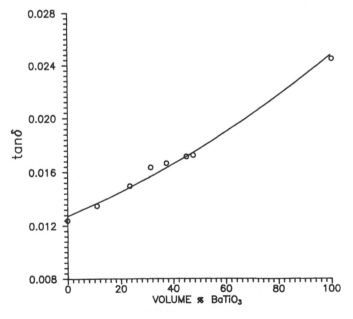

Figure 8 Tan δ in dependence on volume percent BaTiO$_3$ of PMMA/BaTiO$_3$ composite [33].

b. *Temperature Dependences of* ε, *tan* δ, *and DC Conductivity for Composites*
Temperature dependence of permittivity ε at various volume fractions of BaTiO₃ in PMMA/BaTiO₃ composites are shown in Figure 9.

The electrical conductivity of these composites measured at room temperature varies between 2.7×10^{-13} $\Omega^{-1}m^{-1}$ for PMMA and 1.6×10^{-11} $\Omega^{-1}m^{-1}$ for a composite containing 57.5 vol% of BaTiO₃, the measuring field being 1.5 MV/m. For the pure BaTiO₃ ceramics the conductivity was 4×10^{-10} $\Omega^{-1}m^{-1}$. Hence it appears clear that the electrical conductivity of PMMA at room temperature is about three orders lower than that of barium titanate, but its rise with temperature is more impetuous. Both for the pure components and for their mixtures, ln γ is an approximately linear function of $1/T$ (Fig. 10), which is evident with regard to the activation character of the electrical conductivity. The apparent activation energies obtained from the relevant graphs vary in dependence on the mixture composition, as is shown in Table 1. As is seen, PMMA has much lower values of permittivity than barium titanate, whereas its activation energy is larger. By mixing these materials in different proportions it is possible within certain limits to obtain materials of different values of ε, γ, and U, which determine electret, piezoelectric, and pyroelectric properties of PMMA/BaTiO₃ composites.

2 PMMA/PZT Composites (Polar-Polymer/Antiferroelectric Ceramic Systems)

Similar to PMMA/BaTiO₃ systems, composite samples of PMMA/PZT were prepared by polymerization of MMA where fine particles of modified PZT (S-I) and PZT (K-2) ceramics were dispersed. The piezoelectric ceramic PZT of type S-I and type K-2 were produced by Radioceramic Concern CERAD in Warsaw, Poland. The dependence of permittivity ε on the volume fraction of PZT in composites of PMMA/PZT is shown in

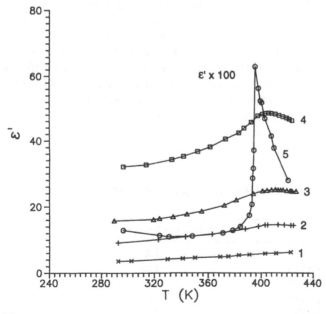

Figure 9 Temperature dependence of permittivity ε′ at various volume fractions of BaTiO₃ in PMMA/BaTiO₃ composites [33]: 1, 0; 2, 0.25; 3, 0.37; 4, 0.57; 5, 1 volume fraction of BaTiO₃.

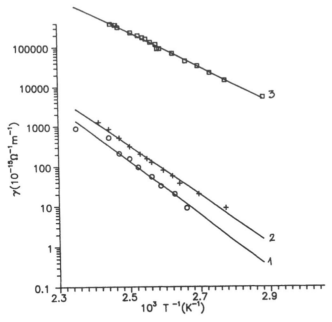

Figure 10 Temperature dependence of electrical conductivity γ [9]: curve 1, pure PMMA; curve 3, BaTiO$_3$; curve 2, composite containing 0.32 BaTiO$_3$.

Figure 11 (experimental data are denoted as circles). In this figure also are compared with experimental data the Maxwell-Wagner, Lichtenecker, and modified Lichtenecker mixing rules. As in the PMMA/BaTiO$_3$ system, the permittivity ϵ predicted by the modified Lichtenecker mixing rule for PMMA/PZT composite found relatively good agreement with observed values. In Figure 12 we present tan δ for composite PMMA/PZT in dependence on volume fraction of PZT. By comparison of the results presented in Figures 8 and 12, we see that the behavior of PMMA/BaTiO$_3$ composites is completely different from PMMA/PZT composites. This can be caused by the fact that tan δ (BaTiO$_3$)/tan δ (PMMA) > 1, and tan δ (PZT)/tan δ (PMMA) < 1. Temperature dependence of permittivity ϵ and tan δ of PMMA/PZT composites for various volume fractions of PZT are shown in Figure 13 and Figure 14, respectively.

Table 1 Apparent Activation Energy for PMMA/BaTiO$_3$ Composites with Various Volume Fractions of BaTiO$_3$ [9]

Composite	Activation Energy, U (eV)
PMMA	1.8
PMMA + 0.15 BaTiO$_3$	1.58
PMMA + 0.26 BaTiO$_3$	1.45
PMMA + 0.32 BaTiO$_3$	1.30
PMMA + 0.45 BaTiO$_3$	1.03
BaTiO$_3$	0.87

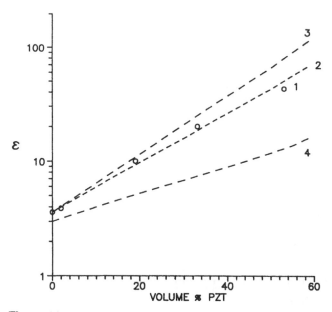

Figure 11 Dependence of permittivity ε on volume fraction of PZT in PMMA/PZT composite [111]: 1, experimental data; 2, modified Lichtenecker rule [Eq. (9)]; 3, Lichtenecker rule [Eq. (5)]; 4, Maxwell-Wagner rule [Eq. (7)].

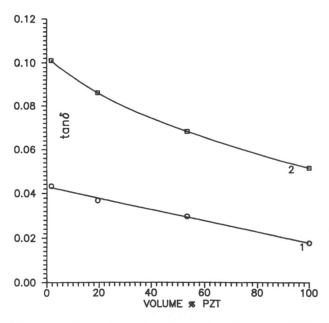

Figure 12 Dependence of tan δ on volume fraction of PZT in PMMA/PZT composite; curve 1, at room temperature; curve 2, at T_M.

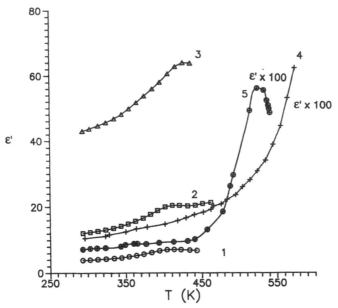

Figure 13 Dependence of permittivity ϵ' on temperature for PMMA/PZT composites: curve 1, 0.02 PZT; curve 2, 0.19 PZT; curve 3, 0.54 PZT; curve 4, PZT (S-1); curve 5, PZT (K-2, CERAD).

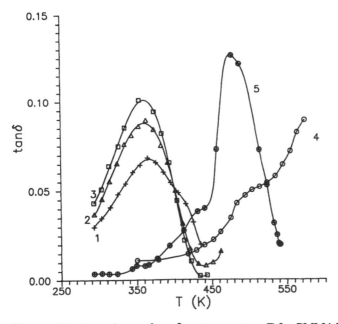

Figure 14 Dependence of tan δ on temperature T for PMMA/PZT composites: Curve 1, 0.54 PZT; curve 2, 0.19 PZT; curve 3, 0.02 PZT; curve 4, PZT (S-1, CERAD); curve 5, PZT (K-2, CERAD).

3. PVDF/PZT Composites (Ferropolymer/Antiferroelectric Ceramic Systems)

Another example of a multiphase polymer-ceramic material is PVDV/PZT 0-3 composite. The method of preparation and the piezoelectric properties of PVDF/PZT composite were described by Furukawa et al. [12–14]. Some results of this paper are presented in another section of this chapter. Here we limit our interest to the permittivity ϵ in dependence on volume fraction PZT in composite PVDF/PZT which is presented in Figure 15. The samples were prepared in the following way. PVDF pellets (produced by Nitrogeneous Concern, Tarnów, Poland) and modified PZT (PP-6CM) ceramic powders (produced by Radioceramic Concern CERAD, Warsaw, Poland) were mixed at 455 K using a hot-press method and pressed into about a 200-μm- to 300-μm-thick film at this temperature [110]. As is seen in this figure, the value of dielectric constant for PMMA is much lower than that for PVDF. However, at about 20 vol% fraction of PZT in the composites PVDF/PZT and PMMA/PZT [110,111], these values are the same. For higher contents of PZT up to about 60 vol%, the values of the dielectric constants for PMMA/PZT are larger than for PVDF/PZT. This fact may be due to the various packing densities of PZT inclusions in polymer matrix from different methods of preparation. The samples of PMMA/PZT composite were prepared by polymerization of MMA in which fine particles of BaTiO$_3$ were dispersed, whereas PVDF/PZT films were obtained by hot-pressing of PZT ceramic and PVDF powders.

4. Nonpolar Polymer/Ferro- and Antiferroceramic Composite Films

Five different composite films of polypropylene (PP)/ceramic (both ferro- and antiferroelectric) materials were prepared by introducing fine-grain ceramic powder in the matrix of polypropylene [38]. The results of these investigations are presented in Figures

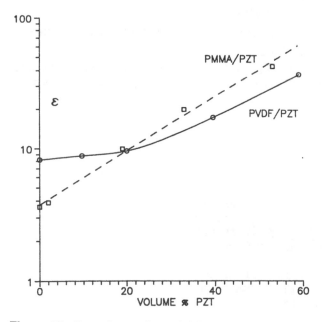

Figure 15 Dependence of permittivity ϵ on PZT fraction in composites PVDF/PZT [110] and PMMA/PZT [111].

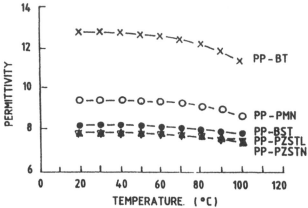

Figure 16 Temperature dependence of permittivity of the composites. Volume fraction of ceramics: 30%. (After Ref. 38.)

16 and 17, and in Tables 2 and 3. Figures 16 and 17 show the nature of permittivities and dissipation factors of the composites as a function of temperature, respectively. It is noticeable that the composite PP-PMN shows the highest loss factor (Fig. 17) but not the highest permittivity (Fig. 16) as expected, although PMN possesses the highest permittivity among the materials selected. Moreover, the composites containing BT and BST, which were produced by a well-known coprecipitation method and possessing ultrafine particle size, appear to have higher dielectric strength than the other composites [38]. Besides, the present results by Das Gupta and Shuren [38] indicate that the ferroelectric ceramics are more effective than the antiferroelectric ceramics in enhancing the energy storage capabilities of the PP-ceramic composites.

5. Temperature and Frequency Dependence of Dielectric Parameters of Some Other 0-3 Composites

Dielectric dispersion of PPN-P(VDF/TFE) (0.98/0.02) composites obtained by hot-pressing of PPN ceramic and PVDF powder were performed by Hiczer et al. [8,112].

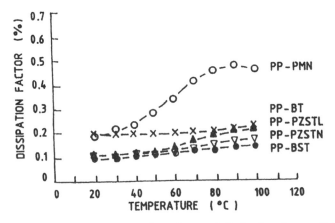

Figure 17 Temperature dependence of dissipation factors. Volume fraction of ceramics: 30%. (After Ref. 38.)

Table 2 Ceramic Materials for Composites

Code	Specifications
BT	$BaTiO_3$ doped by Nb_2O_5, Sb_2O_3
BST	$(Ba,Sr)TiO_3$ system
PMN	Modified $Pb(Mg,Nb)O_3$ system
PZSTN	$Pb(Zr,Sn,Ti)O_3$ doped by Nb_2O_5
PZSTL	$Pb(Zr,Sn,Ti)O_3$ doped by La_2O_3

Source: After Ref. 38.

Composite samples were prepared from $Pb_{0.9}Ba_{0.1}(Zr_{0.5}\ Ti_{0.5})O_3$ modified with 1 wt% of Nb_2O_5 (PPN) ceramic powder (produced by Radioceramic Concern CERAD, Warsaw, Poland) and poly(fluorovinylidene fluoride/tetrafluoroethylene) P(VDF/TFE) copolymer powder of 0.98 molar fraction of vinelidene fluoride (produced by Nitregeneous Concern, Tarnów, Poland). Temperature dependence of permittivity ϵ and tan δ for PPN-P(VDF/TFE) composite with 70 wt% PPN at various frequencies are shown in Figure 18. It should be pointed out that dielectric losses of these composites are lower than those for the polymers. Frequency dependence of dielectric constant ϵ and tan δ of PPN-P(VDF/TFE) composite measured at room temperature is shown in Figure 19. Dielectric dispersion was observed in this composite, with the maximum of absorption in the range 10 MHz.

Previously, the dielectric constants ϵ'/ϵ_0 and ϵ''/ϵ_0 were measured by Furukawa et al. [12] for other 0-3 composites, i.e., a PVA/PZT system with 8 vol% fraction of PZT. In accordance with opinion of Furukawa et al. [12], the monotonous increase of ϵ'' with decreasing frequency suggests that the DC conductivity presumably due to ionic impurities is considerably high at these temperatures (Fig. 20). The first increase of ϵ' observed at low temperature is ascribed to the accumulation of ionic impurities at the interface of PVA polymer and PZT ceramic powder [12]. It is important for electret behavior of the polymer-ceramic composite. The second increase of ϵ' at higher temperatures is ascribed to the electrode polarization [12].

Lastly, we present in Figures 21 and 22 the dependence of dielectric constant and tan δ on volume fraction of ferroelectric inclusion in a few matrixes, for comparison.

With regard to experimental data presented in Figures 21 and 22, we see that it is impossible to describe the dielectric constant (Fig. 21) and tan δ (Fig. 22) by the same mixing rules. Theoretical approaches to the linear response of two-phase systems, even in the simplest noncoupled case, are based on various models of mixing rules [112]. A

Table 3 Dielectric Strengths of Composites

Composite	Dielectric strength (kV/cm)
PP-BT	800
PP-BST	750
PP-PMN	700
PP-PZSTN	650
PP-PZSTL	400

Source: After Ref. 38.

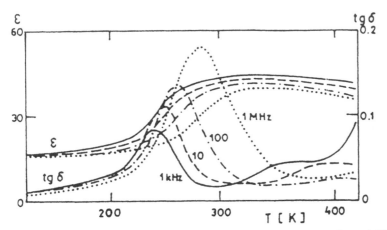

Figure 18 Dielectric dispersion and absorption of 0-3 composites of $(Pb,Ba)(Zr,Ti)_3$: $Nb_2O_5/$ P(VDF/TFE) in a wide temperature range. (After Refs. 8 and 67.)

few years ago [43,113], theoretical equations for dielectric permittivity and tan δ of composites were presented which generalize the parallel, series, and cubes systems, and this model was certified experimentally for dielectric properties of composites consisting of a polymer and mixed ceramic PZT and $PbTiO_3$ powder as a function of the volume fraction of the powder.

A powerful and versatile variational principle, allowing the estimation of the effective properties of nonlinear heterogeneous systems, has been introduced recently by Ponte Castañeda [26]. In other papers [27,28], this author reviews the variational principle and applies it to determine bounds and estimates for the properties of certain classes of nonlinear composite dielectrics with homogeneous, isotropic phases. However, the polymer-ferroelectric ceramic composites, in general, are anisotropic, particularly after poling. Thus,

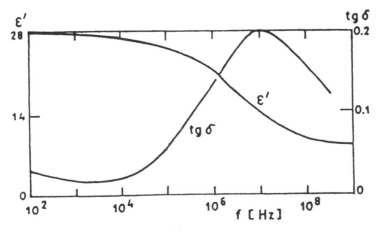

Figure 19 Frequency dependence of dielectric permittivity and tan δ of $(Pb,Ba)(Zr,Ti)O_3:Nb_2O_5/$ P(VDF/TFE). (After Refs. 8 and 67.)

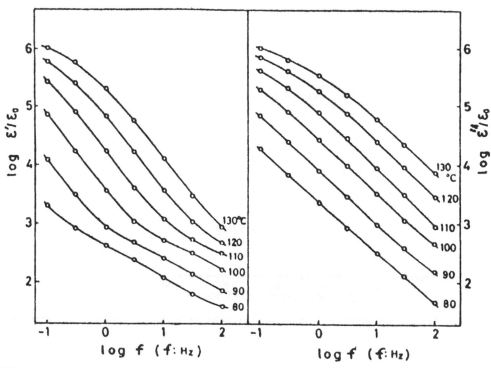

Figure 20 Frequency dependence of ϵ/ϵ_0 for the PVA-PZT system with 0.08 PZT. (After Ref. 12.)

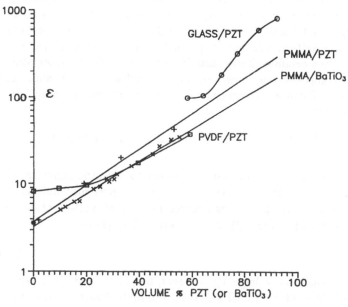

Figure 21 Dependence of permittivity ϵ on piezoelectric ceramics fraction for various 0-3 composites.

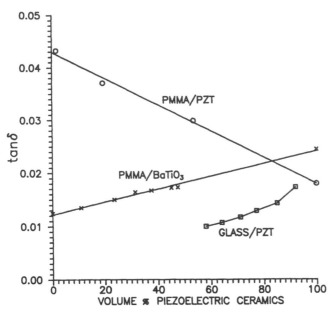

Figure 22   Dependence of tan δ on piezoelectric ceramics fraction for various 0-3 composites.

the bounds and estimations described in this paper cannot be fully applied for comparison between theoretical and experimental results.

Furukawa et al. [12] and Das Gupta and Abdullah [113] have studied the dielectric dispersion in 0-3 polymer-ceramic composite. These authors found the dielectric absorption to be dominated by the polymer, whereas the ceramic phase was found to contribute to the electrical conductivity; Sinha et al. [57,114] reported, however, the major contribution to the conductivity of 0-3 composites to be due to the polymer rather than to the ceramic. In our opinion, it is self-evident that for parallel composite systems, DC conduction, in general, determines ceramic phases, and for series ones, the polymer phase. The 0-3 polymer-ceramic composites are, however, statistical (chaotic) composites, and their DC conductivity and tan δ can be determined mainly in an experimental way (Figs. 7, 8, 12, and 22, for example).

6. Steady-State Current–Voltage Characteristics of PMMA/BaTiO₃ Composites

The problem of conductivity dependence of volume fraction ferroelectric ceramics in a polymer matrix is complicated by the fact that the charge transport mechanisms are different for the polymer and ceramic phases. In general, the current–voltage characteristics of PMMA/BaTiO₃ composites (Fig. 23) can be expressed by [111]

$$j = gV_p^n \tag{10}$$

where g is independent of polarizing voltage V_p; however, this factor and the exponent n are dependent on the BaTiO₃ content in PMMA. As is seen in the figure, n varies from 0.15 to 2 in dependence on BaTiO₃ content in PMMA/BaTiO₃ composites. If $n < 1$, the current–voltage (subohmic) characteristics of these composites are connected with the

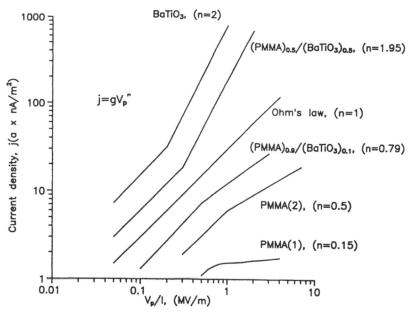

Figure 23 Current–voltage characteristics of PMMA/BaTiO₃ composites at polarizing temperature T_p = 353 K [111]. For PMMA(1), T_p = 293 K. Polarizing time t_p = 3 h.

diffusion model of charge transport; while if $n > 1$ these j–V characteristics (superohmic)
are due to space charge-limited current (SCLC).

C. The 3-1 Type of Epoxy Resin/PZT Ceramic Composite

We present in Figure 24 the dielectric constants of 3-1 type epoxy/PZT composites as a
function of volume percent PZT (as has been described by Newnham et al. [16]). In
conformity with the opinion of the authors of the cited paper, the dielectric constant of
a 3-1 type of composite can be estimated from a model based on two capacitors connected
in parallel [see Eq. (1) for n = 1]. It is seen in this figure that the value of ϵ varies
linearly with the volume fraction of PZT and agrees well with the theoretical one [Eq.
(1) at n = 1].

IV. ELECTRET PROPERTIES OF POLYMER-CERAMIC SYSTEMS

It is well known that phenomenological theories of electret effect are based on the Gross
two-charge theory [115], according to which, in dielectrics which are exposed to the
influence of an electric field, there occurs an orientation of dipoles and an accumulation
of the charges on the macroscopic heterogeneity of the material. The surface charges
which are due to these processes have opposite signs to those of polarizing electrodes
(the surface charges cause heterocharge). Besides, according to Gross, in the air layer
between the dielectric and the electrodes there occur local electrical breakdowns, as a
result of which, on the surface of the electret, these charges appear with a sign which
agrees with those of the electrodes (the surface charges cause homocharge). The effective
charge, which has been measured after removing the electrodes, is an algebraic sum of

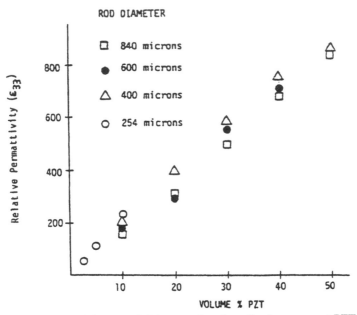

Figure 24   Relative permittivity as a function of volume percent PZT. (After Ref. 16.)

both charges. In the initial stage of the "life" of the electret, heterocharge prevails, but after time τ_1, which is called the time of polarity reversal, effective charge accepts the sign of homocharge.

The first composite electrets were made by Japanese physicist Mototaro Eguchi from carnauba wax and resin with the addition of beeswax [116]. Thermoelectrets which exhibit heterocharge just after poling may also be made from polymers with dipole moments, e.g., PMMA, polyvinyl chloride (PVC), and series of other dipolar polymers [9,76,117–120].

Attempts have also been made to form electrets from inorganic materials (see, for example, Ref. 117). Systematic research on the formation and properties of electrets obtained from perovskite polycrystals as well as from other inorganic materials were made by Gubkin and Skanavi [121]. The perovskites, such as calcium titanate (CaTiO$_3$), magnesium titanate (MgTiO$_3$), barium titanate (BaTiO$_3$), strontium titanate (SrTiO$_3$), and bismuth titanate (BiTiO$_3$), can form homopolar thermoelectrets under electric fields in the range 1–2 MV/m. The stability of electrets made from ceramic materials is low. Such low stability of electrets obtained from ceramic materials is connected with the relatively low resistivity of these materials. In order to heighten the so-called lifetime of the electret, polymer-ceramic composites were made from different materials [9,17–19,31,32,111].

Today, the term electret is associated first of all with a thin polyethylene (PE) or polytetrafluoroethylene (PTFE) [119], containing an implanted and highly persistent space charge. Such electrets have wide and diversified applications [118,119]; however, their piezoelectric parameters are low. With regard to piezoelectric properties of electrets, different polymer-ferroelectric ceramic materials have been made [4,12,16].

In accordance with Furukawa and well-known opinions of other authors [4,9,122], the poling process in a polymer-ceramic composite is rather complex. As is seen from experiments described in Section III, ferroelectric ceramics have much larger permittivities than polymers, so the polarizing field is greatly reduced in the ceramic phase, and it seems impossible to pole the composite dielectrically [122]. However, we can pole it by using the conventional poling procedure which consists of applying an electric field E_p at a temperature T_p for a period t_p. In order to recommend typical poling conditions, the investigations of surface charge density in dependence on E_p, T_p, and t_p for each composite are necessary. However, for theoretical predictions of optimal condiction for electret properties, thermally stimulated discharge (TSD) analysis of the materials is also necessary. Some aspects of these subjects are described in the following subsections.

A. Current TSD of Polymers That Contain Additives

Previously, Van Turnhout studied the increase in the heterocharge of PMMA upon adding highly polar materials such as TiO_2 and SnO_2 [123, p. 244]. We have used the TSD method in order to determine the effect of low-molecular additives (H_2O and $BaTiO_3$) on the dielectric relaxation processes of polarized PMMA [124–126]. From TSD current analysis it was possible to state that such parameters as intensity (j_m), TSD peak area (σ), maximum temperature (T_M), and activation energy (U) of so-called β', α, and ρ relaxations undergo evident changes in dependence on low-molecular additives content. In addition to the effect of additives in changing the temperature of α relaxation, the results obtained, presented in Figures 25–29, lead to the following conclusions: (a) The space charge ρ peak is enlarged. It has been confirmed by Van Turnhout [123] that the additives raise the number of charge carriers that can be frozen-in. (b) The parameters

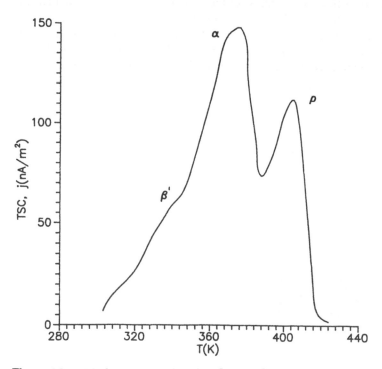

Figure 25 TSC thermogram of PMMA [124,125].

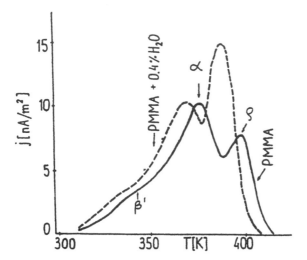

Figure 26 Current density (j) thermograms of polarized PMMA, and PMMA + 0.4 wt% H_2O samples [124]. Polarizing conditions: E_p = 0.5 MV/m, T_p = 403 K. Heating rate during TSD: b = 3 K/min.

of β' relaxation are changed because of the formation of clusters of ester group additives. The plasticizing effect of H_2O in lowering the temperature of α relaxation in PMMA is shown in Figure 26; however, heights of this peak are unchanging. This was to be expected, because the α peak is connected with the dipolar relaxation, and dipolar moments of H_2O and ester groups in PMMA are comparable.

The influence of $BaTiO_3$ inclusions on the low temperature of relaxation processes in this polymer is different. First, in the TSC spectrum of $BaTiO_3$ ceramics, the highest peak appears at the Curie temperature (T_c) of this material. This temperature lies between

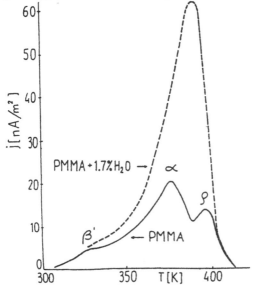

Figure 27 Same as Figure 26, except for PMMA, and PMMA + 1.7 wt% H_2O. E_p = 1 MV/m.

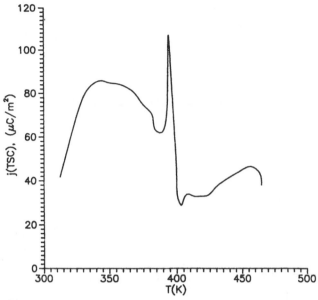

Figure 28 Current density (j) thermograms of polarized BaTiO$_3$ [111]. Conditions: E_p = 0.25 MV/m, T_p = 403 K, t_p = 0.5 h.

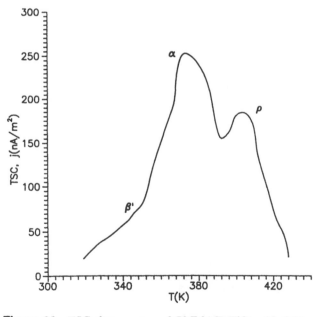

Figure 29 TSC thermogram of PMMA/BaTiO$_3$ with 0.03 volume fraction of BaTiO$_3$ [111]. Conditions: E_p = 5 MV/m, T_p = 393 K, t_p = 2 h, b = 2 K/min.

the temperatures of the α and ρ peaks, so the peak connected with spontaneous polarization of BaTiO$_3$ influences an apparent α peak of PMMA/BaTiO$_3$ composite. However, as is seen in Figures 27 and 30, mainly the space-charge ρ peak is enlarged by the H$_2$O or BaTiO$_3$ additives, due to charge storage at interfaces between different phases of the composite.

B. Time-Temperature Dependence of Surface Charge Density for PMMA, BaTiO$_3$, and Composite PMMA/BaTiO$_3$

On the basis of Debye theory and TSD spectrum analysis, the dependence of the heterocharge on polarizing temperature T_p can be made more clear, as follows: Assume a time dependence of the dipole polarization $P(t)$ with a single relaxation time $\tau(T)$. This builds up according to the well-known equation

$$\frac{dP(t)}{dt} + \frac{P(t)}{\tau(T_p)} = \frac{P_0}{\tau(T_p)} \tag{11}$$

where P_0 equals $N\mu^2 E_p/3kT_p$. Integrating Eq. (11) yields [127]

$$P(t_p) = P_0 \left\{ 1 - \exp\left[-\frac{t_p}{\tau(T_p)} \right] \right\} \tag{12}$$

where $\tau(T_p) = \tau_0 \exp(U/kT_p)$.

Approximating Eq. (12) for polarizing time $t_p \gg \tau(T_p)$ gives

$$P(t_p) = P_0(T_p) \tag{13}$$

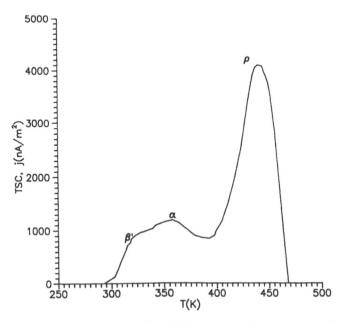

Figure 30 The same as Figure 29, except for the composite with 0.48 volume fraction of BaTiO$_3$ [111].

We note that the inequality $t_p \gg \tau (T_p)$ is likely to hold in our experiments only if $T_p \gg T_g$. However, approximating Eq. (12) for polarizing time $t_p < \tau(T_p)$ gives

$$P(t_p) = \frac{P_0 t_p}{\tau(T_p)} = \frac{\epsilon_0(\epsilon_s - \epsilon_\infty)E_p t_p}{\tau_0} \exp\left(-\frac{U}{kT_p}\right) \tag{14}$$

Thus, in the absence of a homocharge we have for the initial heterocharge density

$$\sigma_0 = P(t_p) = C \exp\left(-\frac{U}{kT_p}\right) \tag{15}$$

where $C = \epsilon_0(\epsilon_s - \epsilon_\infty)E_p t_p/\tau_0$. Obviously, C in Eq. (15) is temperature dependent in accordance with Eq. (12).

We note that inequality $t_p < \tau(T_p)$ is likely to hold in our experiments when the polarization temperature T_p is sufficiently below the glass transition temperature T_g (ca. 105°C) of PMMA. However, if the polarizing temperature increases from room temperature T_r to final temperature T_f, the relation between t_p and $\tau(T_p)$ changes from $t_p < \tau(T_r)$ by $t_p = \tau(T_M)$ up to $t_p \gg \tau(T_f)$. At this condition the polarizing temperature dependence of the dipole polarization fulfills Eq. (12), obviously. However, the relation between $P(t_p)$ and T_p is more involved than is indicated by Eq. (12), because the heterocharge of a PMMA thermoelectret is determined by dipole polarizations and space charge polarization with distribution of relaxation time τ_i or activation energy U_i, where i denotes β', α, or ρ relaxation at temperatures above room temperature. If we assume in the simplest case three relaxation processes in PMMA with discrete relaxation times τ_i, the T_p dependence of charge density σ can be expressed as follows:

$$\sigma(T_p) = \sum P_{0i}\left\{1 - \exp\left[-\frac{t_p}{\tau_i(T_p)}\right]\right\} \tag{16}$$

where $\tau_i(T_p) = \tau_{0i} \exp(U_i/kT_p)$.

In order to compare the theory (Eq. 16) with experiments, one should calculate some relaxation parameters on the basis of the thermally stimulated discharge (TSD) method of the materials considered. From the TSC thermogram of PMMA (Fig. 31) and its analysis, it is possible to determine U_i, P_{0i}, and τ_{0i} [124,125]. Applying Eq. (16) and these parameters, the T_p dependence of surface charge density σ for each (β', α, and ρ) peak and the effective (sum) charge density are presented in Figure 32 in comparison with experimental data. As is seen in this figure, the maximal effective charge density can be obtained in PMMA if $T_p > T_g$, where the glass transition temperature for PMMA, T_g, is 378 K. In the case of polarized BaTiO$_3$ ceramics, this condition is fulfilled because maximal σ appears at $T_p \simeq T_c$, which is higher than T_g (Fig. 33). In order to complete Furukawa lists of recommended poling conditions of different two-phase systems [122], the dependence of σ on E_p for the materials should be known.

C. Field Dependence of Surface Charge Density for 0-3 Composites

In 1936, Thiessen, Winkler, and Hermann [128] found that the kind of permanent charge in thermoelectrets of some mixtures depends very much on the polarizing field E_p; electrets formed in low fields showed only heterocharge, while electrets polarized in high fields exhibited heterocharge immediately after formation that decreased over time τ_1 to zero, following which homocharge appeared. The charge became established at a certain level and did not change thereafter for a very long time. Later research [9,128,129]

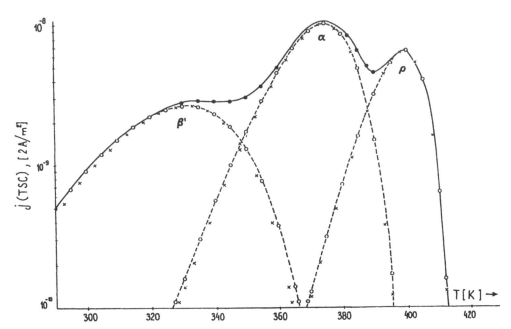

Figure 31   Theoretical fit to β', α, and ρ peaks using Arrhenius ($\circ$) and Eyring ($\times$) relaxations [125]. ———, experimental, ----, synthesis of TSC spectra.

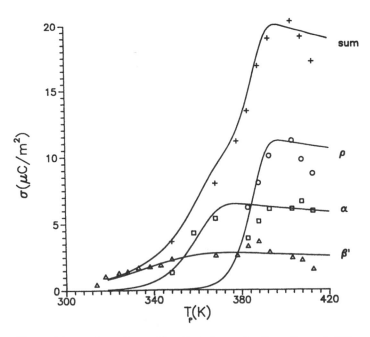

Figure 32 T_p dependence of surface charge density σ for each (β', α, and ρ) peak and effective (sum) charge density. Solid lines, theory; symbols, experimental data.

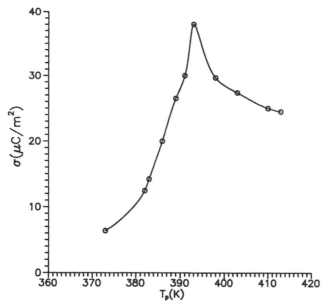

Figure 33 T_p dependence of TSD charges for BaTiO$_3$ [160].

showed that time τ_1 after which heterocharge changes into homocharge depends very strongly on the temperature at which the electrets were stored. In general, both buildup and decay of electret charge are due to a transport mechanism of charge carriers in the material. A kind of mechanism can be established on the basis of a current–voltage characteristic. Generally, for PMMA/BaTiO$_3$ composites, this characteristic can be expressed by Eq. (10),

$$j = gV^n$$

For PMMA and PMMA/BaTiO$_3$ composites with volume fractions of BaTiO$_3$ lower than 0.5, $n < 1$. In this case, the charge transport may be connected with the diffusion mechanism. Besides, if the composite contains more than 0.5 volume fraction BaTiO$_3$ ceramic, then $n > 1$. In this case the charge transport can be described by space charge-limited current (SCLC).

Experimentally (Fig. 23), for PMMA at $T_p = 353$ K and $t_p = 3$ h, with $n = 0.5$, the current–voltage characteristic can be expressed by

$$j = gV_p^{1/2} \tag{17}$$

Applying some equation for the potential distribution as determined by probe technique together with the TSD method [130] and assuming a diffusion model of charge transport, we obtained Eq. (17) analytically [131].

On this basis and with some boundary condition, obtain a theoretical relation between surface charge density σ of a PMMA thermoelectret and polarizing voltage V_p as follows [132]:

$$\sigma(\infty) = a\{bV_p - 2c^{1/2}[(c + V_p)^{1/2} - c^{1/2}]\} \tag{18}$$

where $a = \epsilon_0(\epsilon_s/l + \epsilon_1/l_1)$, $b = [1 + (l\epsilon_1)/(l_1\epsilon_s)]^{-1}$. So, with our assumptions, it is seen [Eq. (18)] that interfacial charge $\sigma(\infty)$ changes nonlinearly with applied voltage V_p. It is this charge that can be frozen in.

As is seen in Figure 34, a plot of $\sigma = f(E_p)$ represents an experimental relation between σ_i and E_p up to 6 MV/m. Approximately, relation (18) may also be applied for PMMA/BaTiO$_3$ and PMMA/PZT composites containing no more than 0.5 volume fraction of ceramic (Figs. 34 and 35).

D. Charge Stability of 0-3 Polymer-Ceramic Composites

In phenomenological theories it is assumed that a dipole component of the heterocharge (σ_f) decays due to heat disorientation according to the rule [133,134]

$$\sigma_f = \sigma_{0f} \exp(-\alpha t) \tag{19}$$

where $\alpha = 1/\tau_f$ is the reciprocal of the time of "frozen" polarization relaxation. However, the ionic (real) component of the electret charge (σ_r) changes by means of conductivity according to Ohm's law [134],

$$-\left(\frac{d\sigma_r}{dt}\right) = \gamma E_i = \beta\sigma \tag{20}$$

where E_i is the internal field of the electret, $\sigma = \sigma_f + \sigma_r$, is the effective value of the surface density of the electrets, and $\beta = 1/\tau_M$ is the reciprocal time of the space charge relaxation. The calculation based on the above assumptions allows us to find for the electret the dependence $\sigma = \sigma(t)$ in the form of a linear combination of exponential functions with different times of relaxation [133,134].

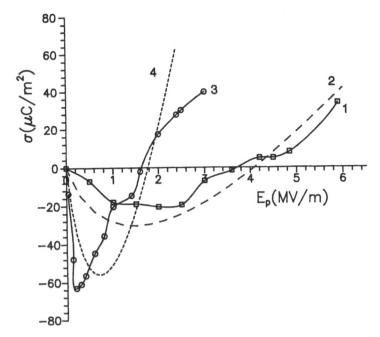

Figure 34 E_p dependence of surface charge density σ for PMMA [1, experimental data; 2, Eq. (18)] and PMMA/BaTiO$_3$ composite with 0.48 BaTiO$_3$ [3, experimental data, 4, Eq. (18)].

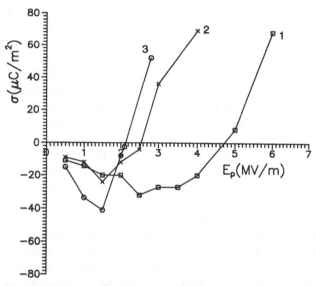

Figure 35 E_p dependence of surface charge density σ for PMMA/PZT composite (1, 0.02; 2, 0.2; 3, 0.5 volume fraction PZT).

The assumption defined by Eq. (20) with reference to a composite dielectric in which there coexist dipole polarization, space charge polarization at interfaces between different phases, and charge injected by the polarizing electrodes, is a barely justified assumption. The resulting current–voltage characteristics for the composites of PMMA/BaTiO$_3$ show an ohmic region preceding the nonohmic region in which the current becomes proportional to the polarizing voltage V_p^n, where $n < 1$ for low volume fraction of BaTiO$_3$ (about 0.3) and $n > 1$ for higher ones [Fig. 23, Eq. (10)].

If $n < 1$, we have to deal with diffusion charge transport. The diffusion transport model in polymers is not fully reasonable due to low carrier mobility. However, for large concentration gradients at the surfaces, which may be found particularly in heteroelectrets made from polar polymers, the neglect of diffusion is not a consistent approximation [123,131].

On the other hand, in BaTiO$_3$ ceramics we have to deal with space charge-limited currents, which also are not linear in relation to the voltage ($n > 1$).

In connection with these facts, the charge decay of electrets made from these composites can be expressed as

$$-\left(\frac{d\sigma_r}{dt}\right) = a\sigma^n \tag{21}$$

A certain nonlinear question in an electret theory has been considered in some papers [135,136]. As follows from these papers, the effective charge density in dependence on t is

$$\sigma = \sigma_f + \sigma_r \simeq \frac{\sigma_0}{[1 + (n-1)a\sigma_0^{(n-1)}t]^{1/(n-1)}} - \sigma_{0f}(1 - e^{-\alpha t}) \tag{22}$$

and then, if $n = 2$

$$\sigma \simeq \frac{\sigma_0}{1 + \sigma_0 at} - \sigma_{0f}(1 - e^{\alpha t}) \tag{23}$$

Comparison of this theoretical result with experimental dependence $\sigma(t)$ for composite PMMA/BaTiO$_3$ is shown in Figure 36. As is seen from this figure, these relations are in good agreement for the initial stage of electret "life."

Time dependence $\sigma(t)$, initial heterocharge, maximal homocharge after change of the polarity, and time of polarity reversal τ_1 for different volume fractions of BaTiO$_3$ in composite PMMA/BaTiO$_3$ are presented in Figures 37–39. The electrets were formed under the conditions $E_p = 0.8$ MV/m, $t_p = 2$ h, and $T_p = 428$ K. As is seen in these figures, both initial heterocharge and maximal homocharge increase, while τ_1 decreases with volume percent BaTiO$_3$.

It was to be expected numerically on the base of Eq. (22) that the time $t = \tau_1$ after which the effective charge, $\sigma = 0$, depends on n, differs for various BaTiO$_3$ contents. At higher polarizing fields ($E_p = 3$ MV/m), time dependence of the effective charge, maximal initial charge, and homocharge after about 2 months are illustrated in Figures 40 and 41. As is seen in Figure 41, the highest electret stability at these conditions can be obtained in composites with 0.3 volume fraction BaTiO$_3$ and 0.7 volume fraction PMMA.

In order to complete the experimental data on this subject, we present in Figure 42 the time dependence of the induction charge density for thermoelectrets with various PZT contents in composite PMMA/PZT. Moreover, in Figures 43 and 44 are presented the dependent TSD charges on volume fractions BaTiO$_3$ and one PZT in the composites PMMA/BaTiO$_3$ and PMMA/PZT, respectively [111].

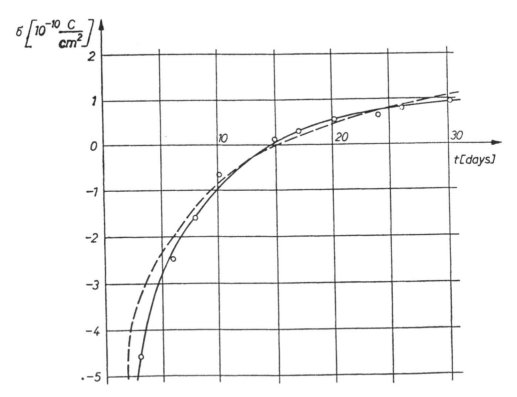

Figure 36 Dependence of surface charge density on time for PMMA/BaTiO$_3$ composite [135,136]. Solid line, experimental curve; dashed line, curve resulting from Eq. (22) at $n = 2$. Conditions: $E_p = 0.8$ MV/m, $T_p = 428$ K, $t_p = 2$ h.

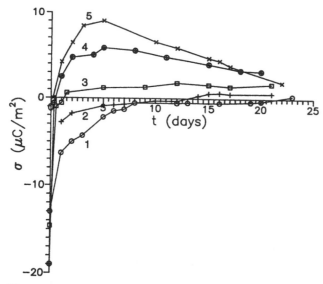

Figure 37 Time dependence of surface charge density σ of PMMA/BaTiO₃ composite with various volume fractions of BaTiO₃. Curves 1, 2, 3, 4, and 5 relate to samples with 0, 0.1, 0.29, 0.48, and 0.56 volume fraction of BaTiO₃ [9]. (Polarizing conditions: see Fig. 36.)

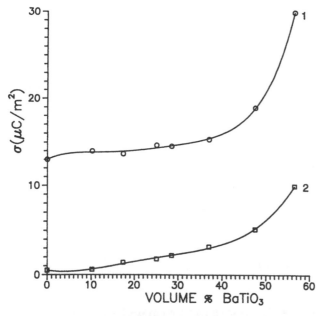

Figure 38 Dependence of (1) the initial values of heterocharge surface density (−σ) and (2) the maximum values of homocharge density (+σ) on the percentage of BaTiO₃ in PMMA/BaTiO₃ composite [9]. (Polarizing conditions: see Fig. 36.)

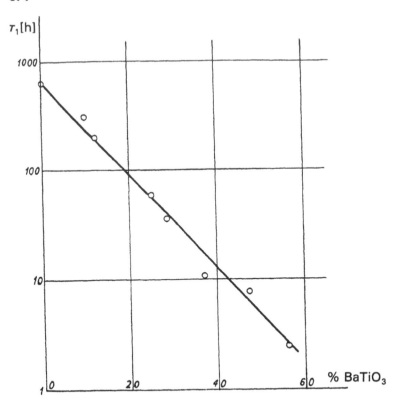

Figure 39 Dependence of polarity reversal time (τ_1) on content of BaTiO$_3$ in PMMA/BaTiO$_3$ composite [9]. (Polarizing conditions: see Fig. 36.)

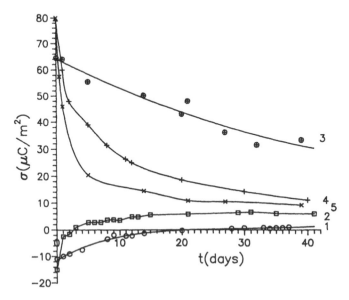

Figure 40 Dependence of surface charge density σ on time for PMMA/BaTiO$_3$ composites with various volume fractions of BaTiO$_3$ [161]: curve 1, 0; curve 2, 0.15; curve 3, 0.28; curve 4, 0.45; curve 5, 0.57 volume fraction of BaTiO$_3$. (Polarizing conditions: same as Fig. 36, except E_p = 3 MV/m.)

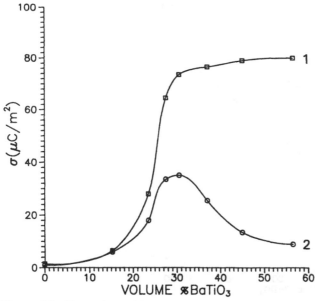

Figure 41 Homocharge (σ) dependence on volume fraction of BaTiO$_3$ in PMMA/BaTiO$_3$ composite [161]: curve 1, initial values; curve 2, after about 2 months.

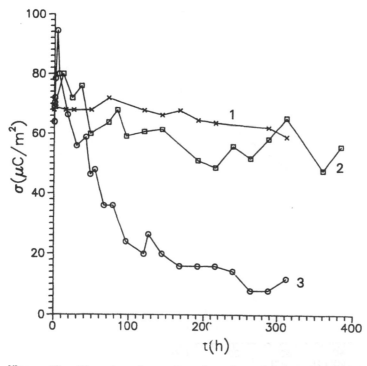

Figure 42 Time dependence of surface charge density σ for PMMA/PZT composite with 1, −0.02; 2, −0.2; 3, 0.5 volume fraction of PZT. Polarizing conditions: E_p = 3 MV/m, T_p = 423 K, t_p = 0.5 h [111].

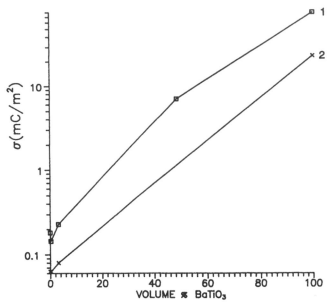

Figure 43 Dependence of TSD charges on volume fraction of BaTiO₃ in PMMA/BaTiO₃ composite. Conditions: E_p = 3 MV/m, T_p = 423 K, t_p = 0.5 h, b = 2 K/min [111]. (Initial values, curve 1; values after 15 days, curve 2.)

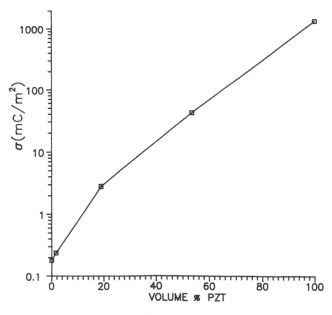

Figure 44 Dependence of TSD charges on volume fraction of PZT in PMMA/PZT composite [111]. For conditions, see Figure 43.

E. Electret Properties of Series Multilayered Systems

The TSC diagrams of polarized films are characterized by the occurrence of different discharge current peaks in the temperature range between room temperature and 420 K (Figs. 45 and 46 for the components and laminated systems, respectively). Heights and positions of these peaks are listed in Table 4. TSD charges as calculated from the TSC peak area of the a-b-a, a-c-a, and a-d-a laminated systems are approximately 69, 18, and 21 mC/m^2 after 10 days and 3, 3.5, and 2.8 mC/m^2 after 100 days from poling.

The induction charge densities of one-sided metallized films are equal to 32, 10, and 16 μC/m^2, respectively (homoelectrets). On the basis of the above data (Figs. 45, 46, and Table 4), we came to the conclusion that the experimentally observed different relaxations of multilayered systems cannot be interpreted simply as a result of the superposition of individual relaxation processes, such as dipolar, spontaneous, or space charge polarizations. An additional effect originates from static contacts between different ma-

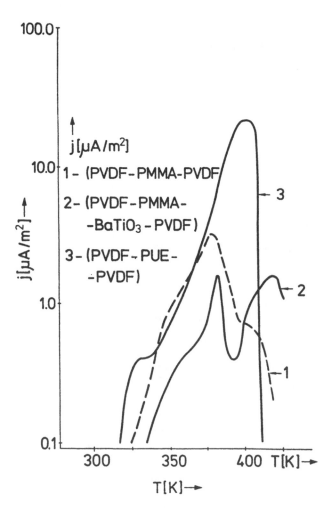

Figure 45 TSC peaks of components of different laminates (after 10 days) [19]. Conditions: E_p = 10 MV/m, T_p = 375 K, t_p = 1 h, b = 2 K/min.

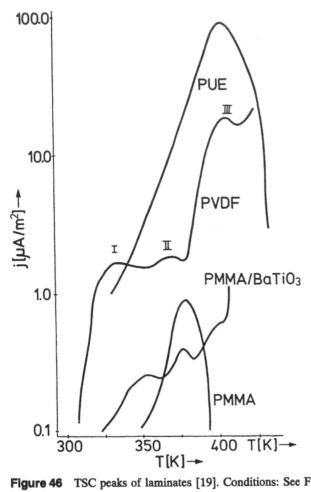

Figure 46 TSC peaks of laminates [19]. Conditions: See Figure 45.

Table 4 Data from TSC Peaks Obtained After 10 and 100 (*) Days

Material	Peak I		Peak II		Peak III	
	j_M ($\mu A/m^2$)	T_M (K)	j_M ($\mu A/m^2$)	T_M (K)	j_M ($\mu A/m^2$)	T_M (K)
PVDF (a)	1.70	335	1.8	371	18.0	408
PUE (b)					92.0	405
PMMA (c)			0.9	380		
PMMA/BaTiO$_3$ (d)	0.25	355	0.4	377	0.6	406
a-b-a system	0.60	330			65.0	408
a-b-a system *	0.40	331			21.5	403
a-c-a system	0.95	348	4.1	380	6.0	400
a-c-a system *	0.80	348	3.2	380	0.7	401
a-d-a system	0.50	333	0.7	365	14.7	412
a-d-a system *	0.50	368	1.6	383	1.6	420

terials (equilibrium triboelectric effect), but it can be seen that the internal layers determine the main shape of TSC thermograms of the laminates.

The degree of crystallinity of PVDF polymer film is 45%. The TSC diagram (Fig. 45) seems to indicate that all crystalline forms, i.e., the γ, α, and β phases, appear in semicrystalline PVDF film obtained by the hot-pressure method. This thermogram shows three well-known peaks: the intermediate α_i peak (peak I in Fig. 45 and in Table 4) connected with the γ and α forms; the α_c peak (peak II), which is probably related to dipole relaxation in the crystalline β phase; and the ρ peak (peak III), which is due to space charge relaxation in the region between the crystalline and amorphous phases (see TSD of PVDF by Van Turnhout in [118]).

As is seen in Figure 45 and Table 4, the space charge ρ peak appears mainly in the TSC thermograms of PUE and PVDF films and to some degree in PMMA/BaTiO$_3$ composite film. However, the TSC spectrum of PMMA film polarized below the glass transition temperature shows only the α peak.

The location and shape of the current peaks are quite different for the various components (Fig. 45) and their laminated systems (Fig. 46 and Table 4). The results seem to indicate that the relaxation of space charge polarization is predominant in the polarized a-b-a system, while in the a-c-a laminate, containing PMMA, the main peak appears from dipole reorientation. In the case of a-d-a system containing PMMA/BaTiO$_3$ composite, both the relaxation of dipolar (or spontaneous) polarization and relaxation of the space charge polarization can have comparable significance.

We assume that all the above-mentioned relaxation processes to some degree determine the piezoelectricity of the considered materials; however, the exact nature of the correlation has yet to be ascertained [19].

V. PIEZO- AND PYROELECTRIC PROPERTIES OF DIFFERENT MULTIPHASE SYSTEMS

A. General Remarks on the Piezo- and Pyroelectric Properties of Dielectrics

Since the dielectric variables (displacement D and field E), mechanical variables (stress X, strain S), and thermal variables (temperature T, entropy S_e) are interrelated [2b,118,119,122], the polarization P can be changed not only due to changes in the external electric field but also as a result of changes in other external parameters, such as mechanical stress X and temperature T, for example.

If the polarization of a dielectric changes under the influence of an external stress X, such material is called a piezoelectric and the piezoelectric d constant is defined by

$$d = \frac{\partial D}{\partial X} = \frac{\partial S}{\partial E} \tag{24}$$

Combinations of all variables (D, E, X, S) give rise to four constants, d, e, g, and h.

Some piezoelectric materials exhibit dielectric polarization without external stress and an electric field and, moreover, this polarization is a function of all external parameters, viz., electric field strength E, stress X, and temperature T. Such materials are referred to as pyroelectrics. In this case the most frequently used constant is the pyroelectric constant p, defined by a change D (or P) caused by a change in T:

$$p = \left(\frac{\partial P}{\partial T}\right)_{E,X} \tag{25}$$

As is known, the piezoelectrics are dielectrics which crystallize in systems without a center of symmetry, and pyroelectrics are crystals with polar axes. In the case of biological and composite materials, another kind of symmetry has to be considered, i.e., point groups of the texture made up of crystalline aggregates randomly arranged on a plane but exhibiting some ordering in the direction to it [2b,119]. Of the seven point groups established by Shubnikov [108] for textures, two (∞ and ∞m) meet the conditions necessary for pyroelectric properties to occur, for they have appropriately infinitely multiple rotation axes at the intersection of an infinite number of mirror planes [109,119].

The expected components of the piezoelectric and pyroelectric tensors for composites of polymer-ceramic films are [122,118,119].

$$d_{ij} = \begin{pmatrix} 0 & 0 & 0 & 0 & d_{15} & 0 \\ 0 & 0 & 0 & d_{15} & 0 & 0 \\ d_{31} & d_{32} & d_{33} & 0 & 0 & 0 \end{pmatrix} \tag{26}$$

$$p_i = \begin{pmatrix} 0 \\ 0 \\ p_3 \end{pmatrix} \tag{27}$$

Here the 3 axis is taken in the poling direction, which is along the film thickness.

As is known, the piezoelectric effect was discovered in 1880 by the brothers Jacques and Pierre Curie, and ever since that discovery the effect has been intensively studied in various materials. The first investigations of the piezoelectric properties of polymers, such as cellulose, were made by Brain [137], and the earliest theoretical discussion of the piezoelectric properties of various polymeric materials can be found in the works of Rez [138] and Gubkin et al. [139]. (For recapitulation of the literature on this subject, see, for example, Refs. 2a, 2b, 3, 4, 118, and 119]).

Such polymers as PVDF, in particular, have wide applications [3]; sometimes its properties are advantageous for some reasons, e.g., low electric permittivity and small thickness, but limit its application in other devices. When, for example, hydrophones, which are electroacoustic transducers used in a water environment, because of the low transducer capacity made of PVDF, amplifiers should be placed very near. Moreover, voltage sensitivity in the open system given in dB in the relation $1/\mu$Pa determined by a product $g_h x_2$ (where g_h indicates hydrostatic piezoelectric voltage coefficient and x_2 is transducer thickness), is low for transducers made of PVDF films.

Similarly, piezoelectric ceramic converters to be applied in ultrasonic medical diagnostics should exhibit large electromechanical feedback coefficients giving high sensitivity and acoustic impedance adjusted to the human body. PVDF has in fact low acoustic impedance (about 4 Mrayl) and high g_h coefficient in a wide range of frequencies; however, its low electromechanical coupling constant (0.2) and low electric permittivity (about 10) limit its application in the transduction of ultrasounds. Similarly, piezoelectric ceramics that are widespread in common use, e.g., PZT ceramics of a morphotropic bound around, have for some reasons unfavorable properties in hydrophones and ultrasonic transducers for medical diagnosis. The figure of merit for the hydrophones is determined by the product $d_h g_h$, where d_h indicates hydrostatic piezoelectric coefficient and $g_h = d_h/\epsilon_0\epsilon$ indicates piezoelectric hydrostatic voltage coefficient. Piezoelectric ceramic PZT has a low coefficient $d_h = d_{33} + 2d_{31}$, since d_{33} and d_{31} have opposite signs. Moreover, high electric permittivity (about 1800) lowers the d_h value, whereas high ceramic density (7900 kg/m^3) makes the impedance adjustment for water difficult. There are, however, piezoelectric ceramics, such as lead methanioban and modified ceramics of lead titanate, that have coefficient $d_h g_h$, but that also have high density and acoustic impedance. High acoustic impedance, which for example, for ceramic PZT equals about 30 Mrayl, makes its applications in ultrasonic transducers for medical diagnosis difficult (acoustic impedancy for leather is ~1.5 Mrayl), though it has a high coefficient of electromechanical coupling (0.5).

The analysis presented above was made by Hilczer and Maiłecki [4,119], and it is clear from this that single-phase piezoelectric systems, both ceramics and piezopolymers, do not fulfill all the requirements to be applied in hydrophones and ultrasonic transducers for medical diagnosis. They can be fulfilled by multiphase system composite materials consisting of piezoelectric ceramics and a polymer. Properties of the composites depend on the properties of particular phases, the volume fractions of the phases, and the means of their connectivity.

The origins of piezoelectricity of multiphase systems have been classified in previous reviews [10,11] by Hayakawa and Wada. In these reviews they have suggested three fundamental mechanisms of piezoelectricity of polymers, i.e., piezoelectricity related to internal deformation (mechanism A), piezoelectricity due to strain dependence of spontaneous polarization (mechanism B), and piezoelectricity related to film nonhomogeneities and stored electric charge (mechanism C). It is generally known also that piezoelectricity is due to strain dependence of frozen dipole polarization of polar polymer films. Complementary classifications are given by Wada [140] and Furukawa [122].

B. Piezoelectric Properties of 0-3 Composites

In this subsection we compare theories for a two-phase system with the different 0-3 types of polymer-ceramic composites.

1. Mixing Rules for Piezoelectric Constants

The piezoelectric constants d, e, g, and h are defined as follows [2b,12,141,142]:

$$d = \left(\frac{D}{X}\right)_E = \left(\frac{S}{E}\right)_X \tag{28}$$

$$e = \left(\frac{D}{S}\right)_E = -\left(\frac{X}{E}\right)_S \tag{29}$$

$$g = \left(\frac{E}{X}\right)_D = \left(\frac{S}{D}\right)_X \tag{30}$$

$$h = -\left(\frac{E}{S}\right)_D = -\left(\frac{X}{D}\right)_S \tag{31}$$

where X is the stress, S is the strain, E is the electric field, and D is the electric displacement.

Furukawa et al. [12] have considered a two-phase system composed of a nonpiezoelectric continuous phase (phase 1) and a piezoelectric spherical phase (phase 2). Assuming that the dielectric and elastic constants of a ceramics are usually much larger than those of polymers, the piezoelectric constants according to Furukawa et al. theory [12] are approximated as

$$d = \frac{15v_2}{(2 + 3v_2)(1 - v_2)} \frac{\epsilon_1}{\epsilon_2} d_2 \qquad (32)$$

$$e = \frac{15v_2}{2(1 - v_2)^2} \frac{\epsilon_1}{\epsilon_2} \frac{c_1}{c_2} e_2 \qquad (33)$$

$$g = \frac{15v_2}{(1 + 2v_2)(2 + 3v_2)} g_2 \qquad (34)$$

$$h = \frac{15v_2}{2(1 + 2v_2)(1 - v_2)} \frac{c_1}{c_2} h_2 \qquad (35)$$

$$c = \frac{2 + 3v_2}{2(1 - v_2)} c_1 \qquad (36)$$

$$\epsilon = \frac{1 + 2v_2}{1 - v_2} \epsilon_1 \qquad (37)$$

If the properties of the ceramic phase (phase 2) are independent of frequency, d and ϵ should have the same frequency dependence as ϵ_1 (retardional), and h and c the same as c_1 (relaxational). Furthermore, g should be independent of frequency [12]. (The c, c_1, and c_2 are elastic constants.)

The frequency dependence of ϵ at high temperatures is considered in detail by Furukawa et al. [12]. However, Eqs. (32)–(37), explicating the relation between piezoelectric constants and electric polarizing field E_p and temperature T_p, are not given.

Wada and Hayakawa [11] considered the piezoelectric properties of polymer films in which fine particles of ferroelectric materials were dispersed. If the particle is assumed to be rigid and further $\epsilon_2 \gg \epsilon_1$, the constant e becomes of the form

$$e = v_2 \frac{3P_s}{\epsilon_2} \epsilon_1 \left(\frac{\kappa_1}{\epsilon_1} + m \right) \qquad (38)$$

where P_s, κ, and m are the spontaneous polarization, electrostriction constant, and Poisson's ratio, respectively. It is to be emphasized from this theory that the piezoelectricity or pyroelectricity due to heterogeneity does not require any piezoelectric or pyroelectric phase in the material, and they arise from coupling of the heterogeneity of matrix to the antisymmetrical charge distribution [11].

2. Dependence of Piezoelectric Constants on Volume Fraction of Ceramics

We compare the piezoelectric constant with the above mixing rules for different 0-3 composites. Figures 47, 48, and 49 plot the maximum values of d_{31} and d_{33} [12,122,110] on volume fractions v_2 of PZT ceramics in composites PVDF/PZT, respectively. The dashed lines in these figures represent the predicted d constants for the PVDF/PZT systems obtained by Eq. (32). As is seen, the observed values are two to three times higher for d_{31} and two to three times lower for d_{33} than predicted. In the opinion of authors of the paper [12], these discrepancies may suggest the deviation of the samples used from the ideal two-phase system with spherical dispersions.

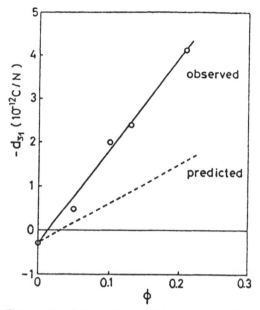

Figure 47 Φ dependence of the d constant for the PVDF-PZT system. (In this figure $\Phi = v_2$). (After Ref. 12.)

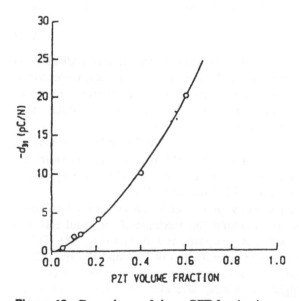

Figure 48 Dependence of d_{31} on PZT fraction in composite. (After Ref. 122.)

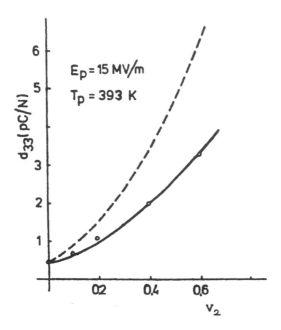

Figure 49 Dependence of d_{33} on PZT volume fraction (v_2) in composite. (After Ref. 110.)

Obviously, the d of polymer-ceramic composites depends on the volume fraction of the ceramic phase, which is limited by the closest packing of spherical particles, if the ceramic powder is dispersed in polymer matrix.

For comparison, in Figure 50 we present the dependence of d_{33} and d_h on the volume fraction of BaTiO$_3$ in composite PMMA/BaTiO$_3$. We note that for 0-3 polymer-ceramic composites, where $d_{31} = d_{32}$, the hydrostatic piezoelectric constant d_h is expressed as

$$d_h = d_{33} + 2d_{31} \tag{39}$$

The dependence of d_{31} on volume percent BaTiO$_3$ for composite PMMA/BaTiO$_3$ is shown in Figure 51, both for initial and stabilized values of this quantity, respectively. The shapes of the curves representing these dependences are quite different. From this figure it appears that the agreement between experimental and theoretical results may depend on the storage time after which the measurements were made.

Other piezoelectric parameters, such as the electromechanical coupling constant k_{33}, and the piezoelectric constants g_{33}, g_h, e_{33}, and e_h for PMMA/BaTiO$_3$ composite, are shown in Figures 52–54. The predicted values in these figures are illustrated by dashed lines. Also, for this composite $d_h g_h$ is presented in Figure 55 [111].

As is seen from the above data, the maximal piezoelectric constants of 0-3 polymer-ceramic composites are relatively low. For example, the maximum d_{31} is about 1/10 that of PZT, even though 60% volume fraction in composite is achieved (Fig. 48). This is due primarily to the substantial differences in permittivities of PZT and PVDF [122].

Skiner et al. suggested [143] that composites in which both ceramics and polymer are continuous are favored for higher activities, although they require elaborate processing.

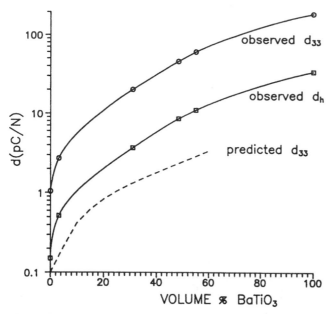

Figure 50 Dependence of d constant on BaTiO$_3$ fraction in composite [111].

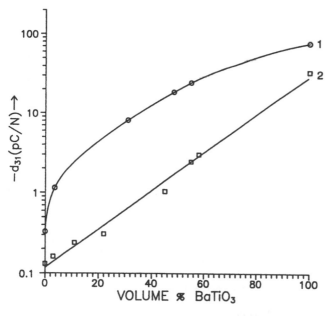

Figure 51 Dependence of d_{31} constant on BaTiO$_3$ fraction in composite [111]: initial values (curve 1) and stabilized values (curve 2).

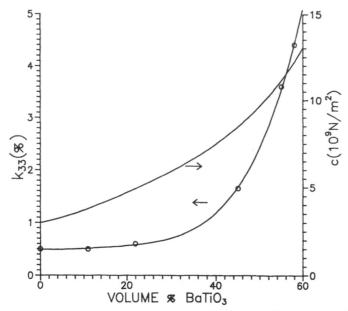

Figure 52 Dependence of electromechanical coupling constant k_{33} and elastic constant c on BaTiO$_3$ volume fraction in composite [111].

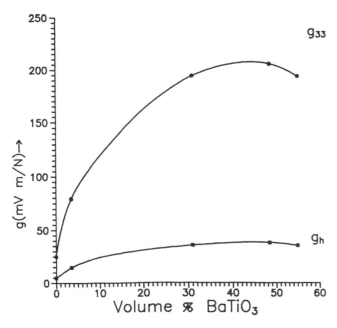

Figure 53 Dependence of g_{33} and g_h on BaTiO$_3$ volume fraction in composite [111].

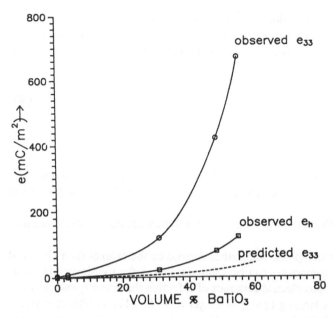

Figure 54 Dependence of piezoelectric e constants on BaTiO$_3$ composite [111].

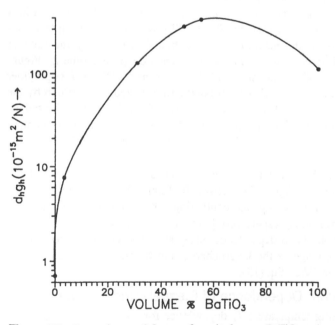

Figure 55 Dependence of figure of merit $d_h g_h$ on BaTiO$_3$ volume fraction in composite [111].

3. Polarizing Field Characteristics of Piezoelectricity

On the basis of thermodynamic considerations, Broadhurst and Mopsik [143a,143b] obtained the following expressions for piezo- and pyroelectric coefficients of film electrets with polarization P:

$$d = \frac{-\epsilon_\infty}{3\epsilon_0} \beta_T P \tag{40}$$

$$p = \frac{-\epsilon_\infty}{3\epsilon_0} \alpha P \tag{41}$$

where β_T is isothermal compressibility, and α is the thermal expansion coefficient. The polarization P and dielectric constant ϵ_∞ are dependent on temperature and polarizing field, obviously.

Poling and temperature characteristics of different polymer-ceramic systems are presented in this subsection.

The poling conditions for two-phase composites depend on the ratio of their electrical conductivities γ_1/γ_2. The 0-3 composites can be approximated to the two-layer Maxwell-Wagner model, consisting of piezoelectric ceramics of effective thickness l_1 and polymer film with thickness l_2. If the polarizing time t_p is longer than the Maxwell-Wagner relaxation time τ_M, the voltage is divided according to the conductances of the layers and the voltage V_1 across the ceramic layer attains its final value:

$$\frac{V_1}{V_p} = \frac{1}{1 + (\gamma_1/\gamma_2)(l_2/l_1)} \tag{42}$$

where V_p is the applied voltage. For polymer-ceramic composites, where $\gamma_2 < \gamma_1$, the ratio V_1/V_p is low. One way of improving the poling conditions of the 0-3 composites is by lowering the conductivity in the polymer phase [4,123].

Figure 56 shows the E_p dependence of the d_{31} constant for samples with various volume fractions of PZT ceramic in a composite PVDF/PZT system. The absolute values of d increase with E_p; however, the maximum of this field is limited by the field of electrical breakdown, E_B. Moreover, the result for d depends on poling time t_p. Figure 57 shows the switching process of a composite of PZT ceramic and VDF/TrFE copolymer at 100°C and 12.5 MV/m [122]. Here a change in polarization is again monitored by the piezoelectric constant. We find that it takes a long time (>1000 s) to reverse the polarization, but PZT in the composite is fully poled toward the reverse direction. As is mentioned by Furukawa [122], in this case t_p is given by the relaxation time inherent to the Maxwell-Wagner effect.

The dependence of d_{33} on E_p at different polarizing times t_p for the composite PZT/Eccogel(1365-0) are presented in Figure 58 [4,144]. In Figure 59 are represented the dependence of d_{33} on E_p for an 0-3 composite containing 60% volume fraction PZT, in PVDF/PZT and PE/PZT composites, respectively [145]. However, the piezoelectric constants can also change the polarity in dependence on E_p as is shown in Figures 60 and 61, and this corresponds to change of the heterocharge into homocharge in accordance with the discussion in Section IV.C, Eq. (18).

4. Polarizing Temperature Dependence of Piezoelectric Constants

Figure 62 shows the polarizing temperature T_p dependence for PZT/PE and PZT/PVDF composites [145]. The maximal values of d_{33} reported in this paper are not very high in

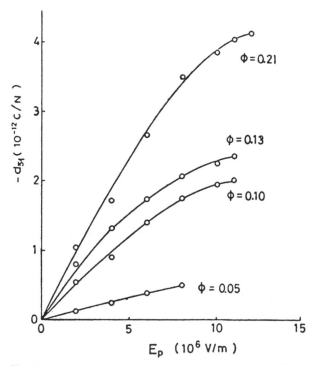

Figure 56 Poling characteristics for PVDF-PZT system. (In this figure $v_2 = \Phi$). (After Ref. 12.)

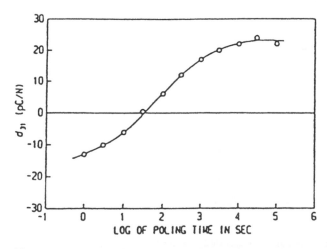

Figure 57 Polarization reversal in PVDF/PZT composite. (After Ref. 122.)

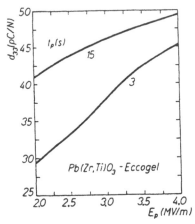

Figure 58 Poling characteristics for PZT-Eccogel system at different polarizing times t_p. (After Ref. 144.)

comparison with other data on this subject. Essential here, however, is the character of the dependence of d_{33} on T_p. As Eq. (12) or (16) predicts, it is similar to the T_p dependence of the TSD charge (remnant polarization P).

On the other hand, as has been shown by Furukawa et al. [12], after poling, the temperature dependence of d for a PVDF/PZT composite with 0.21% PZT is similar to the temperature dependence of the dielectric constant. It was to be expected that the piezoelectric constants are dependent on T_p and T, since these parameters are determined by the dielectric constant, the elastic constant, and the remnant polarization, which depends on T_p and T. We note that d_{33} and the figure of merit $d_h g_h$ are dependent on the pressure during preparation of composite. This dependence is shown in Figure 63 [146].

5. Frequency Dependence of Piezoelectric Constant

The piezoelectric constant d was measured by Furukawa et al. [12] for a PVA/PZT system with 8 vol% PZT in the frequency range from 0.1 to 30 Hz at temperatures between 80

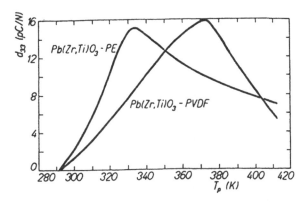

Figure 59 Dependence of d_{33} on polarizing temperature T_p for PZT-PVDF and PZT-PE systems. (After Ref. 145.)

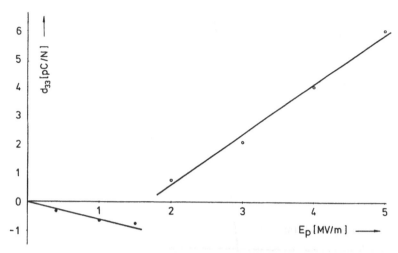

Figure 60   Poling characteristic for PMMA/BaTiO₃ system [163].

and 130°C. The sample was poled prior to measurement. From results of this paper, it appears that d exhibits retardional and relaxional behaviors. Figure 64 plots d' and d'' at 80°C on a complex plane. Using the instantaneous value $d(\infty)$ and the equilibrium value $d(0)$, they obtain the relative relaxation strength for d by

$$\Delta_r d = \frac{d(\infty) - d(0)}{d(\infty)} \tag{43}$$

The authors of this paper find relatively good agreement between observation and prediction for $\Delta_r d$ and the relaxation time τ of this composite. However, Cole-Cole presentation of complex electric permittivity for some composites with higher volume percent PZT suggest not only a single relaxation process but multiple ones [8,67].

C.   Piezoelectricity of 1-3 Composites

Theoretical study of the propagation of the elastic wave in 1-3 composites [83,84,147–149] enable one to indicate how electromechanical properties of piezoelectric composites depend on the properties of the component phases and the volume fraction of the piezoelectric ceramics. The dependence between electromechanical coupling factor (k) and the acoustic impedance (Z) can be determined theoretically for composites of different ceramic contents and to find a compromise between increasing k and Z while increasing ceramic contents. High values of coefficient k and small acoustic impedance Z are required in applications for ultrasonic transducers.

Composites which will be applied in hydrophones should possess, besides low acoustic impedance, high hydrostatic values of piezoelectric coefficients d_h and g_h, since the figure of merit is a product,

$$d_h g_h = \frac{(d_{33} + 2d_{31})^2}{\epsilon \epsilon_0} \tag{44}$$

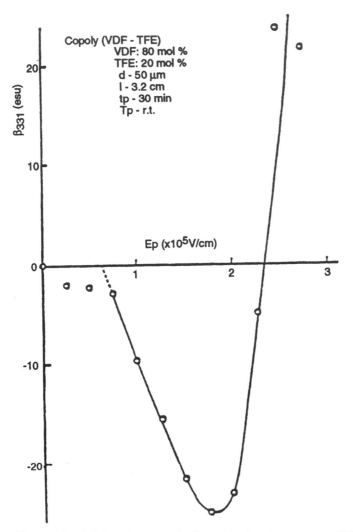

Figure 61 Relations between the β_{331} piezoelectric constant and E_p for copoly(VDF/TFE) (80/20) film. (After Ref. 162.)

The analysis of this problem leads to the conclusion [4] that the composite materials to be used in hydrophones should be prepared in such a way as to minimize the values of the coefficient d_{31} and permittivity ϵ and to maintain the greatest value of the coefficient d_{33}. One can find some information in the literature about piezoelectric composites of various ways of joining phases [2,4,12–16].

D. Piezoelectricity of Laminated Polymer-Ceramic Films

The nature of piezoelectricity in multiphase systems depends on the physical and chemical properties of the components. Greavers et al. [150] and Hayakawa and Wada [10,11] have shown that most heterogeneous polymer films can have piezoelectric properties to

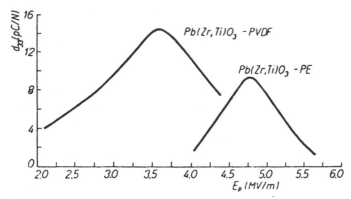

Figure 62 Dependence of d_{33} on polarizing field E_p for PZT-PVDF and PZT-PE systems. (After Ref. 145.)

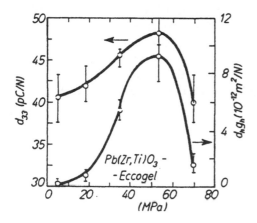

Figure 63 Dependence of d_{33} and $d_h g_h$ on pressure for PZT-Eccogel composite. (After Ref. 144.)

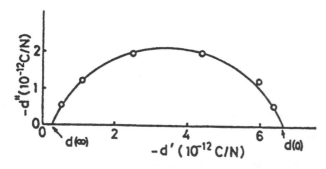

Figure 64 Plot of d' and d'' at 80°C for the PVA-PZT system with $\Phi = 0.08$. (After Ref. 12.)

some degree even if they are not polarized. This may be related to heterogeneity and embedded charges accidentally present in the film [10,150] (machanism C).

In a previous paper [17] we have shown experimentally that uncharged laminate PET-PMMA/BaTiO₃-PET exhibits piezoelectric properties. This laminated system of three-layered structure of different materials fulfills the heterogeneity condition; moreover, the charges can be stored at the interfaces of the layers as a result of the equilibrium triboelectric electrification and some effects caused by partially oriented PET films (mechanism C). However, we note that even PMMA/BaTiO₃ composite as internal layer of the laminate shows piezoelectricity to some extent, due to the spontaneous polarization of barium titanate particles (mechanism B), and the dipole polarization of PMMA (mechanism D).

Similar piezoelectric laminates have been prepared by using other materials [82,151], i.e., polyester-urethane elastomer (PUE) and PMMA/PUE, as an internal layer in the PET-PUE-PET and PET-PMMA/PUE-PET systems. These laminates, both uncharged and charged by an external electric field at higher temperature, can have interesting electret and piezoelectric properties [151]. However, exact consideration of the complex physical nature of piezoelectricity in such systems is an open problem.

In this section we show how the TSD method can be used to establish which of the mechanisms A to D plays the principal role in piezoelectricity of PVDF-PUE-PVDF, PVDF-PMMA-PVDF, and PVDF-PMMA/BaTiO₃-PVDF laminates.

The piezoelectric strain constants d_{33} of the a-b-a, a-c-a, and a-d-a systems were measured at room temperature for about 100 days after poling. The time dependence for these constants are presented in Figure 65. It is seen in this figure that the initial values of d_{33} are 3, 4, and 20 pC/N for the a-c-a, a-b-a, and a-d-a systems, respectively.

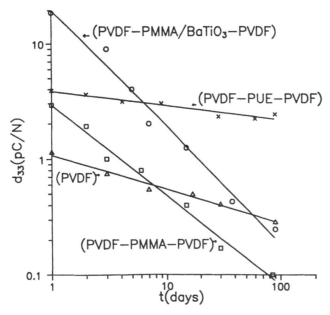

Figure 65 Time dependence of piezoelectric strain constant $d_{33}(t)$ of laminated films. Polarizing conditions: E_p = 10 MV/m, T_p = 375 K, t_p = 1 h [19].

To compare the experimental data with phenomenological results of Wada and Hayakawa [11], we applied their expression for the piezoelectric constant e_{33} or $d_{33} = e_{33}/c_{33}$. This expression for the (a-b-a) system is given as

$$e = \frac{(\kappa_a/\epsilon_a - \kappa_b/\epsilon_b) + (m_a - m_b)}{\epsilon_a \epsilon_b [\Phi/\epsilon_a - (1 - \Phi)/\epsilon_b]^2} \Phi(1 - \Phi)\sigma \tag{45}$$

where Φ is the thickness fraction of film a and σ is the surface charge density which is stored on the interfaces of the system. The maximum value of σ is limited by the breakdown field E_B of films b [11],

$$\sigma_M = \epsilon_0 \epsilon_b E_B \tag{46}$$

where ϵ_0 is the vacuum permittivity.

If we assume for the PVDF-PUE-PVDF system that the dielectric constants $\epsilon_a = 10$ and $\epsilon_b = 6$, the Poisson ratio $m_a = m_b = 0.35$ [10], the electrostriction constants $\kappa_a = -3$ [10] and $\kappa_b = -3.37$, the elastic constant $c = 0.1$ GPa, and $E_B = 18$ MV/m, we get $d \simeq 0.6$ pC/N.

It is seen that the piezoelectric strain constant d calculated from Eqs. (45) and (46) is about one order of magnitude smaller than the experimentally determined d_{33} constant.

Moreover, under conditions when the polarizing field $E_p < E_B$, Eq. (46) cannot be applied. We assume that in this case σ can be evaluated by applying Van Turnhout's equation [123], which appears from consideration of the Maxwell-Wagner effect in nonpolar laminates. This equation for laminated polymer films of three-layered structure can be modified as

$$\sigma = \frac{1 - \epsilon_b \gamma_a/\epsilon_a \gamma_b}{1 + l_b \gamma_a/2l_a \gamma_b} \epsilon_0 \epsilon_a V_p/2l_a \tag{47}$$

where l and γ with subscripts (a, b), are, respectively, the thickness and electrical conduction of the layers, whereas V_p is the polarizing voltage. In particular, Eq. (47) can be applied if the space charge polarization has the predominant role in comparison with dipolar polarization. We assume that under polarizing conditions $\epsilon_a = 13$, $\epsilon_b = 7$, $\gamma_a = 1 \times 10^{-11} \ \Omega^{-1} m^{-1}$, and $\gamma_b = 10\gamma_a$.

The calculations based on Eqs. (45) and (47) and the above data give $d \simeq 1.42$ pC/N. It is interesting that this result is nearer to the experimental value than the previous one, which derives from Eqs. (45) and (46). We note that the best agreement of experimental and calculated results can be obtained if we replace the σ charge by the TSD charge in Eq. (45). In this case the d obtained theoretically (1.9 pC/N) corresponds quite well to that obtained experimentally (3 pC/N).

Similar comparisons for a few other laminated polymer films are listed in Table 5.

In Figure 66 are presented the time dependence of d_{33} for PET-PMMA/BaTiO$_3$-PET laminates [17].

The experimental and theoretical analysis presented above seems to indicate that the initial piezoelectric constant d_{33} is highest for the PVDF-PMMA/BaTiO$_3$-PVDF system; however, better piezoelectric stability can be obtained in polarized PVDF-PUE-PVDF laminates. We note that in the TSC spectrum of this laminate, a very large ρ peak is observed, so that the mechanism of piezoelectricity of the system may be due to heterogeneity and embedded charges at the interfaces.

Table 5 Experimental and Theoretical Values of d_{33} [111]

Laminate	(A) d_{33} (pC/N)	(B) d_{33} (pC/N)	(C) d_{33} (pC/N)	(D) d_{33} (pC/N)	(E) d_{33} (pC/N)
PVDF-PUE-PVDF	2.95	2.25	0.21	22.29	0.57
PVDF-PMMA-PVDF	0.49	0.10	0.005	0.13	0.013
PVDF-PMMA/BaTiO₃-PVDF	2.10	0.22	0.04	0.91	0.003
PTFE-PUE-PTFE	2.20	5.50	2.70	—	1.42

(A) after 10 days; (B), after 100 days; (C), calculated from Eqs. (45) and (46); (D), calculated from Eq. (45) and TSD charges; (E), calculated from Eqs. (45) and (47).

E. Pyroelectric Properties of Multiphase Systems

Pyroelectric materials have been extensively investigated recently for use in integrated imaging devices such as remote sensing, biomedical thermography, gas detection, alarms, etc. [2a,64,152,153].

The pyroelectric coefficients of the polymer-ceramic composites are large compared to polymers, and the relative permittivities are small compared to ceramics [64]. Therefore the figures of merit are enhanced over conventional single-phase polymer materials. (In certain pyroelectric systems a useful figure of merit is p/ϵ, where p is the pyroelectric coefficient and ϵ is the electric permittivity [16]).

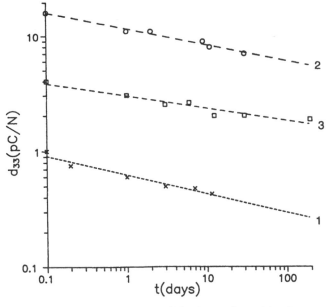

Figure 66 Time dependence of d_{33} for uncharged laminates PET-PMMA/BaTiO₃-PET with 0 (curve 1), 0.03 (curve 2), and 0.22 (curve 3) volume fraction of BaTiO₃ [17].

1. Methods for Measuring the Pyroelectric Coefficient of Composites

The most usual way to measure the pyroelectric coefficient is by the conventional quasi-static method [48,154,155]. Since the polarization P measure is surface charge density $\sigma = Q/A$ (Q denotes electric charge and A is the surface of the dielectric film), Eq. (25) can be rewritten as follows:

$$p = \left(\frac{\partial \sigma}{\partial T}\right)_{E=0} = \frac{\partial \sigma}{\partial t}\frac{dt}{dT} = j\frac{dt}{dT} \tag{48}$$

where j denotes the current density in short-circuited systems induced by a change in temperature at some heating rate dT/dt. Thus the pyroelectric coefficient of a polymer film can be determined by a conventional method [2b,154], analyzing the electric response of a film heated at constant rate dT/dt [155]. However, for multiphase materials the total current is usually made up of two nondistinguishable contributions, such as the irreversible depolarization current due to space charge relaxation and the true reversible pyroelectric current [48,75].

Another possibility is to use a dynamic method, which consists of studying the electric response after the film has been stimulated with heat pulses produced by a modulated light beam [156].

Recently, Dias et al. [48] suggested that a periodic temperature oscillation of a ceramic-polymer composite sample would permit measurement of its true pyroelectric coefficient without the need to anneal the sample.

Both the above-mentioned methods are used in investigations of the pyroelectric properties of polymer-ceramic composites. We will compare the pyroelectric properties of a few polymer-ceramic composites as determined by different methods.

2. Pyroelectricity of PMMA/BaTiO₃, PMMA/PZT, and PVDF/BaTiO₃ Composites

Thermally stimulated current at the first heating of the composite samples is due to space charge, dipolar charge, and spontaneous polarization relaxations. The TSC curves for these composites are given in Section IV.A. These curves do not obviously represent the pyroelectric effect. If we heat the discharged sample for a second time, then the current connected with the pyroelectric effect appears. On the base of pyrocurrent, the pyroelectric coefficient as a function of temperature can be determined. The results for PMMA/BaTiO₃ and PMMA/PZT are shown in Figures 67 and 68. In Figures 69 and 70 we present the dependence of the pyroelectric coefficient on the volume fraction of BaTiO₃ and PZT in the composites, respectively.

By means of this conventional method together with infrared spectroscopy, the pyroelectric behavior in PVDF/BaTiO₃ has been investigated by Muralidhar and Pillai [6]. Composite samples were prepared by grinding PVDF powder and fine powder of BaTiO₃ ceramics in different weight proportions. The temperature dependence of the pyroelectric constants for the samples with various volume fractions of BaTiO₃ in the PVDF/BaTiO₃ system are shown in Figure 71. The authors suggest that this composite material is expected to enhance the pyroelectricity of PVDF by becoming polar in nature.

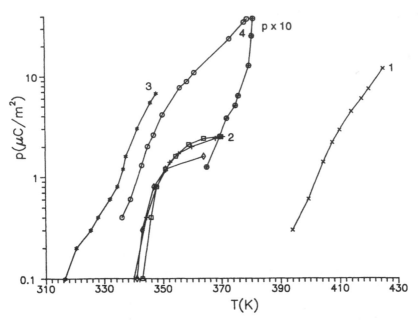

Figure 67 Temperature dependence of pyroelectric coefficient p for PMMA/BaTiO$_3$ composite with various volume fractions of BaTiO$_3$: 0 (curve 1), 0.03–0.05 (curve 2), 0.48 (curve 3), and BaTiO$_3$ (curves 4) [111].

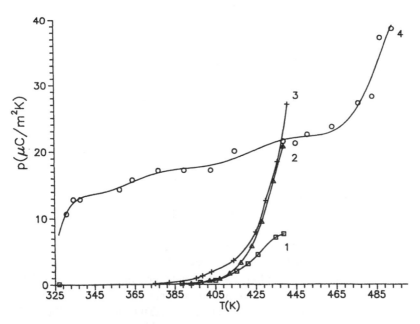

Figure 68 Temperature dependence of pyroelectric coefficient p for PMMA/PZT composite with various volume fractions of PZT: 0.02 (curve 1), 0.2 (curve 2), 0.54 (curve 3), and PZT (curve 4).

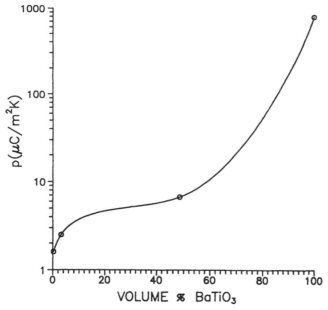

Figure 69 Dependence of pyroelectric coefficient p on BaTiO$_3$ fraction in PMMA/BaTiO$_3$ composite [111].

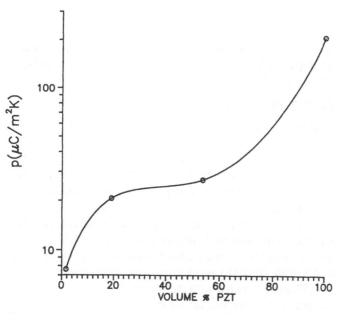

Figure 70 Dependence of pyroelectric coefficient p on PZT fraction in composite PMMA/PZT [111].

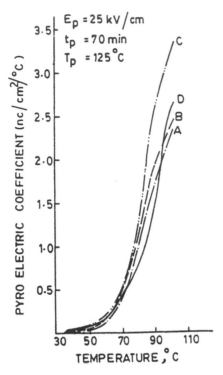

Figure 71 Pyroelectric behavior of PVDF/BaTiO$_3$ composite with various weight fractions of BaTiO$_3$: 0 (curve A), 0.1 (curve B), 0.3 (curve C), 0.5 (curve D), 0.7 (curve E), and 0.9 (curve F). (After Ref. 6.)

It should be noted that some parasitic current can influence the typical pyroelectric current, which is probably accompanied by chemical changes such as oxidation, disproportionation, monomer conversion, pyrolysis, etc. [123].

3. Pyroelectricity of P(VDF/TrFE)/PbTiO$_3$

A measurement method that permits determination of the true pyroelectric current is to sinusoidally modulate the temperature of the sample while recording the current (or voltage) at the same frequency as that of the temperature oscillation [157,158]. This idea has been developed by Dias et al. [48]. These authors suggest that a periodic temperature oscillation of a ceramic-polymer composite sample would permit measurement of its true pyroelectric coefficient without the need to anneal the sample. The temperature distribution inside the composite sample, the magnitude of the pyroelectric current, and their variation with frequency have been simulated with theory presented by these authors. Pyroelectric coefficients measured using this technique have been made on a poled composite of calcium-modified lead titanate (PbTiO$_3$) and a copolymer of vinylidene fluoride and trifluoroethylene P(VDF/TeFE) and compared with the results of direct measurements before and after annealing, in order to test the validity of the method.

After authors of this interest paper [48], we show in Figure 72 the pyroelectric coefficient measured with the static and dynamic methods before and after poling.

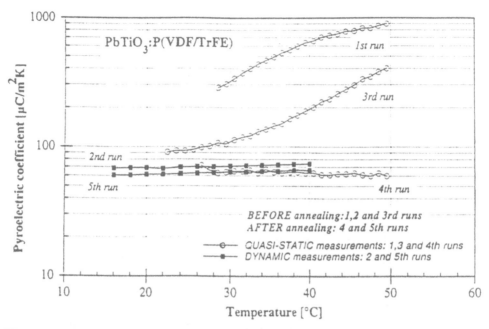

Figure 72 Pyroelectric coefficient of a poled composite sample of lead titanate:P(VDF/TrFE) (60:40 vol%) measured before and after annealing by the dynamic temperature method and comparison with measurements by the direct method. The sample was poled with a field of 30 KV/mm at $T = 102°C$ for 2.6 h. (After Ref. 48.)

VI. SUMMARY AND CONCLUSIONS

In this chapter, all the basic aspects concerning the polymer-ferroelectric ceramics have been discussed, i.e., dielectric behavior, electret, piezo-, and pyroelectric properties, both theoretically and experimentally.

Different mixing rules for electric permittivity have been compared with experimental data for various types of polymer-ceramic composites. Moreover, temperature and frequency dependence of electric permittivity and the dissipation of these composites have been also considered.

Charge transport, charge distribution, and charge storage in some polymeric and polymer-ceramic materials have been analyzed on the basis of the TSD method and some complementary methods.

Piezo- and pyroelectric properties in dependence on volume fraction of the ceramic phase and some external parameters (viz., electric field strength and temperature) have been considered for different types of polymer-ceramic composites.

The experimental and theoretical analysis presented in this chapter seems to indicate that the initial piezoelectric constants are highest for multiphase systems containing piezoelectric ceramics. However, better piezoelectric stability can be obtained in some multiphase polymer systems without any ferroelectric ceramics.

The general notes about piezoelectric composites which we present below were made by Hilczer and Mañecki [4].

Table 6 Piezoelectric Properties of Composites

Material	ϵ	d_{33} (pC/N)	g_h (mVm/N)	d_h (pC/N)	$g_h d_h$ (10^{-15} m²/N)	Ref.
Ceramic BaTiO$_3$	1,200	190		34		4
Ceramic PZT	1,800	450	2.5	40	100	41
PVDF	12	20	100	11	1,100	41
PMMA	3	0.3				76
0-3 BaTiO$_3$/PMMA	40	50	35	10	350	111
0-3 PZT/rubber	40	60	40	15	600	41
0-3 PbTiO$_3$/rubber	40	30	100	35	3,500	41
0-3 PbTiO$_3$/chloroprene rubber	45	65	100	40	4,000	40
0-3 PbTiO$_3$/Eccogel	48	60	96	42	4,032	164
0-3 (PbTiO$_3$+BiFeO$_3$)/Eccogel	40	55	90	30	2,700	24
1-3 PZT/Spurrs resin	54	150	56	27	1,536	87
1-3 PZT/polyurethane	40	170	56	20	1,120	24
1-3-0 PZT/polyurethane	41	180	210	73	15,330	24
3-3 Coral structure	50	100	140	36	5,040	165
3-3 PZT/BURPS rubber	450	200	45	180	8,100	92
3-1 Honeycomb structure of PZT	400	350	30	100	3,000	95
3-1-0 Perforated PZT-epoxy resin-air	650	430	35	200	7,000	86
3-2 Perforated PZT-epoxy resin	375	350	60	200	12,000	86
3-2-0 Performed PZT-epoxy resin-air	360	370	70	220	15,000	86

Source: Based on Ref. 4.

Multiphase materials such as polymer-ceramic composites combine properties of component phases and have useful parameters that cannot be obtained in homogeneous materials. Not only are the properties of separate phases included in a composite, but also the means of their connectivity determines these properties, since they determine charge transport, charge storage, and the distribution of electric field and mechanical stress [4,130–132].

Symmetry of composite materials is defined by the principle of symmetry of super-position [23]: Composite materials will have only those elements of symmetry that are common to component phases and are determined by their geometric arrangement. According to Neumann's law, this determines physical properties of composites, since the elements of symmetry of each of its properties must contain elements of point groups of the composites. An example of a 3-1 composite can be given in which ceramic PZT with the structure of a honeycomb with a square aperture is filled by polymer [94,95]. This composite has a tetragonal structure with quadruple axes (4/mm) [23]. After polarization of this material in the direction of the axes [94], the symmetry of the composite changes for 4 mm, but after polarization across these axes, for 2 mm. Cross-polarized composites with honeycomb structures have very high figures of merit, $d_h g_h$ (Table 6). Figures of merit are also high for 1-3-0 composites consisting of piezoelectric ceramic rods distributed in foam polymer (the third composite is air), but their properties change with pressure and this limits their usage. However, there are so-called cut composites in hydrophone applications [24,89,159]. Between them the best are composites made directly from partially cut, previously polarized ceramic PZT, covered by brazen plate and inserted in an aluminum capsule [24]. The 1-3 composites consisting of PZT ceramic rods which are inserted in a polymer exhibit large electromechanical feedback coefficients and low acoustic impedance, which is why they are applied in medical diagnosis. The easiest composites in mass production with a way of phase merging 0-3 have large values of factor g_h, but the value of piezoelectric factor d_h is low.

The work described in this chapter lead us to the conclusion that the study of electrets in polymer-ferroelectric ceramic composites is a fascinating field of basic and applied research.

REFERENCES

1. F. Jona and G. Shirane, *Ferroelectric Crystals*, Oxford, London, 1962.
2. (a) S. B. Lang, Guide to the literature of piezoelectricity and pyroelectricity, *Ferroelectrics* *139*:141 (1993). (b) S. B. Lang, *Sourcebook of Pyroelectricity*, Gordon & Breach, London, 1974.
3. (a) H. Kawai, The piezoelectricity of poly(vinylidene fluoride), *Jpn. J. Appl. Phys.* *8*:975 (1969). (b) H. S. Nalwa, Recent developments in ferroelectric polymers, *JMS—Rev. Macromol. Chem. Phys.* *C31(4)*, 341–432 (1991).
4. B. Hilczer and J. Maitecki, *Electrets and Piezopolymers*, PWN, Warsaw, 1992 (in Polish), chaps. 3 and 7.
5. T. T. Wang, J. M. Herbert, and A. M. Glass (eds.), *The Applications of Ferroelectric Polymers*, Blacke, Glasgow and London, 1988.
6. C. Muralidhar and P. K. C. Pillai, Pyroelectric behaviour in barium titanate/polyvinylidene fluoride (PVDF) composite, *Proc. 5th Int. Symp. Electrets*, Heidelberg, 1985, pp. 865–870.
7. R. E. Newnham, Composite Electroceramics, *Ferroelectrics* *68*:1 (1986).
8. J. Wolak, Dielectric behaviour of 03-type piezoelectric composites, *IEEE Trans. Elect. Insul.* *28*:116 (1993).

9. K. Mazur, J. Handerek, and T. Piech, The electret effect in a mixture of polymethacrylate and barium titanate, *Acta Phys. Polon. A37*:31 (1970).

10. R. Hayakawa and Y. Wada, Piezoelectricity and related properties of polymer films. *Advances in Polymer Science* (H. J. Cantow et al., eds.), Springer-Verlag, Berlin, 1973, pp. 1–55.

11. Y. Wada and R. Hayakawa, Piezoelectricity and pyroelectricity of polymers, *Jpn. J. Appl. Phys. 15*:2041 (1976).

12. T. Furukawa, K. Ishida, and E. Fukada, Piezoelectric properties in the composite systems of polymers and PZT ceramics. *J. Appl. Phys. 50*:4904 (1979).

13. T. Furukawa, K. Fujino, and E. Fukada, Electromechanical properties in the composites of epoxy resin and PZT ceramics. *Jpn. J. Appl. Phys. 15*:2119 (1976).

14. T. Furukawa and E. Fukada, Piezoelectric properties of the composite epoxy-resin/PZT-ceramic system, *J. Appl. Phys. 16*:453 (1977).

15. R. E. Newnham, D. P. Skinner, and L. E. Cross, Connectivity and piezoelectric-pyroelectric composites, *Mat. Res. Bull. 13*:525 (1978).

16. R. E. Newnham, L. J. Bowen, K. A. Klicker, and L. E. Cross, Composite piezoelectric transducer, *Mat. Eng. 2*:93 (1980).

17. K. Mazur, Polarization effect in uncharged laminates, *Proc. ICSD '89*, Trondheim, Norway, 1989, pp. 437–441.

18. K. Mazur, The way of preparation of multilayered piezoelectrics, Polish Patent PL 156183 (1987).

19. K. Mazur, Piezoelectricity of PVDF/PUE, PVDF/PMMA, and PVDF/(PMMA+BaTiO₃), *Proc. 7th Int. Symp. Electrets*, Berlin, 1991; pp. 512–517; *IEEE Trans. Elect. Insul. 27*: 782 (1992).

20. S. Ahmed and F. R. Jones, A review of particulare reinforcement theories for polymer composites, *J. Mat. Sci. 25*:4933 (1990).

21. H. Zewdie and F. Brouers, Theory of ferroelectric polymer-ceramic composites, *J. Appl. Phys. 68*:713 (1990).

22. J. van Suchtelen, Product properties: a new application f composite materials, *Philips Res. Rep. 27*:28 (1972).

23. M. J. Haun and R. E. Newnham, An experimental and theoretical study of 1.3 and 1.3.0 piezoelectric PZT polymer composites for hydrophone applications, *Ferroelectrics 68*:123–129 (1986).

24. T. R. Gururaja, A. Safari, R. E. Neunham, and L. E. Cross, Piezoelectric ceramic-polymer composites for transducter application, *Electronic Ceramics* (L. M. Levinson, ed.), Marcel Dekker, New York, 1987, p. 92.

25. P. Sheng, Microstructures and physical properties of composites, *Homogenization and Effective Moduli of Materials and Media* (J. L. Ericksen et al., eds.), Springer-Verlag, New York, 1986, p. 196.

26. P. Ponte Castañeda, A new variational principle and its application to nonlinear heterogeneous systems, *SIAM J. Appl. Math. 52*:1321 (1992).

27. P. Ponte Castañeda, Bounds and estimates for the properties of nonlinear heterogeneous systems, *Phil. Trans. R. Soc. Lond. A 340*:531 (1992).

28. P. Ponte Castañeda, G. deBotton, and G. LI, Effective properties of nonlinear inhomogeneous dielectrics, *Phys. Rev. B 46*:4387 (1992).

29. J. van den Boomgaard, D. R. Terrell, R. A. J. Born, and H. F. J. I. Giller, An in situ grown eutectic magnetoelectric composite material, Part I: composition and unidirectional solidification, *J. Mat. Sci. 9*:1705 (1974).

30. A. M. J. G. van Run, D. R. Terrell, and H. Scholing, An in situ grown eutectic magnetoelectric composite material, Part II: physical properties, *J. Mat. Sci. 9*:1710 (1974).

31. K. Mazur, Dielectric properties of BaTiO₃ and polymerized methyl-methacrylate mixture, *Schedule of International Conference, Physics of Ferroelectrics*, Poznań, Poland, 1965, p. 6.

32. K. Mazur, Dielectric and electret properties of PMMA/BaTiO$_3$ composite (in Polish), *Z.N. WSP in Katowice, Phys. Sec. 33* (1968).

33. K. Mazur, The mechanism of build-up of homocharge in the electrets of PMMA/BaTiO$_3$ composite, Ph.D. thesis, Silesian University, Katowice, Poland, 1968.

34. C. M. Dhar and P. K. C. Pillai, Thermocompensated capacitor with barium titanate-polyvinylidene composite, *J. Mat. Sci. Lett. 6*:33 (1987).

35. M. Amin, L. S. Balloomal, K. A. Darwish, H. Osman, and B. Kamal, Pyroelectricity in Rubber Composite Films, *Ferroelectrics 81*:381 (1988).

36. M. Amin, H. Osman, L. Baloomal, K. A. Darwish, and B. Kamal, Electrical properties of acrylonitrile-butadiene rubber-barium titanate composites, *Ferroelectrics 81*:387 (1988).

37. A. Govindan, A. K. Tripathi, T. C. Goel, and P. K. C. Pillai, Pyroelectric and piezoelectric studies on BaTiO$_3$:silica glass composite, *Proc. ISE-7*, Berlin, 1991, pp. 524–529.

38. D. K. Das-Gupta and Zhang Shuren, Non-polar polymer/ferro and antiferroelectric composite films for high energy storage capacitors, *Ferroelectrics 134*:71 (1992).

39. T. Yamamoto, K. Urabe, and H. Banno, BaTiO$_3$ particle-size dependence of ferroelectricity in BaTiO$_3$/polymer composites, *Jpn. J. Appl. Phys. 32*:4272 (1993).

40. H. Banno, S. Saito, Piezoelectric properties of composites of synthetic rubber and PbTiO$_3$ or PZT, *Jpn. J. Appl. Phys. Suppl. 22*-2:67 (1983).

41. R. E. Neunham, A. Safari, J. Giniewicz, and B. H. Fox, Composite piezoelectric sensors, *Ferroelectrics 60*:15 (1984).

42. K. F. Schoch, D. P. Partlow, and W. R. Krauze, Assessment of degree of poling in 0-3 piezoelectric composites by x-ray methods, *Ferroelectrics 77*:39 (1988).

43. H. Banno and K. Ogura, Dielectric and piezoelectric properties of a flexible composite consisting of polymer and mixed ceramic powder of PZT and PbTiO$_3$, *Ferroelectrics 95*:171 (1989).

44. R. P. Tandon, R. Singh, R. D. P. Sinha, and S. Chandra, Dielectric and piezoelectric behaviour of lead titanate/polymer composite, *Ferroelectrics 120*:293 (1991).

45. L. Pardo, B. Jimenez, L. Calzada, C. Alemany, and F. Carmona, Preparation and piezoelectric properties of boron glass and Ca-modified PbTiO$_3$ composites, *Ferroelectrics 127*:179 (1992).

46. I. Taguchi, A. Pignolet, L. Wang, M. Proctor, F. Levy, and P. E. Schmid, Raman scattering from PbTiO$_3$ thin films prepared on silicon substrates by radio frequency and thermal treatment, *J. Appl. Phys. 73*:394 (1993).

47. C. S. Hwang and H. J. Kim, Deposition and characterization of PbTiO$_3$ thin films on silicon wafers using metalorganic sources, *J. Electron. Mat. 22*:7 (1993).

48. C. Dias, M. Simon, R. Quad, and D. K. Das-Gupta, Measurement of the pyroelectric coefficient in composite using a temperature-modulated excitation, *J. Phys. D. Appl. Phys. 26*:106 (1993).

49. B. D. Qu, W. L. Zhong, K. M. Wang, and W. Z. Li, Ion-beam-assisted of ferroelectric PbTiO$_3$ films, *J. Appl. Phys. 74*:4 (1993).

50. M. Miyashita, K. Takano, and T. Toda, Preparation and properties of PZT ceramics with ladder type structure, *Ferroelectrics 28*:397 (1980).

51. T. Yamada, T. Ueda, and T. Kitayama, Piezoelectricity of a high-content lead zirconate titanate/polymer composite, *J. Appl. Phys. 53*:4328 (1982).

52. T. R. Gururaja, Piezoelectric composite materials for ultrasonic transducer applications, Ph.D. thesis, Pennsylvania State University, University Park, PA, 1983.

53. T. Furukawa, K. Suzuki, and M. Date, Switching process in composite systems of PZT ceramics and polymer, *Ferroelectrics 68*:33 (1986).

54. D. A. Hutchins, H. D. Mair, P. A. Puhach, and A. J. Osei, Continuous-wave pressure fields of ultrasonic transducers, *J. Accoust. Soc. Am. 80*:1 (1986).

55. Y. Higashihata, T. Yagi, and J. Sako, Piezoelectric properties and applications in the composite system of vinylidene fluoride and trifluoroethylene copolymer and PZT ceramics, *Ferroelectrics 68*:63 (1986).

56. K. Kikuchi, Composite piezoelectric material, Japanese Patent, Kokai Tokyo Koho JP G1, 141, 188 (June 28, 1986).

57. D. Singha, C. Muralidhar, and P. K. C. Pillai, Dielectric behaviour in lead circonate titanate (PZT) polyvinylidene fluoride (PVDF) composite, *Proc. ICSD '86*, Erlangen, Germany, 1986, pp. 227–231.

58. H. Takeuchi and C. Nakaya, PZT/polymer composites for medical ultrasonic probes, *Ferroelectrics 68*:53 (1986).

59. D. Singha and P. K.C. Pillai, Hysteresis behaviour in lead zirconate titanate/polyvinylidene fluoride composites, *Ferroelectrics 76*:459 (1987).

60. Yu. I. Goltzov and V. E. Yurkevich, The preparation and properties of ferroelectric-composites possessing glass-like matrix, *Ferroelectrics 129*:67 (1992).

61. S. Sherit, H. D. Wiederick, B. K. Mukherjee, and S. E. Prasad, Stress isolation PZT-air composites, *Ferroelectrics 132*:61 (1992).

62. K. Lubitz, A. Wolff, G. Preu, R. Stoll, and B. Schulmeyer, New piezoelectric composites for ultrasonic transducters, *Ferroelectrics 133*:21 (1992).

63. S. Sherrit, H. D. Wiederick, B. K. Mukherjee, and S. E. Prasad, 0-3 piezoelectric-glass composites, *Ferroelectrics 134*:65 (1992).

64. C. E. Murphy, T. Richardson, and G. G. Roberts, Thin-film pyroelectric inorganic/organic composites, *Ferroelectrics 134*:189 (1992).

65. W. B. Harrison and S. T. Liu, Pyroelectric properties of flexible PZT composites, *Ferroelectrics 27*:125 (1980).

66. A. M. Varaprasad and K. Uchino, Dielectric relaxation studies in some polymer-PZT composites, *Ferroelectrics Lett. 7*:55 (1987).

67. B. Hilczer, J. Kuiłek, and J. Wolak, Dielectric behavior of 0-3 type piezoelectric composites, *Proc. 7*, Berlin, 1991, pp. 407–414.

68. I. R. Giniewicz, (Pb, Bi) (Ti, Fe) O₃ polymer 0-3 composite materials for Hydrophone Applications, Pennsylvania State University, University Park, PA 1985.

69. I. M. Tawlar, Ya. Yagik, N. Lal, and K. K. Nagpaul, Field-induced thermally stimulated relaxation of polyvinylchloride, *J. Mat. Sci. Lett. 5*:1084 (1986).

70. H. Takeushi, S. Jyomura, C. Nakaya, New piezoelectric materials for ultrasonic transducers, *Jpn. J. Appl. Phys. Suppl. 24-2*:36 (1985).

71. L. Pardo and J. Mendiola, Piezoelectric properties of modified lead calcium titanate and epoxy resin composites, *Ferroelectrics 81*:397 (1988).

72. Y. H. Lee, M. J. Haun, A. Safari, and R. N. Newnham, Preparation of PbTiO₃ powder for flexible 0-3 piezoelectric composite, *IEEE Trans. Elect. Insul. EL-21*:318 (1986).

73. J. Paletto, G. Grange, R. Goutte, and L. Eyraud, A study of the dielectric properties of powder BaTiO₃, *J. Phys. D: Appl. Phys. 7*:78 (1974).

74. J. Del Cerro, M. Mundi, C. Gallardo, J. M. Criado, F. J. Gotor, and A. Bhalla, Sintering temperature influence on phase stability in barium titanate ceramics with very small grain size, *Ferroelectrics 127*:59 (1992).

75. M. J. Abdullah, D. K. Das-Gupta, Dielectric and pyroelectric properties of polymer-ceramic composites, *Ferroelectrics 76*:393 (1987).

76. G. A. Lushcheikin, *Polymeric Electrets*, Izd. Khimija, 1976 (in Russian).

77. G. A. Lushcheikin, Polymer and composite piezoelectrics, *Bull. Acad. Sci. USSR Ser. Phys. 51*:2273 (1987).

78. J. Glatz-Reichenbach, F. Epple, and K. Dransfeld, The ferroelectric switching time in thin VDR-TrFE copolymers films, *Ferroelectrics 127*:13 (1992).

79. M. Schenk, S. Bauer, T. Lessle, and B. Ploss, Dielectric spectroscopy on P(VDF-TrFE), *Ferroelectrics 127*:215 (1992).

80. K. Ogura, M. Ogawa, K. Ohya, and H. Banno, Receiving characteristics of d_{31}-zero piezo-rubber hydrophone, *Jpn. J. Appl. Phys. 32*:2304 (1993).

81. W. P. Krug, Orientation on piezoelectric tartaric acid composites, *Jpn. J. Appl. Phys. Suppl.* *24-2*:871 (1985).

82. K. Mazur, Electret and piezoelectric properties in the laminated systems of PTFE and PUE polymer films, *Proc. ICSD '92*, Sestri Levante, Italy, 1992, pp. 280–284.

83. Q. M. Zhang, Wenwu Cao, H. Wang, and L. E. Cross, Characterisation of the performance of 1-3 type piezocomposites for low-frequency applications, *J. Appl. Phys.* *73*:1403 (1993).

84. Q. M. Zhang, W. Cao, H. Wang, and L. E. Cross, Strain profile and piezoelectric performance of piezocomposites with 2-2 and 1-3 connectivities, Material Research Laboratory, The Pennsylvania State University, University Park, PA.

85. W. B. Harrison, Flexibility of piezoelectric organic composite, Proceedings of Workshop on Sonar Transducer Materials, Naval Res. Lab., Washington, DC, 1977.

86. A. Safari, Perforated PZT-polymer composites with 3-1 and 3-2 connectivity for hydrophone applications, Ph.D. thesis, Pennsylvania State University, University Park, PA, 1983.

87. K. A. Klicker, J. V. Bigger, and R. E. Newnham, Composite of PZT and epoxy for hydrostatic transducers applications, *J. Am. Ceramics Soc.* *64*:5 (1981).

88. E. C. Galgoci and J. Runt, Interfacial Adhesion in PZT-epoxy composites, *Ferroelectrics* *70*:205 (1986).

89. H. P. Savakus, K. A. Klicker, and R. E. Newnham, PZT-epoxy piezoelectric transducers: Simplified fabrication procedure, *Mat. Res. Bull.* *16*:677 (1981).

90. A. Halliyal, A. Safari, A. S. Bhalla, R. E. Newnham, and L. E. Cross, Grain-oriented glass-ceramics for piezoelectric devices, *J. Am. Ceramics Soc.* *67*:331 (1984).

91. R. Y. Tyng, A. Halliyal, and A. S. Bhalla, New materials for hydrophone applications: Single crystals and polar glass ceramics, *Jpn. J. Appl. Phys. Suppl.* *24-2*:982 (1985).

92. K. Rittenmayer, T. Shrout, W. A. Schulce, and R. E. Newnham, Piezoelectric 3-3 composites, *Ferroelectrics* *41*:189 (1982).

93. K. Hikita, K. M. Nishioka, and M. Ono, Piezoelectric properties of porous PZT and composite with silicon rubber, *Ferroelectrics* *49*:265 (1983).

94. T. R. Shrout, L. Y. Bowen, and W. A. Schulze, Extruded PZT-polymer composites for electromechanical transducer applications, *Mat. Res. Bull.* *15*:1371 (1980).

95. R. Sockel and R. C. Hughes, Numerical analysis of transient photoconductivity in insulators, *J. Appl. Phys.* *53*:7414 (1982).

96. B. Tareev, *Physics of Dielectric Materials*, Mir Publishers, Moscow, 1979, pp. 116–125.

97. A. von Hippel, *Dielectric and Waves*, Part II, Sec. 31 (1959).

98. H. Banno, Theoretical equations for dielectric, piezoelectric, and elastic properties of flexible composites consisting of polymer and ceramic powder of two different materials, *Ferroelectrics* *95*:111 (1989).

99. J. C. Maxwell, *Electricity and Magnetism*, Clarendon Press, Oxford, 1892, p. 452.

100. K. W. Wagner, *Die Isolierstoffe der Elektrotechnik*, Vol. 1 (H. Schering, ed.), Springer, Berlin, 1924.

100a. L. K. H. van Beek, Dielectric behaviour of heterogeneous systems, *Progress in Dielectrics*, *Vol. 7* (J. B. Birks, ed.), Heywood Books, London, 1967, p. 69.

100b. C. F. J. Böttcher and P. Bordewijk, Heterogeneous mixtures, *Theory of Electric Polarization*, Vol. II (C. F. J. Böttchers and P. Bordewijk, eds.), Elsevier, Amsterdam, 1978, p. 476.

101. R. W. Sillars, The properties of a dielectric containing semiconducting particles of various shapes, *J. Inst. Elect. Engs.* *80*:378 (1937).

102. R. G. Barrera, G. Monsivais, and W. L. Mochán, Renormalized polarizability in the Maxwell Garnet theory, *Phys. Rev. B* *38*:5371 (1988).

103. P. Sheng and Z. Chen, Local-field in random dielectric media, *Phys. Rev. Lett.* *60*:227 (1988).

104. R. G. Barrera, P. Villaseñor-González, W. L. Mochán, M. del Castillo-Mussot, and G. Monsivais, Effect of the dispersion of sizes in the dielectric response of composites, *Phys. Rev. B* *39*:3522 (1989).

105. F. Claro and F. Brouers, Dielectric anomaly: The role of multipolar interactions, *Phys. Rev. B 40*:3261 (1989).

106. W. F. Brown (ed.), *Dielectrics*, Section 53, Springer-Verlag, Berlin, 1956.

107. D. A. G. Bruggeman, Berechnung verschiedener physikalischer Konstanten von heterogenenen Substanzen. I. Dielektrizitatkonstanten und Leitfahiggheiten der Mischkorper aus isotropen Substanzen, *Ann. Phys. (Leipzig) 24*:636–679 (1935).

108. J. L. Ericksen, D. Kinderlehrer, R. Kohn, and J.-L. Lions, *Homogenization and Effective Moduli of Materials and Media*, Springer-Verlag, New York, 1986.

109. A. W. Shubnikov, I. S Zheludev, and W. P. Konstantinova, *Investigation of Piezoelectric Textures*, Moscow, 1955.

110. M. Olszowy, Electret and piezoelectric properties of PVDF/PZT composites, *Z.N. WSI Zielona Gora, 94*:39 (1990).

111. K. Mazur, a) Dielectric, electret, piezo-, and polyelectric properties of BaTio$_3$/PMMA and PZT/PMMA composites (Parts I and II), Z. N. Nr 106, Zielona Gora, Poland, *Fiz.-Chem. 6-7, 5* (1994). b) Piezo- and pyroelectricity of BaTiO$_3$/PMMA and PZT/PMMA composites, *Proc. ISE 8*, Paris, 1994, pp. 754–759.

112. H. Banno, Theoretical equations for dielectric and piezoelectric properties of ferroelectric composites based on modified cubes model, *Jpn. J. Appl. Phys. Suppl. 24-2*:445 (1985).

113. D. K. Das Gupta and M. J. Abdullah, Dielectric and pyroelectric properties of polymer/ceramic composites, *J. Mat. Sci. Lett. 7*:167 (1988).

114. D. Sinha and P. K. C. Pillai, The conductivity behavior in lead zirconate titanate polyvinylidene fluoride composites, *Appl. Phys. 64(5)*:2571 (1988).

115. B. Gross, On permanent charges in solid dielectrics. II. Surface charges and transient currents in carnauba wax, *J. Chem. Phys. 17*:866 (1949).

116. M. Eguchi, Further researches on permanently polarized dielectric, *Proc. Phys.-Math. Soc. Jpn. 2*:169 (1920).

117. B. Gross, *Charge Storage in Solid Dielectrics, A Bibliographical Review on the Electret and Related Effects*, Elsevier, Amsterdam, 1964.

118. G. M. Sessler (ed.), *Electrets*, Springer-Verlag, Berlin, 1987.

119. B. Hilczer and J. Maiłecki, (a) *Elektrety*, PWN, Warsaw, 1980 (in Polish); (b) *Electrets*, PWN, Warsaw, Elsevier, Amsterdam, 1986.

120. B. Hilczer, B. Biłaszczyk, and S. Goderska, Investigation of the properties of carnauba wax and polymethyl methacrylate electrets, *Fizyka Diel. i Radiosp. 4*:79 (1968) (in Polish).

121. A. N. Gubkin and G. I. Skanavi, Preparation and properties of new electrets obtained from inorganic dielectrics, *Izv. AN SSSR, Ser. Fiz. 22*:330 (1958).

122. T. Furukawa, Piezoelectricity and pyroelectricity in polymers, *Proc. ISE 6*, Oxford, 1988, pp. 182–193.

123. J. van Turnhout, *Thermally Stimulated Discharge of Polymer Electrets*, Elsevier, Amsterdam, 1975.

124. K. Mazur, The influence of water concentration on the dielectric relaxation processes of PMMA, *Int. Symp. Polymer Materials*, Donostia-San Sebastian, 1987, Session no. 7.

125. K. Mazur, Effect of absorbed water on some properties of PMMA, *ZN WSI Zielona Góra 94*:69 (1992).

126. K. Mazur, PMMA as a component of composite and multilayered electrets, *ZN WSI Zielona Góra 86*:25 (1988).

127. K. Mazur, On the relation between the effective charge of carnauba wax electrets and the polarization temperature, *J. Electrostatics 3*:327 (1977).

128. P. A. Thiessen, A. Winkel, and K. Herrmann, Elektrische Nachwirkungen im erstarren Dielektrikum, *Physik. Z. 37*:511 (1936).

129. B. Gross, Experiments on electrets, *Phys. Rev. 66*:26 (1944).

130. K. Mazur, Measurement of the potential distribution in the electrets by probes together with TSD method, *Proc. ISH-87*, Braunschweig, 1987, 23.14.

131. K. Mazur, Sub-ohmic current-voltage characteristic of PMMA at temperatures below the glass transition temperature, *Proc. ICSD '86*, Erlangen, 1986, pp. 401–405.

132. K. Mazur, Current-voltage characteristic and the relation between surface charge density of PMMA thermoelectrets and polarizing voltage, *Proc. ISE 5*, Heidelberg, 1985, pp. 271–276.

133. W. F. G. Swann, Fundamentals in the behavior of electrets, *J. Franklin Inst. 255*:513 (1953).

134. A. N. Gubkin, The phenomenologic theory of electrets, *Zh. tekh. Fiz.*, 27:1954 (1957).

135. K. Mazur, On a certain non-linear question in a electret phenomenological theory, *Acta Phys. Polon. A46*:497 (1974).

136. K. Mazur, On homocharge in the electrets, *ZN WSI Zielona Góra 22*:115 (1974) (in Polish).

137. K. R. Brain, Investigation of piezoelectric effect with dielectrics, *Proc. Phys. Soc. (Lond.) 36*:81 (1924).

138. I. S. Rez, High-molecular piezoelectrics, *Kristallografija 6* (1961) (in Russian).

139. A. N. Gubkin and W. S. Sorokin, Piezoeffect in electrets, *Izv. AN SSSR, Ser. Fiz. 24*:246 (1960).

140. Y. Wada, Ferro-, piezo- and pyroelectricity, *Proc. ISE 5*, Heidelberg, 1985, pp. 851–856.

141. J. F. Nye, *Physical Properties of Crystals: Their Representation by Tensors and Matrices*, Clarendon Press, Oxford, 1957, p. 110.

142. International Electrotechnical Commission, *Guide to Dynamic Measurements of Piezoelectric Ceramics*, Publication 483, 1976.

143. D. P. Skinner, R. E. Newnham, and L. E. Cross, Flexible composite transducers, *Mat. Res. Bull. 13*:599 (1978).

143a. G. M. Broadhurst, C. C. Malmberg, F. I. Mopsik, and W. P. Haris, Piezo and pyroelectricity in polymer electrets, *Electrets, Charge Storage and Charge Transport in Dielectrics* (M. M. Perlman, ed.), The Electrochemical Society, Princeton, NJ, 1973, p. 492.

143b. F. I. Mopsik and M. G. Broadhurst, Molecular dipole electrets, *J. Appl. Phys. 46*:4204 (1975).

144. S. Sa-Gong, A. Safari, S. J. Jang, and R. E. Newnham, Poling flexible piezoelectric composites, *Ferroelectrics Lett. 5*:131 (1986).

145. M. G. Shakhtakhtinsky, M. A. Kurbanov, B. A. Guseinov, Ju. N. Gazaryan, M. M. Kuliev, and A. Q. Guliev, Pyroelectric and piezoelectric effect in polymer composites, *Izv. AN Azerb. SSR, Ser. Fiz.—Mech. Mat.*, 66–71 (1985).

146. R. K. Sadhir and H. E. Saunders, Plasma processes for electrical and electronics applications, *IEEE Elect. Insul. Mag. 2(6)*:8 (1986).

147. B. A. Auld, Y. A. Shui, and Y. Wang, Elastic wave propagation in three-dimensional periodic composite materials, *J. de Phys. 45*, Suppl. 4:C5–159 (1984).

148. B. A. Auld and Y. Wang, Acoustic wave vibrations in periodic composite plates, *Proc. IEEE Ultrasonic Symp.*, Dallas, TX, 1984.

149. W. A. Smith, A. Shaulov, and B. A. Auld, Tailoring the properties of composite piezoelectric materials for medical ultrasonic transducers, *Proc. IEEE Ultrasonics Symp.*, 1985, p. 642.

150. R. W. Greavers, E. P. Fowler, A. Goodings, and D. R. Lambs, The direct piezoelectric effect in extruded polyethylene, *J. Mat. Sci. 9*:1602 (1974).

151. K. Mazur, Polyester-urethane elastomer as a component of piezoelectric laminates, *Proc. IRC90*, Paris, 1990, pp. 357–358.

152. M. Okuyama and Y. Hamakawa, PbTiO₃ ferroelectric thin films and their pyroelectric application, *Ferroelectrics 118*:261 (1991).

153. T. Kamitsubara, M. Komabayashi, and K. Murase, Pyroelectric element coated with insulator and manufacture thereof (Patent), Jpn. Kokai Tokkyo Koho JP 04063481 (Issued 28 Feb. 1992).

154. R. L. Byer and C. B. Roundy, Detector nsec response-time-pyroelectric coefficient, *Ferroelectrics 3*:333 (1972).

155. H. Burkard and G. Pfister, Reversible pyroelectricity and inverse piezoelectricity in poly(vinylidene fluoride), *J. Appl. Phys.* *45*:3360 (1974).

156. J. H. McFee, J. G. Bergman, and G. R. Crane, Pyroelectric and non-linear optical properties of poled poly(vinylidene fluoride) films, *Ferroelectrics 3*:305 (1972).

157. N. P. Hartley, P. T. Squire, and E. H. Putley, A new method of measuring pyroelectric coefficients, *J. Phys. E: Sci. Instrum. 5*:787 (1972).

158. a) L. E. Garn and E. J. Sharp, Use of low-frequency sinusoidal temperature waves to separate pyroelectric currents from nonpyroelectric currents. Part I: Theory, *J. Appl. Phys. 53*: 8974 (1982). b) E. J. Sharp and L. E. Garn, Use of low-frequency sinusoidal temperature waves to separate pyroelectric currents from nonpyroelectric currents. Part II: Experiment, *J. Appl. Phys. 53*:8980 (1982).

159. N. M. Shorrocks, M. E. Brown, R. W. Whaitmore, and F. M. Ainger, Piezoelectric composites for underwater transducers, *Ferroelectrics 54*:215 (1984).

160. K. Mazur, TSD current of polarized $BaTiO_3$, *ZN WSI Zielona Gora 31*:55 (1974) (in Polish).

161. K. Mazur, Preparation of electrets, *ZN WSI Zielona Gora 12*:99 (1973) (in Polish).

162. T. Takamatsu, R. Wen Tian, and H. Sasabe, Bending piezoelectricity of polymer electrets, *Proc. I SE 5*, Heidelberg, 1985, pp. 942–946.

163. K. Mazur, Electret and piezoelectric property of multiphase dielectrics, *ZN WSI Zielona Góra 90*:5 (1989).

164. D. L. Monroe, J. B. Blum, and A. Safari, Sol-gel derived $PbTiO_3$ polymer piezoelectric composites, *Ferroelectrics Lett. 5*:39 (1986).

165. D. P. Skinner, R. E. Newnham, and L. E. Cross, *Mat. Res. Bull. 13*:599 (1978) (after Ref. 4).

12
Nonlinear Optical Properties of Ferroelectric Polymers

Toshiyuki Watanabe and Seizo Miyata
Tokyo University of Agriculture and Technology, Tokyo, Japan

Hari Singh Nalwa
Hitachi Research Laboratory, Hitachi, Ltd., Hitachi City, Ibaraki, Japan

I. INTRODUCTION

The field of nonlinear optics became of considerable interest in 1990s with the development of promising organic molecular and polymeric materials [1–4]. Nonlinear optical materials are expected to have a wide range of applications in photonic technology, including harmonic generators, optical computing, telecommunications, laser lithography, image processing, switching, sensors, and overall photonic transmission. Organic materials that have been investigated for nonlinear optics can be summarized into several different categories such as single crystals, guest-host systems, Langmuir-Blodgett (LB) films, polymers such as NLO-dye grafted polymers and polar polymers, self-assembled systems, and liquid crystalline materials [5–8]. Recently a great interest in nonlinear optical organometallic compounds has been seen because the metal-to-ligand bonding can be utilized in optimizing photonic functions [6,8]. In particular, organic polymers seem more promising due to their low costs of fabrication and greater compatibility with desired substrates as well as tremendous possibilities for tailoring structures with high mechanical strength and environmental stability. An excellent review dealing with the second-order nonlinear optical polymeric materials has been recently published by Burland et al. [9]. A wide variety of polar polymers or more specifically piezoelectric polymers [10–12] is also available nowadays that can be used for second harmonic generation and electrooptic effects. Polar polymers are important elements in the field of nonlinear optics, as they show temporal stability of second-order NLO effects. Second-order nonlinear optical properties of various ferroelectric polymers such as poly(vinylidene fluoride) (PVDF) and its copolymers, vinylidene cyanide copolymers, polyureas, and ferroelectric liquid crystalline polymers will be discussed here.

II. NONLINEAR OPTICAL EFFECTS

When the electromagnetic field of a laser beam is illuminated on an atom or a molecule, the induced electrical polarization $P_I(\omega_1)$ in a microscopic medium is expressed by the following equation [1–8]:

$$
\begin{aligned}
P_I(\omega_1) = &\sum_j \alpha_{ij}(-\omega_1;\ \omega_2)E_j(\omega_2) \\
&+ \sum_{jk} \beta_{ijk}(-\omega_1;\ \omega_2,\ \omega_3)E_j(\omega_2)E_k(\omega_3) \\
&+ \sum_{jkl} \gamma_{ijkl}(-\omega_1;\ \omega_2,\ \omega_3,\ \omega_4)E_j(\omega_2)E_k(\omega_3)E_l(\omega_4) + \ldots
\end{aligned}
\tag{1}
$$

Where ω_1 is the laser frequency along the ith molecular axis; α, β, and γ are the linear polarizability, first hyperpolarizability, and second hyperpolarizability, respectively; and E_j refers to the electric field component along the jth direction. The macroscopic polarization in the bulk media induced under high electromagnetic fields can be expressed in a power series as

$$
\begin{aligned}
P_I(\omega_1) = &\sum_j \chi_{IJ}^{(1)}(-\omega_1;\ \omega_2)E_J(\omega_2) \\
&+ \sum_{JK} \chi_{IJK}^{(2)}(-\omega_1;\ \omega_2,\ \omega_3)E_J(\omega_2)E_K(\omega_3) \\
&+ \sum_{JKL} \chi_{IJKL}^{(3)}(-\omega_1;\ \omega_2,\ \omega_3,\ \omega_4)E_J(\omega_2)E_K(\omega_3)E_L(\omega_4) + \ldots
\end{aligned}
\tag{2}
$$

Here $\chi^{(1)}$, is the linear susceptibility, while $\chi^{(2)}$ and $\chi^{(3)}$ are the second- and third-order nonlinear optical susceptibilities, respectively. The $\chi^{(1)}$ is related directly to the dielectric constant and refractive index of a material. Based on the symmetry, the $\chi^{(2)}$ effects appear only in a noncentrosymmetric medium where $\chi^{(3)}$ effects have no such restrictions. Ferroelectric polymers have advantages compared with other organic materials as they are noncentrosymmetric materials with large spontaneous polarization.

The macroscopic $\chi^{(2)}$ is related to the microscopic nonlinearity β by the following expression [13]:

$$
\chi_{IJK}^{(2)}(-\omega_1;\ \omega_2,\ \omega_3) = Nf_I(\omega_1)f_J(\omega_2)f_K(\omega_3)\langle\beta_{ijk}(-\omega_1;\ \omega_2,\ \omega_3)\rangle_{IJK}
\tag{3}
$$

here β_{ijk} is an average overall orientation of chromophores, N is the number density of chromophores, and the Lorentz-Lorenz local field factor f can be given by

$$
f(\omega) = \frac{\epsilon_\omega + 2}{3}
\tag{4}
$$

where ϵ is the dielectric constant. A general trend of $\chi^{(2)}$ can be estimated once the β values are known. For SHG where $\omega_2 = \omega_3 \equiv \omega$, the nonlinear optical coefficient d_{ijk} is related to second-order NLO susceptibility $\chi^{(2)}$ as follows:

$$
\chi_{ijk}^{(2)}(-2\omega:\omega,\omega) = 2d_{ijk}(-2\omega:\omega,\omega)
\tag{5}
$$

The d_{ijk} coefficient is also related to the electrooptic coefficient r_{ijk} as follows using the two-level model [9]:

$$r_{IJK}(-\omega; \omega, 0) = -\frac{4d_{KIJ}}{n_I^2(\omega)n_J^2(\omega)} \frac{f_{II}^{\omega}f_{JJ}^{\omega}f_{KK}^0}{f_{KK}^{2\omega'}f_{II}^{\omega'}f_{JJ}^{\omega'}}$$

$$\times \frac{(3\omega_0^2 - \omega^2)(\omega_0^2 - \omega'^2)(\omega_0^2 - 4\omega'^2)}{3\omega_0^2(\omega_0^2 - \omega^2)} \quad (6)$$

Here ω_0 is the frequency of the first strongly absorbing electronic transition in the molecule, and ω' and ω are the fundamental wavelengths in second harmonic generation and for electrooptic coefficient measurements, respectively. The electrooptic effect (Pockels effect) is related to the corresponding second-order NLO susceptibility and by knowing the SHG coefficients, one can also estimate the electrooptic coefficients.

III. MEASUREMENT TECHNIQUE FOR SECOND HARMONIC GENERATION

Maker-fringe second harmonic generation technique is the most common measurement method for evaluating NLO coefficients of poled polymer systems including ferroelectric polymers. The second harmonic power $P(2\omega)$ generated by a single-mode Gaussian beam of power $P(\omega)$ incident on a plane-parallel slab of a NLO crystal is given by [14]

$$P(2\omega) = \frac{128\pi^2\omega^2L^2P(\omega)^2d_{ijk}^2}{c^3w_0^2n(\omega)^2n(2\omega)} \frac{\sin^2(L\,\Delta k/2)}{(L\,\Delta k/2)^2} \quad \text{(in cgs units)} \quad (7)$$

where ω is the angular frequency of the fundamental wave, w_0 is the spot radius of the fundamental beam, d_{ijk} is the nonlinear optical (NLO) coefficient, $n(\omega)$ and $n(2\omega)$ are the refractive indices of the material at the fundamental and the SH wavelengths, respectively, c is the velocity of light in vacuum, $\Delta k = 2k(\omega) - k(2\omega)$ is the phase mismatch between wave vectors at the harmonic and fundamental wavelengths, and L is the optical path length of material. The above expression indicates that SH power undergoes periodic oscillation as a function of thickness and is called Maker fringe when Δk is not zero. This period of oscillation is called as the coherence length (l_c) and is associated with NLO coefficients. The coherence length for normal incidence can be written as

$$l_c = \frac{\lambda_1}{4[n(2\omega) - n(\omega)]} \quad (8)$$

Here λ_1 is the free-space wavelength of the fundamental wave and $n(\omega)$ and $n(2\omega)$ depend on the polarization direction of the fundamental and second harmonic waves, respectively.

Figure 1 shows the experimental setup for Maker fringe. The laser beam reflected from a glass surface is incident on an attenuator consisting of a $\lambda/2$ wave plate. In order to increase the second harmonic power, the laser beam is focused by the lenses F1 and F3. Two Glan-Thomson polarizes or half-wave plates are used to polarize the beam in such a way that the electric field is horizontal or vertical. The rotation speed must be considered, since a thick sample or short-coherence material drastically changes the second harmonic power as a function of rotation angle. Both reference and signal are detected using a boxcar integrator and recorded. The components of the second-order po-

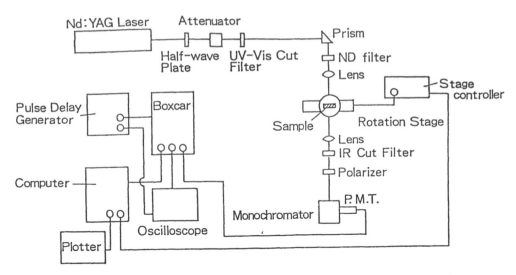

Figure 1 Schematic representation of experimental setup for second harmonic generation.

larization at the second harmonic frequency (2ω) can be written in matrix form as follows:

$$\begin{pmatrix} P_x \\ P_y \\ P_z \end{pmatrix} = \epsilon_0 \begin{pmatrix} d_{11} & d_{12} & d_{13} & d_{14} & d_{15} & d_{16} \\ d_{21} & d_{22} & d_{23} & d_{24} & d_{25} & d_{26} \\ d_{31} & d_{32} & d_{33} & d_{34} & d_{35} & d_{36} \end{pmatrix} \begin{pmatrix} E_x^2 \\ E_y^2 \\ E_z^2 \\ 2E_yE_z \\ 2E_xE_z \\ 2E_xE_y \end{pmatrix} \tag{9}$$

If there is no absorption at fundamental (ω) and second harmonic wavelengths, the following NLO coefficients are obtained according to Kleinman's symmetry:

$$d_{12} = d_{26} \qquad d_{13} = d_{35} \qquad d_{14} = d_{25} = d_{36} \tag{10}$$
$$d_{15} = d_{31} \qquad d_{16} = d_{21} \qquad d_{23} = d_{24} \qquad d_{24} = d_{32}$$

Because the NLO coefficient (d) is a polar tensor of odd rank, its elements are identically zero in a centrosymmetric medium. The number of nonvanishing elements of a third-rank tensor are obtained from the point-group symmetry of a crystal, hence they are reduced to only a few independent elements but have the same form in a given point-group symmetry. Since this chapter deals with SHG activity in poled polymers which have an $mm2$ point-group symmetry, the NLO coefficients of this class are d_{15}, d_{24}, d_{31}, d_{32}, and d_{33}. Far from the resonance the Kleinman symmetry yields $d_{15} = d_{31}$ and $d_{31} = d_{33}$ are approximately equal to $\frac{1}{3}$ in most cases, though in some cases deviation from this ratio has been observed. The NLO coefficients of poled polymer systems are determined by comparing the second harmonic intensity of the polymer with a reference material, often quartz crystal, which has a $d_{11} = 0.40$ pm/V at 1.064 μm.

IV. FERROELECTRIC POLYMER SYSTEMS

A. Poly(vinylidene Fluoride)

Poly(vinylidene fluoride) (PVDF) is a highly crystalline polymer and shows strong pyroelectric, piezoelectric, and ferroelectric behaviors. Bergman et al. [15,16] reported second harmonic generation in PVDF thin films for the first time. The strong piezoelectric effect in PVDF arises from permanently oriented dipoles induced in the film by stretching and poling at elevated temperatures, which also indicates that PVDF has a nocentrosymmetric structure. Because PVDF lacks a center of symmetry, it exhibits both piezoelectricity and second harmonic generation (SHG). PVDF shows birefringence. For PVDF films, the $(n_y - n_z)$ were 0.018 and 0.019 for biaxial stretched 0.4- and 0.8-mm-thick film, respectively, and 0.0 for 2.0- and 5.0-mm-thick films. On the other hand, the $(n_x - n_z)$ were 0.011 for biaxial stretched 0.4- and 0.8-mm-film and 0.013 for 2.0- and 0.0 for 5.0-mm-thick films. Here the refractive index n_z corresponds to the film normal or polar axis, and n_x and n_y correspond to the orthogonal directions in the film plane having the smallest and largest indices, respectively. The refractive index of 1.425 was measured by an Abbe refractometer at 632.8 nm. The SHG polarization in PVDF agrees with $mm2$ symmetry and can be given by following equations [16]:

$$p_x = 2d_{15}E_xE_z \tag{11}$$

$$p_y = 2d_{24}E_yE_z \tag{12}$$

$$p_z = d_{31}E_x^2 + d_{32}E_y^2 + d_{33}E_z^2 \tag{13}$$

There are three independent, second-order, nonlinear coefficients d_{31}, d_{32}, and d_{33}. The NLO coefficients were obtained by comparing with SHG intensity of PVDF films with that of quartz crystal. When the fundamental beam is polarized in the z direction, only d_{33} can be measured. The measured NLO coefficients of PVDF were d_{33}(PVDF) = $2d_{31}$(PVDF) = d_{11}(quartz), and d_{32}(PVDF) = 0.

Some PVDF films showed much higher SHG power for d_{31} coupling than d_{33} coupling because the coherence length l_{31} was larger. The coherence length l_{33} was found to depend on the dispersion of PVDF, while l_{31} depended on both the birefringence and the dispersion of the film. The l_{31} of PVDF film depends on the drawing ratio and varies up to 100 μm.

Broussoux and Micheron [17] reported electrooptic and elastooptic effects in stretched PVDF. The linear EO coefficients in PVDF were measured at room temperature and constant stress at the He-Ne laser wavelength. The refractive indices $n_1 = 1.444$, $n_2 = 1.436$, and $n_3 = 1.425$ and electrooptic coefficients $r_{51} = 0.10$ pm/V, $r_{42} = 0.21$ pm/V were measured. The secondary electrooptic effect estimated for the longitudinal configuration was 2.5% of the measured free electrooptic coefficient. The quadratic electrooptic coefficients at constant strain and stress were $g_{44} = 0.02$ m^4/C^2 and $g_{13} = 0.01$ m^4/C^2.

B. Copolymers of Poly(vinylidene Fluoride)

Like PVDF, copolymers of vinylidene fluoride with trifluoroethylene and tetrafluoroethylene also exhibit strong pyroelectric, piezoelectric, and ferroelectric effects. Robin et al. [18] reported SHG properties of vinylidene fluoride/trifluoroethylene P(VDF/TrFE) copolymer having composition of VDF : TrFE = 65 : 35 mol%. No SHG was noticed for unpoled films, but after electrical poling, a d_{33} coefficient of 0.6 pm/V was determined at room temperature. The d_{33} coefficient was found to decrease with increasing temperature and vanish at the Curie temperature. This also indicates that SHG is related to the remanant polarization. The d_{33} coefficient also depends strongly on the poling field and is proportional to the polarization of the copolymer film. Legrand et al. [19] observed no SHG in the paraelectric phase of a VDF/TrFE copolymer having a composition of 70/30 mol%. The hysteresis loop of SHG intensity recorded at room temperature showed saturation behavior and an identical magnitude of SHG intensities for positive (+) and negative (−) polarizations. The SHG intensity was found to depend on poling field, polarization also changed with electric field, and a ferroelectric hysteresis loop was observed by the SHG measurements. The SHG properties of P(VDF/TrFE) was measured by Sato and Gamo [20]. The P(VDF/TrFE) copolymer has a composition (VDF:TrFE = 78:22 in mol%), Curie temperature = 120°C, thickness = 200 μm. For the poling process, a high voltage of 10 kV/mm was applied along the direction perpendicular to the film surface for 30 min at 100°C, just below the Curie temperature. The P(VDF/TrFE) copolymer exhibits ferroelectric phase without stretching, while the stretching is very important for PVDF. The SHG signal was measured as a function of input beam peak power and showed a quadratic dependence as a function of input power. The P(VDF/TrFE) copolymer behaves much more like a single crystal and belongs to point-symmetry class $mm2$, the same as PVDF. The nonlinear polarization in P(VDF/TrFE) can be described [20], similar to that of equations (11), (12) and (13) used for PVDF. For the unstretched copolymer sample, the NLO coefficients $d_{15} = d_{24}$ and $d_{31} = d_{32}$. Sato and Gamo measured only the SHG signal and coherent length and did not estimate the NLO coefficients. The SHG signal was found to oscillate periodically with a period of about 1.6 mm. The SHG power in the P(VDF/TrFE) copolymer was found to be comparable to or slightly larger than that of PVDF homopolymer. The maximum conversion efficiency was of the order of 10^{-3}. Pantelis et al. [21] reported d_{33} of P(VDF/TrFE) as 1.71 pm/V. Broussoux et al. [22] reported that P(VDF/TrFE) copolymer with composition of 75:25 mol% showed d_{33} of 1 ± 0.5 pm/V. The average electrooptic coefficient $\langle\gamma\rangle$ measured by the surface-plasmon method was 16 pm/V, and the electrooptic part of this coefficient was between 1 and 3 pm/V, similar to that obtained from the d_{33} value. Dumont and Levy [23] measured electrooptic coefficients of P(VDF/TrFE) copolymer using an attenuated total reflection technique. Various measured coefficients were $\chi^{(2)} = 16$ pm/V, $\chi^{(2)}$piezo = 17.3 pm/V, and $\chi^{(2)}$eo = 1.3 pm/V. Berry et al. [24] reported an electrooptic coefficient of 15 pm/V, refractive index = 1.4, and optical loss of 4 dB/cm at 633 nm for P(VDF-TrFE) copolymer. Both SHG and electrooptic coefficients of copolymers depend on the copolymer composition and measurement conditions, leading to some differences in these parameters.

C. Composites and Blends of Ferroelectric Polymers

Hill et al. [25] reported SHG properties of poled PVDF copolymer guest/host composite. The copolymer of VDF and TrFE in the molar ratio of 7:3 was used as the host matrix. The guest NLO chromophore was 4-(4'-cyanophenylazo)NN-bis-(methoxycarbonylmethyl)-aniline (2). The NLO measurements were done on guest-host films kept for 48 h after poling. The host copolymer contained 0–10% by weight of the guest molecule, and the NLO coefficient of the composites was found to be a linear function of the guest concentration. The composite showed some unusual behavior at low concentrations of guest molecule, which caused a change in d_{33} from negative to positive sign; there may have been a small negative contribution to the optical nonlinearity from the host copolymer caused by back-donation of an electron from the fluorine atom to an antibonding s orbital upon electronic excitation of the host polymer chain. A d_{33} coefficient of 2.6 pm/ V was estimated for the guest-host system at 1.064 μm. Figure 2 shows the concentration dependence of the d_{33} coefficient for chromophore (2). The optical loss of P(VDF/TrFE) was measured spectrophotometrically. Waveguide structures composed of the composite showed optical loss of 5 dB/cm at 633 nm. At 1.3- and 1.55-μm wavelengths there were two optical windows, which are important for telecommunications. The bulk losses were significantly lower, of the order of 1.5 dB/cm. The main source of loss in the visible region was from scattering and absorption.

Pentalis et al. [21] used another NLO guest dye for frequency doubling. In this study, P(VDF-TrFE) with guest chromophores was dissolved in acetone up to 15% by weight and thin films onto glass were cast at 30°C under dry nitrogen. The refractive index and dispersive properties of the guest-hosts were obtained by an Abbe refractometer. The effect of thermal annealing on SHG properties of composites containing 2% of chromophore (1) was also investigated. The samples thermally annealed before poling showed 1.4 times SHG efficiency than nonannealed guest-host systems. The substitution of 5% of chromophore (2) by 5% of 4-amino-NN-bis-(methylcarbonylmethyl)-4'-nitrostilbene increased SHG efficiency by an average ratio of 1.3 in fixed 45° angle measurements. The increment in SHG was due to the large product of dipole moment and first hyperpolarizability of the stilbene derivative rather than the aminoazo compound. The d_{33} coefficient for composite having 10% of the aminoazo compound was 2.55 pm/V. More details of second-order nonlinear optical properties of PVDF and its copolymers as hosts can be found in a recently published monograph [26].

(1)   (2)   (3)

Tsutsumi et al. [27,28] reported SHG from oriented β-crystallite dipoles in blends of 75 mol% vinylidene fluoride and 25 mol% trifluoroethylene P(VDF/TrFE) copolymer and poly(methyl methacrylate) (PMMA). The blends of P(VDF/TrFE)/PMMA and P(VDF/TrFE)/PMMA/P(MMA-co-MMA-DR1) with low DR1 contents were optically transparent from the near infrared to ultraviolet regions. The blends of P(VDF/TrFE) and PMMA and DANS were obtained by dissolving in acetone, and the films were melt-processed at 200°C. These blends were annealed at 100°C for 30 min. Table 1 lists the refractive indices and d_{33} coefficients of various blends. Figure 4 shows the temperature dependence of d_{33} coefficient for 90/10 P(VDF-TrFE)/PMMA blend. A gradual decrease of d_{33} for 90/10 P(VDF-TrFE)/PMMA blend between 40 and 60°C occurs due to the molecular relaxation of the amorphous dipoles around its T_g. Figure 4 shows the tem-

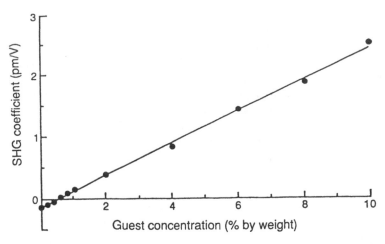

Figure 2 Concentration dependence of d_{33} coefficient for P(VDF/TrFE) copolymer. (After Ref. 25.)

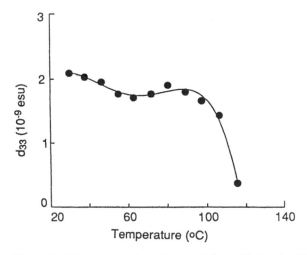

Figure 3 Temperature dependence of d_{33} coefficient for 90/10 P(VDF-TrFE)/PMMA blend. (After Ref. 27.)

Table 1 Refractive Indices and d_{33} Coefficients of Blends

Copolymer blend	Refractive index		d_{33} (pm /V)
	1.064 μm	0.530 μm	
P(VDF/TrFE)	1.420	1.453	1.38
P(VDF/TrFE)/PMMA	1.406	1.427	0.96
P(VDF/TrFE)/PMMA/ P(MMA-co-MMA-DR1)	1.427	1.440	0.38

Source: After Ref. 27.

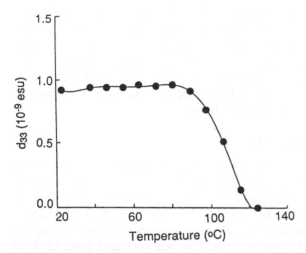

Figure 4 Temperature dependence of d_{33} coefficient for 90/10 P(VDF-TrFE)/P(MMA-co-MMA-DR1) blend. (After Ref. 27.)

perature dependence of d_{33} coefficient for 90/10 P(VDF-TrFE)/P(MMA-co-MMA-DR1) blend. The d_{33} of P(VDF/TrFE)/PMMA/P(MMA-co-MMA-DR1) was thermally stable at T_g. The NLO coefficients of poled polymers are not stable after poling and show orientational relaxation with time. The ferroelectric phase in P(VDF/TrFE) provides the possibility to prevent chromophore reorientation by internal electric field. When the blend films are poled at room temperature, both the guest molecules and the carbon–fluorine bonds of the polymer partially align in the field. The carbon–fluorine groups on the

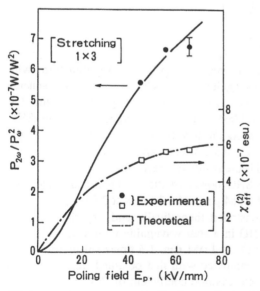

Figure 5 SHG power and effective nonlinear optical susceptibility $c(2)$ as a function of poling field in VDCN/VAc copolymer. (After Ref. 29.)

copolymer chains which have crystalline regions retain their alignment after the field is removed. This causes a net dipole ordering of the crystallites and imparts piezoelectric properties to the whole film which are stable over a period of years after an initial decay in the first day. This remanent internal fields of ferroelectric copolymers cause the guest molecules to retain their dipolar alignment and orientational noncentrosymmetry. These studies demonstrate that guest-host systems comprised of ferroelectric host matrix can play an important role in stabilizing the SHG efficiency.

D. Vinylidene Cyanide Copolymers

Cyanopolymers also show very interesting piezoelectric, pyroelectric, and ferroelectric properties attributed to the large $C=N$ dipole moments in the molecules. Azumai et al. [29] reported frequency doubling in vinylidene cyanide/vinyl acetate P(VDCN/VAc) copolymer thin films. Poling of thin films was done at 182°C for 2 h under electric fields. After cooling the sample to room temperature, the stretching process was also applied parallel to the film surface. Second harmonic generation was evaluated using Er:YAG 2.94-μm laser radiation. The SHG signal of the P(VDCN/VAc) varied periodically on the optical path length and was independent of the strength of the applied poling field. Figure 5 shows the SHG power and effective second-order NLO susceptibility as a function of the poling field. The SHG intensity increases remarkably with increasing poling field. The SHG power was enhanced by 20% with a threefold stretching. The SHG power of copolymer was significantly enhanced by both poling and stretching. The refractive indices of copolymer were $n_\omega = 1.434$ and $n_{2\omega} = 1.446$. The P(VDCN/VAc) copolymer exhibited an effective $\chi^{(2)}$ of 234 pm/V, which is larger by a factor of 3.9 compared to lithium niobate crystal. The conversion efficiency of copolymer was 10^{-4}, relatively smaller, due to the incomplete phase matching.

Azumai et al. [30] reported SHG enhancement in P(VDCN/VAc) thin films with both symmetrical and asymmetrical slab waveguides in the Cerenkov radiation scheme at the fundamental wavelengths of 1.064 and 2.94 μm. Effective $\chi^{(2)}$ of 41.4 and 234 pm/V were measured for the copolymer at 1.064 and 2.94 μm, respectively. The SHG power was found to be enhanced by decreasing the film thickness (d); for example, the SHG power increased by a factor of 10 at $d = 0.5$ μm compared with $d = 5$ μm, which was fivefold larger than the bulk scheme. Theoretically, the SHG power will be increased by a factor of 50 with $d = 0.5$ μm. The SHG power of the asymmetrical guide with 5-μm thickness was enhanced five times at 1.064 μm and 1.3 times at 2.94 μm compared to the bulk. Azumai and Sato [31] demonstrated theoretically and experimentally an improvement of the Cerenkov radiative SHG in a slab waveguide with a periodic nonlinear optical susceptibility in P(VDCN/VAc). At 1.064 μm, 1.5 times enhancement was obtained from Cerenkov radiative scheme compared with that obtained from a uniform scheme. Dumonat et al. [32] reported the electrooptic coefficient of P(VDCN/VAc) copolymer as 1.2 pm/V. Broussoux et al. [33] reported the d_{33} coefficient of 0.4 pm/V at 1.064 μm after poling 160°C.

Kishimoto et al. [34] reported SHG properties of vinylidene cyanide (VDCN) and vinyl acetate (VAc) copolymers with and without nonlinear optical chromophore. The VDCN/VAc backbone was selected because it offers excellent optical transparency, high glass transition temperature to help stabilize chromophore alignment, and ease of functionization and processing. SHG properties of VDCN and p-hydroxy vinyl benzoate copolymers having 4-nitro-4'-oxobiphenyl (BP) and 4-nitro-4'-oxo-stilbene (ST) and referred as P(VDCN/VBZ-ST) and P(VDCN/VBZ-BP) were measured at 1.064 μm. Thin films of these copolymers were corona poled for 0.5 h at poling temperature $T_p \geq T_g$. The d_{33} coefficient for these copolymers varied with the poling temperature, and the

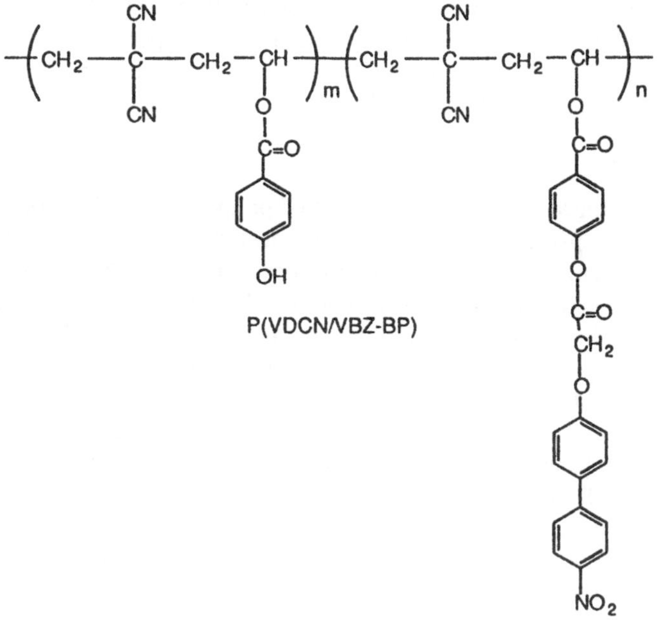

P(VDCN/VBZ-BP)

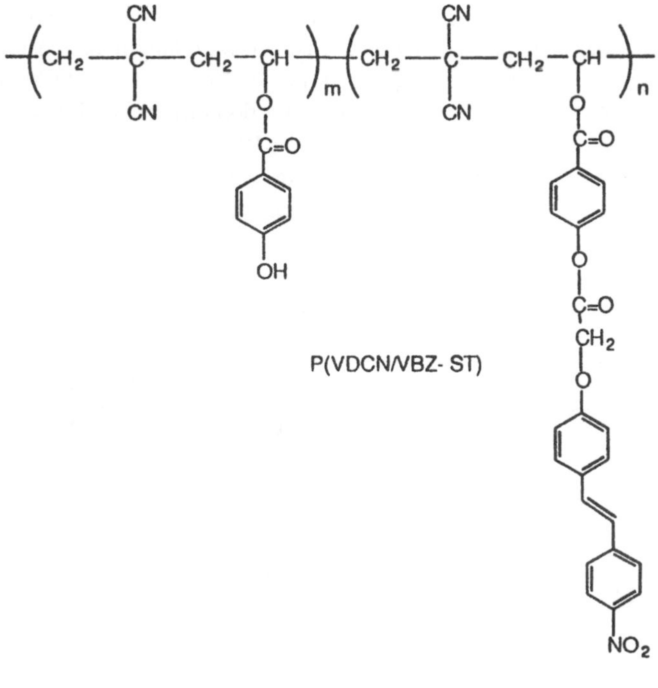

P(VDCN/VBZ- ST)

largest d_{33} values were measured at $T_p - T_g = 10°C$ for P(VDCN/VAc) and 20°C for P(VDCN/VBZ-ST) and P(VDCN/VBZ-BP). The refractive indexes, T_g, absorption cutoff wavelength, and Maker fringe-measured d_{33} coefficients of these three copolymers are listed in Table 2. Remarkably large d_{33} coefficients were obtained by functionization of polymers with NLO chromophore and, in particular, P(VDCN/VBZ-ST) exhibited the largest optical nonlinearity. The large NLO susceptibility of VDCN/VAc copolymer results from the large dipole moment of the C—CN in the vinylidene cyanide segment and larger space provided by the vinyl acetate segment, which facilitates easy rotation of CN dipoles under high electric fields. In P(VDCN/VBZ-ST) and P(VDCN/VBZ-BP) copolymers, the pendant 4-nitro-4'-oxostilbene chromophore in P(VDCNVBZ-ST) and 4-nitro-4'-oxobiphenyl chromophore in P(VDCN/VBZ-ST) are responsible for large d_{33} coefficients, whereas the vinylidene cyanide segment can also provide an additional benefit in enlarging SHG through a cooperative effect.

E. Polyureas

Polyureas have emerged as an important class of piezoelectric and ferroelectric materials. Polyureas can be prepared either by solution polymerization or by vapor deposition polymerization. Several groups have focused their studies on polyureas because they provide ease of synthesis, processing, and chemical modification. The SHG of aromatic polyureas with and without NLO chromophore was reported by Nalwa et al. [35–39]. For SHG measurements, thin films of polyureas were spin-coated onto conducting glass slides at 80°C because this leads to better optical-quality thin films. The corona poling of polyurea thin films spin-coated on soda lime glass (under vacuum coated by aluminum) was done at 130°C.

The spin-coated films show an absorption peak at 253 nm and have a cutoff wavelength at around 307 nm (Fig. 6). Optical transmission for polyurea films was greater than 99% from 350 nm to 2500 nm. Interestingly, PU1 is the first organic NLO polymer to be optically transparent at such low wavelengths. PU2, which has a nitro group, shows an absorption maximum at 378 nm and has a cutoff wavelength at 460 nm. With the introduction of a nitro group at the pendant phenyl ring, polyurea with urea main-chain and side-chain chromophores can be obtained and a significant difference in the absorption spectral features can be seen since the cutoff wavelength increases by about 150 nm. This indicates that optical transparency of polyureas can be tailored by chemical modification. Likely, PU3 has a nitro group and a donating methyl group at the pendant benzene ring, which shifts cutoff wavelength to 510 nm.

The Maker-fringe polyurea thin films were measured at 1.064 μm as functions of the angle of incidence. The SH signal increases symmetrically as the angle of incidence in-

Table 2 Refractive Indices, T_g, Absorption Cutoff Wavelengths and d_{33} Coefficients of Cyanopolymers

Cyanopolymer	T_g (°C)	Refractive index	λ_{cutoff} (nm)	d_{33} (pm/V)
P(VDCN/VAc)	180	1.484	330	8
P(VDCN/VBZ-ST)	190	1.635	520	120
P(VDCN/VBZ-BP)	160	1.598	460	60

Source: After Ref. 34.

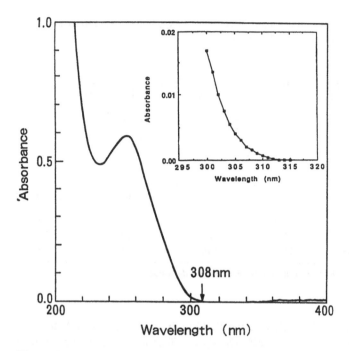

Figure 6 Optical transparency of PU1 showing a cutoff wavelength around 310 nm, the shortest ever reported for a NLO polymer. (After Ref. 37.)

creases. The SHG coefficients of PU samples were evaluated by comparing the SH signal with the SH signal of blank quartz substrate where a $d_{11} = 0.4$ pm/V for the quartz plate was used as the reference. Table 3 lists the d_{33} coefficients of poled polyurea thin films. PU1 shows a d_{33} coefficient of 5.5 pm/V. The d_{31} coefficient was estimated to be 1.67 pm/V. The electrooptical coefficient r_{33} of PU1 was calculated from the simple relationship

$$r_{33} = \frac{4d_{33}}{n^4} \tag{17}$$

where the refractive index (n) of PU1 is 1.577 at 632.8 nm. This yields an r_{33} coefficient of 3.56 pm/V. The NLO efficiency of PU1 is about four times greater than that of urea single crystal ($d_{33} = 1.4$ pm/V), showing a significant increment in the NLO efficiency while retaining the optical transparency of urea NLO moieties. PU2, which possesses large hyperpolarizability NLO chromophores, shows substantial enhancement in NLO coefficient. The d_{33} coefficient of PU2 was estimated as 8.37 pm/V, and the d_{31} coefficient of 2.93 pm/V assuming that the refractive index of PU2 is the same as that of the PU1.

Figure 7 shows the temperature dependence of SHG intensity as a function of the poling temperature at 6 kV recorded in in-situ SHG measurements. The SHG intensity starts increasing after 120°C. Near the glass transition temperature (T_g) of 180°C, the enhanced orientation of NLO moieties leads to a many-fold increase in SH intensity over that observed at room temperature. Figure 8 shows the poling field dependence of the d_{33} coefficient at a constant temperature of 130°C. These data were also recorded in in-situ SHG measurements where the PU1 sample was held at a fixed angle of 45°. The d_{33} coefficient increased with increasing poling voltage and started saturating at 8 kV. The d_{33} coefficient at the highest poling voltage (8–10 kV) was 5.5 pm/V.

Table 3 Chemical Structures, Absorption Cutoff Wavelengths (λ_c), Initial Decomposition Temperatures (IDTs), and d_{33} Coefficients of Polyureas[a]

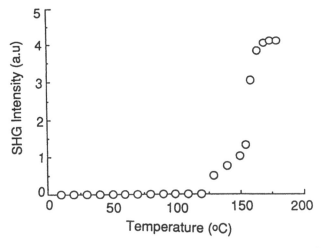

Polyureas	A	X	IDT (°C)	λ_c (nm)	d_{33} (pm/V)	r_{33} (pm/V)
PU1	H	H	300	307	5.5	3.6
PU2	NO$_2$	H	220	460	8.4	5.4
PU3	NO$_2$	CH$_3$	200	510	11.8	7.6

[a]Electrooptic r_{33} coefficients were calculated from the SHG measured d_{33} values.
Source: After Refs. 35–39.

Figure 9 shows the temporal stability of the SHG activity of PU1 at ambient conditions, after the poling field was switched off. The d_{33} coefficient decreased slightly over a period of 60 h and showed no further decay up to 1000 h, where d_{33} stabilized at 5 pm/V. The PU1 retained 90% of the initial d_{33} value. Long-term stability of the NLO coefficient has been a major problem, as its magnitude decreases substantially due to orientational relaxations. The majority of NLO dye grafted polymers relaxed to 40–80% of their initial d_{33} value over a short period, while the decay of d_{33} coefficient in PU1 was significantly less. The large NLO activity and stability of PU1 is associated with hydrogen-bond formation, which plays a significant role in dipolar alignment. Especially in PU1, a net dipole moment appears as a consequence of hydrogen bonding when all

Figure 7 Temperature dependence of SH intensity of PU1 (in situ SH intensity measurements). Poling conditions: applied voltage = 8 kV; incident angle = 45°; incident polarization (S, P) and output polarization (P). (After Ref. 39.)

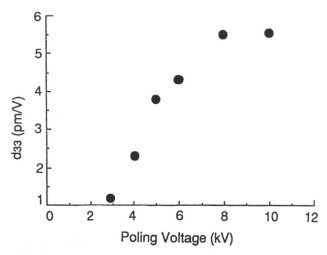

Figure 8 Change of d_{33} coefficient of PU1 polyurea as a function of applied electric field. (After Ref. 39.)

the dipoles are aligned in the same direction, which provides poling-induced NLO stability. The hydrogen-bonding mechanism of optical nonlinearity in PU1 may be considered similar to that in the polyamides such as odd nylons. Because polyureas are piezoelectric materials, they inherently show SHG activity. PU1 also shows interesting ferroelectric and pyroelectric properties [40], as discussed in the chapter dealing with polyureas and polythioureas. A waveguide mode structure for a 0.2477-μm-thick PU1 film coated on a quartz substrate as a function of the waveguide was investigated. The unpoled PU1 has a refractive index of 1.577 at 632.8 nm. The cutoff waveguide widths

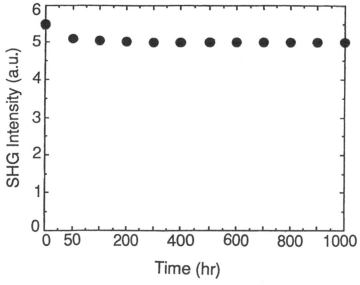

Figure 9 Temporal stability curve of SH intensity for PU1 film at room temperature after poling field was switched off. (After Ref. 39.)

of a planar waveguide were 0.18 and 0.23 μm for the TE and TM modes, respectively. The phase-matching possibilities can also be determined from an intersection of the appropriate fundamental and harmonic curves. The SHG properties of polyurea thin films formed by vapor deposition polymerization were reported by Kajikawa et al. [41]. The chemical structure is shown below. Thin films corona poled at 180°C showed d_{33} of 1.65 pm/V, which was stable indefinitely at room temperature after a slight initial decay in the first week. Polyurea backbone offers a lot of possibilities of chemical modifications to tailoring high-performance NLO materials with desired optical, NLO, and electrooptical properties.

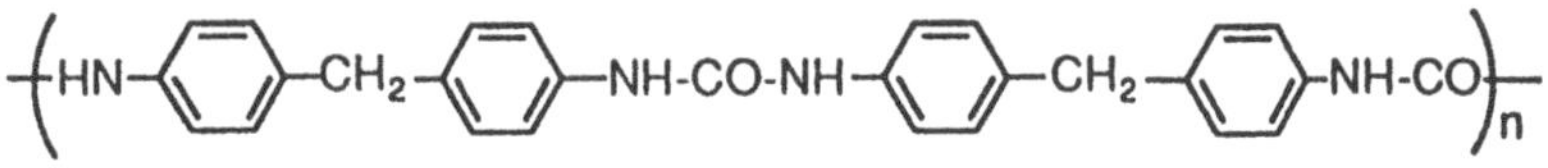

F. Ferroelectric Liquid Crystalline Polymers

Recently, ferroelectric liquid crystalline (FLC) polymers have been considered another important class of nonlinear optical materials. There are only a few reports on the second harmonic generation of FLC polymers. Ozaki et al. [42] reported the SHG properties of a FLC polymer having a siloxane backbone with the chiral mesogen in the side chain. The angular dependence of the SHG intensity of a 12-μm-thick cell in the SM A phase showed no SH signal because it has a centrosymmetric structure. Also, the angular dependence of the SHG intensity in the chiral smectic C (Sm C*) phase showed no SHG in the absence of an electric field. Under the influence of an electric field, the SHG intensity increased in the Sm C* phase depending on the rotating angle. The FLC polymer attains a noncentrosymmetric structure and dipole moments are oriented in the direction of the twofold axis normal to the molecular long axis and parallel to the smectic layer.

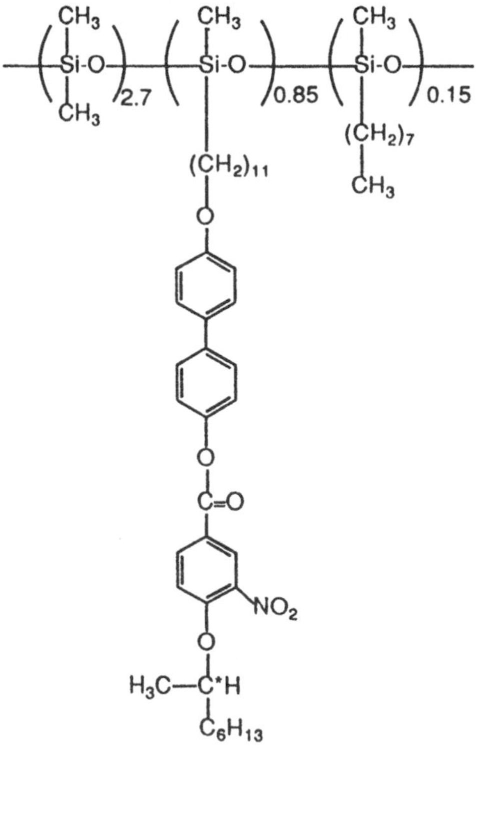

The relative SHG measurement to $LiNbO_3$ yields the effective NLO coefficient of 0.02 pm/V under phase-matching conditions. Type I phase matching was observed. The siloxane polymer has a large spontaneous polarization of 100 nC/cm^2 at $T_c - T = 20°C$ due to the large dipole moment of the nitro group in the side-chain chiral mesogen. The nitro group in the ortho position to the chiral tail has a dominant dipole moment pointing in the direction of the twofold axis. The nitro group is responsible for large hyperpolarizability. The low concentration of the side-chain mesogen as well as poor homeotropic alignment may be responsible for lower SHG activity.

Coles et al. [43] reported SHG in a FLC side-chain copolymer that has a polysiloxane backbone with 30% of the sites functionalized with electroactive side groups. The copolymer has an achiral nonmesogenic chromophore substituted directly onto the backbone next to the chiral mesogen to avoid demixing and NLO chromophore colinear with the ferroelectric dipole. The SHG of the copolymer depends on temperature, which coincides with the phase behavior. This copolymer has a maximum tilt angle of 24° and spontaneous polarization of 21 nC/cm^2. The S_{C*}-to-S_A transition occurs at 75°. The copolymer shows a SHG efficiency equal to that of quartz. Ferroelectric and NLO properties were investigated by varying the ratio of mesogen to chromophore content. Ferroelectric properties were found to be suppressed by increasing chromophore content, and no switching was observed for more than 50% content.

In summary, ferroelectric polymers such as poly(vinylidene fluoride), copolymers of vinylidene fluoride with trifluoroethylene, vinylidene cyanide copolymers, polyureas, and ferroelectric liquid crystalline polymers show interesting second-order, nonlinear optical properties. In addition, ferroelectric polymers can be used host matrix for guest-host poled polymer systems, which show good temporal stability. In particular, ferroelectric polymers seem to have somewhat better SHG stability as a function of time than NLO chromophore grafted conventional polymers. Ferroelectric polymers seem promising NLO materials for emerging photonic technologies.

REFERENCES

1. D. S. Chemla and J. Zyss (eds.), *Nonlinear Optical Properties of Organic Molecules and Crystals*, Academic Press, New York, 1987.
2. S. R. Marder, J. E. Sohn and G. D. Stucky (eds.), Materials for nonlinear optics: Chemical perspectives, *Am. Chem. Soc. Symp. Ser. 455* (1991).
3. J. Zyss (ed.), *Molecular Nonlinear Optics*, Academic Press, New York, 1994.
4. L. A. Hornak (ed.), *Polymers for Lightwave and Integrated Optics*, Marcel Dekker, New York, 1992.
5. H. S. Nalwa, T. Watanabe, and S. Miyata, Optical second-harmonic generation in organic molecular and polymeric materials: measurement techniques and materials, *Progress in Photochemistry and Photophysics*, Vol. 5 (J. F. Rabek, ed.), CRC Press, Boca Raton, FL, 1992, chap. 4, pp. 103–185.
6. H. S. Nalwa, Organometallic materials for nonlinear optics, *Appl. Organometal. Chem. 5*:349 (1991).
7. H. S. Nalwa, Organic materials for third-order nonlinear optics, *Adv. Mat. 5*:341 (1993).
8. H. S. Nalwa and A. Kakuta, Organometallic Langmuir-Blodgett films for electronics and photonics, *Appl. Organometal. Chem. 6*:645 (1992).
9. D. M. Burland, R. D. Miller, and C. A. Walsh, Second-order nonlinearity in poled-polymer systems, *Chem. Rev. 94*:31 (1994).
10. T. T. Wang, J. M. Herbert, and A. M. Glass (eds.), *The Applications of Ferroelectric Polymers*, Blackie, Glasgow, 1988.
11. H. S. Nalwa, Recent developments in ferroelectric polymers, *J. Macromol. Sci.: Rev. Macromol. Chem. Phys. 29*:341 (1991).
12. R. G. Kepler and R. A. Anderson, Ferroelectric polymers, *Adv. Phys. 41*:1 (1992).
13. K. Singer, M. Kuzyk, and J. Sohn, Second-order nonlinear optical process in orientationally ordered materials: Relationship between molecular and macroscopic properties, *J. Opt. Soc. Am. B 4*:968 (1987).
14. P. D. Maker, R. W. Terhune, M. Nisenoff, and C. M. Savage, Effect of dispersion and focusing on the production of optical harmonics, *Phys. Rev. Lett. 8*:21 (1962).
15. J. G. Bergman, Jr., J. H. McFee, and G. R. Crane, Pyroelectricity and optical second harmonic generation in polyvinylidene fluoride films, *Appl. Phys. Lett. 18*:203 (1971).
16. J. H. McFee, J. G. Bergman, and G. R. Crane, Pyroelectric and nonlinear optical properties of poled polyvinylidene fluoride films, *Ferroelectrics 3*:305 (1972).
17. D. Broussoux and F. Micheron, Electro-optic and elasto-optic effects in polyvinylidene fluoride, *J. Appl. Phys. 51*:2020 (1980).
18. P. Rabin, E. Chastaing, D. Broussour, J. Raffy, and J. P. Pocholle, Optical second harmonic generation in copolymer P(VDF/TrFE), *Ferroelectrics 94*:133 (1989).
19. J. F. Legrand, J. Lajzerowicz, B. Berge, P. Delzenne, E. Macchi, C. B. Leonard, A. Wicker, and J. K. Kruger, Ferroelectricity in PVF$_2$ based copolymer, *Ferroelectrics 92*:267 (1984).
20. H. Sato and H. Gamo, New SHG observation in vinylidene fluoride/trifluoroethylene copolymer film using a pulsed YAG laser, *Jpn. J. Appl. Phys. 25*:L990 (1986).
21. P. Pantelis, J. Hill, and P. L. Davies, Poled copoly(vinylidene fluoride-trifluoroethylene) as a host for nonlinear optical molecules, *Nonlinear Optical and Electroactive Polymers*, American Chemical Society, Washington, DC, 1987.
22. C. Broussoux, S. Esselin, P. LeBarny, J. P. Pocholle, and P. Robin, Nonlinear optics in doped amorphous polymers, *Nonlinear Optics of Organics and Semiconductors* (T. Kobayashi, ed.), Springer-Verlag, Berlin, Heidelberg, 1989, p. 126.

23. M. Dumont and Y. Levy, Measurement of electrooptic properties of organic thin films by attenuated total reflection, *Nonlinear Optics of Organics and Semiconductors* (T. Kobayashi, ed.), Springer-Verlag, Berlin, Heidelberg, 1989, p. 256.

24. M. H. Berry, D. M. Gookin, and E. W. Jacobs, Nonlinear optical properties of materials, *Tech. Digest Ser.* 9:121 (1988).

25. J. R. Hill, P. L. Dunn, G. J. Davies, S. N. Oliver, P. Pantelis, and J. D. Rush, Efficient frequency doubling in a poled PVDF copolymer guest/host composite, *Electron Lett.* 23:700 (1987).

26. P. Pantelis and J. R. Hill, Guest-host polymer systems for second-order optical nonlinearities, *Polymers for Lightwave and Integrated Optics* (L. A. Hornak, ed.), Marcel Dekker, New York, 1992, chap. 13, p. 343.

27. N. Tsutsumi, T. Ono, and T. Kiyotsukuri, Internal electric field and second harmonic generation in the blends of vinylidene fluoride-trifluoroethylene copolymer and poly(methyl methacrylate) with a pendant nonlinear optical dye, *Macromolecules* 26:5447 (1993).

28. N. Tsutsumi, I. Fujii, Y. Ueda, and T. Kiyotsukuri, SHG from NLO dyes in ferroelectric polymer, *Polymer Preprints Jpn.* (1993).

29. Y. Azumai, H. Sato, and I. Seo, Second-harmonic generation of Er:YAG 2.94 μm laser radiation using an organic vinylidene cyanide/vinyl acetate thin film, *Opt. Lett.* 15:932 (1990).

30. Y. Azumai, I. Seo, and H. Sato, Enhanced second harmonic generation with Cerenkov radiation scheme in organic film slab-guide at IR line, *IEEE J. Quantum Electronics* 28:231 (1992).

31. Y. Azumai and H. Sato, Improvement of the Cerenkov radiative second harmonic generation in the slab waveguide with a periodic nonlinear optical susceptibility, *Jpn. J. Appl. Phys.* 32:800 (1993).

32. M. Dumont, Y. Levy, and D. Morichere, Electro-optic organic waveguides: Optical characterization, *Organic Molecules for Nonlinear Optics and Photonics* (J. Messier et al., eds.), Kluwer, The Netherlands, 1991, p. 461.

33. D. E. Broussoux, S. Chastaing, O. Esselin, R. Le Barny, Y. Robin, J. P. Pocholle, and J. Raffy, Organic materials for nonlinear optics, *Rev. Tech. Thomson-CSF 20-12*:151 (1989).

34. M. Kishimoto, D. Zou, and I. Seo, New polymers for nonlinear optics, nonlinear guided-wave phenomena, *1991 Tech. Digest Ser. 15*:82 (1991).

35. H. S. Nalwa, T. Watanabe, A. Kakuta, A. Mukoh, and S. Miyata, N-Phenylated aromatic polyureas: A new class of nonlinear optical materials exhibiting large second harmonic generation and u.v. transparency, *Polymer 34*:657 (1993).

36. H. S. Nalwa, T. Watanabe, A. Kakuta, A. Mukoh, and S. Miyata, Aromatic polyureas: A new class of nonlinear optical polymers with large second harmonic generation, *Electronic Lett.* 28:1409 (1992).

37. H. S. Nalwa, T. Watanabe, A. Kakuta, A. Mukoh, and S. Miyata, Aromatic polyurea exhibiting large second harmonic generation and optical transparency down to 300 nm, *Appl. Phys. Lett. 62*:3223 (1993).

38. H. S. Nalwa, T. Watanabe, A. Kakuta, A. Mukoh, and S. Miyata, Aromatic polyurea exhibiting large second harmonic generation and UV transparency, *Synth. Metals 57*:3895 (1993).

39. H. S. Nalwa, T. Watanabe, A. Kakuta, and S. Miyata, (1993). Aromatic polyureas: a new class of chromophore main-chain polymers for second-order nonlinear optics, *Nonlinear Optics*, 8:157(1994).

40. S. Tasaka, K. Ohishi, H. S. Nalwa, T. Watanabe, and S. Miyata, Ferroelectric polarization reversal stability by hydrogen bonding in N-phenylated aromatic polyurea, *Polymer J.* 26:505 (1994).

41. K. Kajikawa, H. Nagamori, H. Takezoe, A. Fukada, S. Ukishima, Y. Takahashi, M. Iijima, and E. Fukada, *Jpn. J. Appl. Phys.* 30:1737 (1991).

42. M. Ozaki, M. Utsumi, K. Yoshino, and K. Skarp, Second harmonic generation in ferroelectric liquid crystalline polymer, *Jpn. J. Appl. Phys.* 32:L852 (1993).

43. H. J. Coles, M. Redmond, O. Mondain-Monval, E. Wischerhoff, and R. Zentel, A study of second harmonic generation in side-chain ferroelectric liquid crystalline copolymers, *Proc. Fourth Int. Conf. on Ferroelectric Liquid Crystals*, Tokyo, September 28–October 1, 1993, p. 83.

13
Dielectric Properties of Ferroelectric Polymers

Péter Hedvig

Plastics Research and Development, Ltd., Budapest, Hungary

I. INTRODUCTION

Dielectric spectroscopy has been and is being used to study multiple transitions in polymers [1,2]. Liberation of motion of polar groups in the polymer chain, or attached to it, results in changes of the dielectric permittivity and loss. For usual homopolymers, copolymers, and blends, the dielectric transitions are correlated with those detected by dynamic mechanical relaxation, thermomechanical analysis (TMA), differential scanning calorimetry (DSC), electron spin resonance (ESR), and nuclear magnetic resonance (NMR) methods [3]. For ferroelectric polymeric compounds and composites this correlation is not straightforward.

The dielectric technique has been used mostly to study polymers which are good electrical insulators, which includes most polymers. From the 1950s on, however, a great many such polymers have been synthesized which are semiconductors and exhibit peculiar electric behavior [4].

Also rather early, it was realized that practically all polar polymers can be prepared in such a state where polarization of the polar groups is fixed, i.e., a relatively large part of the dipole moments are aligned along the polarizing electric field and remain so, even if the field is removed. Also, there are some other ways to obtain permanent polarization and to create polymer electrets involving storage of detrapped or injected charges or those created by high-energy irradiation [5].

In electrets the polarization is not quite stable; it decays with time, although this decay may be very slow. In ferroelectric polymers, however, stable dipole-oriented structures are formed in which the orientation can be reversed by applying an external electric field opposite to the direction of polarization. This problem, together with other pyroelectric, piezoelectric, and ferroelectric, properties, is discussed elsewhere in this volume.

In ferroelectric polymers, permanent polarization is an intrinsic behavior depending on the crystal structure. Thus, permanent polarization can be achieved by ensuring proper crystallization or recrystallization conditions. For comprehensive recent reviews about structure and properties of ferroelectric polymers, see Nalwa [6] and Kepler and Anderson [7].

The purpose of this chapter is to discuss the different dielectric polarization mechanisms in polymers with special emphasis on ferroelectric ones, and to critically review the most important experimental methods. Interpretation of the dielectric spectra is discussed for some selected, representative, polymeric compounds and composites.

II. POLYMER TYPES AND STRUCTURES

A. General Considerations

Polymers are macromolecules built by repetition of monomer units to form long chains which behave thermodynamically differently from small molecules. A polymer molecule may have, for example, side chains or groups which may exhibit thermal motion almost independent of that of the main chain. A methyl group attached to a polymer chain, for example, may rotate freely down to very low temperatures without much perturbation in the motion of the main chain. This does not mean that a side-group motion can be regarded as absolutely independent of the motion of the main chain, but it does mean that it usually acts as a relatively small perturbation. A similar principle is used in interpreting infrared spectra.

Parts of the long polymer molecules may thus be considered thermodynamically separate [8], although interacting, subsystems. Besides side groups, some segments of the macromolecule may also be considered as thermodynamical subsystems because they may change configuration without changing bond lengths and angles. The various possibilities for intramolecular mobilities in solid polymers are discussed in detail by Cowie [9].

In the case of block copolymers, chain segments of another type of monomer are built into the main chain, forming a separate thermodynamical subsystem. In random copolymers, on the other hand, the second kind of monomer unit is built into the main chain randomly and, thus, forms a joint subsystem with the main chain. This, by upsetting the regularity, can affect crystallization processes considerably.

Evidently there are some cases where the thermal motion of a side chain has a significant effect on that of the main chain. An example of this is polymethyl methacrylate, in which the ester side-group rotation is rather strongly coupled to the thermal motion of the main chain. By liberation of the rotation of the ester side group, the motion of the main chain is also affected (see Ref. 8, p. 122).

In considering dielectric behavior, consequently, the chemical and physical structure of the macromolecules must be known. This can be realized by combining various techniques. In order to define the thermodynamical subsystems, the mobility of groups, side chains, and, even parts of the main chain have to be considered (intramolecular mobility). This is also done by a combination of various methods, such as mechanical relaxation, Fourier transform infrared spectroscopy (FTIR), high-resolution magnetic resonance spectroscopy (HRNMR), and in the solid state, cross-polarized magic angle spinning (CPMAS), high-resolution magnetic resonance, wide-line and spin-echo (pulsed) nuclear magnetic resonance, electron spin resonance (ESR), time-resolved fluorescence, polarized luminescence, scattering techniques, and some others. (See Refs. 3 and 10 for details.)

B. Amorphous Polymers

In general, molecules in solids tend to arrange so that the symmetry of their molecular electronic orbitals is copied in a macroscopic scale. Thus, the expected structure is always crystalline. In the case of the very long polymer molecules built by monomer units of the order of 10^4, this is, however, seldom realized accurately because of steric and thermodynamic reasons.

The amorphous phase is also an extremity for polymers because the macromolecules tend to form aggregates or microcrystallites. Indeed, X-ray, electron, and neutron scattering experiments show that solid polymers usually exhibit aggregate structure. Thus, even if there is no long-range symmetry in an amorphous polymer, its structure is by no means homogeneous. The configuration of the macromolecules in amorphous solids is very often considered as being similar to that in solutions, i.e., random coils or globules. This structure in dilute solution is rather well understood, but extrapolation of these properties to the solid state has not been verified experimentally. The concept that the amorphous state is like a supercooled liquid is an oversimplification.

An amorphous polymer is in a glassy (vitreous) state below its glass transition temperature and in a rubbery (viscoelastic) state above it. In non-cross-linked polymers the viscoelastic behavior is attributed to chain entanglements.

Unlike crystalline melting, the glass transition temperature is a relaxation transition. This means that it is dependent on the effective frequency of the measurement. This frequency is found by dynamic mechanical (DMA), dielectric relaxation, and pulsed nuclear magnetic methods. Quasi-static methods, such as dilatometry, differential scanning calorimetry (DSC), and thermomechanical analysis (TMA), show that the effective frequency depends on the rate of temperature scan. This is one of the reasons why the glass transition temperatures reported for various amorphous materials appear so diverse.

Polymers in the glassy state are seldom in thermodynamical equilibrium. By cooling an amorphous polymer down from above its glass transition temperature, a thermodynamically nonequilibrium state is formed, which, usually very slowly, relaxes toward equilibrium. This kind of relaxation process is referred to as volume-enthalpy or thermodynamical relaxation because it involves a decrease of the specific volume and enthalpy. The volume change can be measured dilatometrically, the enthalpy change calorimetrically. Volume relaxation is described quantitatively by the Kovacs [11] equation as

$$\frac{dw}{dt} = \frac{-w}{\tau(w)} \tag{1}$$

where $w = [v(t) - v(\infty)]/v(\infty)$, $v(t)$ being the specific volume and $\tau(w)$ the relaxation time of volume relaxation, which is dependent on w, the variable describing the deviation of the system from thermal equilibrium.

Equation (1) can be solved numerically and, thus, experimentally observed isothermal time-dependent volume changes occurring after quenching an amorphous polymer from above the glass transition temperature to temperatures below it can be described quantitatively.

Thermodynamical relaxation has been found to affect the mechanical, dielectric, and thermal properties of polymers appreciably [12].

C. Polycrystalline Polymers

Polymers are rather difficult to prepare in the single-crystalline state. Polymeric single crystals grown by delicate techniques always show an amorphous halo in their wide-angle X-ray diffraction spectra (WAX), indicating that the long-range symmetry is not perfect.

In polycrystalline polymers there is always an amorphous part and a crystalline one consisting of small-sized crystallites. By X-ray diffraction techniques it is possible to reveal the symmetry of the crystallites and also to measure their average size. Crystallites also tend to aggregate to form spherulites, sphere-shaped formations, containing ordered crystallites and disordered amorphous parts. Spherulites may be very large, reaching directly visible measures.

An amorphous halo in WAX may originate from imperfections of the crystal lattice (defects) and not necessarily from a separate amorphous phase. The amorphous part in a crystalline polymer is, however, usually considered as a separate phase. According to this two-phase model, polycrystalline polymers are regarded as composites of crystallites embedded in an amorphous medium.

The crystalline-to-amorphous ratio can be determined by wide-angle X-ray scattering, by density measurements, and also by differential scanning calorimetry (DSC).

From the viewpoint of dielectric properties it is concluded that polymeric solids, whether they are amorphous or crystalline, exhibit heterogeneous phase structure.

D. Orientation

The physical structure of solid polymers changes by stretching. This change is manifested at different structural levels. Aggregates, e.g., spherulites, would be deformed to spheroids. This process can be studied by small-angle light scattering (SALS). The next level of orientation is that of the crystallites which may be located inside the spherulites. This process may be studied by wide-angle X-ray scattering (WAX). The Debye-Scherrer diffractions pattern measured by WAX for an assembly of randomly oriented crystallites are concentric rings. By orientation the rings become dashed depending on the extent of orientation. The next architectural level is change of the lammellar structure inside the crystallites. Crystal lamellae tend to align perpendicularly to the direction of stretching, thus the molecular segments are aligned along the orientation direction [13].

Stretching may change the polymorphic crystalline structure, as in the case of poly(vinylidene fluoride) (PVDF) when the monoclinic, paraelectric, α phase is transformed to the orthorombic, ferroelectric, β phase by stretching or hot rolling [14].

At high deformations fibrillous structure may be formed. The change of the lamellar structure by stretching can be studied by small-angle X-ray scattering (SAX) and by electron microscopy. The effect of stretching goes down to molecular, or even intramolecular, levels, involving orientation of parts (segments) of polymer molecules in the crystalline as well as in the amorphous parts.

Molecular orientation is usually studied by birefringence measurements, i.e., by measuring the index of refraction of light in directions parallel $(n_\parallel)$ and perpendicular $(n_\perp)$ to the direction of stretching. The orientation is characterized by the difference $\Delta_n = n_\parallel - n_\perp$. Polymer molecules, evidently, may have anisotropic refractive indices, but, by random arrangement, the refractive index difference is $\Delta_n = 0$.

Another widely used method for studying molecular orientation in polycrystalline polymers is the infrared dichroism technique. In the infrared spectra it is practically always possible to find spectrum bands which originate from groups located in the amorphous part and those located in the crystalline one. Dichroism means the ratio of the absorption coefficient of the light polarized along the orientational axis to that polarized in the perpendicular direction. This ratio, referred to as the dichroic ratio, δ, is used to characterize molecular orientation. If the polymer is considered as an assembly of perfectly oriented chains in a volume fraction v_f and the rest is randomly oriented, the dichroic ratio is expressed as

$$\delta = \frac{1 + (1/3)(\delta_0 - 1)(1 + 2v_f)}{1 + (1/3)(\delta_0 - 1)(1 - v_f)} \tag{2}$$

where δ_0 is the dichroic ratio corresponding to perfect orientation ($v_f = 1$).

Orientation in polymers is generally characterized by the orientation function,

$$f\alpha = \frac{\langle 3 \cos^2 \alpha \rangle - 1}{2} \tag{3}$$

where $\langle \ \rangle$ means averaging the angles α of the units (crystallites, intramolecular groups) with respect to the direction of orientation. In polycrystalline polymers the orientation function can be determined for the crystalline as well as for the amorphous parts. Besides infrared dichroism, orientation in the amorphous part can also be determined by measuring the mechanical Young modulus at sonic frequencies. The sonic modulus G is expressed in terms of the orientation function of the amorphous part f_a and that of the crystalline part f_c as

$$\frac{1}{G_{un}} - \frac{1}{G_{or}} = \frac{\beta f_c}{G_{oc}} + \frac{(1 - \beta)f_a}{G_{oa}} \tag{4}$$

where β is the crystalline to amorphous ratio, G_{un} and G_{or} are the moduli of unoriented and oriented polymers, respectively, and G_{oc} and G_{oa} are the intrinsic transversal moduli of the completely oriented crystalline and amorphous parts, respectively.

Polymers may exhibit liquid crystalline, ordered structures similar to those of small molecular liquid crystals. As polymer chains are very long, containing about 10^4–10^5 monomer units, only parts of them or side groups can be ordered in nematic, cholesteric, and different kinds of smectic structures. As in small-molecular nematic liquid crystals, in nematic polymer structures groups of cylindrical-like symmetry are arranged statistically with their long axes along a certain direction.

This arrangement fluctuates because of thermal motion. The order parameter in such systems is characterized by an orientation function such as Eq. (3). The cholesteric structure is a twisted nematic one consisting of nematic layers in which the average orientation is turned by an angle up to 2π. Angles 0 and 2π are, evidently, equivalent. The distance between these equivalent layers is called "pitch." It determines the wavelength at which the reflexion of circularly polarized light is maximum.

In the smectic-type arrangement the molecules are packed in layers with their long axes parallel to each other and perpendicular to the layer plane. According to the symmetry of the ends (or folds) in the layer plane, several smectic types may be distinguished. Also, in some cases the molecules are not arranged perpendicularly to the layer plane, but remain parallel to each other. The angle of molecular orientation with respect to the layer plane may change from one layer to another, resulting in a twisted smectic structure.

The macroscopic order in liquid crystalline materials, called texture, is very variable. It can be studied by microscope using polarized light. Samples for dielectric study prepared from solution are, correspondingly, expected to exhibit highly heterogeneous structure.

E. Composites

Polymer composites are multiphase materials containing, usually, inorganic fillers or reinforcing materials embedded in an amorphous or polycrystalline matrix. The dielectric properties of the inclusions are, usually, very different from those of the matrix. For ferroelectric applications, inorganic ferroelectric materials (e.g., ceramics) are often used as fillers.

The mechanical, thermal, and electrical properties of polymers are changed drastically by the presence of fillers or reinforcing materials. Mechanical properties are usually improved [15]. Electrical conductivities may become very high when the inclusion is conductive and when its volume fraction is higher than a certain (percolation) limit. The dielectric permittivities may become as high as several thousands, indicating nondipolar origin of the polarization, discussed in Section III.D in this chapter.

III. POLARIZATION MECHANISMS

A. Electrical Moments

Electrical moments are used to characterize the potential of an arbitrary charge distribution. In general, this potential is expressed as a series,

$$V(d) = \sum_{i=0}^{\infty} \frac{\mu_i}{d^{i+1}} \tag{5}$$

where d is the distance from the charge distribution, where the potential V is considered,

$$\mu_i = er^i P_i(\cos \phi) \tag{6}$$

is the ith electrical moment of the charge distribution, e is the total net charge of the system, r and ϕ are the polar coordinates of the individual charges inside the distribution, and $P_i(\cos \phi)$ are the Legendre polynomials: $P_0 = 1$, $P_1 = \cos \phi$, $P_2 = (3 \cos^2 \phi)/2$, $P_3 = (5 \cos^3 \phi - 3 \cos \phi)/2$. An iterative formula for calculating higher-order Legendre functions is [16]

$$P_n(z) = \left(\frac{1}{n}\right)[(2n - 1)zP_{n-1}(z) - (n - 1)P_{n-2}(z)] \tag{7}$$

with $z = \cos \phi$.

According to Eqs. (5) and (6), μ_0 corresponds to a monopole (point charge), μ_1 to a dipole (two charges of opposite sign separated by a distance r), and μ_2 to a quadrupole formed by two dipoles of opposite direction.

The conventional unit of the dipole moment is the debye, which is equal to 3.3×10^{-30} Asm in SI units. In this chapter the debye unit will be used for simplicity.

The higher-order moments are usually not considered in dielectric spectroscopy because, usually, they make a small contribution to the macroscopic properties. They may become important by considering internal fields generated by the individual charge distributions.

B. Quantum Aspects of Polarization

A molecule, or a part (group) of a long polymer chain, consists of a set of positively charged nuclei and negatively charged electron density distributed along the molecule. The electron charge density at different sites of the molecule or group can be calculated by quantum chemical methods [17]. Thus, the moments of Eq. (6) can be interpreted at a molecular or even in a supermolecular level.

In polymers, usually, permanent dipole moments of groups or bonds are considered by interpretation of dielectric properties. Formation of permanent dipoles by changing the chemical structure is illustrated by the example of chlorobenzene. In benzene the π-electron density distribution is even along the ring, so the permanent dipole moment is zero. By substituting one of the H atoms by chlorine, the symmetry of the charge distribution is upset. The coulombic integral of the benzene π-electron system is modified to

$$I = I(\text{benzene}) + \alpha K(\text{benzene})$$

where K(benzene) is the exchange integral, and α is a semiempirical parameter for describing the inductive effect of the substituent to the electronegativity of the ring carbons. Using $\alpha = 1.2$, the π-electron density of chlorobenzene is obtained as [17]

where the numbers denote charge densities in units of the electron charge (rounded), and $(-)$ is the negative charge center of the ring, $(+)$ means the positive charge at the chlorine nucleus.

By considering changes of exchange and overlap integrals too, with a little more refined calculation the dipole moment of chlorobenzene was calculated as 0.45 debye. The experimental value is 0.43 debye.

In such quantum chemical considerations, introduction of a foreign atom into a molecule is treated as a perturbation. In the case of long polymer molecules, such a perturbation affects the electronic states of the near neighborhood mainly and not those of the whole molecule, unless it is highly conjugated.

In conjugated polymer molecules—for example, in polyenes—the π-electron orbitals are delocalized over a large part of the molecule or, in some cases, over several molecules. In such cases a perturbation may affect the electron densities at atoms far away from the source of perturbation. Not only the electron densities are affected by the perturbation, but the bond orders, bond lengths, polarizibilities, and free valences are also changed [17]. Thus, the change of the electronic molecular orbitals affects the configuration of the positively charged nuclei too, although to a relatively small extent. In some cases, however, the arrangement of the nuclear configuration may change appreciably by application of external electric fields. Not only the orientation of the dipolar groups, but the crystalline-to-amorphous ratio or the crystal symmetry may be changed.

The effect of an external electric field on a polymer can also be treated as a perturbation which changes the molecular orbitals, and in this way upsets the symmetry of the charge distribution, resulting in formation of induced dipole moments. In polymeric solids, instead of molecular orbitals, collective crystal orbitals (excitons) are considered [18] even in the amorphous phase, where there is no long-range symmetry of location of nuclei. There are various approaches for describing exciton states from the localized Fermi excitons to the most delocalized Wannier excitons. Since polymers do not exhibit rigorous symmetry, not even in single crystals, an exciton picture lying between the Fermi and Wannier limits is used. This is applicable even to aggregates which exhibit no translational symmetry [18].

The dielectric polarization in such systems is treated in terms of exciton states interacting with the external electric field. Although such a treatment can be applied quantitatively only to single crystals of known symmetry, the concept is very useful in understanding the origin of the internal field in dielectrics. A brilliant treatment of this problem may be found in Ref. 19. In the self-consistent field (SCF) quantum treatment, the effective field acting on a charge configuration is composed of the external vacuum field E (V/m) plus all the fields generated by the other charge distributions, which can be calculated by averaging the local fields over the unit cell of the crystal or aggregate. The local field is calculated by considering the exciton orbitals and the effects of charge distribution of ionic cores at the nuclei. Atomic polarizabilities α in a solid are affected by this local field, resulting in the same dependence on the polarizability of the isolated atom α_0 as is obtained by the classical Lorentz procedure:

$$\alpha = \frac{\alpha_0}{1 - (4\pi/3)\,\rho\alpha_0} \tag{8}$$

where ρ is the density of the polarizable units ("oscillators").

The conceptional difference between the classical and quantum description is that in the quantum view the polarizability of the atoms in a solid is changed by the presence of neighbors through the exciton orbitals, while in the classical view the atomic polarization is considered as that of the isolated atom and the change is interpreted as being a result of the Lorentz field created whenever an external field is applied.

The effect of an external electric field on a dielectric material (vacuum field, E, in V/m) is described by the displacement field $D = \epsilon_0 \epsilon E$, where ϵ_0 is the vacuum permittivity (8.854×10^{-12} AS/Vm) and ϵ is the relative permittivity (dimensionless). The polarization field P is defined as

$$P = D - \epsilon_0 E = \left(1 - \frac{1}{\epsilon}\right)D \tag{9}$$

The polarization field P arises from perturbation of the electronic orbitals which can respond to very high excitation frequencies and also that of nuclei, i.e., ion cores, which can respond to excitation frequencies in the order of the frequencies of lattice vibrations.

In the high-frequency (nonresonant optical) limit, in Eq. (9) ϵ is substituted for by $\epsilon(\infty)$, while at low frequencies

$$P(\omega) = \epsilon_0 \left[\frac{1}{\epsilon(\infty)} - \frac{1}{\epsilon}\right]D(\omega) \tag{10}$$

In the general treatment the polarization field P is expressed by considering the nuclear charge distributions and the electron density distributions determined by the exciton orbitals in a crystal. In such a system the positively charged nuclei exhibit vibrational motion which is not much influenced by the external field. The exciton orbitals are, evidently, highly perturbed, resulting in development of a polarization field. The change of the exciton orbitals, however, affects the arrangement of the nuclei of the crystal or aggregate. Thus, there is a mutual interaction between the electronic exciton states and the vibrating nuclear arrangement (phonons). The coupled phonon-exciton is referred to as polaron (Ref. 19, p. 280). The development of polarization field P as a result of a source field D may be described for a simple model [19] by an equation of motion such as

$$\frac{d^2P}{dt^2} + \omega^2 P = \left(\frac{1}{\mu_t}\right)D \tag{11}$$

where ω is the angular frequency of the lattice vibration, and $\mu_t = (4\pi/\omega^2)[1/\epsilon(\infty) - 1/\epsilon]^{-1}$ is the transition moment corresponding to $dP/dt = 0$.

Equation (11) is derived from the Hamiltonian based on the interaction energy density (DE, J/m^3) of an electron in the polarization field [19].

For localized orbitals the development of the polarization in time is very fast, and the corresponding response frequencies lie in the optical range. In isotropic nonpolar materials, the short-time (high-frequency, unrelaxed) polarization is due to deformation of the electron density distribution by the external electric field. The unrelaxed permittivity is related to the refractive index of light (n) as $\epsilon_u = n^2$. In nonisotropic materials ϵ and n are tensors.

The electronic polarizibilities with short response times are, evidently, dependent on the polarizibilities of groups of the polymer molecule and, correspondingly, mainly affect the optical properties. This very important effect is beyond the scope of the present review, and is discussed elsewhere in this volume. For delocalized orbitals, as in highly conjugated polymers, however, the delocalized π-electron orbitals may result in much longer response times and rather high induced moments. To illustrate the effect of delocalization, in the molecular scale for simplicity, Figure 1 shows schematically the change of the π-electron energy levels as the length (order) of the polyenes ($-CH=CH-$)$_n$ is increased. It is seen that the energy gap between the highest bonding and the lowest antibonding energy levels decreases to become small for high polyenes, i.e., by increasing delocalization of the π-electron orbitals. Since in odd-n polyenes the nonbonding orbital is occupied by a single electron, these materials are free radicals and the part of the polymer molecule in which the polyene segment is located becomes semiconductive and very easily polarizible. In even polyenes the nonbonding orbital is empty and, correspondingly, the energy gap does not vanish when the length is increased infinitesimally. In crystallites, or even in aggregates, π-type exciton orbitals may be formed, extending the delocalization to macroscopic measures. This results in an increase of polarizibility and electrical conductivity up to a level approaching that of metals. For example, the conductivity of polyacetylene ($-CH=CH-$)$_n$ with odd n value may reach the level of 10^3 S/m, which is more than that of graphite [20].

In odd polyenes, the sequence of conjugation may be interrupted at some places. These kinds of defects are referred to as solitons [21]. Defects are also formed by introducing electron-donor (Li, Na, K) or -acceptor (Cl, Rr, I) dopants to influence electric and magnetic properties. Cyclic, highly conjugated structures may be prepared by pyrolysis of polyacrylonitrile, $CH_2-CH-C\equiv N$, for example, resulting in an increase of

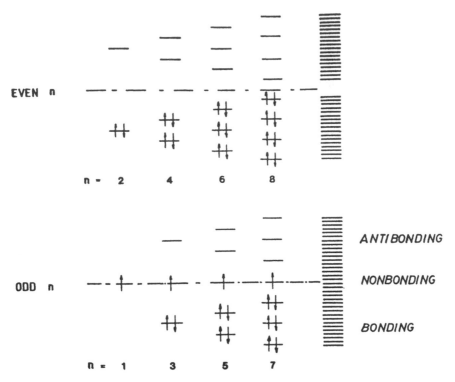

Figure 1   Electronic energy levels of even and odd polyenes by increasing sequence length.

conductivity from about 10^{-12} to 10^2 S/m. The products are paramagnetic, exhibiting a single, Lorentzian-shaped electron spin resonance line indicating overlap of the unpaired electron orbitals.

C. Orientational Dipole Polarization

As discussed in the previous section, the response of the system of charges representing a molecule, group, or even a crystallite to the action of an external force field may be treated as a perturbation, and quantum mechanical methods may be used. In this way the response of the dielectric material to a time-dependent (e.g., periodic) force field can be determined in terms of general complex susceptibilities, which are related directly to the complex permittivity tensor (Ref. 22, pp. 178–201).

The problem may be very much simplified in such cases when the electronic configuration of some polar groups in the polymer is not changed by the external electric field—just the orientation of their permanent dipole moments is affected. Permanent dipoles may be located at different sites of the polymer molecule, forming separate thermodynamical subsystems. It is possible to apply straightforward linear response theory to describe their interaction with the external field. This treatment is also based on the principles of nonequilibrium statistical thermodynamics, as the more rigorous quantum treatment is. As elements of a subsystem, such groups are considered in which the configuration is not changed during thermal motion (cf. Ref. 2, pp. 31–43, and Ref. 22, pp. 193–201).

The dielectric permittivity of such a subsystem can be expressed in terms of a macroscopic response function $\Phi(t)$ representing the response of the subsystem to a step-function external electric field. The subsystem is regarded as being in thermodynamical equilibrium before application of the field step, which upsets the equilibrium until a new one is reached. Upon removing the external field the subsystem usually relaxes to equilibrium again, unless this relaxation is hindered by some structural means.

According to the Kubo formalism of nonequilibrium statistical thermodynamics [2,22], the response $R(t)$ of a statistical system to any time-dependent external action $A(t)$ can be generally expressed as

$$R(t) = N \int_{u=0}^{t} \Phi(t - u)\left[\frac{dA(u)}{du}\right] du \qquad (12)$$

where Φ is the response function and N is a normalization factor.

Equation (12) is very widely applicable. As an action $A(t)$, mechanical, electrical, or magnetic force fields may be considered. Even the response of the polymer to a temperature jump can be treated this way. As a response $R(t)$, the mechanical compliance or modulus may be used. In the dielectric case, the external electric field in the classical meaning may be used as the action, that is, $A(t) = E(t)$ in V/m. As a response, the dielectric polarization field $P(t)$, expressed in terms of the permittivity by Eq. (9), or the displacement $D(t) = \epsilon_0 \epsilon(t) E(t)$ may be used. Substituting these functions into Eq. (12), integrating by parts and considering the limiting values, one obtains for the time-dependent dielectric permittivity:

$$\epsilon(t) = \epsilon(0) + \left[\frac{\epsilon(\infty) - \epsilon(0)}{E(t)}\right] \int_{u=0}^{t} \left[\frac{d\Phi(t - u)}{du}\right] E(u) \, du \qquad (13)$$

The conditions for applicability of this treatment are that the response function should not depend explicitly on the driving force field nor on previous actions. For dipole polarization in polar polymers these conditions are usually fulfilled, but not necessarily for ferroelectrets, especially the condition of the independence of the history. Such a problem also occurs in the case of thermodynamical (volume, enthalpy) relaxation [11], discussed in the previous section, when thermal history affects mechanical and dielectric relaxation. This problem can be treated by the linear response method by introducing an effective time instead of real time in considering the continuous shift of the relaxation time distribution during the course of volume relaxation [11]. To this author's knowledge, no systematic studies have been performed yet to investigate the effect of thermal and electrical history on dielectric relaxation in ferroelectric polymers.

The form of the response function Φ may be derived from molecular models or from other considerations. The simplest function, introduced by Debye, is $\Phi(t) = 1 - \exp(-t/\tau)$, where the parameter τ is referred to as the relaxation time of the assembly of dipoles.

By substituting a periodic external field,

$$E(t) = E_0 \exp(i\omega t)$$

into Eq. (13), for the single relaxation time approximation, the frequency-dependent Debye equations are obtained for the real (ϵ') and imaginary (ϵ'') parts of the complex permittivity as

$$\epsilon'(\omega) = \epsilon_u + (\epsilon_r - \epsilon_u)f'(\omega) \tag{14}$$

$$\epsilon''(\omega) = (\epsilon_r - \epsilon_u)f''(\omega) \tag{15}$$

with

$$f'(\omega) = [(1 + \omega^2\tau^2)]^{-1} \tag{16}$$

$$f''(\omega) = \omega\tau f'(\omega) \tag{17}$$

The relaxed permittivity ϵ_r is the low-frequency limiting value, also referred to as "static" permittivity, while ϵ_u is the high (optical) frequency (unrelaxed) limit.

$\epsilon_r - \epsilon_u$ is referred to as the relaxation strength of the dielectric transition; it depends on the dipole concentration, the dipole moment of the group considered, and the strength of the dipole–dipole interaction. For weak interactions the relaxation strength is expressed by the Fröhlich-Kirkwood equation (see, e.g., Ref. 1, pp. 82–102) as

$$\epsilon_r - \epsilon_u = \frac{\epsilon_r(\epsilon_u + 2)^2}{2\epsilon_r + \epsilon_u} \frac{cg\mu^2 N_\mu}{T} \tag{18}$$

where μ is the group dipole moment (debye), g is the Kirkwood reduction factor, which characterizes the dipole–dipole interactions, N_μ is the dipole concentration in moles per liter, c is a factor having a value for the units used here of $c = 6.1 \times 10^6$ Km/A^{2}s^2 mol.

In ferroelectrets Eq. (18) is applicable if the Kirkwood reduction factor g is modified considering the symmetry of the system where the dipolar groups are located.

The single-relaxation-time approximation corresponds to an assembly of like, and mutually not interacting, dipoles. This is, evidently, never realized. It is a common practice to regard the assembly of real, mutually interacting dipoles by assuming a distribution of relaxation times and approximating the actual response function by a series of simple exponentials of different τ values. By increasing the number of terms of this series, a relaxation distribution function $F(\tau)$ is defined that characterizes a dielectric transition [1]. Although this treatment seems attractive, the number of parameters in such a series, which have to be used to describe an experimentally measured response function, is too high to be able to interpret molecular-scale models. Thus, it is more practical to use response functions which have fewer parameters. Quite a number of such functions have been recommended based on molecular or other (statistical, many-body interaction) considerations.

The dielectric relaxation time distribution function $F(\tau)$ can be calculated from any kind of response function $\Phi(t)$ as

$$F(\log \tau) = \sum c_i \left[\frac{d^i\Phi(\log t)}{d (\log t)^i} \right] \tag{19}$$

where the coefficients c_i in order of i are: 0.34, -2.41×10^{-2}, 5.02×10^{-3}, -3.35×10^{-4}, 2.79×10^{-5}, and -7.75×10^{-7} [23]. Thus, it is still useful to characterize a dielectric transition in terms of relaxation time distribution, even if it is not constructed as a series of simple exponentials.

A two-parameter response function based on consideration of diffusion and damping processes within single polymer chains developed by Gény and Monnerie (GM) [24] is

$$\Phi(t) = \exp\left(\frac{t}{\tau}\right) \exp\left(\frac{-t}{\rho}\right) \text{erfc}\left(\frac{t}{\rho}\right)^{1/2} \tag{20}$$

where erfc is the complementary error function, defined as

$$\text{erfc}(x) = 1 - \int_{u=x}^{\infty} \exp(-u^2)\, du \tag{21}$$

which may be calculated numerically. Parameters τ and ρ are related to orientational damping and diffusion processes within the polymer chains.

The GM response function was successfully applied to describe dielectric spectra of polymer solutions, those of unsaturated polyester resins [25], and even pulsed nuclear magnetic resonance results [26].

Another widely used semiempirical response function is

$$\Phi(t) = 1 - \exp\left[-\left(\frac{t}{\tau}\right)^n\right] \tag{22}$$

where τ is a relaxation time-like parameter and the exponent n characterizes the shape of the response function and, correspondingly, that of the relaxation time distribution function whenever that representation is used. The experimental value of n in Eq. (22) is usually between about 0.4 and 0.8 for polymers. $n = 1$ corresponds to the single-relaxation-time approximation.

The response function of Eq. (22) was recommended a long time ago by Kohlrausch and applied much later to describe dielectric response by Williams and Watts [27]. It will be referred to in this contribution as the Kohlrausch function. It is preferred by the author because it is useful for describing dielectric, mechanical, magnetic, and thermodynamic relaxation processes with just two parameters.

The GM function has explicit forms for the real and imaginary parts f' and f'', but the Kohlrausch function does not. In this case the integral of Eq. (13), which is essentially a Laplace transformation of the derivative of the response function, is calculated numerically. First the current response to a step voltage is measured and the derivative of the Kohlrausch function is fitted to it, then the Laplace transformation is performed.

The time-dependent current density (A/m^2) is expressed as

$$j(t) = j(\infty) + [j(0) - j(\infty)]\frac{dD(t)}{dt} \tag{23}$$

where $j(\infty)$ corresponds to continuous charge transport. The ohmic conductivity corresponding to this charge transport is $\sigma_0 = j(\infty)/E$ (S/m), where E (V/m) is the applied electric field. $j(0) - j(\infty)$ in Eq. (23) is the instantaneous displacement current density, which is related to the high-frequency (unrelaxed) dielectric permittivity ϵ_u. The derivative of the displacement $D(t)$ is proportional to that of the response function $\Phi(t)$.

For illustration, Figure 2 shows the real f' and imaginary f'' components obtained by numerical transformation of the derivative of Kohlrausch functions with the same τ but different exponent (n) values. It is seen that the transition broadens by decreasing the value of the exponent and so does the corresponding relaxation time distribution.

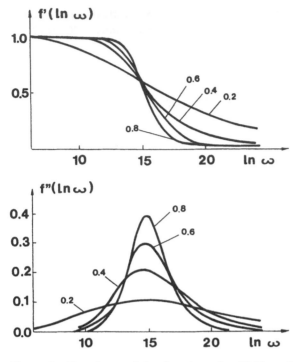

Figure 2 Transforms of the time-dependent Kohlrausch response function to frequency-dependent real (f') and imaginary (f'') parts.

The parameters of the response function are dependent on temperature and pressure. The temperature dependence will be discussed here for the Kohlrausch function only. In this function, parameter τ determines the angular frequency where the dielectric loss and the relaxation time distribution calculated from Eq. (19) are near to maximum. The temperature dependence of this parameter for such dipolar groups, the thermal motion of which do not influence the main structure appreciably, is

$$\tau(T) = \tau(T_0) \exp\left[\left(\frac{H}{R}\right)\left(\frac{1}{T} - \frac{1}{T_0}\right)\right] \tag{24}$$

where T_0 is an arbitrary reference temperature, H is the enthalpy of the transition, and R is the gas constant. This, Arrhenius-like temperature dependence is observed for dielectric polarization of not too bulky dipolar side groups, the rotation of which would not disturb the whole structural arrangement appreciably.

For transitions which involve rearrangement of the structure, such as the glass transitions, the temperature dependence of parameter τ in the Kohlrausch equation is

$$\tau(T) = \tau(T_0) \exp\left[\frac{A(T - T_0)}{B + (T - T_0)}\right] \tag{25}$$

where A and B are the Williams-Landel-Ferry (WLF) constants (See, e.g., Ref. 2, p. 93).

In order to perform the simplified linear response treatment, polymers are characterized in terms of polarity of bonds appearing in their main chains, side chains, or attached to the chains. Polarity of a bond means the dipole moment associated with it, not considering the environment. Thus, the C–H bond is considered as nonpolar, and the C–Cl, C–F, and C–N bonds are considered as polar. This is, of course, an approximation because the environment may have significant effects, but it is very useful to develop a picture about the expected dielectric behavior. The most important bond moments occurring in polymers are listed in Ref. 2 (p. 22). On the basis of this concept, polymers may be classified according to the position of polar groups in their chains.

1. Nonpolar Polymers

Some polymers have absolutely no polar groups in their main chains nor in side groups. A trivial example is polyethylene, which contains C–H bonds and some branches consisting of CH_3 and CH_2 groups only, which are nonpolar. The reason why polyethylene exhibits some dielectric activity is that the polymer is always somewhat oxidized, which results in the appearance of some carbonyl and carboxyl groups attached to its main chain. These groups are polar because the C=O and C–O bonds have dipole moments. These groups are attached directly to the main chain and, correspondingly, follow its thermal motion.

2. Polymers Having Polar Bonds in Their Main Chains

In many kinds of polymers, for example, in oxide polymers, polymethylene oxides of the general composition

$$[(-(CH_2)_m-O-]_n$$

saturated polyesters such as polymethylene terephthalates,

with $m = 2$–10, polar groups are built into the main chains. Such polymers are polycrystalline, and typically exhibit broad, low-temperature (around 180 K) dielectric γ transitions associated with local motion of short segments in the main chain, an α_a transition associated with the glass transition of the amorphous part, and a high-temperature (320–330 K) α_c transition attributed to motion of disordered chain fragments located between the crystal lamellae.

3. Polymers Having Polar Groups Attached Directly to Their Main Chains

The subsystem of dipoles in these polymers is strongly coupled to the main chain. Thus the main dielectric transition is expected, and found, at the glass transition temperature, where large parts of the chain become mobile. Secondary transitions assigned to intramolecular mobility of short chain segments are usually smeared out to form a broad secondary transition band, as in the case of poly(vinyl chloride), where a very strong dielectric transition of relaxation strength of about 10 is observed with WLF-type temperature dependence at about 360 K at 1 kHz, and a very weak, broad transition in the temperature range from about 260 to 350 K exhibiting Arrhenius-like temperature dependence with an activation enthalpy of 64 kJ/mol.

By introducing another chlorine atom symmetrically to form poly(vinylidene chloride), the glass transition is shifted to a lower temperature and the relaxation strength is decreased. A similar tendency can be observed for poly(vinyl fluoride) and poly(vinylidene fluoride). In these polymers the relaxation strength increases abruptly by passing through the glass transition temperature. This indicates a change of the effective dipole moment concentration.

These polymers easily form ferroelectric structures by alignment of dipolar groups in a whole crystallite in a suitable polymorphic symmetry. The most thoroughly studied of these polymers are poly(vinylidene fluoride) and its copolymers. The properties of these ferroelectric structures are discussed elsewhere in this volume. The dielectric transitions of some representative halogen polymers are illustrated schematically in Figure 3.

4. Polymers with Flexible Polar Side Groups

In a large class of polymers the main chain is nonpolar but such polar side groups are attached, which can easily change their dipole moment direction, especially by rotation.

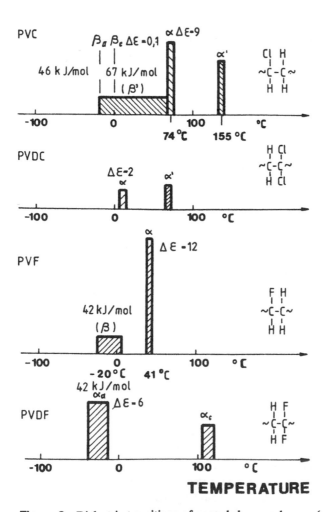

Figure 3 Dielectric transitions of some halogen polymers (schematic).

Poly(alkyl acrylates) and methacrylates belong to this class, with a general formula of

$$
\begin{array}{c}
R_1 \\
| \\
-C\!-\!C\!-\!C- \\
| \\
O\!=\!C\!-\!O\!-\!R_2
\end{array}
$$

where R_1 may be H (acrylates), CH_3 (methacrylates), or a longer alkyl group, while R_2 may be normal, iso-alkyl, cyclohexyl, or even a polar chloro-alkyl group. The ester side group has a group moment of 1.8 debye, so even if both groups R_1 and R_2 are nonpolar, the rotation of the ester side group about the C—C bond, which links it to the main chain, produces dielectric orientational polarization. Group R_2 may also be polar, thus, another rotation about the C—O axis results in dielectric activity originating from the corresponding thermodynamical subsystem. The dielectric transitions in these polymers have been thoroughly studied, as have the transitions assigned by introducing various (e.g., chlorinated) groups at different places of the molecule. The results are summarized in Ref. 2 (pp. 207–219). Here only a schematic representation is presented of the position of the dielectric transitions of poly(methyl acrylates) and methacrylates in Figure 4, indicating shifts caused by increasing the length of the nonpolar R_2 group. The apparent activation energies of the transitions are also indicated in Figure 4.

The thermodynamical subsystem representing long-range or local mobilities of the main chain is, evidently, coupled to those corresponding to mobilities of the side groups;

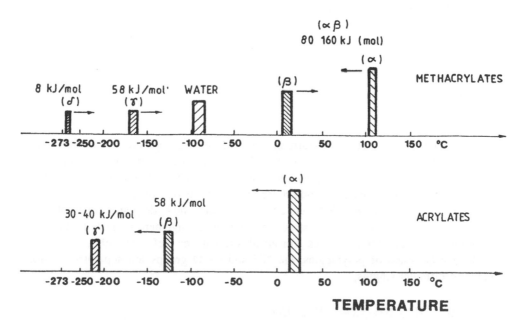

Figure 4 Dielectric transitions in acrylates and methacrylates (schematic). Arrows indicate shifts caused by alkyl substitution at the ester groups.

thus, at the glass transition temperature range a dielectric transition is observed, although the main chain is nonpolar. This dielectric transition is, however, weaker than that corresponding to liberation of the rotation of the main chain. For the mechanical transitions and for those measured by the pulsed NMR technique, the reverse is true. By substituting R_2 with alkyl groups of increasing length, the dielectric α peaks corresponding to large-scale main-chain motion increase and shift to lower temperatures, while the β peaks corresponding to side-group rotation decrease until the two transitions merge. This behavior is explained satisfactorily by considering the hindered rotation of the side group in a potential barrier model. For details, see Ref. 2 (p. 216).

The flexible side groups of poly(alkyl acrylates) have been used to prepare ferroelectric liquid crystalline structures [28].

5. Polymers with Hydrogen-Bond Chains

In certain groups of polymers intermolecular hydrogen bonds are present which may form hydrogen-bond chains, e.g., sequences of hydrogen bonds across the main hydrocarbon chains (see Ref. 2, p. 243). In certain polyamides and polyurethanes the chain conformation may be favorable for formation of such chains by interaction of C=O and N—H bonds. The hydrogen-bond chains are interrupted by defects caused by unfavorable configurations of groups.

The intermolecular interaction in hydrogen-bond chain polymers is rather high, because the hydrogen-bond is much stronger than the van der Waals interaction. Also, the dipole moment of a hydrogen bond is rather high, and there is a possibility of proton transfer along the hydrogen-bond chain. Depending on the chemical structure, hydrogen-bond dipoles may form ferroelectric structures (Ref. 6, p. 402). The general chemical structure of the repetition unit of a class of polyamides prepared by condensation of ω-amino acids is

$$-\underset{\underset{H}{|}}{N}-(CH_2)_n-\overset{\overset{O}{\parallel}}{C}-$$

It is seen that the hydrogen-bond-forming NH and O=C groups are separated by alkyl chains of length according to the value of n. If n is odd, the number of carbon atoms in the repetition unit is even. These polyamides are conventionally referred to as even ones. The polyamides corresponding to odd n are referred to as odd ones.

Depending on the composition and crystal symmetry the hydrogen-bond dipoles may be oriented to form a ferroelectric structure. Such an arrangement is more probable for odd polyamides than for even ones. For an illustration, see Ref. 6 (p. 403).

In another class of polyamides, the NH and C=O groups are separated by alkyl chains of different lengths,

$$-NH-(CH_2)_n-NH-CO-(CH_2)_m-CO-$$

These polyamides are prepared by condensation of diamines with dicarboxylic acids.

The general structure of polyurethanes is

$$-N-(CH_2)_n-N-\overset{\overset{\displaystyle O}{\|}}{C}-O-(CH_2)_m-O-\overset{\overset{\displaystyle O}{\|}}{C}-$$
$$\quad\ |\qquad\qquad\ |$$
$$\quad\ H\qquad\qquad H$$

In these polymers, hydrogen-bond chains may also be formed by NH $\cdots$ O=C interactions.

Figure 5 shows schematically the main transitions in polyamides and polyurethanes. The highest temperature transition below crystalline melting (α) is attributed to mobility in the amorphous phase, since it decreases with increasing crystalline-to-amorphous ratio, exhibits WLF-type temperature dependence, and is shifted to lower temperatures by increasing sorbed water content (cf. Ref. 1, pp. 480–494). The lowest temperature (γ) transition is attributed to local motions of the alkyl parts of the main chain in the amorphous as well as in the crystalline phase. Transition β is independent of the crystalline-to-amorphous ratio but decreases by decreasing amount of sorbed water in the polymer. The β process is attributed to motion of water molecules coupled to the carbonyl groups of the polymer by hydrogen bonding. (Ref. 1, p. 496.)

The dielectric spectra of polyamides are considerably obscured by the high proton-conductivity background, which is observed even in carefully dried samples. The situation is even more complicated by the presence of sorbed water, which makes the polymer dielectrically very heterogeneous.

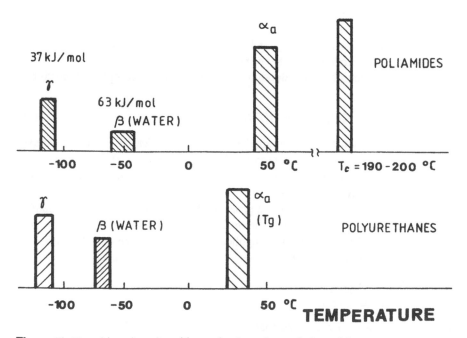

Figure 5 Transitions in polyamides and polyurethanes (schematic).

D. Interfacial Polarization

As was discussed in Section II, polymeric solids exhibit heterogeneous structure. In semicrystalline polymers there are interfaces between the surface of the crystallites and the surrounding amorphous material. Even within the crystallites there are interfaces between the lamellae. In amorphous polymers, aggregates are usually formed. The presence of conjugated segments or areas in the polymer, as discussed in Section III.B, also makes the polymeric solid highly heterogeneous as a dielectric material.

At the interface of two dielectric materials having different permittivities and/or conductivities, charges are accumulated according to simple electrodynamic considerations. Thus, a layer of dipoles induced by the external electric field is formed at the interface, resulting in an increase of the total polarization field, i.e., an increase of the dielectric permittivity, often up to very high values.

If, for example, highly conductive inclusions are present in a nonpolar polymer matrix, the dielectric permittivity may reach a level of several thousand as a result of the interfacial polarization at the interface of the inclusions. On the other hand, if conjugated parts are formed in the molecular or supermolecular measures, as discussed in Section III.B, the resulting polarization will be a MWS one. This means that the very high "giant" or "nomadic" polarization considered by Pohl [29] is, in fact, a MWS polarization caused by the dielectric heterogeneity introduced by the highly conjugated parts exhibiting high polarizability and conductivity. Thus, there is a "giant" polarization whenever the conjugation reaches macroscopic measures, but the very high permittivities actually observed are due to the MWS polarization at the interface of the highly conductive conjugated parts and the less conductive matrix.

Interfacial polarization can be treated quantitatively if the volume fraction of the inclusion to the matrix and the shape of the inclusion is known. This is possible even if both phases exhibit dielectric dispersion. For a recent analysis, see Ref. 30.

In general, the complex permittivity $\bar{\epsilon}$ of a two phase system consisting of a homogeneous phase and inclusions can be expressed as a function of the individual permittivities ϵ_1, ϵ_2, volume fractions v_1, v_2, and a factor A characterizing the shape of the inclusion. Several such functions have been introduced [30].

In these arguments the ohmic conductivity is included in the complex permittivities as

$$\epsilon(\omega, T) = \epsilon'(\omega, T) - i[\epsilon''(\omega, T) + \sigma(T)/\epsilon_0\omega] \qquad (26)$$

where ω is the angular frequency of the applied electric field (2π Hz), T is the temperature (K), ϵ_0 is the vacuum permittivity, and σ is the conductivity (S/m) arising from charge transport, which is usually not dependent on the angular frequency and increases exponentially by increasing temperature.

In this chapter only one of the many approaches for describing interfacial polarization is discussed: the one which seems to be useful for application to ferroelectric systems. The most straightforward and easily applicable function is the one developed by Wagner and Sillars (cited in Ref. 32):

$$\bar{\epsilon} = \frac{(1 - A)\epsilon_1 v_1 + \epsilon_2(v_2 + v_1 A)}{\epsilon_1 + Av_1(\epsilon_2 - \epsilon_1)} \qquad (27)$$

By simulations it is easy to recognize that, even when neither of the phases exhibits dielectric dispersion, the two-phase system may show a Debye-like response to external electric fields. This is illustrated in Figure 6, where the real and imaginary parts of the complex permittivity are plotted against frequency for such a system which has a non-polar matrix (e.g., polyethylene) and a somewhat more conductive sphere-shaped inclusion with $\epsilon_1' = 2$, $\epsilon_1'' = 10^{-4}$, $\sigma_1 = 10^{-20}$ S/m, $\epsilon_2' = 10$, $\epsilon_2'' = 1$, $\sigma_2 = 10^{12}$ S/m. The volume fraction of the inclusion is changed as shown in the figure. In the present scale the change of the dielectric permittivity and conductivity of the matrix polyethylene is negligible. The inclusion has no dielectric dispersion but has a relatively large (temperature-dependent) conductivity. As a result of the change of the interfacial polarization, a transition is observed in the dielectric spectrum. Such a transition will be referred to here as a Maxwell-Wagner-Sillars (MWS) transition, and has nothing to do with the phase transition of the polymeric system, but originates from its electrical heterogeneity. For details for inclusions of different shape, see Ref. 32.

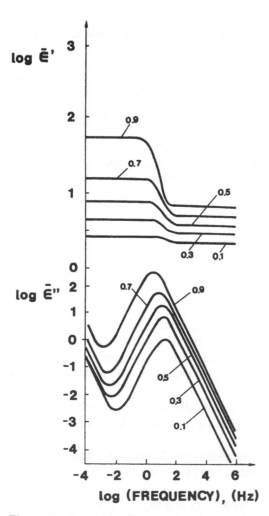

Figure 6 A transitionlike response of a nonpolar low-conductivity polymer mixed with a higher-permittivity, medium-conductivity, nondispersive filler. Simulation by Eq. (27).

In the case of polymer blends and composites, one or both phases may be dispersive. In such systems, in addition to the dielectric transitions, separate MWS transitions may be detected, or some transitions may be distorted, even obscured, by the MWS effect. This problem has recently been thoroughly analyzed and checked experimentally [30].

IV. EXPERIMENTAL ASPECTS

A. Time-Domain Techniques

The most straightforward method for measuring dielectric permittivities and losses is to apply a voltage step to the sample and measure the current response. According to Eq. (23), the measured time-dependent current is

$$I(t) = I(\infty) + [I(0) - I(\infty)] \frac{d\Phi(t)}{dt} \tag{28}$$

where Φ is the response function.

The ohmic conductivity σ_0 is calculated from the current extrapolated to long times $I(\infty)$ as

$$\sigma_0 = \left(\frac{\epsilon_0}{C_0}\right) \frac{I(\infty)}{V} \tag{29}$$

where $C_0 = \epsilon_0(A/d)$ is the geometric capacitance (F), A is the area of the electrode (m^2), ϵ_0 is the vacuum permittivity, and V is the applied voltage (V).

For numerical transformation of the real and imaginary parts of the complex permittivity, only that part of the current is considered which arises from polarization, i.e., $\bar{I}(t) = I(t) - I(\infty)$. Using this current, the numerical transformation formulas are [31]

$$\epsilon'(\omega) - \epsilon_u = \left(\frac{t}{C_0 V}\right) \left[\int_0^t \bar{I}(u)\, du + \sum_{i=1}^4 a_i \bar{I}(4it) \right] \tag{30}$$

$$\epsilon''(\omega) = \left(\frac{t}{C_0 V}\right) \left[b_0 \bar{I}\left(\frac{t}{2}\right) + \sum_{i=1}^4 b_i \bar{I}(2it) \right] \tag{31}$$

$$\omega = \frac{1}{t}$$

where V is the voltage (V), $C_0 = 0.8854$, A/d is the geometric capacitance in pF when the area of the electrode A is in square centimeters, and the distance d is in millimeters. t is time (s), ω is the angular frequency (2π Hz), and a_i and b_i are numerical coefficients: $a_1 = -2.61 \times 10^5$, $a_2 = -1.232 \times 10^6$, $b_1 = 314$, $b_2 = 2.378 \times 10^3$, and $b_3 = -1.3 \times 10^3$ [31].

This simple step-voltage technique is easily applicable at low frequencies from about 10^{-5} to 100 kHz, and probably even higher by using recently available high-speed data collecting facilities.

A generally applicable more accurate transformation formula of a time-dependent response function $\phi(t)$ to a frequency-dependent complex function $f(i\omega)$ is

$$f(i\omega) = \frac{dt\, \exp(i\omega\, dt/2)}{2 \sin(\omega\, dt/2)} \sum_{k=0}^n \{\phi(k\, dt) - \phi[(k - 1)\, dt]\} \exp(i\omega k\, dt) \tag{32}$$

where dt is the sampling interval and n is the total number of sample points.

Attempts have been made to use this method in the gigahertz range by using very-high-speed switches and waveguides. The method is referred to as time-domain reflectometry [34], because for calculating the permittivity the complex reflection coefficient of the sample located at an end of a waveguide is used. The technique is by no means an easy one, because there are limitations to the sample thickness and because spurious reflections occur. Recently, this technique has been used to study dielectric spectra of polymer coatings in the frequency range from 10^7 to 10^9 Hz [34].

B. Frequency-Domain Techniques

The most widely used method for measuring frequency-dependent dielectric permittivities is to apply a periodic electric voltage to the sample, $V(i\omega) = V_0 \exp(i\omega t)$, and measure the in-phase and out-of-phase components of the current $I(i\omega)$. The results are usually expressed in terms of complex admittance Y, defined as $Y(i\omega) = I(i\omega)/V(i\omega)$.

The real and imaginary parts of the complex admittance are related directly to the parts of the complex permittivity as

$$Y'(\omega) = \omega C_0 \epsilon''(\omega) + \left(\frac{C_0}{\epsilon_0}\right)\sigma(T) \tag{33}$$

$$Y''(\omega) = \omega C_0 \epsilon'(\omega) \tag{34}$$

where C_0 is the geometric capacitance and σ is the ohmic conductivity.

It is seen that with this technique the loss arising from the ohmic conductivity cannot be separated from dielectric loss. Instead of ϵ' and $\sigma(\omega, T)$, ϵ' and the loss tangent, defined as $\tan \delta = \epsilon''/\epsilon'$, is very often used. The value actually measured is $\tan \delta = Y'/Y''$, which means that the ohmic conductivity is mixed up with the dielectric permittivity and loss.

Since in ferroelectric polymers ohmic conductivity may be frequency dependent, the $\tan \delta$ representation is not recommended. The dielectric transition may be represented by the frequency-dependent permittivity related to the imaginary part of the actually measured admittance (Y'') and by the frequency-dependent AC conductivity related to the real part Y' as

$$\sigma(\omega, T) = \sigma_0(T) + \omega \epsilon_0 \epsilon''(\omega, T) \tag{35}$$

where the ohmic conductivity σ_0 usually, but not always, depends on the temperature only.

It is common practice to define the dielectric transition frequency by the maxima of the $\epsilon''(\omega)$ or $\tan \delta(\omega)$ curves, although these are somewhat shifted with respect to each other. Neither of these methods is correct, because the relaxation time distribution is usually nonsymmetric. The transition frequency may be defined as that corresponding to the derivative of the real part ϵ' by log(frequency). According to Eq. (19), this corresponds to the reciprocal of the maximum of the relaxation time distribution. The first derivative is usually sufficient. ϵ' is affected by the ohmic conductivity only when interfacial polarization is dominant in the frequency range studied.

C. Temperature-Domain Techniques

As discussed in Section IV.B, the parameters of the response function, e.g., τ and n in the Kohlrausch function [Eq. (22)], are dependent on the temperature. This is shown for

τ by Eqs. (24) and (25). The exponent n may be considered in many cases as being temperature independent. This means that the response function $\Phi(t)$ and its transformed real and imaginary parts f' and f'' are also temperature dependent. Besides, according to the Fröhlich-Kirkwood equation, the relaxation strength $\epsilon_r - \epsilon_u$ exhibits a $1/T$ dependence according to Eq. (18). The ohmic part of the conductivity for usual polymers is exponentially increasing with increasing temperature as

$$\sigma_0(T) = \sigma_\infty \exp\left(\frac{-E_c}{RT}\right) \tag{36}$$

where E_c is the activation energy of charge transport.

Despite these apparent complications, a very common practice is to measure dielectric transitions by scanning the temperature at constant frequency or changing the frequency in steps while the temperature is scanned or changed stepwise. By studying multiple transitions, this method has an advantage that the results can be easily compared with dilatometric, scanning calorimetric, pulsed nuclear magnetic resonance, and thermomechanical results which can only be obtained by temperature scan.

A typical, simple temperature-domain spectrum band is shown in Figure 7, where the dielectric glass transition is shown for a conventional plasticized poly(vinyl chloride) compound in ϵ' and σ representation. In this compound the ohmic conductivity (10^{-11}–10^{-12} S/m) is not too large to obscure the dielectric loss [$\epsilon''(t)$] peak. The exponentially

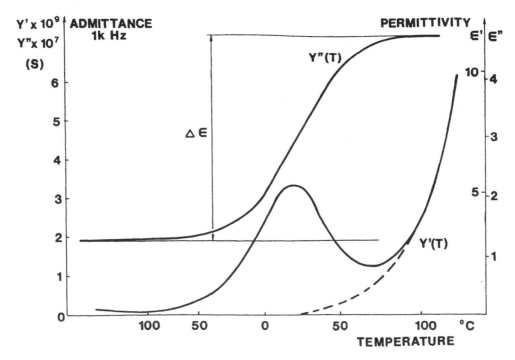

Figure 7 A typical temperature-domain dielectric transition for a plasticized poly(vinyl chloride) in ϵ', σ representation showing admixture of conductivity with loss.

increasing ohmic conductivity is also visible, and by extrapolation according to Eq. (36) it can be subtracted from the total AC conductivity $\sigma(\omega, T)$ to obtain the part representing the dielectric loss ϵ''. The activation energy of charge transport appears in this case as 48 kJ/mol. The $1/T$ decrease of the unrelaxed permittivity appearing at the high-temperature side is observed too. At higher conductivities, and in the presence of strong interfacial polarization, the dielectric loss peak may be totally obscured. In such cases the transition may be represented by the relaxation time distribution function calculated according to Eq. (19) from the real permittivity (ϵ')-versus-temperature curve, which can usually be detected even if the conductivity background is high. Instead of frequency, the derivation of ϵ' is performed by using temperature as an independent variable. The first derivative $d\epsilon'/dT$ is usually sufficient to characterize the dielectric transition, as in the frequency domain.

A more refined analysis is possible by introducing the temperature dependence of the parameters of the response function into the transformed equations of $\epsilon'(\omega, T)$ and $\sigma(\omega, t)$ and performing an iterative fitting procedure. With this method, closely located dielectric transitions can be separated.

The temperature-domain analog of the time-domain method is the thermally stimulated depolarization (TSD) technique, which involves measurement of the (usually) short-circuit current by heating up a previously polarized sample at a constant rate. Provided that the permanent polarization originates from dipole polarization, the response functions of Eq. (20) or Eq. (22) may be used in describing the current density measured as a function of the temperature, taking into account the temperature dependency of the parameters of the response function.

In the simplified case, when the exponent n of the Kohlrausch function [Eq. (22)] is considered as independent of temperature, the time-dependent response function can be transformed into the temperature scale by replacing the real time t with the effective time λ, defined as

$$\lambda(t) = \int_{u=0}^{t} \frac{du}{\tau[T(u)]} \tag{37}$$

where the relaxation time-like parameter of the Kohlrausch function τ is implicitly dependent on the variable of integration u through the temperature T.

The condition of this transformation is that the shape of the response function should not depend on the temperature, i.e., in the Kohlrausch function the exponent n should be temperature independent. In the relaxation time distribution representation this means that the shape of the distribution should be independent of temperature, i.e., the distribution is just shifted along the relaxation time axis. This condition is also referred to as the time–temperature superposition principle in classical rheology. The method of effective time is widely used in temperature-stimulated techniques such as the quasi-static thermomechanical method, thermoluminescence, differential scanning calorimetry, and even considering relaxation processes running simultaneously with thermodynamical (volume) relaxation, as was mentioned in Section II.B.

In TSD the sample is first heated up to the poling temperature under an external electric field E. The correspondingly developing polarization is

$$P(\lambda) = \epsilon_0(\epsilon_r - \epsilon_u)E\Phi(\lambda) \tag{38}$$

At a constant rate of heating, Eq. (37) (the effective time) simplifies to

$$\lambda = \left(\frac{1}{r}\right) \int_{u=0}^{T} \frac{du}{\tau(u)} \tag{39}$$

where r is the rate of heating (K/s).

The current density measured in the heating period is

$$j(T) = \frac{dP(T)}{dT} + \sigma_0(T)E \tag{40}$$

where σ_0 is the ohmic conductivity.

At the final temperature the sample is stored for a while until a final polarization P_f is reached, then the sample is cooled down at a constant rate to a temperature where it is stored again to reach a polarization P_F. The degree of polarization may be characterized by a factor Γ, defined as the ratio of polarization P_F to $\epsilon_0 E(\epsilon_r - \epsilon_u - 1)$, which is considered the highest polarization obtainable for the given transition.

With subsequent heating at a constant rate, the short-circuit current density generated by the depolarization of a dipolar group represented by a response function Φ is

$$j(\lambda) = \Gamma \epsilon_0 \, \Delta \epsilon \, E \, \frac{d\Phi}{d\lambda} \tag{41}$$

where $\Delta \epsilon = \epsilon_r - \epsilon_u$ is the relaxation strength, which depends on the dipole moment concentration according to Eq. (18).

The appearance of several TSD current peaks in experimental TSD spectra may originate from the presence of several types of dipolar groups characterized by different response functions, from depolarization of interfacial polarization, from space charges formed at the electrodes, or from recombination of trapped charge carriers [33]. Consequently, in order to assign the experimentally observed TSD peaks several techniques are combined. Simple transformation of a TSD current peak to the frequency domain is by no means recommended without careful consideration, taking into account results obtained by thermomechanical, DSC, NMR, and other techniques. Even if a TSD peak appears at such a temperature, where a transition is detected by other methods, the relaxation strength obtained by straightforward transformation to the frequency domain usually appears too high as a result of the interfacial polarization, which is practically always present in polymeric solids. Especially in the case of polymer electrets, which are dielectrically essentially heterogeneous systems, TSD is not the best method for detecting and assigning multiple transitions.

As a simple example, Figure 8 shows the TSD spectrum of a blend of plasticized poly(vinyl chloride) with polyethylene. At the actual sensitivity level of the measurement, polyethylene alone shows no dispersion, and the plasticized PVC matrix has a glass transition temperature at 70°C at the effective frequency corresponding to the rate of heating of the TSD measurement, verified by various other techniques. The TSD current peak appearing at 10°C is due to the interfacial polarization, and has nothing to do with intramolecular mobilities. According to Eq. (27) and to more detailed studies [30], the exponentially increasing ohmic conductivity by temperature-domain measurements induces very high interfacial polarization in heterogeneous systems no matter in which phase this increase occurs.

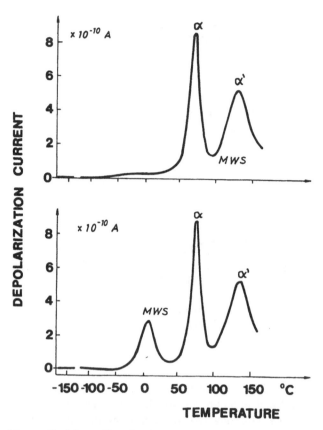

Figure 8 Thermally stimulated depolarization in poly(vinyl chloride) (PVC) and in its blend with polyethylene (PVC + PE).

D. Technical Considerations

Although there are various excellent dielectric spectrometers available commercially, when one meets a special problem, as in study of electrets, some "homemade" solutions are necessary. In investigating ferroelectric polymers the possibility of applying a relatively high bias voltage in frequency-domain measurements is necessary. This voltage may be used to perform time-domain experiments in the same sample holder, if it is constructed accordingly. Also, TSD experiments may be performed by using the same sample holder. The control and data collecting facilities have to be flexible enough to be changed easily, if necessary. Using available parts, it is not difficult to set up a combined dielectric spectrometer which meets the special requirements for studying ferroelectric polymers.

Figure 9 shows a simplified block scheme of an automated frequency- and/or temperature-domain dielectric spectrometer. As a generator, usually a digital frequency synthesizer is used in the frequency range from 1 Hz (or lower) up to about 1 GHz. For detectors, usually automated bridges or impedance analyzers are used, which are able to separate the in-phase and out-of-phase components of the complex admittance. Thus the signals going to the data acquisition unit are directly proportional to the real and imagi-

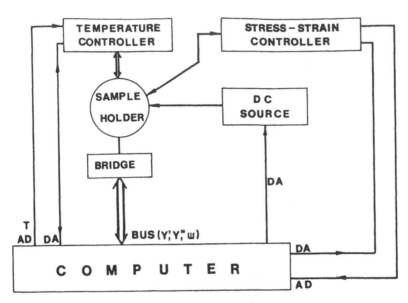

Figure 9   Scheme of a dielectric spectrometer to study ferroelectric polymers. AD and DA mean analog-to-digital and digital-to-analog channels, respectively.

nary parts of the complex permittivity. The data collecting device is a microcomputer, equipped with analog-to-digital (AD) and digital-to-analog (DA) converters. The temperature controller may be driven by the DA converter, while through the AD converter the frequency, temperature, Y', and Y'' signals, and, in some cases, the signal corresponding to the change of the sample thickness are collected. The data collecting unit for frequency-domain dielectric spectroscopy does not need to be a high-speed one. For data collection frequencies the widely available 100-kHz 16-bit AD controllers are sufficient.

The most delicate part of the spectrometer is the sample holder. It should be easily thermostatted from about 100 to about 800 K, and dry inert atmosphere in the sampling area is necessary. Very important is to avoid development of a temperature gradient across the sample thickness during the course of the temperature scan. Application of high scanning rates may cause serious errors, even with thin samples. The electrodes should preferably be made in contact with the sample by a pneumatic device by a command from the microcomputer. Since ferroelectric polymers are usually piezoelectric, the actual force and deformation should be controlled and detected during the course of the measurement. The electrodes (preferably gold) should be vacuum deposited onto the sample to avoid interfacial polarization. The electrodes of the sample holder act only to ensure contact with the vacuum-deposited electrode. Samples prepared by compression of powders, unfortunately, always produce high MWS effects, which may obscure dielectric transitions, especially at low frequencies. To study ferroelectrets, a DC bias voltage is also needed. This voltage may be used for poling samples for thermally stimulated depolarization (TSD) studies. In this case, instead of the bridge or impedance analyzer, a high-gain, high-stability current amplifier is used.

All these parts are available commercially, together with the microcomputer needed to control the measurement and for data collection.

V. EXAMPLES

A. Liquid Crystalline Polymers

Although the chemical structure of liquid crystalline polymers is rather complex, one of them is selected as a first illustrative example for dielectric studies because its structure is very well defined.

In liquid crystalline side-chain polymers (see Ref. 7, p. 49), there is a backbone of long nonpolar chains to which polar mesogenic groups are attached by a relatively short, flexible, nonpolar side chain. This makes it possible for the polar mesogenic groups to arrange in liquid crystalline layers of different types depending on the method of preparation.

As an example, the polymer synthesized and studied by Pfeiffer et al. [35] is considered. This polymer is an isotropic fluid above 115°C. By cooling it down in the dielectric cell very slowly under a magnetic field of 1.2 T, an oriented smectic A* (SmA*) phase is formed, where the dipolar mesogenic groups are aligned along the direction of the magnetic field. Upon removing the magnetic field, the alignment is restored. By cooling the fluid below 85°C, a smectic C* (SmC*) phase is formed consisting of layers of groups tilted by an angle α, which can be measured with a polarization microscope. The thickness of the layers was measured by SAX. This way, arrays of dipolar groups arranged in layers and oriented at a known tilt angle with respect to the layer surface are obtained. By application of a periodic electric field a dipole orientational polarization is induced with a relaxation time depending on the mobility of the mesogenic groups within the layers. One of the processes considered for low-molecular liquid crystals of similar symmetry [35] is referred to as the "soft mode." It is attributed to change of the tilt angle α. The other mode, referred to as the Goldstone mode [35], is attributed to a precessionlike motion of the long axis of the group. The Goldstone-mode motion is influenced by external DC field, while the soft mode is not.

Figure 10 shows the dielectric spectrum of the ferroelectric liquid crystalline polymer after Pfeiffer et al. [35] in the temperature range of the smectic A* phase and also that in the smectic C* phase. Just the imaginary component is shown for simplicity, as a function of the frequency. The rather intense relaxation transition observed around 100 Hz in the SmA* region is attributed to the soft mode. It is thought to be associated with tilt angle movement, which becomes more and more intense as one approaches the SmA*–SmC* transition temperature (82°C).

The high-frequency relaxation observed between 10 and 100 kHz in the SmC* phase is attributed to rotation of the mesogenic groups about their long axis, coupled to the motion of the main chain. This process is referred to as the molecular mode. The soft-mode relaxation strength, $\epsilon_r - \epsilon_u$, is found to exhibit a sharp peak at about 85°C starting from a value of 2.5 in the SmC phase, reaching a value of 4.5, from where it drops steeply almost to zero. The sharp maximum of the relaxation strength observed near 85°C corresponds to the destruction of the ferroelectric alignment; i.e., this is the Curie temperature. The steep decrease of the relaxation strength above 85°C is due to the reduction of the effective dipole moment.

B. Halogen Polymers

Halogen polymers have halogen atoms coupled directly to their nonpolar main chain. Main representatives are poly(vinyl chloride), poly(vinylidene chloride), poly(vinyl flu-

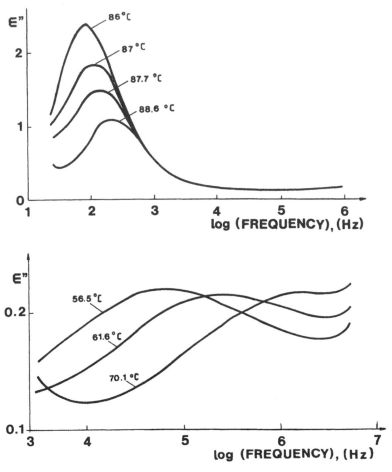

Figure 10 Frequency-domain dielectric spectra at different temperatures of a ferroelectric liquid crystalline polymer synthesized by Pfeiffer et al. (Reproduced from Ref. 35 by permission of Gordon and Breach Science Publishers, S.A.)

oride), poly(vinylidene fluoride), poly(trifloro ethylene), poly(chlorotrifluoroethylene), poly(tetrafluoroethylene), and poly(vinyl bromide).

Poly(vinylidene fluoride), $(CH_2CF_2)_n$, is a polycrystalline polymer with repetition units n of the order of 30,000 and a crystalline-to-amorphous ratio of about 0.6. The crystalline part may exist in several polymorphic modifications depending on the method of preparation. Details of its structure are discussed elsewhere in this volume.

By simply quenching from melt, the α form (crystal form II) is formed, which is not ferroelectric. In this polymer three dielectric transitions may be detected, the β transition at about $-60°C$, the α_a transition near $-10°C$, and the α_c transition near $150°C$. The β transition exhibits an Arrhenius-like relaxation behavior with an activation enthalpy of 50 kJ/mol; it is attributed to local motions in the amorphous as well as in the crystalline state. The α_a transition has a WLF-type relaxation behavior; it is attributed to the glass transition of the amorphous part. The α_c transition is attributed to motions in the crystalline phase, but most probably not within the lamellae but in the partially disordered interlamellar space. According to this view, the mobile parts are the chain folds located between the lamellae. Their motion, coupled to motions within the lamellae, would be responsible for the α_a transition. As an illustrative example, the pressure dependence of the frequency-domain dielectric α_c transition will be discussed here after Miyamoto [36]. By application of hydrostatic pressure, the dielectric spectrum band is shifted to higher frequencies without significant change of shape. This is shown in Figure 11, by Miyamoto [37], by plots of ϵ' and ϵ'' as a function of log frequency. It is worth noting that the unrelaxed permittivity in these experiments appears to be extremely high, of the order of 10. This indicates the presence of interfacial polarization caused by the

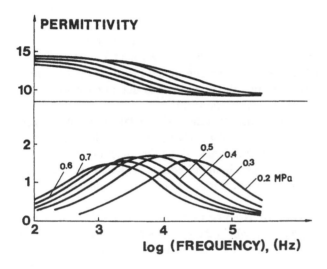

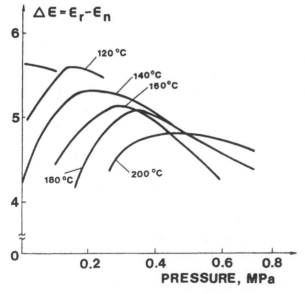

Figure 11 Frequency-domain dielectric transition in β-poly(vinylidene fluoride) at 180°C at different pressures indicated in MPa (a) and the pressure dependence of the relaxation strength at different temperatures (b). (Redrawn from Ref. 37 by permission of Butterworth Heinemann, Ltd.)

heterogeneity of the lamellar structure in such conditions when both phases are dispersive. Such a situation can be handled by using a proper mixture formula [32] whenever the volume fractions of the phases are known. Miyamoto used a correction to reduce MWS effect caused by the relatively high ohmic conduction, at low frequencies, but did not consider the problem of high unrelaxed permittivities observed.

The mean relaxation time associated with the maxima of the loss curves (or the parameter τ of the Kohlrausch function) depends on the pressure p according to

$$\tau(p) = \tau(p_0) \exp(H_p p) \tag{42}$$

where H_p is the activation enthalpy of the pressure and p_0 is an arbitrary reference pressure. By increasing the temperature, this activation enthalpy is somewhat decreased and the τ values decrease by decades.

Miyamoto also observed that the relaxation strengths measured at different temperatures exhibit maxima as a function of pressure. This is shown in Figure 11b. This behavior could be interpreted qualitatively by a two-site model based on the site theory of Hoffman (see Ref. 2, p. 41), assuming that the α_c process is due to changes of conformation from $TGT\overline{G}$ to $\overline{G}TGT$ in the α-crystalline form.

Miyamoto and co-workers [37] also studied the effect of orientation on the dielectric α_c transition of the α-polymorphic crystal form of poly(vinylidene fluoride). The relaxation strength of the transition was found to be highest along the direction of orientation and to decrease down to near zero in the perpendicular direction. With random orientation the relaxation strength is about 4, along the orientation about 22, perpendicularly to the orientation direction near zero.

From the dielectric studies, Miyamoto assumed that there were conformational defects in the chains in the crystalline phase in which the all-trans modification is the more stable one. All-trans modification is ferroelectric (β form). Now it is quite clear that the all-trans ferroelectric β form can be formed only if the defect concentration is higher than about 12% [7]. Defects can be introduced easily by copolymerization. This is why vinylidene fluoride is copolymerized with other monomers such as trifluoroethylene. In this way defects can be introduced into chains in variable amounts in order to help formation of the ferroelectric β form.

At the Curie temperature T_c, the dielectric relaxation strength increases to a maximum value. By scanning the temperature downward, this maximum is reached at much lower temperature. T_c is known as a first-order transition with enthalpy change; where an endothermic peak is observed by differential scanning calorimetry [7], Glatz-Breitenbach et al. [38] studied the dielectric transition at T_c in thin films of a vinylidene fluoride-trifluoroethylene copolymer of composition 75:25. The permittivity showed no dispersion in the frequency range from 20 to 100 kHz in the temperature range between 30 and 150°C, where crystalline melting occurs. With increasing temperature the permittivity (ϵ_r) reaches a maximum at 130°C, while with cooling the maximum is observed at 85°C. The film thickness was varied from 150 to 0.2 μm. With decreasing film thickness by ϵ_r maximum also decreases, as shown in Figure 12. The reduction of the ϵ_r(max) with decreasing thickness is explained by the reduced mobility, when the film thickness approaches the order of the lamellar thickness. Since the $\epsilon(T)$ curves show no dispersion and the ϵ values are very high, the dielectric transition can hardly be associated with a relaxation due to mobility of dipoles. The transitions connected with mobility are probably at lower temperatures. The change of the relaxation strength of the dielectric transition associated with the ferroelectric–paraelectric phase transition in liquid crystals (see,

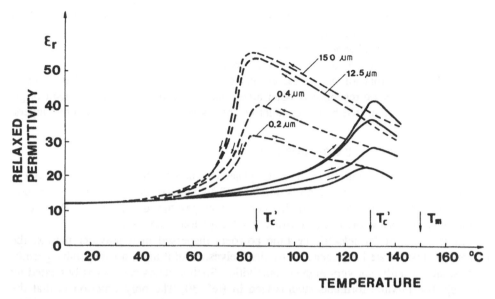

Figure 12 Effect of sample thickness on the dielectric transition of a vinylidene fluoride-trifluo-roethylene copolymer at 1–100 kHz. T_m is the melting temperature; T_c is the Curie temperature. (Reproduced from Ref. 38 by permission of Gordon and Breach Science Publishers, S.A.)

e.g., Ref. 35) is about 10 times smaller than that observed by the thinnest film of the copolymer. The interplanar spacing determined by SAX in these polymers is of the order of 0.4–0.5 nm (see Ref. 7, p. 27); the long period is of the order of 15 nm. The smallest film thickness measured by Glatz-Breitenbach et al. is 200 nm, thus enough interfaces are still present in the film to produce MWS polarization. ϵ_r is large in the ferroelectric state inside the lamellae, but small in the disordered interlamellar space. At the transition, ϵ_r in the lamellae drops suddenly, resulting in change of the overall $\bar{\epsilon}_r$ value [cf. Eq. (27)]. This might be an explanation of the very high ϵ_r values observed even in the 200-nm-thick films.

C. Highly Conjugated Polymers

As was mentioned in Section III.B, in highly conjugated polymers the conjugation may reach supermolecular measures, resulting in high conductivity and polarizability. The dielectric properties of such polymers will be discussed by the example of metal phthal-ocyanines showing ferroelectric behavior in experiments of Nalwa et al. [39]. Nalwa synthesized a series of metal phthalocyanines with different metal atoms such as Fe, Co, Ni, and Cu and studied their conductivities an dielectric permittivities. These polymers are paramagnetic and exhibit exponentially temperature-dependent ohmic conductivities of the order of 10^{-8} S/m. There is not much difference of the conductivities of the monomeric and polymeric phthalocyanines. On the other hand, by thermal treatment of the polymers the conjugation may be extended and the conductivity increased to the level of 1 S/m.

The dielectric permittivities at kilohertz frequencies are very high, reaching a level of 10,000. The permittivity increases with increasing temperature and drops steeply near 120°C, where spontaneous current generation is observed without external field. Upon cooling, the sign of this current is not reversed as in normal pyroelectric materials. With increasing frequency the permittivity decreases, but the temperature corresponding to the maximum does not change. The room-temperature dielectric permittivity of Cu-phthalocyanine was found to decrease from about 1000 to 10 at frequencies from 100 Hz to 100 kHz. From all these characteristics, Nalwa et al. concluded that in these polymers the dielectric response is determined mainly by the MWS mechanism originating from the conjugated highly conductive areas separated by less conductive layers.

D. Polymer-Ceramic Composites

As was shown in Section III.D, in heterogeneous materials the interfacial (Maxwell-Wagner-Sillars, MWS) polarization is dominant. This mechanism usually results in high overall permittivities and losses. The frequency or temperature dependence of the overall permittivity may show the characteristics of a Debye-like response.

In the case of ferroelectric ceramic powders dispersed in a polymer matrix, the ceramic inclusions are heterogeneous by themselves, exhibit high, nondispersive permittivities, and, usually, not very high conductivities. Such systems can be easily treated by the dielectric mixture formulas summarized in Ref. 30. The only criterion is that the volume fraction of the inclusions is known and also something about their shape. The size distribution is not important. The dielectric dispersion of the composite is determined by that of the matrix, which can be measured separately. The usual effect of the high-permittivity, nondispersive, ceramic filler is to raise the permittivity level of the dispersion bands. Sometimes new MWS transitions are created, which are not connected to molecular mobilities of the matrix.

Unfortunately, although the importance of interfacial polarization in such systems is generally realized, the results are not analyzed quantitatively by using dielectric mixture formulas.

As an illustrative example of the dielectric behavior of polymer-ceramic composites, the work of J. Wolak [40] is discussed here in some detail. Wolak studied the dielectric spectra of a vinylidene fluoride-tetrafluoroethylene copolymer (PVDF$_2$-TFE) filled with a ferroelectric ceramic Pb$_{0.9}$Ba$_{0.1}$(Zr$_{0.5}$Ti$_{0.5}$)O$_3$ + 1% Nb$_2$O$_5$ in a weight factor of 0.7. The grain size of the inclusion was 2.5 μm; the shape was not reported. The room-temperature permittivity (real part) was 1660, and tan δ = 0.018 at 1 kHz. The Curie temperature of the ceramic was 536 K. Unfortunately, the temperature dependence of tan δ was not measured, so the contribution of the ohmic conductivity is not known. Samples were prepared by hot-pressing of the powder mixture; vacuum evaporated Al electrodes were used. The dielectric spectrum of the unloaded copolymer was also studied. Frequency-as well as temperature-domain dielectric studies were performed in the frequency range from 20 Hz up to 1 GHz, in the temperature range from 110 to 420 K, using automated techniques.

Figure 13 shows a frequency-domain spectrum of the composite measured by Wolak in ϵ' and tan δ representation, and also a series of temperature-domain spectra measured at increasing frequencies. The main dielectric transition observed near 250 K is associated with the glass transition of the amorphous phase of the matrix. The smaller tan δ peak occurring at low frequencies is interpreted as being a crystalline transition in the matrix. This is observed in the pure matrix as well. The transition observed near 300 K at 1 MHz, shown in Figure 13, has the peculiarity that the relaxation strength decreases with increasing frequency, while the tan δ peak increases. It is difficult to understand this,

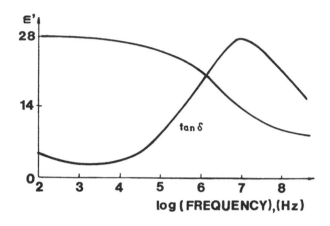

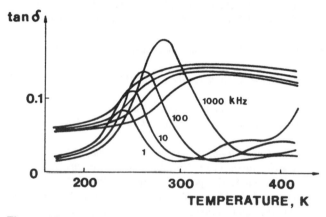

Figure 13 Frequency-domain (a) and temperature-domain (b) dielectric transitions of a ferroelectric ceramic-polymer composite. (Redrawn from Ref. 40 by permission.)

because in tan δ the ohmic part is also included but it is usually independent of the
frequency. The opposite tendency of the change of the relaxation strength and the tan δ
peak, and the area below the tan δ-versus-temperature curves, would indicate that the
ohmic conductivity of either the matrix or the inclusion increases with frequency.

Wolak fitted the frequency-domain spectrum of Figure 13 to the Davidson-Cole
response function (Ref. 1, p. 120), but it can be fitted to the Kohlrausch function as well.
The temperature shift of the tan δ maxima are found to be non-Arrhenius-like; in the
glass transition a WLF dependence is expected [Eq. (25)]. By replotting Wolak's data to
the WLF system, good agreement is obtained.

REFERENCES

1. N. G. McCrum, B. E. Read, and G. Williams, *Anelastic and Dielectric Effects in Polymeric
   Solids*, John Wiley, Bristol, U.K., 1967, p. 617.
2. P. Hedvig, *Dielectric Spectroscopy of Polymers*, Adam Hilger, Bristol, U.K., 1977, p. 431.
3. P. Hedvig, Combined relaxation spectroscopy of polymers, *J. Polymer Sci. Macromol. Rev.*
   15:375 (1980).

4. F. Guttman and L. E. Lyons, *Organic Semiconductors*, John Wiley, New York, 1967.
5. G. M. Sessler, Polymeric electrets, *Electronic Properties of Polymers*, (J. Mort and G. Pfister, eds.), John Wiley, New York, 1982, pp. 59–100.
6. H. S. Nalwa, Recent developments in ferroelectric polymers, *Rev. Macromol. Chem. Phys. C31(4):*341 (1991).
7. R. G. Kepler and R. A. Anderson, Ferroelectric polymers, *Adv. Phys. 41:*1 (1992).
8. N. Baur, *Einführung in die Thermodynamik der irreversiblen Prozesse*, Wissenschaftliche Buchgesellshaft, Darmstadt, 1984, pp. 72–140.
9. J. M. G. Cowie, Relaxation processes in the glassy state: Molecular aspects, *J. Macromol. Sci. Phys. B18:*569 (1980).
10. W. Klöpfer, *Introduction to Polymer Spectroscopy*, Springer-Verlag, Berlin, 1984, pp. 53–61.
11. A. J. Kovacs, J. J. Aklonis, J. M. Hutchinson, and A. Ramos, Isobaric volume and enthalpy recovery of glasses, *J. Polymer Sci. Polymer Phys. Ed. 17:*1097 (1979).
12. L. C. E. Struik, *Physical Aging in Amorphous Polymers and Other Materials*, Elsevier, Amsterdam, 1978, p. 229.
13. R. J. Samuels, *Structured Polymer Properties*, John Wiley, New York, 1974, pp. 1–113.
14. A. J. Lovinger, *Developments in Crystalline Polymers 1* (D. C. Basset, ed.), Applied Science, London, 1982, p. 195.
15. L. E. Nielsen, *Mechanical properties of polymers and composites*, Vol. I, Marcel Dekker, New York, 1962, pp. 379–503.
16. P. R. Bevington, *Data Reduction and Error Analysis for the Physical Sciences*, McGraw-Hill, New York, 1969, p. 149.
17. P. Hedvig, *Experimental Quantum Chemistry*, Academic Press, New York, 1975, p. 533.
18. N. F. Mott and E. A. Davies, *Electronic Processes in Non Crystalline Materials*, Clarendon Press, Oxford, 1971.
19. S. A. Rice and J. Jortner, *Physics and Chemistry of the Organic Solid State*, Vol. 3, (D. Fox, M. M. Labes, and A. Weissberger, eds.), Interscience, New York, 1967, pp. 201–489.
20. J. Tsukamoto, Recent advances in highly conductive polyacetylene, *Adv. Phys. 41:*509 (1992).
21. D. Baeriswyl, G. Harbeke, H. Kiess, and W. Meyer, Conducting polymers: Polyacetylene, *Electronic Properties of Polymers* (J. Mort and G. Pfister, eds.), John Wiley, New York, 1982, pp. 276–295.
22. G. M. Bartenev and Yu. V. Zelenev (eds.), *Relaxation Phenomena in Polymers*, John Wiley, New York, 1974, pp. 113–123, 178–201.
23. Yu. Zelenev and A. P. Molotkov, *Vysokomolekularnie Soed. 6:*1426 (1964) (in Russian); see P. Hedvig, Characterization of polymers by dielectric spectroscopy, *Applied Polymer Analysis and Characterization*, Vol. II (J. Mitchell, Jr., ed.), Hanser, Munich-New York, 1992, p. 135.
24. F. Gény and L. Monnerie, Dynamics of macromolecular chains V. Interpretation of the dielectric relaxation data, *J. Polymer Sci. Polymer Phys. Ed. 17:*147 (1979).
25. Z. Jelcic, P. Hedvig, F. Ranogajec, and I. Dvornik, Study of crosslinking of unsaturated polyester resins by relaxation methods. *Angew. Makromol. Chem. 130:*21 (1985).
26. B. I. Hunt and G. Powles, Nuclear spin relaxation and a model for molecular reorientation in supercooled liquids and glasses, *Proc. Phys. Soc. 88:*513 (1966).
27. G. Williams and D. C. Watts, Multiple dielectric relaxation processes in amorphous polymers as a function of frequency, temperature and applied pressure, *Dielectric Properties of Polymers* (F. E. Karasz, ed.), Plenum Press, New York-London, 1972, pp. 17–44.
28. G. Scherowsky, Fast switching ferroelectric liquid crystalline polymers containing one or two centers of chirality in the side chain, *Polymer Adv. Technol. 3:*219–229 (1992).
29. H. A. Pohl, Superdielectrics polymers, *IEEE Trans. Elect. Insul. EI-25:*683 (1986).
30. G. Bánhegyi, P. Hedvig, Z. S. Petrovic, and F. E. Karasz, Applied dielectric spectroscopy of polymeric composites, *Polymer Plastics Technol. Eng. 30:*183.
31. J. van Turnhout, *Thermally Stimulated Discharge of Polymer Electrets*, Elsevier, Amsterdam, 1974.

32. G. Bánhegyi, Numerical analysis of complex dielectric mixture formulae, *Colloid Polymer Sci. 266*:11 (1988).

33. G. M. Sessler, Polymeric electrets, *Electrical Properties of Polymers* (D. A. Seanor, ed.), Academic Press, New York, 1982, pp. 242–279.

34. D. J. Eadline and H. Leidheiser, Jr. High frequency time-domain spectrometer for determination of water in polymer coatings on metal substrates, *Rev. Sci. Instrum. 56*:1432 (1985).

35. M. Pfeiffer, L. A. Bresnev, W. Haase, G. Scherowski, K. Kühnpast, and K. Jungbauer, Dielectric and electro-optic properties of a switchable ferroelectric liquid crystalline side-chain polymer, *Mol. Cryst. Liq. Cryst. 214*:125 (1992).

36. Y. Miyamoto, Dielectric relaxation and the molecular motion of poly(vinylidene fluoride) crystal form II under high pressure, *Polymer 25*:63 (1984).

37. Y. Miyamoto, H. Miyaji, and K. Asai, Dielectric relaxation in crystal form II of poly(vinylidene fluoride), *J. Polymer Sci. Polymer Phys. Ed. 18*:597 (1980).

38. J. Glatz-Reichenbach, Li-Jie, D. Schilling, E. Schreck, and K. Dransfeld, Dielectric and piezoelectric properties of very thin films of VDF-TrFE copolymers, *Ferroelectrics 109*:309 (1990).

39. H. S. Nalwa, L. T. Dalton, and P. Vasudevan, Dielectric properties of copper-phthalocyanine polymer, *Eur. Polymer J. 21*:943 (1985).

40. J. Wolak, Dielectric behavior of O3-type piezoelectric composites, *IEEE Trans. Elect. Insul. 28*:116 (1993).

14
Pyroelectric Applications

Elso Yamaka
Tsukuba College of Technology, Tsukuba, Japan

I. INTRODUCTION

Surface charges due to internal electric polarization of a material which has a definite-axial crystal symmetry cannot be observed when the material is at thermal equilibrium with its surroundings, because stray charges are attracted and trapped on its surfaces and neutralize surface charges. However, if the temperature of the material is changed within a short time by an external source such as conduction of heat or incidence of radiation, its internal polarization will be changed and the resulting surface charges can be measured by an external electric circuit before they are neutralized by other stray charges. It is obvious that devices using this pyroelectric effect have the following characteristics: electrical response to rate of temperature change, capacitive element, response independent of wavelength of absorbed radiation, and cooling not required. Therefore an uncooled pyroelectric radiation detector using an element of high pyroelectric coefficient is very useful for detection of weak incident radiation. At present the main practical use of pyroelectric devices is in various applications of incident radiation detection.

It is clear that the pyroelectric detector described above belongs to a family of so-called thermal detectors, since its operating principle depends on both thermal and electrical phenomena. It can be used without cooling and has a response that is independent of the wavelength of the incident radiation as well as other thermal detectors. However, the behavior of the pyroelectric detector is different from that of other thermal detectors such as thermistor bolometers, thermocouples, and thermopiles, because it behaves essentially with pure capacitance due to its high resistivity. In particular, it is necessary to notice that, from its operating principle, it does not give a response to continuous radiation, but only to time-dependent radiation, and hence a light chopper must be used for continuous or very slowly varying radiation. However, this restriction does not mean that it is inferior to other radiation detectors, because radiation detectors are generally used with a light chopper and corresponding narrow-band amplifier in order to reduce the detector noise, especially in the case of weak radiation.

Conversely, the pyroelectric detector has a benefit in being able to detect only the time-dependent radiation in the background of continuous radiation without the light chopper in several special applications. For example, an intruder moving in the optical field of a pyroelectric detector can be easily monitored by detecting the sudden, small change of infrared (IR) radiation intensity, when the temperature of the intruder is either higher or lower than that of the background, because the IR radiation emitted from the intruder has an intensity that depends on his temperature. Hence the pyroelectric detector is utilized in low-cost, passive intruder monitoring appliances as well as in door-opening sensors. Application to a fire alarm system is another example of detecting time-dependent radiation.

Thermal detectors including pyroelectric detectors have a small detectivity and slow response, compared with so-called photon detectors. Therefore, from their benefit of noncooling, thermal detectors are used mainly in the IR region, where photon detectors are inconvenient to use, although they are sometimes used for radiation in other wavelength regions, such as the microwave region and even for high-energy particles, if they can absorb sufficient thermal energy from the radiation and particles.

Pyroelectric polymers have additional features as pyroelectric devices, such as large surface area, thin films, low thermal diffusion, small dielectric constant, flexible for conformity to curved surface and low cost, though their pyroelectric coefficients are generally smaller than other materials. Pyroelectric devices using polymer elements with these features may have a large number of applications which are not necessarily restricted to IR radiation detector.

In the following, the operating performances and typical structures of pyroelectric single-detector, one- and two-dimensional array detectors are discussed, with special attention to PVDF, P(VDF/TrFE), P(VDCN/VAc), and polyurea polymers, in Sections II, III, and IV. In Sections V and VI, other applications of pyroelectric devices using polymer element are introduced.

II. SINGLE-POINT DETECTORS

A. Responsivity [1]

The detector responsivity can be described by a combination of thermal and electrical processes, as follows. In the thermal process, when a thin pyroelectric element is held at its rim in position by supports of low thermal conductance in an evacuated constant-temperature enclosure of temperature T K, the thermal conductance, coupling the element to its surroundings, is determined mainly by the radiative exchanges between the element and the enclosure. If the temperature of the element rises uniformly to $T + \theta_\omega$ by absorbing the incident radiation $I_\omega e^{i\omega t}$ of angular frequency ω, the heat flow from the element to its surroundings is given by $G\theta_\omega$. The temperature of the element is determined by solving the energy balance equation

$$I = H\left(\frac{d\theta}{dt}\right) + G\theta$$

where G is the thermal conductance and H is the heat capacity, and hence θ_ω is expressed by the equation

$$\theta_\omega = I_\omega(G^2 + \omega^2 H^2)^{-1/2} = \left(\frac{I_\omega}{G}\right)(1 + \omega^2\tau_T^2)^{-1/2}$$

where $\tau_T\ (= H/G)$ is the thermal time constant.

It should be pointed out that the simple calculation described above assumes uniform absorption of the incident radiation throughout the element. Since, in general, the absorption of polymers in the IR region depends strongly on the specified wavelength, θ_ω will be quite nonuniform through the element. When the front surface of the element is coated with a thin IR-absorbing layer such as gold black or Ni-Cr film in order to absorb whole incident radiation, the heat conduction from the coated layer to the pyroelectric element causes a decrease of θ_ω from the surface of the element, especially at high ω, so that θ_ω will have a more complicated form, though this is not discussed here in detail.

By the electrical process in the detector element of pyroelectric coefficient p, θ_ω produces an alternating charge $p\theta_\omega$, and the element becomes equivalent to a current generator $i_p = \omega p\theta_\omega$ with a parallel capacitor and resistor. When the element is connected across an input of a high-impedance amplifier, whose input is represented by a parallel combination of a capacitor and resistor, the voltage V applied to the input of the amplifier is given as

$$V = i_p R(1 + \omega^2\tau_E^{-2})^{-1/2}$$

where $\tau_E\ (= RC)$, R and C are electrical time constant, equivalent resistance and equivalent capacity for the combination of the element and input circuit, respectively. Then the voltage responsivity is expressed as

$$\mathcal{R}_V = V/I_\omega = \left(\frac{\omega pR}{G}\right)(1 + \omega^2\tau_T^2)^{-1/2}(1 + \omega^2\tau_T^2)^{-1/2}$$

It is clear that $\mathcal{R}_V$ has a symmetric frequency dependence on τ_E and τ_T, and is zero when $\omega = 0$. In the low-frequency range where $\omega \ll \tau_T^{-1}$, $\mathcal{R}_V$ is equal to $\omega pR/G$ and increases with ω, while in the high-frequency range where $\omega \gg \tau_T^{-1}$, τ_E^{-1}, it is equal to $p/\omega CH$ and decreases with ω^{-1}. In the medium-frequency range between τ_T^{-1} and τ_E^{-1}, $\mathcal{R}_V$ is frequency independent, either p/GC for $\tau_E^{-1} < \tau_T^{-1}$ or pR/H for $\tau_E^{-1} > \tau_T^{-1}$. Thus for the pyroelectric detector it is desirable to have large p, R, and small G, C, H for large responsivity.

It is interesting that τ_E can be easily reduced by several orders from the highest value by adding a small resistor parallel to the input resistor of the amplifier, while τ_T cannot be changed. Figure 1 shows the schematic frequency dependence of $\mathcal{R}_V$ for $\tau_T = 1$ s and $C = 10$ pF with R values between 10^{12} and 10^7 Ω. It can be seen that the pyroelectric detector can have a wide frequency response with a reduced value of $\mathcal{R}_V$, if a resistor of low value is used in the input of the amplifier. This is caused by its capacitive nature and gives a remarkable characteristic that distinguishes it from other thermal detectors.

Also, it is worth noting that the heat capacity of the detector element can be reduced by using thinner film, and then the maximum value of $\mathcal{R}_V$ in the frequency range between τ_T^{-1} and τ_E^{-1} can be increased with $1/H$, though τ_E^{-1} decreases by increased C and the bandwidth of the detector decreases under the condition $\tau_T^{-1} < \tau_E^{-1}$. However, the frequency dependence of $\mathcal{R}_V$ in the high-frequency range follows the theoretical form ω^{-1}

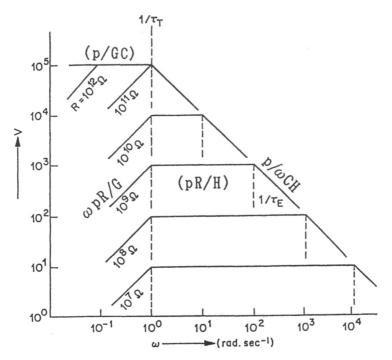

Figure 1 Schematic dependence of pyroelectric detector on frequency for $\tau_T = 1$ s, $C = 10$ pF, and $R = 10^7$–10^{12} Ω, omitting smoothly curved shapes at frequencies equal to $1/\tau_T$ and $1/\tau_E$ on each curve for simplicity.

in the pyroelectric detector; this dependence can be easily compensated by a suitable electrical circuit, e.g., an operational amplifier which has a small noise figure, so that a fabrication method of thin film is important for the pyroelectric detector with high, wide band responsivity. Here it should be pointed out that polymer materials are generally preferred to other materials from the standpoint of very thin film fabrication, especially in a silicon integrated circuit, which is discussed in a later section.

It should be pointed out that the pyroelectric coefficient p is generally not constant but is temperature dependent even at room temperature, especially at temperatures near and below the Curie temperature, T_c, as shown in a later section for the case of a P(VDF/TrFE) copolymer. Hence a detector element operating at room temperature has to use a pyroelectric material which has a small temperature coefficient dp/dT resulting from high T_c, for quantitative measurement of the incident radiation.

B. Detector Noise [1,2]

Detector noise is an important factor for detector performance, because the signal-to-noise ratio, S/N, cannot be improved by an electrical amplifier. Therefore, smaller noise voltage V_n is desirable as well as larger responsivity $\mathcal{R}_V$. Pyroelectric detectors have various noise sources, some of which come from the detector itself and others of which come from associated input amplifier:

Radiation or temperature noise voltage. Incident radiation from the surroundings at temperature T K has random-fluctuation noise $(4kT^2G)^{1/2}$, and causes noise voltage $V_{n,T}$ through the detector responsivity $\mathscr{R}_V$. Thus this noise is given by

$$V_{n,T} = \mathscr{R}_V(4kT^2G)^{1/2}$$
$$= (4kT^2G)^{1/2}(\omega pRG^{-1})(1 + \omega^2\tau_T^2)^{-1/2}(1 + \omega^2\tau_E^2)^{-1/2}$$

Of course $V_{n,T}$ has the same frequency dependence as $\mathscr{R}_V$.

Johnson noise voltage. This noise voltage is due to the equivalent resistor R of the detector and amplifier, and is given by

$$V_{n,J} = (4kTR)^{1/2}(1 + \omega^2\tau_E^2)^{-1/2}$$

This noise voltage decreases as ω^{-1} in the high frequency range.

Tan δ noise voltage $V_{n,D}$. This noise is associated with the dielectric loss tan δ of the pyroelectric material and is given by

$$V_{n,D} = (4kT \tan δ)^{1/2}(\omega C)^{-1/2}$$

where C is the equivalent capacity of the detector and input amplifier. This has a frequency dependence of $\omega^{-1/2}$.

Amplifier noise voltage $V_{A,V}$ and $V_{A,I}$. These are associated with the voltage noise generator in series and the current noise generator in parallel to the input amplifier.

The square of the total noise voltage is equal to the summation of the squares of each noise voltage component. Each component of noise voltage has a different frequency dependence, and its magnitude varies with the detector element and input amplifier, so that optimum design of the detector element and input amplifier is necessary to achieve larger detectivity in a specified frequency range. $V_{n,T}$, $V_{n,J}$, $V_{n,D}$ are generally larger than $V_{A,V}$ and $V_{A,I}$ in the lower-frequency range where pyroelectric detectors give their best performance.

Among various noise voltages mentioned above, $V_{n,D}$ may change in a wide range, depending on the element fabrication technique, especially for ceramics and polymers having composite components, and it will frequently become the dominant noise source in the frequency range from 1 to 100 Hz. The dielectric losses of elements having composite components are, in general, larger than those of single crystals, so special care must be taken to minimize dielectric losses in fabrication.

C. Detector Structure

Essential points to be taken into account for designing the structure of a pyroelectric single detector are to obtain maximum absorption of incident radiation, maximum electric signal, and minimum electrical noise. A typical structure of a single detector is shown in Figure 2, for which several comments are described in the following, especially in the case of a detector using polymer film.

In the IR range the absorption coefficients of various polymer materials have, in general, not flat but rather complicated shapes and are small, so that thin polymer film almost cannot absorb the incident radiation. Then an IR-absorbing layer such as gold black, which can absorb the radiation almost perfectly throughout the whole IR range,

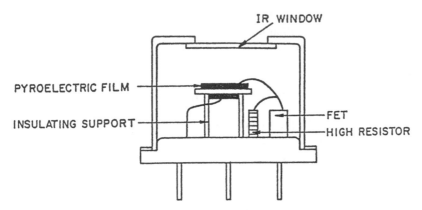

Figure 2 Schematic structure of single pyroelectric IR radiation detector.

has to be deposited on the front surface of the detector. The layer thickness must be determined from the standpoint of the optical absorption of the radiation, its thermal capacity, and heat conduction to the pyroelectric material. Sometimes the responsivity of the detector may have a tendency to drop in the high-frequency region in relation to the thermal diffusion time of the absorbed energy from the absorbing layer to the pyroelectric element. If a NiCr thin front-electrode layer is designed to have a maximum absorption and a back-electrode metal layer is designed for perfect reflection to the leakage of the radiation through the pyroelectric material with minimum thickness, better performance in the high-frequency range can be obtained without an IR-absorbing layer, because of the lack of a thermal diffusion mechanism.

The rim of the detector element is, in general, mounted on a thermally insulating cylinder to reduce the thermal capacity and heat conductance. A special technique is needed to keep the detector element flat in the case of a thin polymer film. The detector mount is encapsulated by a metal case for electrical shielding. The size of a window plate using IR-transparent material, such as Si, Ge, or CaF_2, should be as small as possible to avoid any stray radiation. A partial reflection of incident radiation due to relatively large refractive indexes of these window materials can be decreased by using an anti-reflecting coating on both surfaces of the window plate. These coatings can also electrically shield the inside of the detector due to their high conductivity. The detector mounting case is generally evacuated in order to avoid microphonic vibration and heat conduction through the surrounding air, and also to reduce the electrical surface leakage of the detector element and amplifier. In this case it is pointed out that the heat conduction of the detector is mainly radiation exchange between the detector element and its surroundings.

As was pointed in the previous section, the electric impedance of the pyroelectric element is very high, e.g., 10^{11} Ω, so that a high input-impedance amplifier is necessary to be enclosed near the detector element as shown Figure 2 and its output impedance should be low. A junction field-effect transistor (JFET) or an operational amplifier with a JFET input is usually used as the enclosed amplifier, because the JFET has a very high input impedance and generates only small voltage and current noise.

D. Polymers as Detector Materials

Physical properties of pyroelectric materials required for IR detectors are discussed in the following.

1. IR Absorption

Pyroelectric materials, in general, do not have a constant absorption coefficient, but a wavelength-dependent value in the IR region. Polymers especially have complicated structures of absorption curves compared to other pyroelectric materials, so the IR-absorbing layer must be deposited on the front surface for wavelength-independent responsivity. This structure may cause the frequency dependence of the polymer detector to be modified from the simple theory for even thin film, because polymers have small heat conduction.

2. Pyroelectric Coefficient p

Since detector responsivity $\mathcal{R}_V$ is proportional to p from the detection mechanism described in Section II.4, large p values are required. Spontaneous electric polarization P_s of ferroelectric materials decreases toward the Curie temperature T_c and vanishes over T_c, so ferroelectric materials may have large p value which correspond to their temperature coefficients dP_s/dT below T_c. At present, all pyroelectric materials used as IR detectors belong to the class of ferroelectrics.

3. Curie Temperature T_c

The pyroelectric detector has to be used below T_c, and the ferroelectric element has to be repoled when its temperature rises above T_c. Therefore, T_c of the detector element is preferred to be much over room temperature, e.g., 100°C. Higher T_c is especially important for quantitative detection of incident radiation, because p around room temperature as well as dielectric constant for low T_c materials such as P(VDF/TrFE) copolymer [3] is not constant but strongly temperature dependent, as shown in Figure 3.

4. Dielectric Constant ϵ

Small ϵ value of the detector material is preferred, because small electric capacity of the detector corresponding to small ϵ value improves the performance of the detector, especially in the higher-frequency range, as discussed in Section II.A.

5. Specific Heat C_s

Small C_s of the element material is favored, since small heat capacity related to small C_s gives better performance in the lower-frequency range, as also discussed in Section II.A. However, C_s values of the element materials are almost equal, so the heat capacity of the detector depends mainly on the thickness of the element as well as extra heat capacities of other materials such as the IR-absorbing layer and any supporting substrate. Thus the fabricating technique of the detector becomes more important in the case of a thin film element.

6. Tan δ

The main noise source of a pyroelectric detector with a carefully designed input amplifier comes from the tan δ of the element material itself. It may have divergent values depending on the fabricating methods, especially in the case of thin films.

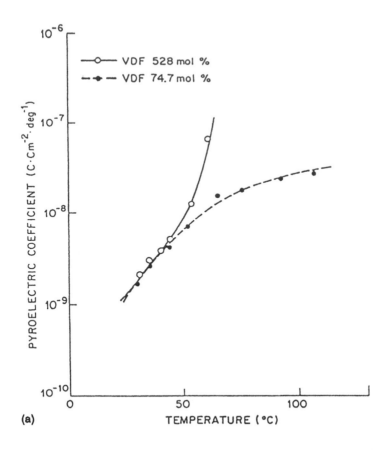

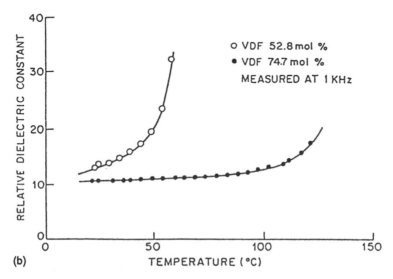

Figure 3 Temperature dependence of (a) pyroelectric coefficient and (b) dielectric constant of P(VDF/TrFE) copolymers of VDF 52.8 and 74.7% [3].

The physical properties of pyroelectric materials used for detectors, including candidate materials, are shown in Table 1. For polymers, their p values are generally one order smaller than for other materials, but they have smaller ϵ values, so their material figures of merit p/ϵ and $p/\epsilon \, C_s$ are not so small as for other materials. In addition, polymers have several other benefits: Films thinner than several micrometers can easily be fabricated, their cost is attractively low, they are flexible to conform a curved surface, and so on. Thus it can be said that polymers are better materials for low-cost single detectors or specially designed detectors.

E. Detectors Using Polymer Film

1. General

In 1969, Kawai [6] pointed out that poly(vinylidene fluoride) (PVDF) has the largest piezoelectric coefficient among various polymers. In 1970, stimulated by Kawai's report, two groups, one [7,8] in the United States and the other [9,10] in Japan, began independently to study the pyroelectric properties of PVDF film which had been produced as a capacitor dielectric by Kureha Engineering Company in Japan in order to use it as IR detector element. At the beginning of their studies the p and tan δ were so poor that the responsivity and detectivity values of PVDF detectors [11] were inferior to those of detectors using TGS monocrystal and PbTiO$_3$ ceramics. While the material properties of PVDF film have been gradually improved since then, many other kinds of pyroelectric polymers have been studied to date. However, PVDF is still one of the best polymers for the pyroelectric detector, as shown in Table 1.

An example of typical performance of a PVDF single detector is shown in Figure 4 [12], in which the same, unirradiated PVDF element is connected in series and reverse to the 30-μm, 7.4-cm-diameter detecting element, in order to cancel the effects of the temperature change and mechanical vibration of the detector enclosure case. Unfortunately, at present, single detectors using polymer films are generally inferior to detectors using other pyroelectric materials due to their low p values, so they are used mainly as low-cost Ir detectors in simple household equipment such as intruder monitors and door-opening sensors, except specially designed detectors described in the following.

2. Detectors Using Large, Flexible Polymer Films

Large IR pyroelectric detectors of special forms such as hemisphere, cone, or cylinder can be easily fabricated by using polymer films. An example of such a special detector is shown in Figure 5 [13], which is used to measure directional-hemispherical spectral reflectances in the IR region. It is a cavity-shaped pyroelectric detector that itself collects the radiation reflected by the test sample, thereby obviating the need for an intermediate collector such as an integrating sphere or concave mirror. The detector is made in a cone shape from an electrically polarized PVDF film coated with a gold-black absorbing layer on its inner surface and backed with brass shim on its outer surface in order to provide mechanical strength. A reflectometer using this detector has been used with a Fourier spectrometer to measure spectral reflectance over the wavelength range 5–30 μm.

Another example is a large, electrically calibrated radiation detector using PVDF film, as shown in Figure 6 [14]. A circular thin PVDF film 10–20 μm thick is clamped taut between a pair of rigid rings with the center region away from any heat sink in order to have extended low-frequency response for accurate calibration. On the front and back

Table 1 Physical Properties of Various Pyroelectric Materials

Material		T_c (°C)	p (10^{-8} C/cm²-K)	ϵ	C_s (J/cm³-K)	$\epsilon^{-1}C_s^{-1}p$ (10^{-10} C/cm-J)	$C_s^{-1}p$ (10^{-8} C/cm-J)	Reference
TGS	Single crystal	49	3.4	50	2.5	2.8	1.4	1
LiTaO₃	Single crystal	618	1.7	43	3.2	1.2	0.53	2
PbTiO₃	Ceramic	490	3.0	200	3.2	0.47	0.93	
PVDF	Polymer	—	0.10	5	2.3	0.87	0.04	
PVDF	Polymer	(~150)	0.2	11	2.3	0.79	0.087	
P(VDF/TrFE)	Copolymer VDF 52.8%	70	0.1	12	(2.3)	(0.36)	(0.04)	3
P(VDF/TrFE)	Copolymer VDF 74.7%	125	0.17	11	(2.3)	(0.67)	(0.073)	3
P(VDCN/VAc)	Alternating copolymer	—	(~0.2)	4.5	(2.3)	(~2.1)	(~0.08)	4
POLYUREA	Polymer	200	0.18	4.0	(2.3)	(1.8)	(0.078)	5

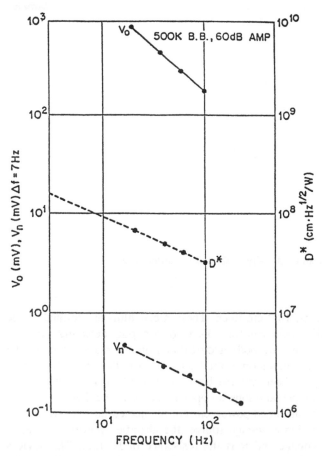

Figure 4 Signal output V_0, noise level V_n, ($\Delta f = 7$ Hz), and detector detectivity D^* of PVDF IR detector versus frequency [12].

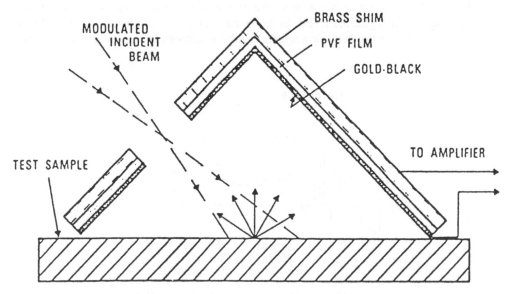

Figure 5 Schematic structure of a cavity-shaped pyroelectric radiation detector [13].

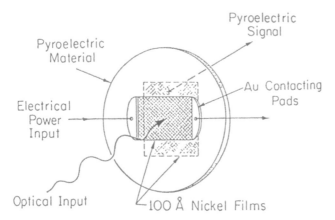

Figure 6 Schematic structure of an electrically calibrated radiation detector [14].

sides are evaporated thin Ni films, 10^{-6} cm thick. These metal films are formed in the shape of rectangles and carefully positioned for a 1-cm² overlapping detecting area.

Calibration of these devices involves both electrical and optical measurements. The electrical measurements relate the calibration to the basic power units. The optical measurements are necessary simply to determine the fraction of incident power that is absorbed by using a standard technique, and thus no optical source is required.

To determine the response to absorbed power, current is injected into the Ni film on the front face of the detector. Using a voltage source, the absorbed power is equal to V^2/R, where V is the applied voltage and R is the resistance of Ni film. The devices cannot respond to unmodulated signals and thus the calibrating voltage must be modulated. This modulation of the applied voltage may lead to capacitively coupled signals that may add to or subtract from the pyroelectric signal and produce an error in the calibration. To avoid this error, a balancing circuit is used that reduces the capacitively coupled signals to negligible levels. In the electrical calibration circuit shown in Figure 7 [14], the output of the detector amplifier can be monitored by applying a square-wave

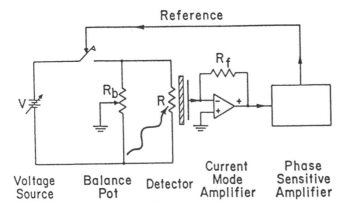

Figure 7 Electrical calibration circuit for the detector in Figure 6 [14].

voltage to the Ni film, and the balance potentiometer of the figure can be dialed to square up the signal and eliminate the spikes due to the capacitively coupled currents.

3. Detectors Using Very Thin Polymer Films

Recently, in order to improve the frequency performance of pyroelectric detectors, thin films, especially of polymers, have been studied for fabrication on a silicon substrate or the input gate of a Si integrated device by using various methods such as a spinner method [3,15], a vapor deposition and polymerization method [16], and an electrospray method [17]. The forming temperatures of thin pyroelectric polymer films are relatively low, so they do not damage Si integrated devices. It is important to fabricate the integrated detectors using thin polymer films with corresponding signal processing electronics (for example, Si CCD devices) to uncooled, one- and two-dimensional IR imaging array detectors, as described in the following sections, since other pyroelectric oxide films, e.g., of $PbTiO_3$, require high fabricating temperatures which might seriously damage the Si integrated device.

Reported performance of pyroelectric IR single detectors integrated on Si FETs [18], shown schematically in Figure 8, for PVDF thin film (1–2 μm) deposited by the electrospray method, compared with uniaxially stretched PVDF sheet film (9 μm), are shown in Figures 9a and 9b for $\mathcal{R}_V$ and V_n, respectively. These data for $\mathcal{R}_V$ and V_n seem to be difficult to use for detailed analysis with the simplified model described in Sections II.A and II.B, because the complicated structure with hole arrays for etching of Si substrate and Si_3N_4 membrane supporter with back Al electrode in Figure 8 needs more detailed analysis. Although the electrospray film gave smaller $\mathcal{R}_V$ than the uniaxially stretched film in Figure 9a, it might be expected that further improvement of the fabricating technique for the electrospray film can give higher $\mathcal{R}_V$ in a specified frequency range.

III. TWO-DIMENSIONAL IMAGE DEVICES

A. Effect of Thermal Diffusion in Imaging Target Plane

While a photon detection mechanism using semiconducting material is very useful for two-dimensional imaging devices as well as single detectors due to its high response and speed in the visible region, it is inferior to the thermal detection mechanism in the IR region because of the need for a cooling system, which is very complicated for imaging devices. Thus it is desirable to use a thermal mechanism for uncooled thermal imaging

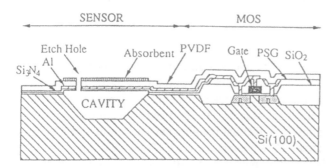

Figure 8 Schematic cross-section structure of PVDF thin-film single IR detector deposited on silicon substrate with a readout circuit by electrospray method. Absorbent (gold black) layer, not shown in this figure, is connected to silicon substrate as an upper electrode [18].

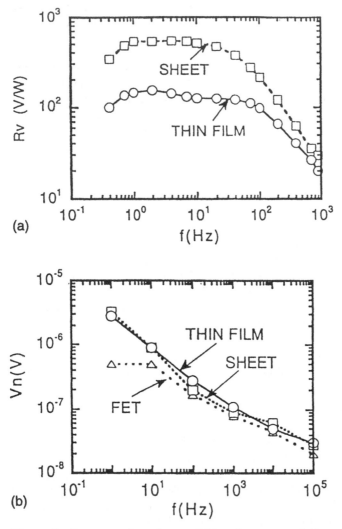

Figure 9 Frequency dependence of (a) voltage responsivity and (b) noise voltage of PVDF thin-film (1–2 μm) single IR detector shown in Figure 8 versus frequency PVDF sheet (9 μm) detector [18].

devices using IR radiation. Among various thermal detection mechanisms, pyroelectric detection is easy to use because of the electrical charges produced by IR radiation.

Electrical signals of imaging devices are generally read out in a time-sequential form to an external amplifier by an electrical scanning method over imaging pixels. If the output signal has a TV format, a low-cost thermal imaging system can be made by using the imaging device with commercial TV equipment, e.g., VTR and TV display.

Since pyroelectric materials including polymers usually have very high resistivities, there is no degradation of image quality caused by electrical charge spreading. However, if the pyroelectric material has a large thermal diffusion in an imaging target plane, the temperature distribution in that plane will be smeared out in a short time and details of

Table 2 Pyroelectric and Thermal Properties of TGS, PbTiO$_3$, and PVDF [9]

Property	Material		
	TGS	PbTiO$_3$	PVDF
Curie temp. T_c (°C)	49	470	120
Pyro. coeff. p(C/cm^2°C)	3.5×10^{-8}	3×10^{-8}	0.24×10^{-8}
Dielectric const. ϵ	42	200	11
Specific heat C_s (J/cm^c °C)	2.5	3.2	2.5
Heat cond. K (w/cm^{c}C)	6.4×10^{-3}	32×10^{-3}	1.3×10^{-3}
Heat diffusion coeff. κ (cm^2/s)	2.6×10^{-3}	9.9×10^{-3}	0.53×10^{-3}

the imaging pattern will be lost. Since the thermal diffusion coefficients κ of pyroelectric polymers are much smaller than those of other pyroelectric materials, as shown in Table 2, imaging devices using polymer films will be expected to give a better thermal image. Besides this advantage, thin, large, and flexible film of pyroelectric polymers is easily fabricated. Thus pyroelectric polymer films are suitable for thermal imaging devices as well as single detectors.

The degradation of the thermal image due to thermal diffusion in the target plane is expressed by a thermal modulation transfer function (MTF). The thermal MTF is given by the following formula for a light chopping mode D_c or a camera panning mode D_p:

$$D_c = \left(\frac{f}{\pi^2 \kappa n^2}\right) \tanh\left(\frac{\pi^2 \kappa n^2}{f}\right)$$

$$D_p = \left[1 + 2\left(\frac{\pi \kappa n}{V}\right)^2\right]^{-1/2}$$

where n is the spacial frequency (line pairs/cm) and V is the panning velocity (cm/s) on the target plane. Assuming that the thermal diffusion takes place only in the pyroelectric target plane, thermal MTF curves versus n through materials listed in Table 2 are shown in Figure 10 [19]. It is clear that PVDF film has the best performance due to its small value of κ. However, in an actual device it is expected that IR-absorbing and electrode layers deposited on the pyroelectric target plane will reduce the thermal MTF values shown in Figure 10 in the higher-n region. Therefore these layers should be as thin as possible in order to obtain better images.

B. Pyroelectric Vidicon Tubes

The structure of a pyroelectric vidicon tube is quite similar to that of a visible vidicon tube. It consists of a pyroelectric target plane and an electron gun, as shown in Figure 11 [19,20]. The main differences from a visible vidicon tube are a window plate of IR-transparent material, e.g., Ge or CaF$_2$ instead of glass, and pyroelectric target film instead of a photon-detecting layer that is sensitive to visible light.

When a chopped IR image is focused on the front surface of the target plane by an IR optical lens system, a two-dimensional distribution of temperature change is built in the target plane. This results in positive and negative charge distributions during each chopper open–close period, which correspond to the thermal images developed on the back, unelectroded surface of the target plane through the pyroelectric effect.

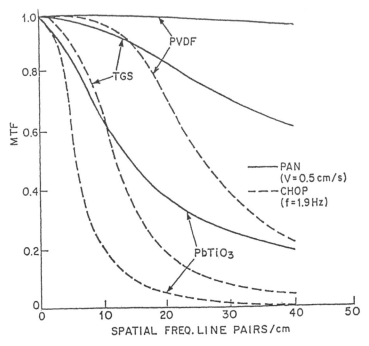

Figure 10 Theoretical thermal MTF curves of pyroelectric materials of TGS, PbTiO₃, and PVDF for light chopping and camera panning modes [19].

Although the positive charges can be read out by the scanning electron beam and become a time-sequential signal, the negative charges cannot be read out by the scanning electron beam and accumulate gradually on the back surface, causing the electron beam do cut off. In order to maintain continuous operation, bias charges are required to cancel these accumulated negative charges. One method is to fill the tube with low-pressure gas, which can be ionized by the scanning electron beam and which neutralizes negative

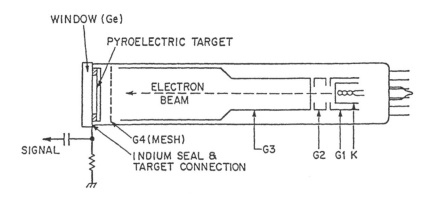

Figure 11 Schematic structure of pyroelectric vidicon tube [19].

charges. The other method is to use secondary electron emission from the target plane, which is generated by bombardment with an electron beam of high enough energy to produce a secondary electron yield greater than 1 during the horizontal flyback time of the scanning beam. Both methods can be used, but the former method has the possibility of shortening tube life through damage to the cathode, while electron beam bombardment does not damage polymer targets such as PVDF, so that the latter method is preferred for pyroelectric vidicons.

By this bias charging technique, positive and negative electric image patterns can be alternatively obtained through the load resistor in Figure 11 for the open and closed periods of the light chopper, which rotates synchronized with TV frames. By using a substracting method between positive and negative image patterns, an improved thermal image pattern can be obtained, because the signal intensity becomes twice, while the noise unrelated to the TV frame can be canceled.

Pyroelectric vidicon tubes are very easy to fabricate compared with the solid-state image devices described in the following section, because the tube's target plane can be easily mounted with its rim on the supporting cylinder and sealed with the electron gun in the glass tube. The electron beam is easily scanned by a deflection coil held separately from the vidicon tube. However, the electric voltage required to operate the vidicon tube is too high for other transistorized circuits, so it is inconvenient to use in a portable, uncooled thermal imaging camera.

Figure 12 shows a view of a car parking area using a Matsushita Research Institute (Tokyo) pyroelectric vidicon camera [21] in which an uniaxially stretched PVDF film (6 μm) is used.

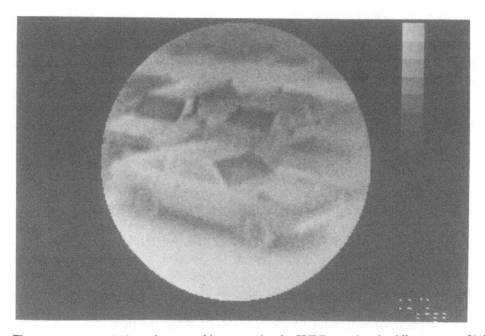

Figure 12 Thermal view of a car parking area taken by PVDF pyroelectric vidicon camera [21]. Courtesy of Matsushita Research Institute, Tokyo.

C. Pyroelectric Solid-State Imaging Devices

Pyroelectric solid-state thermal imaging devices are more desirable than the visicon tube described in Section III.B, because these devices can be operated with low voltage and small power and may have longer lives. This is important, especially for a portable, uncooled IR imaging camera using a transistorized electrical circuit.

Two methods of transfer of pyroelectric signal charges from the target plane to the input of the electronic scanner circuit, e.g., a charge-coupled device (CCD), have been studied. One method uses one soft In metal pad per pixel, which connects the input gate of each pixel of the scanner both mechanically and electrically to the corresponding pixel of the pyroelectric plate. In this type of device, the pyroelectric plate and the scanner are fabricated separately and then combined, so it is possible to use a uniaxially stretched plate of PVDF with a large pyroelectric coefficient. However, this is a rather difficult technique to use for a large number of pixels such as in a TV format, resulting in nonuniformity of thermal capacity and heat conduction.

The other method is based on direct deposition of the pyroelectric thin layer on an FET amplifier and electronic scanning circuit, without connecting pads. However, if the film deposition requires high temperatures as in the case of $PbTiO_3$, it may seriously damage the electronic circuit. On the other hand, various polymers can be used in fabrication at much lower temperatures, so the direct deposition method could be applied to polymer film. Indeed, various methods such as an electrospray method [17] for PVDF, a spinner method [3] for P(VDF/TrFE), and a vapor deposition and polymerization method [16] for polyurea, described in Section II for the single-element detector, are also expected to be useful in fabricating one- and two-dimensional array detectors on silicon integrated circuits. Figure 13 [22] shows a schematic cross section and top views of one pixel of an IR imaging device using P(VDF/TrFE) thin film fabricated on a silicon FET and CCD by the spinner method.

When the common front electrode on the polymer film is connected to the Si substrate, the back electrode of each pixel gives positive and negative charges alternatively to the input gate of each pixel of the scanner for open and closed periods of the light chopper. Since electronic scanners, such as CCDs, can read out only one type of charge, the other type must be removed in order to maintain continuous operation. One method of charge removal is to use a bias charge generated by visible light irradiation, as shown in Figure 13 [3]. If a pinhole is made in the poly-Si rod of each pixel and continuous visible light is introduced through the hole to the Si, the *p-n* junction of that pixel is controlled to fill half of the charge storage capacity with positive charges. The positive charges increased and decreased with pyroelectric charges will be transferred to an output through the scanner for the open and closed periods of the light chopper. Thus continuous operation is possible using this bias method. The other method of supplying bias charges is to use a combination of a bias cell and transfer gate for each pixel and to fill the corresponding storage cell with bias positive charge through the transfer gate to cancel negative pyroelectric charge. This bias cell method requires a large modification from the visible CCD, and the electric circuit is more complicated because of the additional timing pulses needed for transfer of the bias charges from the bias cell to the signal storage cell in each TV frame.

At a depth equal to thermal diffusion $(2K/\omega C)^{1/2}$ below the IR absorbing layer, the thermal amplitude will be reduced to $1/e$ of its value at the absorbing layer. If the thickness of the pyroelectric film is greater than this length for the chopping period, the

(a)

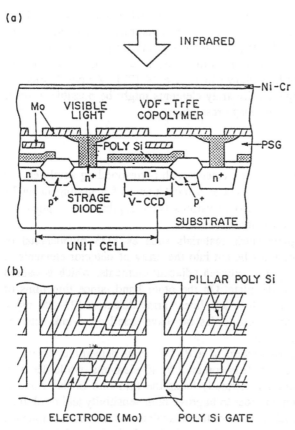

Figure 13 Schematic diagrams of (a) cross-section and (b) top views of one pixel of one- or two-dimensional imaging devices using P(DVF/TrFE) thin film fabricated by the spinner method. Ni-Cr electrode is connected to silicon substrate, not shown in the figure [22].

scanner sustrate gives no effect. However, in the panning mode of camera operation, there is a possibility that the thickness will be less than the diffusion length. In this case it will be better to remove the substrate under the pyroelectric film in order to reduce the heat capacity and increase the signal, e.g., by using Si anisotropic etching through the small pinholes in the pyroelectric film, as shown in Figure 8 [18].

Though several fundamental researches of pyroelectric solid-state imaging devices have been published in journals, as described in this section, performance of the device has not yet been reported, except for Thomson CSF's preliminary data sheets of IR solid-state imaging devices using pyroelectric copolymer, having pixel size of 81 μm $\times$ 81 μm, pixel number 128 $\times$ 128, sensitivity 11.3 mV/K, spectral range 8–14 μm, noise level 7 mV, NETD ($f = 1$, $T_B = 300$ K, $T_{op} = 0.8$) 0.62 K.

D. Improvement of Two-Dimensional Pyroelectric Targets

Pyroelectric vidicon tubes using flat PVDF film has not enough responsivity as the IR imaging device, as shown in Section III.B, compared with a pyroelectric vidicon tubes having a reticulated TGS single crystal target, which have better image resolution.

If an improved target having an aligned structure of PbTiO$_3$ ceramic rods of large pyroelectric coefficient embedded in polymer binder of low thermal conductivity can be fabricated, IR imaging devices using this target will have greater responsivity and better thermal MTF, in the solid-state device as well as the vidicon tube. An example of the aligned structure of high-density rods is shown in Figure 14 [22]. Various fabrication methods for the thin target with high-density array elements might be possible in order to produce a large responsivity and high image resolution.

IV. LINEAR ARRAY DETECTORS

Linear array detectors for a small number of elements with corresponding amplifiers are useful in various special fields. They are used at a focal plane of a grating in an IR spectroanalyzer for measuring simultaneously at various wavelengths, or of optical reflecting lens in a passive intruder-monitoring system.

In IR array detectors using pyroelectric materials such as TGS monocrystal or PbTiO$_3$ ceramics, the detecting plate must be cut into the array of detector elements in order to reduce the cross-talk of response between adjacent elements, which is caused mainly by high thermal diffusion in the plate. On the other hand, since thin films of pyroelectric polymers have low thermal diffusion, linear array detectors [23–25] using polymer film with an array of separated electrodes can have small cross-talk between adjacent elements without cutting the film.

An example of a schematic cross section of a PVDF array detector is shown in Figure 15 [23], in which the back electrode is common to 40 elements, while the front electrodes are separated. Surrounding air thermally loads the thin film in this case, so the array detector is sealed in vacuum in order to improve both sensitivity and resolution. FET input amplifiers connected to each element of the array detector and the multiplexing electronics for electrically scanning output are also packed with the array detector in a shielding metal case to reduce electrically induced noise. The properties of the linear array detector (40 elements, 3 mm long, with 0.1-mm spacing, thin PVDF film of 0.5–0.8

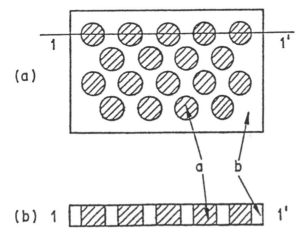

Figure 14 Schematic structure of one- or two-dimensional array detector with pyroelectric ceramic rods embedded in polymer film: (a) cross-section and (b) top views, in which a is pyroelectric ceramics and b is polymer [22].

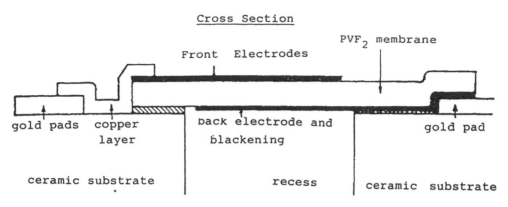

Figure 15   Schematic diagram of one pixel of PVDF array detector [23].

μm thickness) have been reported: responsivity (500 K, 10 Hz, no window loss) of 10,000 V/W, D^* (500 K, 10 Hz, no window loss, 1–30 Hz bandwidth) of 1.5×10^9 cm/Hz/W and resolution of < 30% thermal cross-talk at adjacent detectors at 20 Hz.

It is important, especially for home use, that the array detector can be manufactured at substantially lower cost with PVDF film than with other materials, because of its simple structure and low material cost.

V.   HEAT FLOW DETECTION

Since the pyroelectric detector gives an electrical signal with temperature change, it can be used to measure the temperature change of a solid specimen in intimate contact with the detector, or of a noncontacted specimen by heat flow through the surrounding air, as described in the following.

A.   Photopyroelectric Detection

When monochromatic, sinusoidally modulated light irradiates a specimen to be studied, modulated thermal energy, which is converted from optical energy due to light absorption in the specimen, flows into a pyroelectric detector in intimate contact with the specimen, and the electrical signal produced by the pyroelectric effect is proportional to the light absorption of the specimen. Hence this new, simple spectroscopic technique [26,27] can be applied to detect optical absorption and nonradiative energy conversion processes in condensed-phase matter, and is named photopyroelectric spectroscopy (PPES) [27]. The in-situ study of physicochemical processes performed by PPES may involve complex sample geometries which cannot be easily handled by the conventional photoacoustic spectroscopy (PAS) method used intensively in recent years, so PPES can have distinct advantages over PAS. Figure 16 [27] shows the PPES experimental apparatus using both a broad-band Xe lamp and a He-Ne laser source.

Although any kind of pyroelectric material can be used in PPES, the pyroelectric polymers such as PVDF film have the great advantage because their thin, flexible properties make them easy to use as well as low in cost. An example of a PPES cell using PVDF film is shown in Figure 17 [27].

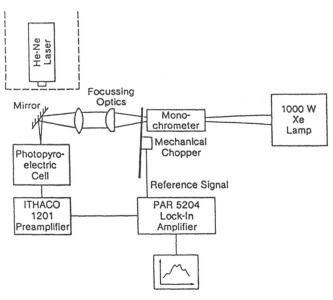

Figure 16 Photopyroelectric experimental apparatus. Both broad-band Xe lamp and He-Ne laser sources are shown [27].

Because of simplicity of the PPES method, shown in Figures 15 and 16, and high sensitivity, this method has already been applied in various fields such as thermal wave scanning microscopy [28], thermal diffusivity measurement [29], phase-transition phase study [30], and so on. In all cases the pyroelectric device for each design uses PVDF film.

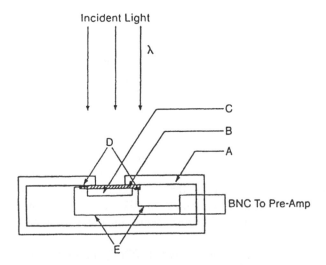

Figure 17 Photopyroelectric cell with a 0.5-in. × 0.25-in. PVDF sample holder: A, Teflon cell; B, 25-μm-thick PVDF film; C, Teflon support; D, solderable copper foil electrodes with conductive adhesive; E, silver wire [27].

B. Surrounding Temperature Detection

Even when the pyroelectric detector cannot contact the specimen directly, the temperature change of the specimen can be measured by heat conduction through surrounding liquid or gas between the specimen and detector. The most simple example is a nonradiative fire-alarm device. Since the pyroelectric detector is very sensitive to the temperature of the surroundings, the small thermal energy generated in the specimen can be detected if a cell containing the specimen, detector, and surroundings is designed to minimize the thermal energy leakage from the cell to the outside.

An example of this type of cell is shown in Figure 18 [31] for optical fiber absorption loss measurement. A 500-mm length of PVDF tube is coated with electrically conductive paint and inserted into a tightly fitting Cu tube. A 50-μm-diameter metal wire runs along the inside wall of the PVDF tube. The optical fiber is aligned centrally within the tube. The water in the tube conducts the heat produced in the fiber to the PVDF detector and also acts as an inner conductor for the accumulation of pyroelectric charge. This charge is measured by a charge amplifier, capable of measuring 1 pC, through a low-noise coaxial cable from the BNC. Radiation of 1.06 μm from a Nd:YAG laser is admitted and radiation absorbed by the fiber is converted into heat, which is conducted through the water into the PVDF tube. This technique can measure the fiber absorption loss in only 50 mm in length, with the same accuracy obtained by the standard calorimetric method.

The PVDF cylinder used in Figure 18 was made using small-bore, thick-walled pyroelectric tubing produced continuously from an 18-mm screw extruder and oriented along its axis in a hot air-heated region between two caterpiller haul-offs. It is obvious that such machine work of the cylinder-type detector cannot be applied using pyroelectric materials other than polymers.

VI. SPECIAL APPLICATIONS

A number of pyroelectric devices have been proposed to use the additional features of polymeric materials, and some have undergone further development. Several examples are introduced in the following.

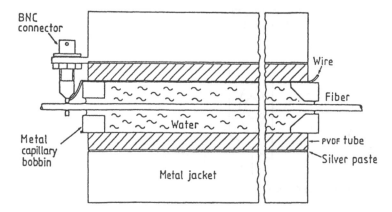

Figure 18 Schematic diagram of pyroelectric detection cell used for optical fiber absorption loss [31].

A. Pyroelectric Copying Devices

Conventional xerographic devices use the photoconducting effect, in which the electrical conductivity of the photoconductor is changed by exposure to light. Since the photoresponse usually depends on wavelength, copying response is not flat for documents having various colors. If the pyroelectric effect is used instead of the photoconductive phenomenon, the light response of the copying device is wavelength independent.

Electrostatic copying devices based on the pyroelectric effect were proposed by the BTL group [32,33] early in the 1970s. When a thin plate of a pyroelectric material is exposed to a thermal image, it develops a charge distribution corresponding to the thermal image; indeed, this is the basis of operation of the pyroelectric vidicon described in Section III. The charges persist for a length of time comparable to the electric time constant RC. Since this electrostatic image produces electric fields at the surface of the pyroelectric material, a conventional electrostatically charged ink will be attached to those regions of (relative) opposite charge. This results in a visual image, which in turn may be readily transferred to paper.

Figure 19 [32] is a schematic diagram of a pyroelectric copying device, in which a pyroelectric element is used instead of the photoconductive semiconductor element in a conventional xerographic device. Light from an intense source is focused by a first lens and passes through the document to be copied. The light image from the document is focused by a second lens onto the pyroelectric element, which is composed of a light-absorption layer, a metal-conducting layer and a pyroelectric layer.

It is clear that polymer films having large surface area and low thermal diffusion are preferred to crystalline materials for this device, and also that further development is required for practical use.

B. Thermal Energy-to-Electricity Conversion

Thermal energy-to-electricity converters had been proposed using the pyroelectric effect, i.e., the thermodielectric converter [34] and the heat-to-electricity converter [35], in which the conversion efficiencies are expected to increase for a drastic change of permittivity or D-E (displacement versus applied electric field) of ferroelectric materials around the

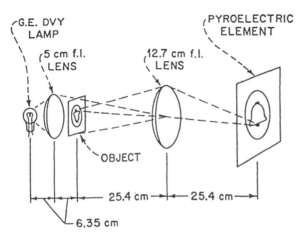

Figure 19 Pictorial view of the method used to demonstrate a pyroelectric copying device [32].

Curie temperature. Since ferroelectric polymers such as PVDF and P(VDF/TrFE) have characteristics of low Curie temperature and high applicable voltage as well as large surface area, they might be useful for thermal energy-to-electricity converters.

Direct conversion of thermal energy to electric power may be described in terms of *D-E* behavior of a ferroelectric material [35]. For any cycle process, the area on the *D-E* diagrams represents the electrical work, where the direction of path of the *D-E* curve (clockwise or counterclockwise) determines whether electrical energy is produced or dissipated.

Figure 20 [35] shows an overlay of *D-E* curves of PZST ceramics at two different temperatures of ferroelectric and paraelectric phases. The clockwise cycling process between two temperatures is obtained as follows. Starting in the upper right-hand corner at high voltage and low temperature, the ferroelectrics is discharged as it is heated. After reaching a high temperature, the ferroelectric is further discharged by reducing the externally applied voltage. The ferroelectric is then recharged at low voltage while cooling. After reaching low temperature, the ferroelectric is further recharged by increasing the externally applied voltage. Thus, the electrical cycle may be executed in a clockwise manner, producing the electrical energy of the shaded area in Figure 20.

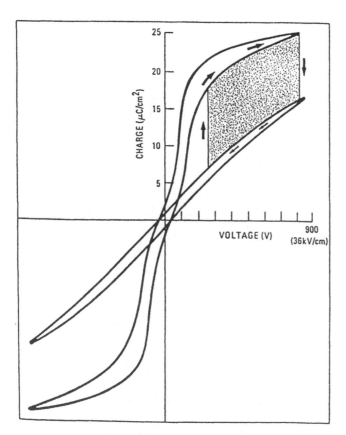

Figure 20 Electrical energy production cycle may be realized by approximately phased changes in temperature and applied electric field. The shaded area is equal to the electrical energy per cycle which may be produced by $Pb_{0.99}Nb_{0.02}[Zr_{0.68},Sn_{0.25},Ti_{0.07}]_{0.98}O_3$ [35].

The dielectric strength and remanent polarization are important factors for high conversion efficiency, because the electric output energy density scales as the product of the electric field and spontaneous polarization, as seen in Figure 20. P(VDF/TrFE) has a lower Curie temperature and much higher dielectric strength than those of PZST ceramics, so a great improvement might be expected for the copolymer.

The experimental equipment [35] for P(VDF/TrFE) copolymer was constructed by the combination of the controlling system of the test copolymer and the managing system of the heat flow, in which the spiral stack including the copolymer specimen was used. As shown from the top view in Figure 21, the spiral was wound around a central support rod and consists of alternate layers of plastic sheet and nylon separator screen. The copolymer specimen was wrapped into the stack. The nylon separator screen allows heat exchange fluid (a high-dielectric-strength silicone) to flow over the surface of the plastic sheet and specimen.

Figure 21 Schematic top view of a stack for pyroelectric conversion cycle of P(DVF/TrFE) copolymer [35].

From the experiment using the equipment described above, it was found that standard isothermal D-E hysteresis loop measurements were not necessarily accurate predictors of pyroelectric conversion performance for this material. Conduction effects were found to obscure the observation of conversion cycles in most cases for the materials available at that time. In spite of these difficulties, a conversion cycle was measured whose output electric energy density was 30 mJ/cm^3 for P(VDF/TrFE) of 73% VDF, where, unfortunately, the 73/27 material could be studied only in a temperature range substantially below its Curie point. It is now well established that the largest pyroelectric conversion cycles will be obtained near and above the Curie point of a given material. It is therefore expected that much larger cycles will be obtained without conduction effect once resins are available with increased resistivity (about 10^{15} Ω-cm) above their Curie temperatures, and great improvement of the output density might be achieved.

VII. CONCLUSIONS

When the pyroelectric properties of PVDF film was first studied in 1969 [7–10], the application of pyroelectric devices using monocrystalline TGS, LiTaO$_3$ and polycrystalline PbTiO$_3$, and PZT materials applied to IR radiation detection had already been established in various fields involving space, engineering, and home use. PVDF film improved for p and tan δ values could be utilized instead of these materials as the low-cost, single IR detector for home use, though performance has been inferior to other pyroelectric detectors because of the small p value.

On the other hand, by using the advantages of its low thermal diffusivity κ and commercially available thin film, the pyroelectric vidicon tube using PVDF film was developed in 1977 [19,20], in which a reticulation technique for the target in the vidicon could be avoided in contrast with TGS and PbTiO$_3$ tubes. The responsivity of the PVDF tube was also lower than those of TGS and PbTiO$_3$ tubes, though its spatial resolution was of the same order.

The additional advantages of PVDF film opened up a number of other applications besides IR radiation detection, some representatives of which have been described. At present, almost all pyroelectric devices using polymer film are handmade, with a special design for each purpose, and so many studies have been reported every year that all reports except those cited in previous sections are omitted. One can find them in "Literature Guides to Pyroelectricity" and "Bibliographies on Piezoelectricity and Pyroelectricity of Polymers," published until early 1990 in *Ferroelectrics* by S. B. Lang. A list of all published "Literature Guides to Pyroelectricity" appears in *Ferroelectrics 103*:91 (1990). A list of all published "Bibliographies on Piezoelectricity and Pyroelectricity of Polymers" is given in *Ferroelectrics 103*:219 (1990). The successor of these lists is the series of "Guides to the Literature of Piezoelectricity and Pyroelectricity," by S. B. Lang, for which the following have so far been issued: Guide 1, *Ferroelectrics 116*:277 (1991); Guide 2, *Ferroelectrics 119*:137 (1991); Guide 3, *Ferroelectrics 123*:69 (1991); Guide 4, *Ferroelectrics 129*:175 (1992).

The application of pyroelectric devices using polymer materials requires the preferred properties among their features for the specified application field. Table 3 summarizes the preferred properties in order of importance and the target values for practical use, which might serve as a guide to further improvement of available polymers and development of new polymers.

Table 3 Preferred Properties and Target Values in Various Fields of Use for
Pyroelectric Polymers

Field of use	Preferred properties	Target values
A. Radiation detector		
1. Single detector	High p value	$2 \times 1(2 \times 10^{-8})$ C/cm^{-2}-K
	Relatively high T_c	$100°C < T_c < 200°C$
	Thin film	$0.3 \ \mu m < t < 10 \ \mu m$
	Small area	$d = 1.5-5$ mm
	Small tan δ	tan $\delta = \sim 0.001$
	Small dielectric const.	$\epsilon = \sim 5$
2. One-dimensional detector	High p value	2×10^{-8} C/cm^2-K
	Relatively high T_c	$100°C < T_c < 200°C$
	Thin film	$0.3 \ \mu m < t < 10 \ \mu m$
	Small tan δ	tan $\delta = \sim 0.001$
	Low thermal diff. κ	$\kappa = 0.5 \times 10^{-3}$ cm^2/s
3. Two-dimensional	High p value	2×10^{-8} C/cm^2-K
	Relatively high T_c	$100°C < T_c < 200°C$
	Thin film	$0.5 \ \mu m < t < 10 \ \mu m$
	Medium size	$d = \sim 15-25$ mm
	Small tan δ	tan $\delta = \sim 0.001$
	Low thermal diff. κ	$\kappa = 0.5 \times 10^{-3}$ cm^2/s
4. Curved-shaped detector	High p value	$1 \times 1 \ (1 \times 10^{-8})$ C/cm^2-K
	Flexible to conform to curved surface, adhesive to surface	
B. Nonradiative device		
1. Heat flow detector	High p value	1×10^{-8} C/cm^2-K
	Flexible to conform to curved surface	
	Large area to be possibly thick	$t = 1-30 \ \mu m$
C. Heat-to-electricity converter		
	Low T_c	$\sim 100°C$
	High resistivity around T_c	10^{15} Ω-cm
	Flexible to conform to curved surface	
	Stable to heat cycles	
	Large size	
	Thick, stable	

REFERENCES

1. E. H. Putley, The pyroelectric detector, *Semiconductors and Semimetals* (R. K. Willardson and A. C. Beer, eds.), Academic Press, New York, 1970, p. 259.
2. E. Yamaka, Pyroelectric infrared detector, *Natl. Tech. Rep. 18*:141 (1972) (in Japanese).
3. E. Yamaka, Pyroelectric sensor using vinylidene fluoride-trifluoroethylene copolymer film, *Ferroelectrics 57*:337 (1984).

4. S. Miyata, M. Yoshikawa, S. Sakata, and M. Ko, Piezoelectricity revealed in the copolymer of vinylidene cyanide and vinyl acetate, *Polymer J. 12*:857 (1980).
5. X.-S. Wang, M. Iijima, Y. Takahashi, and E. Fukada, Dependence of piezoelectric and pyroelectric activities of aromatic polyurea thin films on monomer composition ratio, *Jpn. J. Appl. Phys. 32* (Part 1): 2768 (1993).
6. H. Kawai, The piezoelectricity of poly(vinylidene fluoride), *Jpn. J. Appl. Phys. 8*:975 (1969).
7. J. B. Bergman, J. H. McFee, and G. R. Grane, Pyroelectricity and optical second harmonics generation in polyvinylidene fluoride films, *Appl. Phys. Lett. 18*:203 (1971).
8. A. M. Glass, J. H. McFee, and J. B. Bergman, Jr., Pyroelectric properties of polyvinylidene fluoride and its use for infrared detection, *J. Appl. Phys. 42*:5219 (1971).
9. E. Yamaka, M. Matsumoto, and C. Hayashi, Pyroelectric detector using polymer, *32th Meeting Appl. Phys.*, 1971, p. 272 (in Japanese).
10. E. Yamaka, Infrared intensity detector using a pyroelectric polymer, U.S. Patent 3,707,695 26 (Dec. 1972).
11. R. J. Phelan, Jr., R. J. Mahler, and A. R. Cook, High D^* pyroelectric polyvinylfluoride detectors, *Appl. Phys. Lett. 19*:337 (1971).
12. K. Shigiyama, Pyroelectric IR detector using PVDF, *Natl. Tech. Rep. 26*:517 (1980) (in Japanese).
13. W. R. Blevin and J. Geist, Infrared reflectometry with a cavity-shaped pyroelectric detector, *Appl. Opt. 13*:2212 (1974).
14. R. J. Phelan, Jr., and A. R. Cook, Electrically calibrated pyroelectric optical-radiation detector, *Appl. Opt. 12*:2494 (1973).
15. A. Lee, A. S. Fiorillo, J. van Der Spiegel, P. E. Bloomfield, J. Dao, and P. Dario, Design and fabrication of a silicon-P(DVF-TrFE) piezoelectric sensor, *Thin Solid Films 181*:245 (1989).
16. Y. Takahashi, M. Iijima, and E. Fukada, Pyroelectricity in poled thin films of aromatic polyurea prepared in vapor deposition on polymerization, *J. Appl. Phys. 28*:L2245 (1989).
17. R. Asahi, O. Tabata, M. Mochizuki, J. Sakata, and S. Sugiyama, Pyroelectric IR sensor using PVDF thin film deposited by electro-spray method, *Tech. Dig. 9th Sensor Symp.*, 1990, p. 79.
18. R. Aashi, S. Sakata, O. Tabata, M. Mochizuki, S. Sugiyama, and Y. Taga, Integrated pyroelectric infrared sensor using PVDF thin film deposited by electro-spray method, *Proc. Int. Transducer '93*, Yokohama, 1993, p. 656.
19. E. Yamaka and A. Teranishi, Pyroelectric vidicon with PVDF film and its application, *Proc. 1st Meeting Feeroelectric Materials and Applications*, Kyoto, 1977, p. 258.
20. Y. Hatanaka, S. Okamoto, and R. Nishida, Pyroelectric vidicon with a sensitive PVF_2 film, *Proc. 1st Meeting Ferroelectric Materials and Applications*, 1977, p. 251.
21. A. Kaneko, T. Tanabe, T. Fumoto, K. Tomii, and J. Nishida, Development of high sensitive pyroelectric vidicon, *J. Inst. TV Eng. Jpn. 40*:878 (1986) (in Japanese).
22. E. Yamaka, Recent infrared study in Japan, *SPIE 1157, Infrared Technology XV*:286 (1988).
23. U. Korn, Z. Rav-Noy, and Shtrikman, Pyroelectric PVF_2 infrared detector arrays, *Appl. Opt. 20*:1980 (1981).
24. E. Bunnuel, D. Esteve, J. Farre, V. V. Pham, and J. J. Simmone, Performance of a pyroelectric PVDF detector extension to a linear array as a image detector, *Proc. SPIE 702*:35 (1986).
25. A. Reynes, D. Esteve, J. Farre, V. V. Pham, and J. J. Simonne, Elaboration and properties of a linear pyroelectric PVDF focal plane array, *Proc. SPIE 865*:40 (1988).
26. H. Coufal, Photothermal spectroscopy using a pyroelectric thin-film detector, *Appl. Phys. Lett. 44*:59 (1984).
27. A. Mandelis, Frequency-domain photopyroelectric spectroscopy of condensed phases (PPES): A new, simple and powerful spectroscopic technique, *Chem. Phys. Lett. 108*:388 (1984).
28. I. F. Faria, Jr., C. C. Ghizoni, and L. C. M. Miranda, Photopyroelectric scanning microscopy, *Appl. Phys. Lett. 47*:1154 (1985).
29. S. B. Lang, Technique for the measurement of thermal diffusivity based on the laser intensity modulation method (LIMM), *Ferroelectrics 93*:87 (1989).

30. A. Mandelis, F. Care, K. K. Chan, and L. C. M. Miranda, Photopyroelectric detection of phase transitions in solids, *Appl. Phys. A 38*:117 (1985).
31. R. Kashyap and P. Pantelis, Optical fibre absorption loss measurement using a pyroelectric poly(vinylidene fluoride) tube, *J. Phys. D 18*:1709 (1985).
32. J. B. Bergman, G. R. Crane, A. A. Ballman, and H. M. O'Byran, Jr., Pyroelectric copying process, *Appl. Phys. Lett. 21*:497 (1972).
33. J. B. Bergman, Jr., and G. R. Crane, Pyroelectric copying device, U.S. Patent 3,824,098 (July 16, 1974).
34. C. F. Pulvari and F. J. Garcia, Conversion of solar to electrical energy utilizing the thermodielectric effect, *Ferroelectrics 22*:769 (1978).
35. R. B. Olsen, D. A. Bruno, J. M. Briscoe, and E. W. Jacobs, Pyroelectric conversion cycle of vinylidene fluoride-trifluoroethylene copolymer, *J. Appl. Phys. 57*:5036 (1985).

15
Electromechanical Applications

Iwao Seo and Dechun Zou*
Mitsubishi Chemical Corporation, Ibaraki, Japan

I. INTRODUCTION

Basic theories of pyro-, piezo-, and ferroelectric properties and their related organic polymer materials have been described by other authors. In this chapter, we will focus mainly on some applications based on the electromechanical properties of piezoelectric polymer materials.

Four typical classes of polymeric piezoelectrics are shown in Table 1. The first group is optically active polymers. Most of the polymers in this group are biological materials, e.g., cellulose derivatives such as cellulose triacetate, cellulose diacetate, and cyanoethyl cellulose, proteins such as collagen and keratin, as well as synthetic polypeptides such as poly-γ-methyl-glutamate (PMG) and poly-γ-benzyl-glutamate (PBG). The origin of piezoelectricity in these polymers is ascribed to the internal rotation of dipoles related to asymmetric carbon atoms which give the polymer optical activity. The piezoelectric effect in these polymers can be observed when the polymer film is oriented uniaxially. It is found that the piezoelectric constant d_{14} in these polymers is not zero, which means that the shear stress in the plane of orientation causes electric polarization perpendicular to the plane of stress. Piezoelectric effect has also been observed in synthetic optically active polymers such as polypropylene oxide (PPO) and poly-β-hydroxybutyrate (PHB). The second group of piezoelectric polymers is poled polar polymers (electret polymers). Poly-vinyl chloride (PVC), polyvinyl fluoride (PVF), polyacrylonitrile (PAN), odd nylons such as nylon-11, and copolymers of vinylidene cyanide and vinyl acetate [P(VDCN/VAc)] are examples belonging to this group. Piezoelectricity in these polymers is caused by orientation of dipoles in the polymer chains which are frozen in during the poling process.

---

*Present affiliation: Tokyo University of Agriculture and Technology, Tokyo, Japan.

Table 1 Classification of Piezoelectric Polymers

Category	Material	Origin of piezoelectricity	d_{14} or d_{31} (pC/N)
Optically active polymers	Polysaccharides Fibrous proteines Polypeptides Poly-β-hydroxybutylate Polypropylene oxide	Internal rotation of dipoles Optical activity Uniaxially oriented state	$d_{14} < 4$
Poled polar polymers (electret polymers)	Polyvinyl chloride Polyacrylonitrile Polyvinyl fluoride .... Vinylidene cyanide copolymers Odd nylons Polyurea	Polarization of oriented molecular dipoles "Frozen-in" polarization	$d_{31} < 3$ $d_{31} < 10$
Ferroelectric polymers	Poly(vinylidene fluoride) Vinylidene fluoride copolymers	Spontaneous polarization of polar crystal	$d_{31} < 40$
Ceramic-polymer composites	PVDF/PZT POM/PZT Rubber/PZT,PbTiO₃ Epoxy/PZT	Ferroelectric ceramics embedded in matrix polymer	$d_{31} < 40$

In the case of amorphous polymers such as PVC, PAN, and P(VDCN/VAc), the poling electric field is applied at a temperature around the glass transition temperature and the preferential alignment of dipoles is frozen in by cooling the poled film to room temperature before removing the poling field. In the case of crystalline polymers such as PVF and nylon-11, the polarization resides in a crystalline phase where poling fields orient chains in directions close to the electric field. The third group is ferroelectric polymers, in which an oriented polar domain gives piezoelectric effect. Poly(vinylidene fluoride) (PVDF) and the copolymer of vinylidene fluoride (VDF) with trifluoroethylene (TrFE) or tetrafluoroethylene (TFE) are well known as ferroelectric polymers [1–6]; they were the first and also the only group have been completely verified to be ferroelectric polymers. The presence of spontaneous polarization in a polar crystal phase (polar domain) is the origin of piezoelectricity. In recent years, several other polymers have also been reported to show some ferroelectric behaviors. These polymers are copolymers of vinylidene cyanide and fatty acid vinyl ester derivatives [7–9], the polyamide (odd nylon) [10] family, and polyurea [11]. In all of the above three groups, the origin of piezoelectricity is basically from the polymer molecules; we call them pure polymer piezoelectric systems. In these kinds of piezoelectrics, both the film-forming ability and piezoelectric properties are from polymer molecules. The polymer needs to have not only good film-forming ability, but also polar function groups in its molecules. Compared with above pure polymer systems, there is another kind of polymeric piezoelectric called polymer/ferroelectric ceramic composite, which is listed in the fourth rank in Table 1. The most often used ferroelectric ceramics are lead titanate zirconate (PZT) and $PbTiO_3$ [4–6]. In a polymer/ceramic composite system, the film-forming ability is from matrix polymer materials and piezoelectric activity arises from the intrinsic piezoelectricity of the ceramics. The merit of composite piezoelectrics is that the physical properties can be controlled through the choice of matrix polymers and ferroelectric ceramics, as well as fabrication and poling procedures. They have a combination of high piezoelectric properties of ferroelectric ceramics and film-forming ability of matrix polymers. In this way it is possible to create high-performance piezoelectric films such as these with high thermal stability and sensitivity, which is very difficult to achieve in a pure polymer system because of the thermal relaxation of oriented dipoles.

Table 2 shows the physical properties of commercially available piezoelectric polymers compared with some typical ferroelectric materials. The piezoelectric constants d_{31} of all the polymers are relatively small compared with the best ferroelectric materials, but this is not a drawback because the dielectric constants are also much smaller. Thus the voltage generated per unit stress, which is called g_{31}, is larger than that of ferroelectric ceramic material. That is, polymer piezoelectric film has higher sensitivity to force (stress) than do ferroelectric ceramics, so the electromechanical coupling factor (transducer capacity) is better than that of ferroelectric materials. The density of polymer is lower than that of ceramics and quartz, which is generally a plus factor. Furthermore, these polymers have a low acoustic impedance, resulting in a desirable impedance match to that of water, human tissue, and adhesive materials. In addition to these advantages, these polymer are flexible, lightweight, tough, and readily manufactured into sheet form in a continuous roll process or into complex shapes for specific applications. The main disadvantages of piezoelectric polymers are their relatively poor stability. One is the dimension stability, due in part to mechanical relaxation of residual stresses induced in the fabricating process, and another is thermal relaxation, which leads to relaxation of piezoelectricity. The melting or softening points of polymers are lower than those of ceramics, and the available

Table 2 Comparison of Piezoelectric Properties of Polymers and Nonorganic materials

Materials	Density ρ (10^3 kg/m³)	Modulus C_{11} (10^9 N/m²)	Dielectric constant ϵ_r	Piezoelectric constants				Coupling factor k_{31}	Maximum available temperature (°C)
				d_{31} (10^{-12} C/N)	e_{31} (10^{-2} C/m²)	g_{31} (10^{-3} V·m/N)	h_{31} (10^7 V/m)		
PVDF	1.78	3.0	13	20	6.0	174	53	0.10	80
P(VDF/TrFE) (VDF=55%)	1.90	1.2	18	25	3.0	160	19	0.07	70
P(VDF/TrFE) (VDF=75%)	1.88	2.0	10	10	2.0	110	22	0.05	100
P(VDCN/VAc)	1.20	4.5	4.5	6	2.7	169	76	0.06	160
PVDF/PZT	5.3	3.0	120	20	6.0	19	6	0.07	100
Rubber/PZT	5.6	0.04	55	35	1.4	72	11	0.01	100
POM/PZT	4.5	2.0	95	17	3.4	20	4	0.08	140
Quartz	2.65	77.2	4.5	2	15.4	50	387	0.09	573
PZT	7.5	83.3	1200	110	920	10	87	0.31	250

temperature for long-term use is presently limited to below about 100°C for most polymers.

Figure 1 shows the basic principle of the two-port piezoelectric transducer [2]. The general transducer has an electrical port and a mechanical port, where the parameters at the mechanical port are force (stress tensors) or elastic displacement (strain tensors), and the parameters at the electrical port are voltage, current, and surface charge. The transducer can be driven from either port with another port driving a load. The basic equations of the transducer for a finite element of material are given by

$$F = AH + BQ \tag{1}$$
$$V = CH + DQ \tag{2}$$

where F is force, H is the elastic displacement, Q is the electric charge, and V is the electrical voltage. A, B, C, and D are device constants determined by the piezoelectric constants of the transducer material and structure factors of the device.

A device based on the mechanoelectrical properties converts a mechanical input signal such as force (stress) or elastic displacement (strain) to an electrical output signal. These devices are often used as sensors to detect displacement, stress, vibration, and sound. Typical sensors in practical use are hydrophones, blood pulse counters, blood pressure meters, pressure sensors, flow meters, acceleration sensors, shock sensors, vibration sensors, touch sensors, microphones, antinoise sensors, and keyboards. Devices belonging to the second group are the reverse. In these devices, the input is an electrical signal, and the output is a mechanical signal such as force, displacement, vibration, or sound. These devices are often used in fields associated with position control, acoustic systems, or actuators, for example, headphones, loudspeakers, ultrasonic driving motors, ultrasonic transducers, and bimorph displays.

In recent years, many kinds of piezoelectric devices from organic polymer materials have been developed and widely used as industrial as well as medical instruments [2,14,15,151]. Table 3 lists some typical applications of ferroelectric polymers. These applications can be grouped into sensors, medical instrumentation, robotics, optical de-

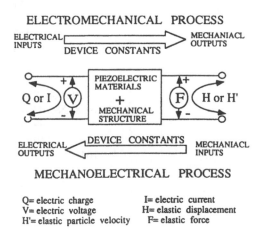

Figure 1 The basic working principle of two-port piezoelectric transducer. (From Ref. 2.)

Table 3 General Applications of Ferroelectric Polymers

Application	References
1. Sensors	
Accelerometers	20,57,146
Flow meters	58
Stress and strain gauges	55
Switching or information storage devices	56
Shock-wave sensors	59
Ballistic transducers (impact detection for shooting)	60
Impact sensors	61–63,147
Underwater shock-wave sensors	64
Coin sensors	32
2. Medical instrumentation	149
Artificial sensitive skin	79
Tactile sensors	83
Measurement of tongue shape	81
Pulse-wave monitors	80,82,84
Fetal heart sound detectors	76
Catheter-tip pressure transducers	77,78
Phonocardiographic transducers	80
Korotkov's sound detectors	80
3. Robotics	
Matrix tables for object recognition	74
Soft-touch position sensors	72
Tactile force sensors	73,75
4. Optical devices	
Optical fiber modulators	131–133
Variable mirrors (deformable mirrors)	136,137
Light modulators	88,138
Fiber optical switches	139
Electric field sensors	130,134,135
5. Computer and office automation	
Keyboards	14,15,28,66–70
Telephone dials	66
Tablets	71
6. Energy conversion	
Physiological power supplies	143
Electric power generation (sea waves)	143
Electric energy storage	33
7. Ultrasonic or underwater transducers	
Ultrasonic bulk waves	19,40,41,85,86
High-frequency transducers	85,87,88,103,141
Nondestructive testing transducers	89–91
PVDF MOSFETs	110,111
SAW devices	104–109,125,140
Hydrophones	
General	21,112–114
Miniature	92,115,116
Membrane	93–96
Needle	97,117
Flexural disk	98

Table 3 Continued

Application	References
Sphere	120
Tubular or cylindrical	99,100
Cable	65,118,119
Variable-focus transducers	101
Confocal transducers	84,86
Pressure transducers	102
Linear array transducers	121–124,148
Scanning acoustic microscopes	103,126,127
Flexible ultrasound transducers	128
Annular array transducers	129
Optoacoustic devices	88
8. Audio	
Microphones	
General	12,13,14
Noise cancellation	15,16,17
Lightweight	18,19
Membrane	13
Water and moisture resistant	18,20–22
Telephone transmitters	12,27,28
Headphones and earphones	12,13
Loudspeaker and tweeters	13,24
Flat-type loudspeakers	26
Digital speakers	31
Transparent speakers	29,30
Phonograph cartridges	13,25,20
AE microphones	23
Guitar pickups	65
Electric pianos	65
9. Actuators	
General	34–37
Air flow fans	41–43,46,47
Micromanipulator	14,44,45
Wind electrical generators	38
Tent and balloon transducers	39,40
Bowtie transducers	49
Light shutters	48,142
Vibrators for marine fouling prevention	50
Traveling micromechanisms	51,52
Display devices	42,53,54

vices, computer and office automation, energy conversions, ultrasonic and underwater transducers, electroacoustic transducers, and acutators. In this chapter we will focus mainly on some applications based on the electromechanical properties of piezoelectric polymers. In particular, electroacoustic transducers, ultrasonic transducers, and actuators will be described in detail.

II. ELECTROACOUSTIC TRANSDUCERS

We will mainly introduce electroacoustic transducers made of piezoelectric polymer films. In the audiofrequency range, when an alternating voltage is applied to electrodes attached to the two surfaces of piezoelectric polymer film, the film vibrates in a transverse direction. If the frequency of applied voltage is high enough, the film vibrates in the thickness direction at high frequency, and then ultrasonic waves can be produced. Electroacoustic transducers such as headphones, speakers, and ultrasonic generators all work on this principle. In these kinds of devices, the dominant parameters are piezoelectric constant d_{31} and Young's modulus of the piezoelectric film. Large piezoelectric constant can introduce large displacement (deflection), and high Young's modulus can give a strong output force even when the same drive voltage.

A. Speakers and Headphones

The first commercial application of piezoelectric polymer film was audio transducers such as high-frequency loudspeakers (tweeters) and stereophonic headphones. Piezoelectric polymer films are very suitable for audio transducers with a wide frequency range because the material can be formed into thin film (1–300 μm) with low mechanical stiffness. Audio transducers using piezoelectric polymer film are based on the transverse piezoelectric effect. The film vibrates in the transverse direction, and coupling to the surrounding medium is weak. Better coupling can be achieved by the conversion of transverse motion into pulsating motion by using cylindrical elements. A perfect omnidirectional tweeter utilizing a piezoelectric PVDF film was constructed by the Pioneer Electronic Corporation in 1974. Omnidirectional patterns are realized almost up to 20 kHz. These cylindrical tweeters were mounted on the top of surface of the associated loudspeaker cabinet to improve the sound quality in the high-frequency range. The use of high-frequency (5 kHz and up) tweeters reinforces the upper frequency spectrum of conventional speakers in high-fidelity audio systems. Figure 2 shows one example. On top of the speakers there are cylindrical tweeters made of PVDF piezoelectric film [13,15]. Headphones have also been commercialized. As shown in Figure 3, the structure of the headphones is quite simple [13]. This type is designed as a lightweight and thin headphone. These can be used without setup transformers and exhibit excellent frequency response with low distortion. In addition, many other audio transducers with piezoelectric polymer films have been devised and described in the literature [2,13–15]. In the following section, we will describe several types of speakers fabricated from piezoelectric polymer films.

1. Paper Speakers, Transparent Speakers

By using polymeric piezoelectric film, loudspeakers can be made as thin as a paper sheet (<100 μm). Figure 4 shows a commercialized paperlike speaker (paper speaker) made of composite piezoelectric film of PZT ceramic powder and polyoximethyl [PZT/POM]. The paper speaker is one of the typical applications using the thin film-forming ability of polymeric material which is impossible for ceramic materials with their large piezoelectric constants. For the first step in fabricating a poled composite piezoelectric film, ferroelectric ceramic filler such as PZT is mixed with melting polymer by hot roller, then pressed and passed through the gap of the rollers to form a thin sheet (about 100 μm). The second step is thermal poling. Generally, metallic electrodes are first attached onto both surfaces of the film by vacuum evaporation or other methods such as printing

Figure 2 PVDF tweeter (on the top of speaker box) used to improve the sound quality of traditional speaker in high frequency area. (From Ref. 15.)

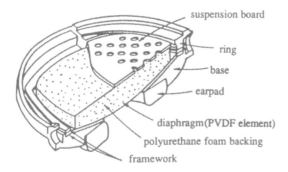

Figure 3 A cross-sectional view of PVDF headphone. (From Ref. 13.)

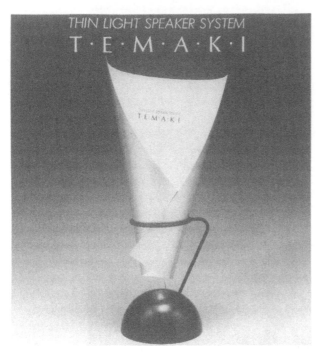

Figure 4   Paper speaker made of composite piezoelectric film (PZT/POM).

conductive paste, after which the film is subjected to an electric field of about 300 kV/ cm at 110°C for about 1 h. At this stage the poled film is ready to be used to fabricate the paper speaker. Figure 5 shows the frequency response of the paper speaker when the sound detector distance is 50 cm and the drive voltage is 20 V. Over a wide frequency range from 600 Hz to 1600 Hz, the speaker shows a flat sound pressure property, indicating its suitability to be used as speaker material.

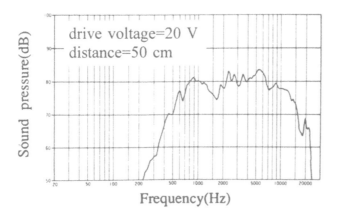

Figure 5   Frequency feature of speaker made of composite piezoelectric film (PZT/POM).

For some practical applications, it is desirable to making a transparent speaker by using poled transparent polymer film. In this case it is necessary to have both transparent and conducting electrodes. This can be done by using indium tin oxide (ITO) or conducting polymer. Figure 6 shows a transparent speaker made of poled P(VDCN/VAc) film [29,30]. The electrodes on both sides are indium tin oxide (ITO) deposited by vacuum evaporation. Generally, the sound quality is greatly influenced by the conductivity of electrodes. When ITO is used as the electrodes, its thickness should be large enough to obtain a sound quality similar to that of metallic electrode speakers. If the electrodes are conducting polymer, an all-polymer speaker can be realized. This is not easy, however, because high-conductive polymers generally have strong absorption bands in the visible to near-infrared region, so visible light cannot pass through even if the film is only several micrometers thick. Recently, a new method for preparing both transparent and conducting polypyrrole film was developed [144,145]. One example is the preparation of poly(vinyl alcohol) composite film by chemical vapor deposition (CVD) technique [145]. It is similar to many other conducting polymers in that higher conductivity leads to lower transmittance. High-quality transparent and conducting film can be obtained only by proper controlling of polymerization time and concentration of oxidant.

Speakers made of piezoelectric polymers have the advantages of typical polymers. In particular, the thickness may lead to very promising applications such as paperlike speakers suitable for the future's thin televisions, decoration, and interiors. Generally, the cathode-ray tube (CRT) TV is bulky, heavy, and expensive. In recent years, liquid crystal display (LCD) TV has been developed, and has become very common in the market. The size of the LCD TV is comparatively thin and light, and there should be some kind of speaker suitable for such a thin and light TV. This can be realized by using paper speakers. However, the disadvantage of paper speakers is that it is difficult to get as

Figure 6 Transparency speaker made of P(VDCN/VAc) film. (From Ref. 30.)

high-quality low-frequency sound as from other types of speakers at present. In order to overcome this problem, some improvements such as using the resonance effect between the film and the supporting frame, have been proposed so that by appropriate design, high-quality low-frequency loudspeakers may be possible.

2. Digital Speakers and Headphones

Another promising application of piezoelectric polymers to the audio field is utilization for digital-type speakers and headphones. Digital speakers can change a digital signal to an acoustic analog output directly [31]. Figure 7 shows the basic structure and working principle of a digital speaker. The digital headphone is composed of a piezoelectric PVDF film and an acoustic desampling acoustic filter. Seven pairs of circular aluminum electrodes with the same center are attached on one side of the film; the area of each circular

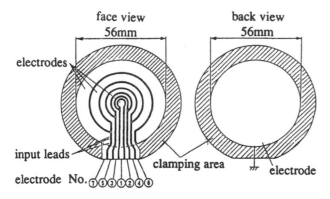

(a) Face- and back-views of a digital speaker[31]

Area: $S_1 = S_0$, $S_2 = 2S_0$, $S_3 = 4S_0$, $S_4 = 8S_0$,...., $S_n = 2^{n-1}S_0$

digital signal	active electrode No.	active area	volume
00000000	00000000	0	0
00000001	00000001	S_0	V_0
00000010	00000020	$2S_0$	$2V_0$
00000011	00000021	$1S_0 + 2S_0$	$3V_0$
00000100	00000300	$4S_0$	$4V_0$
00000101	00000301	$1S_0 + 4S_0$	$5V_0$
00000110	00000320	$2S_0 + 4S_0$	$6V_0$
00000111	00000321	$1S_0 + 2S_0 + 4S_0$	$7V_0$
00001000	00004000	$8S_0$	$8V_0$
00001001	00004001	$1S_0 + 4S_0$	$9V_0$
00001010	00004020	$2S_0 + 8S_0$	$10V_0$

(b) Relationship between the digital
signal and output sound volume

Figure 7 Principle of digital speaker. (a) Face- and back-views of a digital speaker (Ref. 31) and (b) relationship between the digital signal and output sound volume.

electrode is two times larger than the next one from inside to outside. The opposite side of the film is completely covered with aluminum by vacuum evaporation. By using this type of electrode structure, the strength of sound produced from the headphone is determined by the electrodes to which the drive voltage (digital signal) is applied. Each electrode corresponds to one bit in the digital control system, so that, if there were enough control bits, it would be possible to reproduce sound precisely through a digital signal. A speaker with 8 bits was developed, but its bit number was only half of the needed value, so the dynamic properties of the reproduced sound was not very good. More bit electrodes are needed to improve sound quality.

III. ULTRASONIC TRANSDUCERS

One of the most important and promising applications of piezoelectric polymers is ultrasonic transducers. As shown in Figure 8, various ultrasonic transducer have been developed which cover a wide frequency range from 10 kHz to 10 GHz.

Polymer materials are very soft, light, pliable, and easily fabricated into large sheets and cut or bent into complex shapes. They are also mechanically lossy, with very flat response over a wide range of frequency. All these make polymer materials the best candidates for vital information-converting materials or human body-suitable materials. Table 4 compares some related physical properties of piezoelectric polymers with these of ferroelectric ceramics used widely in ultrasonic transducers. For ultrasonic transducer applications, polymer materials have special advantages compared with ceramic materials:

1. The acoustic impedance of piezoelectric polymers is very close to that of the human body and water, so that when these soft materials are used, more complete energy

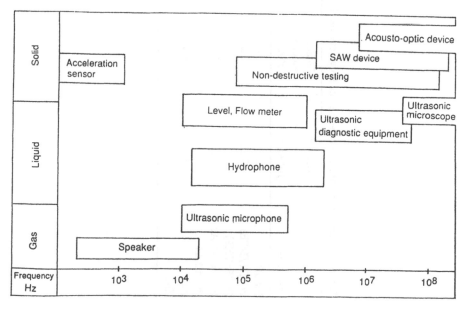

Figure 8 Applications of piezoelectric polymers at various frequencies.

Table 4 Electromechanical Properties of Polymers and Nonorganic Materials for Ultrasonic Transducers

Materials	Density ρ (10^3 kg/m^3)	Sound velocity (10^3 m/s)	Coupling factor k_{31}	Dielectric constant ϵ_r	Dielectric loss tan δ_m	Mechanical loss tan δe	Acoustic impedence Z (10^6 kg/m^{2}s)
PVDF	1.78	3.0	0.20	6.2	0.31	0.13	4.0
P(VDF/TrFE) (VDF=75%)	1.88	2.0	0.30	6.0	0.15	0.05	4.5
P(VDCN/VAc)	1.20	4.5	0.28	4.2	0.05	0.02	3.0
POM/PZT	4.5	2.0	0.25	45	0.05	0.15	8.0
PZT	7.5	83.3	0.64	1200		0.002	30
ZnO	5.7		0.28	8.84			36

transfer occurs into the transducer and there are fewer reflections and distortion of the wave pattern at the transducer interface.

2. The large mechanical loss of polymers makes it easy to get a short pulse with high distance resolution.
3. The ease of fabricating into any shape makes it possible to focus the ultrasonic wave to one point without the help of a lens. High position and direction resolution can be achieved. In many cases, the use of a focused transducer provides an increase in resolution, as the acoustic power is concentrated at the focal point.
4. The ease of producing area film makes large-area transducers possible.

Another merit of polymer piezoelectrics when used as ultrasonic transducer material is their thin film-forming ability. This is a very important factor for the selection of ultrasonic materials. As shown in Figure 8, it is evident that the working frequency of an ultrasonic transducer is very high, for example, from several kilohertz to several gigahertz. This means the film needs to be very thin, in some cases even thinner than several micrometers. It is not easy for ceramic materials to form such a thin film, but it can be easily done using polymer materials. The main disadvantages of polymer trans-ducers are their relatively low piezoelectric constant and relatively poor dimensional stability.

Figure 9 shows some commercialized ultrasonic transducers for industrial measure-ments. The available frequency for these polymer transducers ranges from several MHz to 50 MHz. Figure 10 gives a comparison of the reflect echo pulses detected by a

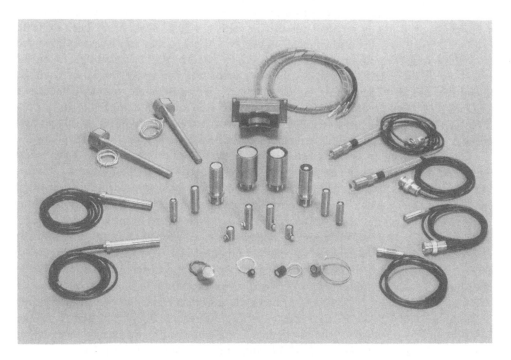

Figure 9   Various types of ultrasonic transducers made of piezoelectric polymer film for industrial measurement.

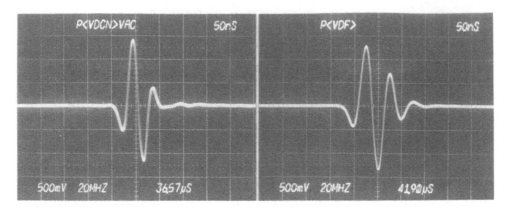

Figure 10   Pulse echo waveforms of PVDF and P(VDCN/VAc) transducers obtained in water.

P(VDCN/VAc) transducer and a PVDF transducer. Both gave a short pulse (wide range) compared with a traditional PZT transducer. This means that, if the transducer is made of polymer material, high distance resolution can be achieved. Figure 11 gives a comparison of the measured results of a model defect embedded inside steel plate by using a P(VDCN/VAc) transducer and a ceramic transducer. At a diameter of 0.5 mm, a defect existing at a depth of 1.75 mm from the surface can be correctly detected using the P(VDCN/VAc) transducer. Undrawn cast P(VDCN/VAc) film has a large electromechanical coupling constant kt in the thickness direction; sound velocity in this direction is 2500 m/s. These make it possible to make a 1-GHz ultrasonic wave generator in which a thickness of 1 μm is required for the piezoelectric film. This is not very difficult, because 0.5-μm to several micrometer P(VDCN/VAc) thin film can be easily prepared by casting or spin-coating technique. If the P(VDCN/VAc) film is casted directly on the sound-absorbing material (substrate) and then poled, the applicable frequency could be higher because of the absence of an adhesive layer. It is possible to detect a defect (hole size is 8 μm) embedded in the side of hot-pressed aluminum oxide plate by using a 50-MHz P(VDCN/VAc) transducer. A P(VDCN/VAc) transducer working at higher frequency, used for detecting microscopic defects in ceramics, has also been developed.

Real-time imaging systems using linear array transducers are widely accepted in ultrasonic diagnosis. Some ultrasonic transducers composed of piezoelectric polymer films are shown in Figure 12. Figure 12a is a 10-MHz linear array transducer for medical applications. Mechanical damage-protecting coating and acoustic impedance-matching layers are coated on the surface. The curved surface is designed to focus the wave beam. Sixty-four array elements are arranged on the surface. This type of transducer is specially developed to be used to observe the surface organisms of the human body. Figure 13 shows some examples of imagery photographs of breast cancer taken by a 10-MHz transducer. The cancer zone can be clearly distinguished from normal tissue [150].

Figure 12b shows a new type of ultrasonic transducer made of piezoelectric composite film, POM/PZT [128]. This transducer is an ultrasonic doppler for fetal heart-rate monitoring. A comb-shaped gold electrode is formed on the film by vacuum evaporation. The size of the transducer is 3.5 × 14 cm², and it is loaded by a 6-Vp-p, 1.5-MHz alternating current and receives doppler signals. This transducer can catch fetal heart rate continuously.

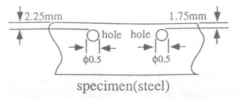

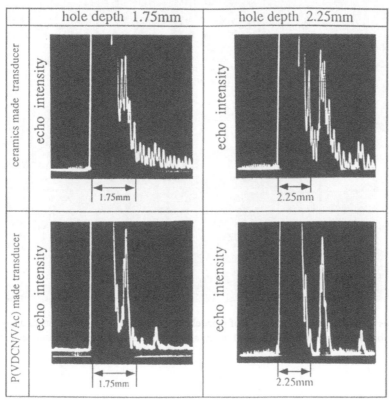

Figure 11 Reflected pulse echo from model defects detected by polymer transducer and ceramic transducer.

Figure 12c shows a flexible ultrasonic transducer used in detecting the liquid level of a tank or a tube. The flexible transducer made of piezoelectric composite material can be engaged with the circumference of the container in face-to-face relation to the work uniformly along the entire effective surface of the transducer. It is possible to transmit and receive signals efficiently so that measurement of the liquid level can be achieved even if the container is metallic.

Another interesting application is an array device in which the piezoelectric polymer is combined with integrated circuit technology for use in medical imaging [110,111]. The array is composed of a PVDF film bonded by an epoxy layer to the surface of a MOSFET wafer. A linear 34-element receiving transducer array has been built with a bandwidth of 6 MHz and a dynamic range over 70 dB.

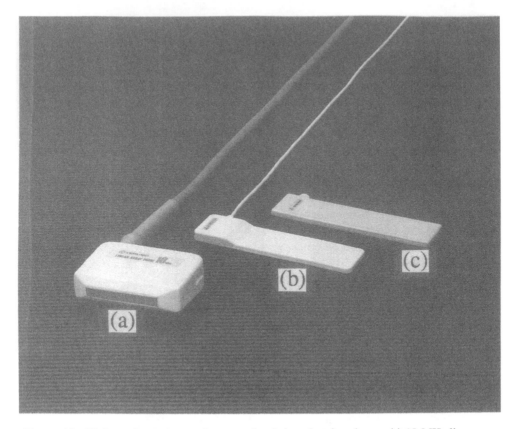

Figure 12 Various ultrasonic transducers made of piezoelectric polymer. (a) 10-MHz linear array transducer, (b) flexible doppler transducer for fetal heart, and (c) flexible ultrasonic transducer for liquid level meter.

IV. ACTUATORS

The basic structure used to obtain a force or elastic displacement from an electrical input is the bimorph, as shown in Figure 14. Figure 14a is the most simple and basic bimorph structure, which is also called the transverse effect bimorph. Figure 14b is the structure of a longitudinal-effect bimorph which can produce a much stronger force than the transverse effect bimorph can. Figure 14c is a multilayer-type transverse-effect bimorph designed to generate greater force than the monolayer type. We introduce some basic theories on bimorph structures before discussing practical applications.

A. Fundamental Properties of Bimorphs

Figure 14a shows a two-layer bimorph cantilever structure (transverse-effect bimorph structure). Two piezoelectric polymer films with the same polarization direction are attached altogether by adhesive. One end of the piece is fixed and the other is free. When electric fields with opposite directions (to the polarization direction) are applied to them, one layer tends to elongate and the other layer to shrink. The bimorph deflects because of the opposite stresses, and a large elastic displacement can be obtained at the free end. Let the radius of curvature of the deflected bimorph shown in Figure 14a be R_0, the

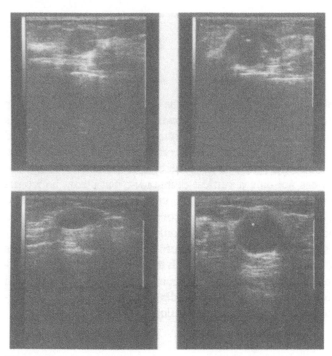

Figure 13 Various ultrasound images of the breast obtained with a 10-MHz linear array transducer.

displacement at the free end $X_{F=0}$, the generated maximum force $F_{X=0}$, the fundamental resonance frequency f_0, and the reserved elastic energy U_M. The relations among these parameters and the drive voltage can then be described by the following equations:

$$R_0 = \frac{2}{3} \frac{t^2}{V} \frac{1}{d_{31}} \tag{3}$$

$$X_{F=0} \approx \frac{L^2}{2R_0} = \frac{3}{4} d_{31} \frac{L^2}{t^2} V \tag{4}$$

$$F_{X=0} = \frac{3}{2} d_{31} C_{11} \frac{wt}{L} V \tag{5}$$

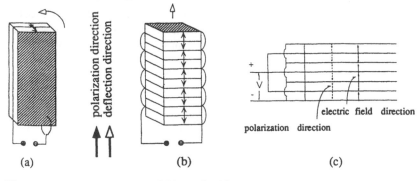

Figure 14 Basic structure of bimorph. (a) Transverse effect bimorph, (b) longitudinal effect bimorph, and (c) multilayer type bimorph.

$$f_0 = \frac{t}{2\pi} \left(\frac{1.875}{L}\right)^2 \sqrt{\frac{C_{11}}{3\rho}} \tag{6}$$

$$U_M = \frac{9}{16} d_{31}{}^2 C_{11} \frac{Lw}{t} V^2 \tag{7}$$

where t is the thickness of the piezoelectric film, L is the length of the bimorph, w is the width, ρ is the density, d_{31} is a piezoelectric strain tensor component, C_{11} is Young's modulus in the stretch direction, and V is the driving voltage. As seen in Eqs. (4) and (5), the transverse-effect bimorph, by using a thin film, produces a large displacement, making it very suitable for utilization where large deflection is needed. It should be noticed that the displacement has not only a vertical component, but also a horizontal component. Generally, the horizontal component is much smaller than the vertical component; in many cases, its effect can be neglected. For control purposes, however, both the horizontal component and the vertical component should be taken into account.

The two-layer bimorph structure often produces insufficient bending force for uses where strong force is needed. One approach to dealing with this problem is the multilayer-type transverse-effect bimorph structure as shown in Figure 14c. This type of bimorph structure can generate large force without sacrificing bending amplitude. In a multilayer-type transverse effect bimorph structure, we can rewrite Eqs. (3)–(6) as follows:

$$R_0 = \frac{2}{3} \frac{Nt^2}{V} \frac{1}{d_{31}} \tag{8}$$

$$X_{F=0} \approx \frac{L^2}{2R_0} = \frac{3}{4} d_{31} \frac{L^2}{Nt^2} V \tag{9}$$

$$F_{X=0} = \frac{3}{2} d_{31} C_{11} \frac{wtN^2}{L} V \tag{10}$$

$$f_0 = \frac{Nt}{2\pi} \left(\frac{1.875}{L}\right)^2 \sqrt{\frac{C_{11}}{3\rho}} \tag{11}$$

where $2N$ is the total number of superposed layers. From these equations, we can see that the generated maximum force $F_{X=0}$ at the free end is proportional to N^2, the deflection is inversely proportional to N, and the resonance frequency is linearly proportional to N.

Figure 14b shows the structure of a multilayer-type longitudinal-effect bimorph. Electrodes are attached on both sides of the film to form a parallel circuit. In this structure, the displacement $X_{F=0}$ in thickness direction, the generated force $F_{X=0}$, the resonance frequency, f_0, and the reserved mechanical energy U_m can be expressed as follows:

$$X_{F=0} = d_{33} \frac{L}{t} V \tag{12}$$

$$F_{X=0} = 2w d_{33} C_{33} V \tag{13}$$

$$f_0 = \frac{1}{2L} \sqrt{\frac{C_{33}}{\rho}} \tag{14}$$

$$U_m = d_{33}{}^2 C_{33} \frac{Lw}{t} V^2 \tag{15}$$

where d_{33} is the piezoelectric constant and C_{33} is Young's modulus in the thickness direction. The longitudinal-effect bimorph can create very strong force, and it has high energy-converting efficiency and good durability. Moreover, it is possible to use it to drive a device with very high frequency due to its high resonant frequency. Small displacement may be the disadvantage of this structure.

The bimorph can be also considered as a condenser, so the input (reserved) electrical energy (U_E) can be calculated from Eq. 16 [12]:

$$U_E = \frac{1}{2} CV^2 \tag{16}$$

where C is the electrical capacitance of the bimorph. Hence the energy-converting efficiency (η) of the device can be expressed by

$$\eta = \frac{U_M}{U_E} \tag{17}$$

Figures 15–21 show some experimental results on various types of bimorphs made of typical piezoelectric polymers, PVDF and PZT/POM. Figure 15 shows the relationship between the deflection and drive voltage in the two-layer-type bimorph ($N = 1$, curve A) and miltilayer-type bimorph ($N = 5$, curve B) made of PVDF film [35]. The dotted line is the calculated result using Eq. (4) and Eq. (9), and marks indicate the experimental results. The theoretical and experimental results are in a good agreement. The deflection in the two-layer-type bimorph ($N = 1$, length $L = 10$ mm or 20 mm, thickness $t = 100$ μm, and width $w = 10$ mm) made of PZT/POM composite film as a function of applied voltage is shown in Figure 16. The theoretical curves from Eq. (4) are also shown in the figure. The theoretical curves are in good agreement with experimental results. As shown in Eq. (4), a large displacement is favored by the use of a longer bimorph.

Figure 17 compares the observed and calculated force of the two-layer bimorph and multilayer bimorph as a function of applied voltage. The solid lines show the calculated results from Eq. (5) and Eq. (10). It is clear that the multilayer structure can deliver a relatively large bending force at low applied voltage.

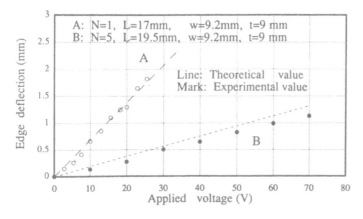

Figure 15 Bimorph voltage deflection characteristics for PVDF film (d_{31}=22pC/N). (From Ref. 35.)

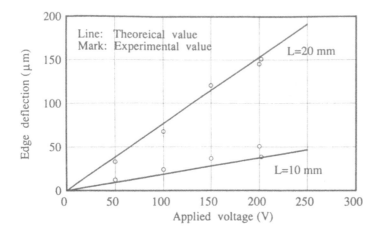

Figure 16 Bimorph voltage deflection characteristics for PZT/POM composite piezoelectric film (d_{31}=18pC/N).

Figure 18 shows the curvature radius R_0 of the PVDF bimorph (t = 9 μm) as a function of applied voltage [53]. It gives good agreement to the theoretical calculation when the applied voltage is low, but if the applied voltage is over 300 V (limiting voltage is 500 kV/cm), the slope of R_0 against the applied voltage becomes smaller, and when the voltage reaches 450 V, R_0 begins to decrease. This decrease of R_0 in the high-voltage range may be due to the polarization reverse of the polar domain. It means that this kind of device is suitable for applied voltages lower than 500 kV/cm.

Generally, the free end reaches its static state through a dynamic process when a voltage is applied to a bimorph. Figure 19 shows typical vibration behavior of a PZT/POM bimorph before the deflection reaches a stable value. When DC voltage is applied to the sample, a vibration occurs, and then gradually decays due to mechanical loss and air resistance. The vibration is closely related to the resonance behavior of the bimorph

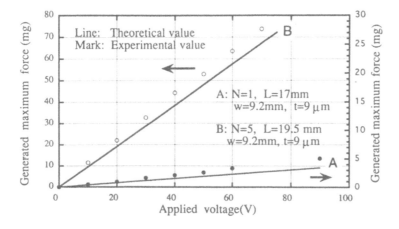

Figure 17 Bimorph maximum force characteristics for PVDF film (d_{31}=22pC/N). (From Ref. 35.)

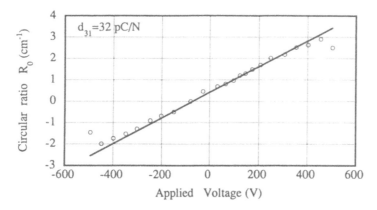

Figure 18 Radius as a function of applied voltage of a bimorph composed of 9 μm-thick PVDF layers. (From Ref. 53.)

structure; in other words, we can calculate the resonant frequency from Figure 19 by using Eq. (6).

Some attention should be paid to the influence of the adhesive layer on the output behavior of bimorphs when the adhesive layer is thick. Generally, when the adhesive layer is thinner than several micrometers, its influence on the deflection can be neglected, but when the adhesive layer is thicker than 10 μm, its influence should be taken into account in device design.

One disadvantage of bimorphs made of polymer materials compared with those made of ceramic materials is that the polymer bimorph has larger hysteresis and zero-point shift. Figure 20 shows one typical example of the hysteresis for a bimorph made of PZT/

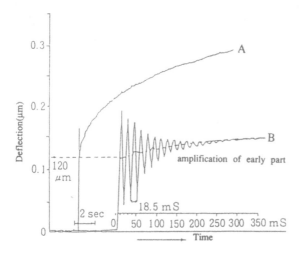

Figure 19 Vibration behavior of the free end in bimorph before reaching its stable state. A) transient response to square-wave voltage, B) magnifying the rising part of A.

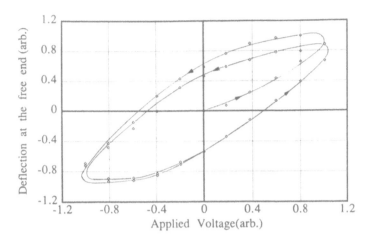

Figure 20 Residual displacement characteristic of PZT/POM bimorph. (L=3 cm, t=100 μm, d_{31}=18.3pC/N).

POM composite film. Hysteresis gives poor data reproducibility, which makes it difficult to use the bimorph in fields that require position control systems. Composite piezoelectric films have many advantages over polymer or ceramic piezoelectric materials, such as high thermal resistance and little polarization relaxation, but unfortunately the hysteresis is generally also much more larger than that of other materials. The large hysteresis and zero-point shift in composite piezoelectric films can be considered to be due to the plastic deformation and the existence of microcracks at the boundary between the polymer matrix and ceramic particles.

B. Displays

Display devices can be realizable by using the movement of bimorph elements. Figure 21 shows one segmental structure for a slotted-mask-type display [42,53]. A small, light-weight paper tip is fixed perpendicular to the free end of the bimorph, and the paper piece can be seen through the slit window. The width and height of the element are 1.5

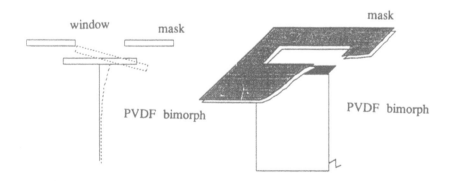

Figure 21 Basic structure of bimorph type display. (From Ref. 53.)

cm and 3.5 cm, respectively; the size of the indicating plate is 0.6×1.5 cm^2. The driving voltage is ± 120 V; power consumed is 2.2×10^{-8} W/cm^2 at the stable state, and 2.4×10^{-5} W/cm^2 during the [on] or [off] process. The switching speed is 21 ms.

Bimorphs used as binders to shut off light or large-scale indicating devices have been developed [31,48]. In this structure, a lot of bimorph elements (2.5×10 cm) are arranged in a distance of 2 cm. When the applied voltage is zero, the bimorphs open to the same plane and light can pass through. When a driving voltage ($+170$ V) is applied, they close (responding time = 21 ms) and act like a curtain. When each bimorph element is controlled selectively, it acts like a display.

C. Displacement-Generating Systems

A structure which can give parallel motion under a pulse driving voltage can be realized by using bimorph structure. Typical applications in medicine and biology are micromanupulators and microneedles. Figures 22a and 22b show the structure of traveling micromachines made of PVDF bimorphs [45,51,52]. The radius of the arch is 1 cm, and it is made of two pieces of PVDF films (thickness 30 μm) attached by epoxy resin. The polarization direction in the two pieces is the same. Figure 22a is a pair-arch-type model, in which the front structure and back structure in the forward movement have symmetry; if there is no unbalancing weight, the machine cannot walk. Figure 22b is a triangle-type structure. Here the front structure and back structure in forward movement do not have symmetry, so if a driving voltage is applied, the machine can walk without the help of an unbalancing weight. It is possible to control the walking state by regulating the width and legnth of A and B at the two sides to change their resonance frequency. By the selection of resonance frequency f_A or f_B, walk-forward or back movement can be

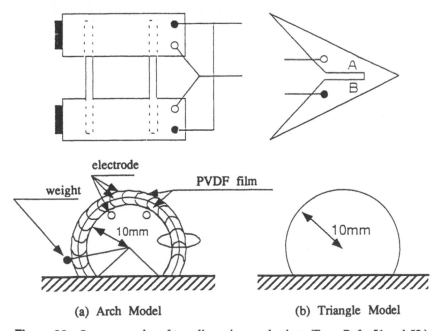

(a) Arch Model   (b) Triangle Model

Figure 22 Some examples of traveling micromechanism. (From Refs. 51 and 52.)

achieved. Figure 23 shows the structure of a microneedle [45], which can be used to penetrate bacteria. The sensitivity of the needle end is 1.5×10^{-6} N, and 80-μm parallel motion (displacement) can be obtained with a driving voltage of 300 V. Figure 24 shows the structure of a composite-type micromanupulator. This type can operate over an area of 330 μm $\times$ 500 μm with the same driving voltage (300 V) [2,146].

D. Light-Control Systems

In optical communication systems, it is frequently necessary to switch the active fiber. Figure 25 shows the principle of a light switch structure made of polymer bimorph and optical fiber [139]. By using this structure, little electrical energy is consumed and very high response can be achieved. For example, in the system shown, where the distance moved was 3 mm and the driving voltage was 3–5 V, the power consumed was several microwatts. Figure 26 shows the principle of a light intensity-controlled limit switch consisting of a photoconductor coupled to a PVDF bimorph [142]. Conducting paste is applied and the transparent electrode (ITO) is evaporated on one side of the PVDF film. The applied voltage on the piezoelectric film is proportional to the strength of the incident light, so the displacement is also proportional to it. It is possible to control a camera lens, or use the structure as a light shutter, based on this principle. Other applications associated with light control, such as a laser wave front surface repairing system by controlling the refractive index of the PVDF attached on the mirror, and three-dimensional picture-producing systems, have also been proposed [136,137].

E. Transportation and Flow Systems

Bimorphs can be used not only as display elements but also to transmit substance such as powder, liquid, or gas from one place to another. In this section, some applications of bimorph devices to moving liquid and gas are introduced. In this kind of application,

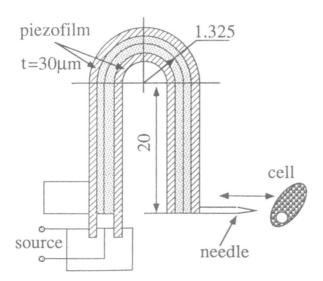

Figure 23   Micromanipulator using PVDF bimorph. (From Ref. 45.)

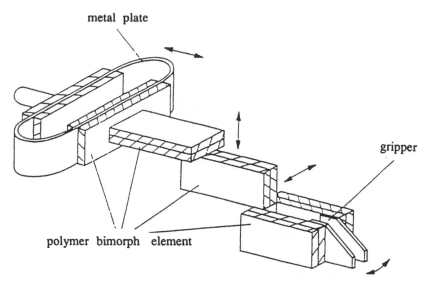

Figure 24 Composite micromanipulator. (From Ref. 2.)

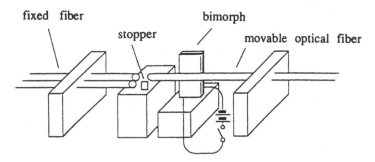

Figure 25 Optical switch system. (From Ref. 139.)

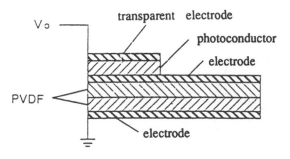

Figure 26 Light shutter. (From Ref. 142.)

many bimorph elements have to be connected in a tail-to-head form, and driven by a pulse voltage with proper time delay for neighboring bimorphs.

1. Powder and Liquid Transportation Systems

In Figure 27, many bimorph elements are connected head to tail. When DC voltage is applied from left to right with proper delay time, the substance on the bimorph surface can be transmitted from A to B. This means that it is possible to create a rotationless transmitting system or microtransmitting machine. Figure 28 shows the principle of a transmitting tube in which many bimorph elements are attached on the surface of a soft tube. When an alternative voltage is applied on the bimorph with proper delay time, the liquid or powder inside the tube can be driven from one side to another. Figure 29 shows the basic structure of a micropump and micropipet, which can be used to drive gas or liquid with a driving speed of 12.5 ml/min at a driving voltage of 200 V [44].

2. Air and Gas Flow-Generating Systems

Bimorph structure can be also applied to air and gas flow-generating systems. The flow generated by a two-layer bimorph is not very strong, but it can be improved by using a multilayer structure. If the device is working at its resonant frequency, very strong air flow can be produced [46,47]. Figure 30 shows a schematic diagram of the measurement for air flow generated by a bimorph fan made of PVDF films [41,43]. The width of the bimorph was 10 cm; the layer number was 3 (pairs). Air flow and flow speed at the resonance frequency are shown in Figure 31. The flow speed decreases with the increase of bimorph length, but the flow changes little Properties of different kinds of bimorph fans are shown in Table 5. Power of 14 mW (driving voltage is ±150 V, resonance frequency is 13 Hz) is necessary to generate air flow of 1.4 m/s. When two sets of such fans were used to cool a television set (heat radiating plate), the inside temperature decreased by 5°C. If this work was done by a motor-type fan, about 10 times more energy would be necessary. This means a possibility of using bimorph fans to cool transistor circuits.

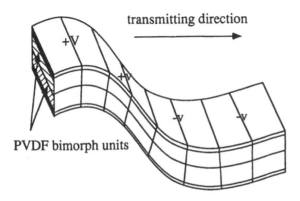

Figure 27 Microtransmitting belt composed of bimorph units.

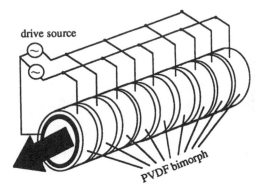

Figure 28   Transmitting tube.

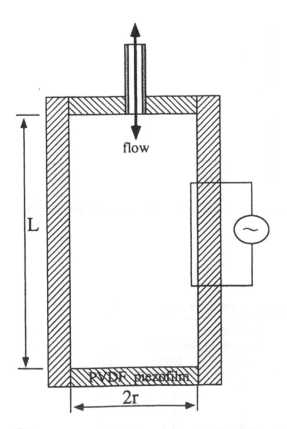

Figure 29   Liquid driving pipette. d_{31}=20 pC/N, r=1.25 cm, L=8 cm. (From Ref. 44.)

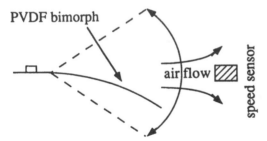

Figure 30 Schematic diagram of measurement for the air flow generated by a PVDF bimorph. (From Ref. 43.)

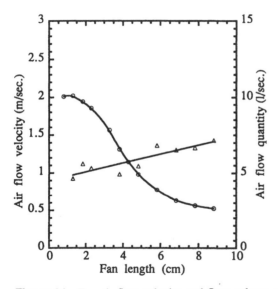

Figure 31 Fan air flow velocity and flow volume as a function of fan length. (From Ref. 43.)

Table 5 Properties of bimorph fan (thickness = 9 μm, drive voltage = ±150 V)

Total number of layers	Bimorph length (cm)	Resonance frequency (Hz)	Width of bimorph (cm)	Amplitude (cm)	Air flow velocity (m/s)	Air flow quantity (l/s)
6	1.55	60	10	2.4	2.0	4.8
6	3.7	12	10	3.7	1.3	4.8
6	5.7	4.3	10	8.7	0.8	6.7
8	3.8	13	9.5	5.0	1.4	6.8

REFERENCES

1. T. Furukawa, Ferroelectric properties of vinylidene fluoride copolymers, *Phase Transitions 18*:143 (1989).
2. T. T. Wang, J. M. Herbort, and A. M. Glass (eds.), *The Applications of Ferroelectric Polymers*, Blackie, Glasgow, 1988.
3. T. Furukawa, Piezoelectricity and pyroelectricity in polymers, *IEEE Trans. Elec. Insul. 24*: 375 (1989).
4. R. E. Newnham, L. J. Bowen, K. A. Klicker, and L. E. Cross, Composite piezoelectric transducers, *Materials Eng. 112*:93 (1980).
5. T. Yamada, T. Ueda, and T. Kitayama, Piezoelectricity of a high-content lead zirconate titanate/polymer composite, *J. Appl. Phys. 53*:4328 (1982).
6. H. Banno, Recent developments of piezoelectric ceramic products and composites of synthetic rubber and piezoelectric particles, *Ferroelectrics 50*:3 (1983).
7. S. Tasaka, K. Miyasato, M. Yoshikawa, S. Miyata, and M. Ko, Piezoelectricity and remanent polarization in vinylidene cyanide/vinyle acetate copolymer, *Ferroelectrics 57*:267 (1984).
8. I. Seo and K. Nakajima, Development of piezoelectric polymer in cyanide family. *Nikkei New Materials*, No. 6, p. 50 (February 24, 1986) (in Japanese).
9. D. Zou, S. Iwasaki, T. Tsutsui, S. Saito, M. Kishimoto, and I. Seo, Anomaly in dielectric relaxation in alternating copolymers of vinylidene cyanide and fatty acid vinyl ester, *Polymer 3*:1888 (1990).
10. J. I. Schcinbeim, J. W. Lee, and B. A. Newman, Ferroelectric polarization mechanisms in Nylon 11, *Macromolecules 25*:3729 (1992).
11. Y. Takahashi, M. Ijima, and E. Fukada, Pyroelectricity in poled thin films of aromatic polyurea prepared by vapor deposition polymerization, *Jpn. J. Appl. Phys. 28*:L2245 (1989).
12. F. Micheron, P. Ravinet, D. Guillou, and C. Claudepierre, Dome-shaped piezopolymer electroacoustic transducers, *Ferroelectrics 51*:803 (1983).
13. M. Tamura, T. Yamaguchi, T. Oyaba, and T. Yoshimi, Electroacoustic transducers with piezoelectric high polymer films, *J. Audio Eng. Soc. 23*:21 (1975).
14. N. Murayama and H. Obara, Piezoelectric polymers and their applications, *Jpn. J. Appl. Phys. 22* (S22-3):3 (1983).
15. H. Hayashi, Applications of piezoelectric organic polymer materials, *Nikkei Electronics*, No. 180, p. 68 (February 20, 1978) (in Japanese).
16. J. F. Sear and R. Carpenter, Noise cancelling microphone using a piezoelectric plastics transducing element, *Elect. Lett. 11*:532 (1975).
17. H. Naono, T. Gotoh, M. Matsumoto, and S. Ibaraki, Design of an electroacoustic transducer using piezoelectric polymer film, Preprint of 58th AES Conv., p. 1271, 1977.
18. H. R. Gallantree, Review of transducer applications of polyvinylidene fluoride, *IEE Proc. 130*, Part T:219 (1983).
19. H. R. Gallantree and R. M. Quilliam, Polarization poly(vinylidene fluoride)—Its application to pyroelectric and piezoelectric devices, *Marconi Rev. 39*:189 (1976).
20. T. Yamamoto and M. Tamura, Application of piezoelectric polymers to audio transducers, *J. Acoust. Soc. Am. 60*:S46 (1976).
21. J. H. Lai, S. A. Jenekhe, R. J. Jensen, and M. Royer, Polymer in electronics áT, *Solid State Technol. 27*:165 (1984).
22. J. V. Chatigny, Piezo film yields novel transducers, *Electron Week 57*:74 (1984).
23. M. Onoe, Piezoelectric polymer film transducer for detection of acoustic emission, *Jpn. J. Appl. Phys. 20*(S20-4):125 (1981).
24. M. Tamura, K. Ogasawara, N. Ono, and S. Hagiwara, Piezoelectricity in uniaxially stretched poly(vinylidene fluoride), *J. Appl. Phys. 45*:3768 (1974).
25. M. Tamura, K. Iwama, and T. Yoshimi, Electro-acoustic transducers with piezoelectric high polymer films, *J. Acoust. Soc. Jpn. 31*:414 (1975) (in Japanese).

26. J. Ohga, A flat piezoelectric polymer film loudspeaker as a multi resonance system, *J. Acous. Soc. Jpn.* 4:113 (1983).

27. S. Shirai, T. Yamada, and J. Ohga, A bimorph microphone using composite piezoelectric polymer, *J. Acoust. Soc. Jpn.* 4:2 (1983).

28. I. Namiki, K. Sugiyama, T. Kitayama, and T. Ueda, Piezoelectric key board electric design condition, *IEEE Trans. Components Hybrids Manuf. Technol.* 4:304 (1981).

29. A. Techagumpuch, H. S. Nalwa, and S. Miyata, in *Electroresponsive Molecular and Polymeric Systems* (T. A. Skotheim, ed.), Marcel Dekker, New York, 1991, p. 284.

30. S. Miyata and K. Nakajima, Film speaker, *Trans. Elect. Commun. Jpn.* 72:676 (1989) (in Japanese).

31. S. Sakai, N. Kyouno, and S. Fujiwara, Digital-to-analog conversion by piezoelectric headphone, Preprint of 78th AES Conv., 1985, p. 2231.

32. G. R. Crane, Poly(vinylidene fluoride) used for piezoelectric coin sensors, *IEEE Trans. Sonics and Ultrasonics* SU-25:393 (1978).

33. F. Micheron, Ferroelectric polymers and applications, *Ferroelectrics* 53:139 (1984).

34. I. Seo, Piezoelectric polymer actuators, *J. Jpn. Soc. Precision Eng.* 53:689 (1987) (in Japanese).

35. M. Toda and Osaka, Electromotional devices using PVF2 multilayers bimorph, *Trans. IECE Jpn.* E61:507 (1978).

36. M. Toda, Design of piezoelectric polymer motional devices with various structures, *Trans. IECE Jpn.* E61:513 (1978).

37. M. Marcus, Performance characteristics of piezoelectric polymer flexure mode device, *Ferroelectrics* 57:203 (1984).

38. V. H. Schmitt, M. Klakken, and H. Darejeh, PVF2 bimorphs as active elements in wind generators, *Ferroelectrics* 51:765 (1983).

39. J. G. Linvill, PVF2 modes, measurement and devices, *Ferroelectrics* 28:29 (1980).

40. H. Sussner and K. Dransfeld, Der piezoelektrische effekt in polyvinyliden fluorid unt seine anwendingen, *Colloid Polym. Sci.* 257, 591 (1981) (in German).

41. P. Th. A. Klaase, Low-frequency and ultrasonic applications on piezoelectric PVDF films, *Ferroelectrics* 60:215 (1984).

42. M. Toda, Voltage-induced large amplitude bending device—PVF2 bimorph—Its properties, *Ferroelectrics* 32:127 (1984).

43. M. Toda and S. Osaka, Vibrational fan using the piezoelectric polymer PVF2, *Proc. IEEE* 67:1171 (1979).

44. Y. Umetani, Principle of a piezoelectric micromanipulator with tactile sensitivity, *Proc. 8th ISIR*, 1980, p. 406.

45. Y. Umetani and H. Suzuki, Piezoelectric micromanipulator in multidegrees of freedom with tactile sensibility, *Proc. 10th ISIR*, 1980, p. 571.

46. M. Toda, Theory of air flow generation by a resonant type PVF2 bimorph cantilever vibrator, *Ferroelectrics* 22:911 (1979).

47. M. Toda, High field dielectric loss of PVF2 and the electromechanical conversion efficiency of a PVF2 fan, *Ferroelectrics* 22:919 (1979).

48. M. Toda and S. Osaka, Large area electronically controllable light shutter array using PVF2 bimorph fans, *Ferroelectrics* 23:121 (1980).

49. D. H. Dameron and J. G. Linvill, Cylindrical PVF2 electromechanical transducers, *Sensors and Actuators* 2, 73 (1981/1982).

50. M. Latour and P. V. Murphy, Application of PVF2 transducers as piezoelectric vibrators for marine fouling prevention, *Ferroelectrics* 32:33 (1981).

51. T. Hayashi, Micro mechanism, *Inst. Elect. Commun. Eng., Paper of Tech. Group on Electro Acoust.* US84-8:25 (1984) (in Japanese).

52. T. Hayashi and K. Yoshida, Traveling micromechanism, *Proc. 8th World Congress on T.M.M.*, 1991, p. 59.

53. M. Toda and S. Osaka, Application of PVF2 bimorph cantilever elements to display devices, *Proc. SID 19*:35 (1978).

54. M. Toda, S. Osaka, and S. Tosima, Large area display element using PVF2 bimorph with double-support structure, *Ferroelectrics 23*:115 (1980).

55. A. S. DeReggi and S. Edelman, Piezoelectric polymer membrane stress gage, U.S. Patent, 4166229 (Aug. 28, 1979).

56. H. Sussner and K. Dransfeld, Importance of the metal polymer interface for the piezoelectricity of polyvinilidene fluroide, *J. Polymer Sci., 16*:529 (1978).

57. Y. Obara, M. Murayama, K. Koumoto, and H. Yanagida, PT-polymer piezoelectric composites: A design for an acceleration sensor, *Sensors and Actuators A36*:121 (1993).

58. A. Obara, K. Kato, B. Honda, and H. Yamazaki, Hydrodynamic oscillator type flow meter using piezoelectric film, *Sensor Technol. 5*:65 (1985).

59. G. Heine, Shock wave sensors, *Ferroelectrics 75*:357 (1987).

60. F. Bauer, PVF2 polymers: Ferroelectric polarization and piezoelectric properties under dynamic pressure and shock wave action, *Ferroelectrics 49*:231 (1983).

61. A. S. DeReggi, Piezoelectric polymer transducer for impact pressure measurement, Natl. Bur. Std. Rep. NBSIR 75-740, July 1975.

62. T. Jingu, H. Matsumoto, K. Nezu, and K. Sakamoto, Experimental consideration of the technological issues in measuring the impact load by various dynamic response transducers, *Trans. Jpn. Soc. Mech. Eng. A54*:506 (1988) (in Japanese).

63. I. Seo, Impact sensor using piezoelectric polymer, *Polymer Prepr. Jpn. 26*:1558 (1977) (in Japanese).

64. S. W. Meeks and R. Y. Ting, The evaluation of PVF2 for underwater shock-wave sensor application, *J. Acoust. Soc. Am. 75*:1010 (1984).

65. H. Banno and H. Sofue, Properties of piezoelectric composite materials and electroacoustic applications, *The Plastics 32*:27 (1986) (in Japanese).

66. G. T. Pearman, J. L. Hokason, and T. R. Meeker, Design and evaluation of a contactless piezoelectric keyboard using PVDF as the active element, *Ferroelectrics 28*:311 (1980).

67. N. Murayama, K. Nakamura, H. Obara, and M. Segawa, The strong piezoelectricity in polyvinylidene fluoride, *Ultrasonics 14*:15 (1976).

68. P. Dario, D. DeRossi, C. Giannotti, F. Vivaldi, and P. C. Pinotti, Ferroelectric polymer tactile for prostheses, *Ferroelectrics 60*:199 (1984).

69. R. Takano, 300 bit/s keyboard printer terminal for public data network, Jpn. Telecommun. Rev., October 1977, p. 318.

70. J. Sako, Piezoelectric and pyroelectric sensors, *Eng. Materials 30*:71 (1982).

71. I. Namiki, Y. Shiratori, K. Sugiyama, and T. Kitayama, Piezoelectric tablet Kanji input equipment, *Trans. Elect. Commun. Jpn. J64-A*:682 (1981) (in Japanese).

72. M. Toda, PVF2 piezoelectric bimorph devices for sensing presence and position of other objects, *IEEE Trans. Electron Devices ED-26*:815 (1979).

73. A. A. Schoenberg, D. M. Sullivan, C. D. Barker, H. E. Bouth, and C. Galway, Ultrasonic PVF2 transducers for sensing tactile force, *Ferroelectrics 60*:239 (1984).

74. P. Dario and D. DeRossi, Piezoelectric polymer: New sensor materials for robotic applications, *Proc. 13th ISIR*, 1983, p. 14.

75. P. Dario and D. DeRossi, Tactile sensors and the gripping challenge, *IEEE Spectrum 22*:46 (1985).

76. K. Kobayashi and T. Yasuda, An application of PVDF-film to medical transducers, *Ferroelectrics 32*:181 (1981).

77. P. Dario, R. Bedini, and D. DeRossi, A piezoelectric polymer catheter tip dynamic pressure transducer, *Digest of Papers BioMed 80*:229 (1980).

78. K. Tamiya, M. Sugawara, Y. Sakurai, T. Tomita, I. Seo, T. Aramaki, and Y. Ejiri, Trial manufacture of catheter tip manometer using piezoelectric material, *Jpn. Soc. Med. Elect. Bio. Eng. 15*:470 (1977) (in Japanese).

79. P. Dazio, R. Bardelli, D. DeRossi, L. R. Wang, and P. C. Pinotti, Touch-sensitive polymer skin using piezoelectric to recognize orientation of objects, *Sensor Rev* 2(4):194 (1982).

80. H. Araki, I. Ibe, K. Suzuki, and T. Ohno, High molecular piezoelectric transducer for medical electric instruments, *Inst. Elect. Commun. Eng., Paper Tech. Group Electro Acoustics EA76-31*:25 (1976) (in Japanese).

81. M. Matsumura and K. Fujii, Measurement of tongue shape by using piezoelectric film, *Inst. Elect. Commun. Eng., Paper Tech. Group Electro Acoustics MBE83-16*:77 (1983) (in Japanese).

82. W. Nitsche and R. Thunker, Application of the piezo-electric effect in measuring the arterial pressure pulse, *Ferroelectrics 75*:381 (1987).

83. S. B. Lang, B. D. Sollish, M. Moshitzhy, and E. H. Frei, Model of a PVDF piezoelectric transducer for use in biomedical studies, *Ferroelectrics 24*:289 (1980).

84. H. Sussner, The piezoelectric polymer PVF2 and its applications, *Proc. 1979 IEEE Ultrasonics Symp.*, 1979, p. 491.

85. H. Sussner, D. Michas, A. Assfalg, S. Hunklinger, and K. Dransfeld, Piezoelectric effect in polyvinylidene fluoride at high frequency, *Phys. Lett. 45A*:475 (1973).

86. N. Chubachi and T. Sannomiya, Confocal pair of concave transducers mode of PVF2 piezoelectric films, *Jpn. J. Appl. Phys. 16*:2259 (1977).

87. H. R. Gallantree, Ultrasonic applications of PVDF transducers, *Marconi Rev. 45*:49 (1982).

88. H. Ohigashi, R. Shigenari, and M. Yokota, Light modulation by ultrasonic waves from piezoelectric polyvinylidene fluoride films, *Jpn. J. Appl. Phys. 14*:1085 (1975).

89. Y. Tomikawa, M. Ohki, H. Yamada, and M. Onoe, *Jpn. J. Appl. Phys. 22*(S22-3):133 (1983).

90. A. S. DeReggi, S. C. Roth, J. M. Kenney, S. Edelman, and G. R. Harris, Piezoelectric polymer probe for ultrasonic applications, *J. Acoust. Soc. Am. 69*:853 (1981).

91. W. H. Chen, H. J. Shaw, D. G. Weinstein, and L. T. Zitelli, PVF2 transducers for NDE, *Proc. 1978 IEEE Ultrasonics Symp.*, 1978, p. 780.

92. R. A. Lewin, Miniature piezoelectric polymer ultrasonic hydrophone probes, *Ultrasonics 19*: 213 (1981).

93. D. R. Bacon, Characteristic of a PVDF membrane hydrophone for use in the range 1–100 MHz, *IEEE Trans. Sonics and Ultrasonics SU-29*:18 (982).

94. G. R. Harris, Sensitivity considerations for PVDF hydrophones using the spot-poled membrane design, *IEEE Trans. Sonics and Ultrasonics SU-29*:370 (1982).

95. R. C. Preston, D. R. Bacon, A. J. Livett, and K. Ragendom, PVDF membrane hydrophone performance properties and their relevance to the measurement of the acoustic output of medical ultrasonic equipment, *J. Phys. E. Sci. Instrum. 16*:786 (1983).

96. G. R. Harris, E. F. Carome, and H. D. Dardy, An analysis of pulsed ultrasonic fields as measured by PVDF spot-poled membrane hydrophone, *IEEE Trans. Sonics and Ultrasonics, SU-30*:295 (1983).

97. M. Platte, A polyvinylidene fluoride needle hydrophone for ultrasonic applications, *Ultrasonics 23*:113 (1984).

98. T. D. Sullivan and J. M. Powers, Piezoelectric polymer flexural disk hydrophone, *J. Acoust. Soc. Am. 63*:1316 (1978).

99. A. S. DeReggi, *Ferroelectrics 60*:83 (1984).

100. T. A. Henriquez, Application of tubular PVDF to shock resistant hydrophones, *Ferroelectrics 50*:365 (1983).

101. H. J. Shaw, D. Weinstein, L. T. Zitelli, C. W. Frank, R. C. DeMattei, and K. Fesler, PVF2 transducers, *Proc. 1980 IEEE Ultrasonics Symp.*, 1980, p. 927.

102. A. S. DeReggi, Transduction phenomena in ferroelectric polymers and their role in pressure transducers. *Ferroelectrics 50*:347 (1983).

103. H. Ohigashi, K. Koyama, S. Takashi, A. Ishizaki, and Y. Maida, Piezoelectric properties of P(VDF-TrFE) thin films at low temperatures and their application to ultrasonic transducers

for scanning acoustic microscopy operating in a wide temperature range, *Jpn. J. Appl. Phys. 28*(S28-1):66 (1989).

104. R. S. Wagers, SAW transduction on silicon substrates with PVF2 films, *Proc. 1979 Ultra-sonics Symp.*, 1979, p. 645.

105. F. Mattioco, E. Dieulesaint, and D. Royer, PVF2 transducers for Rayleigh waves, *Electronics Lett. 16*:250 (1980).

106. E. Dieulesaint, F. Mattiocco, and D. Roger, Exeitation et detection d'ondes de rayleigh a l'aide d'une feuille de polymere piezoelectrique, *C. R. Acad. Sci. Paris 287*:B-171 (1978).

107. R. S. Wager, Low-temperature Lamb wave propagation in PVF2, *J. Appl. Phys. 51*:5797 (1980).

108. E. Carome, K. Fesler, H. J. Shaw, D. Weinstein, and L. T. Zitelli, Poly(vinylidene fluoride) surface wave transducers, *Proc. 1979 IEEE Ultrasonics Symp.*, 1979, p. 641.

109. P. Merilainen and M. Luukkala, Convolution by membrane waves, *Electron. Lett. 14*:745 (1978).

110. R. G. Swartz and J. D. Plummer, Integrated silicon-PVF2 acoustic arrays, *IEEE Trans. Electron Devices ED-26*:1921 (1979).

111. R. G. Swartz and J. D. Plummer, On the generation of high-frequency acoustic energy with polyvinylidene fluoride, *IEEE Trans. Sonics and Ultrasonics SU-27*:295 (1980).

112. R. Y. Ting, Evaluation of new piezoelectric composite materials for hydrophone applications, *Ferroelectrics 67*:143 (1986).

113. B. Woodward, The suitability of polyvinylidene fluoride as an underwater transducer material, *Acustica 37*:264 (1977).

114. F. Micheron and C. Lemonon, Moulded piezoelectric transducers using polar polymers, *J. Acoust. Soc. Am. 64*:1720 (1978).

115. K. C. Shotton, D. R. Bacon, and R. M. Quilliam, A PVDF membrane hydrophone for operation in the range 0.5 MHz to 15 MHz, *Ultrasonics 17*:123 (1980).

116. K. Koyama, T. Eda, H. Suzuki, Y. Wada, and O. Ishizuka, The sensitivity of miniature hydrophones made of P(VDF-TrFE) thin film, *Acoustical Imaging 17*:579 (1989).

117. S. Tsuchiya, T. Sato, K. Koyama, S. Ikeda, and Y. Wada, Application of piezoelectric film of vinylidene fluoride-trifluoroethylene copolymer to a high sensitive miniature hydrophone, *Jpn. J. Appl. Phys. 26*(S26-1):183 (1987).

118. H. Mikami and S. Sugata, *Proc. Acoust. Soc. Jpn.*, Paper no. 2-7-9, March 1982, p. 721 (in Japanese).

119. K. Okada, N. Wakita, S. Saito, and K. Ohya, *Proc. Acoust. Soc. Jpn.*, Paper no. 3-7-3, March 1982, p. 643 (in Japanese).

120. Y. Saito, K. Koyama, S. Ikeda, and Y. Wada, *Rep. Prog. Polymer Phys. Jpn. 30*:263 (1987).

121. K. Kimura, N. Hashimoto, and H. Ohigashi, Performance of a linear array transducer of vinylidene fluoride trifluoroethylene copolymer, *IEEE Trans. Sonics and Ultrasonics SU-32*:566 (1985).

122. B. Grang, A linear monolithic receiving array, *Acoustical Imaging 12*:307 (1982).

123. S. Takebayashi, K. Matsui, Y. Onohara, and H. Hidai, Sonography for early diagnosis of enlarged parathyroid glands in patients with secondary hyperparathyroidism, *A.J.R. 148*:911 (1987).

124. A. A. Shaulov and W. A. Smith, Ultrasonic transducer arrays made from composite piezoelectric materials, *Proc. 1985 IEEE Ultrasonics Symp.*, 1985, p. 648.

125. J. Fraser, B. T. Khuri-Yakub, and G. S. Kino, The design on efficient broadband wedge transducer, *Appl. Phys. Lett. 32*:698 (1978).

126. H. Ohigashi, K. Koyama, S. Takahashi, Y. Wada, Y. Maida, R. Suganuma, and T. Jindo, Ferroelectric polymer transducers for high-resolution scanning acoustic microscopy, *Acoustical Imaging 16*:521 (1988).

127. H. Ohigashi, S. Takahashi, Y. Tasaki, and G. R. Li, Piezoelectric properties of ferroelectric polymers at low temperature, *Proc. 1990 IEEE Ultrasonics Symp.*, 1990, p. 753.

128. T. Terao, K. Sumimoto, and T. Yu Lin, A new CTG equipped with real time autoanalizing system and flexible ultrasound transducer, *Recent Advances in Perinatology* (K. Maeda, K. Okuyama, and Y. Takeda, eds.), Elsevier, Amsterdam, 1986, p. 149.

129. N. Hashimoto, T. Miya, K. Yoneya, A., Ando, and H. Ohigashi, High resolution ultrasonic imaging using large aperture annular array transducer of P(VDF-TrFE) copolymer, *Acoustical Imaging 17*:561 (1989).

130. M. Imai, H. Tanizawa, Y. Ohtuska, Y. Takase, and A. Odajima, Piezoelectric copolymer jacketed single-mode fibers of electric-field, *J. Appl. Phys. 60*:1916 (1975).

131. E. F. Carome and K. P. Koo, PVF2 phase shifters and modulators for fiber optic sensor systems, *Proc. 1980 IEEE Ultrasonics Symp.*, 1980, p. 710.

132. K. P. Koo and E. F. Carome, Frequency mixing in fiber-optic interferometer systems, *Electronics Lett. 17*:380 (1981).

133. J. Jarzynski, Frequency response of a single-mode optical fiber phase modulator utilizing a piezoelectric phase jacket, *J. Appl. Phys. 55*:3243 (1984).

134. M. D. Mermelstein, Optical-fiber copolymer-film electric field sensor, *Appl. Opt. 22*:1006 (1983).

135. L. J. Donalds, W. G. French, W. C. Mitchell, R. M. Swinehart, and T. Wei, Electric Field sensitive optical fiber using piezoelectric polymer coating, *Electronics Lett. 18*:327 (1982).

136. S. A. Kokorowski, Analisis of adaptive optical elements made from piezoelectric bimorphs, *J. Opt. Soc. Am. 69*:181 (1979).

137. T. Sato, H. Ishida, and O. Ikeda, Adaptive PVDF piezoelectric deformable mirror system, *Appl. Opt. 19*: 1430 (1980).

138. M. Kimura, H. Takahashi, and T. Miyamoto, Light modulation using piezoelectric films with waveguides, *Electronics Lett. 20*:772 (1984).

139. T. Ebato, T. Kajiwara, and S. Kobayashi, Mechanical fiber optic switching using PVDF bimorph, *Electronics Lett. 16*:829 (1980).

140. K. Toda and K. Ikenohira, High temperature-dependent performance of a polyvinylidene fluoride Lamb wave device, *J. Appl. Phys. 45*:3768 (1974).

141. H. Ohigashi, Electromechanical properties of polarized polyvinylidene fluoride films as studied by the piezoelectric resonance method, *J. Appl. Phys. 47*:949 (1976).

142. Optical to mechanical transducer using photoconductor piezoelectric polymer sandwich, *Research Disclosure*, p. 117 (March 1980).

143. E. Hausler and L. Stein, Hydromechanical and physiological mechanical-to-electrical power converter with PVDF film, *Ferroelectrics 75*:363 (1987).

144. M. Kobayashi, N. Colaneri, M. Boysel, F. Wudl, and A. J. Heeger, The electronic and electrochemical properties of poly(isothianophthene), *J. Chem. Phys. 82*:5717 (1985).

145. T. Ojio and S. Miyata, High transparent and conducting polypyrrole-poly(vinyl alcohol) composite film prepared by gas state polymerization, *Polymer J. 18*:95 (1986).

146. F. Yokosuka and T. Tomita, Acceleration sensor, *Transistor Technol.*, No. 11, p. 475 (1986).

147. I. Seo, Impact sensor using piezoelectric polymer, 142th Report on organic materials for information science, 1984, p. 125 (in Japanese).

148. H. Yagami, Y. Ishitsuka, T. Fujii, M. Sasaki, H. Takami, and I. Seo, Development of a 10 MHz linear array transducer of piezoelectric polymer composite, *J. Ultrasonic Med. Soc.*, Preprint, p. 583 (1986).

149. P. M. Galletti, D. DeRossi, and A. S. DeReggi (eds.), *Medical Applications of Piezoelectric Polymers*, Gordon & Breach, New York, 1988.

16
Transduction Applications

Thomas R. Howarth and Kurt M. Rittenmyer
Naval Research Laboratory, Orlando, Florida

I. INTRODUCTION

The use of ferroelectric polymers in transduction applications has had an impact on the development of advanced specialized devices. Many of the original devices were oriented toward military applications such as large-area sensor arrays and surveillance systems for underwater detection. Although some of these systems have been implemented, later emphasis has been toward commercial transduction application in the fields of geophysical exploration, biomedicine, robotics, and measurement instrumentation.

This chapter reviews some of the more significant ferroelectric polymer transduction applications over the past decade. The features and advantages of ferroelectric polymer as an active transducer element will be discussed as they pertain to specific applications. Previous books have discussed earlier ferroelectric polymer developments to the extent that an effort to review the technology of these prototype devices would be counterproductive. The interested reader is urged to refer to [1] for an in-depth discussion of the original and early (1970s) ferroelectric polymer transducer prototypes and concepts. During that time frame, the selection of ferroelectric polymer materials was also limited. As the selection of shapes and properties improved, more advanced devices have emerged.

This chapter begins with a presentation of the state of the art of current underwater acoustic devices featuring ferroelectric polymers. These prototypes include a variety of polymer material types and geometries as deemed appropriate in order to solve specific application interests. Included are plates, multistacked plates and disks, cylinders, and coaxial lines. Each of these transducers was developed for specific sensing of either underwater acoustic pressure or shock wave.

Another strong area of technology development has been in the field of ultrasonics, with specific application to biomedical uses. There is a wide variety of devices that have featured ferroelectric polymers for this area. Included are needle hydrophones and probes, wide-band ultrasonic transmitters, acoustic imaging systems, noninvasive sensing of arterial pressure contours, and posture rehabilitation mapping sensor platforms. Each of these transducers will be discussed with respect to how the ferroelectric polymer is configured within the transducer to address a specific biomedical application.

Robotics has been a field of recent rapid growth. Ferroelectric polymers are finding significant transduction applications for tactile sensor arrays. These arrays are finding direct implementation in robotic force and pressure detectors, multiplexing sensor array schemes, object imaging, and shear stress detection. Also included within the robotics framework is the use of ferroelectric polymer in active vibration-control systems. The inherent lightweight, conformable features of the polymer have offered advanced sensing capabilities without introducing field disturbance effects. This benefit results in active control systems that can sense a variation of a vibration or acoustic field and thus feed back into an electronic switching network for a structure to react to the variation in a prescribed manner.

The final category to be discussed is a general one which will be denoted measurement instrumentation. Included within this is accelerometers, strain and stress gauges, acoustic emission sensors, fiber optic phase modulators, gas emission detectors, and external pressure field flow detection by use of surface acoustic wave (SAW) ferroelectric polymer transducer configurations.

This area does not address the wide variety of devices that feature ferroelectric polymers as detectors based on their inherent pyroelectric effect. Instead, the focus of this chapter is to present recent (within the last decade) poly(vinylidene fluoride) (PVDF) transducer applications that feature their ferroelectric effect as the preferred active material property.

II. UNDERWATER ACOUSTIC HYDROPHONES

One of the original applications of ferroelectric polymers was in the field of underwater acoustic transduction. The properties offered by ferroelectric polymers for this application include the following: ease of fabrication of both large- and small-cross-sectional-area elements, high elasticity, low density, and acoustic impedance properties of the same order of magnitude as the underwater medium. This last property permits an underwater acoustic hydrophone to be placed within a sound field and sense the acoustic pressure within that field without disturbing the field because of its own acoustic presence. The high elasticity further enhances ferroelectric polymer hydrophones because of low transient ringing within the hydrophone device.

These positive traits come at costs in terms of dielectric and piezoelectric constants when compared with traditional piezoceramics (lead zirconate titanates and modified lead titanates). However, the low material properties can be partially design compensated for in many applications.

In an application in which ferroelectric polymer is featured as the active transduction material, one of the piezoelectric constants is governing. Which piezoelectric constant is dominant is dependent on the application, transducer structure, operating frequency band, and environmental exposure conditions. For low-frequency hydrophone applications, the hydrostatic piezoelectric voltage constant g_{3h} is of main interest, while at higher frequencies the piezoelectric voltage constant g_{33} is dominant. For strain gauges, the piezoelectric field constant h_{31} is of concern. In high-frequency thickness-mode-driven transmitters, the piezoelectric charge constant d_{33} is of interest, while at low-frequency broadband transmitting applications, the d_{3h} is dominant [2]. The peizoelectric voltage constants important in underwater towed arrays for sensing are both g_{31} and g_{32}. The uniqueness of these two constants means that the individual hydrophone elements can be oriented to minimize cable strum and acceleration while maximizing the acoustic pressure reception normal to the towed array.

A. Hydrophone Basics

The simplest underwater PVDF hydrophone is one in which a single electroded sheet of the ferroelectric polymer has electrical leads attached across its thickness direction and is encapsulated within a waterproofing elastomer. This "hydrophone" senses the acoustic pressure as it impinges upon the ferroelectric sheet. At frequencies where the size of the hydrophone is small compared to the acoustic wavelength, the acoustic pressure is equal on all of the three axes of the hydrophone. The output signal of such a hydrophone is a function of the hydrostatic piezoelectric voltage constant g_{3h}. A hydrostatic (also known as the "volume" mode) operating hydrophone refers to the configuration of a hydrophone that is completely exposed to the acoustic field on all surfaces. The hydrostatic-mode piezoelectric voltage constant is defined as the superposition of the modes acting on all three axes,

$$g_{3h} = g_{31} + g_{32} + g_{33} \tag{1}$$

where the g_{3h} means that the electrical excitation is occurring along the 3-axis and the h means that the constant is indicative of the hydrostatic mode. For piezoceramics and for ferroelectric copolymers, the 31 and 32 modes are equal to each other, which means that the hydrostatic constant is often presented in the following form:

$$g_{3h} = 2g_{31} + g_{33}. \tag{2}$$

In lead zirconate titanate piezoceramics the sign of the g_{31} and g_{32} are negative while that of the g_{33} is positive. This results in values of g_{3h} that are small when compared to the pure (3,3)-mode excitation. For this reason, lead zirconate titanate piezoceramics are typically poor hydrostatic-mode hydrophones. Typically, the transduction configuration is such that acoustic pressure-release material is included in the design to isolate a hydrophone material in terms of a single excitation mode in order to take advantage of the higher piezoelectric properties of the pure modes such as the (3,3) mode.

For piezoelectric homopolymer there is a difference in the (3,1) and (3,2) modes because of the stretching and poling processes involved during the fabrication. This means that ferroelectric homopolymers have a relatively high and useful hydrostatic receiving sensitivity. Because of this, ferroelectric homopolymer hydrophones have found favor in hydrostatic-mode applications.

The free field voltage (receiving) sensitivity (FFVS) can be defined as

$$\text{FFVS} = 20 \log m_0 - 120 \text{ (dB) re } 1\text{V/}\mu\text{Pa @ face} \tag{3}$$

where m_0 has been defined as the open-circuit voltage sensitivity, given as

$$m_0 = g_{3j}t \text{ (V/Pa)} \tag{4}$$

with j indicating the mechanical excitation direction axis and t the thickness of the piezoelectric polymer along the 3-axis.

B. Planar Hydrophones

The simple single-sheet hydrostatic-mode sensor configuration was a primary focus of early hydrophone designs. During the early to middle 1980s, this type of design was improved upon by placing a stiffener either behind a single sheet or between two sheets of ferroelectric polymers. The reason for the stiffener addition to the hydrostatic transducer design was to improve the low-frequency response. By placing the piezoelectric polymer onto a backing plate, the inherent material low-frequency resonance frequencies (extensional modes) can be scaled to other frequencies in the spectrum. This is because the stiffened effect of the backing plate serves to clamp the naturally occurring extensional modes of the ferroelectric polymer in such a manner that the extensional mode may often be proportionally scaled to other regions of the frequency spectrum of less interest.

It has since been noted that the presence of the stiffening plates actually induces the hydrophones to operate in a thickness (3,3) mode as opposed to a hydrostatic mode [3]. This may be attributed to the fact that the stiffener layer laterally clamps the active ferroelectric polymer layer such that there is no excitation of the (3,1) and (3,2) modes. The first stiffening layers were thick plates of epoxy-loaded graphite and aluminum. Later it was found that electroplating thick copper-plated PC board laminates onto the ferroelectric polymer was effective as a combination stiffener layer and electrode.

The existence of the low-frequency extensional modes has been an area of much interest. It has been shown [4] that the presence of the low-frequency extensional modes is an indication that the multilayered and stiffened ferroelectric polymer hydrophone does not behave in a strict thickness mode. Figure 1 shows the low-frequency extensional modes of two layers of ferroelectric rectangular plates mounted mechanically in series and electrically in parallel. Figure 1a presents the plot for a two-layer hydrostatic hydrophone design where the mechanical stretch axis is aligned in parallel, while Figure 1b shows the resulting plot for a perpendicular alignment scheme. A drawing of how the actual three extensional (contour) modes occur is given in Figure 2.

An investigation featuring an analytical model with experimental results [4] concluded that an eight-layer (where the copper electrodes are each counted as an individual layer), rectangular ferroelectric polymer sensor arrangement can be placed in a cross-axis orientation to negate out two of the three electrically coupled, low-frequency, contour extensional modes (modes 1 and 2 as drawn in Fig. 2). This configuration represents a promising ferroelectric polymer hydrophone design for underwater, low-frequency (and broad-band) hydrostatic receiving applications.

The typical low-cost, large-area hydrophone design features two PVDF sheets bonded together with stiffening plates and a polymer encapsulant as shown in Figure 3. Other schemes often used include using a centrally mounted stiffening layer between the sheets. The stiffening layer has the dual advantage of pushing the lateral modes out of most low-frequency operating bands and a slight improvement of the hydrostatic sensitivity response (≤ 1 dB) because of the smaller lateral piezoelectric voltage constants [5]. The configuration of electrically connecting the sheets in parallel while mechanically arranging them in series results in a hydrophone that is self-shielded. This means that the stray capacitive coupling losses that may occur between an odd balanced number of layers and hydrophone housings or structures. Other advantages of two-layer designs include higher capacitance with increased energy output. The fact that the hydrophone is large-area results in a sensor that by natural design will area-average the impinging

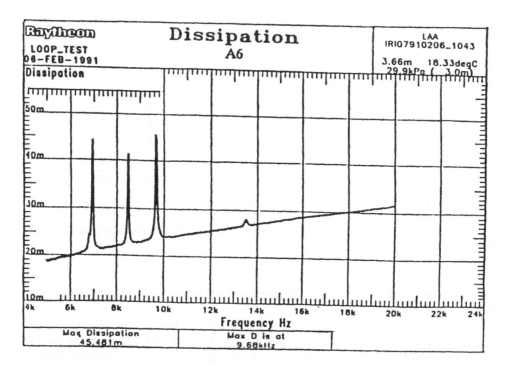

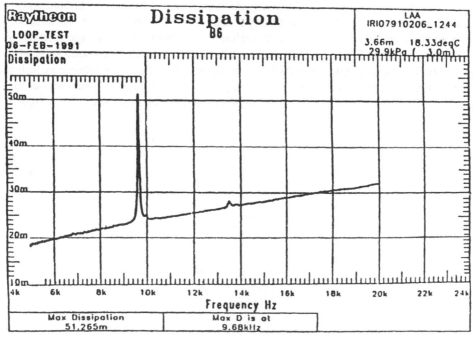

Figure 1 Modal dissipation measurements of a 6-in. × 6-in. PVDF bilaminar plate: (top) parallel orientation and (bottom) cross-axis orientation. (Adapted from Ref. 4.)

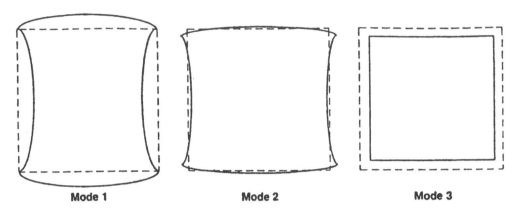

Mode 1   Mode 2   Mode 3

Figure 2 Contour-extensional vibrations of square composite PVDF plates planarly isotropic. (Adapted from Ref. 4.)

acoustic pressure field as opposed to single-point detection in fields where flow or spurious pressure is incumbent.

A recent introduction to the ferroelectric polymer family has been copolymer. This material is typically cast as opposed to stretched and drawn with the homopolymers. The advantage offered by the copolymer process is that the (3,1) and (3,2) piezoelectric and stiffness coefficients are equal. Additionally, the copolymer processing technique allows versatility in selecting the thickness of the polymer layer. This means that the selection of copolymer material can be done with preferred capacitance and voltage sensitivity performance. The copolymer also offers nominally higher piezoelectric coefficients with reduced lateral coupling and higher thickness-mode coupling. However, when comparing the properties of homopolymer and copolymer, there is a noticeable difference in the copolymers' reduced storage temperature capability. Also, the copolymer has a higher

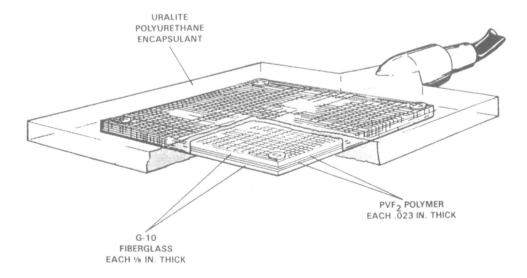

URALITE
POLYURETHANE
ENCAPSULANT

PVF_2 POLYMER
EACH .023 IN. THICK

G-10
FIBERGLASS
EACH 1/8 IN. THICK

Figure 3 Large-area hydrophone. (Adapted from Ref. 5.)

density and thus, larger acoustic impedance (20–60% higher, depending on the selection of electrodes and individual manufacturers' homopolymers).

The copolymer is attractive for low-frequency, hydrostatic-mode hydrophone applications. It is presently being offered commercially in single-sheet and multilayer configurations with aluminum stiffening electrodes [6].

C. Cylindrical Hydrophones and Arrays

Perhaps the simplest cylindrical hydrophone is a cylinder-shaped piezoelectric polymer. Such a product is available as a piezoelectric copolymer hydrophone cylinder for hydrostatic-mode operation [7]. The cylinders are available with nominal outside diameters ranging from 0.64 to 2.54 cm and lengths up to 30 cm. With nominal wall thicknesses of 0.5 to 1.3 mm available, hydrostatic free-field voltage sensitivities (FFVS) of −202 to −194 dB re 1 V/μPa may be obtained. This type of performance and mechanical characteristics are for the active material. For many hydrophone uses, the cylinder must be placed into a housing or an encapsulant for underwater application.

An application that has recently benefitted from the use of ferroelectric polymers is autonomous underwater vehicle (AUV) hydrophones and conformal arrays [8,9]. These sensors need to operate over wide frequency bands. They are also designed to be mounted in or on cylindrical shell structures. The main need for these types of hydrophones and arrays is for obstacle avoidance, navigation, and target detection. To accomplish this it is necessary to maintain systems with high directivity resolution, reduced side-lobe levels, and high element-to-element uniformity. From a manufacturing viewpoint, it is desired to have low cost, ease of fabrication, low weight and volume, and minimal field profile.

To accomplish these needs, piezoelectric polymer is an attractive material. The desired element-to-element uniformity is inherent in the ferroelectric polymers, as is low weight, conformability, and minimal volume occupation. Furthermore, using a modified Chebyshev tapered design as an electrode pattern has been found to produce clean directivity responses with side-lobe levels down more than 30 dB (as opposed to the traditional −13.3 dB side-lobe levels of a rectangular-shaped element) as is shown in Figure 5 [8,9]. The implementation of shaping an element into the modified Chebyshev tapered design is easily and inexpensively accomplished with ferroelectric polymers by using silkscreening (etching) techniques on the electrode pattern that is on top of the active polymer sheet.

Two different cylindrically conformal-type arrays fabricated include a receive armband array featuring 20 hydrophone disks and a toroidal volume search sonar (TVSS) featuring 120 Chebyshev-shaped hydrophones. The armband cylindrical array contained 20 large-aperture elements with 5° beam width per element for 120° of coverage.

The TVSS design consisted of 20 hydrophones etched into the modified Chebyshev tapered design on a single sheet of voided (homopolymer) PVDF as shown in Figure 4. To accomplish the objectives, the aperture width function of the modified Chebyshev tapered design was equivalent to the amplitude shading function.

The final sensor design was double layered to achieve self-shielding and double the capacitance. Figure 5 shows the PVDF sheet conformed directly to the outside of the shell, without any baffle inclusion. This method for fabrication results in minimal assembly labor and accurate element positioning tolerances. The resulting free-field voltage sensitivity response is flat over a wide frequency band, as shown in Figure 6 [9].

In general, the sensing of acoustic pressure consists of two processes, the excitation of the piezoelectric strain in a material by conversion of an external behavior and the

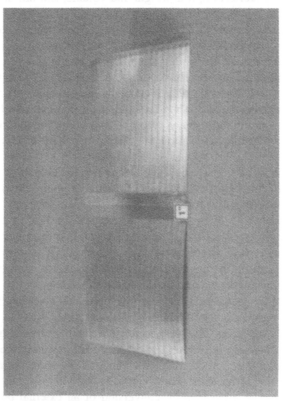

Figure 4 Twenty hydrophones formed on a single sheet of PVDF. (Adapted from Ref. 9.)

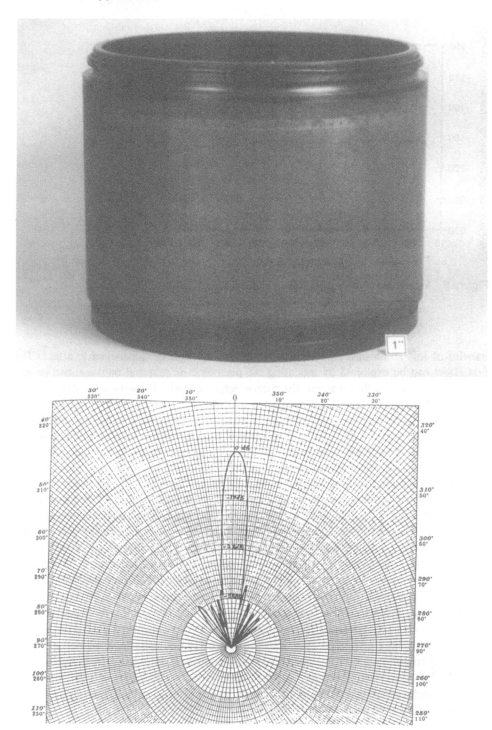

LONGITUDINAL BEAM PATTERN

Figure 5 TVSS cylindrical array and directivity response. (Adapted from Ref. 9.)

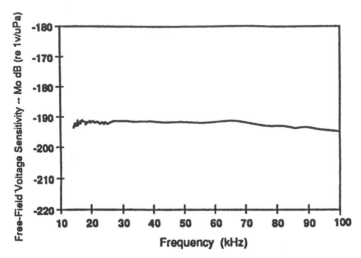

Figure 6   Measured FFVS of the TVSS cylindrical array. (Adapted from Ref. 9.)

transfer of this strain into an electrical waveform through the piezoelectric effect. This first effect can be exploited by mounting the piezoelectric material onto a substrate material. When used in this manner, the piezoelectric material behaves as a strain gauge. Several configurations have been developed that take advantage of this trait [2,3,10,11].

A hydrophone structure that features a compliant substrate results in sensors with high sensitivity gains. The gains are attributable directly to the high compliance of the hydrophone as a structure as opposed to a basic material property enhancement. One design consists of cylindrical structures with slotted acrylic form with piezoelectric polymer wrapped around the structure, as shown in Figure 7. A similar approach has been to use plate-type gradient sensors, where the PVDF behaves as either a strain gauge or a membrane type; two such designs are shown in Figure 8. Comparing the acoustic performance of each of these designs, the sensitivity of the sensor with the piezoelectric polymer placed in a membrane-type arrangement is almost an order of magnitude higher than the design that features PVDF as a strain gauge in tension. The reason for this gain is that the membrane design takes simultaneous advantage of both the excitation of the piezoelectric strain and the direct conversion into an electrical signal [2].

Another cylindrical hydrophone design approach that features piezoelectric polymer as the active sensing element is shown in Figure 9. In this concept, PVDF coaxial cable is wrapped around a compliant cylindrical structure [11]. A theoretical model with supporting experiments of such a concept has suggested that cylindrical hydrophones featuring PVDF cable wound around an air-filled cylinder (with exposed ends) results in enhanced performance as compared to a similar piezoelectric polymer coil around a free flooded cylindrical structure as demonstrated by the FFVS of Figure 10.

D. High-Sensitivity Hydrophones

A hydrophone that has been designed for high-sensitivity gains is one which has been configured for acoustic detection along the g_{31} axis (as opposed to the common g_{33} or

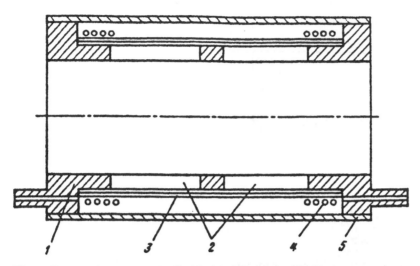

Figure 7 Membrane type of cylindrical PVDF sensor with (1) slotted acrylic form; (2) slots; (3) PVDF wrapped around the form; (4) fiber band for tightening; (5) casing. (Adapted from Ref. 2.)

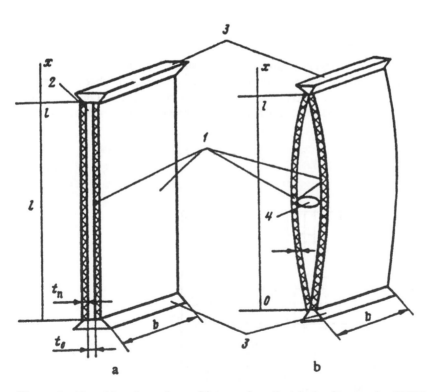

Figure 8 Plate (a) and membrane (b) type of gradient hydrophones using PVDF strain-sensing element with (1) PVDF; (2) acrylic plate; (3) rigid supports; (4) spreader bar. (Adapted from Ref. 2.)

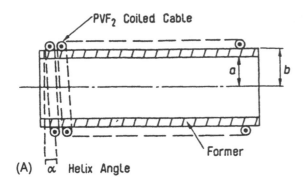

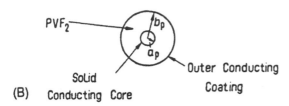

Figure 9 Cross-sectional view of (A) coiled PVDF cable and (B) PVDF cable. (Adapted from Ref. 11.)

g_{3h} excitation modes) [12]. A recent effort has concentrated the acoustic pressure field to a small cross-sectional area within a transducer mounting structuring that features the (3,1) mode. Figure 11 shows a conceptual drawing of ferroelectric polymer mounted between springs along the (3,1) axis in a hydrophone structure. The resulting FFVS is shown in Figure 12.

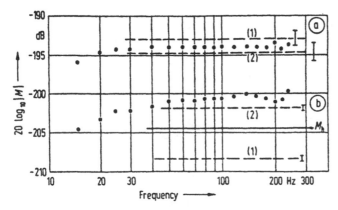

Figure 10 FFVS of coiled PVDF hydrophone for (a) air backed with exposed ends and (b) flooded interior; (•••) is experiment and (————) theory. (Adapted from Ref. 11.)

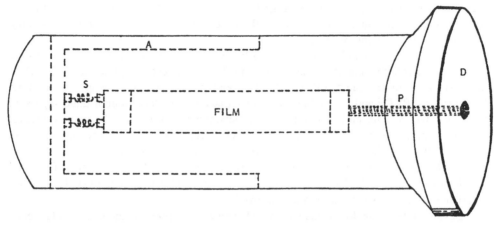

Figure 11   Cross-sectional view of the (3,1) mode transducer. D = diaphragm, P = driver pin, S = spring-loaded screws, A = aluminum frame. (Adapted from Ref. 12.)

E. Shock Wave Sensors

Because of the mechanical robustness of PVDF, it is an attractive transduction material for the measurement and characterization of shock waves. PVDF is desirable because it can withstand the physical demands required for placing such an object within a shock wave field.

Of the present transduction materials used in underwater shock wave sensors, tourmaline has been a dominant choice. It is a naturally occurring piezoelectric crystal that has attractive properties such as mechanical robustness, chemical stability, independent performance as a function of pressure, and relatively good hydrostatic pressure sensitivity response at high frequencies [13]. However, tourmaline is a semiprecious mineral, which

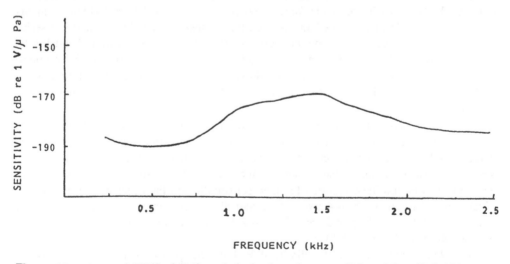

Figure 12   Measured FFVS of (3,1)-mode hydrophone in water. (Adapted from Ref. 12.).

means that it is expensive. It also is rather brittle and is therefore susceptible to chipping, surface defects, and cracks. As a result of these deficiencies, there is an interest in using other active transduction materials.

A study has been conducted comparing the behavior of PVDF-featured sensors with that of the tourmaline-featured sensors for shock wave detection application [13]. Both materials were of the same disk construction and placed into drop-weight calibrators. Both sensors (PVDF and tourmaline) were then placed into a large indoor tank, separated by a distance of 1 m, and exposed to an underwater shock wave. The results suggest that there might be a problem with piezoelectric coefficient uniformity in uniaxially stretched PVDF during shock wave exposure. However, the results also show good agreement with theoretical predicted values for peak pressure and decay times. The final conclusions were that the PVDF could offer improved performance and desirability characteristics if the transducer design is improved.

Besides interest in low-frequency shock wave hydrophones, there have also been investigations of shock wave hydrophones for ultrasonic applications. Focused fields of ultrasonic pulses occurring in lithotripsy can be measured using ferroelectric polymer sensors. A study in a water-filled tank has led to the development of a 25-μm PVDF hydrophone. The sensitivity of this sensor is 20 mV/MPa with a bandwidth of 20 MHz. More than 100,000 focused shock wave pulses of 20 MPa have been recorded successfully without a significant decrease in acoustic sensitivity [14].

F. Towed Arrays

Ferroelectric polymers continue to show excellent application in underwater towed arrays. Because of the ease in forming ferroelectric polymer into a variety of shapes and geometries, PVDF has been featured in towed array applications as planar elements, coaxial lines, and cylindrical elements. The towed array applications have included military and seismic uses.

During the 1980s, Raytheon Company manufactured a selected number of towed array sensors [5]. These sensors consisted of multiple homopolymer PVDF elements connected electrically in parallel to form an extended line sensor. The individual elements were constructed from electroded and bonded PVDF sheets stacked in mechanical series and electrically in parallel to simultaneously boost both the capacitance and sensitivity. The anisotropy of the homopolymer is exploited in these extended sensors by aligning the low (3,2) axis sensitivity along the towed axis. This results in an array that, by design, will suppress flow noise and vibration strum while being most sensitive in the normal direction.

A product known as Vibetek-20 is a ferroelectric polymer sensor cable [15]. This cable has been formed into a hydrophone for towed array applications by wrapping the cable in a helical fashion about a potted cylinder structure. Figure 13 shows a cross-sectional drawing of an extended sensor line featuring the Vibetek-20 PVDF cable. The hydrophone has been reported to show sensitivities on the order of −188 dB re 1 V/μPa, lengths ranging from 4 cm to 1 m, and diameters from 8 to 40 mm. Figure 14 is a representative FFVS plot from 1.5 to 20 kHz. This figure shows a broad resonance with a rolloff beginning at 5 kHz.

In 1990, a product line of cylindrical PVDF hydrophones was introduced to the market for bottom reference and downhole applications in the seismic industry [16]. This

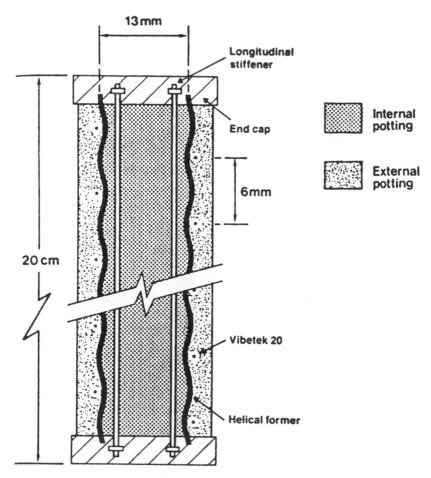

Figure 13 Cross-sectional view of the L29 hydrophone. (Adapted from Ref. 15.)

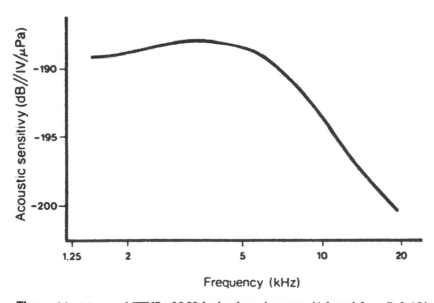

Figure 14 Measured FFVS of L29 hydrophone in water. (Adapted from Ref. 15.)

technology was recently extended into a solid towed array product. Because of the light-weight characteristics of PVDF, the towed array has high buoyancy characteristics without oil filling a hose but instead using a closed-cell foamed polyurethane jacket on the cable itself. The PVDF featured towed array is 20% of the weight and volume of similar oil-filled towed arrays. The reported sensitivity is −188 dB re 1 V/μPa per element over an operating frequency range of 6 to 10,000 Hz.

III. BIOMEDICAL TRANSDUCER APPLICATIONS

The electrical and mechanical properties of piezoelectric polymers make them a possible alternative to ferroelectric ceramics such as lead zirconate titanate. For several reasons, they are attractive for transducer design. The mechanical flexibility and conformability of thin-film PVDF means that it can be configured into a wide range of transducer products. The low acoustic impedance of PVDF is comparable to body tissues, which makes it useful for acoustic imaging applications. Short impulse response and high axial resolution in acoustic imaging systems are possible with PVDF-featured devices because of the robustness and broadband characteristics of the polymer.

Poly(vinylidene fluoride) has become an attractive material for use in ultrasonic biomedical transducers because of its large bandwidth, low acoustic impedance (about 4.5 MRayl, which is fairly close to human tissue), long stability with time, conformability to shape, and low cost. This makes the material intrinsically broadband and permits construction of transducers with short impulse responses without the need for matching layers. The low electromechanical coupling factor is partially offset by the use of time-delay spectrometry [16]. This technique allows a signal-to-noise ratio (SNR) of at least 60–75 dB to be maintained in the frequency range from 1 to 40 MHz.

Over the past decade, frequencies have been pushed from 2.25 MHz up to the 20-MHz range for ophthalmological systems and beyond 30 MHz for acoustic shock waves used to disintegrate kidney stones. Biomedical applications of ferroelectric polymers are focused mainly on the use of their electromechanical transduction properties for the implementation of various kinds of dynamic pressure measurements of sounds, fluid flow, heart rate, and soft tissue contact pressure [17]. The low acoustic impedance and higher mechanical losses, as compared with piezoceramic, implies that the frequency responses of PVDF transducers can be maintained to be quite flat over an extended operating band. With PVDF available in thicknesses ranging from 6 to 110 μm, thickness-mode resonance frequencies between 160 kHz and 10 MHz are achievable. Furthermore, a PVDF hydrophone will not disturb the acoustic field and will register rapid dynamic behavior in pressure as required. Ferroelectric sensors for biomedicine are implemented mainly in the form of external sensors such as flexible bands, wrappings or strips, rigid arrays and platforms, and as implantable sensors used for cardiovascular and pulmonary systems. Disadvantages of featuring piezoelectric polymer as the active transduction component in biomedical applications include lower electromechanical conversion efficiency and an intrinsic lack of a DC electrical response.

Biocompatibility with body tissues is often important and is related to the inertness of the introduced material with both tissue and blood. Water absorption of PVDF is extremely low (less than 0.04% by weight) and by nature is nontoxic; thus it is an attractive material with respect to biocompatibility.

A. Sensors and Probes

Transducers for biomedical application are classified into sensors, acoustic sources, and imaging systems. First, sensors and probes will be discussed. Several types of hydrophones have been devised, including membrane sensors, miniature sensor probes, and needle sensors [17–21].

The basic requirement for an ideal sensor is a small enough size so as to not perturb the acoustic field together with the smallest possible sensing area so as to perform a point measurement with improved angular response and minimal spatial averaging effects. Also, the bandwidth of the sensor should be broadband, with uniform frequency response for precise point measurement within the acoustic field.

1. Probe, Needle, and Membrane Sensors

A sensor consisting of a metal needle point, rounded at its tip and coated with a layer of PVDF, has been recently developed [20]. The main purpose of this probe is the scanning of acoustic fields generated by focusing or plane transducers. The concept is shown diagrammatically in Figure 15.

A similar sensor has also been described in which during poling the field is concentrated at the tip so that the tip is the most active region [17–19]. Only the lower tip region is sensitive. The layer of PVDF at this tip is about 20 μm, which corresponds to a thickness-mode resonance frequency at 27 MHz, which is well above the intended operating frequency band of 1–10 MHz. The needle serves as the mechanical support for the PVDF layer and as one of its electrodes. A vapor-deposited coating of silver or nickel serves as the outer electrode. The inside of the metal pipe is filled with nonconductive epoxy. After poling by the corona point discharge method, the tip of the needle is assumed to be poled to the maximum for PVDF with a g_{33} of 80–120 mV-m/N. The order of magnitude of the probe capacitance is in picoFarads (pF).

The sensitivity of a probe sensor is a fairly strong function of frequency, with sensitivities ranging from -150 to -160 dB re 1 V/μPa at the end of a 40-cm cable over the operating frequency band of 1 to 10 MHz. This is due to the convex curvature and the frequency-dependent sensitivity of the probe design. The convex shape of the PVDF

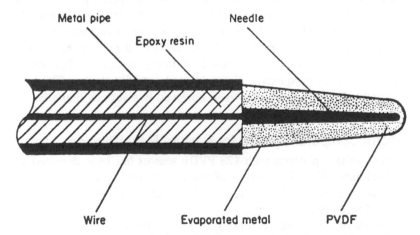

Figure 15 Cross-sectional view of PVDF needle hydrophone. (Adapted from Ref. 20.)

needle probe sensor makes it more suitable for measuring convergent wave fields relative to plane hydrophones. Its curvature makes it largely independent of the angle of incidence of the impinging wave.

Membrane and probe sensors have flatter frequency responses compared with those of needle hydrophones. The membrane hydrophone has a flat response of about -128 dB re 1 V/μPa from 1 to 22 MHz, whereas the probe hydrophone has a sensitivity of -160 dB re 1 V/μPa from 1 to 9 MHz. A major drawback of the membrane sensor is the presence of side lobes due to Lamb wave excitation as well as improved directionality characteristics. Needle sensors have greater applicability in measuring convergent wavefronts in the point of convergence compared with plane transducers and may be used for measuring converging shock waves because of their mechanical stability. Because of the curvature of the sensitive surface, the needle sensor is independent of the angle of incidence in a wide angular range with application in the 1–10 MHz operating frequency range [21].

2. Noninvasive Sensing of Arterial Pressure

Noninvasive cardiopulmonary sensors continually monitor heart and breathing rates. Several kinds of devices have been conceived which allow continuous and comfortable monitoring of the main cardiac and pulmonary parameters. A biaxially drawn PVDF film sensor for peripheral pulse measurement has been realized in a bilaminate configuration [22]. Faithful reproduction of a carotid pulse was accomplished by this sensor, with systolic time intervals easily determined in patients at rest.

A similar sensor was used for ambulatory monitoring of breathing. The sensor is applied to a patient's chest by an elastic band. The piezoelectric polymer sensor becomes charged as the patient's chest cage expands. After suitable filtering, the PVDF signals were compared to those recorded by a pneumotachograph with favorable results. Detection of cardiac sounds were also possible by using a biadhesive film between the skin and sensor. This increases the cardiac signal relative to sounds resulting from friction [23].

Fluid-filled catheters connected to external pressure transducers are the most commonly used method for measuring intravascular pressure. There are several problems involved with this process, including delay in the transmission of high-frequency signals, smoothing of the pressure wave, and vibrations along the catheter. These problems can be addressed by means of catheter-tip piezoelectric polymer transducers [22]. These types of transducer designs avoid the problems of transmitting a pressure wave from the fluid-filled catheter to the external pressure transducer. PVDF has been demonstrated to be a suitable catheter tip transducer because it can be used in various geometric configurations and it has the potential to be an inexpensive disposable and sensitive device. The lack of a truly static response of a ferroelectric polymer, however, is a serious limitation to widespread application.

3. Sensor Platform Computer System for Posture Rehabilitation

A computerized system using a platform with 128 PVDF sensors has been designed as a tool for posture rehabilitation [24]. The purpose of this system is to allow for improved understanding of the mechanisms involved in foot load transfer for diagnosing and programming corrective rehabilitative procedures.

The platform consists of a 128 circular sensing sites (8×16 element configuration matrix) with a diameter of 10 mm for each sensor and spacing between sensors of 20 mm along the platform width and 25 mm along the platform length. Each piezoelectric polymer sensor is 80 μm thick. The sensors are individually connected to a preprocessing unit by coaxial cables. The core processing unit contains an analog multiplexor, a charge amplifier, and an addressing unit. This addressing unit is under computer control to sequentially scan each of the sensors and convert their particular charge into voltage signals. These individualized signals are then input into an A/D converter. The voltage level on each of the sensors is proportional to the exerted force, which is calculated by the computer. Calibration was performed on each of the sensors using a load cell. After processing, the computer outputs a footprint with square areas whose size is proportional to the pressure value on each sensor and, hence, a mapping of the foot load transfer.

B. Wide-Band Acoustic Sources

Commercially available hydrophone probes are generally not calibrated beyond 10 MHz. One reason is the lack of available acoustic sources. Imaging transducers constructed from lead zirconate titanate show sharp resonant characteristics and must use matching layers due to the high acoustic impedance of the ceramic. The matching layers in turn limit the bandwidth of the transducer. The 3-dB bandwidth of a high-quality imaging transducer is about 70% of the fundamental frequency and thus a single transducer cannot provide an output of 1–40 MHz. Furthermore, most available sources are highly resonant devices and cannot be used as wide-band sources. Instead, the National Bureau of Standards (NBS) supplies a series of calibrated sources with resonant frequencies that are approximately 1 MHz apart. Use of such probes can overlook the many rapid variations in frequency response due to mechanical resonances. Most transducers using piezoelectric ceramics, such as lead zirconate titanate, will deliver maximum power when the dimension of the piezoelectric material is equal to one-half of the acoustic wavelength. However, the mismatch between the ceramic and the propagation medium causes a significant part of the acoustic wave to be reflected, which in turn introduces ringing in the transducer response, which decreases the bandwidth.

Recently, piezoelectric polymer sources which can be used at frequencies above 40 MHz have been described [19]. These transmitters are intended for absolute calibration of hydrophone probes. PVDF transmitters have good characteristics for wide-band calibration applications. Their low electromechanical coupling reduces transmitting efficiency; however, if the goal is bandwidth maximization, this is of little consequence. The high mechanical losses of the PVDF material provides fairly flat transmitting voltage response over fairly wide frequency ranges. Such high-frequency sources are used to generate acoustic shock waves with frequency content well beyond 30 MHz. They can be used to noninvasively disintegrate kidney stones with acoustic shock waves as well as for imaging applications.

A 9-μm PVDF film has been bonded directly to the backing material based on analysis using the Krimholz-Leedom-Mathei (KLM) model [19]. The one-dimensional approximation of several layered structures, consisting of the piezoelectric material, the acoustic impedance matching layers, and the propagation medium, can all be accounted for in the model. Additionally, the influence of the electrical matching and the backing material cause an alteration of the resonance frequency. The resonance of a 9-μm-thick PVDF film occurs at approximately 120 MHz, whereas with a 12-MRayl backing the resonance drops to approximately 56 MHz. The thickness of the PVDF material governs the bandwidth of the transducer.

A 9-μm film piezoelectric polymer was used in a transducer, as shown in Figure 16, which resulted in a peak in the transmitting voltage response of 52–62 dB re 1 Pa/V at a distance of 50 cm over a frequency range of 1 to 40 MHz [17]. This compares to a similar piezoelectric ceramic hydrophone backed with a material of an acoustic impedance of 18 MRayl with a peak response of 62 dB re 1 Pa/V but over a much more limited frequency range. The directivity patterns displayed in Figure 17 show a 20-dB decrease for 1° off the acoustic axis at 15 MHz.

Compared with conventional heavily damped piezoelectric ceramic transducers, PVDF transmitters show lower efficiency; however, the lower transmitting efficiency is fully adequate for narrow-band measurement systems such as time-delay spectrometry. Furthermore, the frequency response of the PVDF transmitters well exceeds the 40 MHz tested here and as such provide a very broadband alternative to traditional ceramic transmitters [19].

C. Imaging Systems

The electrical and mechanical properties of piezoelectric polymers make them a suitable alternative to conventional ferroelectric ceramics (such as PZT) in transducer design, particularly as sensing probes. The lower electromechanical coupling factor makes the polymer inferior as a transmitting material and has limited its use as a projector or pulse-echo transduction material.

Piezoelectric polymer has been demonstrated successfully for "near-surface" tissues or organs where the penetration depth is sufficient given the power limitations. Imaging systems reviewed elsewhere in this volume show that the ultrasonic images of a thyroid gland are superior to those obtained from piezoceramic transducers. A 5-MHz P(VDF-

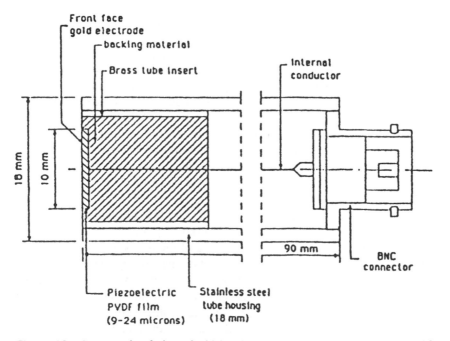

Figure 16 Cross-sectional view of wideband PVDF ultrasonic projector. (Adapted from Ref. 17.)

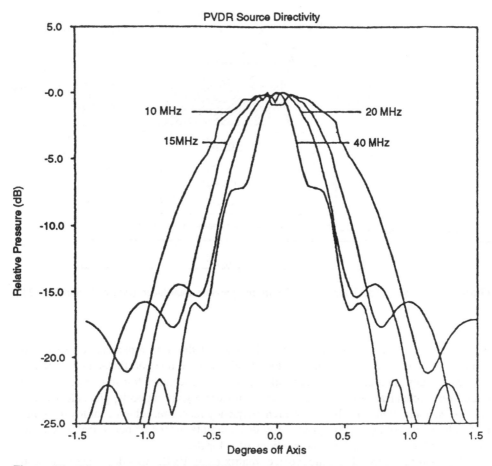

Figure 17 Directivity of wide-band PVDF ultrasonic projector. (Adapted from Ref. 17.)

TrFE) linear transducer array was also constructed from a 64-element arrangement to provide good tomographic images. Annular arrays made from a single sheet of piezo-electric polymer were developed and tested. A 5-MHz annular array transducer with high resolution has been developed with eight annular elements for a total diameter of 70 mm, a lateral resolution of less than 1 mm, and a depth resolution of 0.3 mm (see Chap. 18).

A transducer system for recording arterial pressure contours consists of a thin (30-μm) film of poly(vinylidene) fluoride, which serves as the active element and as the mechanical coupler to the skin [23]. One side of the disk, 2.5 cm in diameter, is cut and bonded to an insulating ring. This assembly is set between the main body of the trans-ducer housing and the transducer cap as shown in Figure 18. The PVDF membrane, which serves as both the active device and as the transducer diaphragm, is coupled directly to the skin. The active element, which is backed by foam, is quite compliant and conforms to the contour of the body. Additional mechanical coupling between a piezoelectric ceramic and a diaphragm of the transducer is unnecessary because of the good acoustic impedance match between PVDF and skin. Comparisons with the trans-ducer and with an indwelling catheter show similar signals for the arterial profiles. The

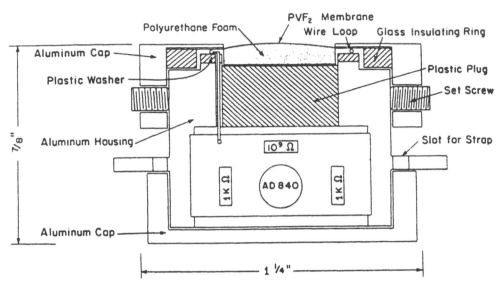

Figure 18 Cross-sectional view of PVDF arterial pressure transducer. (Adapted from Ref. 23.)

PVDF transducer has several advantages. Because of the low impedance of PVDF, the contact component and electrical transduction element are one and the same and thus do not require any mechanical or electrical coupling. The contact transducer element embodies no mechanical inertia, unlike devices with strain-gage supports, piezoelectric crystals, or magnetic fields. It also has a high frequency response, which enables it to make high-fidelity recordings. This system provides a simple means of periodically screening cardiovascular changes in large populations of subjects.

A two-dimensional ultrasonic receiver matrix using PVDF was developed and tested in an ultrasonic transmission camera [25]. This camera produces medical images which are different from the B-scan images found in most camera systems where linear arrays are used. These arrays must be mechanically moved to form an image. This limitation has been overcome by means of a receiver matrix with a high number of receiver elements. The receiver matrix has both high sensitivity and bandwidth along with a large number of small receiver elements. Furthermore, no mechanical movement of the receiver array is required. Figure 19 shows the configuration of the array. The matrix readout is by a stack of 29 thin-film substrates with 128 switchable preamplifiers. One side of the stack forms a total covering area of 96 mm × 26 mm. On the front side of the ceramic substrate are 128 contact areas of size 0.65 mm × 0.75 mm with element-to-element spacing of 0.75 mm. On the flat side there are 128 signal leads, resistors and contact pads for 128 dual-gate MOSFETs, which function as switches. The contact matrix is pressed against a common 25-μm PVDF foil as a common transducer and as a window for the water bath. On the rear side there are 128 switch leads which select one of the 128 transistors per substrate and amplify it. The foil is metalized only on the inner side to contact the water and form a common ground. The alternating charge (AC) charge pattern of the ultrasonic image is capacitively coupled to the contact matrix. The 29 × 128 receiver matrix is read out through 29 signals that are each representative of the average pressure amplitude which has been detected by the respectively addressed matrix

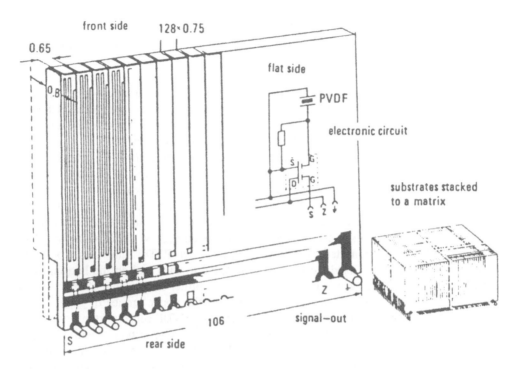

Figure 19 PVDF imaging transducer system on substrate with 128 contact areas. (Adapted from Ref. 25.)

column. These output levels are multiplexed by a 29:1 ratio to form a video signal. With the two-dimensional PVDF receiver matrix, transmission images can be clearly presented.

IV. ROBOTICS

Another emerging field of applications for PVDF sensors is in force contact sensing for robotics. A sense of touch is particularly important to robotics. Current research involves the measurement of contact forces. The mimicking of humanlike spatial resolution and sensitivity, operation with a large bandwidth for fast response, and having a linear response while simultaneously avoiding hysteresis are important guiding design factors, along with cost, compliance, and physical robustness. Piezoelectric polymer possesses several advantages over competing piezoresistive technologies because of lower noise, improved linearity, longer response times, hysteresis, fatigue, and material stiffness.

Presently, most robotic applications require control of position only for assembly purposes. They are not suitable for operations requiring control of contact forces such as mating of parts with close tolerances. Future robotic applications will require sensory capabilities that are functionally equivalent to human senses, including vision and touch. Important to robotic applications are the ability to sense when an object comes in contact with another. Touch sensors are able to provide sensing over an area in which there is spatial resolution, such as the posture rehabilitation sensor described in Section III.A.3.

A. Tactile Sensors

Tactile sensitivity is a complex task, since its purpose is to mimic the sensitivity of human skin in terms of pressure and temperature.

A sensor robot gripper has been which utilizes a PVDF tactile sensor as shown in Figure 20 [26]. This sensor is reported to be fairly rugged, linear, and nonhysteretic. A PVDF film of 110 μm is attached to the bottom of a robot gripper with parallel fingers. An electrode-etched pattern of 128 circular electrodes was disposed in an orthogonal grid on the upper side of the printed circuit board (PCB). Each electrode was connected through a metalized hole to a pad deposited on the bottom side of the PCB. The electrodes are circular with a 1.5 mm diameter. The metallization on the bottom of the PVDF was removed and then the film was fixed to the PCB with a nonconductive adhesive. This arrangement was chosen because it enables the PVDF to operate predominately in the thickness mode when the sensor is subject to contact forces. It also reduces mechanical crosstalk between sensors if the sensing sites had been defined directly on the PVDF film and electrically connected by pathways on the same PVDF film. This would also reduce the site density. Coaxial cables were mounted to each pad on the bottom of the PCB, which were then each led to scanning and amplifier instruments. A rubber sheet with an array of 128 plastic spheres was bonded to the upper surface of the sensor in order to both decouple the sensing sites and shield the PVDF sensors from temperature variations. If the shielding was not present, the strong temperature sensitivity of the sensors due to the large pyroelectricity of PVDF would couple into the signal outputs. Charge amplifiers were used to process signals generated by the piezoelectric polymer, since a charge amplifier output is largely independent of sensor and cable capacitance. The charge of each sensor was connected to the charge amplifier by a multiplexing unit from which it was fed into an analog-to-digital (A/D) converter and then onto a computer for further processing. The unit was tested for sensitivity (20 pC/N), force resolution (0.01 N), linearity (1%), hysteresis (5%), drift (1%), and crosstalk (<2%). Future recommended improvements for the system include separation of the functions of internal grasping from reconstruction of tactile images.

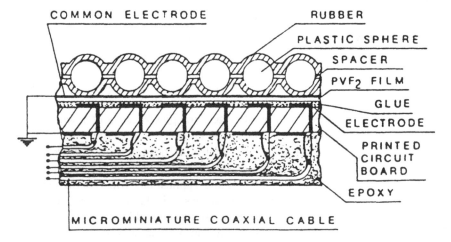

Figure 20   Cross-sectional view of PVDF-based tactile sensor. (Adapted from Ref. 26.)

A touch sensor featuring piezoelectric polymer as the active material has been described [27]. This system employs sensor elements mounted on a printed circuit board with 256 sensors defined on a single PVDF sheet. The shape of an object pressed against the sensor is identified by electronically scanning the 256 sensors and digitally analyzing the pressed and nonpressed sensors. Also described is a tactile sensor which utilizes an ultrasonic time-of-flight method to measure the thickness of a rubber layer that is deformed when touched. The design of this sensor is shown in Figure 21, where PVDF is featured as the active material both in the ultrasonic emitter and in the ultrasonic receiver. The duration of the time of flight of the ultrasonic waves through the compliant layer can be calibrated to provide a measurement of the pressure impinging on the sensor. This technique eliminates the influence of the pyroelectric effect but increases the hysteresis due to the elastomeric layer and requires more complex electronics.

Another similar piezoelectric polymer tactile sensor array system has been described featuring six piezoelectric polymer tactile sensor arrays [28]. One design featured a square pad configuration, while another used stripe electrodes on both upper and lower surfaces with the upper surface being in a direction perpendicular to the lower surface (shown in Fig. 22). A difference in the two configurations is that they featured piezoelectric polymers of different thicknesses. A similarity is that they both used metal evaporation masks and photoetching techniques to apply electrodes of each geometry to the surfaces of the PVDF film. Thermal poling was used to create or increase the piezoelectric activity of the film. Results demonstrated that thicker film produced higher sensitivity, as expected,

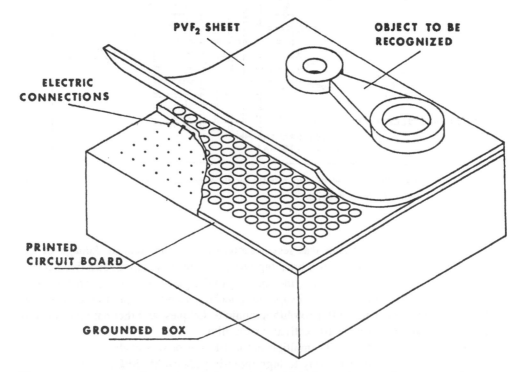

Figure 21 PVDF matrix sensing platform with multiplexing and preamplifiers mounted inside of grounded box. (Adapted from Ref. 27.)

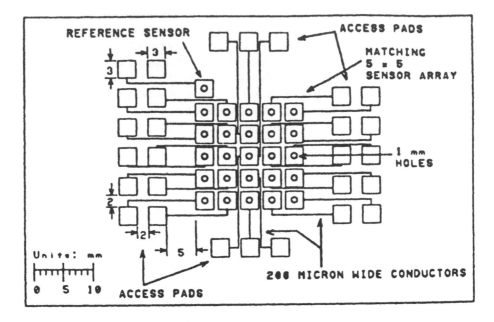

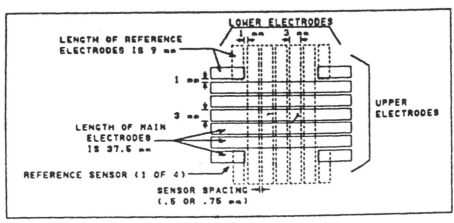

Figure 22 Cross-sectional view of two robotic tactile sensor designs featuring PVDF as the active material. (Adapted from Ref. 28.)

and that the square pad design was less susceptible to failure due to damage of the electrodes. This is sensible because breaking one stripe electrode eliminates five sensors in the striped array design. However, the square pad design requires more wiring and is therefore more difficult to fabricate. Evaporation techniques were found to produce more robust electrodes compared with photolithographic techniques, and thermal repoling was found to degrade the piezoelectric activity of the film.

The performance of a similar multiplexed tactile sensor was evaluated by coupling sensors in a 5 × 5 square pad array to high-input-impedance MOSFET amplifiers [29]. Five different thicknesses of PVDF film were evaluated; films with thicknesses between 25 and 52 μm provided high sensitivity response while being flexible enough to conform to the surface of the underlying integrated circuit. Thinner film (25-μm) was found to provide the quickest response times. A direct current bias voltage of 2.5 V_{DC} was applied to each sensor for 0.1-s prior to testing. This biasing was done to stabilize and improve the uniformity of the response. The output of the sensors were found to be linear with applied force over the range from 0.8 to 76 gmf. Several different shapes of loads were applied to the sensor in order to test their ability to discern circular, toroidal, rectangular, and hexagonal shaped loads. The ability to recognize different shapes was clearly demonstrated.

Cross-talk measurements were compared between loaded and unloaded sensors. Levels of noise of 15 mV were noted on the unloaded sensors, while adjacent loaded sensors had noise of 7.5 V. Thermal insulation was required to prevent the pyroelectric effect from overwhelming the piezoelectric effect utilized during the measurement.

Future work is focusing on using a microthermistor in the array to detect thermal fluctuations and to separate them from the detection of pressure. This would also provide temperature sensing to supplement the pressure-detection capability of the robotic sensor.

B. Active Vibration Control

The use of piezoelectric polymer for active control systems is a recently rapid growing interest. Piezoelectric polymer is attractive for these applications because of its light weight, low profile, conformability, and relative good impedance match with fluids. Although active vibration control systems are important in many different fields, inclusion of this is representative of the broad definition and understanding of robotics. This is because the piezoelectric polymer in the active systems is typically used to sense and characterize a behavior (vibration and/or acoustic) and respond accordingly by transferring the behavior through a processor and into an actuator. In theory, this is similar to a robotic application in which an acoustic pressure or temperature is detected and then processed and output to a prescribed behavior.

One of the first active vibration control systems that investigated the use of PVDF was for the control of vibration decay of a cantilever beam [30]. The attraction of using a PVDF polymer for these studies, and for later research, is the low weight and profile of PVDF. The presence of the PVDF does not change the uncontrolled behavior of the structure, but during the active control stage the PVDF provides an accurate representation of the vibration signal for proper control manipulations.

Later research has concentrated on shaping PVDF into strips and placing them onto structures [31]. The shape and mounting positions of the ferroelectric polymer were made to account for the active modal radiating modes while ignoring other nonradiating vibration behavior. This work has recently been extended to full plate structures and compared with free field microphones for error sensors [32]. It was found in the case for using error information to input into an active electronic controller that two narrow PVDF sensors that are positioned on the plate structure, such that the dominant acoustic radiators, were in contact is a more efficient system than one that uses microphones in the far field. The efficiency was in providing accurate information necessary to reduce the far-field acoustic radiation of the plate structure.

V. MEASUREMENT INSTRUMENTS

PVDF is useful for a variety of measurement instruments including accelerometers, stress gauges, strain gauges, shock gauges, turbulence sensors, fiber optic phase modulators, and nondestructive testing transducers. Its long stability with time, large bandwidth, low weight, conformability, and low cost give it some advantages over conventional piezoelectric or piezoresistive materials. A major question about its use is its reproduction.

A. Accelerometers

Due to its robustness, PVDF has found application as an accelerometer [33,34]. Piezoelectric polymer accelerometers have high sensitivity, low-frequency operation, large dynamic range, excellent linearity, high resonant frequency, low mechanical quality factors (Q_m), low transverse sensitivity, and small temperature dependence.

One accelerometer design consists of a PVDF film placed between metallic electrodes [33]. The lower electrode is attached to a substrate, while the top electrode is attached to a seismic mass. The mass is used to apply a force to the PVDF film as the mass is exposed to an acceleration. In turn, the force on the piezoelectric polymer film induces a prescribed voltage signature on the electrodes that is proportional to the acceleration. The time constant is close to a few tens of microseconds, which is an order of magnitude higher than the measured phenomenon. The range of acceleration that can be measured is controlled by the characteristics of the seismic mass. Care must be taken in bonding the electrodes and masses to avoid stripping the electrodes. "Smart accelerometers" have been designed by bonding a MOS transistor and printed circuit onto the accelerometer to provide a direct corresponding readout (shown in Fig. 23). Results show a linear response of the acceleration up to 50,000 g's.

AMP Corporation offers a family of various commercial accelerometers for applications ranging from general purpose, geoacoustics, machine health monitoring, and suspension control systems [34].

B. Strain Gauges

PVDF has recently found application as a dynamic strain gauge [35]. The distinct advantage of using a piezoelectric polymer, as opposed to the traditional wire and semiconductor resistance gauges, is that PVDF-based gauges are nonparametric transducers because their output is directly proportional to strain. This means that the electronic

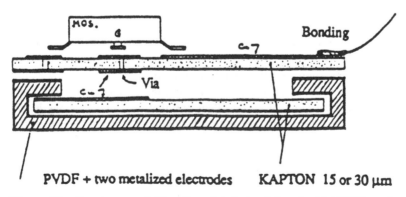

Figure 23 Smart sensor design of a printed circuit board with PVDF accelerometers. (Adapted from Ref. 33.)

components associated with them are considerably simpler. Additionally, the PVDF strain gauges are inherently more robust than the traditional gauges. Furthermore, piezoelectric polymer can be used to measure the strain of objects of almost any shape.

The error of PVDF-based strain gauges is somewhat larger (2%) than that of wire strain gauges; however, the dynamic range, sensitivity and frequency range are superior. An example of a two-layer piezoelectric polymer strain gauge is shown in Figure 24. The sensing element is a folded film with an even number of layers of dimensions 10 mm × 10 mm and a thickness of 300 μm (depending on the number of layers). For the two-layer configuration shown, the inner conductor of the cable is connected to the inner electrode while the shield is connected to the outer layer. The sensitivity of the gauge to strain (γ), in volts per unit strain, may be expressed as

$$\gamma = \frac{U}{S} = \frac{d_{31}E_{11}t}{\epsilon_{33}\epsilon_0} \tag{5}$$

where d_{31} is the piezoelectric charge constant, E_{11} is the induced electric field, t is the piezoelectric film thickness, ϵ_{33} is the permittivity of PVDF, and ϵ_0 is the permittivity of free space (8.8542×10^{-12} F/m). The entire assembly is wrapped with unmetalized polymer film or potted in a sealing compound. The capacitance of the gauge is between 600 and 800 pF. For strain measurement, the gauge is cemented to the vibrating structure with the polarization direction of the film oriented along the vibration. The output is corrected for cable capacitance, and the gauge can be calibrated by direct comparison to a standard strain gauge or measured directly by a laser interferometer after it is attached to the center of a rod that vibrates harmonically along its axis with frequency ω. The frequency range of the gauge is 20 Hz to 100 kHz. The dynamic range of the strain measurement is 10^{-7} to 10^{-2}, which is limited on the lower end by electrical noise and on the upper end by linearity of the film properties and the strength of the cemented joints. The PVDF gauge is operationally linked to a temperature range between -20 and $+70°$C. This limitation is because of vitrification of the film on the lower end and depolarization on the upper end. Measurements show that the sensitivity varied by less than 5% over the operating temperatures of 17–65°C.

Errors caused by strains in directions which differ from the measured axis are reduced by the anisotropy of the PVDF to approximately one order of magnitude less, although they are still distinct enough that they must be accounted for.

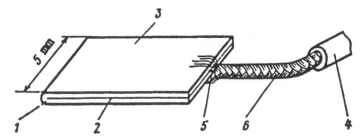

Figure 24 Cross-sectional view of two-layer PVDF strain gauge: (1) PVDF; (2) inner electrode; (3) outer electrode; (4) cable; (5) center conductor; (6) shield. (Adapted from Ref. 35.)

C. Stress and Shock Gauges

PVDF has been studied under uniaxial pressure and shock waves including dielectric hysteresis and d_{33} coefficient [36,37]. The results allow stress gauges to be developed using 35-μm PVDF. The delivered charge is proportional to the applied stress, so that the electric charge plotted as a function of time exhibits the shock wave profile.

Powder-gun plate impact experiments were used to measure the response of a different shock gauge between pressures of 10 and 40 kbar. Measurement of the details of stress pulses has been accomplished by means of Bauer PVDF stress gauges [38]. As shown in Figure 25, a PVDF gauge element is placed directly on the impact face of either a Z-cut sapphire or Z-cut quartz. The impactor projectile, which is shot out of a compressed-gas gun, is usually made from the same material. The target and impactor materials remain elastic up to about 20 and 13 GPa, respectively; the gauge elements are subjected to unusually well-behaved stress pulses. A major requirement for measuring shock compression loading is a high degree of reproduction in the material constants of the sensitive material used in the shock gauges. Gauges have been made from biaxially stretched or transversely stretched films which have been poled by the Bauer method after electroding with vapor-deposited gold [39]. The elements were 3.4 mm in diameter, and connected by leads which were 2 mm wide and 60 mm long by soldering to the plating. Powder-gun plate impact experiments were used to test the gauges between 10 and 40 kbar. The dynamic compressed loading behavior was investigated using step compressional loading waves and shock rarefaction waves induced by a thin projectile. The gauge was embedded in PMMA. The delivered charge of PVDF was seen to exhibit the shock profile in both cases. The ability to produce an undistorted signal and conse-

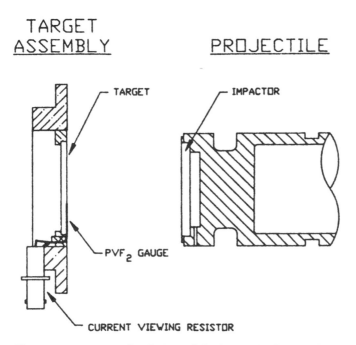

Figure 25 Cross-sectional view of the impact loading configuration showing the Bauer PVDF stress gauge. (Adapted from Ref. 38.)

quently, the maximum measurable stress was found to depend on the degree of poling of the PVDF. At shock pressures of 35 GPa, the response is highly distorted for a film with a polarization of about 8.0 μC/cm^2 but has good fidelity at a stress of 20 GPa. As the degree of polarization is lowered, the ability to detect high stresses without distortion of the pulses is increased. Sensitivity to lower shock pressures is increased with higher poling. The level of shock pressure must be considered in the design of a shock gauge.

D. Ultrasonic Resonators and Transducers

A PVDF transducer has been described which is used as an ultrasonic resonator for the purpose of measuring ultrasonic velocity and absorption in fluids [40]. The PVDF transducers were shown to be quite useful for this application because they can be made in large sizes of diameters near 100 mm. The large radiating surface avoids diffraction effects. Other transducer materials, particularly pure piezoelectric crystals such as quartz, cannot be used practically in such large sizes.

The system described operates in the frequency range from 100 kHz to the low megahertz range. The method supplements acoustic interferometer, pulse methods, and light-scattering techniques for measuring ultrasonic velocities. The apparatus is shown in Figure 26. The thickness of the PVDF film was 30 μm. The fundamental resonance frequency was estimated to be 30 MHz. The overall diameter of the film was 50 mm. The film was sandwiched between two adaptor rings and lightly pressed by a soft spring and a narrow Bakelite ring to prevent waves or wrinkles in the film. The adaptor ring was grounded and a hot lead attached to the other electrode. Two of these transducers were produced and separated in the cavity by 20 mm. Continuous electrical signals drive one of the transducers to excite the cavity into resonance. The amplitude of the sound waves is monitored by the other transducer. The interval between two resonances gives

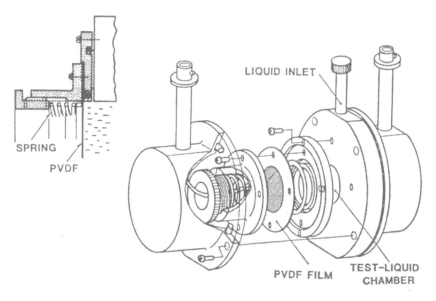

Figure 26 Cross-sectional view of ultrasonic resonator and PVDF transducer assembly. (Adapted from Ref. 40.)

the half-wavelength of sound, and therefore the phase velocity. The width of one of the resonances is related to the acoustic absorption in the liquid. The signal loss in the instrumentation can be calibrated out by comparing it with a standard liquid of known absorption. Tests of ultrasonic velocity and absorption in ethanol were made, and where diffraction effects were avoided, the data were in fair agreement with theoretical values.

Ultrasonic sensors with high resolution in the direction of sound propagation can be used as proximity sensors and distance measurements in the range of several meters. Poly(vinylidene fluoride) provides broadband characteristics which are of great interest due to their good pulse response, as is the case for medical applications.

A cylindrical transducer as well as an array of such transducers was made by wrapping PVDF around a cylindrical foam backing material [41]. The resonance frequency is nearly inversely proportional to the radius. These transducers have a radius of 2.6 mm which makes them suitable for operating over the frequency range of 40–200 kHz. Average values of the transducer sensitivity are 0.2–1 mV/Pa and 20–50 mPa/V in the transmitting mode at a distance of 1 m. Because of the short decay time of the acoustic pulse, PVDF transducers are well accommodated for pulse-echo measurements. For the transducer described here, the minimum measurement distance is about 30 mm. The transducer is easily transformed into an array by cutting segments of PVDF out from the cylindrical shell, which permits a beam to be steered by phase shifting each element. This accomplishes ultrasonic scanning within a limited solid angle.

E. Surface Acoustic Wave Sensors

A surface acoustic wave (SAW) sensor has been researched to detect surface forces and the direction of turbulent flow as a function of position and time on a structure [42]. The sensor is composed of two SAWs with identical center frequency and shear waves experiencing forces in opposite directions. The difference in SAW velocity is proportional to the shear stress associated with the turbulent flow. The direction of fluid flow is determined using an arrangement of three SAW sensors. On one of the interdigital transducers (IDT) mounted as an electrode pattern on one side of the PVDF sheet, no surface force is applied and the SAW velocity is used as a reference. The other two sensors experience both normal and shear stresses caused by the turbulent flow. Opposite ends of the other two sensors are fixed such that when they are subjected to shear and normal stresses, the velocity of one of the SAW increases whereas the velocity of the other sensor decreases. The difference between the mean velocities of the two sensors and the shielded sensor is related to the normal pressure, whereas the difference between the two sensors exposed to the flow is related to the shear stress associated with the flow. The direction of the turbulent flow can be determined by the arrangement of three pairs of SAW sensors in a manner similar to a strain rosette. Different sizes of SAW sensors would enable a wider range of spatial and temporal resolution.

The SAW sensor was composed of a 0.53-mm-thick PVDF film bonded to a Plexiglass substrate with silver IDT electrodes. The sensor is 152 mm × 76 mm × 19 mm, and each electrode of the IDT has 10 fingers. The SAW propagates in the stretched direction of the PVDF film. The center frequency of the IDT was 702 kHz in air with a velocity of 1180 m/s and a wavelength of 1.68 mm. The SAW velocity was measured as a function of normal stress and shear stress. The normal stress was determined by varying the depth of the sensor after the sensor was immersed in oil. The sensor showed a 1.5° phase shift for 60-Pa external pressure field up to a range of 2400 Pa.

The same SAW sensor was exposed to shear stress. The shear stress deforms the layers of the sensors in opposite directions. The layer surface of one sensor is compressed while the other is expanded. The direction of the phase shift was proportional to the magnitude of the shear stress, and the direction of the phase shift was opposite for the two types of shear stress. The phase-shift difference was linear with shear stress. There is 1° of phase shift for 10 Pa of each type of shear stress. The results suggest that the SAW detector is more sensitive to shear stress than to normal stress. Depending on the size of the sensor, it can be used to measure the space- and time-dependent surface forces as well as the direction of turbulent flow (locally and globally).

F. Fiber Optic Phase Modulator

A fiber optic phase modulator using a PVDF-VDF/TFE copolymer to jacket an optical fiber has been designed [43]. Figure 27 shows a schematic view of the configuration.

The copolymer, which does not require stretching prior to poling, was melt extruded onto a single-mode fiber. The fiber has a 4-μm core with an 80-μm cladding. The thickness of the copolymer coating was between 80 and 90 μm. It was annealed to enhanced a degree of crystallinity and poled using the corona technique. Electrodes were evaporated over the outside of the polymer and attached to a glass-epoxy substrate as shown in Figure 27. The optical phase shift was measured using a Mach-Zender interferometer with the optical fiber as one arm. The frequency response of the phase shift was measured and is presented in Figure 28. A flat response of 0.04 rad/V/m is observed and maintained between 10 kHz and 20 MHz. The axial resonance dominates the low-frequency response, while modal contamination from the fiber-jacket composite dominates the response above 2 MHz. The lowest-order radial resonance of the fiber jacket occurs at 7.3 MHz. Improvements in design should eliminate the resonances between the axially constrained region and the radial resonance.

VI. CONCLUSIONS

Ferroelectric polymers offer new avenues for transduction applications. The use of PVDF in fields such as underwater acoustics and sonar, biomedicine, robotics, and measurement instrumentation represent advanced capabilities of present-day standard transducers. Each

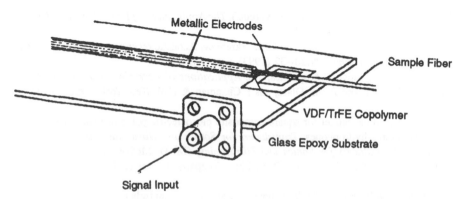

Figure 27   Cross-sectional view of fiber optic phase modulator featuring PVDF attached onto a glass epoxy substrate. (Adapted from Ref. 43.)

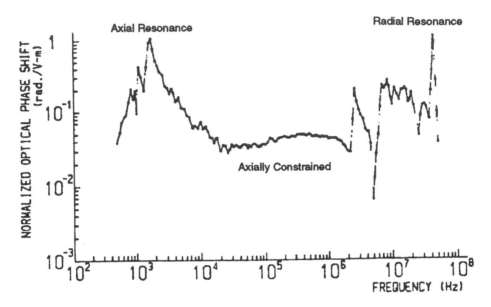

Figure 28 Measured phase shift per applied voltage of device shown in Figure 27, normalized with poled length, as a function of frequency. (Adapted from Ref. 43.)

of the aforementioned applications of this chapter have featured ferroelectric polymer as the active transduction material of choice because it offers the potential for increased capabilities and lower costs. It is believed that future devices will also emerge where the continued evolvement of ferroelectric polymer will also be enhanced.

In conclusion, the applications discussed in this chapter are representative of conceptual technologies from the late 1980s and early 1990s. Many future trends appear to be headed toward higher-frequency applications, composite smart structures, and multi-layer projectors for both high and low frequencies. With rapid advances in the material processing capabilities, new devices will follow.

REFERENCES

1. T. T. Wang, J. M. Herbert, and A. M. Glass (eds.), *The Applications of Ferroelectric Polymers*, Chapman & Hall, New York, 1988.

2. I. P. Golyamina, D. L. Rastorguez, and G. K. Skrebnev, Acoustic receivers using piezoelectric polymer films, *Acoust. Phys. 39*:30–32 (1993).

3. J. M. Powers, Long range hydrophones, *The Applications of Ferroelectric Polymers* (T. T. Wang, J. M. Herbert, and A. M. Glass, eds.), Chapman & Hall, New York, 1988, pp. 118–161.

4. D. Ricketts, Analytical model for square composite piezoelectric polymer plates, with application to large-area hydrophones, *Transducers for Sonics and Ultrasonics* (*Proc. Third Int. Workshop on Transducers for Sonics and Ultrasonics*) (M. D. McCollum, B. F. Hamonic, and O. B. Wilson, eds.), Technomic Publishing Company, Lancaster, PA, 1993, pp. 320–335.

5. R. H. Tancrell, D. T. Wilson, N. T. Dionesotes, and L. C. Kupferberg, PVDF piezoelectric polymer: Processing, properties and applications, *Transducers for Sonics and Ultrasonics* (*Proc. Third Int. Workshop on Transducers for Sonics and Ultrasonics*) (M. D. McCollum,

B. F. Hamonic, and O. B. Wilson, eds.), Technomic Publishing Company, Lancaster, PA, 1993, pp. 103–112.

6. ATOCHEM Sensors, Incorporated (now AMP Corporation), Piezoelectric copolymer hydrophone tile (hydrostatic mode), Product data sheet number 13, Rev. 4, Valley Forge, PA, 3/11/91.

7. ATOCHEM Sensors, Incorporated (now AMP Corporation), Piezoelectric copolymer hydrophone cylinders (hydrostatic mode), Product data sheet number 14, Valley Forge, PA, 4/15/91.

8. W. J. Hughes and C. W. Allen, A shaped PVDF hydrophone for producing low sidelobe patterns, *Proc. 1992 IEEE Symp. on Autonomous Underwater Vehicle Technology*, Ocean Engineering Society of IEEE, Washington, DC, 1992, pp. 219–223.

9. W. J. Hughes, Shaped polyvinylidene fluoride (PVDF) conformal arrays at ARL/PSU, ONR Conf. on Transducer Materials and Transducers, State College, PA, April 11–13, 1994.

10. D. Ricketts, Electroacoustic sensitivity of composite piezoelectric polymer cylinders, *J. Acoust. Soc. Am. 68*:1025–1029 (1980).

11. B. V. Smith, A theoretical model of a cylindrical PVF2 hydrophone, *Acustica 67*:62–65 (1988).

12. J. J. Bhat, T. T. Thomson, and P. R. Saseendran Pillai, Development of (3,1) drive low-frequency piezofilm hydrophones with improved sensitivity, *J. Acoust. Soc. Am. 94*:3053–3056 (1993).

13. P. Leaver, M. J. Cunningham, and B. E. Jones, Piezoelectric polymer pressure sensors, *Sensors and Actuators 12*:225–233 (1987).

14. B. Granz, PVDF hydrophone for the measurement of shock waves, *IEEE Trans. Elect. Insul. 24*:499–502 (1989).

15. D. R. Fox, A low-density extended acoustic sensor for low-frequency arrays, *IEEE J. Oceanic Eng. 13*:291–295 (1988).

16. Innovative Transducers Incorporated, ST-5 solid towed array, Product data sheet, Fort Worth, TX, 1994.

17. P. A. Lewin and M. E. Schafer, Design of piezoelectric polymer transducers for time delay spectrometry applications, *IEEE Ultrasonics Symp. Proc.*, 1987, pp. 721–724.

18. P. A. Lewin and M. E. Schafer, Piezoelectric polymer transducers for ultrasound dosimetry applications, *IEEE Ultrasonics Symp. Proc.*, 1986, pp. 515–518.

19. P. A. Lewin and M. E. Schafer, Wide-band piezoelectric polymer acoustic sources, *IEEE Trans. Ultrasonics, Ferroelectrics, and Frequency Control 35*:175–184 (1988).

20. M. Platte, A polyvinylidene fluoride needle hydrophone for ultrasonic applications, *Ultrasonics*, May 1985, pp. 113–118.

21. K. K. Shung and B. McGuire, Development of ultrasonically marked needle for ultrasonically guided biopsy, *Proc. 1989 IEEE Symp.*, 1989, pp. 119–120.

22. D. De Rossi, P. Dario, C. Marchesi, M. G. Trivella, F. Pedrini, C. Contini, A. S. De Reggi, S. Edelman, and S. Roth, Piezoelectric polymers as peripheral pulse sensors, *Proc. 1981 IEEE Symp. on Computers in Cardiology*, 1982, pp. 111–114.

23. S. G. Karr, T. Karwoski, J. E. Jacobs, and L. F. Mockros, Transducer system for the noninvasive recording of arterial pressure contours, *Ann. Biomed. Eng. 13*:425–442 (1985).

24. A. Starita, A. Battaglini, M. Bergamasco, and P. Dario, A sensorized platform-based, computerized system for posture rehabilitation, *Proc. Ninth Annual IEEE Conf. of the Engineering in Medicine and Biology Society*, 1987, pp. 1779–1780.

25. B. Granz and R. Oppelt, A two dimensional PVDF transducer matrix as a receiver in an ultrasonic transmission camera, *Acoustical Imaging, Volume 15* (H. W. Jones, ed.), Plenum Press, New York, 1986, pp. 213–225.

26. A. S. Fiorillo, P. Dario, and M. Bergamasco, A sensorized robot gripper, *Robot Sensors*, Elsevier, North Holland, 1988, pp. 49–55.

27. P. Dario, C. Domenici, R. Bardelli, D. De Rossi, and P. C. Pinotti, Piezoelectric polymers: New sensor materials for robotic applications, *Proc. 13th Int. Symp. on Industrial Robots and Robots*, Volume 2, Society of Manufacturing Engineers, Dearborn, MI, 1983, pp. 14-34 to 14-49.

28. D. G. Pirolo and E. S. Kolesar, Piezoelectric polymer tactile sensor arrays for robotics, *Proc. IEEE 1989 Natl. Aerospace and Electronics Conf. NAECON 1989*, Dayton, OH, 1989, pp. 1130–1135.

29. E. S. Kolesar, R. R. Reston, D. G. Ford, and R. C. Fitch, Jr., Multiplexed piezoelectric polymer tactile sensor, *J. Robotic Systems 9*:37–63 (1992).

30. T. Bailey and J. E. Hubbard, Jr., Distributed piezoelectric-polymer active vibration control of a cantilever beam, *AIAA J. Guidance and Control 8*:605–611 (1985).

31. C. K. Lee and F. C. Moon, Modal sensors/actuators, *ASME JAM 57*:434–411

32. R. L. Clark and C. R. Fuller, Control of sound radiation with adaptive structures, *J. Intell. Mater. Sys. and Struct. 2*:431–452 (1991).

33. B. Andre, J. Clot, E. Partouche, and J. J. Simonne, Thin film PVDF sensors applied to high acceleration measurements, *Sensors and Actuators A 33*:111–114 (1992).

34. ATOCHEM Sensors, Incorporated (now AMP Corporation), Accelerometers ACH-01, ACH-03-03, ACH-05-01S, ACH-06-03, ACH-LN-15, Product data sheet numbers 17, 23, 36, 38, 50, Valley Forge, PA, 3/91.

35. M. V. Belova, I. P. Golyamina, and D. L. Rastorguez, Piezoelectric-polymer strain gauges, *Pribory i Tekhnika Eksperimenta 6*:160–163 (1988).

36. F. Bauer, Piezoelectric and electric properties of PVF2 polymers under shock wave action: Application to shock transducers, *Shock Waves in Condensed Matter*, J. R. Asay, R. A. Graham, and G. K. Straub, eds.), Elsevier, Amsterdam, 1984, pp. 225–228.

37. F. Bauer, PVF2 polymers: Ferroelectric polarization and piezoelectric properties under dynamic pressure and shock wave action, *Ferroelectrics 49*:231–240 (1983).

38. R. A. Graham, L. M. Lee, and F. Bauer, Response of Bauer piezoelectric polymer stress gauges (PVDF) to shock loading, *Shock Waves in Condensed Matter* (S. C. Schmidt and N. C. Holmes, eds.), Elsevier, Amsterdam, 1987, pp. 619–622.

39. L. M. Lee, W. D. Williams, R. A. Graham, and F. Bauer, Studies of the Bauer piezoelectric polymer gauge (PVF2) under impact loading, *Shock Waves in Condensed Matter* (Y. M. Gupta, ed.), Plenum Press, New York, 1986, pp. 497–502.

40. P. K. Choi and K. Takagi, An attempt at ultrasonic resonator with piezoelectric polymer film, *J. Acoust. Soc. Jpn. 6*:15–19 (1985).

41. F. Harnish, N. Kroemer, and W. Manthey, Ultrasonic transducers with piezoelectric polymer foil, *Sensors and Actuators A 25-27*:549–552 (1991).

42. Y. Roh, Development of local global SAW sensors for measurement of wall shear stress in turbulent flows, Ph.D. thesis, The Pennsylvania State University, 1990.

43. M. Imai, T. Yano, Y. Ohtsuka, K. Motoi, and A. Odajima, Wide-frequency fiber-optic phase modulator using piezoelectric polymer coating, *IEEE Photonics Technol. Lett. 2*:727–729 (1990).

17
Ferroelectric Optical Memory

Munehiro Date
The Institute of Physical and Chemical Research, Saitama, Japan

I. INTRODUCTION

For mass computer memory, it is essential that either much information has to be stored in one material unit, or a large number of elements and selection circuits have to be combined to act as one element that can store much information. As examples of the former type, ultrasonic delay line, magnetic tape, and disk can be mentioned. As for the latter, we can refer to the magnetic core memory that is created by attaching 1-bit ferrite cores at the cross points of a woven insulated wire, and integrated circuit (IC) memory that is built up from two parts: microcapacitors, each storing 1 bit of information corresponding to existing or nonexisting electrostatic charge, with selection circuits incorporated.

While the delay line is designed to store a large amount of information in the time domain, the magnetic tape, the flexible and hard magnetic disks, and the magnetooptical disk [1] are intended to store more information in the space domain of a uniform medium.

Recently, experiments have been carried out to produce a mass memory device applying both time (frequency) and spatial domains. Photochemical hole burning (PHB) technique [2] can be mentioned as an example that is capable of recording multiple information at one point by recording different wavelengths at the same point. As this device has not yet achieved practical usage, the previously designed optical type of memory is the most effective device for mass storage that is available now. It is particularly suitable for application as mass memory because it is capable of dividing the disk space by specifying the light wavelength order.

We can divide currently used optical memories into three groups: "read only" type (RO), "write only" type (WO), and "rewritable" type (RW). The RW type can be further divided into "write after erase" type and "overwrite" type. Compact disks (CD) and laser disks (LD) are typical examples of the RO type, the group that uses dye [3] or other type of pitting [4] is the WO type, while the amorphous-to-crystalline phase-change type [5], the magnetooptical type, and the ferroelectrooptical type we are going to discuss are RW-type memories.

II. THE PRINCIPLE OF THE FERROELECTRIC OPTICAL MEMORY

To construct a multibit memory using ferroelectrics, the polarity of polarization at a selected point on a medium has to be controlled. This can be achieved by two methods, using either a multielectrode or a single electrode. The former requires a matrix array of electrodes as is used in IC memory [6]. In the latter, a point on a fully electroded medium is addressed by irradiating with a focused laser beam. To construct a ferroelectrooptical memory, the latter method was adopted. Ferroelectrooptical memory is a device that records binary information corresponding to the polarization polarity appearing at each point of a thin-film medium. In terms of practical realization, it is necessary to provide methods for arbitrarily setting the polarization polarity of a minute area without affecting the polarization of other areas (writing rule) and for reading the polarity of each minute area (reading rule). In order to achieve this purpose, we can use the facts that the spontaneous polarization of ferroelectrics reverses only when an inverse field is applied over the coercive electric field, the polarization is stable under a field lower than the coercive field, the coercive field decreases with increasing temperature, and the temperature increases without any effect on the other parts outside the given minute area. The first two facts are called the hysteresis property. When increasing the temperature locally at a point of a voltage lower than the one corresponding to the coercive field, the coercive field will decrease only at the given point and the voltage of that point will be higher than the voltage of the coercive field. Consequently, the polarization at the given point will be reversed. Figure 1 illustrates the writing mechanism applied by heating. The applied electric field of reverse direction to the polarization is constant during the writing. By increasing the temperature starting from point A, the coercive field (E_c) decreases under the applied field, and because of this the polarization first decreases and then inverses toward counterpolarity from point A to point B. From point B to point C, by decreasing the temperature, the polarization keeps the new direction, because it is the same as the electric field's direction. Finally, at point C the coercive field regains its initial magnitude and the polarization keeps holding its direction.

A. Writing Method

Considering the writing method applied for the memory device, "write after erase" type and "overwrite" type devices can be distinguished. In the case of "write after erase" type, writing and erasing operations are separated. First there is an erasing step resulting in a polarization of uniform direction on the whole device. After the erasing step, either reversing or skipping corresponding to information 1 and 0 under a constant field is imposed during the writing step. It is important for this type to be capable of verifying the polarization change simultaneously by monitoring the current in writing time due to applying constant field. The total writing time is twice as long as the basic writing time in this case, because of the erasing operation applied before writing. The "overwrite" type sets either plus or minus polarization polarity corresponding to information 0 and 1 by changing the polarity of the applied field independent of the information previously

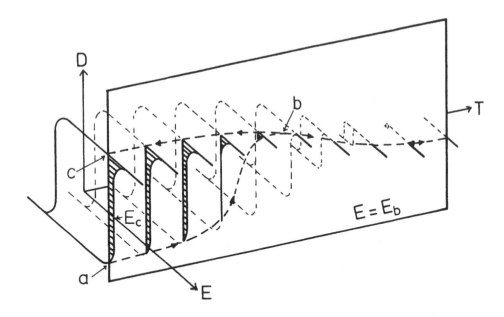

Figure 1 Principle of thermal writing.

written. The total writing time is shorter in this case, because an erasing operation before the writing is not necessary. However, in this case, it is impossible to verify simultaneously by current under the writing step because the reversed or unreversed polarization is related to both the present and the previous state and because the induced current is overlapped by the field change.

B. Reading Method

In the case of magnetooptical memory, we can detect the direction of the magnetization polarity by using the fact that the polarity of the optical polarizing angle of the light varies by changing the magnetization polarity when it passes the magnetic layer. This is called the Faraday effect. However, in the case of vinylidene fluoride and its copolymers, the electric polarization does not affect the optical polarization plane axis. Furthermore, for these polymers, the rotation axis of the polarization and the anisotropy axis of the dielectric constant are parallel, and the polarization reverse does not show any change in optical characteristics. Therefore the optical method cannot be applied in our case. However, it is possible to read the polarization polarity electrically by using the fact that the sign of the pyroelectric constant is related to the sign of the polarization. The polarity of the polarization can be determined from the sign of the detected pyroelectric current generated by heating. By setting electrodes onto the whole surface, it is also possible to read the polarization polarity of the divided part separately by heating it locally by applying irradiating focused light. In Figure 2 we present a read–write cycle explaining the states occurring on the different memory types while carrying out these processes.

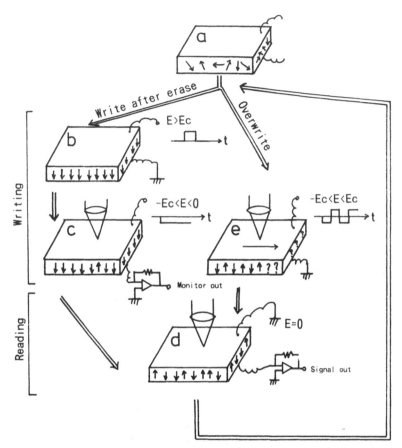

Figure 2 Schematic diagram of the read–write method: (a) initial state showing random polarity; (b) erasing for "write after erase" type, resulting in unidirectional polarity; (c) writing for "write after erase" type, monitoring polarization reversal; (d) reading; (e) writing for "overwrite" type.

C. The Relation Between the Characteristics of the Optical Memory and the Physical Properties

In this section we present a review of the physical parameter dependence of the optical memory characteristics. For memory devices, the most important characteristics can be listed as recording density, writing speed, reading speed, sensitivity, signal-to-noise (S/N) ratio, information retaining property, rewritability, etc.

1. *Recording density.*   Recording density can be determined as the maximum of the following values: independent polarization area, beam diameter of the laser light, area of the polarization reversal generated by laser light, reading area. (All the above-mentioned characteristics are minimum values.)

2. *Independent polarization area used as memory unit.*   This characteristic can be defined as the minimum independent polarization retaining area that is always greater than or equal to the spontaneous polarization domain. If one crystal represents one

domain, the independent polarization area is equal to the crystal size. In the usual case the crystal size is smaller than 1 μm.

3. *Beam diameter.* The beam diameter can be defined as the shortest radius of the laser light, which is nearly equal to the laser light wavelength, because the Airy disk radius defined as including 83.8% of the whole laser light power can be written as 0.6 λ/NA (where λ is the wavelength and NA is the numerical aperture).

4. *Reversal area.* Polarization reversal can be generated by increasing the temperature above the critical temperature by irradiating light. Even tough the light irradiates a given area uniformly, the temperature is lower on the border because of the lateral heat diffusion. Accordingly, due to the fact that we can set the temperature above the critical temperature only at the center of the area, the reversal area may become smaller than the beam diameter.

5. *Reading area.* The lower limit of the reading area can be determined from the light beam diameter. However, it is usually different from this, because by applying pyroelectric reading technique, the output signal is proportional to the value obtained by the multiplication of the amount of temperature change and the heating area. Due to the fact that the temperature change is limited by the temperature at which the polarization information volatilizes, an area larger than the limit value determined in the previous paragraphs has to be applied in order to keep the S/N ratio as high as is required. This will be explained in detail in the paragraph describing the S/N ratio.

6. *The read/write speed.* The writing speed, like the scanning speed, is limited by the time that is needed to heat one point up to the necessary temperature and by the polarization reversal time. For "write after erase" type memory the real polarization reversal of the writing area is allowed to delay during the writing of the next area, because constant field voltage has to be applied during the total writing time. On the contrary, for "overwrite" type memory, we cannot write in the next point before the reversal at the current point has been completed, because a unique field must be provided for each point. Therefore the writing can be carried out in a shorter time in the case of "write after erase" type than in the case of "overwrite" type.

7. *Signal-to-noise ratio.* By applying the pyroelectric readout method, the output signal can be calculated as

$$V_s = \frac{\rho C_v p_y S \, \Delta T}{C} \tag{1}$$

where V_s is the output voltage, ρ is the density, C_v is the heat capacity, p_y is the pyroelectric constant, S is the area storing 1 bit of information, ΔT is the temperature difference generated by light irradiation, and C is the electrode capacity.

The noise signal for FET input can be written as

$$V_n = \sqrt{\left[4kT\left(R_s + \frac{1}{g_m}\right) \Delta f \right]} \tag{2}$$

where V_s is the noise voltage, k is Boltzmann's constant, T is the temperature, R_s is the source resistance, g_m is the mutual conductance of FET of the top stage, and Δf is the frequency bandwidth. As is obvious from this equation, if we apply smaller per bit area and larger electrode area in order to store more information in one electrode, a smaller voltage value will result and due to this the signal-to-noise ratio will worsen. By applying thicker film, the S/N ratio (V_s/V_n) becomes better because the capacity increases; however, it worsens the resolution by diffusing heat spots.

III. MEMORY CHARACTERISTICS OF VINYLIDENE FLUORIDE TRIFLUOROETHYLENE COPOLYMERS

A. Polarization Reversal

Vinylidene fluoride trifluoroethylene (VDF/TrFE) copolymer, as the only stable ferroelectric polymer, appears to be the most suitable material to produce optical disks, because it is easier to provide thin films of polymers than ferroelectric ceramics. In the former case no elongation process for making ferroelectric β-form structure is necessary, as opposed to the case of poly(vinylidene fluoride) (PVDF). Figure 3 depicts the molecular structure of VDF/TrFE copolymer (a) and PVDF (b). The changes of the spontaneous polarization of poly(vinylidene fluoride) and its copolymers are caused by the rotation of the CF_2 functional groups. Figure 4 shows how the polarization, the dielectric constant, and the infrared dichroism according to the CF_2 symmetric stretching mode vibration vary versus the electric field change in PVDF [7]. At the point where the polarization shows a sharp variance versus the field change, the dichroism presents a maximum. From this dichroism change, it is obvious that the CF_2 dipole is parallel to the electric field at high electric field areas, and it is perpendicular to the electric field direction during the polarization change. The dielectric constant increases when the dipole becomes perpen-

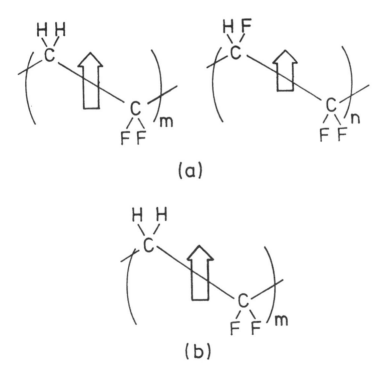

Figure 3   Molecular structure of VDF/TrFE copolymer (a) and PVDF (b).

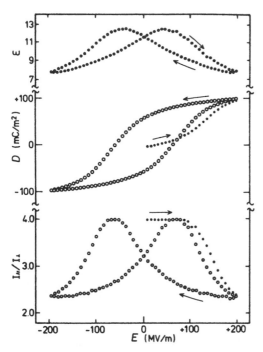

Figure 4 Hysteresis curves at −5°C for the linear dielectric constant ε, the electric displacement D, and the ratio of IR transmitted intensities at 1273 cm^{-1} polarized parallel ($I_{\parallel}$) and perpendicular ($I_{\perp}$) to the stretch direction of the PVDF film when a triangular electric field with a period of 1000 s was cyclically applied. The closed circles show the data when the electric field is initially applied and the open circles depict the data for the tenth cycle of the applied field. (From Ref. 7.)

dicular to the electric field, and decreases when it appears to be parallel. We can easily understand the reason for this change in the dielectric constant if we consider that the change in the electric field-directed component of the polarization that was carried out by the rotation of the dipole perpendicular to the electric field is more significant than the change in the dipole moment itself that is parallel to the electric field.

A typical hysteresis loop of VDF/TrFE (65/35) copolymer is shown in Figure 5, where P_r is the remnant polarization and E_c is the coercive field. The rotation of the dipole created by applying a reverse electric field has a time delay. The polarization response in the case of applying a step electric field is shown in Figure 6 [8]. The time corresponding to the maximum response gradient is called switching time, t_s. Figure 7 presents a decreasing tendency in the switching time by increasing temperature. The switching time can be approximated as inversely proportional to the electric field, and the speed limit can be estimated to be less than 10 ns in order on the basis of Figure 8.

The thinner the ferroelectric layer is, the better it can fulfill the thermal diffusion. If it becomes thinner than 0.1 μm in order, the hysteresis characteristic will worsen [9]. The polarization reverses by applying reverse electric field over the coercive electric field. Figure 9 shows the temperature dependence of the pyroelectric constant P_y of 52/48 copolymer under a bias field [10]. The decrease of the coercive field with increasing

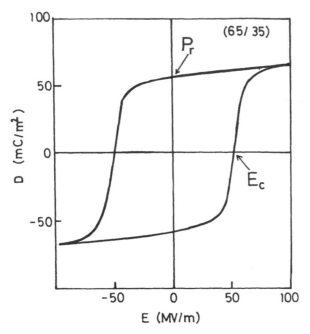

Figure 5 Typical hysteresis loop of VDF/TrFE (65/35) copolymer: P_r, remnant polarization; E_c, coercive field.

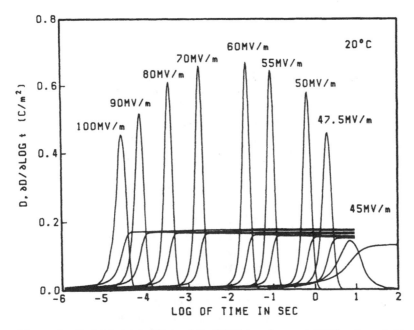

Figure 6 Switching transitions of the 35/65 copolymer at 20°C under a field of 45–100 MV/m. (From Ref. 8.)

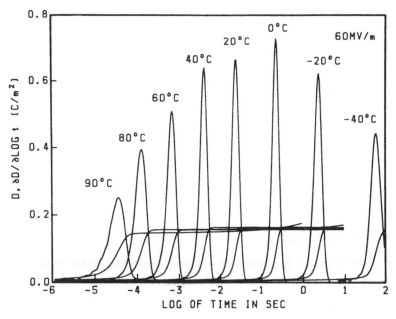

Figure 7 Switching transient of the 65/35 copolymer under a field of 60 MV/m at temperatures between −40 and 90°C. (From Ref. 8.)

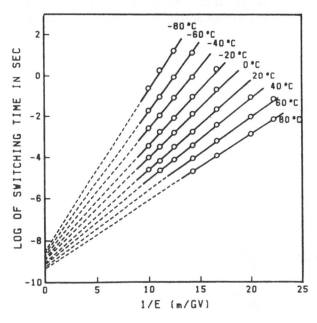

Figure 8 Plots of switching times against the reciprocal of applied field in logarithmic scale. (From Ref. 8.)

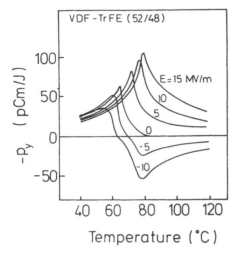

Figure 9 Temperature dependence of the pyroelectric constant P_y under bias electric field. (From Ref. 10.)

temperature can be explained as the temperature increases at the zero cross points of the pyroelectric constant, such as at the polarization-disappearing points during the reversing process, by decreasing the applied inverse electric field.

The temperature dependence curves of the electric field on constant reversal time taken as a parameter are plotted in Figure 10. The writing reverse electric field is necessary to fulfill two conditions. First, it should not cause a change in the polarization of any unselected area; on the other hand, the polarization reversal has to be completed at the heating point in a time necessary to the writing access at one point. The two areas

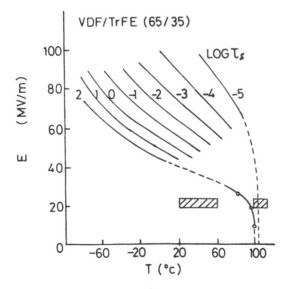

Figure 10 Equipolarization switching time curves on E–T map.

filled by hatching show the parts of fulfilled condition on heating and nonheating state in this figure. The switching time is reasonably fast near the Curie point, even in relatively low fields [11]. The VDF fraction that is suitable for memory is located where the Curie point is higher than the room temperature and lower than the melting point. This sufficient region of VDF fraction from 65 to 75 is shown in Figure 11. Due to the fact that the pyroelectric response time is shorter than 0.1 μs [12], the readout speed appears to be high enough not to be concerned about.

B. Bit Density

If the number of bits stored in one electrode is limited by the signal-to-noise ratio, this value can be estimated by applying Eqs. (1) and (2), assuming that $T = 300$ K, $\Delta f = 1$ MHz, $g_m = 100$ mS, $\rho = 1.8$ g/cm^3, $C_v = 2.4$ J/g-K, $\epsilon = 10$, $p_y = 20$ pC-m/J, $\Delta T = 20°$C, and $d = 1$ μm, total bits per electrode of about 100 million bits can be obtained for $S/N = 1$. However, it has to keep in mind that this has no relations to the electrode area.

C. Dye Doping Effect

In order to reduce the reading and writing laser power and increase the read/write density, it is better to absorb the light into the ferroelectric layer directly, rather than transfer it into heat at the electrode. For this purpose, one side of the electrode is created transparent. Furthermore, we can dope dye into it in order to make the ferroelectric layer more suitable for light absorption. In general, there is a tendency that adding a material such as a dye increases the DC conduction. Because the worsening of the dielectric hysteresis property caused by DC conduction has to be avoided, a nonionic, nonconductive dye is necessary. Figure 12 shows the structure of naphthalocyanine dye that was selected as a proper absorbent for 780-nm laser light and which has low DC conduction property. The absorption characteristic of this dye is shown in Figure 13. Dispersing it in polycarbonate (PC) and vinylchloride/vinylacetate (VC/VAc), a sharp absorption similar to the one that occurs in an organic solvent is shown. However, if we disperse it into VDF/TrFE co-

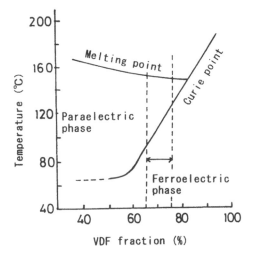

Figure 11 Optimum VDF fraction for memory in phase diagram of VDF/TrFE copolymer.

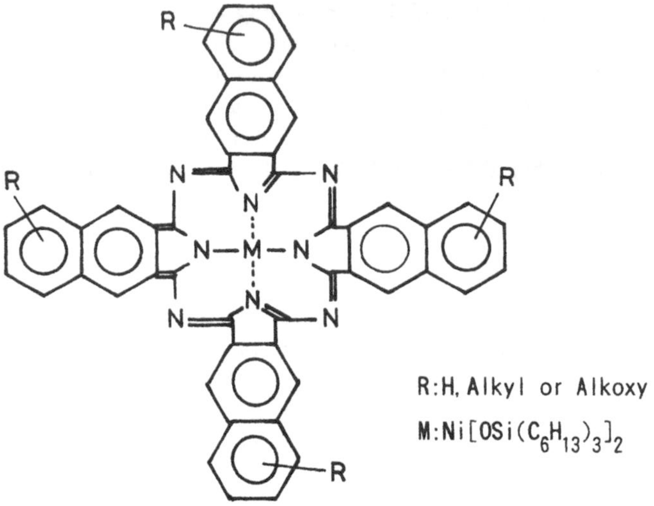

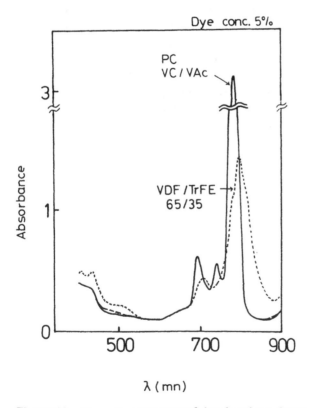

R
N
N—M—N
R:H, Alkyl or Alkoxy
M:Ni[OSi(C6H13)3]2

Figure 12 Molecular structure of naphthalocyanine dye.

Dye conc. 5%
PC
VC / VAc
VDF / TrFE
65/35
Absorbance
3
1
0
500
700
900
λ (mn)

Figure 13 Absorption spectrum of dye-doped copolymer.

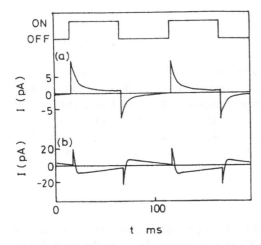

Figure 14 Pyroelectric response of copolymer irradiated from ITO side: (a) containing 1% dye; (b) containing 5% dye. (From Ref. 13.)

polymer, a broad absorption peak is obtained. This case shows that some interaction occurs between the dye and the polymer. The dye doping effect not only increases the light–heat conversion efficiency, it also abnormally changes the time response of the pyroelectric current [13]. Figure 14 depicts the difference in the pyroelectric response between the two cases of adding 5% dye and 1% dye. It was observed that if the dye concentration is larger than 3%, the sign of the pyroelectric current will be inverted after a long period.

IV. PRACTICAL REALIZATION AS AN OPTICAL DISK

A. Medium Structure

Figure 15 illustrates the structure of a medium applied for evaluating memory property through the following experiment [14]. An indium tin oxide (ITO) transparent electrode is set on a glass plate, then 1- to 2-μm-thick VDF/TrFE (65/35) copolymer is coated by the spin-coat method on it from tetrahydrofuran (THF) solution. It is annealed at 145°C for an hour and then an A1 electrode is deposited on the plate.

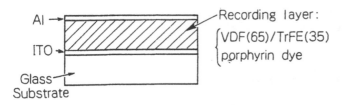

Figure 15 Sample structure. (From Ref. 14.)

B. Measurement System

Figure 16 is a schematic diagram of the measurement system developed to test the writing and reading procedure under the control of a microcomputer system. The reading of the polarization pattern is accomplished by using two modes, one static and another dynamic. The static mode uses a chopped laser beam to evaluate the value of polarization at each point including its sign, though phase-sensitive detection of the pyroelectric current. The dynamic mode is implemented by scanning a continuous laser beam over the medium at a given speed. The resulting pyroelectric signal reflects the spatial change of the polarization. This dynamic characteristic is more intimately related to the practical specification of memory application.

C. Experimental Results

In Figure 17 the stability of polarization of 65/35 copolymer versus the exposing time is illustrated at high temperature. Under 64°C, no polarization decay is observed. Figure 18 shows the stability of polarization against an inverse pulse field of 20 ms. No change can be obtained in the polarization under 51% of the coercive field.

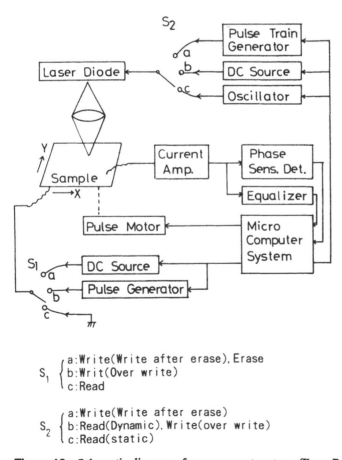

$$S_1 \begin{cases} a:\text{Write(Write after erase), Erase} \\ b:\text{Writ(Over write)} \\ c:\text{Read} \end{cases}$$

$$S_2 \begin{cases} a:\text{Write(Write after erase)} \\ b:\text{Read(Dynamic), Write(over write)} \\ c:\text{Read(static)} \end{cases}$$

Figure 16 Schematic diagram of measurement system. (From Ref. 14.)

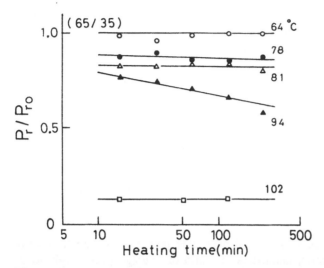

Figure 17 Temperature dependence of the polarization of VDF/TrFE (65/35) copolymer, measured by the pyroelectric measurement at room temperature after an applied high-temperature period.

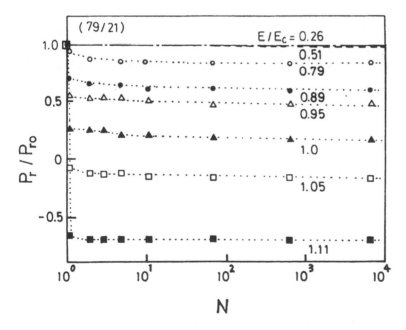

Figure 18 Stability of the polarization against inverse pulse field of 20 ms pulse width.

In Figure 19 the read-by-static-mode polarization pattern in the sample is presented. The height of the pattern corresponds to the value of polarization, while the flat level corresponds to the nonelectrode portion. The initially polarized region is indicated at the bottom part, and the locations where the polarization was reversed during the writing process are represented by the peaks. Each pattern is obtained as a result of the following procedure.

1. A field of 100 MV/m is applied to initialize the memory.
2. A single line is written in term of a sequence of points 5 μm in diameter, using 10-mW laser irradiation for 50 μs in the 20-MV/m reverse field.
3. A field of 100 MV/m is applied to erase the previous pattern.
4. Ten lines are written at varying laser power levels, from 11 mW (left side) to 2 mW (right side) in 1-mW steps.

From these figures, it can be found that 4-mW laser power is enough to achieve reverse polarization. The S/N ratio, defined as the ratio of average peak height to its fluctuation in the pyroelectric signal, is about 30 dB under optimal condition.

The optimum writing conditions of laser power, pulse width, and thickness of ferroelectric layer are shown in the next two figures. Figure 20 shows the dependence of writing ability on the writing laser pulse width monitored by readout signal under a bias field of 25 MV/m, reading power 2 mW, beam diameter 5 μm, and writing power 4, 8, 10, and 12 mW. By applying 10-mW writing power, the writing can be carried out within 100 μs up to the saturation level. In order to decrease the writing time to the 10-μs level, power higher than 12 mW is necessary. This curve corresponds to the polarization switching under low field at high temperature. Figure 21 illustrates the recording power dependence of the reading output level under a bias field of 25 MV/m, recording speed 50 mm/s, and recording frequency 500 Hz. The recording power decreases with decreasing film thickness (d).

Figure 22 depicts the output signal and frequency spectrum that is obtained by dynamic-mode measurement of data consisting of a regular repetition of 0/1 states at a pitch of 20 μm, under the conditions that the laser power is 12 mW and the field strength

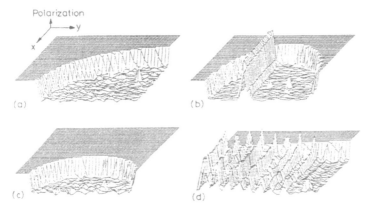

Figure 19 Memory patterns of erased and written states: (a) after erasing; (b) first writing; (c) re-erasing; (d) rewriting.

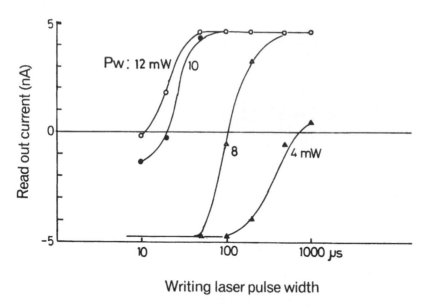

Figure 20 Monitoring the writing level could be achieved by applying different laser pulse widths. Read–write conditions: bias field, −25 MV/m; reading power, 2 mW, beam diameter 5 μm.

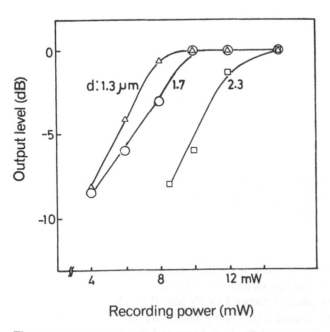

Figure 21 Monitoring the writing level could be achieved by applying different recording powers. (d) Thickness of the ferroelectric layer. Read–write conditions: bias field, −25 MV/m; writing speed, 50 mm/s; writing frequency, 500 Hz; reading speed, 200 mm/s; reading power, 2 mW; beam diameter, 5 μm.

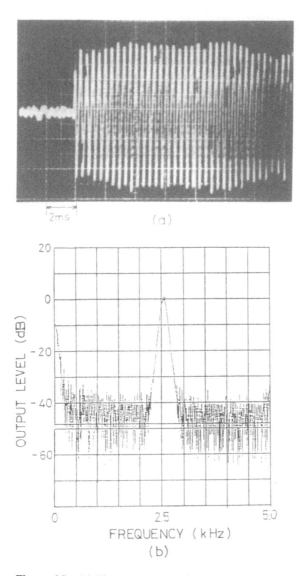

Figure 22 (a) The output signal from dynamic reading. (b) Its frequency spectrum. Read–write conditions: bias field, − 25 MV/m; writing pitch, 20 μm; writing power, 12 mW; reading speed, 100 mm/s; reading power, 2 mW; beam diameter, 5 μm. (From Ref. 14.)

is 25 MV/m. The carrier-to-noise (C/N) ratio of 48 dB can be obtained at a reading speed of 100 mm/s. This value shows that this medium may be suitable for application as an optical disk memory. When the pitch is reduced to 10 μm, the C/N ratio decreases to about 30 dB. We assume that this decrease is caused primarily by reduction of the reading signal, due to the fact that the diameter of the laser beam and the data pitch are nearly equal.

The experimental results described above confirm that the ferroelectric polymer can be applied for read–write memories. Its simple configuration and stainless property in humid conditions make it especially suitable for the mass production of low-cost erasable mass memories.

REFERENCES

1. H. J. Williams, R. C. Sherwood, F. G. Foster, and E. M. Kelly, Magnetic writing on thin films of MnBi, *J. Appl. Phys. 28*:1181 (1957).
2. A. A. Gorokhovskii, R. K. Kaarli, and L. A. Rebane, Hole burning in the contour of a pure electronic line in a Shpolskii system, *JETP Lett. 20*:216 (1974).
3. C. O. Carlson, Helium-neon laser thermal high recording, *154*:1550 (1966).
4. D. Maydan, Micromachining and image recording on thin films by laser beams, *Bell Syst. Tech. J. 50*:1761 (1971).
5. J. Feinleib, J. de Neufville, S. C. Moss, and S. R. Ovshinsky, The amorphous to crystalline phase change, *Appl. Phys. Lett. 18*:254 (1971).
6. S. S. Eaton, D. B. Butler, M. Parris, D. Wilson, and H. McNeillie, A ferroelectric nonvolatile memory, *IEEE Int. Solid-State Circuit Conf.*, 1988, p. 130.
7. T. Takahashi, M. Date, and E. Fukada, Dielectric hysteresis and rotation of dipoles in polyvinylidene fluoride, *Appl. Phys. Lett. 37*:791 (1980).
8. T. Furukawa, M. Date, M. Ohuchi, and A. Chiba, Ferroelectric switching characteristics in a copolymer of vinylidene fluoride and trifluoroethylene, *J. Appl. Phys. 56*:1481 (1984).
9. K. Kimura and H. Ohhigashi, Ferroelectric properties of poly(vinylidenefluoride-trifluoroethylene) copolymer thin film, *Appl. Phys. Lett. 43*:834 (1983).
10. M. Date, Effect of electric field on the phase transition in vinylidene fluoride trifluoroethylene copolymers, *IEEE Trans. Electr. Insul. 21*:539 (1986).
11. Y. Tajitsu, T. Masuda, and T. Furukawa, Switching phenomenon in vinylidene fluoride/trifluoroethylene copolymers near the Curie point, *Jpn. J. Appl. Phys. 26*:1749 (1987).
12. R. G. Kepler and R. A. Anderson, On the origin of pyroelectricity in polyvinylidene fluoride, *J. Appl. Phys. 49*:4918 (1978).
13. M. Date, T. Furukawa, M. Kutani, A. Katashima, Y. Uematsu, and T. Yamaguchi, Effect of dye doping on photo-induced pyroelectric response in thin films of vinylidene fluoride/trifluoroethylene copolymers, *IEEE Trans. Electr. Insul. 27*:777 (1992).
14. M. Date, T. Furukawa, T. Yamaguchi, A. Kojima, and I. Shibata, Opto-ferroelectric memories using vinylidene fluoride and trifluoroethylene copolymers, *IEEE Trans. Electr. Insul. 24*:537 (1989).

18
Biomedical and Robotic Applications of Ferroelectric Polymers

Danilo De Rossi and Elisa Stussi
Centro "E. Piaggio," University of Pisa, Pisa, Italy

Claudio Domenici
C.N.R. Institute of Clinical Physiology, Pisa, Italy

I. INTRODUCTION

Sensing devices and actuators in biomedicine and robotics are a highly specific and peculiar, but very diversified field of application for ferroelectric polymers.

Biomedicine has been relying increasingly on engineering methods and computer-driven instruments both for therapeutics and diagnostics. Sophisticated machines are progressively substituting, where possible, for the classical figure of the physician examining patients with the help of only his senses and experience.

In robotics and teleoperation as well, especially with the realization of tactile sensors as well as micro- and nonmicroactuators and manipulators, the increasing difficulty of the tasks to be performed stimulated the search for new materials and technologies for both sensing and actuation.

In spite of the major progress in the field of microelectronics, which allowed the implementation of powerful electronic processing units and memory and display systems, several limitations in accuracy and reliability still exist for the measuring device as a whole. The transducer often represents the ultimate limit to the performance of the instrument, since the interface between the external environment and the device remains the most delicate point of the sensing (or transduction) chain.

The discovery of ferroelectric properties in several polymers presented an excellent alternative to inorganic ferroelectric materials, which, however, are still used in most applications.

Present research on physical properties, performance, and possible applications of electroactive polymers has concentrated on conducting polymers because of their peculiar electrical properties, but the interest in polymers originally arose from the dielectric ones. The strong piezoelectric effect in poly(vinylidene fluoride) (PVDF or PVF_2) poled under high electric field was observed for the first time by Kawai [1] in 1969, while its pyroelectric properties were discovered in 1971 by Bergman et al. [2] and Nakamura and Wada [3]. Since then, a large variety of new devices has been developed, based on the piezo and pyroelectric effect of such polymers, which often offer better performance in comparison with the corresponding inorganic ones [4]. The peculiar properties of ferroelectric polymer materials found in health care a potentially wide field of application. In fact, the characteristics of excellent coupling in terms of mechanical and acoustical impedance, very large bandwidth, lightness, low cost, and easy conformability to irregular shapes with large transducing areas make polymers interesting materials for the development of medical devices coupled with living tissues and organs. Electromechanical devices making use of the piezoelectric effect could benefit by the availability of flexible membranes, foils, and bars, vibrating at high frequency with relatively high amplitudes or sensing dynamic physiological parameters of thermal or mechanical origin with an intrinsically wide bandwidth, ranging from subhertz to ultrasonic frequencies. The majority of ferroelectric polymer features can also be exploited successfully in robotic applications, where the versatility and excellent electromechanical properties of these materials allow new design solutions.

A large variety of devices for biomedical and robotic applications has been conceived and implemented based on the direct and inverse piezoelectric and pyroelectric effects of ferroelectric polymers. These applications will be illustrated briefly in the following paragraphs, under classification into four groups: sensors, actuators, diagnostic imaging devices, and charge generators.

As a matter of fact, the application of ferroelectric polymers in medicine and robotics has often been limited to the laboratory development stage, and it remains mostly confined to a restricted group of users, leading in just a few cases to an industrial production stage.

Increasing interest is now being given to composite materials [5], while research activity on biomedical and robotic applications of ferroelectric polymers has been slightly decreasing during these last years, as can be seen from the amount of published work.

II. REQUIRED MATERIAL CHARACTERISTICS FOR BIOMEDICAL APPLICATIONS

Different types of artificial materials are used in biomedical applications, their characteristics depending on the kind of function to be performed, on the level of interaction with the biological system, and on its duration in service.

In medical applications, the interaction between material and human body must be taken into account not only in terms of physical transduction properties, but also considering the mutual influence between the biological environment and extraneous material. In fact, if on one hand the harmfulness of the material to be used must be taken into account (contamination, poisoning, low biocompatibility), on the other hand, human body self-defense can be very aggressive.

Sterilization is usually mandatory also for those devices which are external to the human body and intended only for short-term contact.

Biocompatibility with body tissues is often due to high inertness of the introduced materials; even more pressing requirements are needed in the case of blood contact, where problems of coagulation and thrombus formation can arise.

The water absorption of PVDF, measured following ASTM DS70, is extremely low (less than 0.04% by weight). The homopolymer of PVDF is considered nontoxic, and may be used for repeated contact with food (e.g., packaging) in accordance with extraction and washing conditions [6]. The Solvay material PVDF satisfies the Dutch regulations for plastic materials in contact with food. No adversal toxicological or biological response from PVDF when ingested by, inhaled by, or implanted in test animals under controlled conditions has been reported.

Extensive studies of blood–material and tissue–material interactions are not available in the literature, but Mason et al. [7] reported PVDF to have "intermediate" characteristics of blood compatibility as determined through an in vitro test system. Irreversible loss of remanent polarization occurs above 70°C, with a quasi linear decrease in residual pyro- and piezoelectric response approximating 1%/°C [8]. This must be taken into account in device sterilization, which can be achieved by exposure to ethylene oxide, or by irradiation with an electron beam from a radioactive source, with a dose of the order of 2.5 Mrad. PVDF films retain their polarization fully below 2.5-Mrad irradiation from a ^{60}Co source [9]. According to data provided by Dynamit Nobel, PVDF shows no appreciable chemical attack from ethylene oxide up to a temperature of 100°C.

For the incorporation of piezoelectric films in devices, suitable electric connections are established with the film electrodes using silver-filled epoxy or rubber cements, pressure contacts, or other nonsoldering techniques.

Because of the high electrical impedance of piezoelectric materials, electrical insulation is critical, particularly when flexible incapsulants are needed for implantable devices. Although synthetic plastic and rubber encapsulants tend to be permeable to the electrolytes of body fluids, animal waxes such as beewax and spermaceti have proved to be highly effective, although little used. Enger and Simeone [10] reported no loss of electrical output in a piezoelectric ceramic bimorph encapsulated in beeswax after 6 months of immersion in a salt solution.

Satisfactory impermeability has been coupled with excellent mechanical flexibility by soaking sheets of medical-grade silicone rubber with wax [11]. High fatigue resistance and very low fluid permeability have been reported for flexible vapor-deposited thin carbon films. Strain to fracture has been measured in excess of 5%, and no fatigue failure under cyclic loading has been found for low-temperature-pyrolitic carbons [12], which are among the most biocompatible materials presently known.

Preliminary experiments suggest that an adherent vacuum-sputtered carbon film can be deposited on gold-electroded PVDF elements at a temperature below the depolarization temperature. Thus, the combined goals of efficient electrical insulation, impermeability to blood fluids, flexibility and fatigue resistance, and minimum interaction with surroundings cells may be achieved for implantable devices.

III. SENSORS

Biomedical applications of ferroelectric polymers are focused mainly on the use of their electromechanical transduction properties for the implementation of various kinds of sensors, measuring dynamic pressures, sounds, fluid flow, heart rate, and soft tissue contact pressures in different configurations within and outside the human body. Ferroelectric polymers are currently used in several branches of medical diagnostics, follow-up and rehabilitation.

Peculiar properties of ferroelectric polymers, namely, their intrinsic broad bandwidth and high mechanical-to-electrical conversion efficiency, easy conformability to biological structures, low cost, and the almost ideal mechanical impedance matching to human soft tissues, make them attractive as sensor component materials for biomedical measurements. On the other hand, the intrinsic lack of a DC electrical response and the high electrical impedance of ferroelectric sensors limit their field of application.

Ferroelectric sensors for biomedicine are implemented mainly in the form of external sensors (flexible bands, wrappings or strips, rigid arrays, platforms), catheter tips, and implantable sensors used for monitoring mechanical events of the cardiovascular and pulmonary systems. Sensor responses depend on external stimulus characteristics, sensor properties, dimensions, and mounting conditions, requiring appropriate signal elaboration and analysis.

A review of main cardiovascular and pulmonary applications is presented in this section.

A. Noninvasive Cardiopulmonary Sensors

Continuous monitoring of heart and breathing conditions represents a fundamental technique for determining patient cardiopulmonary conditions. Ferroelectric polymer-based external devices allow continuative and comfortable monitoring of the main cardiac and pulmonary physiopathological parameters. Several kinds of devices have been conceived, externally tied to the human body and measuring different pressure and displacement signals.

A biaxially drawn PVDF film-based sensor for peripheral pulse measurement has been realized by De Rossi et al. [13] in a bilaminate configuration, which shows excellent characteristics of immunity from external electrical noise and interference. Faithful reproduction of carotid pulse was achieved by means of this sensor in preliminary clinical experiments; the combination with suitable algorithms for the detection of carotid pulse upstroke and incisura allows easy determination of systolic time intervals in patients at rest.

A similar bilaminate configuration was also used for ambulatory monitoring of breathing [14]. The sensor was applied to the patient's chest by means of an elastic band and provided electric signals related to the expansion of the chest cage during breathing. After suitable filtering, PVDF signals compared favorably to tidal volume recordings picked up from a pneumotacograph. Because of the intrinsic lack of DC response and because filtering occurred during analysis, the transducer was able to record a limited breathing rate range (5–60 beats/min). The detection of cardiac sounds was also possible, with the introduction of a biadhesive film between skin and sensor, which increased the amplitude of the signal related to cardiac sounds and reduced the artifacts due to friction.

External transducers for the recording of peripheral arterial pulses were also realized by Shuford et al. [15]. A transducer system for the noninvasive recording of arterial pressure contours was realized by Karr et al. [16] in a form that resembles a thick wristwatch, permitting the noninvasive recording of the pressure-versus-time arterial profile at any palpable site on the body. The active element in the transducer assembly is a thin (30-μm), uniaxially stretched PVDF film glued to an insulating ring and sandwiched between the main body of the transducer housing and the transducer cap (Fig. 1). The PVDF membrane, which serves as both the active device and the diaphragm of the transducer, is coupled to the skin. Comparisons with records taken with an indwelling catheter indicate congruence of the two signals.

The interest in fetus health monitoring has led to the implementation of recording devices for fetal heart rate and sounds, such as the one described by Kobayashi and

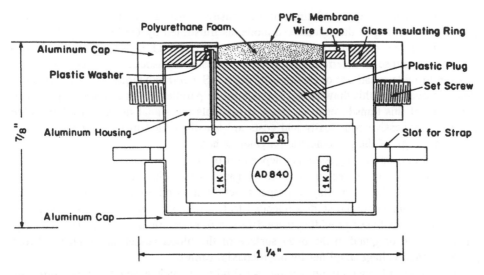

Figure 1 Cross section of the piezoelectric polymer transducer for the recording of arterial pressure contours. (From Ref. 16.)

Yasuda [17]. A fetal phonocardiographic transducer to monitor the fetal heart rate during pregnancy and labor has also been conceived and realized by Steenkeste et al. in a prototype form [18], based on a 190-μm-thick PVDF film, shaped as a portion of a sphere, 40 mm in diameter. The film operates in the transverse mode, and the transducer is connected to an amplifier enclosed in the box that supports the film.

Fundamental diagnostic and prognostic information about cardiac function can be derived from long-term ambulatory monitoring by an electrocardiogram; however, the simultaneous recording of a cardiac performance index of mechanical origin is greatly needed in order to increase the sensitivity, specificity, and reliability of the measurement. For this purpose, the flexibility and low mass of PVDF sensors would, in principle, allow this measurement to be performed with minimum patient discomfort. The major difficulty in the development of such sensors lies in the inaccuracy of sensor positioning and in artifacts due to patient movements. Linear and bidimentional arrays having high sensor density have been developed, and a preliminary assessment of motional artifacts has been made [19].

B. Indwelling and Implantable Transducers

Fluid-filled catheters connected to external pressure transducers still represent the most widely adopted solution for measuring intravascular pressure in the cardiovascular system. Nevertheless, problems such as delay in the transmission of high-frequency signals, smoothing out of the pressure wave or overshooting, and occurrence of vibrations along the length of the catheter limit the performance of these systems. These problems can be overcome by means of catheter-tip transducers, which allow "in loco" measurements, avoiding artifacts due to the transmission of a pressure wave. Ferroelectric polymers proved to be suitable as catheter-tip transducers, since they can be used in different geometrical configurations for disposable, low-cost devices, which also proved to be very robust and sensitive (up to 40 times more sensitive than a strain gauge transducer), especially if working in the extensional mode, as illustrated by Dario et al. [20].

In spite of the ease of construction, robustness, and high sensitivity, the lack of a truly static response of a ferroelectric polymer pressure transducer represents a serious limitation to widespread use of this type of device. Several techniques can be conceived in order to overcome this drawback [21] (such as the parallel use of a strain gauge sensor and the use of this transducer to calibrate at intervals the signal detected by the PVDF sensor), but excessive complexity and costs are usually added to the originally simple and cheap design of the piezoelectric polymer catheter-tip sensor.

Information on patency, eventual deposition of solid materials on the luminal side, flow disturbances, and detection of solid or gaseous particles in the blood stream are important in both the monitoring of proper functioning conditions in extracorporeal circulation lines and in animal studies intended to test new vascular grafts. In its simplest form, a sensor capable of providing these data is a ferroelectric polymer metallized strip, wrapped around or glued to the outer surface of the blood vessel, and connected to a high-impedance voltage amplifier by microcoaxial cables.

Calculations of open-circuit voltage sensitivity of such a sensor to pressure and sounds fluctuations are based on the determination of the circumferential strains in the external surface of a thick walled cylinder under uniform internal pressure. It is assumed that the strains at the outer surface of the cylinder are the same as those existing inside the PVDF strip. Free-field voltage sensitivity was evaluated numerically for a PVDF strip wrapped around a cylinder, with its draw direction oriented circumferentially. In the case of 25-μm-thick PVDF strip wrapped around a polytetrafluoroethylene tube of the type used in extracorporeal circulation lines, the calculated sensitivity was 11 mV/mmHg.

The source time constant (excluding electrode and cable resistance and capacitance) is the Maxwell relaxation time = 500 s, and this determines the theoretical limit for the low frequency roll-off of the sensor. The mechanical resonance frequency of the radial vibration mode determines the high-frequency cutoff, which is tipically in the kilohertz range.

The detection of partial occlusions of the conduit, thrombus formation, pressure fluctuations, and sounds emission has been accomplished in in vitro and in vivo animal experiments by a sensorized vascular graft [22]. The sensors were designed for low-frequency use (0.1–1000 Hz), with special suitability for the main components of the cardiac pulse. Short-term in vivo measurements have been carried out on grafts bearing PVDF sensors implanted in the external carotid arteries of sheep. The implanted grafts detected signals with frequencies as low as that of respiration and also showed changes in signal response to deliberate and to naturally occurring graft occlusions. A similar device was realized by Gupta et al. [23]: Sensors were constructed from ultrathin PVDF with piezoelectric activity and attached with silicone fixative to 6-mm-diameter polytetrafluoroethylene grafts. Ten of these grafts were placed in mongrel dogs as iliofemoral bypass. Real-time data were acquired from the sensors at a rate of 200 Hz, while simultaneous blood flow (using an electromagnetic flow meter) and intraluminal pressure were processed by using separate channels of the same data acquisition board, making simultaneous signal comparison by regression analysis possible. An almost ideal correlation between the two methods was found.

Ultrasounds possess attractive features as a nondestructive technique for biological tissue analysis. Ultrasound scattering and attenuation techniques can be used as methods for analyzing fine tissue structures. Spatial resolution increases with the working frequency of the ultrasonic transducer, while penetration depth decreases, but transducers with very high resonance frequency are not easily available. Ceramic transducers can provide a relatively large-bandwidth operation just by means of multiple matching layers, which add complexity to the transducer. On the other hand, piezoelectric materials can be prepared in thin films with a large bandwidth and good impedance matching to biological tissues to construct high-frequency resonant transducers for detecting trombous formation in implanted vascular grafts [24] as well as for detecting high-frequency pressure fluctuations induced by graft occlusion [22]. Additional information can also be obtained on the performance of prosthetic vascular grafts by monitoring tissue ingrowth on the inner surface of the graft [25]. These phenomena can be detected by ultrasounds transmitted and received by piezoelectric polymeric transducers mounted on the external surface of a graft.

The problem of ill-functioning in implanted prosthetics is fundamental in the monitoring of total artificial heart (TAH) performance, whose reliability is still far from ideal. Major mechanical failures in pneumatically driven TAH occur at the level of the pumping diaphagm, valves, valve holders, and inlet air lines. The left ventricle of a TAH has been provided with three thin ferroelectric polymer sensors to assess the ability of early detection of mechanical failure at the most probable sites of occurrence [26]. A 200-μm-thick ferroelectric ultrasonic transceiver, made of a 40-μm PVDF film with 0.1-μm aluminum electrodes backed by a 150-μm-thick copper reflecting layer, has been glued to the outer surface of the polyurethane ventricle wall and protected with a thin coating of polyurethane, being, in effect, embedded into the wall. The location of the transducer was chosen in order to maximize reflections of ultrasonic waves from the air-driven pumping diaphragm. The ventricle was placed on a mock circulatory system and the transducer pulsed by a commercial pulser-receiver, which provided 10-μs-duration pulses at a repetition rate of 10 kHz. Diaphragm movements were clearly detectable, as was the failure of the flexing diaphragm, consisting of the detachment of one of the leaflets composing the layered diaphragm. Fault detection appeared as an additional echo at the receiver, occurring immediately before the echo generated by the intact intermediate layer. A similar ultrasonic transducer was applied to the external surface of the valve holder, again monitoring the valve disk movement, which was easily detected. Because of the inherently large bandwidth of the transducers, sound signals generated by valve opening and closure were detected by switching the sensor output in the receiving mode to a high-impedance voltage amplifier. Alteration of sound emission in amplitude and spectrum occurred and was detected when partial occlusion of the inflow valve was deliberately created. A third sensor, having the same configuration as the former, was wrapped around the PVC drive line, close to its junction with the ventrical base inlet connector. Sound generated by deliberately created small air leaks was again detected by the FEP acoustic sensor, but poor signal-to-noise ratio of the sensor used in these experiments did not allow unequivocal differentiation of minor air leaks from background noise [27]. Evidence of the in vitro detection of thrombus formation of the inner surface of the TAH was also obtained by ultrasonic measurements.

C. Sensors for Prosthetics, Orthotics, and Rehabilitation Devices

The detection of contact forces and the determination of their time-space distribution at the interface between biological tissues and a sensing plane plays a fundamental role in prosthetics, orthotics, and rehabilitation devices. Sensing platforms are also used in the determination of deambulation problems and weight distributions while walking or running.

High-spatial-density ferroelectric polymer tactile sensor arrays have been developed for measuring the space–time distribution of foot–shoe contact pressure. The measuring platform includes a matrix of sensing sites made of piezoelectric polymers and protected with a sheet of abrasion-resistant rubber. The PVDF film is electroded and bonded to a rigidly supported printed circuit board, on which a multiple pattern of electrodes has been defined by means of a thin layer of nonconductive glue. The platform scanning is performed by an analog multiplexer, and data are provided to a PC in a digitalized form [28].

Foot–ground pressure patterns have also been recorded [29] by means of multisensor piezoelectric polymer insoles made of uniaxially stretched and selectively poled PVDF on which aluminum disks were vacuum evaporated, serving as electrodes for pressure sensors. The sensors were located so as to optimize the response of the five metatarsal transducers, which are particularly critical for a correct gait analysis. The tracks, one for each transducer, converge toward one side of the insole, where contact pads are provided to bond connecting wires, leading to an electronic acquisition unit conveniently attached to the leg of the patient.

An ultrasonic PVDF tactile transducer has also been developed by Shoemberg et al. [30], intended to provide a force feedback during object grasping by quadriplegic patients whose hand control and finger movements are restored by electrical stimulation of forearm muscles. The transducer is made of a four-element sensor array; each PVDF sensor measures the amount of compression of a compliant rubber layer by detecting the time of flight of the ultrasonic pulse transmitted into the layer and reflected, at the rubber–object interface, back to the PVDF element. Normal forces in the range of 0.3 to 12 N exerted on a 4×4 mm area can be measured. A dynamic response up to 10 Hz was demonstrated, and much higher response rates up to 200 Hz were possible.

Slippage control by tactile sensor has been analyzed by Mingrino et al. [31] in order to provide myoelectric hand prostheses with a tactile sensor system to correctly control the grasp of an object. Incipient and actual slippage are detected by sensors located on the fingertips, and this information is used to control the slippage of the object.

D. Robotic Tactile Sensors

The development of dexterous end effectors and tactile sensing systems for robots capable of operating in an unstructured environment is one of the main research topics in the field of robotics and teleoperation. Since the beginning of this kind of studies, the human tactile system has represented and still constitutes the ultimate model, inspiring research and development activity in artificial tactile sensing systems in terms of performance to be implemented and often in terms of device designing as well.

The high degree of dexterity which characterizes grasping and manipulative functions in humans, and the sophisticated capability of recognizing the features of an object, is the result of a powerful sensory motor integration. This is usually known as haptic sensitivity and fully exploits the wealth of information provided by the cutaneous and kinesthetic neural afferent systems. The potential applications of such a powerful exploration tool to artificial devices has increased considerably in the last few years. As suggested by experiments carried out on human subjects, tactile perception is both fast and accurate in exploring the environment, and it is even superior to vision in determining some features of the objects such as texture, compliance, and fine spatial details.

The primary goal of an ideal tactile sensing device is to measure variable contact forces on a sensing area. However, observation of how humans recognize objects suggests that it may be useful to sense other variables besides the forces perpendicular to the sensor surface, such as hardness, contours, texture, and tangential forces for the detection of incipient slippage [32].

Approaching the challenging task of developing artificial sensory systems requires synergistic contributions from a number of disciplines, including materials science, control theory, artificial intelligence, and mathematics.

The simplest form of tactile sensors is represented by contact sensors, which are constituted mainly of microswitching systems, providing binary tactile images of objects. Toda [33] developed a PVDF-based transducer which senses both the presence and the position of an object by soft contact. The transducer is formed by two uniaxially oriented PVDF films, bonded together to form a bimorph configuration. One film is connected to an alternate voltage source and causes bimorph bending. The second bimorph element senses bending strains. The amplitude of oscillations depends on the contact with the object. Calibration of the sensor allows the measurement of object position.

An anthropomorphic PVDF tactile sensor that aims at reproducing most sensing properties of the human fingertip has been developed by Dario and De Rossi [34]. The design can be used in both flat grippers and curved artificial fingers. The sensor, shown in Fig. 2, consists of a deep (dermal) and superficial (epidermal) layer. Each layer has distinct sensing functions: a PVDF film about 100 μm thick at its base mimics human dermis, while a 40-μm-thick film simulates human epidermis. An elastomer layer between the two provides a compliant backing for the epidermal layer. Metal plated on its upper surface, the dermal PVDF film is bonded with nonconductive glue to a supporting printed circuit board. Electric charges are generated by mechanical deformation of the surfaces of the dermal PVDF film, and these charges are coupled capacitively to circular metal electrodes on the circuit board. Electrodes on the top and bottom of the epidermal

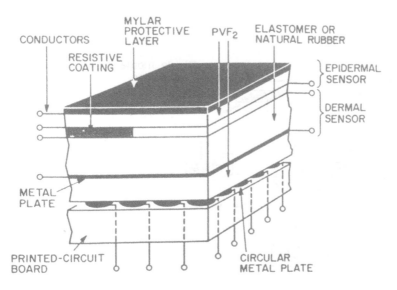

Figure 2 PVDF-based tactile sensor designed to emulate the sensing capabilities of human skin. (From Ref. 34.)

PVDF layer similarly collect the charges generated there, transferring them to conductors. Typical sensitivities are 20 pC/N for the dermal and 200 pC/N for the epidermal layer. Linearity is within about 1% between 0.01 and 2.56 N, peak-to-peak. Some hystheresis was introduced by the intermediate layer.

A multicomponent sensor was also described by De Rossi et al. [35], based on six piezoelectric polymer elements, whose surfaces normal to the z axis (see Fig. 3) are covered with thin metal electrodes. This sensor shows the distinctive feature of sensing each individual component of the stress tensor field generated inside the sensor during contact. The sensing elements are respectively made of: (1) uniaxially oriented PVDF film; (2) uniaxially oriented PVDF positioned with its drawn direction rotated at 90° in the x-y plane with respect to element 1; (3) biaxially oriented PVDF; (4) PVDF element obtained by cutting an uniaxially oriented PVDF thick slab along its thickness, in a direction parallel to the drawn direction; (5) PVDF element obtained by cutting an uniaxially oriented PVDF thick slab along its thickness, in a direction perpendicular to the drawn direction; (6) uniaxially oriented polyhydroxybutirate sample. Transducer dimensioning was performed by a process which combines engineering reasoning and the analytical solution of a simple direct elastic contact problem. A two-dimensional, half-infinite elastic medium in which the sensor is embedded, with a line vertical force acting at the straight boundary, was paradigmatically used. Preliminary experiments with planar and circular indenter loads were performed, being in good agreement with theory.

Further developments of the PVDF-based tactile sensor led to the implementation of a high-resolution tactile sensing device for object fine-form discrimination: A system able to provide robots with tactile sensitivity was realized for continuous scanning of variable contact forces occurring during object exploration and manipulation by means of miniature sensors. An investigation of the expected responses of the sensor has been carried out by modeling the mechanics of object–sensor interaction and finding the stress solutions of a forward elastic problem. The system has been tested by comparing the theoretical results with data measured experimentally from the sensor [36].

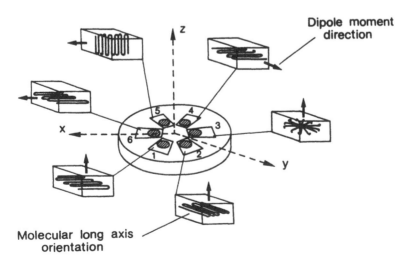

Figure 3   Schematic diagram of the six-element, stress-component-selective tactile sensor.

IV. ACTUATORS

Actuators are based on the inverse piezoelectric effect, converting electrical into mechanical energy. The displacements that can be obtained and the forces that can be exerted are very limited (up to some micrometers per volt, with a maximum of some hundreds of micrometers and a few grams, respectively). Configurations such as bimorphs, multimorphs, tent structures, and push-pull vibrating elements have been devised in order to obtain higher displacements and forces, but the applicability of piezoelectric actuators remains limited to microscale applications, such as instrument manipulation (needles, pipettes, electrodes) for cell biology, or various kinds of manipulations in the fields of microsurgery, micromechanics, and microelectronics.

Umetani and Suzuki [37] described the design and fabrication of some piezoelectric polymer actuators for special applications. In particular, a composite micromanipulator, illustrated in Fig. 4, has been realized using some bimorph PVDF elements. The microactuator assembly has three degrees of freedom and a gripper end, which is able to grasp and position small objects within a volume of $330 \times 500 \times 150 \ \mu m^3$. Such structures, with linear dimensions of few centimeters, can easily be used on the deck of a microscope.

The same authors proposed a microinjector based on a thin PVDF tube acting with a radial contraction for squeezing out its contents in microliter doses. The performance of the device is incompatible with any clinical application (too little volume output at too high voltage), but the principle of using PVDF for drug delivery has been demonstrated. The flexibility of PVDF geometry makes further research possible.

Another multielement actuator has been described by Nevill and Davis [38]. Theirs is made of three unit cells (3.2 mm wide and 70 mm long) based on the competitive deflectional behavior of two alternately oriented PVDF bimorph elements. When a voltage is applied, the cooperative action of bimorph elements produces a curvature of each unit, giving rise to a sinusoidlike configuration. The device yields an isometric force of 0.005 N with a power supply of 400 V and can also be used as a contact sensor. The authors developed the concept of a sensor based on the vibrations induced by sliding

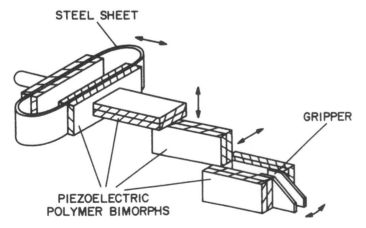

Figure 4 Scheme of mounting of the micromanipulator incorporating four PVDF-based, bimorph-type actuating elements. (From Ref. 37.)

motion across a sample object. This concept is based on the theory that the papillary ridges of the fingertip provide useful information for the identification of objects by touch.

When used as a sensor, the multilayered bimorph element structure would initially be mounted in a curved state, producing an electrical response upon mechanical compression or flattening.

V. DIAGNOSTIC IMAGING

The electrical and mechanical properties of piezoelectric polymers make them interesting also in the development of electroacustic or ultrasonic transducers for medical applications. Comparison of the representative PVDF material characteristics with a conventional ferroelectric ceramic as PZT shows several features of piezoelectric polymers which make them attractive in transducer design (Table 1).

The mechanical flexibility and the easy conformability of the polymer, which can be manufactured in very thin film, allow a wide range of transducer configurations. Moreover, the acoustic impedance of PVDF is much smaller than that of ceramics, being comparable with water and body tissues, and makes the polymer useful in biomedical imaging transducers. In fact, the optimum acoustic matching, coupled with the acoustic absorption characteristics of PVDF (comparable to those of water) allow the design of nonperturbing probes. Furthermore, polymer transducers present broad bandwidth characteristics, which allow a very short-impulse response, corresponding to a high axial resolution in an acoustical imaging system.

Although the low dielectric constant of PVDF (two order of magnitude smaller than that of the ceramics) accounts for the high receiving properties of the polymer, the lower value of the electromechanical coupling factor, k_t, with respect to ceramics (Table 1), means that the polymer is inferior to a ceramic as a transmitting material. However, this disadvantage can be compensated for through the possibility of using a larger driving voltage. The low transmitting efficiency has limited the employment of piezoelectric

Table 1　Selected Comparative Properties of PVDF and PZT Ceramic (Lead Zirconate Titanate)

Properties	PVDF	PZT
Density (10^3 kg/cm^3)	1.78	7.5
Sound velocity, v_3 (km/s)	2.26	4.63
Acoustic impendance, z_0 (10^6 kg/m^2 s)	4.02	34.4
Stiffness constant, c_{33} (10^9 N/m^2)	9.1	159
Coupling factor, k_t	0.20	0.51
Piezoelectric constants:		
$\quad e_{33}$ (C/m^2)	−0.14	15.1
$\quad d_{33}$ (pC/N)	−31	374
$\quad d_{31}$ (pC/N)	27	−123
$\quad d_h$ (pC/N)	−9	32
Dielectric constant, ϵ_3/ϵ_0	6.2	635
Dielectric loss tangent, tan δ_e	0.25	0.004
Mechanical loss, tan δ_m	0.10	0.004
Coercive field (MV/m)	45	0.7

polymers in projectors or pulse-echo transducers in the typical frequency range of the biomedical field (1–15 MHz). However, several applications in medical ultrasonics have been proposed [39].

Piezoelectric polymer hydrophones have been proposed as probes for mapping ultrasonic fields and determining acoustic field parameters in both water and biological media. Among the several PVDF hydrophones developed, the most widely used have been the original hoop membrane configuration (spot-poled membrane supported by means of a hoop) [40] and the needle-type probe [41]. While both can be used to assess ultrasonic field patterns in water, only the needle-type probe can be used for bioacoustic measurements, such as in vitro and in vivo evaluation of tissue properties.

Miniature needle-like hydrophones mounted in a biopsy device have been proposed and tested in vivo for the measurement of the attenuation coefficient of human tissues [42].

Imaging of human tissues and noninvasive assessment of their lesions is the most important diagnostic area of ultrasonics, and PVDF devices have also been exploited in medical ultrasound imaging. Different transducer configurations have been adopted in the realization of the PVDF-based devices.

Although acoustic power limitations restrict the PVDF transducer's penetration depth, polymeric systems have been used successfully in imaging of "near-surface" tissues or organs, such as the thyroid gland, which lie immediately under the skin layer. In particular, the clinical evaluation of the single-element concave transducer fabricated by Ohigashi and co-workers showed that the ultrasonic images of tyroid gland obtained with polymeric transducers are much superior in quality to those obtained with conventional ceramic transducers [43].

Better results have been obtained recently using a derivative of PVDF: the copolymer of vinylidene fluoride and trifluoroethylene, P(VDF-TrFE). This copolymer, which is crystallized from the melt directly into a polar crystal form, can be poled without a stretching process. It exhibits a stronger piezoelectric activity than PVDF [44]. Its electromechanical coupling factor k_t depends on VDF content and preparation conditions, and in the VDF range of about 70 mol%, it becomes quite large (>0.3).

Ultrasound images of the breast, obtained with P(VDF-TrFE) transducers working at 7.5 MHz, have shown how the image quality is improved by the increased sensitivity due to the use of the copolymer, and demonstrated that the capabilities of this kind of transducer can potentially allow imaging and diagnosis of very small lesions (even less than 5 mm in diameter) [45].

Linear and annular transducer arrays are also widely used in ultrasound diagnostics. The linear array currently employed to produce ultrasonic beams focused on selected areas (able to give high-resolution images) are manufactured by cutting a piezoelectric ceramic plate into small separate elements. This means a complicated process of preparation which is not necessary when piezoelectric polymer films are used. In fact, among the advantages of piezoelectric polymers over piezoelectric ceramics, such as their good impulse response, they do not need to be cut into separate elements to reduce cross-talk phenomena.

An array of backing electrodes, realized by photolithography and chemical etching techniques, can be bonded on a polymeric substrate and can lodge a ferroelectric polymer film onto which a front electrode has been previously coated. PVDF or copolymer films can be poled locally through such electrodes, but also a film poled over its whole surface may be used without affecting the distribution of the acoustic field or the cross-talk between adjacent elements.

The disadvantage of this system is that the electric impedance of each element is very high, making electric excitation problematic. This problem can be solved using a matching circuit able to feed the array elements properly.

A 5-MHz P(VDF-TrFE) linear transducer array, made of 64 elements and able to provide good-quality tomographic images, has been realized by Ohigashi and co-workers [46]. Further improvement in image quality can be obtained by increasing the working frequency, and by increasing the number and the spatial frequency of the array elements.

Single-element focusing transducers are able to give good-quality images on their focal zone, but they have limited lateral resolution. This becomes more critical for large-area transducers.

The use of annular array transducers provides high-resolution images over large areas. In fact, concentric annular ring elements driven by a phase-controlled pulse can shift the focal point along the transducer axis.

Annular arrays composed of a single sheet of piezoelectric polymer have been developed and tested in vivo. In particular, a 7.5-MHz annular array transducer with very high resolution has been realized recently [39]. The transducer, based on P(VDF-TrFE) film and composed of eight annular elements of equal areas, has a diameter of 70 mm. In the focal zone (about 150–200 mm from the transducer surface), it has a lateral resolution less than 1 mm, while the depth resolution is 0.3 mm. A clear ultrasound image of the thyroid gland obtained with this transducer is shown in Fig. 5. Ultrasound imaging systems based on these transducers are suitable for examination of breast and abdominal organs. In Japan, P(VDF-TrFE) copolymer transducers are used routinely for examination of breast, thyroid, and eyes in many hospitals.

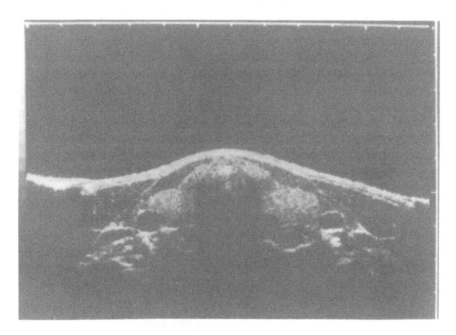

Figure 5 Ultrasound image of the thyroid gland observed with an annular PVDF array transducer. (From Ref. 39.)

The capability of VDF-TrFE copolymer films to be poled directly after casting suggested the possibility of realizing an integrated silicon-piezoelectric copolymer transducer array, by depositing a thin layer of copolymer directly on silicon wafers [47]. Following the original approach proposed by Swartz and Plummer [48], who designed an integrated silicon-PVDF transducer for medical imaging in which a thick film of PVDF was bonded with an epoxy layer to the surface of a silicon wafer, a 1.5-μm thin layer of P(VDF-TrFE) has been obtained by spinning a polymer solution onto a metallized silicon surface. The copolymer was successively poled locally by corona discharge, showing a piezoelectric activity of the same order as conventionally prepared piezoelectric polymers.

A further development indicated that a thin piezoelectric transducer coupled directly to the extended gate of a MOS transistor can be processed into an ultrasonic sensor which can serve as the basis of an ultrasonic imaging array [49]. Preliminary results obtained, showing the ability of the ultrasonic sensor to receive signals in acqueous medium at megahertz frequencies, indicate the feasibility of an integrated ultrasonic array sensor for medical applications.

VI. CHARGE GENERATORS AS TISSUE PROMOTERS

Biological mechanisms of growth and control in body tissues are based mainly on a mutual interaction of chemical, mechanical, and electrical phenomena.

The complex physicochemical processes involved in the response of biological structures to stimuli of different origin are very difficult to analyze in their components, because of the intrinsically multiphase and multicomponent structure of biological tissues. For these reasons, explaining the basic mechanisms governing these biological functions and selectively stimulating individual processes is not an easy task. However, empirical methods of mechanical or electrical origin have often been used in the medical field in order to improve tissue regrowth, as in the case of bone healing. Osteogenesis induced by electric currents has been reported in the last century. In the first attempts the procedures were essentially based on the application of external fields, and the small current passing in the living structures has been shown to be effective in the formation of bone tissue [50].

Remarkable work was carried out in the 1970s, which demonstrated that small direct currents or pulsed electromagnetic fields can stimulate osteogenesis. Some of these works investigated the influence of parameters such as current level and electrode materials on the production of new bone [51,52]. On the other hand, studies devoted to understanding the molecular mechanism responsible for the ability of bone to convert mechanical and electrical signals into regenerating stimuli led to the discovery of mechanoelectrical transduction properties in the dry and wet states.

In particular, piezoelectric properties of bone were reported by Fukada and Yasuda [53], who ascribed their origin to the oriented collagen fibers. The studies on bone and other biopolymers stimulated an interest in piezoelectric properties of synthetic materials such as polypeptides and fluorinated polymers. Although the complexity of the microstructure of the tissues does not allow direct comparison with the piezoelectric behavior of polymers, the same research group implanted piezoelectric synthetic polymer to induce osteogenesis electrically. Small pieces of piezoelectric and electret films where applied onto the femur of living animals, and the formation of callus surrounding the polymer films was observed. In the first experiments, owing to its good biocompatibility, electrically poled Teflon film was used by Fukada and co-workers in rabbits [54]. The result was a growth of callus, surrounding the Teflon film in a bridgelike configuration, which caused the formation of rigid bone. A typical result is shown in Fig. 6.

Inoue et al. [55] used piezoelectric polypeptides instead of electret film to induce bone growth in the femur of rats. Films of oriented poly-γ-methyl-L-glutamate (PMLG) were positioned around the femur by fixing their ends to quadriceps and biceps muscles. In this way the motion of the animal induces dynamic deformations in the PMLG film, producing its piezoelectric response. Bone growth was observed a few weeks after the operation, as demonstrated by X-ray and histological examinations.

At the same time, Suzuki and co-workers implanted films of a highly piezoelectric polymer: thin strips of PVDF were applied to the femur and mandible of a macaque monkey [56]. After 6 weeks, osteogenesis was observed around the film, and a slight remodeling of the cortical bone was evident under it, while the callus remained spongy bone.

More recently, the use of thick monomorph and bimorph PVDF films has been reported [57]. Piezoelectric polymer samples were implanted around the femoral diaphysis of rabbit and a comparative study on effects showed a great amount of callus and an important remodeling of the cortical bone for the bimorph PVDF.

A further improvement in piezoelectrically stimulated bone-healing has been provided by experimental work on the repair of fracture using piezoelectric endomedullary ostheosynthesis [58]. Steinman pins have been purposely prepared to lodge a piezoelectric PVDF tubular layer. After the generation of fracture in rat femurs, the pins were inserted into the medullary cavity of bone.

The radiographic and histomorphologic examinations showed a stronger bridgelike periostal reaction and a faster healing of fractures treated with piezoelectric ostheosynthesis with respect to controls, once again indicating the positive effect in the consolidation of bone fractures of the use of piezoelectric polymers. An example of the comparative radiographic examination is reported in Fig. 7, where the rat on the right of Fig. 7A, treated with piezoelectrically active ostheosynthesis, presents a good fracture repair (Fig. 7B).

The desirable availability of piezoelectric polymers that are bioresorbable in addition to biocompatible would eliminate the need for explantation of the polymeric material at the end of the healing process. A possible explanation of the osteogenetic power of these materials can be found into the piezoelectric polarization induced by muscular activity or minute bone movements as suggested by Fukada [59]. The charges generated can

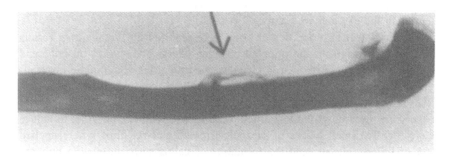

Figure 6   X-ray photograph showing the bone regrowth stimulated by the application of an electret Teflon film around a rabbit femur. (From Ref. 54.)

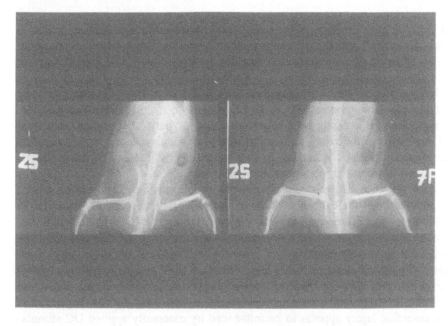

(A)

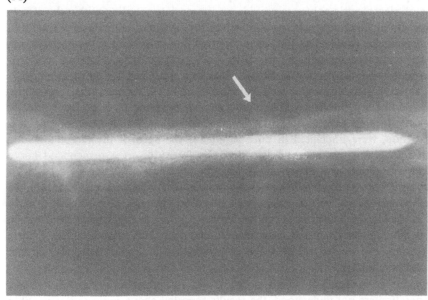

(B)

Figure 7 (A) Comparative X-ray examination of the regenerative effect of piezoelectrically active osteosynthesis: rat treated with poled (right) and unpoled PVDF (left). (B) Detail of the fracture repair of the piezoelectrically treated rat.

produce a flow of ionic currents in the tissue fluid surrounding the implant, which is electrically conductive. These ionic currents might represent a stimulus for cell migration or even differentiation of fibroblasts into osteoblasts, providing a fast proliferation of bone cells.

Electrically active polymers demonstrated their ability to enhance the time course of other in vivo repair processes such as wound healing and, mainly, nerve regeneration. Concerning the latter, it has long been appreciated that the peripheral nervous system is capable of regeneration following transection injury, although the precise mechanisms controlling regeneration are not known. A wide range of physical and chemical factors have been shown to influence the regeneration process.

The use of guidance channels can improve this process, and experiments have been performed to determine the influence of their material properties on the regenerative effects. Valentini et al. [60] reported how the composition and physical characteristics of synthetic materials used as nerve guidance channels can influence the rate and morphology of peripheral nerve regeneration.

Recently, exogenous electrical fields have also been used in affecting neural regeneration. Several groups have reported that applied electric fields influence the extent and direction of neurite outgrowth from neurons cultured in vitro [61]. In vivo regeneration following transection injury appears to be influenced by externally applied DC stimulation [62] and pulsed electromagnetic fields [63].

The use of piezoelectric polymers provide an appropriate coupling within the effects on neurite growth of electromagnetic stimuli and the regenerating function of channelling on severed nerves. Aebischer et al. [64] have reported that peripheral nerve regeneration is significantly enhanced through the use of piezoelectric guidance channels made of PVDF tubes with small diameter. The PVDF tubes were stretched 3.5 times along their long axis at a rate of 1 cm/min and at a temperature of 110°C. A corona poling procedure, based on the use of a cylindrical configuration device, rendered them piezoelectrically active. After sterilization, the piezoelectric channels, of a diameter of about 0.8 mm and a length of 6 mm, were implanted in mice to anchor the ends of the resected sciatic nerve (Fig. 8). A few weeks after implantation, the formation of a regenerated nerve bridging the nerve stumps was observed. The regenerated nerve bundles were larger in section, and their content of myelinated axons was higher with respect to the control unpoled channels (Fig. 9). Nerves regenerated in poled channels also displayed more consistent morphologic characteristics and the highest number of myelinated axons when compared to other materials tested as guidance channels. This suggests that electric charge generation, due to the transient charge production of PVDF, improves axonal elongation and myelination.

Although the mechanisms by which electrical activity influences peripheral nervous system regeneration are not understood, the results reported above indicate the possible helpful clinical use of piezoelectric polymers in the surgical repair of injured or severed peripheral nerves. The piezoelectric characteristics of poled PVDF tubes can obviate the need for external power sources and electrical connections, as requested in regenerating systems based on the application of electromagnetic fields or direct currents.

Moreover, as in the case of piezoelectrically stimulated bone healing, the availability of piezoelectric synthetic polymers that are both biocompatible and bioresorbable would render this technique very attractive for human application.

In addition, in the field of nerve regeneration, research is also in progress toward the development of connectors capable of interfacing the peripheral nervous system to

1. CREATION OF 3–4 mm NERVE DEFICIT

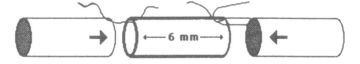

2. PLACEMENT OF 6 mm GUIDANCE CHANNEL w/ 10–0 SUTURES

3. NERVE-CHANNEL RELATIONSHIP AND GAP DISTANCE

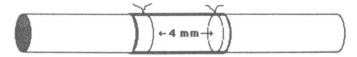

Figure 8   Scheme of the surgical implantation of PVDF guidance channels.

artificial devices. For example, Cocco et al. [65] proposed a methodology for in vitro stimulation of neuro cells, cultured directly on a piezoelectric polymer film, in which electric charges might promote nervous differentiation. The use of active PVDF, as substrate for cell culture, has been shown to improve the cell differentiation process. In particular, neurites with pronounced length (up to 100 mm) are obtained in the "stimulated" samples. More recently, Valentini et al. [66] carried out a study aimed at determining whether the addition of charged surface groups to the surfaces of fluorinated ethylenepropylene (FEP) and PVDF would modify the influence of bulk electrical charges on cultured neurons. Mouse neuroblastoma were cultured on electrically charged and

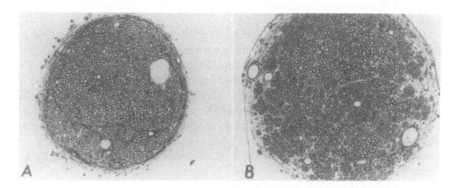

Figure 9   Toluidine blue-stained transverse sections of regenerated nerves at the midpoint of the guidance channels at 12 weeks after implant of channels: A, unpoled PVDF channel; B, poled PVDF channel. (From Ref. 64.)

uncharged FEP and PVDF substrates with covalently modified surfaces containing hydroxyl and amine groups. Surface chemical modification was performed on the entire surface or in discrete striped regions. Neuroblastoma cell cultures on electrically active FEP and PVDF showed greater levels of differentiation than cells on electrically neutral substrates. Investigations on the presence of NH_2 groups were also carried out. In summary, bulk electrical charges proved to be more important than surface charges in stimulating neuroblastoma cell differentiation. Surface groups serve to modulate neuronal morphology and confer specific attachment-promoting properties in serum-containing media.

Finally, another in vivo application of piezoelectric polymers, used as tissue healing promoters, has been proposed recently by Okoshi et al. [67]. Microporous vascular prostheses were fabricated with P(VDF-TrFE) copolymer by a spray phase-inversion technique. The grafts, polarized with a corona poling device like that used for the PVDF channels, were implanted in the infrarenal aorta of rats. The combination of piezoelectric properties and the microporous wall structure of the grafts may enhance antithrombogenic phenomena and accelerate the development of a stable nonthrombogenic neointima. Preliminary results showed that the use of piezoelectric P(VDF-TrFE) vascular grafts can help in the investigation of the relationships of charged surfaces to thrombus formation and vascular healing.

REFERENCES

1. H. Kawai, The piezoelectricity of poly(vinylidene fluoride), *Jpn. J. Appl. Phys.* 8:975 (1969).
2. J. G. Bergman, J. H. McFee, and G. R. Crane, Pyroelectric and optical second harmonic generation in poly(vinylidene fluoride) films, *Appl. Phys. Lett.* 18:203 (1971).
3. K. Nakamura and Y. Wada, Piezoelectricity, pyroelectricity, and the electrostriction constant of poly(vinylidene fluoride), *J. Polymer Sci. A-2, 9*:161 (1971).
4. D. De Rossi and P. Dario, Biomedical applications of piezoelectric and pyroelectric polymers, *Ferroelectrics 49*:49 (1983).
5. T. Bui, H. L. W. Chan, and J. Unsworth, Specific acoustic impedances of piezoelectric ceramic and polymer composites used in medical applications, *J. Acoust. Soc. Am.* 83:2416 (1988).
6. U.S. Code of Federal Regulations, Title 21, Food and Drugs, chap. F, p. 121.
7. R. G. Mason, D. E. Scarborough, S. R. Saba, K. M. Brinkhous, L. D. Ikenberry, J. J. Kearney, and H. G. Clark, Thrombogenicity of some biomedical materials: platelet-interface reactions, *J. Biomed. Mat. Res.* 3:615 (1969).
8. D. De Rossi, A. S. De Reggi, M. G. Broadhurst, S. C. Roth, and G. T. Davis, Method of evaluating the thermal stability of the pyroelectric properties of polyvinylidene fluoride: Effect of poling temperature and field, *J. Appl. Phys.* 53:6520 (1982).
9. T. T. Wang, Aging behavior of piezoelectric poly(vinylidene fluoride) films irradiated by γ rays, *J. Polymer Sci., Polymer Lett. Ed. 19*:289 (1981); Properties of piezoelectric poly(vinylidene fluoride) films irradiated by γ-rays, *Ferroelectrics 41*:213 (1982).
10. C. C. Enger and F. A. Simeone, Biologically energized cardiac pacemaker: In vivo experience with dogs, *Nature 218*:180 (1968).
11. C. C. Enger and F. W. Rhinelander, Implantable impermeable flexible encapsultants for artificial organs, *J. Biomed. Mat. Res. Symp. 1*:149 (1971).
12. J. C. Bockros, Carbon biomedical devices, *Carbon 15*:355 (1977).
13. D. De Rossi, P. Dario, C. Marchesi, M. G. Trivella, F. Pedrini, C. Contini, A. S. De Reggi, S. Edelman, and S. C. Roth, Piezoelectric polymers as peripheral pulse sensors, *Computers in Cardiology (Proc., IEEE Cat. N.81CH, 1750-9)*, p. 111 (1981).

14. G. Kraft, M. Raciti, R. Francesconi, P. Pisani, C. Carpeggiani, and M. Emdin, Ambulatory recording of respiration. Preliminary experience with a piezoelectric transducer, *J. Ambulatory Monitoring 5*:79 (1992).

15. R. J. Shuford, A. F. Wilde, J. J. Ricca, and G. R. Thomas, Characterization and piezoelectric activity of stretched and poled poly(vinylidene fluoride). I. Effect of draw ratio and poling conditions, *Polymer Eng. Sci. 16*:25 (1976).

16. S. G. Karr, T. Karwoski, J. E. Jacobs, and L. F. Mockros, Transducer system for the noninvasive recording of arterial pressure contours, *Ann. Biomed. Eng. 13*:425 (1985).

17. K. Kobayashi and T. Yasuda, An application of PVDF-film to medical transducers, *Ferroelectrics 32*:181 (1981).

18. F. Steenkeste, Y. Moschetto, M. Boniface, P. Ravinet, and F. Micheron, An application of PVF$_2$ to fetal phonocardiographic transducers, *Ferroelectrics 60*:193 (1984).

19. D. De Rossi, G. Kraft, R. Francesconi, M. Bertoncini, and C. Contini, Ferroelectric polymers technology and sensors for non-invasive cardiovascular measurements, *Proc. XIII Congress European Society for Non-Invasive Cardiovascular Dynamics*, Brescia, Italy, 1985, p. 32.

20. P. Dario, D. De Rossi, R. Bedini, R. Francesconi, and M. G. Trivella, PVF$_2$ catheter-tip transducers for pressure, sound and flow, *Ferroelectrics 60*:149 (1984).

21. A. S. De Reggi, Transduction phenomena in ferroelectric polymers and their role in biomedical applications, *Ferroelectrics 60*:83 (1984).

22. P. D. Richardson, P. M. Galletti, and P. Dario, PVF$_2$ sensors for monitoring pulse and turbulence in prosthetic vascular grafts, *Ferroelectrics 60*:175 (1984).

23. S. K. Gupta, A. M. Dietzek, F. J. Veith, H. B. Kram, and K. R. Wengerter, Use of a piezoelectric film sensor for monitoring vascular grafts, *Am. J. Surgery 160*:182 (1990).

24. P. Dario, P. D. Richardson, and P. M. Galletti, Monitoring of prosthetic vascular grafts using piezoelectric polymer sensors, *Trans. Am. Soc. Artificial Intern. Organs 29*:318 (1983).

25. P. Dario, P. D. Richardson, M. Bertoncini, D. De Rossi, L. A. Trudell, and P. M. Galletti, Prosthetic vascular graft monitoring by ultrasound using piezoelectric polymers, *Trans. Am. Soc. Artificial Intern. Organs 30*:645 (1984).

26. Yu Long Sheng, D. De Rossi, P. Dario, and P. M. Galletti, Indwelling acoustic sensor for early detection of total artificial heart failure, *Life Support Systems 4*:239 (1986).

27. D. De Rossi, Design criteria for ferroelectric polymer sensors and their applications to artificial organ technology, *Trans. Am. Soc. Artificial Intern. Organs 32*:697 (1986).

28. A. Starita, F. Basta, B. Carbone, P. Dario, D. De Rossi, and C. C. Perfetti, A computerized platform for the analysis of spatio-temporal foot-ground pressure patterns, *Med. Biol. Eng. & Comp. 23* (Suppl. Pt. 1): 420 (1985).

29. A. Pedotti, R. Assente, G. Fusi, D. De Rossi, P. Dario, and C. Domenici, Multisensor piezoelectric polymer insole for pedobarography, *Ferroelectrics 60*:163 (1984).

30. A. A. Shoemberg, D. M. Sullivan, C. D. Baker, M. E. Booth, and C. Galway, Ultrasonic PVF$_2$ transducers for sensing tactile force, *Ferroelectrics 60*:239 (1984).

31. A. Mingrino, P. Dario, and A. Sabatini, A hand prosthesis with slippage control by tactile sensors, *15th Annual Int. Conf. IEEE Med. Biol. Soc.*, San Diego, CA, 1993, p. 1276.

32. D. De Rossi, Artificial tactile sensing and haptic perception, *Meas. Sci. Technol.*, 2:1003 (1991).

33. M. Toda, A PVF$_2$ piezoelectric bimorph device for sensing presence and position of other objects, *IEEE Trans. Electron. Devices ED-26*:815 (1979).

34. P. Dario and D. De Rossi, Tactile sensors and the gripping challenge, *IEEE Spectrum 22*:46 (1985).

35. D. De Rossi, A. Nannini, and C. Domenici, Biomimetic tactile sensor with stress-component discrimination capability, *J. Mol. Electron. 3*:173 (1987).

36. D. De Rossi, G. Canepa, G. Magenes, F. Germagnoli, A. Caiti, and T. Parisini, Skin-like tactile sensor arrays for contact stress field extraction, *Mat. Sci. Eng. C 1*:23 (1993).

37. Y. Umetani and H. Suzuki, Piezoelectric micro-manipulator in multi-degrees of freedom with tactile sensibility, *Proc. 10th Int. Symp. on Industrial Robots*, Milano, Italy, 1980, p. 571.

38. G. E. Nevill and A. F. Davis, The potential of corrugated PVDF bimorphs for actuation and sensing, *Proc. 1st World Conf. on Robotics Research*, Bethlehem, PA, 1984, MS84-491.

39. H. Ohigashi, Ultrasonic transducers in the megaherz range, *The Applications of Ferroelectric Polymers* (T. T. Wang, J. M. Herbert, and A. M. Glass, eds.), Blackie & Son, London, 1988, p. 237.

40. A. S. De Reggi, S. C. Roth, J. M. Kenney, S. Edelman, and G. R. Harris, Piezoelectric polymer probe for ultrasonic applications, *J. Acoust. Soc. Am. 69*:853 (1981).

41. P. A. Lewin, Miniature piezoelectric polymer ultrasonic hydrophone probes, *Ultrasonics 19*: 213 (1981).

42. P. A. Lewin, Miniature piezoelectric polymer hydrophones in biomedical ultrasonics, *Ferroelectrics 60*:127 (1984).

43. M. Kobayashi, E. Nishihara, M. Maruyama, J. Morita, R. Omoto, M. Suzuki, T. Miya, and H. Ohigashi, High resolution ultrasound imaging with 7.5 MHz new polymer transducer, *Ultrasound Med. Biol. 8* (Suppl. 1): 99 (1982).

44. H. Ohigashi and K. Koga, Ferroelectric copolymers of vinylidene fluoride and trifluoroethylene with a large electromechanical coupling factor, *Jpn. J. Appl. Phys. 21*:L445 (1982).

45. H. Ohigashi, K. Koga, M. Suzuki, T. Nakanishi, K. Kimura, and N. Hashimoto, Piezoelectric and ferroelectric properties of P(VDF-TrFE) copolymers and their application to ultrasonic transducers, *Ferroelectrics 60*:263 (1984).

46. K. Kimura, N. Hashimoto, and H. Hoigashi, Performance of a linear array transducer of vinylidene fluoride trifluoroethylene copolymer, *IEEE Trans. Sonics Ultras. SU-32*:566 (1985).

47. A. Fiorillo, P. Dario, J. Van der Spiegel, C. Domenici, and J. Foo, Spinned P(VDF-TrFE) copolymer layer for a silicon-piezoelectric integrated US transducer, *Proc. IEEE 1987 Ultras. Symp.*, Denver, CO, 1987, p. 667.

48. R. G. Swartz and J. D. Plummer, Integrated silicon-PVDF acoustic transducer arrays, *IEEE Trans. Electron. Devices ED-26*:1921 (1979).

49. A. S. Fiorillo, J. Van der Spiegel, P. E. Bloomfield, and D. Esmail-Zandi, A P(VDF-TrFE)-based integrated ultrasonic transducer, *Sensors and Actuators A22*:719 (1990).

50. I. Yasuda, K. Noguchi, and T. Sata, Dynamic callus and electric callus, *Proc. J. Bone Joint Surg. 37*:1992 (1955).

51. Z. B. Friedenberg, J. D. Roberts, N. H. Didizian, and C. T. Brighton, Stimulation of fracture healing by direct current in the rabbit fibula, *J. Bone Joint Surg. 53A*:1400 (1971).

52. J. A. Spadaro, Electrically stimulated bone growth in animals and man, *Ann. N.Y. Acad. Sci. 238*:564 (1974).

53. E. Fukada and I. Yasuda, On the piezoelectric effect in bone, *J. Phys. Soc. Jpn. 12*:1158 (1957).

54. E. Fukada, T. Takamatsu, and I. Yasuda, Callus formation by electret, *Jpn. J. Appl. Phys. 14*: 2079 (1975).

55. S. Inoue, T. Ohashi, E. Fukada, and T. Ashihara, Electric stimulation of osteogenesis in the rat: Amperage of three different stimulation methods, *Electrical Properties of Bone and Cartilage* (C. T. Brighton, J. Black, and S. R. Pollack, eds.), Grune & Stratton, New York, 1979, p. 199.

56. H. Suzuki, H. Sasakura, H. Uchida, T. Fukuhara, Y. Togawa, H. Takahashi, and T. Konno, *Jpn. Comm. Elect. Enhanc. Bone Heal.*, Abst., 1976, p. 11.

57. J. J. Ficat, G. Escourrou, M. J. Fauran, R. Durroux, P. Ficat, C. Lacabanne, and F. Micheron, Osteogenesis induced by bimorph polyvinylidene fluoride films, *Ferroelectrics 51*:121 (1983).

58. F. Carlucci, P. Puntoni, M. Arispici, M. Cecconi, C. Domenici, P. Dario, and D. De Rossi, Prime esperienze sull'impiego di un polimero piezoelettrico (PVF$_2$) nella riparazione di fratture sperimentali nel ratto, *Ann. Fac. Med. Vet. (Pisa, I) 37*:47 (1984).

59. E. Fukada, Piezoelectricity of bone and osteogenesis by piezoelectric films, *Mechanisms of Growth Control* (R. O. Becker, ed.), Charles C Thomas, Springfield, IL, 1981, p. 192.

60. R. F. Valentini, P. Aebischer, S. R. Winn, S. K. Kunz, and P. M. Galletti, Peripheral nerve regeneration through guidance channels: The effect of channel composition, *Life Support Systems* 4:392 (1986).

61. N. B. Patel and M. M. Poo, Orientation of neurite outgrowth by extracellular electric fields, *J. Neurosci.* 2:483 (1982).

62. W. A. Nix and H. C. Hopf, Electrical stimulation of regenerating nerve and its effect on motor recovery, *Brain Res.* 272:21 (1983).

63. A. R. M. Raji and R. E. M. Bowden, Effects of high peak pulsed electromagnetic field on the degeneration and regeneration of the common peroneal nerve in rats, *J. Bone Joint Surg.* 65:478 (1983).

64. P. Aebischer, R. F. Valentini, P. Dario, C. Domenici, and P. M. Galletti, Piezoelectric guidance channels enhance regeneration in the mouse sciatic nerve after axotomy, *Brain Res.* 436:165 (1987).

65. M. Cocco, M. Toro, P. Migliorini, P. Bongioanni, A. M. Sabatini, P. Dario, R. Sacchetti, and H. Schmidl, Development of an implantable neural connector for limb prostheses: In vitro investigations, *Proc. 6th Mediterranean Conf. Med. Biol. Eng.*, Capri, Italy, 1992, p. 101.

66. R. F. Valentini, T. G. Vargo, J. A. Gardella, Jr., and P. Aebischer, Patterned neuronal attachment and outgrowth on surface modified, electrically charged fluoropolymer substrates, *J. Biomater. Sci. Polymer Ed.* 5:13 (1993).

67. T. Okoshi, H. Chen, G. Soldani, P. M. Galletti, and M. Goddard, Microporous small diameter PVDF-TrFE vascular grafts fabricated by a spray phase inversion technique, *Trans. Am. Soc. Artificial Intern. Organs* 38:201 (1992).

19
Applications of Ferroelectric Liquid Crystalline Polymers

Rudolf Kiefer

Fraunhofer-Institut für Angewandte Festkörperphysik, Freiburg, Germany

I. INTRODUCTION

In 1974 R. B. Meyer predicted ferroelectricity in chiral, smectic $C(S_C{}^*)$ liquid crystals by pure symmetry considerations, and his French co-workers gave first experimental evidence of this phenomenon [1]. Since then the interest in this field has grown steadily, and research activity emphasized the understanding of the underlying basic physical phenomena and synthesis of new low-molar-mass compounds exhibiting a broad ferroelectric phase. It was finally the work by Clark and Lagerwall in 1980 on a fast electrooptic effect in a surface-stabilized ferroelectric liquid crystal (SSFLC) structure [2] which opened up the promising possibility of technical applications of $S_C{}^*$ ferroelectric liquid crystals ($S_C{}^*$ FLCs) in high-information display devices. Through this work both basic and applied research were stimulated and accelerated tremendously as the SSFLC structure showed much faster switching times and bistability behavior in comparison with commonly used nematic liquid crystal displays.

At about the same period of time the first side-chain liquid crystalline polymers (SCLCP) were synthesized by Finkelmann et al. [3]. In these comblike polymers, mesogenic side chains are covalently bonded via flexible spacer units to the flexible polymer backbone. In contrast to low-molar-mass liquid crystals (LMM LCs), the most essential feature of SCLCPs is their glassy state, which allows a uniaxially ordered state to be frozen in by cooling below the glass transition temperature. Due to this unique additional property it may be possible to cover new fields of application such as nonlinear optical and optical data storage devices [4]. But these nematic or smectic A SCLCPs suffered from their slow switching time due to their high rotational viscosity.

It was therefore natural to combine the advantageous mechanical properties of side-chain liquid crystalline polymer materials with the fast switching characteristics of LMM

S_C^* FLCs to obtain a mechanically more sturdy but still electrically switchable film. This was achieved for the first time in 1984 by Shibaev et al. [5].

As with their LMM counterparts, S_C^* FLC side-chain polymers show piezoelectric, pyroelectric, and nonlinear optical effects, but because of their inherent glassy phase, a nonvanishing spontaneous polarization can be frozen in. This makes it possible to envisage new fields of applications but requires the development of novel materials which have to be tailored for the physical effect applied. Ideally, S_C^* FLC side-chain polymers can meet this requirements, as multifunctional properties can be induced by using different side-chain moieties chemically bonded on the polymer backbone. Apart from the large flexibility in molecular engineering, the processability of these materials should be easier than that of LMM FLCs. Although the number of new materials is now expanding enormously, the whole field of S_C^* FLCPs is still in its infancy.

In order to access the entire potential of S_C^* FLCPs it is very helpful to review the field of LMM S_C^* FLCs, which is better investigated and understood. This will provide the basis from which we can take full advantage when we deal with S_C^* FLC side-chain polymers.

II. PHYSICAL PROPERTIES, ELECTROOPTIC EFFECTS AND APPLICATIONS OF LMM S_C^* FLCs

A. The Ferroelectric Smectic Phases

Liquid crystals can exhibit a variety of mesophases which are distinguished with respect to the directional and positional ordering of their rod-shaped molecules. In the less ordered, so-called nematic phase, the long molecular axes possess an average preference direction which is given by the unit vector **n**, the director (Fig. 1a). If the molecules are chiral, i.e., lack mirror symmetry, a cholesteric (ch) phase is formed whereby the director undergoes a helical distortion as shown in Fig. 1b. In contrast to the nematic phase, in the higher-ordered smectic phases the constituent achiral molecules have directional ordering and are in addition packed in layers. If the average direction of the long molecular axes is perpendicular to the layer planes, and positional ordering within the layer planes is completely absent, one arrives at the S_A phase (Fig. 1c). In the S_C phase the long molecular axes are tilted with respect to the layer normal. For both the perpendicular and the tilted smectic phases, there exist more ordered smectic phases which are characterized by different positional ordering of the molecules within the smectic layers (S_B, S_E for the nontilted phases and S_I, S_F, S_G, S_H, S_J, S_K for the tilted ones). Tilted phases built up with *chiral* molecules give rise to chiral smectic phases, denoted by an asterisk, which all can be expected to be ferroelectric by symmetry considerations (S_C^*, S_I^*, S_F^*, S_G^*, S_H^*, S_J^*, S_K^*) [1]. In the following we will deal only with the least ordered, chiral, tilted S_C^* phase, which seems to be the most promising one for applications owing to its fast electrooptic response time.

B. Structure of the S_C^* Phase

In the S_C^* phase the constituent rod-shaped molecules are arranged in layers, and long-range positional ordering of the centers of gravity of the molecules in the smectic layer planes does not appear. Furthermore, the long axes of the constituent molecules are tilted on average by an angle θ with respect to the normal direction **z** to the layer planes. The

(a)

(c)

(b)

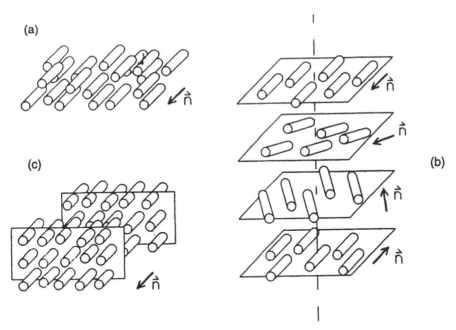

Figure 1 Schematic representation of (a) nematic, (b) cholesteric, (c) smectic A.

chirality of the molecules induces a helical arrangement of the tilted director on moving from layer to layer with a helical periodicity p, which is typically large compared to the layer spacing $l(p/l \approx 10^3$; see Fig. 2).

The appearance of ferroelectricity in the $S_C{}^*$ phase (or, more generally, in chiral, tilted smectic phases) can be explained by considering the local symmetry of the phase in question (see Fig. 3) [1]. A single S_C layer consisting of *achiral* molecules possesses a twofold axis of rotation (C_2 symmetry) and a mirror symmetry in the *x-z* plane. This local C_2 symmetry is based on the fundamental assumption that the director **n** is invariant against sign reversal: **n** and −**n** are equivalent. This means that a ferroelectric polarization along **n** may not appear, as so far it has not been observed. Recently, however, first evidence of such a phase, probably breaking the C_2 symmetry of the achiral S_C phase, has been found [6]. It can be shown that the existence of spontaneous polarization is related to the symmetry properties of the $S_C{}^*$ phase.

For this purpose we assume an arbitrary polarization vector $\mathbf{P}_S$ and apply the C_2 symmetry operation (180°C rotation around the *y* axis) according to Fig. 3:

$$\mathbf{P}_S = \begin{bmatrix} P_x \\ P_y \\ P_z \end{bmatrix} \xrightarrow{C_2} \begin{bmatrix} -P_x \\ P_y \\ -P_z \end{bmatrix} \longrightarrow \mathbf{P}_S = \begin{bmatrix} 0 \\ P_y \\ 0 \end{bmatrix}$$

We realize that P_S can have only a nonvanishing component in the *y* direction. Next, we look for the consequences of the mirror symmetry of the S_C phase in the *x-z* plane:

$$\mathbf{P}_S = \begin{bmatrix} 0 \\ P_y \\ 0 \end{bmatrix} \xrightarrow[\text{symmetry}]{\text{mirror}} \begin{bmatrix} 0 \\ -P_y \\ 0 \end{bmatrix} \longrightarrow \mathbf{P}_S = \begin{bmatrix} 0 \\ 0 \\ 0 \end{bmatrix}$$

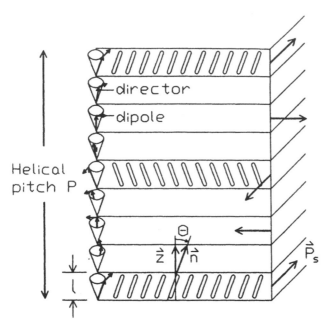

Figure 2 Structure of the S_C^* phase: p = helical periodicity (pitch); l = smectic layer thickness; θ = tilt angle between the layer normal z and the director n; P_S = local spontaneous polarization. In the picture P_S is chosen positive.

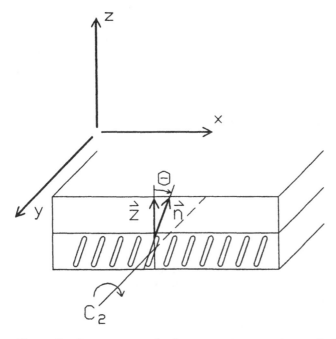

Figure 3 C_2 symmetry and mirror symmetry operation applied to a single smectic C layer.

This symmetry operation results in $P_S = 0$ and the conclusion is that an S_C phase cannot be a ferroelectric one. However, the mirror symmetry of the S_C phase is spontaneously broken when it is composed of chiral molecules. Then the mirror plane is lost and only the C_2 symmetry is preserved locally in each smectic layer. Consequently, the C_2 axis is the polar axis, which admits the existence of a spontaneous polarization along it. For the same symmetry reasons, ferroelectricity may also exist in higher-ordered tilted smectics composed of chiral molecules.

However, as a secondary and necessary prerequisite for the appearance of spontaneous polarization, the molecules must possess a dipole component perpendicular to the long molecular axis (as illustrated schematically in Fig. 2), and the rotation around the long axis is no longer free but is asymmetrically hindered in such a way that on average the lateral dipoles of all molecules spend a longer time in pointing along the y (or $-y$) direction than in all other azimuthal directions.

In general, the spontaneous polarization in the S_C* phase is about two orders of magnitude lower than one could expect if the dipoles were to add to full strength. Depending on the molecular structure of the FLC, the magnitude of polarization varies typically between 2 nC/cm^2 and 500 nC/cm^2. In contrast to solid-state ferroelectrics, the existence of spontaneous polarization in the S_C* phase is not the result of dipole–dipole interactions but is due to a secondary effect of the intermolecular forces of the optically active molecules, as is characteristic of *improper* ferroelectrics.

Within one smectic layer P_S points along the y direction, which is perpendicular to both the director n and the smectic layer normal z (Fig. 3):

$$P_s = P_0(z \times n) = P_0 \sin \theta \cdot y \tag{1}$$

The sign of P_0 is an intrinsic property for a given material, and by convention P_S is called positive if z, n, and P_S form a right-handed system. Optical antipodes always show opposite signs of polarization.

Due to the helical structure, the director n spirals on a cone when moving perpendicular to the planes of the smectic layers (see Fig. 2). Consequently, the local polarization vector P_S also rotates around the helix axis in phase with the change in the azimuthal angle of the director. The resultant macroscopic polarization of a sufficiently bulky sample therefore averages to zero by symmetry reasons.

There is one pecularity concerning the helix structure which should be mentioned. In contrast to the cholesteric helix, the chiral smectic C helix contains a spontaneous bend in addition to the spontaneous twist [7]. This leads to a local flexoelectric polarization which adds up to the ferroelectric polarization [8]. Beresnev et al. [9] have claimed that the latter contribution clearly dominates, but the flexoelectric term may become significant for materials with a short helix period $p < 1$ μm.

To prove the existence of spontaneous polarization P_S, one has to unwind the helix by an electric field, by applying a shearing force parallel to the layers [10] or by suppressing the helix by surface interactions using a thin surface-stabilized FLC structure [2]. The magnitude of P_S can be measured by applying a sine-shaped electric field with the conventional Sawyer-Tower capacitance bridge method [11], with the polarization reversal method using triangular waves [12a], or square waves [12b], or by a pyroelectric technique [13].

Employing a DC electric field induces a S_C* monodomain, thereby orienting all dipoles in the same direction and creating a structure as shown in Fig. 4. Reversing the electric field direction switches the polarization vector in the opposite direction, and the

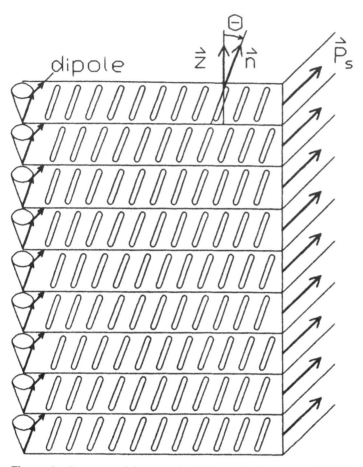

Figure 4 Structure of the smectic C* phase in the unwound helix-free state.

director **n** by twice the optical tilt angle. In this situation the optical tilt angle θ can be also measured by determining the minima of transmission between crossed polarizers.

Due to Eq. (1), P_S and θ show similar temperature behavior, depending on whether the phase transition into the above-lying phase is of first order or of second order. The phase transition S_C^*-ch is always first order, and an abrupt change of P_S and θ is observed when the transition temperature is passed from above. Within the S_C^* phase, both parameters increase weakly when the temperature is further reduced, and the optical tilt angle usually saturates rapidly to a value of approximately 45°. Materials that exhibit an S_C^*-S_A phase sequence usually show a continuous increase of P_S and θ by cooling below the transition temperature, because this type of phase transition is usually second order. However, within the S_C^* phase, P_S and θ depend more strongly on temperature than in the case of the S_C^*-ch phase sequence, and θ saturates more slowly to a value of about 25°.

Common solid ferroelectrics are generally defined as materials showing a zero-field spontaneous polarization which can be reversed by an external electric field, and then they exhibit a hysteresis behavior. In the absence of aligning forces as imposed by electric fields or by boundary surfaces, thick samples of chiral, tilted S_C^*, S_I^*, or S_F^* materials possess a helical superstructure which averages the spontaneous polarization macroscopically to zero and forbids the appearance of hysteresis. Nevertheless, in the presence of an AC electric field the helical structure is unwound, and these materials reveal an apparent polarization hysteresis if the frequency of the AC field is higher than or comparable to the inverse relaxation time of the helix. At static driving conditions the hysteresis disappears completely. Meyer pointed out that it is always possible to obtain helix-free (unwound) materials with zero-field spontaneous polarization by mixing different compounds with right-handed and left-handed helices, where the twists but not the dipole moments are compensated at a certain concentration [1]. First evidence of nonhelical ferroelectric liquid crystals has been given by Beresnev et al. [14]. On the other hand, in the higher-ordered ferroelectric phases S_G^*, S_H^*, S_J^*, and S_K^*, the helical ordering is suppressed by the long-range positional ordering of the molecules within the layer planes, creating a zero-field polarization and a distinct hysteresis [15].

C. Electrooptic Effects

In principle, ferroelectric switching between the two states of polarization can be achieved by employing a DC electric field, thereby unwinding the helical arrangement of the molecules. However, if the electric field is switched off, the uniform aligned monodomain will relax slowly into the helical structure.

In 1980 a much more brilliant approach was described by Clark and Lagerwall in which the ferroelectric character of the S_C^* phase is optimally exploited [2]. They proposed the so-called surface-stabilized ferroelectric liquid crystal (SSFLC) structure, which has the unique combined properties of fast response speed (tens of microseconds) *and* bistability. The originally suggested SSFLC structure is based on the following ideas:

The formation of the helix is suppressed in the fieldless state by choosing a very thin FLC layer thickness $d < 2$ μm and appropriate boundary conditions. In general, d should be smaller than the helical pitch p: $d < p$.

A homogeneous planar surface alignment is imposed by shearing the sample (or by appropriate alignment layers), whereby the smectic layers develop perpendicular to the two confining glass plates, forming the so-called bookshelf structure.

As a result of the added constraint enforced by the boundary conditions, the conical degeneracy of the director is removed and *only two* molecular orientations are possible which are uniform from one plate to the other and have an angular difference of nearly twice the S_C^* tilt angle θ, as illustrated schematically in Fig. 5. These two directions, termed n_1 and n_2, form energetically equivalent domains with their corresponding polarization vectors pointing either upward or downward along the normal of the glass substrates. By applying an electric field perpendicular to the glass plates, this twofold degeneracy is eliminated and the director orientation becomes favored in which P_S aligns parallel to the electric field. Upon switching off the field this state is preserved by surface interactions until a reverse electric field is applied. This then makes the polarization vector swing around by 180°, thereby forcing the director to rotate on the cone until the net change of orientation is nearly twice the S_C^* tilt angle. The memory capability holds for both states, i.e., the SSFLC structure, is *bistable* and can be switched by voltage pulses. Furthermore, the switching effect possesses a *dynamic threshold* and, due to the strong ferroelectric torque ($P_S \times E$) between the electric field E and P_S, it exhibits *high-speed*

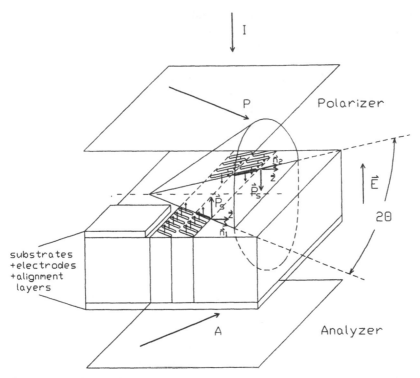

Figure 5 Surface-stabilized ferroelectric LC structure (SSFLC) with the two possible bistable orientations and the corresponding electrooptic effect according to Ref. 2.

response (some 10 μs) compared with the slowly responding nematic LCs (some 10 ms). The combination of high speed, bistability, and threshold makes the SSFLC effect in particularly suitable for passive matrix-addressed displays with high information content.

In its simplest geometry the electrooptic effect is achieved by placing the birefringent SSFLC structure between crossed polarizers and aligning the director n_1, for example, parallel to the direction of one polarizer, leading to maximum extinction of the transmitted light. Upon switching, the optical axis n can be rotated over a net angle 2θ pointing along the n_2 direction within the substrate plane. In this case the light becomes in general elliptically polarized by the SSFLC structure and the fraction of transmitted light in the bright state is given by

$$I = I_0 \sin^2(2\phi) \sin^2\left(\frac{\pi d \, \Delta n}{\lambda}\right) \tag{2}$$

Here I_0 is transmission for parallel polarizers, d is the cell thickness, ϕ is the angle between the polarization direction and the director n_2, Δn is the refractive index anisotropy, and λ is the vacuum optical wavelength. Maximum transmission and contrast are achievable for

$$\phi = 2\theta = 45° \qquad \theta = 22.5°$$

and $$d \cdot \Delta n = (2k + 1) \cdot \frac{\lambda}{2} \qquad k = 0, 1, 2, \ldots \tag{3a}$$

The required tilt angle of 22.5° can be conveniently fulfilled by FLC materials exhibiting an S_C^*–S_A phase transition. The transmission I as a function of $d \cdot \Delta n/\lambda$ according to Eq. (2) for $\phi = 45°$ is plotted in Fig. 6. The bars mark the range of the visible spectrum for the indicated value of Δnd. One clearly recognizes that in the first maximum ($k = 0$) the transmission is less sensitive to wavelength than for k values ≥ 1. In order to obtain colorless black and white states in this birefringent-type cell, the optimal choice will therefore be

$$d \cdot \Delta n = \frac{\lambda}{2} \simeq 0.28 \tag{3b}$$

Since currently available FLC materials possess a Δn of approximately 0.14, which results in a thickness d of the FLC layer of 2 μm. Fortunately, this value is quite compatible with the required thin FLC layer thickness for realizing the helix-free SSFLC structure.

In its ideal state fulfilling the conditions given in equation (3) with $k = 0$, such a SSFLC device is a switchable $\lambda/2$ wave plate as it rotates the optical axis by 45° and the polarization of the incident light by 90°. It can be used as a display, spatial light modulator, shutter, real-time mask, or in a linear array such as a printhead.

In the past, alignment of FLCs had been one of the key problems to be solved. Clark and Lagerwall originally used a flow alignment procedure whereby the FLC, after being introduced between the two substrates, is flow aligned by shearing the top substrate relative to the bottom substrate. Nowadays, however, as with nematic LCs, usually a thin polymeric alignment layer is taken which is spin-coated onto both substrates and then

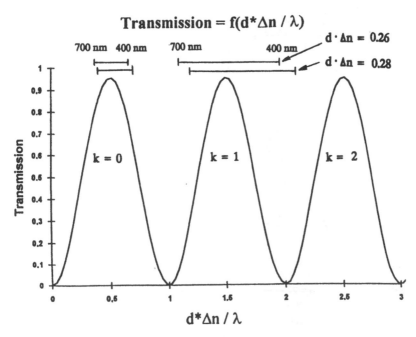

Figure 6 Transmission I according to Eq. (2) as a function of $d^*\Delta n/\lambda$. The bars mark the range of the visible spectrum (400–700 nm) for the indicated value of $d \cdot \Delta n$ (0.26, 0.28) in the first ($k = 0$) or second ($k = 1$) order of birefringence.

both are unidirectionally rubbed. Next, the two substrate plates are assembled using an appropriate spacer technology and the ready cell is filled with the FLC in the isotropic phase. Thereafter the FLC cell is slowly cooled into the S_C^* phase to establish the required SSFLC structure. Another technique uses an oblique evaporated SiO_x layer instead of a polymeric layer. However, only the former method seems suitable for production use. The alignment mechanism is very complex and is still not well understood and is thus a matter of trial and error.

The review concerning the SSFLC structure, as described above, represents an idealization. X-ray studies on planar aligned cells in the S_C^* phase with materials possessing the ch-S_A-S_C^* phase sequence have revealed that the smectic layers do not form a bookshelf structure but are inclined by an angle δ with the bounding plate normal, forming the so-called chevron structure as shown schematically in Fig. 7a [16]. In the above-lying smectic A phase the layers in fact stand perpendicular to the confined plates if the molecules at the boundaries are oriented parallel to the surfaces. Cooling from the S_A to the S_C^* phase leads to a layer shrinkage as the molecules tilt away from the layer normal by the tilt angle θ. Since the layers are not free to slide along the surfaces to remove layer defects caused by the layer shrinkage (strong anchoring), and the number of layers is conserved at the S_A–S_C^* phase transition, the formation of the chevron structure can be explained in a quite natural way.

If the long molecular axes are aligned parallel (or approximately parallel) to the substrate surface, the layers may tilt in either direction at the S_A–S_C^* transition and two

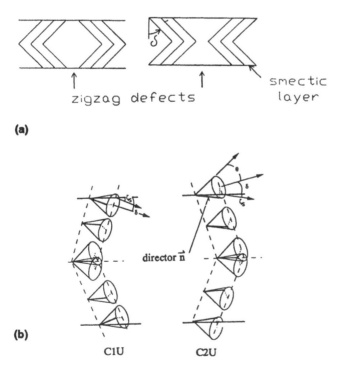

Figure 7   (a) Chevron structure and "zigzag" defects in the S_C^* phase. (b) The types of chevrons which may occur in thin SSFLC cells. C1U and C2U both have uniform director structure and occur preferably for high-pretilt (C1U) and low-pretilt angle (C2U). (From Ref. 31.)

energetically equivalent types of chevrons may form as shown in Fig. 7a. As a consequence, two new types of defects can develop at the boundary between chevrons of opposite inclination. These defects are called "zigzags." Their main characteristics were reviewed by Fukuda et al. [17], and their location is indicated schematically in Fig. 7a. Because they degrade the optical uniformity of the display, they should be avoided.

Contrary to the mild shearing procedure used originally [2], the alignment of FLCs is today commonly obtained by a thin buffed polymeric alignment layer. This method actually induces a small inclination (pretilt angle ζ_s) of the FLC molecules relative to the substrate plane and the buffing direction is parallel to the smectic layer normal (or director $\mathbf{n}$) in the S_A phase. Owing to this pretilt angle, one type of chevron may be preferred and the occurrence of the zigzags can be kept at a minimum if the rubbing directions on the upper and lower boundary plates are parallel assembled [18]. However, the display manufacturer aims to select one particular type of chevron.

Which type of chevron actually appears can be controlled by carefully selecting a particular pretilt angle ζ_s for a given tilt angle θ and layer tilt angle δ. In essence, high values of pretilt angles bias the so-called C1U chevron, whereas the C2U chevron is preferred for medium (low) values of pretilt [19]. A high pretilt angle (> 15°) can be realized by a thin SiO_x layer oblique evaporated at an angle of 85° relative to the substrate normal [20] or with a buffed high-pretilt polyimide [21,22]. In both cases zigzag-free alignment of the FLCs is obtained. Figure 7b shows both types of chevrons including the director structure for one polarization state. Both the C1U and the C2U states possess uniform director structure and are therefore very suitable for achieving full extinction between crossed polarizers. Only the C1U state, however, shows a large memory switching angle, which guarantees high contrast. Depending on the cell thickness, the director–surface interaction, and the inherent bend of the smectic C* phase, so-called twisted (also called splayed) director structures may occur [17]. These states are characterized by low contrast and should be avoided as they show no light extinction between crossed polarizers, because the incident light is partially guided through the twisted structure. They can be suppressed by choosing a small cell gap a low polarization P_S and a high pretilt angle [23].

If zigzag defects already exist, they can be removed by an adequate AC electric field treatment which causes a change from the chevron structure to a slightly bent layer structure [24,25] which is termed "quasi-bookshelf." Alternatively, a quasi-bookshelf alignment has been obtained for special classes of FLC materials [26]. To summarize, there are at present four different optical structures which avoid the occurrence of zigzag defects and are considered for application:

High-pretilt C1U chevron state
Low-pretilt C2U chevron state
Quasi-bookshelf by electrical field treatment
Quasi-bookshelf by choice of appropriate FLC materials

We will now discuss the dynamic aspects of SSFLC structures. Neglecting elastic and dielectric torques, a simplified model of the optical switching process gives the switching time τ of the director as

$$\tau = \frac{\gamma \sin^2 \theta}{P_S \cdot E} = \frac{\gamma_c}{P_S \cdot E} \tag{4}$$

where γ is the rotational viscosity for switching motion on the cone and $\mathbf{E}$ is the applied electric field. The quantity $\gamma_c = \gamma \cdot \sin^2 \theta$ is the rotational viscosity for the motion of the projection of the director on the smectic layer plane. The optical switching time τ is usually given for a change in transmitted intensity from 10% to 90% of the maximum intensity.

Fast switching times can be achieved by FLC materials possessing a low viscosity γ_c and a high polarization value P_S. Using high P_S compounds, however, leads to two problems. First, the electric current i which flows through the sample cell during the switching process is roughly proportional to P_S/τ and, from Eq. (4), this results in $i \simeq P_S^2/\gamma_c$. As high currents should be avoided, it is obvious that fast switching and small currents are preferably obtained with low rotational viscosities rather than large polarization values. Second and more significant, high P_S values generate large internal electric fields, which make the bistability deteriorate and give rise to reverse switching and ghost picture phenomena in FLC displays [27,28]. For both these reasons, small P_S materials (<15 nC/cm^2) are preferably used in high-information FLC displays (see, for example, Ref. 23), and the switching times cannot be shortened simply by high P_S values.

With regards to addressing of the SSFLC structure, there is another peculiarity. Instead of a static threshold voltage for switching, a dynamical threshold condition exists, which depends on voltage amplitude V and pulse length τ_L required for latching into the opposite optical state of a single domain [29]:

$$V \cdot \tau_L = A = \text{const.} \tag{5}$$

with $\tau_L = k \cdot \tau$ and τ according to Eq. (4). One should bear in mind that τ describes the bulk switching behavior, whereas τ_L is strongly influenced by surface interaction controlling the memorization process. k depends in a complicated way on the anchoring conditions, the voltage V, and the addressing scheme. For high electric fields and weak surface interactions, $\tau_L \approx \tau$ is approximately true [29].

Equation (5) is fulfilled for the high-voltage region (>5 V/μm), and the critical pulse area $\tau_L \cdot V$ depends on the cell thickness d. If the applied value $\tau_L \cdot V$ is below a certain constant value called A, monostable switching happens while the above latching occurs.

However, as the ferroelectric torque is proportional to $P_S \cdot E$, and the dielectric torques act proportionally to E^2, the influence of the latter can become significant, particularly in the case of low-P_S materials. Then, as a consequence, the τ_L–V characteristic deviates from the relation predicted by Eq. (5), and results in a slowing down of the switching speed and the observation of a minimum response time in the corresponding τ_L–V curve as shown schematically in Fig. 8 [30]. In the past a negative dielectric anisotropy $\Delta\epsilon = \epsilon_3 - \epsilon_1$ was suggested as the cause of this effect, but now it seems clear that the dielectric biaxiality $\delta\epsilon = \epsilon_2 - \epsilon_1$ often dominates over the influence of $\Delta\epsilon$ ([31] and references therein). Here ϵ_3 is the permittivity along $\mathbf{n}$, and ϵ_2, the permittivity along the $\mathbf{P}_S$ direction, is usually larger than ϵ_3 and ϵ_1. Therefore, an additional applied high-frequency AC electric field which couples to the dielectric biaxiality but not to P_S tends to stabilize the two fully switched ferroelectric states. This can be exploited for contrast improvement if the C2U chevrons are used. As the corresponding memory switching angle is inherently low, a superposed AC field which couples to $\delta\epsilon$ can maintain fully switched states with large extinction angle. Operating in the τ_L–V minimum according to Fig. 8 with AC field stabilization, the C2U chevrons are preferable to the C1U chevrons as they allow shorter pulse widths for memory switching [32].

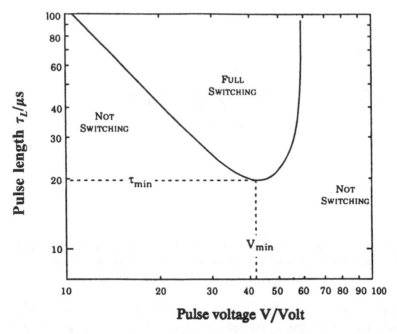

Figure 8 Typical pulse width versus pulse voltage curve for an FLC material with low P_S. (From Ref. 31.)

The problem of alignment in the S_C^* phase is closely connected with the type of electrooptic effect which is applied. In the birefringence mode as illustrated in Fig. 5, one ideally uses an FLC material with the phase sequence isotropic-ch-S_A-S_C^* with a switching angle $2\theta = 45°$. Using the same geometry as in the birefringent-type cell, alternatively a FLC-guest host type of display can be built by doping the FLC with appropriate dichroic dyes (Hitachi [33]). The dye molecules align with their long molecular axes approximately parallel to the FLC director. They absorb a maximum amount of light if the incident light is polarized parallel to the long axis of the dye molecule, while the absorption shows a minimum when the dye molecules are oriented perpendicular to the polarization direction of the incident light (see Fig. 9). This type of display therefore requires one polarizer, and the optimum switching angle 2θ must be $90°$, i.e., twice the birefringent case. The latter requirement can only be met by FLC materials with the isotropic-cholesteric-S_C^* phase succession, avoiding the S_A phase. But these materials have alignment problems if rubbed polymer alignment layers are used. Due to the strong nematic surface anchoring, smectic layers are generated whose layer normals point at two equivalent directions $\pm\theta$ from the buffing direction [29]. In this case, the two ferroelectric states correspond to the same molecular orientation, thus making the device useless. A switchable monodomain can be created by the application of a DC electric field during the cooling from the cholesteric into the S_C^* phase, or by choosing different surface treatments for the two substrate plates [34,35]. Furthermore, in order to obtain acceptable absorbance, thicker cells (about $4-5$ μm) are needed, which require specially designed long-pitch mixtures, avoiding the appearance of twisted states. On the other hand, a larger cell thickness variation can be tolerated. Besides the birefringent mode and the guest

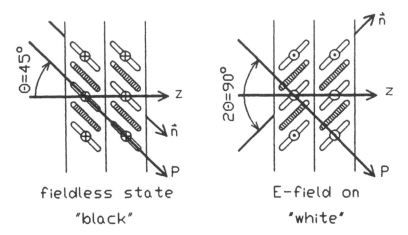

Figure 9 Top view of a guest host-type FLC display. The dye molecules are schematized by hatched rods and P_s points upward ("white") or downward ("black") to the drawing plane.

host mode of SSFLC structure, as discussed above, a number of additional FLC device configurations, particularly operating in the reflective mode, have been proposed [36], which could be usable for electrooptic applications with S_C^* FLC polymers.

On the basis of the SSFLC effect, a number of FLC display prototypes have been developed [23,37–40]. The first multicolor 12-in.-diagonal FLC display with 639 × 400 pixels and video capability ($33\frac{1}{2}$ Hz) was presented by Toshiba [37]. Matsushita has demonstrated a high-resolution 38-in.-diagonal, 8-color projection display with 2000 × 2000 pixels [38]. With a line addressing time of 240 µs, it did not have video capability.

In the ambitious JOERS/ALVEY Project, STC, Thorn EMI, RSRE, BDH, and the University of Hull have joined to push forward the development of ferroelectric displays. A 10.4-in.-diagonal, VGA display with 8 colors and 10-Hz frame rate for laptop applications was presented [39]. To establish this technology the C2U chevron state was selected and a high-frequency AC stabilization operating in the τ–V minimum was used [31].

The most remarkable R&D effort has been initiated by Canon, Inc. [19,23,40], which presently represents the state of the art of FLC display technology in many respects. The newest development is a 15-in.-diagonal display which is available as a monochrome or digital color version with 1280 × 1024 pixels. The panel, which is intended for personal computers and workstations, works with a 70-µs line address time, a 14-Hz frame frequency, and shows high image quality and good readability up to large viewing angles. In the color version, each pixel consists of the subpixels red, green, blue, and white. The monochrome display contains pixels which are subdivided into two subpixels, allowing at least 4 gray levels. Canon used a polyimide alignment layer with a high pretilt angle (>18°) in combination with a 1.5-µm cell spacing to suppress twisted states and to obtain the uniform C1U chevron state without zigzag defects [23]. However, this technology still has difficulties, which have to be overcome in order to produce competitive products. The most serious problems are:

Cell technology
Defect-free smectic C* alignment and their mechanical and thermal shock sensitivity
Gray-scale capability

Bistability and threshold voltage of the SSFLC structure allow time multiplex addressing. In this driving scheme, the image information is composed of m columns and n rows requiring $m + n$ connections. The frame is built up first by simultaneously feeding the data voltages into all column electrodes and addressing a single row electrode. Afterward, the data voltages for the next line are set and the corresponding row is addressed. In this way, the whole frame is built up line by line and for a given frame rate. Due to the memory feature of the SSFLC effect, the number of lines which can be shown is limited by the response time of the FLC (for a 25-Hz frame rate and 625 lines, a switching time of 64 μs/line is required). Nowadays the commercially available broad-range FLC mixtures have achieved switching times of about 40 μs at room temperature for electric field strengths of 15 V/μm and moderate P_S values.

Due to the inherent bistability of the SSFLC structure, intermediate states, which are necessary for gray scales, are not straightforward to implement. Digital gray-shade techniques rely on the spatial subdivision of pixels into subpixels or the temporal subdivision of the frame time into subframes. In the former method, the subpixels are addressed individually, which increases the number of connections and sets limits on the achievable spatial resolution. The latter technique requires shorter line address times. Thus the switching speed of the FLC material has to be increased with the number of gray levels. Often a combination of these two "dither" techniques is applied [41]. An analogous gray shade can be established by multidomain switching [42]. If an SSFLC structure is addressed by a voltage pulse of definite pulse length and amplitude, and afterward electrically isolated, only a part of the pixel area can be switched, which is determined by the amount of charge which is transferred to the pixel during selection. A multidomain structure appears which remains stable. It is possible to stabilize the domain growth by controlling the amount of charge and thus obtain a continuous gradation. A drawback to this technique is the necessity of a thin-film transistor (TFT) per pixel in order to isolate each pixel electrically.

According to Eq. (5), the dynamic threshhold depends on the cell thickness d. This requires a very precise 1–2 μm cell technology where only a small thickness variation can be tolerated. Canon has developed a cell technology with a thickness tolerance of 0.05 μm [43].

The most severe problem which has delayed the commercialization of SSFLC displays seems to be the sensitivity of the smectic alignment against mechanical shock and pressure. As a consequence, mechanical load irreversibly damages the smectic ordering and creates defects which cannot be repaired in a simple way. This is the point where S_C* FLC polymers can be superior to LMM FLCs, as they should be more resistive to mechanical stress and pressure.

Another FLC application is line-at-a-time image generation on photosensitive media for nonimpact printing. A single-line array of FLC light valves is imaged onto the rotating photoconductor drum, producing a charge pattern line by line. The first operating print head was realized by Hitachi in 1985 [44]. This FLC device consisted of 2048 light shutters in parallel, had a response time of 250 μs per line and a resolution of 10 lines/mm. As a result, 12 A4 pages could be printed per minute. New challenges are color, gray scale, high resolution, and multiple-line arrays with multiplex addressing. Recently, a full-color printing head with 640 dots and 6 lines for the colors red, green, and blue has been presented [45]. This device is able to generate 32 gray levels in each RGB color and uses a 1:6 multiplex driving scheme. FLC print heads have no moving parts, and no complicated mechanics are involved. The writing speeds achievable with FLCs can be compared to commercial xerographic printers or laser printers.

A new field of growing interest is in optical computing, taking advantage of highly parallel optical architectures where information is encoded into optical "images" [46–49]. A fundamental component in these systems is a spatial light modulator (SLM), which can produce a two-dimensional modulation pattern on the cross section of a light beam by electrical or optical addressing. Such a device can be used in optical image processing systems (optical pattern recognition, optical correlation, spacial filtering) [46–49], or in optical neuro computing [48]. The *electrically addressed* versions have some similarity with the flat-panel displays described above, but the speed requirements for optical computing are more stringent. Due to the sequential line-at-a-time addressing, electrically controlled FLC SLCs are limited in their frame rate. In *photoaddressed* devices, frame rates equal to the switching speed can potentially be attained by optically addressing all picture elements *in parallel* via a photoconductive sensor (see Fig. 10). A frame rate of 1 μs is expected [47], and with an image consisting of 1000×1000 elements (bits), this would result in a bit processing rate of about 10^{12}/s. An SLC device based on FLCs has a good chance to compete with other currently used inorganic materials such as lithium niobate and PLZT [47]. While these materials have intrinsic switching times much faster than that of optimized FLC SLC devices, the power dissipation limitations prevent conventional materials from coming close to their intrinsic switching times. Only the FLC comes close to the microsecond frame target, because a large birefringence change can be managed in a small-sized pixel and, most important, with a very low switching energy [47]. Besides, the inherent bistability of FLCs allows binary optical memory. Since the individual picture elements of a SSFLC structure can be switched between two orthogonal polarization states of light (half-wave plate), polarization-based binary logic can be implemented, whereby light losses are noncritical [48]. In the optically addressed FLC SLM the incident intensity pattern is absorbed in the photoconductor and converted into a spatially varying charge pattern. This in turn controls and switches the FLC layer and modulates the read beam (see Fig. 10), which is operated in the reflective mode. For

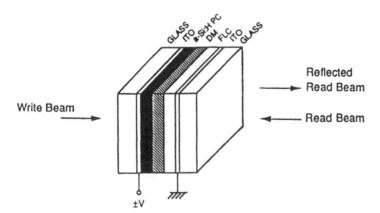

Figure 10   The schematic sandwich structure of the optically addressed FLC-SLM (spatial light modulator) operating in the reflective mode. The dielectric mirror (DM) prevents the incident read light from illuminating the photoconductor (a-Si:H PC). The read beam is linearly polarized.

the photosensor a *p-i-n* photodiode using hydrogenated amorphous silicon (a-Si:H) shows a sufficiently fast photoresponse (~100 ns) [50].

The whole field of optically addressed SLMs is at an early stage of research, and at present much effort is being made to improve the performance of these devices [51].

Besides the Clark-Lagerwall effect, some further electrooptic effects have been discovered which are discussed in particular for applications where gray scale is required.

The distorted helix ferroelectric (DHF) effect is based on FLC compounds with a very short pitch (0.2–0.5 μm) [52–55]. In suitable cells (cell gap much larger than the pitch and weak surface interaction), the smectic layers form a bookshelf arrangement and the helix is not suppressed as in the SSFLC structure, but is fully developed. The director spirals around an axis perpendicular to the smectic planes (*z* axis). If the helix period becomes smaller than the wavelength of visible light, the local refractive index ellipsoid is spatially averaged over the helix, resulting in an apparent birefringence $\langle n_e \rangle - \langle n_0 \rangle = \langle \Delta n \rangle$ which is smaller than in the case of the SSFLC structure. In its fieldless state the apparent optical axis $\langle n_e \rangle$ points along the helix axis direction *z* and, by applying a weak electric field, the helix is easily distorted in such a way that, depending on the polarity of the electric field, the azimuthal directions with cos φ = −1 (*E* < 0) or cos φ = 1 (*E* > 0) are preferred because they have the favorable orientation of P_S to the applied electric field direction **E** (see Fig. 11a). This distortion results in a rotation of the index ellipsoid as well as in an increase in the apparent birefringence (see Fig. 11b). Between crossed polarizers the DHF effect shows a *linear* electrooptic effect *without* an inherent *bistability*. As a consequence, this effect has gray-scale capability and operates at much lower voltages than the SSFLC effect, as the onset of helix unwinding must be effectively prevented. Nevertheless, short response times in the range of 30 μs with driving voltage of ±5 V have been achieved [56].

A DHF cell can be operated in two different modes. In the asymmetric driving mode the polarizer is set at an angle of 22.5° to the helix axis direction. To obtain maximum contrast, the maximum apparent rotation angle $\langle \theta \rangle$ of the optical axis, which corresponds to the tilt angle, should be ±22.5°. To turn the cell on or off, a positive or negative AC voltage is always required. In the preferential symmetrical driving mode the helix axis is parallel to the entrance polarizer and in the powerless state the transmission is zero. The transmission increases for positive and negative voltages, but for optimum transmission a larger switching angle of ±45° is required.

A drawback of the DHF effect is that it requires active matrix addressing, as it has no inherent threshold voltage, which is a prerequisite for multiplexing. On the other hand, the electrooptical performance should be less sensitive to alignment and layer thickness variations than in the case of the SSFLC effect. Most recently, it has been claimed that alignment defects induced by mechanical stress can be recovered by AC voltage treatments [56]. If active matrix addressing is acceptable and gray scale is required, the DHF effect can be employed, in principle, for all applications discussed so far for the SSFLC structure.

More recently, Fünfschilling et al. have described another new *short-pitch* but *bistable* ferroelectric S_C^* (SBF) configuration, which requires materials with a short pitch ($p \leq 0.4$ μm) and large spontaneous polarization $P_S > 60$ nC/cm^2 [57,58]. The probable SBF structure, which is not yet completely clear, is modified by an electric field to an "inverted chevron structure," which appears between crossed polarizers as a pattern of fine black-and-white stripes if one polarizer is set parallel to the optical axis within one stripe. Contrary to the SSFLC structure, the chevrons are arranged in the substrate plane

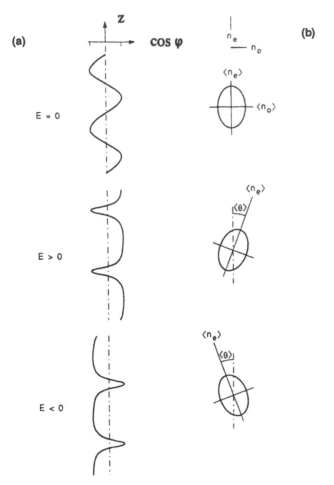

Figure 11 Distorted helix ferroelectric (DHF) effect (from Ref. 54). The effect of moderate electric field strength E on the apparent direction $\langle n_e \rangle$ of the optic axis is shown schematically. Influence of E on: (a) the azimuthal direction ϕ of the tilt along the helix axis z; (b) averaged indicatrix, averaged optic axis $\langle n_e \rangle$, and apparent tilt $\langle \theta \rangle$ along the helix axis with $\langle \Delta n \rangle \langle n_e - n_0 \rangle$.

(see Fig. 12), and conventional rubbing techniques are used for alignment. The direction of these stripes is roughly parallel to the rubbing direction. The stripes prevent helix formation in short-pitch mixtures. Due to the chevrons, a periodic modulation of the refractive index perpendicular to the lines occurs. As in the SSFLC structure, the electrooptic effect of the SBF texture is due to a rotation of the refractive index ellipsoid in the plane of the cell by twice the tilt angle θ within each stripe. Therefore thin cells with $d \cdot \Delta n = \lambda/2$ and a switching angle of $2\theta = 45°$ are required; however, it is difficult to obtain a good black state as there is no unique director orientation in the different stripes (see Fig. 12). Therefore, contrast is limited by the residial transmission in the off state. When the voltage across the cell is removed, the switched states persist within a memory

bookshelf (S_A or S_C)   SBF with stripes

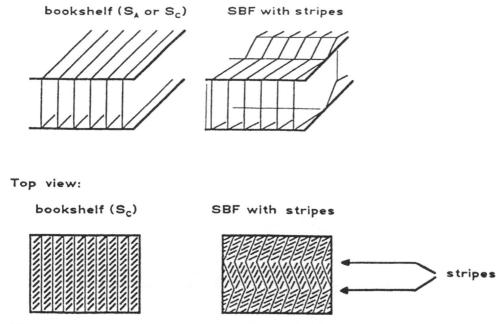

Top view:

bookshelf (S_C)   SBF with stripes

stripes

Figure 12 Layer orientation in the SSFLC (left) and SBF (right) configuration. (From Ref. 58.)

time of several hundred milliseconds. Thereafter helix formation occurs. The attractive features of the SBE effect are its short response time of 25 μs with fields of 15 V/μm, bistability (for several hundred milliseconds), multiplexability, and the possibility to generate gray levels.

Another submicrosecond electrooptic switching effect in S_C^* materials was discovered by Garoff and Meyer in the above-lying nonferroelectric S_A phase [59,60]. Tilted smectic phases composed of appropriate chiral molecules with lateral dipoles are known to be ferroelectric whereby the existing spontaneous polarization is linearly coupled to the tilt angle. In orthogonal smectic phases such as S_A built up from chiral molecules, this coupling induces a molecular tilt $\theta(E)$ relative to the smectic layer normal on applying an external electric field E parallel to the smectic layers (See Fig. 13). This *electroclinic* effect or *soft-mode* effect (SMFLC) is a kind of inverse piezoelectric effect in crystalline phases and is even observable in the S_A phase when no lower-lying S_C^* phase exists. This effect can be observed in very thin samples between crossed polarizers with the smectic layers perpendicular to the glass plates in the same bookshelf geometry typical for SSFLC cells. The magnitude of the induced tilt angle $\theta(E)$ is proportional to the applied field strength E for small tilt angles and depends on the sign of E:

$$\theta = e_c E \tag{6}$$

In order to induce large tilt angles, materials with large electroclinic coefficients e_c are required. The best compounds which meet this requirements are ferroelectric compounds with high spontaneous polarization [61]. The coefficient e_c is temperature dependent and increases when the temperature approaches the S_C^*–S_A phase transition from above. First commercial SMFLC mixtures which can work near room temperature are

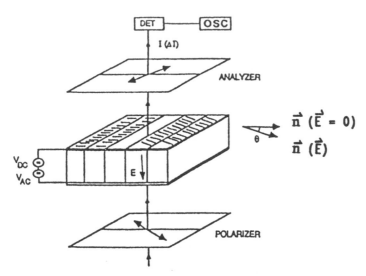

Figure 13   Basis geometry of the electroclinic effect in the smectic A* phase. The electric field induces a tilt $\theta(E)$ of the optical axis $\hat{\mathbf{n}}(\vec{E})$ in the substrate plane. (From Ref. 64.)

now available. Tilt angles of 10–15° and optical switching times below 10 μs [62] have been achieved at field strengths of 20 V/μm and at 5–10°C above the $S_C{}^*$–S_A phase transition. The response time is independent from the applied field strength but shows a strong temperature dependence near the $S_C{}^*$–S_A transition. Close to the phase transition, response times of 200 ns have been reported. Like the SSFLC cell, a SMFLC device acts as a retardation plate with a field-controllable optical axis. It can be operated in several modes, and its transmitted intensity is also given by Eq. (2). The most severe drawback of the electroclinic effect is its considerably lower tilt angle θ, and in particular the strong temperature dependence of θ, which limits the achievable maximum transmission and contrast ratio of an SMFLC device to a narrow temperature range. If the entrance polarizer is set at an angle $\phi = 22.5°$ to the **n** ($E = 0$) direction, the intensity variation is linear in a limited voltage range and will reach a maximum value but the contrast becomes low. If instead a modulation around a high extinction and a maximum contrast ratio is desirable, a crossed polarizer setting with $\phi = \pm\theta$ (E) should be used [63]. In the latter case the response is nonlinear. Finally, with a polarizer setting symmetrical with regard to both tilted states ($\phi = 0$), the electrooptic effect is insensitive to the polarity of the applied voltage but the contrast is still a maximum and the optical modulation frequency is 2ω if a field of frequency ω is applied.

SMFLC devices are not suitable for multiplex addressing, as they lack bistability and an optical threshold. They are high-speed devices with capacity of gray scale and continuous color switching [64]. Besides, the alignment problem seems to be not so serious for the orthogonal smectic phases. Just as for SSFLCs, one would like to use the SMFLC device as a switchable half-wave plate according to Eq. (2), which requires a cell thickness d of about 2 μm. For electroclinic devices, higher-order solutions ($d \cdot \Delta n = \frac{3}{2}\lambda$, which corresponds to $d = 6$ μm, etc.) are conceivable, as helix suppression by surface interaction is not required.

Concerning the main difficulty with SMFLCs, the small tilt angle, the Chalmers group has proposed the use of double or multiple electroclinic cells in special constellations to realize electrically tunable high-speed color filters [64]. With two electroclinic cells in series, the $\lambda/2$ condition can be fulfilled by an induced tilt angle of $11.25°$. Such materials are already available, but one serious problem is the strong temperature dependence of the induced tilt angle. This electrooptic effect is the fastest which has been found in liquid crystals to date. Despite its high application potential (shutter arrays, SLM, etc.), the material situation has to be improved by developing broad-range smectic A* mixtures. The potential and the limitations of this effect for device applications have been reviewed recently [65].

A further very promising alternative electrooptic effect to the SSFLC mode has been found by the Fukuda group [66] with the discovery of tristable switching in antiferroelectric liquid crystals (AFLCs). In the corresponding so-called S_{CA}^* phase, the molecules in the neighboring smectic layers are tilted in opposite directions and their corresponding dipole moments perpendicular to the molecular long axes cancel out within two adjacent layers. In thick samples both tilt directions spiral around the layer normal with a phase difference π and form, respectively, a helical structure as sketched schematically in Fig. 14b. The structure can be regarded as two S_C^* helices which mesh into each other, but the helix period is only half as long as the pitch in the S_C^* phase. In thin samples (2 μm thick), the helix is unwound (see Fig. 14c).

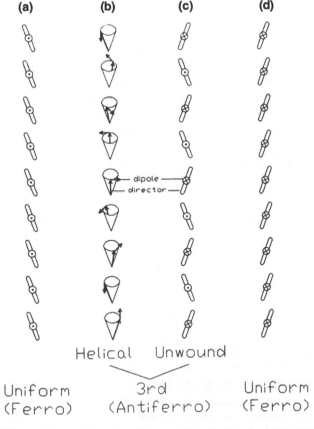

Figure 14 Antiferroelectric state (third state) in the helical configuration (thick sample, Fig. 14b) and the unwound state (thin sample, Fig. 14c). For comparison, the two uniform ferroelectric states with upward and downward polarization are shown (Figs. 14a and 14d).

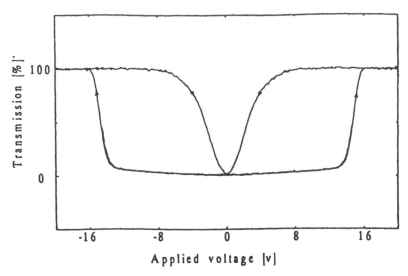

Figure 15 Tristable switching, DC threshold, and double hysteresis in the antiferroelectric phase (S_{CA}^{*}).

This unwound state is spontaneously antiferroelectric and stable (it is also called third state), and appears dark between crossed polarizers if the polarization direction of the incident light is parallel (or perpendicular) to the smectic layer normal. A sufficiently strong electric field induces the phase transition from the S_{CA}^{*} phase to the ferroelectric S_{C}^{*} phase and, depending on the polarity of the applied field, one of the two uniform ferroelectric states is created (Fig. 14a or 14d) whereby the director is tilted from the layer normal by the optical tilt angles $\pm\theta$. Both ferroelectric states will therefore appear bright. By removing the electric field, both uniform ferroelectric states relax into the stable antiferroelectric structure. The "tristable" switching effect in AFLCs can be exploited in a display [67–69] which possesses the following characteristic features [70]:

In its fieldless, stable, antiferroelectric state, no spontaneous polarization exists which collects ions and induces undesirable ghost pictures. The spontaneous polarization appears only when necessary.

The switching effect is characterized by simultaneous director switching and layer switching. An AC electric field induces reversible layer switching between the bookshelf (or chevron) structure in the uniform aligned FLC state and quasi-bookshelf structure in the AFLC state. This produces an internal shear in the cell which promotes self-alignment recovery from damage caused by mechanical shocks. A similar recovery effect by reversible layer deformation has been reported for SSFLC structures [70].

The switching effect has a sharp DC threshold and distinctive hysteresis which allows for multiplex addressing (see Fig. 15). Strictly speaking, the uniform ferroelectric states are not stable as in the case of the SSFLC structure, but they can be memorized by a DC bias voltage. As the hysteresis is symmetric with respect to positive and negative voltages, the polarity of the bias voltage as well as the addressing pulses can be changed alternately. Moreover, recently, analog gray-scale capability by multidomain switching has been demonstrated [69].

A disadvantage is that, for maximum brightness according to Eq. (2), an angle $\phi = \theta = 45°$ is required, which cannot be met by AFLC materials with the phase sequence S_C^*-S_A-ch-isotropic. The switching from the ferroelectric to the antiferroelectric state is a field independent, passive, and slower process, but it may be accelerated by a bias voltage with reversed polarity [71].

Based on already available AFLC mixtures, Nippondenso has recently presented a video-capable, 6-in., full-color display with 320×200 pixels, possessing a line address time of 64 μs, analog gray-scale capability, and a very wide viewing range [69]. First spatial light modulators with AFLC materials have also been demonstrated [72].

D. Nonlinear Optical (NLO) Effects

If a dielectric material is exposed to a strong electromagnetic field, it shows nonlinear optical behavior. The macroscopic induced polarization $P_i(\omega)$ can be expanded in a power series of the optical field $\hat{E}$ [73]:

$$P_i = \chi_{ij}^{(1)}E_j + \chi_{ijk}^{(2)}E_jE_k + \chi_{ijkl}^{(3)}E_jE_kE_l \cdots \tag{7}$$

The indices i, j, k, l denote the Cartesian indices, and the usual linear response of the medium is described by the dielectric susceptibility $\chi_{ij}^{(1)}$. The second-order or third-order processes are taken into account by the susceptibilities $\chi_{ijk}^{(2)}$ and $\chi_{ijkl}^{(3)}$. Similarly, the polarization μ_I induced in a molecule can be expressed by [73]

$$\mu_I = \alpha_{IJ}F_J + \beta_{IJK}F_JF_K + \gamma_{IJKL}F_JF_KF_L \tag{8}$$

where the indices I, J, K, L are related to a molecular coordinate system and $\hat{F}$ is the local electric field seen by the molecule. The tensor α denotes the linear molecular polarizability, and the tensors β and γ are the molecular hyperpolarizabilities of second and third order. Nonlinear optical responses give rise to a series of phenomena such as second or third harmonic generation, Pockels effect, optical bistability, optical amplification and optical-phase conjugation [64], which are interesting from the application point of view. Second harmonic generation (SHG) and Pockels effect are most frequently investigated.

A prerequisite for highly efficient optical second harmonic generation is a large second-order susceptibility $\chi^{(2)}$. This requires

Molecules with a large molecular hyperpolarizability β
A noncentrosymmetric ordered crystal structure which prevents averaging out of the molecular β on a macroscopic scale

Organic materials such as liquid crystals and their corresponding polymers are good candidates for large second-order effects if they contain the strongly polarizable conjugated π-electron system and donor–acceptor groups which provide the necessary polar asymmetry [4,74]. Furthermore the individual molecules should be aligned in such a way that the molecular nonlinearities add constructively as much as possible. An FLC material with an unwound helical structure may be a promising approach, as it inherently contains the required noncentrosymmetric polar order and, due to its large birefringence anisotropy, offers the possibility of phase matching. Since the polar axis (C_2 axis) of FLCs is perpendicular to the long molecular axis, the donor–acceptor system has to be aligned and fixed in the C_2 direction. In general, an FLC with a large spontaneous polarization also has a large $\chi^{(2)}$ [73]. SHG with FLCs has been proven for the first time by Vtyurin et al. [75]. Due to optical interference phenomena, the efficiency of FLCs for SHG depends on the experimental geometry chosen. Efficient energy conversion occurs if the phase-matching condition for the refractive index $n(\omega) = n(2\omega)$ is fulfilled. This can be achieved by rotating the sample with respect to the incident beam (see Fig. 16 [76]) or by varying the temperature of the sample and exploiting the strong temperature dependence of the molecular tilt angle. Thick FLC cells (>10 μm) with homeotropic alignment are commonly used, and the helix is unwound by applying a strong DC field within the substrate plane. Unwound FLC layers have C_2 symmetry, and the nonlinear susceptibility tensor $\chi^{(2)}$ has four nonzero components. If the polar axis (C_2) is chosen parallel to the y direction (see Fig. 16) and, as is quite common, the contracted d_{ij} tensor is used instead of $\chi_{ijk}^{(2)}$, one arrives at [77]

$$d = \begin{bmatrix} — & — & — & d_{14} & — & d_{16} \\ d_{21} & d_{22} & d_{23} & — & d_{25} & — \\ — & — & — & d_{34} & — & d_{36} \end{bmatrix} \tag{9}$$

with

$$d_{y,xz} = d_{14} = d_{25} = d_{36}$$
$$d_{y,zz} = d_{23} = d_{34}$$
$$d_{y,xx} = d_{16} = d_{21}$$
$$d_{y,yy} = d_{22}$$

Up to now, however, the SHG efficiency of the FLCs investigated [77–79] was less than that of state-of-the-art inorganic crystals such as lithium niobate with d_{33} (LiNbO$_3$) ≈ 40 pm/V [80]. All d_{ij} values have been measured for one FLC material by the Boulder group [79], with the largest coefficient $d_{22} = 0.6$ pm/V. Most recently, however, Schmitt et al. have achieved a new record value of $d_{22} = 5$ pm/V for a newly designed FLC material [81]. Due to the existence of a glass transition T_g, it was possible to "freeze

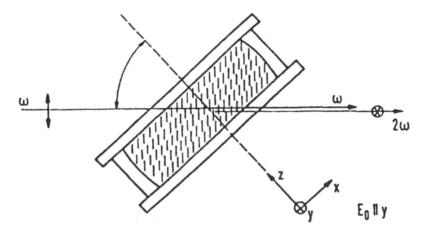

Figure 16 Experimental geometry for SHG (second harmonic generation) in the homeotropic unwound S_C^* phase. Phase matching can be realized by rotation of the sample around the y axis. (From Ref. 73.)

in'' the SHG active configuration by cooling the material from the S_C^* phase below T_g, and to evaluate all components of the d tensor.

It is to be expected that FLC polymers might exhibit all the same NLO properties as LMM FLCs. Additionally, however, they have three advantages:

1. The NLO active ordering can be frozen in in the glassy phase.
2. In the fabrication of thin planar waveguiding structures, FLC polymers may be easier to process, for example, by spin-coating processes.
3. They offer higher flexibility in molecular engineering. Different side-chain moieties can be incorporated into copolymers to design materials which combine NLO activity, spontaneous polarization, and broad room-temperature S_C^* phase.

Besides a high second-order NLO coefficient, low absorption and low scattering losses are prerequisites in order to make these materials applicable in waveguiding devices. A damping constant of 10 dB/cm for light propagation in a 6-μm-thick homogeneous planar-aligned cell has been determined for the newly developed FLC material [81]. This is still quite large compared with 0.1 dB/cm for devices fabricated of Ti: LiNbO$_3$, a current NLO device material [82].

A fast linear electrooptic light modulation effect, which can be employed in waveguiding structures, is the Pockels effect. In contrast to ferroelectric switching in SSFLC structures, the Pockels effect is basically a pure NLO effect, whereby a small refractive index change Δn_e is induced along the optical axis by an applied electric field E_y, which acts along the polar axis [83]:

$$\Delta n_e = \frac{n_e^3 r_{32} E_y}{2} \tag{10}$$

The Pockels coefficient r_{32} has its origin in $\chi^{(2)}$, and is proportional to d_{23}. Most recently, a value of $r_{32} \approx 3$ pm/V has been reported for the commercially available S_C^* FLC material SCE9 [83], and it was shown that Δn_e is of the order of $10^{-4}-10^{-5}$ of the refractive index change caused by ferroelectric switching of the long molecular axes. This small refractive index shift can be exploited in modulating the light transmission through a waveguiding structure. At room temperature the authors achieved a 10-MHz modulation, which is two orders of magnitude faster than the ferroelectric switching mode.

E. Piezoelectricity

As a center of symmetry is lacking, ferroelectric S_C^* liquid crystals are piezoelectric and show mechanoelectrical [10,84,85] as well as electromechanical effects [86–88]. A mechanoelectrical response was observed in thick homeotropic samples [10] as well as in planar samples [85]. In both cases a shear flow applied parallel to the smectic layers (y direction) disturbs the helix configuration, and induces an electric polarization $P_{in,x}$ perpendicular to the flow direction and the layer normal (see Fig. 17a). The mechanism resulting in the observed effect was explained as follows. Shear flow distorts the initial helical director structure, and due to the coupling between the polarization and director, the net polarization changes and an electric current flows through the sample. The shear flow-induced voltage depends linearly on the shear velocity [10].

The inverse of the above-described effects is called electromechanical effects. In this case an alternating electric field induces either a flow or a smectic layer compression

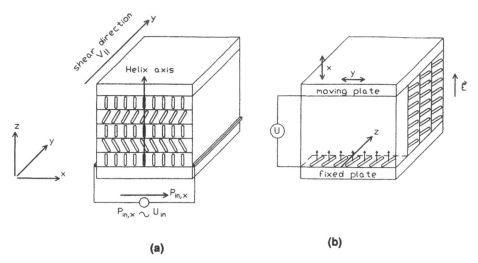

Figure 17 Experimental geometry for detecting: (a) Mechanoelectrical effects [10]; (b) electro-mechanical effects [89].

which results in a mechanical displacement or vibration of the cell substrates. Linear electromechanical effects (exitation frequency corresponds to the vibration frequency) were first observed up to few kilohertz in the $S_C{}^*$ phase of FLCs by Jákli et al. [86]. Although a complete theoretical description of the electromechanical (EM) effects is still lacking, two different basic mechanisms have been identified [89] which act simultaneously and induce predominantly horizontal (y) and vertical (x) vibrations, respectively, when the electric field is applied in the x direction (normal to the glass plate) and the smectic layers form a bookshelf (or chevron) structure (see Fig. 17b).

The first mechanism is the so-called Goldstone mode, which corresponds to collective rotation of the director field on the cone with the cone angle 2θ around the layer normal (z axis) without a change of layer thickness. If the helix is unwound, this mode is identical to the electrooptic switching effect as discussed in Section II.C and illustrated in Fig. 5. Due to the viscous coupling, the director rotation induces flow parallel to the layer planes, which results predominantly in a horizontal force on the top substrate plate (y displacement). In the experiments the upper plate could move freely with respect to the lower one. This effect is not a true piezoelectric effect, as viscous flow behavior dominates rather than the elastic properties as in the case of solid crystals. Consequently it is not a static effect; i.e., a displacement is not observable when a DC voltage is applied [87]. The corresponding EM response disappears in the S_A phase.

The second mechanism corresponds to the electroclinic effect, where a director tilt is induced due to the coupling between polarization and tilt. This, in turn, leads to layer compression and a flow which causes a displacement in the x direction. This response can be regarded as a true piezoelectric effect which is still recognizable in the S_A phase [89] and should be observable with a static electric field.

The observed EM effects depend very strongly on the virgin texture and the director alignment of the smectic samples (helix developed, unwound, etc.). The linear effect is proportional to the applied voltage. Nonlinear EM response (with respect to SHG) has also been observed and studied by Jákli et al. [90].

The EM response is quite audible, as it works in the acoustic frequency range and could be of interest for practical applications (speakers, electromechanical transducer, piezosensors, microphones). Even ultrasonic generation has been observed [91]. FLC polymers show qualitatively very similar piezoelectric effects as LMM FLCs, which will be discussed in Section VII.

F. Pyroelectricity

FLC's possess a temperature-dependent spontaneous polarization and therefore they are pyroelectric because a change of temperature results in a flow of charge to and from the surfaces through the FLC sample. Pyroelectricity in FLCs was first observed in 1976 [92]. The pyroelectric response can be measured by the dynamic Chynoweth technique [93]. In this method the FLC cell is periodically heated by a chopped light source. The AC pyroelectric current $i(\omega)$ generated in the FLC cell can be detected with a lock-in amplifier and is given by

$$i(\omega) = \frac{dQ}{dt} = F \cdot \frac{dP_S}{dt} = F \cdot \frac{dP_S}{dT} \cdot \frac{dT}{dt} = F \cdot p \cdot \frac{dT}{dt} \qquad (11)$$

with

$$p = \frac{dP_S}{dT}$$

where p is the pyroelectric coefficient, which should not be confused with the helical period P, F is the electroded area of the sample, dT/dt is the heating rate of the FLC sample, and dQ is the amount of charge changed by a temperature variation dT in the time interval dt. Alternatively, the pyroelectric coefficient can be evaluated by measurement of the pyroelectric response to heat pulses [13] or by differentiation of the temperature-dependent $P_S(T)$ curve.

Recently, the potential of LMM FLCs as pyroelectric infrared detectors has been investigated by Glass et al. [94]. As expected, the pyroelectric coefficient reaches a maximum value just below the phase transition from the S_A phase into the S_C^* phase. Despite the low P_S values of the FLC materials used, the figures of merit compare well with ferroelectric solid materials commonly used for infrared detection. It is to be expected that FLC polymers should have figures of merit of the same order of magnitude. The discussion will be continued and extended, going into more detail, in Section VIII.

III. PHYSICAL PROPERTIES OF S_C^* FLC POLYMERS

We begin this section with a résumé of the available LMM FLC materials. Due to the intense R&D efforts to commercialize FLC displays during the last 5 years, there has been a strong demand for suitable FLC mixtures. Chemists have responded to this challenge and have designed materials with properties adapted and optimized for the electrooptic effect used. For this purpose the following strategy is pursued:

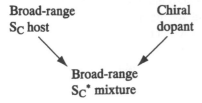

First an S_C host mixture is developed which has a broad S_C phase range and is optimized mainly with respect to a low viscosity. Next the ferroelectric properties are induced by adding suitable chiral dopants to the S_C host mixture. For the potential application of the SSFLC mode the following material requirements have to be fulfilled:

Use of stable compounds
Broad $S_C{}^*$ temperature range
Phase sequence $S_C{}^*$-S_A-Ch-I which is favorable for the quality of the alignment
A long or, better, a compensated helical pitch in the cholesteric and the $S_C{}^*$ phase
A tilt angle θ close to 22.5° (birefringence mode) or 45° (guest host mode)
An optical path difference $d \cdot \Delta n = 0.28$, which results in $\Delta n = 0.14$ for a 2-μm cell technology
A low rotational viscosity γ_C
A low spontaneous polarization P_s (< 10 nC/cm^2)
A large positive dielectric biaxiality $\delta\epsilon$ if AC stabilization is required

In the past quite acceptable FLC mixtures with low P_s values and switching times of about 40 μs ($E = 15$ V/μm, 20°C, $P_s = 20$ nC/cm^2) has been developed for commercial use. Not so much research effort has stressed tailoring mixtures for the application of the electroclinic effect, the antiferroelectric switching effect, or the DHF effect. Nevertheless, some corresponding mixtures are available.

The situation is quite different in the case of the FLC polymers. In 1984 Shibaev was successful in synthesizing the first FLC polymer containing an $S_C{}^*$ phase [5]. In the beginning the main impact was to copy LMM FLCs by covalently bonding familiar mesogenic monomer structures to a polymer backbone and seeing how the ferroelectric properties are transferred to the polymer [95–97]. Later, attempts were made to vary more systematically the polymer backbone (acrylates, metacrylates, polysiloxanes, and so on), the spacer length, and the mesogenic side chain, mainly in order to study their influence on the mesomorphic properties [98–101]. In the meantime new and widely ranging molecular architectures such as combined $S_C{}^*$ FLC polymers (mesogenic groups in the main chain as well as in the side groups) [102] as well as cross-linked $S_C{}^*$ FLC polymers (so-called elastomers) have been introduced [103]. These $S_C{}^*$ elastomers have the exciting new feature that they can be oriented by small mechanical strains. For higher mechanical fields the helical structure according to Fig. 2 can be *reversibly* unwound, giving rise to the appearance of spontaneous polarization [103] (Fig. 4).

Unfortunately, however, for the most part, of the FLC polymer materials published so far, only a phase identification has been possible; many of the physical parameters relevant for application are still unknown at this time [104]. One reason for this might be the fact that with the conventional alignment methods used for nematic LCs it is generally more difficult to obtain homogeneous aligned, unwound, FLC polymer samples with high optical quality than in the case of LMM FLCs. This has impeded and delayed the determination of physical parameters. The second point is that the physical properties depend on the average molecular weight M_n and its distribution (dispersity) D. As a consequence, for $S_C{}^*$ FLC-polymers with identical molecular structure but different values of M_n and D, large differences in the phase-transition temperatures and physical properties have been found. This makes it difficult to find relations between the molecular structure and physical properties.

In the past, research activity on FLC polymers has been directed toward:

Molecular tailoring of materials with fast switching times for electrooptic applications. This development has been initiated by the important work of Uchida et al. [105]

Development of materials which take advantage of the polymer specific properties (glassy phase) in combination with a high spontaneous polarization for NLO or piezoelectric applications.

Now we will focus our attention predominantly on properties involved in electrooptic applications, and will cover NLO, piezoelectric, and pyroelectric properties in detail in later sections.

First, the temperature range of the S_C^* phase of FLC polymers is in general broader and the transition temperatures higher than in the case of the corresponding LMM FLCs. This has been demonstrated with an acrylate (metacrylate) backbone [95,97], with a siloxane main chain [106], and most recently with a polyoxyethylene backbone [107]. Polymerization seems to be effective in stabilizing the S_C^* phase range. Besides the molecular weight M_n, the dispersity D in particular has a significant influence on the S_C^* phase range, as well as the physical properties which are involved in ferroelectric switching [108].

According to Refs. 95 and 97, the spontaneous polarization decreases by an order of magnitude in going from LMM FLCs to the metacrylate FLCP. Shibaev used a material with a high M_n and a large dispersity D. Recently, Sekiya et al. [107] observed no significant difference in the P_S values of LMM FLCs and of polyoxyethylene FLC polymers for $M_n \approx 5000$. Even an increase of P_S for acrylate polymers ($M_n > 38,000$) in comparison to the corresponding monomers has been reported in [109]. Siemensmeyer et al. [108] have shown that for $M_n < 10,000$ g/mol, the influence of the dispersity D on P_S and θ is predominant over that of M_n. With increasing D, the S_C^* phase range, P_S, and θ decrease and finally disappear for large D, whereas the influence of M_n seems to be less pronounced. But for very large M_n values (15,000–240,000) Scherowsky found a distinct reduction of the polarization with increasing molecular weight [110].

Figure 18 shows the comparison of the temperature dependence of P_S for a polymer material which possesses a second-order S_A-S_C^* phase sequence (FLC 112) or a first-order isotropic-S_C^* phase transition (FLC 120). In accordance with LMM FLCs, the temperature dependence near the phase transition is gentle in the case of FLC 112 but strong in the case of FLC 120. Within the S_C^* phase the former exhibits a stronger dependence on temperature whereas the latter saturates more rapidly.

The tilt angle θ is approximately proportional to P_S. Indeed, the temperature dependence is very similar for P_S and θ, as represented in Fig. 18 for the polymer FLC 112.

Perhaps the most significant parameters for electrooptic applications are the optical response time τ and the rotational viscosity γ_c. It was a milestone in the development of FLC polymers when Uchida et al. [105] reported on low-temperature acrylate FLC polymers with fast optical response times τ of 4 ms near room temperature (32°C, $E = 20$ V/μm, $M_n = 3300$, $P_S = 3$nC/cm^2). But no tilt angles were given. Besides, in this remarkable paper the authors gave first experimental evidence that the optical response increases strongly with the molecular weight but saturates above $M_n = 15,000$. From this it was concluded that low molecular weights are desirable for achieving fast switching times. A number of papers have reported on attempts to achieve faster response by using siloxanes as polymer backbone [106,111–114]. Dilution of the mesogenic groups on the main chain considerably reduces the rotational viscosity, but also the bulk viscosity [112,113]. Such diluted polymers with molecular weights M_n of about 2000 show broad

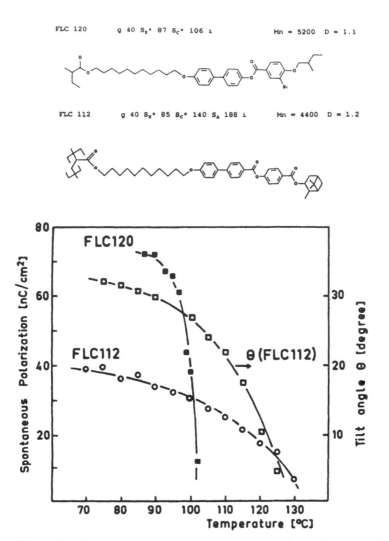

Figure 18 Temperature dependence of P_s for the polymer FLC 120 (FLC 112), possessing an S_C^*–isotropic (S_C^*–S_A) phase transition. In addition, the temperature dependence of the tilt angle θ of FLC 112 is depicted on the right vertical scale from Ref. 120.

S_C^* phase ranges near room temperature, fast electrical response times of about 12 ms (32°C, $P_s = 120$ nC/cm², $E = 6$ V/μm), and large tilt angles of about 30° [112,113]. Another quite attractive feature from the application point of view is the very weak temperature dependence of the tilt angle in a broad temperature range. Even faster optical response at room temperature has been reported most recently in [107] for acrylate polymers. But the corresponding tilt angles are very low, suggesting that electroclinic switching is involved. At even lower temperatures (5°C), where indeed ferroelectric switching occurs, the response time increases considerably and achieves some seconds.

The reduction of the optical switching time τ with decreasing molecular weight was confirmed for the FLC polymer PO [107] and is shown in Fig. 19. However, despite

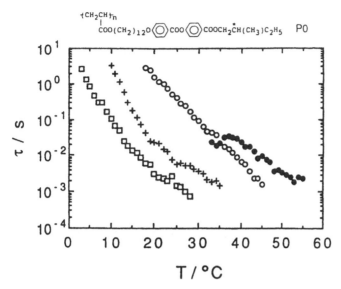

Figure 19 Dependence of optical switching time τ on temperature T for the FLC polymer PO and M_n = 3100 (□), 4800 (+), 5900 (○) and 19000 (●). (From Ref. 107.)

much effort up to now, the optical switching times of FLC polymers are still two to three orders of magnitude slower than that of corresponding LMM FLCs. This was impressively demonstrated by Sekiya et al. on FLC polyoxyethylenes and the corresponding FLC monomers [107] and can be easily explained according to Eq. (4) by a corresponding increase of the rotational viscosity γ_c, which is 10^2–10^3 times larger in the case of FLCPs than for LMM FLCs (see Fig. 20 [107]). Additionally, the temperature dependence of γ_c is stronger in the case of FLCPs.

Very little information exists about further physical parameters such as the helical pitch p, the dielectric biaxiality $\delta\epsilon$, the dielectric anisotropy $\Delta\epsilon$, and the optical birefringence Δn of FLC polymers.

In just one recent paper the temperature dependence of the helical pitch of a polyacrylate FLC polymer has been reported by Endo et al. [116]. Near the $S_C{}^*$–S_A phase transition the helical pitch scarcely changed with temperature, but the pitch increased with increasing molecular weight. The authors succeeded in synthesizing an FLC copolymer with a compensated helical pitch. These copolymers contain two different kinds of chiral side chain units, which are supposed to generate respectively different-handed helical structures. This result is important for realizing bistabile SSFLC structures with FLC polymers. As in LMM FLCs, the pitch p can be determined in thick cells (>20μm), where a typical fan-shaped texture appears with equidistant lines caused by the helical structure.

First dielectric measurements on FLC polymers for the different relaxation processes were reported [117,118], but values of the dielectric anisotropy have not been reported yet. Evidence has been given that collective and molecular dynamic processes in FLC polymeric systems are comparable with LMM FLCs [117].

At present the most serious problem which has to be tackled seems to be alignment of FLC polymers. Many authors have reported on physical measurements on nonperfectly oriented samples. Therefore many results should be used with care. The general feeling

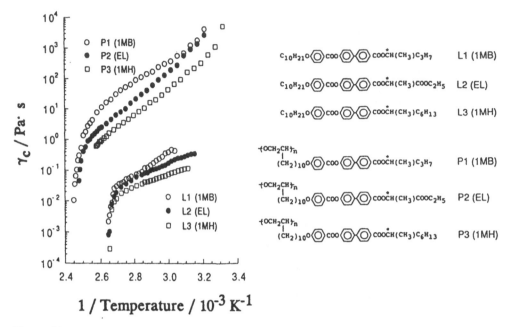

Figure 20 Rotational viscosity γ_c versus the reziproke temperature T of the LMM FLCs L1, L2, L3 and the corresponding polyoxyethylene FLC polymers P1, P2, P3 with $M_n \approx 5000$. (From Ref. 107.)

is that it is more difficult to achieve optically clear and homogeneous aligned samples of FLC polymers than for LMM FLCs. However, so far only the conventional orientation techniques for LMM FLCs, such as polymeric alignment layers, shearing procedures, or electric (magnetic) fields, have been employed. In the field of LMM FLCs it has been attempted to clarify the alignment mechanism and to find correlations between the physical properties of the orientation layer and the quality of the alignment. Many excellent works have covered this field, and we would like to refer to the paper by Myrvold [119] and references therein. So far, however, nobody has tried to investigate or even to understand alignment of FLC polymers on the same level as for LMM FLCs.

Nevertheless, in the last years much progress in alignment has been achieved by using commercially available, low-pretilt polymeric alignment materials. Some polyacrylate FLC polymers which possess a S_A-S_C* phase sequence but lack a cholesteric phase could be oriented, and the desired bistable, defect-free SSFLC structure has been implemented in 2-μm-thick samples with high optical quality by using buffed polyimide layers with low pretilt and by applying the common alignment procedures of LMM FLCs. Contrasts up to 200:1 in the powered states were obtained [120]. The two optical states showed distinct black–white appearance. Therefore a uniform director structure within the smectic layers can be supposed, but in the absence of electric field these structures relax into states with lower contrast. Besides, the actual layer structure (chevron or quasi-bookshelf or other) is still an open question.

Recently, the successful alignment of siloxane-based FLC polymer by nylon-brushed surfaces has been reported [121]. As a new feature, during the slow cooling process from the isotropic to the S_C* phase simultaneously a slight shear along the direction of rubbing was applied to support the orientation process. No zigzag walls were observed.

As will be discussed later in Section IV, the Idemitsu group [115] succeeded in creating an SSFLC structure with an FLC polymer even without an alignment layer by just using an appropriate shearing procedure [115]. This result gave a first hint that alignment and processing might be easier for FLC polymers than for LMM FLCs. But zigzag defects were observed and first X-ray studies [115] indeed revealed a chevron structure which is very similar but not as simple as that of LMM FLCs.

Besides the ambitious efforts to arrive at well-aligned SSFLC structures, some work has been dedicated to other electrooptic effects as are well known with LMM FLCs, such as the electroclinic (SMFLC), the antiferroelectric (AFLC), and the deformed helix ferroelectric effect (DHF).

The electroclinic effect has been observed in a variety of FLC polymer materials [122–124]. The effect shows qualitatively the same behavior as in LMM FLCs. Naciri et al. [106] investigated S_C^* copolymers based on polysiloxanes (M_n = 2000) with an S_A phase lying between 136°C and 150°C. At 137°C, 1° above the S_C^*–S_A phase transition and for E = 10 V/μm, they observed very large induced tilt angles up to 18° with corresponding response times of about 100 μs. For higher temperatures an even faster response was achieved, but with smaller induced tilt angles as shown in Fig. 21. From these results one can conclude that the electroclinic coefficients $e_c = \theta/E$ of FLC polymers according to Eq. (6) are quite comparable to the case of LMM FLCs, but the response again is distinctly slower due to increased viscosity of the FLC polymers and the temperatures at which the electroclinic effect works are still too high. Clearly there is a demand for ferroelectric polymers with a broad room-temperature S_A phase.

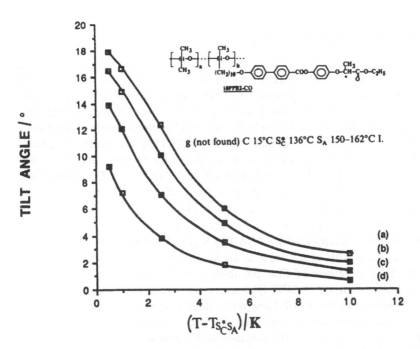

Figure 21 Temperature dependence of the induced tilt angle (electroclinic effect) at different voltages (a) 100 V; (b) 75 V; (c) 50 V; (d) 25 V, in the S_A phase of the copolymer 10 PPB 2-CO as shown in the figure ($M_n \approx$ 2000 from Ref. 106).

Evidence for antiferroelectric behavior of FLC polymers has also been given by some authors [125–128]. The existence of a DC double hysteresis has been found, and birefringence measurements are in accordance with an unwound antiferroelectric structure according to Fig. 14c [126]. Some care has to be taken in the assignment of the antiferroelectric phase to assure that the optical effect is not caused by a transition from a fieldless helical structure to a deformed or unwound ferroelectric structure. This question may be decided by analyzing the polarization reversal current for the transition between the unwound antiferroelectric state and the two uniform ferroelectric states and accompanying optical studies. *Two* characteristic current peaks have to be observed under an applied triangular wave within one half-period. Whether this phase is identical to the $S_{CA}*$ phase of LMM FLCs, or whether polymeric effects give rise to a different antiferroelectric phase in polymeric systems, is still open for discussion. In one case [127] the authors have not excluded that at low voltages a DHF effect occurs.

In conclusion, on the basis of the existing experimental results, some of the physical properties of FLC polymers and LMM FLCs seem to be similar, but the dynamic properties are quite different. Moreover the characterization of the FLCPs is far from being complete and has to be extended and improved. To attain these goals the problems of smectic layer alignment have to be solved and better understood, and a precise knowledge of the smectic layer structure as well as the actual director structure as in the case of LMM FLCs is required. This endeavor should be accompanied by simultaneously widening the material reservoir. This will give us the possibility to realize new polymer-specific effects which can occur due to the presence of the polymer backbone and the bonding of the mesogenic moiety on the main chain. The same is true for the electrooptic effects observed so far.

IV. ELECTROOPTIC APPLICATIONS

When electrooptic applications of FLC polymers are discussed, they are always compared with their LMM FLC counterparts, which set the standard. However, the major problem of SSFLC display prototypes developed so far, which has prevented their commercial exploitation, is the considerable mechanical and thermal shock sensitivity. To overcome these latter problems the idea of a very simple flexible FLC polymer display with plastic substrates has been introduced by the Idemitsu workers [107,115]. Due to the polymer's inherent high flow viscosity, such a device can be produced by an easy fabrication process. In the first step a 2-μm-thick FLC polymer film is coated onto an ITO-coated plastic substrate, and then laminated with a second ITO-coated substrate. Next the sandwich structure undergoes a bending procedure whereby the FLC polymer molecules suffer a shear stress, and align spontaneously and uniformly [107] in a similar way as shown originally by Clark and Lagerwall with LMM FLCs [2]. Thus satisfactory molecular alignment can be achieved without requiring an alignment layer [115]. The authors claim that this method allows alignment of FLC polymers even if the cholesteric phase does not appear in the phase sequence. A SSFLC structure was established which exhibits an electrooptic effect similar to the Clark-Lagerwall effect observed on thin LMM FLC layers. As a further consequence of its high viscosity, such a plastic cell does not require a spacer to control layer thickness, as the polymer itself maintains its film thickness. In addition, this device is flexible, and in particular should show an enhanced alignment stability against mechanical shock. A reflective static driven display (15 cm × 40 cm) was fabricated by using the process described above [115]. A photograph of this display is reproduced in Fig. 22. Most recently, a dynamic driven, 96×288 pixel matrix display with 12 cm $\times$ 36 cm size and a response time of 5 ms/line (2-Hz frame rate) was presented [129]. A contrast ratio of 8 was achieved. Most material properties relevant for application, such as P_S and θ, can be adapted to the display requirements by molecular engineering. However, at present the main disadvantage of the FLC polymer materials is slow optical switching time due to the high rotational viscosity γ_c. Both are typically two to three orders of magnitude larger than those of their LMM FLC counterparts.

Even when using FLC polymers with low molecular weight ($M_n \approx 2000$), high electric field strengths ($E = 15$ V/μm), and medium polarization values ($P_S = 100$ nC/cm^2), the fastest optical response times τ observed so far amount to tens of milliseconds at room temperature. At lower temperatures the response times increase dramatically and arrive at some seconds near 0°C [107]. The unfavorable strong temperature dependence of τ may be improved by employing FLC polymers with lower glass transition temperatures. Indeed, Blatter et al. [130] reported a reduced temperature dependence of switching times of FLC polysiloxanes by using glass transition temperatures as low as -100°C.

Besides the work of Idemitsu, well-aligned 2-μm-thick FLC polymer cells with few defects and high black-and-white contrast was achieved by using conventional low-pretilt polyimid alignment layers [120]. This suggests the establishment of a uniform director structure, but the precise director structure is still unknown. In order to achieve a high-performance electrooptic FLC polymer device, the same requirements as discussed in Sections II and III for the Clark-Lagerwall effect with LMM FLCs have to be fulfilled. These demands are independent of the cell technology (conventional glass cell or flexible plastic cell).

There is a wide field of applications for electrooptic FLC polymer devices. They may be especially applicable as relatively simple, matrix addressed, flexible displays for slowly changing information content (for example, alphanumeric data), and for small or large display areas. Of particular interest are display applications where the unique memory capability of the FLC polymer can be fully exploited. Working in the reflective mode, such a device would not consume any power during times when no information change is required. Such a 180×32 pixel display on the basis of LMM FLCs was demonstrated in 1991 by GEC-Marconi for use as an autonomous electronic label [131]. An FLC polymer-based display, however, would benefit from improved mechanical ruggedness and easy fabrication process. Even high-resolution matrix-addressed displays can be made

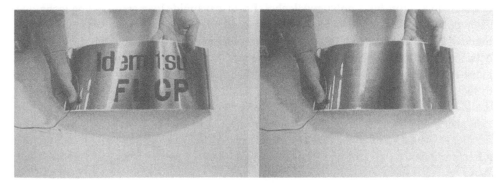

Figure 22 Photograph of the 15 cm $\times$ 40 cm flexible, reflective FLC polymer display in the on state. (From Ref. 115.)

if a slow frame repetition time can be accepted. For example, with a line address time of 10 ms and 200 lines, a frame time of 2 s is achievable. Certainly there is no doubt that FLC polymers cannot replace LMM FLCs in high-resolution displays intended for workstations or TV, as it cannot be expected that the optical response will be fast enough. However, perhaps FLC polymers may be useful as components blended with LMM FLCs to improve the mechanical and thermal shock behavior.

Even so, a lot of research work has been carried out to reduce further the optical response times of the FLC polymer materials. There is a distinct trend toward synthesizing materials with a low degree of polymerization (for example, oligomers), as the rotational viscosity is lowered with decreasing molecular weight of the FLC polymer. Recently, with new FLC oligome sogens (P_S = 80 nC/cm^2) response times of 10–20 ms have been obtained at room temperature [120]. Moreover, dilution of the mesogenic groups and/or chemically bonding molecular moieties which promote low viscosity may be a promising way to arrive at a faster response. Recently, Coles [132] presented a new FLC material. A short, flexible siloxane moiety is incorporated in an LMM mesogenic structure, thus leading to a new class of S_C* compounds that show switching times as fast as LMM FLCs but still possesses polymerlike properties. However, these compounds are no longer polymer materials.

An additional way to increase the switching speed would be to use materials with high P_S. As already mentioned in Section II.C, a high P_S makes the bistability deteriorate and gives rise to image sticking. A large memory time requires a material with low P_S/γ_c. In the case of FLC polymers, higher P_S values should be acceptable, as the related rotational viscosities are higher too. In addition, good bistability is favored if alignment layers are as thin as possible [133].

Besides displays, other electrooptic applications are conceivable—for example, a high-resolution matrix-addressed light valve in which an arbitrary diaphragm shape can be programmed, switched into an optically transparent state, and memorized without the need of further power consumption. Such a device could be extremely useful in optical imaging systems or, if the response speed of an S_C* FLC polymer can be accepted, in more sophisticated parallel optical data processing systems.

Another application of SSFLCs is the control of light in an optical waveguide for data processing and communication. Recently, Ozaki et al. proposed an LMM FLC waveguide composite device [134] which allows fast electrooptic modulation in the microsecond range as shown in Fig. 23a. The lower ITO-coated glass substrate (Pyrex) is coated with a 1-μm-thick polyvinylalcohol (PVA) waveguide by a spin-coating technique. The upper glass plate is also coated with ITO and then with a thin polymer film (polyimide or PVDF) to establish the homogeneous alignment of the FLC which is sandwiched between the PVDF and the waveguide. The passive polymer waveguide has refractive index n_p, and $n_\parallel$ and $n_\perp$ are the refractive indices of the FLC parallel and perpendicular to the molecular axis, respectively. θ is the angle between the propagation direction of the incident light and the long molecular axis. We assume that a monodomain, termed "on state" in Fig. 23b, is established under a DC bias field. If light with polarization parallel to the substrate plane (TE mode) propagates in the waveguide, it sees an effective refractive index n_e, given by

$$n_e = \frac{n_\parallel n_\perp}{\sqrt{n_\perp^2 \sin^2 \theta + n_\parallel^2 \cos^2 \theta}} \qquad (12)$$

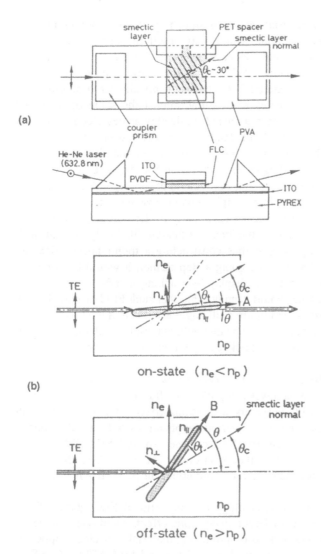

Figure 23 (a) Schematic representation of electrooptic switching in a polymer waveguide by using a FLC material. (b) Top view of the experimental configuration and relevant physical parameters for achieving waveguide switching. (From Ref. 134.)

When θ is smaller than a critical angle θ_c determined by substituting n_e in Eq. (13) by n_p, $n_e < n_p$ holds, and total internal reflection occurs at the FLC–waveguide interface. High transmission through the waveguide is observed. If the polarity of the DC bias is changed, the molecules are reorientated in such a way that $\theta > \theta_c$ and $n_e > n_p$ results (off state in Fig. 23b). Consequently, the light leaves the waveguide and the transmission decreases. It is attractive to think of a full polymer device where the FLC is replaced by an FLC polymer. If the application is not limited by its slower time, the benefits are easier processing and enhanced mechanical stability.

New multifunctional polymer materials can be designed by chemically bonding various side groups on the polymer backbone, which may result in abundance of novel applications. Using this molecular concept, Scherowsky [135] has synthesized the first colored ferroelectric side-chain copolymer. Aside from the chiral mesogenic moiety, which guarantees the S_C^* phase, these materials contain a chemically fixed dichroic dye as a second side chain. The dye molecules will assume almost the same orientational order as the mesogenic moieties. As a result, a guest host display as described in Section II.C but based on FLC polymers can be built. High dye concentrations are achievable without any solubility problems because molecular demixing and crystallization is effectively suppressed by the chemical bonding of the dye on the backbone. Using only *one polarizer*, such a display would allow switching between an absorbing colored state and a nonabsorbing transparent white state. In principle, several dye moieties which cover the whole visible spectral range can be attached chemically to the polymer backbone to make black-and-white displays. To combine the idea of a flexible FLC polymer film with this new class of colored FLC copolymers, thus constructing a monochrome, flexible, large-area display is an obvious step. Finally, going a step further, it would be attractive to make a multicolor, flexible FLC polymer display by stacking a red-, a blue-, and a green-colored FLC polymer film. One should keep in mind that such FLC display devices based on the absorptive guest host mode require a tilt angle $\theta = 45°$ to achieve maximum contrast of the display. This requirement cannot be met with materials available at present. On the other hand, the variation of the cell thickness is not so critical when the display operates in the absorptive mode.

Further research work has been devoted to creating new FLC polymers with multifunctional properties. Recently, Scherowsky reported on fluorescent FLC polymers [136]. The endeavor to produce FLC polymers with strong SHG activity will be dealt with in Section VI. So far we have discussed various applications with an electrooptic effect analogous to the SSFLC Clark-Lagerwall effect observed on LMM FLCs. Many of these applications are conceivable with other electrooptic effects such as the antiferroelectric, electroclinic, or deformed helix ferroelectric effect, in particular if gray scale is required.

Furthermore, flexible polymer film technology or conventional FLC glass cell assembly may be used to exploit all the different electrooptic effects mentioned above.

Antiferroelectric polymers would indeed be promising materials for electrooptic applications if they show the same advantageous characteristics as LMM AFLCs (see Section II.C):

In the fieldless state $P_S = 0$, which avoids image sticking.

In the powered state a high P_S can be used to speed up the optical response.

An AC electric field induces a director switching and additionally a reversible layer switching which promotes the self-alignment recovery from damage caused by mechanical shocks.

Due to its characteristic electrooptic response (double hysteresis), analog gray scale might be possible.

As a drawback, only the antiferroelectric state is stable and does not need a power supply, whereas the ferroelectric state must be maintained by a DC bias voltage. Moreover, to achieve full contrast, a tilt angle of 45° is required, and this can only be met by materials possessing the S_{CA}^*-ch-i phase sequence. It is not known whether the shear-induced orientation process of FLC polymers as demonstrated by Idemitsu [115] also works with compounds without S_A phase. Evidence for the existence of an antiferro-electric phase in SCLC polymers has been given by some authors [125–128].

Primarily due to its fast response, the electroclinic switching effect may become important for light shutter applications even with FLC polymers. However, these sub-millisecond response speeds have always been obtained in the S_A phase at elevated temperatures. In comparison to the other electrooptic effects, the electroclinic effect suffers particularly from its small induced tilt angle θ and above all from the strong temperature dependence of θ (see Fig. 21). Recently, an S_A material with a strongly reduced temperature dependence of θ has been discovered [137]. The authors believe that the absence of a tilted phase below the smectic A phase is the cause for this behavior. Perhaps this observation may be a stimulus for the synthesis of such nonferroelectric S_A polymer materials.

In conclusion, we can state that everything is in an early stage. With some exceptions, most of the current available S_C* FLC polymer materials can be operated only at elevated temperatures. Polymer compounds which possess at room temperature a S_C*, S_A, or S_{CA}* phase, as is vitally important for applications, are hardly available. Moreover, the characterization of the materials is still insufficient, and the electrooptic effects involved are far from being well understood. Nevertheless, a prototype of a large, flexible FLC polymer display has already been presented which demonstrated the ease and promise of this new technology. As there is large flexibility in molecular design, these materials might be further improved and adapted to the electrooptic effect which is used. Therefore, application of FLC polymers can be expected, in particular for display or imaging electrooptic devices which do not require predominantly fast response times but which take advantage of the mechanical ruggedness of this material, the ease of processing, and the memory capability to save energy consumption.

V. THERMOOPTIC APPLICATIONS; OPTICAL DATA STORAGE

So far we have discussed applications where the most essential feature of side-chain liquid crystalline polymers—its glassy state—has not been exploited. Because of this unique property, uniaxially ordered or disordered structures can be frozen in and stored for long periods by cooling the sample below the glass transition temperature [138,139]. This makes such materials suitable as optical data storage media. In the past, extensive work has been done on nematic, cholesteric, and S_A SCLC polymers or copolymers to evaluate the potential of these materials as media for erasable optical data storage (ODS) or storage displays [140,141]. System aspects have also been considered, and we would like to refer to the excellent review article written by McArdle, which summarizes the state of the art of optical storage with SCLC polymers [4]. A review which includes competitive storage technologies was given by Kämpf [142].

Depending on the initial condition and the final liquid crystalline structures which are written in, optical contrast can be generated in a number of ways. We will describe the procedure usually employed, which is sketched schematically in Fig. 24. Utilizing an S_A SCLC polymer, one starts with an optically clear, uniaxially aligned, homeotropic structure whose optical axis stands perpendicular to the glass plates. The polymer material is heated up to the isotropic phase by a short laser pulse, whereby the orientational order is lost. Subsequently cooling down into the glassy state freezes in the disordered, strongly light-scattering state. Optical contrast results from the difference in transparency or light scattering between the locally homeotropic and the disordered structure. Selective erasure

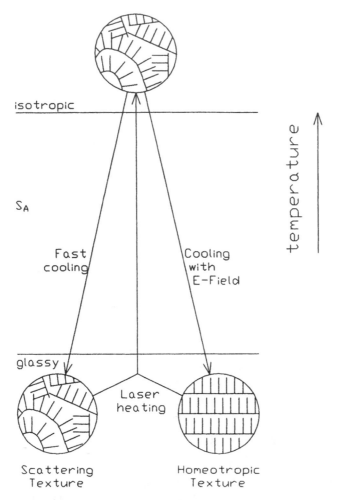

Figure 24   Optical information storage in S_A SCLC polymers.

is provided by repeating the write process but applying simultaneously an electric field during cooling. If the dielectric anisotropy $\Delta\epsilon$ is positive, this reestablishes the homeotropic texture. On the basis of the above-described procedure, the potential of side-chain LC polymers as analog optical stores was demonstrated [140] by using a scanning laser writing system.

Other variants have been proposed which start with different initial condition. For example, Schmidt [143] used an inherently colored nematic SCLC polymer with a few percent dichroic dye covalently bonded on the polymer backbone. Starting with an initially homogeneous, planar-aligned sample (optical axis parallel to the substrate plane) has potential advantages. First, improved recording sensitivity can be achieved, as the laser energy can be coupled very efficiently into the recording medium if the polarization of the writing light coincides with the direction where the dye absorption has a maximum. Indeed, with such dyed SCLC polymer films with a thickness between 1 and 2 μm, recording sensitivities of 0.18–0.4 nJ/μm^2 have been achieved [141]. These values can compete easily with those of other materials proposed for reversible optical data storage. Second, as the optical contrast results from the difference between a scattering structure (isotropic) and a planar (uniaxial) structure, optical contrast can be read out alternatively by detecting the birefringence or the dichroism of the written spots. In particular, the birefringence mode seems to be preferable concerning the attainable signal/noise ratio [141]. As a drawback of the homogeneous planar-aligned sample, selective erasure requires a polymer material with $\Delta\epsilon < 0$. Beside the high sensitivity, a remarkable high spatial resolution of <1 μm was measured [141]. Studies of the dynamic response during the writing process revealed that spot information is completed in a time $t < 1$ μs [141]. Retrieval of data which were written in with a frequency of 1 MHz was clearly demonstrated. But the most serious limitation for the employment of SCLC polymers in optical data storage seems to be the slow response time for local erasure, which is in the millisecond range [140,144].

One might speculate that $S_C{}^*$ FLC polymers could be promising candidates as media for reversible optical data storage, as the response for optical switching is typically two orders of magnitude faster than that of the S_A (or nematic) polymers discussed so far. But some care has to be taken, as the dynamic behavior depends on the optical states which are involved. Two possible cases will be sketched here, which are shown schematically in Fig. 25. The first case is very similar to the one described above for the S_A

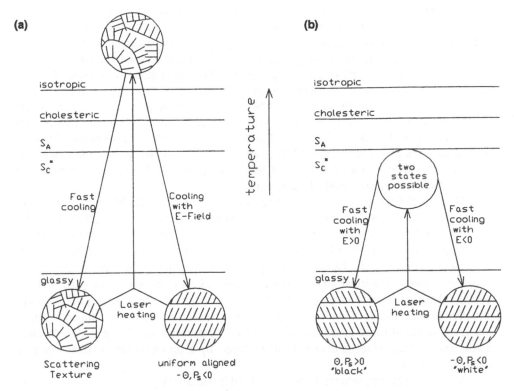

Figure 25 Optical information storage in $S_C{}^*$ FLC polymers. Two different operation modes are sketched: (a) using a scattering texture and a uniform aligned SSFLC state; (b) using the two uniform aligned SSFLC structures with P_s pointing in the upward or downward direction.

SCLC polymer. First, using a thin FLC polymer layer, a uniform aligned SSFLC structure is established in the S_C^* phase. This structure appears optically clear, possesses a definite polarization P_S, and can be transformed in the glassy state by cooling below T_g as shown in Fig. 25a ($P_S < 0$). Information can be written in by two different ways. In the first case the laser pulse heats up the sample into the isotropic phase and induces a scattering texture which is frozen in by the following cooling process. Thus writing time should be comparable as in the case of the ODS based on an S_A material. For selective erasure the sample is heated up again by a laser pulse to the isotropic phase but simultaneously an electric field is applied to assist the uniform alignment during the cooling process through the S_C^* phase. So far, however, experimental evidence that in this case selective erasure occurs more rapidly has not been given.

An alternative operation mode takes advantage of the existence of two uniform aligned states in an SSFLC structure (Fig. 25b). The same initial optical state is chosen as in the operation mode as sketched in Fig. 25a ($P_S < 0$). However, in the writing mode the laser pulse should heat up the sample just into the S_C^* phase, close below the S_C^*– S_A phase transition. A simultaneously applied electric field reverses the polarization vector in the heated spot and twists the optical axis by twice the tilt angle θ. If the sample is placed between crossed polarizers, $\theta = 22.5°$ and $d \cdot \Delta n = \lambda/2$ are chosen, the written spots which are frozen in again in the glassy phase act as $\lambda/2$ wave plates and turn the polarization of the read light by 90°. In this case the writing time will be of the order of the optical switching time of the S_C^* FLC polymer, which is typically two to three orders of magnitude faster than that of nematic (S_A) SCLC polymers [144]. As switching in the millisecond range has been observed with the latter material, optical response of the order of 10 µs should be achievable with FLC polymers, in particular if materials with a high S_C^*–S_A phase-transition temperature (>150°C) and a high spontaneous polarization ($P_S > 400$ nC/cm^2) are chosen and high electric field strengths are applied. Initially, however, experience must be gained on adequate materials. One should keep in mind that in this operation mode there should be no difference between the writing and the local erasure procedure. Only the polarity of the electric field is changed. This lets the optical axis switch back to the initial direction. Consequently, the response times for writing and selective erasure should be nearly the same. As two polarization states are involved, this type of storage device defines a binary storage. Even faster response can be expected if the electroclinic effect is exploited.

The sensitivity is expected to be comparably as high as for S_A SCLCPs if a colored S_C^* polymer film is used. In this case the wavelength of maximum absorption of the dye must be matched to the wavelength of the writing laser. As already mentioned, the stored information can be retrieved by optical means (birefringence, dichroism, scattering), but also by electrical means. Optical techniques are very efficient because of the strong coupling between electrical and optical properties of FLC materials. In particular, recording the birefringence change seems to be a very sensitive technique [141]. Nondestructive electrical readout exploits the piezo- or pyroelectric effects; i.e., the entire memory is uniformly stressed or heated and the piezo- or pyroelectric charge which develops across each storage pixel is detected, for example, by a tip electrode or other electrical techniques [142].

Another technique whereby the two possible electrical polarization states can be recorded as well as read by means of a transistor was described recently [145]. A non-volatile 64K ferroelectric random-access memory (RAM) based on PZT as ferroelectric storage medium in combination with the common semiconductor circuits was developed

[145]. However, in principle, other ferroelectric materials such as S_C^* FLC polymers should be usable.

In summary, we have tried to give an insight into thermooptical addressed optical data storage with S_A SCLC polymers and particularly with S_C^* FLC polymers, but experimental evidence for the latter material regarding its suitability is still so far completely lacking. Therefore, first of all, novel S_C^* FLC polymers have to be tailored to meet the main requirements for optical data storage:

Glass transition temperature T_g much higher than room temperature, i.e., $T_g > 60°C$
Phase sequence S_C^*-S_A-ch-i for attaining good molecular alignment
High P_s (500 nC/cm^2), low γ_c, and high phase-transition temperature S_C^*–S_A ($\geq 150°C$)
    to obtain a fast optical response
Use of an FLC copolymer which contains a few percent of a covalently bonded dichroic
    dye for absorption of laser light
$\theta = 22.5°$, $d \cdot \Delta n = \lambda/2$ to achieve maximum optical contrast

Regarding device design, considerable flexibility is possible, as the FLC polymer film can be coated on a single substrate or filled between two supporting substrates as is conventionally done with displays. Even a flexible optical storage film might be conceivable. This has been realized on the basis of nematic or S_A SCLC polymers with a single plastic substrate by Bowry et al. [146], and with S_C^* FLC polymers and two plastic substrates by Yuasu et al. [115].

The fabrication of thin, defect-free aligned FLC polymer films is one of the key problems to be solved, as higher optical quality is required for ODS than for display devices. Next, figures of merit relevant for ODS have to be evaluated by experiments. Presumably, response speed could be most critical and decisive for the application of FLC polymers in ODS.

VI. NLO APPLICATIONS

The interest in organic polymer materials for application in nonlinear optics is based mainly on three important properties [147]:

1. These materials are compatible with semiconductor technology. Low-cost thin films or waveguiding structures of polymers can be fabricated easily by familiar spin-coating processes which meet the requirements of integrated nonlinear optics.
2. Polymers and, in particular, side-chain polymers, liquid crystalline or amorphous ones, offer a high potential for molecular engineering.
3. Organic polymer materials have the advantage of possessing a high optical damage threshold and high nonlinear coefficients. For applications mostly discussed, such as second harmonic generation or nonlinear electrooptic light modulation by the Pockels effect, a large noncentrosymmetric order is requested on a macroscopic scale.

Up to now three classes of nonlinear optical polymers have been investigated:

1. Side-chain liquid crystalline polymers (SCLC polymer: nematic, S_A, or amorphous)
2. Ferroelectric polymers
3. Amorphous polymers

So far SCLCPs with a nematic or an S_A phase have been mainly used, as these materials possess an inherent uniaxial macroscopic order. However, to establish the required noncentrosymmetric structure, these materials have to undergo a poling procedure in a large DC applied field for some hours. One promising alternative is to use an FLC polymer which combines the inherent *polar* symmetry of an LMM FLC and the advantageous properties of polymers as described above. The NLO-active orientation can be frozen in by cooling below the glass transition temperature. Attempts to obtain such materials have been few so far. FLC polymers have been synthesized preferably for electrooptic display applications. However, NLO-active FLC polymers require a different molecular design. In this case the axis of maximum molecular hyperpolarizability β has to be aligned parallel to the polar axis (C_2 axis), which is perpendicular to the long molecular axis of the mesogenic moiety. NLO chromophores with large β values consist of a donor and an acceptor group, which are linked by aromatic rings. The direction of maximum β points along the donor–acceptor axis, and this axis has to be parallel to the direction of the spontaneous polarization P_S. To our knowledge, Kapitzka et al. [148] were the first to report on effective second harmonic generation with FLC polysiloxanes. Using a material with a promising high $P_S = 205$ nC/cm^2, they did not succeed in determining a d_{22} coefficient due to intensive light scattering. Recently, Coles et al. [149] reported on a new FLC copolymer with a polysiloxane backbone which showed an SHG comparable to quartz, although the experimental conditions were not ideal. However, this approach demonstrated impressively how multifunctional materials can be designed with side-chain liquid crystalline polymers. By employing this principle, three different side chains have been covalently bonded to the backbone. First, a chiral mesogenic moiety with a lateral dipole is introduced, which guarantees the existence of the $S_C{}^*$ phase and its ferroelectric properties. Second, an achiral, nonmesogenic NLO-active chromophor is substituted onto the copolymer backbone close to the chiral mesogen to induce a strong interaction between the NLO-active group and the polar axis. Finally, about 60% of the available sites at the backbone are occupied by nonelectroactive methyl groups, which ensure a reduction of viscosity and improve the alignment behavior. Additional functional side chains may be introduced to influence other physical properties or the phase behavior of the FLC material. The chemical fixation of the chromophor inhibits demixing previously reported for chromophors incorporated in polymer matrices.

At present, NLO-active FLC polymers (or copolymers) are at a very early stage of development, and this demands for compounds with high P_S values because these seem to reveal strong SHG response [73,149]. One serious problem for NLO-active SCLC polymers seems to be alignment, because even with nematic or S_A SCLC polymers, considerable light scattering due to imperfections in sample orientation has been observed. The situation is more difficult with FLC polymers as, in general, it is even more difficult to align these materials. Therefore a trade-off has to be made between high nonlinear optic efficiency and low absorption and scattering losses. Light losses in FLC polymers have not yet been presented, but Schmitt et al. determined a damping constant of 10 dB/cm for light propagation in a 6-μm-thick, homogeneous-aligned cell of a newly designed LMM FLC compound (see Section II.D). This is still quite large compared to the propagation loss of ~0.1 dB/cm for devices fabricated of Ti:LiNbO$_3$, a current material for production of nonlinear optic devices [82].

Thin $S_C{}^*$ polymer layers can be aligned by conventional means (alignment layer, shearing process) and using a supporting electric field to create a monodomain with a well-defined direction of P_S. To achieve good orientation in thicker samples, slightly cross-linked $S_C{}^*$ FLC elastomers may be a more convenient material, as orientation can be attained by mechanical deformation. Next this orientation can be locked in by a second cross-linking reaction which freezes in the NLO-active configuration.

In conclusion, so far only a few cautious attempts have been made to tailor FLC polymers for nonlinear optic applications, mainly for SHG and electrooptic light modulation by the Pockels effect, but corresponding coefficients have not yet been published. However, the high value of $d_{22} = 5$ pm/V observed recently with an LMM FLC is promising, and there is no reason to believe that such values cannot be achieved with FLC polymers. Furthermore, these materials are attractive in that they allow a lot of freedom in the molecular design due to the possibility of incorporating several functional pendant groups into the SCLC polymer. Also, processing of thin FLC polymer films should be comparative to that of S_A SCLC polymers. But at present the most serious problem to be tackled seems to be alignment. Much progress has to be made to improve the optical quality of the samples in order to reduce light absorption and scattering losses. Simultaneously, the development of novel FLC polymers which possess large NLO coefficients has to be pushed.

VII. PIEZOELECTRIC APPLICATIONS

The display of piezoelectricity of S_C^* FLC polymers can be explained by the inherent C_2 symmetry. For this symmetry the piezo tensor has the following form, with eight nonzero coefficients [150]:

$$
\begin{aligned}
d_{ij} &= \begin{pmatrix} 0 & 0 & 0 & d_{14} & d_{15} & 0 \\ 0 & 0 & 0 & d_{24} & d_{25} & 0 \\ d_{31} & d_{32} & d_{33} & 0 & 0 & d_{36} \end{pmatrix} \\
&= \begin{pmatrix} 0 & 0 & 0 & L_S & T_S & 0 \\ 0 & 0 & 0 & T_S & L_S & 0 \\ T & T & L & 0 & 0 & L_S \end{pmatrix}
\end{aligned}
\tag{13}
$$

The piezo coefficients d_{ij} are defined by [150]

$$
d_{ij} = \frac{dP_i}{dX_j}
\tag{14}
$$

where dP_i is the polarization induced in the i direction when mechanical stress dX_j is exerted in the j direction. In Eq. (13) the piezo tensor is given additionally in symbolic form, where T, L denote transverse or longitudinal compression and L_S and T_S denote longitudinal and transverse shear stresses. In particular, d_{33} gives the longitudinal piezo coefficient if the FLC sample is mechanically stressed in the bookshelf geometry perpendicular to the substrate planes, and the polarization induced in the same direction is detected as shown in Fig. 26. So far, to our knowledge, values of d_{33} have not been published either for LMM FLCs nor for FLC polymers. The mechanical induced piezoelectric voltage was measured only for cross-linked S_C^* FLC polymers (S_C^* FLC elastomers) and cholesteric elastomers [151,152], but no d_{33} values were given.

Recently, however, Wagenblast [153] determined for an S_C^* FLC polyacrylate a value of $d_{33} = 0.6$ pC/N. In this experiment a spontaneous polarization was frozen into the glassy state by applying a high DC electric field (20 V/µm) and simultaneously cooling the sample down from the S_C^* phase at room temperature. For the measurement

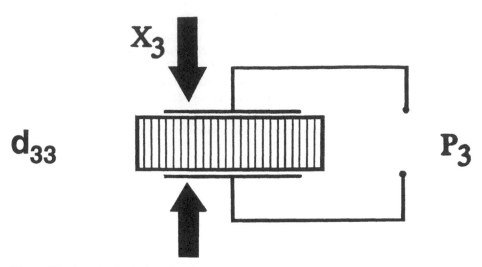

Figure 26 Longitudinal piezoelectric effect: A mechanical stress X_3 along the polar axis induces a corresponding polarization P_3.

of d_{33} the sample, with a thickness of 15 μm, was periodically stressed at a low frequency (1–10 Hz) and the induced surface charge was collected by a charge amplifier. The applied force was measured via a force sensor. In a different study, Hikmet reported on the piezoelectric behavior of a plasticized S_C^* network [154]. First a mixture of a non-reactive chiral molecule with a polymerizable LC diacrylate, which exhibits an S_C phase, was made and oriented by a rubbed nylon orientation layer in 60-μm-thick sample cells. In a second step a highly transparent, uniaxially oriented S_C^* network was established by a photoinduced polymerization of the diacrylates under a DC electric field whereby the chiral molecules were not yet chemically attached to the network. Values of 3.1 pC /N (1.4 pC/N) have been measured for the piezoelectric coefficient parallel (perpendicular) to the molecular orientation, although the spontaneous polarization of this material was very low ($P_S \approx 8$ nC/cm^2). This kind of material belongs to the class of main-chain S_C^* FLC polymers where the liquid crystalline properties are induced by mesogenic moieties incorporated in the polymer main chain.

However, LMM FLCs and FLC polymers are liquidlike even in the glassy state. They cannot resist mechanical stress by elastic forces and, depending on their viscosity, they relax more or less quickly by material flow. On the other hand, corresponding elastomers possess the desired elastic response and behave like soft solids [155]. They can support mechanical stress under equilibrium conditions, and are preferable to non-cross-linked S_C^* FLC polymers for applications in piezo sensors.

Most recently, for such a uniformly aligned, cross-linked S_C^* FLC polysiloxane sample, a maximum piezoelectric coefficient $d_{33} = 6.5$ pC/N was determined [156]. The temperature dependence of d_{33} was measured for a heating and a cooling run and is reproduced in Fig. 27. Apart from the small hysteresis, piezoelectricity is even observed in the nonferroelectric S_A phase and is reversibly obtained when the sample is heated up into the isotropic phase and cooled back to the S_A or S_C^* phase. The maximum d_{33} value is larger than that of quartz ($d_{11} = 2.3$ pC/N) and comes very close to the value $d_{33} = 28$ pC/N for the polymer PVDF, which possesses the highest piezocoefficient measured

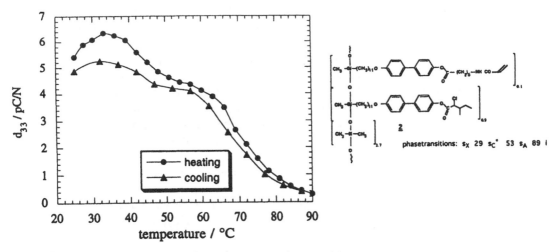

Figure 27 Temperature dependence of the piezocoefficient d_{33} (from Ref. 155) for an S_C^* FLC elastomer as also shown.

for a polymer so far. A comparison of piezoelectric coefficients of some other competitive piezoelectric compounds is given in Table 1.

As this material exhibits a spontaneous polarization $P_S \approx 80$ nC/cm^2 in its non-cross-linked state, further improvement of d_{33} values might be possible in the near future by using high-P_S compounds.

For the fabrication of uniformly aligned samples, two techniques may be applicable. In the first case, as shown in [156], a non-cross-linked sample is aligned in its S_C^* phase by alignment layers (or by a shearing process) and the helix is unwound by a DC electric field which results in a monodomain with maximum polarization. Then the cross-linking process is accomplished while the electric field is maintained. If cross linking is complete, these samples are no longer ferroelectric switchable [156], even with field strengths of $E = 50$ V/μm.

In the second method, first a pre-cross-linking action is done. Then this material can be oriented and the helix can be unwound in the S_C^* phase by a mechanical strain [103]. Full macroscopic polarization will appear if the helix is unwound. However, if the me-

Table 1 Piezoelectric Coefficients d_{ij} of Some Piezopelectric Materials

Material	d_{ij} (pC/N)	
Quarz	2.3	(d_{11})
PZT-4	289	(d_{33})
BaTiO$_3$	190	(d_{33})
Rochelle salts	53	(d_{25})
Nylon-11	0.26	
PVDF	28	(d_{31})
S_C^* polysiloxane [156]	6.5	(d_{33})

chanical stress is removed, the elastomer relaxes back into the disordered state. To avoid this, a second cross-linking action is employed which reversibly locks in the macroscopic polarization even above the glass transition temperature [155]. This shows another advantage of cross-linked materials. As the polarization is already frozen in in the S_C^* phase, there is no need to cool into the glassy state, and other orthogonal smectic phases between the glassy state and the S_C^* phase should not have a detrimental influence on the alignment. Moreover, these materials are in a rubbery state and show a soft Young's modulus; i.e., a weak mechanical stress induces a large strain [151,152,155]. This may become important for a new class of piezoelectric sensors different from classical piezoelectric polymers such as PVDF.

At present we know very little about the piezoelectric behavior of these new materials, and further experiments are required to judge the full potential of these compounds. The frequency dependence of d_{33} of an S_C^* FLC polymer (or elastomer) has not been investigated so far. Other piezoelectric coefficients [see Eq. (13)] have also not yet been determined.

So far, few experimental results have been published on inverse piezoelectric effects, known as electromechanic effects [86]. The linear electromechanical effect in two S_C^* FLC polymers (polysiloxane and polyacrylate) was studied by Jákli and Saupe [157]. The mechanical vibrations of the cover glass were measured in the horizontal direction parallel to the smectic layers (y direction) and perpendicular to the substrate planes (x direction), as shown in Fig. 17b. The electromechanical behavior of these FLC polymers was found to be very similar to that of LMM FLCs. Mechanical vibration has been observed in the frequency range from 100 Hz to about 6 kHz. Linear and quadratic response has been detected. The strength S of the linear electromechanical effect is defined as the maximum obtainable displacement induced by a unit electric field [158]. Typically, S for LMM FLCs varies between 10^{-16} and 10^{-14} m²/V and depends critically on the alignment [158]. For FLC polymers, $S = 3 \times 10^{-16}$ m²/V has been obtained, but one can expect comparable S values if poor alignment of the studied polymers is improved [158]. As in the case of LMM FLCs [89], vibration in the x and y directions can be assigned to two different mechanisms. According to [158], in the kilohertz range the vertical and the horizontal motions have the same order of magnitude of the vibrational amplitude, whereas at lower frequencies preferably horizontal vibration occurs. The vibration in the y direction can be reasoned by the director movement on the cone around the layer normal (Goldstone-mode exitation) which, due to the coupling with the viscous flow, induces material flow in the y direction. This does not represent true piezoelectric response.

On the other hand, the vertical vibration is presumably due to coupling between P_S and the molecular tilt, i.e., it is caused by the electroclinic effect, which leads to a layer compression. This gives rise to an oscillatory material flow within the layers which yields an integrating vertical (x) vibration of the substrates. The vertical vibration can be regarded as a true piezoelectric response.

These electromechanic effects result in sound generation in the audible frequency range up to some kilohertz [87,89]. Higher harmonics may cause ultrasound [91]. Therefore these effects lead to an electroacoustic response which can be exploited in applications such as speakers or headphones. The frequency characteristics and the achievable maximum vibration amplitude are two parameters that are decisive for the application of this material in such devices. As already mentioned, both depend very strongly on the quality of the alignment of the S_C^* FLC polymer, the layer structure (chevron, bookshelf, or other), the existence of zigzag defects in the sample, and the sample cell itself [158]. The frequency dependence of the vertical and horizontal vibrations is still not very well understood. Resonance peaks occur which can be explained by resonance vibrational modes of the glass plates [89] or nonperfect alignment [158,159]. A large disagreement has been observed concerning the vibrational amplitudes caused mainly by misalignment of the sample.

A dispersion of three orders of magnitude in the measured vibrational amplitudes was reported on several samples of the same LMM FLC which were prepared using the same procedure [88]. By using 20-μm thick samples and electric field strengths E of 10 V/μm, a maximum amplitude of 1 μm for the horizontal vibration was detected, which corresponds to a maximum expansion of 5% [88]. According to [158,159], one can expect that FLC polymers should achieve comparable maximum vibrational amplitudes as LMM FLCs. A correlation between the strength of the electromechanical effect and the spontaneous polarization has not yet been found [88].

In conclusion, there is first of all a need for well-aligned samples which provide reproducible electromechanical effects with vibrational amplitudes as large as possible.

Electromechanical effects of S_C^* elastomers should preferably be investigated. As material flow cannot occur in these materials, they may have particular advantages in devices which operate at low frequencies. With a combined main-chain/side-chain S_C^* elastomer, a piezoelectric signal for a 10% thickness variation was reported [160]. If reproducible expansion on this length scale is possible on such materials, they may indeed open a new class of positioning devices, as the current piezo devices achieve expansions of 0.2% at best.

Finally, technological aspects have to be taken into account. Thin films of FLC polymers or elastomers can be easily fabricated by different techniques onto single substrates or between two plastic (or glass) substrates. With thicker samples, alignment problems may occur which can be best solved by the use of S_C^* FLC elastomers which allow alignment by mechanical strain.

To make further progress in this field, precise control of the physical starting conditions of the different piezoelectric effects is necessary and an improved understanding of the underlying mechanisms seems to be of major importance. Therefore the alignment again plays a key role and has to be controlled in order to achieve reproducible results. From the material point of view, cross-linked polymers (S_C^* or cholesteric) are preferable, as their piezoelectric effects result in reversibly elastic deformations (not just in material flow), and spontaneous polarization can be frozen in even in the S_C^* phase [156]. Furthermore, they can be oriented by mechanical forces. Therefore, study of piezoelectric and electromechanic effects of these promising materials seems of particular importance. Thereafter the relevant material properties have to be improved to meet the diverse demands for applications such as piezo stress sensors, microphones, loudspeakers, headphones, or piezo-positing devices. Besides, one should emphasize that electromechanical and its inverse effects may have an impact on the electrooptic characteristics of FLC displays, as material flow can modify the director configuration. Therefore a full understanding of the electromechanical phenomena will also be useful for designing displays with optimum performance.

VIII. PYROELECTRIC APPLICATIONS

Pyroelectric devices are commonly used for the thermal detection of infrared radiation or as temperature-sensitive sensors. In particular, more sophisticated detectors are able to convert infrared images into a spatially replicated charge pattern which can be read using an electron beam or a charge-coupled device (CCD).

The performance of the pyroelectric detector is strongly determined by material properties as well as device parameters. The suitability of FLC polymers for this kind of application has to compete with the as yet most widely used pyroelectric compounds such as triglycine sulfate (TGS), lithium tantalate (LiTaO$_3$), strontium barium niobate (SBN), bariumtitanate (BaTiO$_3$), PVDF, and PVF$_2$.

A pyroelectric detector based on an FLC material has essentially the same sandwiched structure as a display cell. A schematic drawing of a pyroelectric device is shown in Fig. 28. The cell consists of two glass plates separated by polyimide spacers (2 μm thickness). The bottom glass plate is coated with a transparent layer of conductive indium tin oxide (ITO). In contrast to the common display cell, a layer to absorb the incident light power is needed. For this purpose, Glass et al. [94] used a 100-Å-thick conducting Cr-Au layer to achieve 30% absorption of the incident light power. Alternatively, the FLC can be doped with a certain amount of dye or, even better, a colored FLC polymer [135] can be used. Finally, thin polymer layers are spin-coated onto the two glass substrates and rubbed to provide alignment of the FLC molecules.

According to Eq. (11), high detector sensitivity requires a large pyroelectric coefficient p. Its magnitude was determined and compared for the first time for an FLC monomer and an FLC polymer by Koslowskii et al. [95] and later by Shibaev [97] for a polymethacrylate. This comparison is reproduced in Fig. 29 together with the temperature dependence of the spontaneous polarization. The p value of the FLC polymer is about an order of magnitude lower than that of the corresponding FLC monomer, and its temperature dependence is less pronounced, particularly just below the S$_C$*–S$_A$ phase transition. This is due to the weaker temperature dependence and the lower absolute values of P_S in the case of the FLC polymer. From Fig. 29 a typical value of the polymer P6*Cl is $p \approx 0.1$ nC/cm^2 K with $P_S \approx 2$ nC/cm^2. Although quite comparable maximum P_S values can be obtained for polymers and monomers [107], the temperature dependence of P_S is often weaker in the polymer case due to a broader phase range. As a result, the pyroelectric coefficient p is lower in the case of the polymer. To establish full polarization

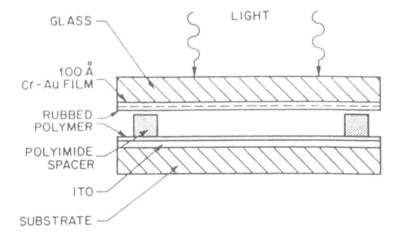

Figure 28 FLC cell for pyroelectric studies. (From Ref. 94.)

(a)   **(b)**

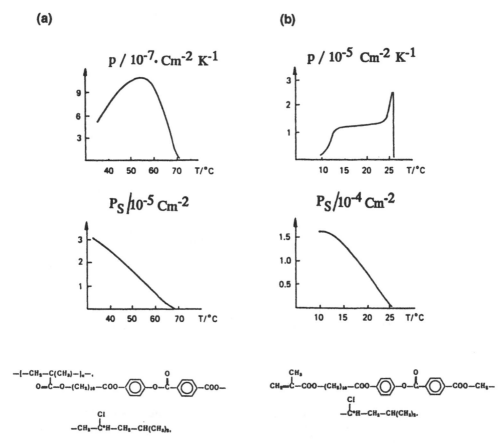

Figure 29 Temperature dependence of the pyroelectric coefficient p and the spontaneous polarization P_S (from Ref. 97) for (a) FLC polymer P6*Cl; (b) LMM M6*Cl.

in thick samples, the helix must be unwound and maintained by a DC electric field [94,161]. This makes the simultaneous application of a DC bias field practical. The use of a thin SSFLC structure also suppresses the appearance of the helix and additionally increases the sensitivity of the pyroelectric device. By using S_C^* FLC polymers with P_S values of 400–500 nC/cm³, it should be possible to obtain high pyroelectric coefficients such as $p \approx 20$ nC/cm² K. Maximum polarization values of 420 nC/cm² have been observed so far on a S_C^* polysiloxane [148], but the p values are not yet known.

Recently, Skarp et al. [162] obtained a maximum p value of about 13 nC/cm² K for a diluted S_C^* polysiloxane (see Fig. 30), which has a P_S of about 100 nC/cm² at a temperature of 95°C. If a small temperature range is acceptable or even desirable for device operation (temperature sensor), further enhancement of p is possible by using materials which have the phase sequence S_C^*-ch-i instead of S_C^*-S_A, as the former material has a stronger temperature dependence of P_S near the (first-order) phase transition.

In conclusion, one can expect that p values quite close to competitive solid-state materials such as TGS ($p = 30$ nC/cm² K) can be obtained close to the S_C^*-S_A (ch) phase transition. In order to exploit the temperature range where the pyroelectric coefficient is

Phase sequence: g 6 S$_X$ 47 C* 98 A* 144 i

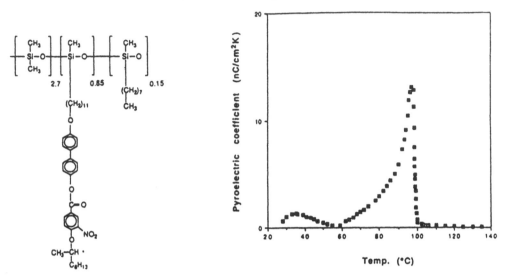

Figure 30 Temperature dependence of the pyroelectric coefficient p for an S$_C$* FLC polysiloxane as shown and according to Ref. 162. The maximum P_S value of this compound is 150 nC/cm^2 (65°C).

a maximum, there is clearly a need for novel S$_C$* polymer materials whose phase transition temperatures to the S$_C$* phase from above lie only slightly above the intended working temperature of the devices.

A prerequisite for achieving a strong pyroelectric response is highly efficient transfer and conversion of the incident light power. Besides the pyroelectric coefficient, this process is determined by the rate dT/dt as described in Eq. (11). dT/dt depends on the cell geometry, the rate of absorbed light, the thermal capacity of the FLC polymer, and the thermal relaxation time with the cell walls.

To achieve optimized performance of a pyroelectric detector, various figures of merits are in common use, depending on whether a high current responsivity, a high voltage responsivity, or a high signal noise/ratio is required. We assume that radiation, modulated at a frequency ω, is incident on the detector in a continuous manner. *If the thermal relaxation is neglected* ($\omega\tau_T \gg 1$, τ_T = thermal relaxation time), then the current responsivity r_i of a detector with area F and thickness d is given by [163]

$$r_i = \frac{i(\omega)}{W_0} = \frac{e \cdot p}{c_p \cdot d} \tag{15}$$

In this equation, W_0 is the maximum incident light power, $i(\omega)$ is the pyroelectric current according to Eq. (11), e is the fraction of energy which is absorbed and converted in the pyroelectric signal, c_p is the specific heat at constant pressure and per cubic centimeter of the pyroelectric material and d is the thickness of the ferroelectric film.

If the detector is connected to a load consisting of a parallel RC circuit (with a load resistance R_L and a load capacity C_L), a voltage responsivity can be defined in a similar manner [163]:

$$r_v = \frac{\Delta V}{W_0} = \frac{e \cdot p}{c_p \cdot F \cdot \omega \cdot \epsilon'} \tag{16}$$

Equation (16) holds only if again $\omega \tau_T \gg 1$, the detector impedance is much smaller than that of the load, and $\omega \tau_E \gg 1$, where $\tau_E = C \cdot R$ is the electrical time constant of the whole RC circuit, ϵ' is the dielectric constant and F is the irradiated detector area. To obtain a maximum signal noise/ratio with the same conditions as given above, a useful figure of merit is the reciprocal of the minimum detectable power $(W_m)^{-1}$ [163]:

$$(W_m)^{-1} = \frac{e \cdot p \cdot \Delta f^{-1/2}}{c_p (4 \ kT)^{1/2} (\sigma \cdot d \cdot F)^{1/2}} \tag{17}$$

where Δf is the bandwidth and σ is the AC conductivity, which is usually dominated by the dielectric loss ϵ'' according to $\sigma = \omega \epsilon''$. From the figures of merits as given in Eqs. (15)–(17), it becomes clear that optimum detector performance involves both the detector geometry (d, F) and the material characteristics $(p, e, \epsilon', \epsilon'', c_p)$.

Dielectric measurements of ϵ' and ϵ'' for an LMM FLC material were also made by Glass et al. [94] and are reproduced in Fig. 31. All data were taken using 2-μm thick cells at 1 kHz measuring frequency for heating and cooling runs. Two features are of interest. First, the dielectric constant ϵ' is relatively small compared to the common solid-state materials TGS, BaTiO$_3$, LiTaO$_3$, and SBN, which have values typically varying from 50 to 1000. Second, ϵ' can be further reduced by a factor of 2 by applying a DC electric field, which results in a suppression of the so-called Goldstone mode. The origin of this mode is a collective motion of the long molecular axes on a cone with a constant cone angle 2θ in each smectic layer which is related to a fluctuation of the corresponding polarization vector. The Goldstone mode is exited by small AC electric fields up to frequencies of about 1 kHz and then contributes to ϵ'. By using higher frequencies or supperposing a DC electric field, the occurrence of this mode can be suppressed. Related to this Goldstone mode is a corresponding dielectric loss ϵ'' as also depicted in Fig. 31. For a zero DC bias voltage, a considerable ϵ'' due to the exitation of the Goldstone mode is observed. At 9 V DC bias, the Goldstone mode is completely suppressed, which results in a dramatic decrease of the dielectric loss.

The Goldstone mode has also been observed with FLC polymers [117], but the corresponding frequency range is shifted about one decade to lower frequencies. Although only a few dielectric measurements on S_C^* side-chain polymers, S_C^* combined side-group main-chain polymers, and S_C^* elastomers exist, these systems seem to behave in a similar way as their LMM counterparts [117]. While the ϵ' values of FLC polymers are comparable than for LMM FLCs [ϵ' (1 kHz) $\approx$ 4] [118], the corresponding dielectric loss values ϵ'' have been found to be larger for polymeric systems [ϵ'' (1 kHz) $\approx$ 0.50] [118]. However, in the case of an S_C^* FLC polytartrate which possesses a high $P_S = 500$ nC/cm^2, a large ϵ' value of $\epsilon' = 25$ was observed, although the high measuring frequency of 1 kHz suggests the suppression of the Goldstone mode [164]. Further work has to be done on the available FLC polymers to confirm these first, very preliminary results.

To evaluate the figures of merit according to Eqs. (15)–(17), we have to take into account the specific heat c_p. For LMM FLCs possessing an S_A–S_C^* phase transition, we

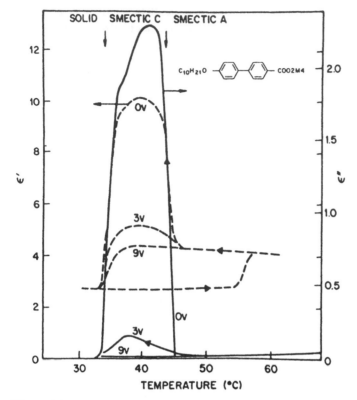

Figure 31 Dielectric constant ϵ' and dielectric loss ϵ'' for a 2-μm thick sample of an LMM FLC measured at 1 kHz according to Ref. 94. Different DC bias electric fields (0 V, 3 V, 9 V) were applied, and heating and cooling were made whose direction is indicated by arrows. Broken lines show ϵ' (left-hand scale); solid lines represent ϵ'' (right-hand scale).

have found c_p values varying between 2 J/cm^3 K and 3.6 J/cm^3 K [165,166] at temperatures some degrees below the phase transition. To our knowledge, at present c_p data of FLC polymers in the $S_C{}^*$ phase have not been reported at all. So, in the following we will take a value $c_v \approx c_p = 7.2$ J/cm^3 K as reported for an S_A SCLC polysiloxane by McArdle [4]. All data and figures of merit relevant for pyroelectric detection with $S_C{}^*$ FLC polymers (and LMM FLCs) and other well-known competitive pyroelectric materials are presented in a comprehensive form in Table 2. The data for the nonliquid crystalline materials have been taken from Ref. 163 and for the LMM FLC from Ref. 94. The p value for the FLC polymer is from Ref. 162, and the values of ϵ' and ϵ'' are from Ref. 118. However, the FLC polymer data have to be used with caution and are very preliminary. They represent an incomplete set of data taken from various materials because at present a complete set of pyroelectric data for one $S_C{}^*$ FLC polymer material is not yet available. These estimated figures of merit for FLC polymers can compete with established pyroelectric materials. In contrast to the latter compounds, FLCs take advantage of their low ϵ' and ϵ'' values.

In the following we will briefly describe dynamic aspects of pyroelectric devices. As with all pyroelectric detectors, thermal coupling of the pyroelectric material with the

Table 2 Properties of Various Pyroelectric Detector Materials and Some Figures of Merit for Their Detector Operation

Material	p (nC/cm^2 K)	ϵ'/ϵ_0	c_p (J/cm^3 K)	ϵ''/ϵ_0	p/c_p (n A cm/W)	$p/c_p\epsilon'$ (V cm^2/J)	$p/c_p(\epsilon'')^{1/2}$ (cm^3/J)$^{1/2}$
Triglycine sulfate [TGS]	30	50	1.7	0.16	17.8	4000	0.149
LiTaO$_3$	19	46	3.19	0.16	6	1470	0.050
Sr$_{1/2}$Ba$_{1/2}$<Nb$_2$O$_6$ (SBN)	60	400	2.34	8	25.6	720	0.030
PLZT	76	1000	2.57	8.8	29.9	340	0.034
PVF$_2$	3	11	2.4	0.25	1.3	1290	0.009
LMM FLC according to [94] [9 V, bias]	0.8	1.7	(2.8)	0.012	0.29	1900	0.009
S$_C$* FLC polymer	13	(4)	(7.2)	(0.50)	1.80	5080	0.009

substrate should be kept as small as possible, as it degrades the response at low frequencies. For low-speed thermal detection a long relaxation time is desirable for optimum device performance. The maximum response is achieved at times shorter than the thermal relaxation time of the device. A pulse response measurement is an adequate technique for determining the thermal relaxation behavior. Unfortunately, such investigations have not yet been done on FLC polymers. Glass et al. [94] have studied the pulse response of three LMM FLCs with 2-μm thick cells. By using optical heating pulses from a YAG: Nd laser with 0.1-μs duration and detecting the pyroelectric current with a boxcar integrator, a rise time $\tau < 7$ μs was observed followed by a relaxation time of about $\tau \approx 80$ μs. This is indeed a very rapid thermal relaxation with the cell walls, but it may be influenced strongly by the metallic layer which acts as a heat sink, transferring the absorbed energy to the neighboring glass substrate.

Alternatively, the FLC can be doped by some percent of a dichroic dye. If the incident radiation power is absorbed more directly by the dye molecules or the pyroelectric material itself (IR absorption bands), a slower thermal relaxation can be expected. As the thermal relaxation time depends in a complicated way on a number of factors such as cell geometry, light-absorbing layer, and thermal conducting coefficients, careful design has to be chosen to produce a device possessing a thermal relaxation as slow as possible.

Another point which has not yet been considered is the infrared absorption of the FLC polymer itself, which influences the spectral response of the pyroelectric device and can limit the use of polymer materials to certain wavelength ranges. With the exception of a strong absorption at 3.0–3.6 μm, which occurs because of excitation of various CH_2 and C–H vibrational modes of the main chain and the mesogenic group, no further absorption is observed in the wavelength range from the visible up to the infrared until 5.5 μm [167]. However, additional absorption bands may appear depending on the class of polymer material which is used.

Concerning questions of fabrication and mechanical strength, a simple pyroelectric device may be based on a thin flexible FLC polymer film with plastic substrates as presented by Idemitsu [107,115] and discussed in Section IV. This would be a very easy and cheap technology, but depending on the wavelength range intended for application, absorption in the plastic substrate has to be taken into account. Thin layers, 1–2 μm thick, can be easily fabricated by spin-coating processes on single substrates or by the current FLC cell technology between two substrates. Great flexibility is offered in detector design, as a patterned charge-coupled device (CCD) could, in principle, be incorporated on one of the cell walls, to record the pyroelectric charge in a pyroelectric imaging device. In contrast to LMM FLCs, corresponding S_C^* FLC polymer (or even elastomer) films are more rugged and should better withstand mechanical shock. We believe that FLC polymers are very promising for pyroelectric applications, but to make them successful, and competitive with conventional pyroelectric materials, the next step has to be taken by the chemists. Novel compounds with high P_s values have to be developed. Thereafter the physical characterization of these new material has to be improved and completed to evaluate the full potential of this new class of pyroelectric materials.

IX. MISCELLANEOUS APPLICATIONS

Recently a new image recording system based on a ferroelectric polymer [poly(VDF)-TFE] was proposed [168], which requires the following steps for transferring an image onto paper (see Fig. 32).

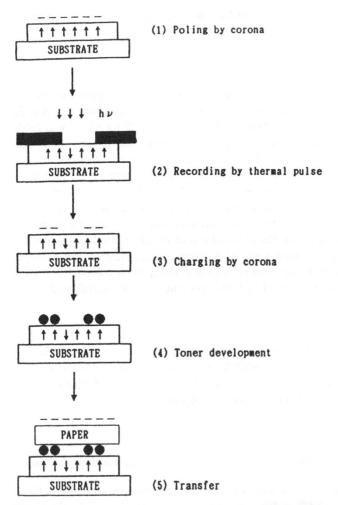

Figure 32 Image recording process with a ferroelectric polymer. (From Ref. 168.)

1. First, the 2-μm thick polymer layer, which is spin-coated on a conductive substrate, is poled by corona pulses.
2. Next, the image is recorded onto the polymer layer by locally changing the state of polarization. As a result, a remanent polarization pattern is built up.
3. The film undergoes a second corona charging.
4. The latent image is developed using a toner.
5. The toner image is transferred onto the paper.

In principle, the procedure, as described in [168], can also be applied to S_C^* FLC polymers. Just one additional alignment layer is required to support uniform orientation of the FLC polymer, which can be spin-coated onto the coated single substrate. For recording, the sample can be heated locally by a laser pulse into a nonferroelectric phase where P_S disappears, or, alternatively, the spontaneous polarization can be locally inverted in the S_C^* phase by a tip electrode. In the case of laser heating, incorporation of a dichroic dye can be helpful to facilitate light absorption.

X. CONCLUSIONS

We have made a careful attempt to give an idea of the potential for application of the new but strongly expanding research area of S_C^* ferroelectric liquid crystalline polymers, which is still in its infancy. Since the synthesis of the first S_C^* FLCP by Shibaev in 1985, enormous progress has been achieved in developing novel materials using various molecular architectures. However, physical characterization is still very fragmentary and incomplete. Nevertheless, in the last few years the possibility of electrooptic, piezoelectrical, and pyroelectrical applications with S_C^* FLC polymers has been demonstrated. The related figures of merit which have been obtained are promising; to develop the full potential of these materials, they have to be well aligned and their physical properties have to be adapted and optimized for the physical effect which is used in the device. This requires new material developments and also better characterization and a deeper understanding of the physical properties and phenomena which are involved. Then great progress and various novel applications of S_C^* FLC polymers can be anticipated.

ACKNOWLEDGMENT

The author acknowledges gratefully the benefits of collaboration with BASF AG, in particular, support and helpful discussions with K. Siemensmeyer and G. Wagenblast. I would also like to thank G. Baur and my colleagues M. Kamm and A. Stieb for fruitful discussions and H. Klausmann for his contribution of figures.

REFERENCES

1. R. B. Meyer, L. Liébert, L. Strzelecki, and P. Keller, Ferroelectric liquid crystals, *J. Physique 36*:L-69 (1975).
2. N. A. Clark and S. T. Lagerwall, Submicrosecond bistable electro-optic switching in liquid crystals, *Appl. Phys. Lett. 36*:899 (1980).
3. H. Finkelmann, H. Ringsdorf, and J. H. Wendorff, Model consideration and examples of enantiotropic liquid crystalline polymers, *Makromol. Chem. 179*:273 (1978).
4. (a) G. R. Möhlmann and C. P. J. M. van der Vorst, Side chain liquid crystal polymers as optically nonlinear media, *Side Chain Liquid Crystal Polymers* (C. B. McArdle, ed.), Blackie, Glasgow and London, 1989, p. 330. (b) C. B. McArdle, The application of side chain liquid crystal polymers in optical data storage, *Side Chain Liquid Crystal Polymers* (C. B. McArdle, ed.) Blackie, Glasgow and London, 1989, p. 357.
5. V. Shibaev, M. Kozlovsky, L. A. Beresnev, L. M. Blinov, and N. A. Platé, Chiral smectics "C" with spontaneous polarization, *Polymer Bull. 12*:299 (1984).
6. F. Tournilhac, L. M. Blinov, J. Simon, and S. V. Yablonsky, Ferroelectric liquid crystals from achiral molecules, *Letters to Nature* (London) 359:621 (1992).
7. S. T. Lagerwall and I. Dahl, Ferroelectric liquid crystals, *Mol. Cryst. Liq. Cryst. 114*:151 (1984).
8. R. B. Meyer, Piezoelectric effects in liquid crystals, *Phys. Rev. Lett. 22*:918 (1969).
9. L. A. Beresnev, V. A. Baikalov, and L. M. Blinov, Flexo- and piezo-electric terms in the spontaneous polarization of ferroelectric liquid crystals, *Ferroelectrics 58*:245 (1984).
10. P. Pieranski, E. Guyon, and P. Keller, Shear flow induced polarization in ferroelectric smectics C, *J. Physique 36*:1005 (1975).

11. C. B. Sawyer and C. H. Tower, Rochelle salt as a dielectric, *Phys. Rev. 35*:269 (1930).

12. (a) K. Miyasato, S. Abe, H. Takezoe, A. Fukuda, and E. Kuze, Direct method with triangular waves for measuring spontaneous polarization in ferroelectric liquid crystals, *Jpn. J. Appl. Phys. 22*:L-661 (1983). (b) Ph. Martinot-Lagarde, Direct electrical measurement of the permanent polarization of a ferroelectric chiral smectic C liquid crystal, *J. Phys. (Lett.) 38*:L17 (1977).

13. L. M. Blinov, L. A. Beresnev, N. M. Shtykov, and Z. M. Elashvili, Pyroelectric properties of chiral smectic phases, *J. Phys.* (Paris) *40*:C3-269 (1979).

14. L. A. Beresnev, V. A. Baikalov, L. M. Blinov, E. P. Pozhidaev, and G. V. Purvanetskas, First nonhelicoidal ferroelectric liquid crystal, *JETP Lett. 33*:536 (1981).

15. J. W. Goodby, Properties and structures of ferroelectric liquid crystals, *Ferroelectric Liquid Crystals* (G. W. Taylor, ed.), Gordon & Breach, New York, 1991, p. 80.

16. T. P. Rieker, N. A. Clark, G. S. Smith, D. S. Parmar, E. B. Sirota, and C. R. Safinya, "Chevron" local layer structure in surface-stabilized ferroelectric smectic-C cells, *Phys. Rev. Lett. 59*:2658 (1987).

17. A. Fukuda, Y. Ouchi, H. Arai, H. Takano, K. Ishikawa, and H. Takezoe, Complexities in the structure of ferroelectric liquid crystal cells. The chevron structure and twisted states. *Liq. Cryst. 5*:1055 (1989).

18. C. Bowry, A. Mosley, B. M. Nicholes, F. LeTexier, J. Ph. Boyer, J. P. Le Pesant, J. F. Clerc, J. Dijon, and C. Ebel, Alignment treatments for surface stabilised ferroelectric liquid crystals, *Ferroelectrics 85*:31 (1988).

19. J. Kanbe, H. Inoue, A. Mizutome, Y. Hanguu, K. Katagiri and S. Yoshihara, High resolution, large area FLC display with high graphic performance, *Ferroelectrics 114*:3 (1991).

20. T. Uemura, N. Ohba, N. Wakita, H. Ohnishi, and I. Ota, Alignment of chiral smectic C liquid crystal by oblique evaporation method, *Proc. 6th Int. Displ. Res. Conf.*, "*Japan Display '86*," Tokyo, 1986, pp. 464–467.

21. N. Yamamoto, Y. Yamada, K. Mori, H. Orihara, and Y. Ishibashi, Ferroelectric liquid crystal display with high contrast ratio, *Jpn. J. Appl. Phys. 28*:524 (1989).

22. Y. S. Negi, Y. Suzuki, T. Hagiwara, I. Kawamura, N. Yamamoto, K. Mori, Y. Yamada, M. Kakimoto, and Y. Imai, Surface alignment of ferroelectric liquid crystals using polyimide, polyamide-imide and polyamide layers and their effect on pre-tilt angle, *Liq. Cryst. 13*:153 (1993).

23. A. Tsuboyama, Y. Hanyu, S. Yoshihara, and J. Kanbe, Characteristics of the large size, high resolution FLCD, *Proc. 12th Int. Displ. Res. Conf.* "*Japan Display '92*," Hiroshima, Japan, 1992, pp. 53–56.

24. Y. Sato, T. Tanaka, H. Kobayashi, K. Aoki, H. Watanabe, H. Takeshita, Y. Ouchi, H. Takezoe, and A. Fukuda, High quality ferroelectric liquid crystal display with quasi-bookshelf layer structure, *Jpn. J. Appl. Phys. 28*:L483 (1989).

25. W. Hartmann, Uniform SSFLC director pattern switching, *Ferroelectrics 85*:67 (1988).

26. (a) A. Mochizuki, T. Yoshihara, M. Iwasaki, M. Nakatsuka, Y. Takanishi, Y. Ouchi, H. Takezoe, and A. Fukuda, Electro-optical switching of bookshelf layer structure S_C^* cells aligned with a rubbed polymer film, *Proc. 9th Int. Disp. Res. Conf.* "*Japan Display '89*," Kyoto, Japan, 1989, pp. 30–33. (b) A. Mochizuki, K. Motoyoshi, M. Nakatsuka, M. Satoh, and K. Yoshio, Improved electrooptical properties of the naphtalene base FLC by tolane FLCs, *Proc. 12th Int. Displ. Res. Conf.*, "*Japan Display '92*," Hiroshima, Japan, 1992, pp. 575–578.

27. Y. Inaba, K. Katagiri, H. Inoue, J. Kanbe, S. Yoshihara, and S. Iijima, Essential factors in high-duty FLC matrix display, *Ferroelectrics 85*:255 (1988).

28. C. Escher, H.-R. Dübal, T. Harada, G. Illian, M. Murakami, and D. Ohlendorf, The SSFLC switching behaviour in view of chevron layer geometry and ionic charges, *Ferroelectrics 113*:269 (1991).

29. J. Dijon, Ferroelectric LCDs, *Liquid Crystals, Applications and Uses* (B. Bahadur, ed.), World Scientific Publishing, Singapore, 1990, vol. 1, p. 305.

30. H. Orihara, K. Nakamura, Y. Ishibashi, Y. Yamada, N. Yamamoto, and M. Yamawaki, Anomalous switching behaviour of a ferroelectric liquid crystal with negative dielectric anisotropy, *Jpn. J. Appl. Phys. 25*:L839 (1986).

31. J. C. Jones, M. J. Towler, and J. R. Hughes, Fast, high-contrast ferroelectric liquid crystal displays and the role of dielectric biaxiality, *Displays 14*:86 (1993).

32. M. Koden, H. Katsuse, N. Itoh, T. Kaneko, K. Tamai, H. Takeda, M. Shiomi, N. Numao, M. Kido, M. Matsuki, S. Miyashi, and T. Wada, Ferroelectric liquid crystal device using the τ-V_{Min} mode, *Proc. 4th Int. Conf. on FLC*, Tokyo, 1993, pp. 369–370.

33. M. Isogai, K. Kondo, T. Kitamura, A. Mukoh, Y. Nagae, and H. Kawakami, Bistable ferroelectric liquid crystal display device using conventional thickness cell, *Proc. 6th Int. Disp. Res. Conf. "Japan Display '86,"* Tokyo, 1986, pp. 472–474.

34. T. Hatano, K. Yamamoto, H. Takezoe, and A. Fukuda, Alignment controls and switching characteristics in a ferroelectric liquid crystal with the phase sequence of N*-S_C*, *Jpn. J. Appl. Phys. 25*:1762 (1986).

35. J. S. Patel and J. W. Goodby, Alignment of liquid crystals which exhibit cholesteric to smectic C* phase transitions, *J. Appl. Phys. 59*:2355 (1986).

36. (a) S. T. Lagerwall, J. Wahl, and N. A. Clark, Ferroelectric liquid crystals for displays, *Proc. 5th Int. Disp. Res. Conf. 85*, San Diego, CA, 1985, pp. 213–221. (b) J. S. Patel and J. W. Goodby, Properties and applications of ferroelectric liquid crystals, *Opt. Eng. 26*:373 (1987).

37. S. Matsumoto, H. Hatoh, and A. Murayama, Matrix liquid-crystal display device technologies, *Liq. Cryst. 5*:1345 (1989).

38. Y. Iwai, N. Wakita, T. Uemura, S. Fujiwara, Y. Gohara, S. Kimura, Y. Matsumoto, Y. Miyatake, T. Tsuda, Y. Horio, and I. Ota, Multi-color, high resolution projection display with ferroelectric liquid-crystal light-valves with 4M pixels, *Proc. 9th Int. Disp. Res. Conf. "Japan Display '89,"* Kyoto, Japan, 1989, pp. 180–183.

39. P. W. Ross, K. Alexander, L. G. Banks, A. N. Carrington, L. K. M. Chan, D. J. Gibbons, R. L. Hedgley, N. Lui, M. J. Naylor, B. Needham, N. E. Riby, P. W. H. Surguy, A. W. Vaidya, and J. C. White, Color digital ferroelectric LCDs for laptop applications, *Proc. SID Int. Symp. Dig. of Techn. Pap.*, Boston, MA, 1992, pp. 217–220.

40. H. Inoue, A. Mizutome, S. Yoshihara, J. Kanbe, and S. Fizima, High resolution, large area FLC display with high graphic performance, *Proc. Int. Symp. of FLCs 1989*, Göteborg, Sweden, 1989, O 12.

41. P. W. Ross, 720 × 400 matrix ferroelectric display operating at video frame rate, *Proc. Int. Disp. Res. Conf. 1988*, San Diego, CA, 1988, pp. 185–190.

42. W. J. A. M. Hartmann, Ferroelectric liquid crystal video display, *Proc. Int. Disp. Res. Conf. 1988*, San Diego, CA, 1988, pp. 191–194.

43. Canon, Inc., Ferroelectric liquid-crystal display product development, *Displays 1*:175 (1991).

44. T. Umeda, Y. Hori, and A. Mukoh, Print head with ferroelectric liquid-crystal light-shutter array, *Proc. SID Int. Symp. Dig. of Techn. Pap.*, Orlando, FL, 1985, pp. 373–376.

45. M. Matsunga, S. Masubuchi, and S. Takahashi, A full-color ferroelectric liquid-crystal shutter array, *Proc. SID Int. Symp. Dig. of Techn. Pap.*, Boston, MA, 1992, pp. 737–740.

46. K. M. Johnson, M. A. Handschy, and L. A. Pagano-Stauffer, Optical computing and image processing with ferroelectric liquid crystals, *Opt. Eng. 26*:385 (1987).

47. M. A. Handschy, K. M. Johnson, G. Moddel, and L. A. Pagano-Stauffer, Electro-optic applications of ferroelectric liquid crystals to optical computing, *Ferroelectrics 85*:279 (1988).

48. K. M. Johnson and G. Moddel, Motivations for using ferroelectric liquid crystal spatial light modulators in neurocomputing, *Appl. Opt. 28*:4888 (1989).

49. D. Armitage, J. I. Thackara, and W. D. Eades, Ferroelectric liquid-crystal devices and optical processing, *Ferroelectrics 85*:291 (1988).

50. G. Moddel, K. M. Johnson, W. Li, R. A. Rice, L. A. Pagano-Stauffer, and M. A. Handschy, High-speed binary optically addressed spatial light modulator, *Appl. Phys. Lett. 55*:537 (1989).

51. T. Kurokawa and S. Fukuchima, Ferroelectric liquid crystal spatial light modulators and their applications, *Proc. 4th Int. Conf. on FLC*, Tokyo, 1993, pp. 69–70.

52. B. I. Ostrowski and V. G. Chigrinov, Linear electrooptic effect in chiral smectic C liquid crystals, *Kristallografiya 25*:560 (1980).

53. B. I. Ostrowski, A. Z. Rabinovich, and V. G. Chigrinov, Behaviour of ferroelectric smectic liquid crystals in electric field, *Advances in Liquid Crystal Research and Applications* (L. Bata, ed.), Pergamon, Oxford, Adademiai Kiado, Budapest, 1980, p. 469.

54. (a) L. A. Beresnev, V. G. Chigrinov, D. Dergachev, E. P. Pashidaev, J. Fünfschilling, and M. Schadt, Electrooptic effect based on the deformation of the helix of a ferroelectric smectic C* liquid crystal, *Proc. 12th Int. Liq. Cryst. Conf.*, Freiburg, Germany, 1988, p. 282. (b) L. A. Beresnev, L. M. Blinov, and D. I. Dergachev, Electro-optical response of a thin layer of a ferroelectric liquid crystal with a small pitch and high spontaneous polarization, *Ferroelectrics 85*:173 (1988).

55. J. Fünfschilling and M. Schadt, Fast responsing and highly multiplexible distorted helix ferroelectric liquid-crystal displays, *J. Appl. Phys. 66*(8):3877 (1989).

56. J. Fünfschilling and M. Schadt, High speed deformed helix ferroelectric (DHF)-liquid crystal displays with video potential, *Proc. 13th Int. Disp. Res. Conf.*, Strasbourg, France, 1993, pp. 63–66.

57. J. Fünfschilling and M. Schadt, Short-pitch bistable ferroelectric LCDs, *Proc. SID Int. Symp. Dig. of Tech. Pap.*, Las Vegas, NV, 1990, Vol. 21, p. 106.

58. J. Fünfschilling and M. Schadt, New short-pitch bistable ferroelectric (SBF) liquid crystal displays, *Jpn. J. Appl. Phys. 30*:741 (1991).

59. S. Garoff and R. B. Meyer, Electroclinic effect at the A-C phase change in a chiral smectic liquid crystal*, *Phys. Rev. Lett. 38*:848 (1977).

60. S. Garoff and R. B. Meyer, Electroclinic effect at the A-C phase change in a chiral smectic liquid crystal, *Phys. Rev. A 19*:338 (1979).

61. S. Nishiyama, Y. Ouchi, H. Takezoe, and A. Fukuda, Giant electroclinic effect in chiral smectic A phase of ferroelectric liquid crystals, *Jpn. J. Appl. Phys. 26*:L1787 (1987).

62. G. Andersson, I. Dahl, L. Komitov, M. Matuszczyk, S. T. Lagerwall, K. Skarp, B. Stebler, D. Coates, M. Chambers, and D. M. Walba, Smectic A* materials with 11.25 degrees induced tilt, *Ferroelectrics 114*:137 (1991).

63. G. Andersson, I. Dahl, P. Keller, W. Kuczynski, S. T. Lagerwall, K. Skarp, and B. Stebler, Submicrosecond electro-optic switching in the liquid-crystal smectic A phase: The soft-mode ferroelectric effect, *Appl. Phys. Lett. 51*:640 (1987).

64. G. Andersson, I. Dahl, L. Komitov, S. T. Lagerwall, K. Skarp, and B. Stebler, Applications of the soft-mode ferroelectric effect, *Proc. 18. Freiburger Arbeitstagung*, Freiburg, Germany, 1989.

65. A. B. Davey and W. A. Crossland, Potential and limitations of the electroclinic effect in device applications, *Ferroelectrics 114*:101 (1991).

66. A. D. L. Chandani, E. Gorecka, Y. Ouchi, H. Takezoe, and A. Fukuda, Antiferroelectric chiral smectic phases responsible for the tristable switching in MHPOBC, *Jpn. J. Appl. Phys. 28*:L1265 (1989).

67. Y. Yamada, N. Yamamoto, K. Mori, K. Nakamura, T. Hagiwara, Y. Suzuki, I. Kawamura, H. Orihara, and Y. Ishibashi, Ferroelectric liquid crystal display using tristable switching, *Jpn. Appl. Phys. 29*:1757 (1990).

68. Y. Yamada, N. Yamamoto, M. Yamawaki, I. Kowamura, and Y. Suzuki, Multi-color video-rate antiferroelectric LCDs with high contrast and wide view, *Proc. 12th Int. Disp. Res. Conf. "Japan Display '92,"* Hiroshima, Japan, 1992, pp. 57–60.

69. N. Yamamoto, N. Koshoubou, K. Mori, K. Nakamura, and Y. Yamada, Full-color antiferroelectric liquid crystal display, *Proc. 4th Int. Conf. on FLC*, Tokyo, 1993, pp. 77–78.

70. K. Itoh, M. Johno, Y. Takanishi, Y. Ouchi, H. Takezoe, and A. Fukuda, Self-recovery from alignment damage under AC fields in antiferroelectric and ferroelectric liquid crystal cells", *Jpn. J. Appl. Phys. 30*:735 (1991).

71. H. Okada, T. Sakurai, T. Katoh, H. Onnagawa, N. Nakatoni, and H. Miyashita, Electro-optic responses of antiferroelectric liquid crystals, *Proc. 12th Int. Disp. Res. Conf. "Japan Display '92,"* Hiroshima, Japan, 1992, pp. 527–529.

72. T. Oyama, T. Masuda, S. Hamado, N. S. Takahashi, S. Kurita, I. Kawamura, and T. Hagiwara, Tristable operation of antiferroelectric liquid crystal light valves optically addressed using amorphous silicon photosensor, *Jpn. J. Appl. Phys. 32*:L668 (1993).

73. I. Drevensek, and R. Blinc, Nonlinear optical properties of ferroelectric liquid crystals, *Cond. Matter News, Vol. 1, Issue 5*:14 (1992).

74. D. J. Williams, Organische polymere und nichtpolymere Materialien mit guten nichtlinearen optischen Eigenschaften, *Angew. Chem. 96*:637 (1984).

75. A. N. Vtyurin, V. P. Ermakov, B. I. Ostrowski, and V. F. Shabanov, Study of optical second harmonic generation in ferroelectric liquid crystal, *Phys. Stat. Sol. (b) 107*:397 (1981).

76. A. Taguchi, Y. Ouchi, H. Takezoe, and A. Fukuda, Angle phase matching in second harmonic generation from a ferroelectric liquid crystal, *Jpn. J. Appl. Phys. 28*:L997 (1989).

77. J. Y. Liu, M. G. Robinson, K. M. Johnson, and D. Doroski, Second-harmonic generation in ferroelectric liquid crystals, *Opt. Lett. 15*:267 (1990).

78. D. M. Walba, M. B. Ros, N. A. Clark, R. Shao, K. M. Johnson, M. G. Robinson, J. Y. Liu, and D. Doroski, An approach to the design of ferroelectric liquid crystals with large second order electronic nonlinear optical susceptibility, *Mol. Cryst. Liq. Cryst. 198*:51 (1991).

79. J. Y. Liu, M. G. Robinson, K. M. Johnson, D. M. Walba, M. B. Ros, N. A. Clark, R. Shao, and D. Doroski, The measurement of second-harmonic generation in novel ferroelectric liquid crystal materials, *J. Appl. Phys. 70*:3426 (1991).

80. R. Lytel, G. F. Lipscomb, J. Thackara, J. Altman, P. Elizonda, M. Stiller, and B. Sullirun, Nonlinear and electro-optic organic devices, *Non-linear Optical Electro-Active Polymers.* Proc. (P. N. Prasad and D. R. Ulrich, eds.), Plenum Press, New York, 1988, p. 415.

81. K. Schmitt, R.-P. Herr, M. Schadt, J. Fünfschilling, R. Buchecker, X. H. Chen, and C. Benecke, Strongly non-linear optical ferroelectric liquid crystals for frequency doubling, *Liq. Cryst. 14*:1735 (1993).

82. K. O. Singer, M. G. Kuzyk, and J. E. Sohn, Orientally ordered electro-optic materials, *Nonlinear Optical Electro-Active Polymers*, Proc. (P. N. Prasad and D. R. Ulrich, eds.), Plenum Press, New York, 1988, p. 189.

83. J. Y. Liu, K. M. Johnson, and M. G. Robinson, Room temperature 10 MHZ electro-optic modulation in ferroelectric liquid crystals, *Appl. Phys. Lett. 62*:934 (1993).

84. L. Bata, A. Buka, N. Éber, A. Jákli, K. Pintér, J. Szabon, and A. Vajda, Properties of a homologous series of ferroelectric liquid crystals, *Mol. Cryst. Liq. Cryst. 151*:47 (1987).

85. A. Jákli and L. Bata, Mechano-electrical effects on planar S_C^* liquid crystals, *Mol. Cryst. Liq. Cryst. 201*:115 (1991).

86. A. Jákli, L. Bata, A. Buka, N. Éber, and I. Jánossoj, New electromechanical effect in chiral smectic C* liquid crystals, *J. Phys. Lett. 46*:L759 (1985).

87. A. Jákli, L. Bata, A. Buka, and N. Éber, Electromechanical effect in S_C^* liquid crystals, *Ferroelectrics 69*:153 (1986).

88. A. Jákli and L. Bata, Resonances in linear electromechanical responses of S_C^* liquid crystal samples, *Ferroelectrics 103*:35 (1990).

89. A. Jákli and A. Saupe, Electromechanical responses of chiral smectic liquid crystals, *Proc. 21. Freiburger Arbeitstagung*, Freiburg, Germany, 1992.

90. A. Jákli and L. Bata, Non-linear electromechanical response of S_C^* liquid crystals, *Liq. Cryst. 7*:105 (1990).

91. F. Gießelmann, I. Dierking, and P. Zugenmaier, Strong electroacoustic effect in ferroelectric liquid crystal cells, *Mol. Cryst. Liq. Cryst. Lett. 8*:105 (1992).

92. L. J. Yu, H. Lee, C. S. Bak, and M. M. Labes, Observation of pyroelectricity in chiral smectic-C and -H liquid crystals, *Phys. Rev. Lett. 36*:388 (1976).

93. A. G. Chynoweth, Dynamic method for measuring the pyroelectric effect with special reference to barium titanate, *J. Appl. Phys. 27*:78 (1956).

94. A. M. Glass, J. S. Patel, J. W. Goodby, D. H. Olson, and J. M. Geary, Pyroelectric detection with smectic liquid crystals, *J. Appl. Phys. 60*:2778 (1986).

95. M. V. Kozlovsky, L. A. Beresnev, S. G. Kononov, V. P. Shibaev, and L. M. Blinov, Spontaneous polarization of a liquid-crystal monomer and its polymer, *Sov. Phys. Solid. State 29*:54 (1987).

96. G. Decobert, J. C. Dubois, S. Esselin, and C. Noel, Some novel smectic C* liquid-crystalline side-chain polymers, *Liq. Cryst. 1*:307 (1986).

97. V. P. Shibaev, M. V. Kozlovsky, N. A. Platé, L. A. Beresnev, and L. M. Blinov, Ferroelectric liquid-crystalline polymethacrylates, *Liq. Cryst. 8*:545 (1990).

98. G. Decobert, F. Soyer, and J. C. Dubois, Chiral liquid crystalline side chain polymers, *Polymer Bull. 14*:179 (1985).

99. S. Esselin, L. Bosio, C. Noel, G. Decobert, and J. C. Dubois, Some novel smectic C* liquid-crystalline side-chain polymers—Polymethacrylates and poly α-chloroacrylates, *Liq. Cryst. 2*:505 (1987).

100. P. Keller, Synthesis and properties of new liquid crystalline side-chain polysiloxanes, *Mol. Cryst. Liq. Cryst. Inc. Nonlin. Opt. 157*:193 (1988).

101. S. Ugiie and K. Iimura, Ferroelectricity of chiral side-chain liquid crystalline polytartrate, *Rep. Prog. in Polymer Phys. Jpn. 32*:351 (1989).

102. R. Zentel, G. Reckert, and B. Reck, New liquid-crystalline polymers with chiral phases, *Liq. Cryst. 2*:83 (1987).

103. (a) R. Zentel, Preliminary communication, *Liq. Cryst. 3*:531 (1988. (b) R. Zentel, Liquid Crystalline Elastomers, *Angew. Chem. Int. Ed. Engl. Adv. Mater. 28*:1407 (1989).

104. P. Le Barny and J. C. Dubois, The chiral smectic C liquid crystal side chain polymers, *Side chain liquid crystal polymers* (C. B. McArdle, ed.), Blackie, Glasgow and London, 1989, p. 130.

105. S. Uchida, K. Morita, K. Miyoshi, K. Hashimoto, and K. Kawasaki, Synthesis of some smectic liquid crystalline polymers and their ferroelectricity, *Mol. Cryst. Liq. Cryst. 155*:93 (1988).

106. J. Naciri, S. Pfeiffer, and R. Shashidhar, Fast switching ferroelectric side-chain liquid-crystalline polymer and copolymer, *Liq. Cryst. 10*:585 (1991).

107. T. Sekiya, K. Yuasa, S. Uchida, S. Hachiya, K. Hashimoto, and K. Kawasaki, Ferroelectric liquid crystalline polymers and related model compounds with a low-moderate degree of polymerization, *Liq. Cryst. 14*:1255 (1993).

108. K. Siemensmeyer, K. H. Etzbach, G. Wagenblast, V. Bach, and P. Delavier, Influence of molecular weight and polydispersity on the physical properties of ferroelectric liquid crystalline polymers, *Proc. 14th Int. Liq. Cryst. Conf.*, Pisa, Italy, 1992, Vol. 1, p. 209.

109. K. Kühnpast, J. Springer, G. Scherowsky, F. Giesselmann, and P. Zugenmaier, Ferroelectric liquid-crystalline side group polymers—Spacer length variation and comparison with the monomers, *Liq. Cryst. 14*:861 (1993).

110. G. Scherowsky, Fast-switching ferroelectric liquid crystalline polymers containing one or two centres of chirality in the side chain, *Polymers for Advanced Technologies* (M. Levin, ed.) John Wiley, New York, 1992, Vol. 3, p. 219.

111. T. Suzuki, T. Okawa, T. Ohnuma, and Y. Sakon, Preparation of ferroelectric liquid-crystalline polysiloxanes and electrooptical measurements, *Makromol. Chem. Rapid. Commun. 9*:755 (1988).

112. M. Dumon, H. T. Nguyen, M. Mauzac, C. Destrade, M. F. Achard, and H. Gasparoux, New ferroelectric liquid crystal polysiloxanes, *Macromolecules 23*:355 (1990).

113. M. Dumon, H. T. Nguyen, M. Mauzac, C. Destrade, and H. Gasparoux, Ferroelectric liquid crystal siloxane homo and copolymers, *Liq. Cryst. 10*:475 (1991).

114. K. Takahashi, S. Matsumoto, S. Tsuru, and F. Yamamoto, Electro-optical effects in new polymeric liquid crystals, *Mol. Cryst. Liq. Cryst. 8*(2):33 (1991).

115. (a) K. Yuasa, S. Uchida, T. Sekiya, K. Hashimoto, and K. Kawasaki, Electro-optical properties of ferroelectric liquid crystalline polymers, *Ferroelectrics 122*:53 (1991). (b) K. Yuasu, S. Uchida, T. Sekya, K. Hashimoto, and K. Kawasaki, Electro-optical properties of ferroelectric liquid crystalline polymers, *Polymers Adv. Technol. 3*:205 (1992).

116. H. Endo, S. Hachiya, S. Uchida, K. Hashimoto, and K. Kawasaki, Helical pitch of ferroelectric liquid-crystalline polymers and copolymers, *Liq. Cryst. 9*:635 (1991).

117. F. Kremer, A. Schönfeld, A. Hofmann, R. Zentel, and H. Poths, Collective and molecular dynamics in ferroelectric liquid crystals: From low molar to polymeric and elastomeric systems, *Polymers Adv. Technol. 3*:249 (1992).

118. M. Pfeiffer, L. A. Beresnev, W. Haase, G. Scherowsky, K. Kühnpast, and J. Jungbauer, Dielectric and electrooptic properties of a switchable ferroelectric liquid crystalline side chain polymer, *Mol. Cryst. Liq. Cryst. 214*:125 (1992).

119. B. O. Myrvold, The relationship between the physical properties of the alignment layer and the quality of SSFLC cells, *Mol. Crystl. Liq. Cryst. 202*:123 (1991).

120. (a) R. Kiefer, G. Baur, P. Delavier, and K. Siemensmeyer, Switching behavior and rotational viscosities of new ferroelectric liquid crystalline oligomers, *Proceed. 4th Int. FLC Conf.* Tokyo, 1993, p. 323. (b) G. Baur, P. Delavier, K.-H. Etzbach, F. Meyer, R. Kiefer, K. Siemensmeyer, and G. Wayenblast, Ferroelectric liquid crystalline oligomesogens: synthesis and properties, *Proceed. Freiburger Arbeitstayung*, Freiburg, Germany, 1994.

121. D. S. Parmar, N. A. Clark, P. Keller, D. M. Walba, and M. D. Wand, Physical properties and alignment of a polymer-monomer ferroelectric liquid crystal mixture, *J. Phys. France 51*: 355 (1990).

122. K. Flatischler, L. Komitov, K. Sharp, and P. Keller, Electroclinic effect in some side-chain polysiloxane liquid crystals, *Mol. Cryst. Liq. Cryst. 209*:109 (1981).

123. H. J. Coles, H. F. Gleeson, G. Scherowsky, and A. Schliwa, Ferroelectric side chain polymer liquid crystals: I. Static and dynamic properties, *Mol. Cryst. Liq. Cryst. Lett. 7*(4):117 (1990).

124. J. Bömelburg, G. Heppke, and J. Hollidt, Electroclinic switching in a tilted smectic phase of a ferroelectric liquid crystalline side chain polymer, *Proc. 20. Freiburger Arbeitstagung*, Freiburg, Germany, 1991.

125. J. Bömelburg, G. Heppke, and J. Hollidt, Evidence for an antiferroelectric smectic phase in a chiral side-chain polymer, *Makrom. Chem. Rapid Commun. 12*:483 (1991).

126. K. Skarp, G. Andersson, F. Gouda, S. T. Lagerwall, H. Poths, and R. Zentel, Antiferroelectric behaviour in a liquid crystalline polymer, *Polymers Adv. Technol. 3*:241 (1992).

127. (a) G. Scherowsky, K. Kühnpast, and J. Springer, Electrooptical investigation on the three switching states of a chiral smectic side group polymer, *Makromol. Chem., Rapid Commun. 12*:381 (1991). (b) F. Giesselmann, P. Zugenmaier, G. Scherowsky, K. Kühnpast, and J. Springer, Three switching states in a chiral smectic side-chain polymer, *Makromol. Chem. Rapid. Commun. 13*:489 (1992).

128. V. Bach, P. Delavier, K.-H. Etzbach, K. Siemensmeyer, and G. Wagenblast, Properties of ferroelectric and antiferroelectric phases in liquid crystalline polymers, *Proc. 22. Freiburger Arbeitstagung*, Freiburg, Germany, 1992.

129. Idemitsu Kosan Co., Ltd., Exhibition 4th INt. FLC Conf., Tokyo, 1993.

130. K. Blatter, P. Hornischfeger, D. Jungbauer, H.-U. Simmrock, and C. Walton, Ferroelektrische Flüssigkristalline Polymere für Display-Anwendungen, *Proc. Makromolekulares Kolloquium*, Freiburg, Germany, 1993, p. 63.

131. A. Mosley, Liquid crystal displays—An overview, *Displays 14*:67 (1993).

132. H. Coles, New low molar mass and polymeric ferroelectric liquid crystals, *Proc. 4th Int. FLC Conf.*, Tokyo, 1993, pp. 63–64, 325–326.

133. T. C. Chieu, Influence of thickness and conductivity of alignment layers on the bistability of ferroelectric liquid-crystalline devices, *J. Appl. Phys. 69*:8399 (1991).

134. (a) M. Ozaki, Y. Sadohara, T. Hadai, and K. Yoshino, Fast optical switching in polymer waveguide using ferroelectric liquid crystal, *Jpn. J. Appl. Phys. 29*:L843 (1990). (b) M.

Ozaki, Y. Sadohara, Y. Uchiyama, M. Utsumi, and K. Yoshino, Electrooptic modulation in the optical waveguide using ferroelectric liquid crystal, *Jpn. J. Appl. Phys. 31*:3189 (1992).

135. G. Scherowsky, A. Beer, and H. J. Coles, A coloured ferroelectric side chain polymer, *Liq. Cryst. 10*:809 (1992).

136. G. Scherowsky, Novel ferroelectric LC polymers and coloured copolymers synthesis and electrooptical properties, *Proc. 4th Int. FLC Conf.*, Tokyo, 1993, pp. 59–60.

137. P. A. Williams, L. Komitov, A. G. Rappaport, B. N. Thomas, N. A. Clark, D. M. Walba, and G. W. Day, Studies of the higher order smectic phase of the large electroclinic effect material W317, *Liq. Cryst. 14*:1095 (1993).

138. V. P. Shibaev, S. G. Kostromin, N. A. Platé, S. A. Ivanov, V. Yu. Vetrov, and I. A. Yakovlev, Thermo-recording on liquid-crystalline polymers with the aid of a laser beam, *Polymer Commun. 24*:364 (1983).

139. H. J. Coles and R. Simon, High-resolution laser-addressed liquid crystal polymer storage displays, *Polymer 26*:1801 (1985).

140. C. B. McArdle, M. G. Clark, C. M. Haws, M. C. K. Wiltshire, A. Parker, G. Nestor, G. W. Gray, D. Lacey, and K. J. Toyne, Laser addressed thermo-optic effect in a novel dyed liquid-crystalline polysiloxane, *Liq. Cryst. 2*:573 (1987).

141. G. Wagenblast, K. Beck, and K.-H. Etzbach, Optical data storage with liquid-crystal/dye copolymers, *Proc. 20. Freiburger Arbeitstagung Flüssigkristalle*, Freiburg, Germany, 1991.

142. G. Kämpf, Polymere als Träger und Speicher von Informationen, *Ber. Bunseng. Phys. Chem. 89*:1179 (1985).

143. H.-W. Schmidt, Dissertation: Synthese, Struktur und Eigenschaften von farbstoff/haltigen flüssigkristallinen Copolymeren, *Doktor der Naturwissenschaften*, Universität Mainz, 1984.

144. (a) R. Kiefer and G. Baur, Fast switching films of nematic side chain copolymers, *Proc. 12th Int. Liq. Cryst. Conf. 1988*, Freiburg, Germany, 1988, P009, p. 121. (b) R. Kiefer and G. Baur, Fast switching films of nematic side chain copolymers, *Liq. Cryst. 5*:1497 (1989). (c) R. Kiefer, F. Windscheid, and G. Baur, Electrooptical behaviour of liquid crystalline side chain copolymers, *Proc. Japandisplay '89*, Kyoto, Japan, 1989, pp. 290–293.

145. H. Weber, Nichtflüchtige Fram-Speicher, *Elektronik Journal 1–2*:56 (1992).

146. C. Bowry, P. Bonnett, and M. G. Clark, A liquid crystal polymer flexible optical storage film, *Proc. Eurodisplay*, Amsterdam, The Netherlands, 1990, pp. 158–161.

147. J.-C. Dubois, P. Le Barny, P. Robin, V. Lemoine, and H. Rajbenbach, Properties of side chain liquid crystal and amorphous polymers—Applications to non-linear optics, *Liq. Cryst. 14*:197 (1993).

148. H. Kapitzka, R. Zentel, R. J. Twieg, C. Nguyen, S. U. Vallerien, F. Kremer, and C. G. Willson, Ferroelectric liquid crystalline polysiloxanes with high spontaneous polarization and possible applications in nonlinear optics, *Adv. Mater. 2*:539 (1990).

149. H. J. Coles, M. Redmond, O. Mondain-Monval, E. Wischerhoff, and R. Zentel, A study of second harmonic generation in side chain ferroelectric liquid crystalline copolymers, *Proc. 4th Int. FLC Conf.*, Tokyo, 1993, pp. 83–84.

150. W. G. Cady, in *Piezoelectricity*, Vol. 1, Dover, New York, 1964.

151. W. Meier and H. Finkelmann, Piezoelectricity of cholesteric elastomers, *Makromol. Chem. Rapid Commun. 11*:599 (1990).

152. S. U. Vallerien, F. Kremer, E. W. Fischer, H. Kapitza, R. Zentel, and H. Poths, Experimental proof of piezoelectricity in cholesteric and chiral smectic C*-phases of LC-elastomers, *Makromol. Chem. Rapid Commun. 11*:593 (1990).

153. G. Wagenblast, private communication, unpublished results.

154. R. A. M. Hikmet, Piezoelectric networks obtained by photopolymerization of liquid crystal molecules, *Macromolecules 25*:5759 (1992).

155. W. Meier and H. Finkelmann, Liquid crystalline elastomers, *Condens. Matter News Vol. 1, Issue 7*:15 (1992).

156. M. Brehmer, R. Zentel, G. Wagenblast, and K. Siemensmeyer, Ferroelectric LC-elastomers, *Makromolekulare Chemie*, in press.

157. A. Jákli and A. Saupe, Linear electromechanical effect in a S_C^* polymer liquid crystal, *Liq. Cryst.* 9:519 (1991).

158. A. Jákli, N. Éber, and L. Bata, Electromechanical effects in ferroelectric liquid crystalline polymers, *Polymers Adv. Technol.* 3:269 (1992).

159. N. Éber, L. Bata, G. Scherowsky, and A. Schliwa, Linear electromechanical effect in a polymeric ferroelectric liquid crystal, *Ferroelectrics 122*:139 (1991).

160. R. Zentel, H. Poths, F. Kremer, A. Schönfeld, D. Jungbauer, R. Twieg, C. G. Willson, and D. Yoon, Polymeric liquid crystals: Structural basis for ferroelectric and nonlinear optical properties, *Polymers Adv. Technol.* 3:211 (1992).

161. L. M. Blinov, V. A. Baikalov, M. I. Barnik, L. A. Beresnev, E. P. Pozhidayev, and S. V. Yablonsky, Experimental techniques for the investigation of ferroelectric liquid crystals, *Liq. Cryst.* 2:121 (1987).

162. K. Skarp, G. Andersson, A. Dahlgren, H. Poths, and R. Zentel, Pyroelectric effect in monomeric and polymeric liquid crystals, *Proc. 14th Int. Liq. Cryst. Conf.*, Pisa, Italy, 1992, Vol. II, p. 905.

163. M. E. Lines and A. M. Glass, Pyroelectric detection, *Principles and Applications of Ferroelectrics and Related Materials*, Clarendon Press, Oxford, 1977, p. 561.

164. S. Ujiie and K. Imura, Ferroelectric liquid-crystalline polytartrate, *Polymer J. 23*:1483 (1991).

165. S. C. Lien, C. C. Huang, and J. W. Goodby, Heat-capacity studies near the smectic-A-smectic-C (-smectic-C*) transition in a racemic (chiral) smectic liquid crystal, *Phys. Rev. A 29*:1371 (1984).

166. M. Meichle and C. W. Garland, Calorimetric study of the smectic-A-smectic-C-phase transition in liquid crystals, *Phys. Rev. A 27*:2624 (1983).

167. R. Zbinden, Characteristic features of polymer spectra, *Infrared Spectroscopy of High polymers*, Academic Press, New York and London, 1964.

168. Y. Guo, K. Hoshino, J. Hanna, and H. Kokado, Image recording with ferroelectric film (I), *Jpn. J. Appl. Phys. 31*:L1432 (1992).

Index